Nissim Sahat

ELECTROMAGNETIC FIELDS:
Sources and Media

ELECTROMAGNETIC FIELDS:
Sources and Media

ALAN M. PORTIS
University of California, Berkeley

JOHN WILEY & SONS
New York · Santa Barbara · Chichester · Brisbane · Toronto

Library of Congress Cataloging in Publication Data

Portis, Alan M., 1926–
    Electromagnetic fields.

    Includes bibliographical references and indexes.
    1. Electromagnetic fields. 2. Solid state physics.
3. Plasma (Ionized gases) I. Title.
QC665.E4P67   537   78-7585
ISBN 0-471-01906-2

Printed in the United States of America

10 9 8 7 6 5 4 3 2 1

PREFACE

Writing this text was begun four years ago in response to a perceived shift in the career goals of undergraduate physics majors. Whereas 10 years ago more than 80 percent of physics majors went on to graduate school in physics, closer to 50 percent now take this option. From surveys conducted at Berkeley, the majority of students who do not continue in physics do work in an applied area, either in graduate school or in industry. And of the students who do go on in physics, nearly half work in solid-state physics, plasma physics, optics, and physical electronics. Thus, more than half of all present undergraduate physics majors will be primarily concerned with the applications of physics.

Although many physics departments do offer courses in solid state and plasma physics as electives, not all departments are able to make such courses available. Nor could all students accommodate such electives in their programs. It seemed that without a major revision in the physics curriculum the need of physics undergraduates to learn something of solid-state and plasma physics and contemporary optics could best be accommodated through the integration of these topics into the electromagnetic fields course. In making this integration I have made a number of subjective judgements that were determined to a very considerable extent by my own background and taste. I was influenced to a very considerable degree by Charles Kittel's highly successful text *Introduction to Solid State Physics* in introducing elementary models as abstractions of real material media. At the same time, I did not want this text to be merely a superficial introduction to solid-state and plasma physics. Thus, the selection of topics has been made within the context of a serious attempt to develop the classical theory of electromagnetic fields from a contemporary point of view. This means that the emphasis has been on macroscopic theory with only as much use made of microscopic models as required for an understanding of the macroscopic behavior.

I was also considerably influenced by E. M. Purcell's magnificent introduction to   v

Electricity and Magnetism in which the special theory of relativity is used in a natural way to develop classical electromagnetic theory. I have tried to adapt Purcell's approach to the needs of the intermediate fields course; and although this text hardly rivals his beautiful exposition of the subject, I tried very seriously to extend his very thoughtful development of this part of physics.

The text begins not unconventionally with electrostatics, the Coulomb and Gauss laws, flux, and potential. The early attention to macroscopic field averages and to polarization as an alternate representation of charge is new to courses at this level. Ellipsoidal geometries are introduced without the need of ellipsoidal coordinates by making use of a theorem of Newton developed for gravitational fields.

The theory of dielectric media is introduced through an analysis of ionic diatomic molecules. The discussion of work and energy is based very carefully on the distinction between *external* charge as an outside agent and *induced* charge as the response of a system to external forces. Diffusion in conducting media is discussed with some care to provide a background both for an analysis of static screening and of longitudinal plasma waves. Dielectric insulators and conducting media are treated from a unified point of view in terms of the dielectric function, an approach characteristic of present analysis in solid-state and plasma physics.

Forces between steady solenoidal currents are shown to arise from the forces between charges at rest. The connection requires the Lorentz transformation of forces together with the not very widely known property that steady solenoidal currents can be synthesized from straight current filaments. Ampère's law and the Biot-Savart law follow directly from this synthesis. The macroscopic magnetic field, magnetization as an alternative description of distributed currents, and ellipsoidal geometries are treated in a way that closely parallels the earlier discussion of electric charge.

The magnetic properties of material media are discussed from the point of view of atomic magnetism and the constancy of angular momentum. This point of view is extended to a discussion of the diamagnetic properties of plasmas and finally to super-conductors, which are treated as macroscopic diamagnets. The London equation is derived from this point of view with application to such topics as the screening of magnetic fields and trapped flux.

The interaction between moving steady solenoidal currents and charges is obtained again from the Lorentz transformation of forces. Just as it was shown earlier that moving charges produce magnetic fields, we find here that moving currents produce electric fields. We introduce Faraday's law as a description of phenomena more general than are predictable from Coulomb's law alone. These two experimental laws together with the Lorentz force statement are taken as the basis of electromagnetic theory. We distinguish between *external* currents and *induced* currents in a careful discussion of magnetic energy that parallels the earlier electric discussion. Ferromagnetic and superconducting materials are treated as examples.

The Maxwell equations are developed from Gauss's law, Faraday's law, and the Lorentz force. The requirement of relativistic invariance yields the Ampère-Maxwell law as a consequence of Gauss's law. Similarly, Faraday's law requires that the magnetic field **B** be solenoidal. Special attention is paid to the linear and angular momentum carried by electromagnetic fields. Only by identifying momentum with fields is it possible to preserve overall momentum conservation where interactions are through electromagnetic forces. Electromagnetic radiation is discussed with particular attention to the multipole radiation of arbitrarily changing charges and currents with synchrotron radiation as an example.

The final chapters treat applications of the theory. We first consider propagation in homogeneous linear media. Magnetooptic effects, both in plasmas and in magnetic insulators, are discussed as examples. Reflection and refraction together with a discussion of surface electromagnetic waves lead to a consideration of bounded waves, both in metallic waveguides and in quartz fibers. The scattering of electromagnetic waves by free charges and then by dielectric bodies leads to a consideration of Fraunhofer diffraction from apertures, gratings, and finally lattices. Rayleigh scattering is treated as is Cherenkov radiation, both in vacuum (from charges moving faster than the speed of light) and in dispersive media. A concluding chapter discusses active media, which exchange energy between *external* sources and propagating electromagnetic fields. As examples, we discuss briefly three-level magnetic systems, the ammonia maser, and the various lasers.

The necessary mathematics is presented in appendices on vector and tensor analysis and on Fourier series and complex variables. Discussions of quantization and relativistic dynamics provide the necessary physics background. Special topics such as equivalent networks, generalized equations, and dispersion relations are available for supplemental treatment. An appendix on the fields of moving charges begins with the fields of an arbitrarily moving charge and, with the use of a formula developed by Feynman, obtains the Coulomb and Biot-Savart expressions for steady charge and current distributions. Although such a result is to be expected, it is instructive to show that not all systems of accelerated charges need radiate.

The chapters are divided into parts and the parts into sequentially numbered sections. Equations through a chapter are numbered sequentially and referred to within a chapter by number as **(15)**. Equations within *other* chapters or appendices are referred to as **(5-15)** or **(D-15)**. Material in other chapters and appendices is referred to by section as Section 8-3 or Section C-5. The titles of sections that contain material that is advanced or of specialized interest are placed in brackets []. I have not found it possible to treat these sections explicitly, even in a full year course.

Important equations are flagged at the left margin and are repeated with equation number in the summary that follows most chapters and appendices. An open flag connotes an equation that is model dependent. A solid flag indicates that the equation is not restricted by the limitations of a model.

Following most sections are exercises that reinforce the material of that section. At the conclusion of each chapter are problems presented in the order in which the material is discussed in the chapter. A special effort has been made to generate problems for the material that is unique to this text, and instructors will want to consult other sources for problems on some of the more standard topics.

I hesitate to advise an instructor concerning the presentation of material or the assignment of problems. I have found it necessary at the beginning of the course to devote a week to vector calculus and tensors. Where a substantial fraction of the students are juniors, a 2-week presentation of vector and tensor analysis would not be excessive. I have found it possible to present the first five chapters in about 8 weeks. Before beginning the discussion of magnetic interactions, I have spent between 1 and 2 weeks on relativity including a discussion of the transformation of forces as presented in Appendix F. Chapters 6 through 11 may be covered in about 8 weeks. I have used the final four chapters together with the appendices on Fourier series and complex variables and dispersion relations in a separate 10-week course in modern optics.

Where a single 15-week term is available for electromagnetic theory and its applications, I recommend that primary emphasis be given to Chapters 1, 2, 4, 6, 9, 10, and 11, together with Appendices B and F. Although a number of recent texts give the instructor the option of teaching electromagnetic theory with or without relativity, I am afraid that my development of the subject leaves the instructor little choice but to discuss relativity together with electromagnetic theory. Although I have made a conscientious effort to produce a readable and teachable text, I am sure that there are many places where improvement is possible, and I welcome suggestions from students and their instructors and other readers as well.

I mentioned the texts by Kittel and Purcell and want to acknowledge my debt both for their published work and for the personal support and encouragement that both have given me. A number of my colleagues at Berkeley have taken an interest in what I was doing and have been generous with their time: J. D. Jackson, E. H. Wichmann, H. H. Bingham, and C. F. McKee, to whom I am particularly indebted for having taught from the manuscript for this text. K. W. Ford at the New Mexico Institute of Technology read through the entire manuscript and has made many valuable suggestions. Final work on the manuscript was completed during the tenure of a visiting professorship at the Victoria University of Wellington. I will always be grateful to my New Zealand colleagues for their hospitality. I also acknowledge the very considerable help that I have received from the some 50 students who have suffered through the development of this text. I particularly want to note the contributions of Allen Olsen and Chenson Chen. At all stages of development, copy has been typed by Linda Billard, who has also overseen the distribution of course material to students. Her commitment to this text has been exceeded only by my own.

Finally, anyone who has written a text knows what an endless task it is and the amount of time that is required. Some of the time would have been spent with colleagues or students, or alone, but much of it comes from family. My wife, Beverly, and our children, Jon, Steve, Lori, and Rick, have been more than understanding, particularly as I have assured them that I do not mean to make this a career. My colleague, Frederick Reif, closed the preface to his text on statistical and thermal physics with the observation that "an author never finishes a book, he merely abandons it." I think I understand that now.

Berkeley, California   Alan M. Portis
January 1978

CONTENTS

Contents

Contents

Contents

Contents

1 SOURCES OF THE ELECTRIC FIELD I

The fact that certain bodies, after being rubbed, appear to attract other bodies, was known to the ancients. In modern times, a great variety of other phenomena have been observed, and have been found to be related to these phenomena of attraction. They have been classed under the name of Electric phenomena, amber, ἠλεκτρον, having been the substance in which they were first described. Other bodies, particularly the lodestone, and pieces of iron and steel which have been subjected to certain processes, have also been long known to exhibit phenomena of action at a distance. These phenomena, with others related to them, were found to differ from the electric phenomena, and have been classed under the name of Magnetic phenomena, the lodestone, μαγνης, being found in the Thessalian Magnesia.

These two classes of phenomena have since been found to be related to each other, and the relations between the various phenomena of both classes, so far as they are known, constitute the science of Electromagnetism.

In the following Treatise I propose to describe the most important of these phenomena, to shew how they may be subjected to measurement, and to trace the mathematical connexions of the quantities measured. Having thus obtained the data for a mathematical theory of electromagnetism, and having shewn how this theory may be applied to the calculation of phenomena, I shall endeavour to place in as clear a light as I can the relations between the mathematical form of this theory and that of the fundamental science of Dynamics, in order that we may be in some degree prepared to determine the kind of dynamical phenomena among which we are to look for illustrations or explanations of the electromagnetic phenomena.

Thus, in 1873 did James Clerk Maxwell introduce his *Treatise on Electricity and Magnetism*,[1] which culminated in his electromagnetic theory of light. It is no exaggeration to say that Maxwell's contribution revolutionized physics and established the foundation for the theory of relativity, which was to follow in 1905.

Maxwell's development of electromagnetism has profoundly affected all subsequent treatments of the subject and this text is no exception. At the same time, a contemporary treatment of electromagnetism is bound to differ in certain respects from that of Maxwell. First, the *Principle of Relativity*, of which Einstein first wrote in his 1905 paper, is now regarded as a fundamental postulate of physics and we discuss the interaction between

---

[1] From *A Treatise on Electricity and Magnetism* by James Clerk Maxwell, 3rd ed., 1891, pp. V–Xi, Oxford University Press.

moving charges from this point of view. Second, we now have microscopic models of condensed matter and in terms of these models may understand the behavior of dielectric and magnetic media as well as plasmas. These phenomena include ferroelectricity and ferromagnetism and the only recently understood superconductivity. Although the justification of these models goes well beyond a classical theory of electromagnetism, we do not hesitate to discuss the response of diverse kinds of media to electric and magnetic fields. Thus, where Maxwell's examples were largely macroscopic—the experiments of Coulomb and Cavendish, of Öersted and Ampère, and of Faraday—ours will be primarily microscopic, taken as they are from plasma physics and solid-state physics.

Our first four chapters generally follow Maxwell's Part I, *Electrostatics*. The fifth chapter, which is concerned with conducting media, corresponds to Maxwell's Part II, which he calls *Electrokinematics*. With Chapter 6 our treatment departs from that of Maxwell and takes the point of view of Einstein. The remaining chapters attempt a synthesis of Maxwell and Einstein while drawing examples from contemporary physics.

We begin with an exposition of classical potential theory as applied to the interaction between electric charges. This subject had its beginnings with Isaac Newton and his development of the theory of gravitational interaction. The subject was applied to electromagnetism by a number of workers in the latter part of the nineteenth century, culminating in James Clerk Maxwell's *Treatise on Electricity and Magnetism*. More recent works[2] have further refined the classical theory.

Some readers may feel impatient that so much formal theory has grown from a relatively simple interaction while others will admire the beauty and elegance of this branch of physics. We have tried to limit our discussion to those topics that will find immediate application in later chapters while at the same time trying to preserve as much as possible the flavor of this early work.

ACTION AT A DISTANCE

The fundamental law of universal gravitation was first postulated by Isaac Newton in 1666 as what we would now call an *action-at-a-distance* theory. We show in Figure 1 a mass m_1 which we regard as the *source* of the action, and a mass m, which is a *test* mass to be acted on. In the notation that we will be using, the force on the test mass is written as:

$$\mathbf{F} = -G \frac{m_1 m}{r_{01}^2} \hat{\mathbf{r}}_{01} \tag{1}$$

where $\mathbf{r}_{01} = \mathbf{r} - \mathbf{r}_1$ and r_{01} is the scalar magnitude of $\mathbf{r}_{01}$. We represent $\hat{\mathbf{r}}_{01}$ as a *dimensionless* vector of unit magnitude parallel to $\mathbf{r}_{01}$. We indicate by **(1)** that the force is along the line of centers of the masses, is proportional to the magnitude of the

[2] O. D. Kellogg, *Foundations of Potential Theory*, J. Springer, 1929. Reprinted by Dover Publications, 1954.
  W. D. MacMillan, *The Theory of the Potential*, McGraw-Hill, 1930. Reprinted by Dover Publications, 1958.

product of the masses, and is inversely proportional to the square of the distance between centers. The constant $G = 6.6720 \pm 0.0041 \times 10^{-11}$ N m^2/kg^2 was first determined in the laboratory by Cavendish in 1798, more than 100 years after Newton's work.

1 **Coulomb's Law.** The behavior of the force between electric charges was established in 1785 by Charles Coulomb with an apparatus very much like that used later by Cavendish for gravitational studies. We write Coulomb's law as the force on a *test* charge q from a *source* charge q_1:

$$\mathbf{F} = \frac{1}{4\pi\epsilon_0} \frac{q_1 q}{r_{01}^2} \hat{\mathbf{r}}_{01}$$
(2)

where $1/4\pi\epsilon_0$ is *numerically* equal to $c^2 \times 10^{-7} = 8.987551679 + 0.000000066 \times 10^9$ and has the unit (Section K-2) farad per metre. Charge is measured in coulombs. (The elementary charge is $e = 1.6021892 \pm 0.0000046 \times 10^{-19}$ coulomb.)

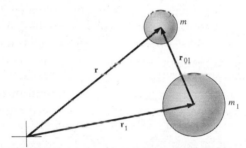

Figure 1   Gravitational attraction. A mass m_1 is regarded as the source of the *action at a distance*. The mass m is the test mass, which is acted on.

Note the similarity between **(2)** and **(1)**. At the same time there are important differences. Gravitational mass is positive and the interaction is always attractive. For electric charge, on the other hand, the interaction between charges of the *same* sign is repulsive while charges of *opposite* sign attract.

We must note that **(2)** describes the interaction between charges at rest in the laboratory. How is the force **F** modified if q moves in the laboratory frame while q_1 remains at rest? We find from experiment that the force on q continues to be quite rigorously given by **(2)** even though the charge velocity may be quite large. Perhaps the most precise experiments involve an attempt to deflect neutral alkali atoms by a uniform electric field,[3] giving an upper limit of 10^{-18} e for the residual charge. Even with the considerable kinetic energy of orbital and subnuclear charges, any residual force is beyond the limits of observation.

It is usually said that electric forces are much stronger than gravitational forces and we may verify this for the interaction between protons. The ratio of electric to

[3] J. C. Zorn, G. E. Chamberlain, and V. W. Hughes, *Phys. Rev.* **129**, 2566 (1963); L. J. Fraser, E. R. Carlson, and V. W. Hughes, *Bull. Am. Phys. Soc.* **13**, 636 (1968).

gravitational force is given by $(e/m_p)^2/4\pi\epsilon_0\,G = 1.2 \times 10^{36}$. The reason we can observe gravitational forces at all is that matter is normally electrically neutral.

Exercises

1. By interchanging the role of source charge and test charge find the force $\mathbf{F}_1$ acting on q_1. Show that $\mathbf{F}_1$ is the negative of **(2)**.
2. Compare the gravitational and electrical forces within the hydrogen atom. Take the separation between electron and proton equal to the Bohr radius, $a_0 = 0.529 \times 10^{-10}$ m. Take for the electron mass $m_e \simeq 9.11 \times 10^{-31}$ kg and for the proton $m_p \simeq 1836.2 m_e \simeq 1.67 \times 10^{-27}$ kg.

ELECTRIC FIELD AND POTENTIAL

If the source charge q_1 **(2)** is moving, this information is communicated to q at finite velocity with the force a complex function of the previous history of q_1. There is no reason to expect that the force on q will even be equal and opposite to the reaction force on q_1 as Newton supposed. For the moment, we can avoid all these difficulties by describing the force on the stationary charge q in terms of an electric *field* $\mathbf{E}$ such that the force on q is *always* given by

$$\blacktriangleright \qquad\qquad \mathbf{F} = q\mathbf{E} \qquad\qquad (3)$$

If q_1 is at rest the field will be given by:

$$\mathbf{E} = \frac{1}{4\pi\epsilon_0}\frac{q_1}{r_{01}{}^2}\hat{\mathbf{r}}_{01} \qquad\qquad (4)$$

but not otherwise. The calculation of the fields of moving charges will occupy us later.

2 **Superposition.** Coulomb's law as given by **(2)** expresses that the force on a test charge is proportional to the amount of source charge and of test charge but does not indicate the force to be expected from an *extended* source distribution of charge. For this we must rely on further experiment, which establishes that electric forces superpose vectorially. For a distribution of charges q_j at positions $\mathbf{r}_j$, the electric field at $\mathbf{r}$ will then be given by:

$$\mathbf{E}(\mathbf{r}) = \frac{1}{4\pi\epsilon_0}\sum_j \frac{q_j}{r_{0j}{}^2}\hat{\mathbf{r}}_{0j} \qquad\qquad (5)$$

We may go to extended charge distributions by identifying q_j with $\rho(\mathbf{r}_j)\,dV_j$ and integrate to obtain:

$$\blacktriangleright \qquad \mathbf{E}(\mathbf{r}) = \frac{1}{4\pi\epsilon_0}\int_{V_1} \frac{\rho(\mathbf{r}_1)}{r_{01}{}^2}\hat{\mathbf{r}}_{01}\,dV_1 \qquad\qquad (6)$$

We emphasize that $\mathbf{r}_{01} = \mathbf{r} - \mathbf{r}_1$ is directed *from* the variable source position $\mathbf{r}_1$ *toward* the field position $\mathbf{r}$.

Exercises

3. Charge is distributed with lineal density λ on the arc of a circle of radius R and subtends an angle 2θ at the center. Find the electric field at the *center*. Imagine now that all the charge were concentrated at a point. Find the position of the charge that gives the same field at the center.
4. Charge is distributed with lineal density λ on a circle of radius R. Find the electric field on the axis of the circle at a distance r from the plane of the circle. Find the position of the charge $q = 2\pi R\lambda$, which gives the same field at r.

3 Work and Potential. The fact that the work done on moving a mass through a gravitational field is independent of the path taken makes it possible to introduce the

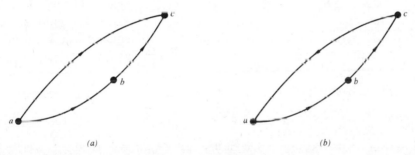

(a)   (b)

Figure 2 Work done in carrying a test charge. (a) The work done in carrying a test charge from a to b and then from b to c is the same as the work done in carrying the charge directly from a to c. (b) The total work done in carrying a test charge around a closed contour must then be zero.

gravitational potential as the gravitational energy per unit mass. In this section we obtain similar results for electric fields and charges. With reference to Figure 2a we show that the amount of work done in carrying a test charge q from a to b and then from b to c along the paths indicated is the same as the work required to carry the charge directly from a to c along another path:

$$W_{ab} + W_{bc} = W_{ac} \tag{7}$$

Without any further discussion, **(7)** implies the existence of an electric potential $\phi(\mathbf{r})$ with:

$$W_{ab} = q(\phi_b - \phi_a) \tag{8}$$
$$W_{bc} = q(\phi_c - \phi_b) \tag{9}$$
$$W_{ac} = q(\phi_c - \phi_a) \tag{10}$$

where ϕ_a, for example, stands for $\phi(\mathbf{r}_a)$. Note that **(8)**, **(9)**, and **(10)** automatically satisfy **(7)**.

To establish **(7)** we compute the work done in carrying q from a to b over the contour C_{ab}:

$$W_{ab} = -q \int_{C_{ab}} \mathbf{E} \cdot d\mathbf{r} \tag{11}$$

We first evaluate **(11)** for the field **(4)** of a charge q_1 at $\mathbf{r}_1$:

$$W_{ab} = -\frac{qq_1}{4\pi\epsilon_0} \int_{C_{ab}} \frac{\hat{\mathbf{r}}_{01} \cdot d\mathbf{r}}{r_{01}^2} \tag{12}$$

As may be seen from Figure 3 we may simplify **(12)** through the relation:

$$\hat{\mathbf{r}}_{01} \cdot d\mathbf{r} = dr_{01} \tag{13}$$

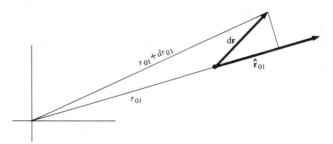

Figure 3   Vector relation between $d\mathbf{r}$ and the unit vector $\hat{\mathbf{r}}_{01}$. For $d\mathbf{r}$ small, dr_{01} is given by **(13)**.

We may then perform the integration in **(12)** to obtain:

$$W_{ab} = \frac{qq_1}{4\pi\epsilon_0} \left(\frac{1}{r_{b1}} - \frac{1}{r_{a1}} \right) \tag{14}$$

For a more general field we write **(5)** in place of **(4)**. The field of each charge may be integrated separately to obtain:

$$W_{ab} = \frac{q}{4\pi\epsilon_0} \sum_j q_j \left(\frac{1}{r_{bj}} - \frac{1}{r_{aj}} \right) \tag{15}$$

Similarly, for the other two segments we must obtain:

$$W_{bc} = \frac{q}{4\pi\epsilon_0} \sum_j q_j \left(\frac{1}{r_{cj}} - \frac{1}{r_{bj}} \right) \tag{16}$$

$$W_{ac} = \frac{1}{4\pi\epsilon_0} \sum_j q_j \left(\frac{1}{r_{cj}} - \frac{1}{r_{aj}} \right) \tag{17}$$

Substituting **(15)** and **(16)** into **(7)** we obtain **(17)**. By **(8)**, **(9)**, and **(10)** the electrostatic potential must be given to within an additive constant by:

$$\phi(\mathbf{r}) = \frac{1}{4\pi\epsilon_0} \sum_j \frac{q_j}{r_{0j}} \qquad (18)$$

For a continuous distribution of charge we write, from **(18)**,

$$\phi(\mathbf{r}) = \frac{1}{4\pi\epsilon_0} \int_{V_1} \frac{\rho(\mathbf{r}_1)}{r_{01}} dV_1 \qquad (19)$$

Finally, we use **(8)** and **(11)** to obtain the field, given the potential. Eliminating W_{ab} between **(8)** and **(11)** and simplifying, we may write:

$$\phi_b = \phi_a - \int_{C_{ab}} \mathbf{E} \cdot d\mathbf{r} \qquad (20)$$

Now, from **(B-29)** we have

$$\phi_b = \phi_a + \int_{C_{ab}} d\mathbf{r} \cdot \mathbf{grad}\ \phi \qquad (21)$$

Since the contour is completely arbitrary we have by comparing **(20)** and **(21)**:

$$\mathbf{E} = -\mathbf{grad}\ \phi \qquad (22)$$

Exercises

5. Prove the identity:

$$d\left(\frac{1}{r_{0j}}\right) = d\mathbf{r} \cdot \mathbf{grad}\ \frac{1}{r_{0j}} = -\frac{d\mathbf{r} \cdot \hat{\mathbf{r}}_{0j}}{r_{0j}^2}$$

6. A disc of radius R carries a charge density σ. Show that the potential at the center of the disc is $\sigma R / 2\epsilon_0$.

7. For $\mathbf{E} = -\nabla\phi$, show that $\mathbf{E}$ is normal to a surface of constant ϕ.

4 Circulation and Curl. The work required to carry q around the contour shown in Figure 2b will be given by:

$$\oint dW = W_{ab} + W_{bc} + W_{ca} \qquad (23)$$

But we have, from **(10)** or **(17)**,

$$W_{ca} = -W_{ac}$$

which gives with **(7)** that the total work done in taking q around a closed path is equal to zero:

$$\oint dW = -q \oint \mathbf{E} \cdot d\mathbf{r} = 0 \tag{24}$$

The circulation of **E** is defined **(B-54)** as:

▶
$$\mathscr{E} = \oint \mathbf{E} \cdot d\mathbf{r} \tag{25}$$

Comparing **(25)** and **(24)** we have:

$$\mathscr{E} = -\frac{1}{q} \oint dW = 0 \tag{26}$$

From **(B-56)** we may write for the component of the curl normal to $\Delta \mathbf{S}$:

$$\hat{\mathbf{n}} \cdot \mathbf{curl}\, \mathbf{E} = \lim \frac{\Delta \mathscr{E}}{\Delta S} \tag{27}$$

But since the right side of **(27)** is identically zero for all $\Delta \mathbf{S}$ we must have:

▶
$$\mathbf{curl}\, \mathbf{E} = \nabla \times \mathbf{E} = 0 \tag{28}$$

Since both **(22)** and **(28)** are consequences of **(7)**, we may expect that they are equivalent statements. To obtain **(28)** directly we simply take the curl of **(22)**:

$$\mathbf{curl}\, \mathbf{E} = \nabla \times \mathbf{E} = \mathbf{curl}\,(\mathbf{grad}\,\phi) = \nabla \times (\nabla\phi) = 0 \tag{29}$$

which is **(B-8)**.

To show that **(28)** implies **(22)** requires Helmholtz's theorem, Section B-15.

Exercise

8. Find the elements of the tensor $\nabla\mathbf{E}$. By using **(28)** show that $\nabla\mathbf{E}$ is a symmetric tensor.

ELECTRIC FLUX

We imagine a source charge q_1 at $\mathbf{r}_1$ and we wish to compute the flux (Section B-6) of $\mathbf{E}$ passing through a surface $d\mathbf{S} = \hat{\mathbf{n}}\, dS$. The differential of flux is given by:

$$d\Psi = \mathbf{E} \cdot d\mathbf{S} = \frac{q_1}{4\pi\epsilon_0} \frac{\hat{\mathbf{f}}_{01} \cdot d\mathbf{S}}{r_{01}{}^2} \tag{30}$$

5 Solid Angle. Let us imagine a sphere of radius R on which we identify a segment of surface ΔS as shown in Figure 4a. By tracing the edge of ΔS with the radius from the origin we trace out a conical region. We define the solid angle enclosed by the conical surface as:

$$\Delta\Omega = \frac{\Delta S}{R^2} \tag{31}$$

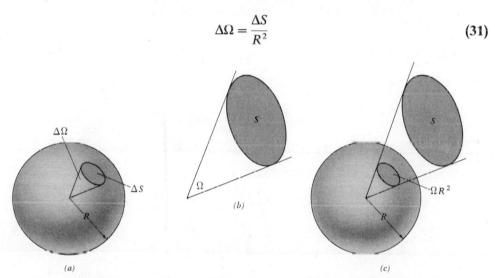

Figure 4   Solid angle Ω. (a) The incremental solid angle $\Delta\Omega$ is related by (31) to the area ΔS on the surface of a sphere of radius R. (b) For some surface S a solid angle Ω is subtended. (c) The angle Ω is obtained from the surface intercepted on a concentric sphere of radius R.

We note that the full solid angle about a point is $4\pi R^2/R^2 = 4\pi$.

To find the solid angle Ω subtended by a surface S as shown in Figure 4b we may find the area ΩR^2 intercepted on a sphere of radius R as shown in Figure 4c. The solid angle subtended by an increment of area dS is then given by:

$$d\Omega = \frac{\cos\theta\, dS}{r^2} = \frac{\hat{\mathbf{r}} \cdot d\mathbf{S}}{r^2} \tag{32}$$

where $\cos\theta\, dS$ is the area projected on a plane normal to $\hat{\mathbf{r}}$. To obtain Ω we integrate (32) over the surface:

$$\Omega = \int_S d\Omega = \int_S \frac{\hat{\mathbf{r}} \cdot d\mathbf{S}}{r^2} \tag{33}$$

As an example of the calculation of solid angle we find the solid angle subtended at a distance L on the axis from a disc of radius R as shown in Figure 5. From **(32)** the increment of solid angle of an area $dS = \rho \, d\phi \, d\rho$ is given by:

$$d\Omega = \frac{\rho \, d\phi \, d\rho \cos \theta}{r^2} = \frac{L \tan \theta \, d\phi \, L \sec^2 \theta \, d\theta \cos \theta}{L^2/\cos^2 \theta}$$

$$= \sin \theta \, d\theta \, d\phi \tag{34}$$

which is the general expression for the increment of solid angle in spherical coordinates as discussed in Section B-12. We integrate over ϕ from 0 to 2π and over θ from 0 to $\tan^{-1} R/L$:

$$\Omega = \int_0^{2\pi} d\phi \int_0^{\tan^{-1} R/L} \sin \theta \, d\theta = -2\pi \cos \theta \Big|_0^{\tan^{-1} R/L}$$

$$= 2\pi \left[1 - \frac{L}{(L^2 + R^2)^{1/2}} \right] \tag{35}$$

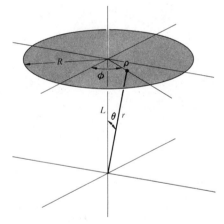

Figure 5   Solid angle subtended by a disc of radius R. The computed result is given by **(35)**.

Exercises

9. Examine **(35)** in the limits $L \ll R$ and $L \gg R$. Interpret these limits.
10. Compute the solid angle subtended at the center by the sides of a cylinder of length $2L$ and radius R. Compare your answer with **(35)** using the fact that the total solid angle must equal 4π.
11. Find the area on a sphere of radius R bounded by the intersection of the sphere and a plane a distance $L \leq R$ from the center of the sphere.

6   Gauss's Law. By comparing **(32)** with **(30)** we see that it is possible to write the contribution to the flux in terms of the solid angle subtended by ΔS at q_1:

$$\Delta \Psi = \frac{q_1}{4\pi\epsilon_0} \Delta\Omega_1 \tag{36}$$

We note that **(36)** is strictly a consequence of the inverse square law of force. If the surface S encloses q_1 as in Figure 6a, the total flux is given by:

$$\Psi = \frac{q_1}{4\pi\epsilon_0} \oint_S d\Omega_1 = \frac{q_1}{4\pi\epsilon_0} \times 4\pi = \frac{q_1}{\epsilon_0} \tag{37}$$

If on the other hand the surface does not enclose q_1, as in Figure 6b the solid angle subtended by the *entire* surface S will be zero:

$$\Psi = \frac{q_1}{4\pi\epsilon_0} \oint_S d\Omega_1 = 0 \tag{38}$$

Now if there is more than one charge within S, or, more generally, if there is distribution of charge, we have, by superposition, *Gauss's law*:

$$\blacktriangleright \qquad \Psi = \oint_S \mathbf{E} \cdot d\mathbf{S} = \frac{1}{\epsilon_0} \sum_j q_j = \frac{1}{\epsilon_0} \int_V \rho(\mathbf{r})\, dV \tag{39}$$

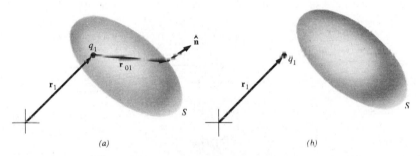

(a) (b)

Figure 6 Intercepted electric flux. (*a*) If the surface S encloses q_1, the total flux is given by **(37)**. (*b*) If the closed surface S does not enclose q_1, then the intercepted flux is zero.

where the volume integration is taken over the entire region enclosed by S. We may apply Gauss's theorem **(B-11)** to the electric field $\mathbf{E}$ to obtain:

$$\Psi = \int_V \mathbf{\nabla} \cdot \mathbf{E}\, dV = \frac{1}{\epsilon_0} \int_V \rho(\mathbf{r})\, dV \tag{40}$$

Since V is a completely arbitrary volume, we must have, at each point,

$$\blacktriangleright \qquad \mathbf{\nabla} \cdot \mathbf{E} = \frac{\rho}{\epsilon_0} \tag{41}$$

that is, the net outgoing flux of $\mathbf{E}$ per unit volume is given by ρ/ϵ_0. Substituting **(22)** into **(41)** we obtain:

$$\mathbf{\nabla} \cdot (\mathbf{\nabla}\phi) = \nabla^2\phi = -\frac{\rho}{\epsilon_0} \tag{42}$$

That is, the Laplacian **(B-64)** of the potential is equal to minus the flux per unit volume, ρ/ϵ_0. This is called *Poisson's equation*. In a charge-free region we have:

$$\nabla^2 \phi = 0 \tag{43}$$

which is called *Laplace's equation*.

Exercises

12. Show that the scalar of the tensor $\nabla \mathbf{E}$ (Section C-6) is equal to ρ/ϵ_0.
13. Find the flux through a circle of radius R from a charge q on the axis of the circle and at a distance L from the center. Compare with **(35)**.
14. A right circular cylinder of length L and radius R has a charge q at its center. Find the flux emerging through the side.

7 **Dirac Delta Function.** We may use **(42)** for discrete charges through the use of the Dirac delta function.[4] For a point charge q at the origin we write:

$$\rho(\mathbf{r}) = q \, \delta(\mathbf{r}) \tag{44}$$

The function $\delta(\mathbf{r})$ is zero except when its argument is zero. At $\mathbf{r} = 0$ the delta function is singular but integrable with:

$$\int_V \delta(\mathbf{r}) \, dV = 1 \tag{45}$$

as long as the origin is included within the volume of integration. From **(18)** the potential of a point charge q at the origin is:

$$\phi = \frac{1}{4\pi\epsilon_0} \frac{q}{r} \tag{46}$$

Substituting **(46)** and **(44)** into **(42)** we obtain the useful result:

$$\nabla \cdot \left(\nabla \frac{1}{r} \right) = \nabla^2 \frac{1}{r} = -4\pi\delta(\mathbf{r}) \tag{47}$$

Another way of visualizing **(47)** is in terms of a velocity field as shown in Section B-10.

---

[4] For a discussion of the use of the Dirac delta function for other charge singularities, see V. Namias, *Am. J. Phys.* **45**, 624 (1977).

8 **Gaussian Surfaces.** The reason that the use of Gauss's law **(39)** offers such a powerful method for finding electric fields is that we are able to incorporate the symmetry of the problem into the choice of the Gaussian surface. We consider several examples:

1. We imagine that the charge density $\rho(\mathbf{r})$ is a function only of r. Then the electric field *must* be radial and will be uniform over a spherical surface centered at the origin as in Figure 7a.

2. For a two-dimensional charge density $\rho(\mathbf{r}) = \rho(r^2 - z^2)$, the electric field will be radial in the xy plane and will be uniform over a cylindrical surface coaxial with the z axis as shown in Figure 7b.

3. For a one-dimensional charge density $\rho(\mathbf{r}) = \rho(x)$, the electric field will be directed along $\hat{\mathbf{x}}$ and will be independent of y and z. The appropriate Gaussian surface for this case is a right cylinder as shown in Figure 7c.

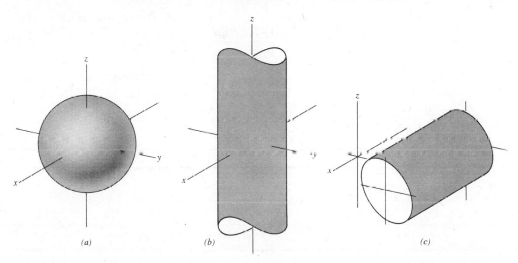

(a)   (b)   (c)

Figure 7 Gaussian surfaces. (*a*) Where the charge density has spherical symmetry, the Gaussian surface is a sphere concentric with the center of symmetry. (*b*) Where the charge density has cylindrical symmetry, the Gaussian surface is a cylinder coaxial with the axis of symmetry. (*c*) Where the charge density is a function of x, the Gaussian surfaces are planes normal to the x axis.

Exercises

15. An infinite circular cylinder of radius R carries a surface charge σ. By use of Gauss's law find the electric field outside. Show that the electric field inside the cylinder is zero. How much work must be done to carry a test charge q from r_1 to r_2, both greater than R?

16. Substitute **(47)** into Gauss's theorem **(B-11)** and show that the theorem is satisfied both for the case where S includes the origin and where the origin is outside S.

17. An infinite plane surface carries a uniform charge density σ. From Gauss's law, find the electric field above and below the surface. Obtain the field by direct integration.

18. Two plane parallel infinite surfaces are charged to $+\sigma$ and $-\sigma$. Find the electric field between and outside the surfaces.

19. A uniformly charged spherical shell contains a small circular hole. Find the electric field at the center of the hole.

9 Spherical Charge Distributions. We wish to obtain expressions for the electric field outside and inside a uniformly charged spherical shell. We first solve the problem by direct integration and then by Gauss's law, the use of which results in a great simplification. Finally, we obtain expressions for the field and potential in the vicinity of a uniformly charged sphere.

We show in Figure 8 a section through a uniformly charged spherical shell for which we want the electric field at the exterior point **r**. We intercept two portions of the shell of area dS_1 and dS_2 with solid angle $d\Omega$ as shown. We write for the contribution to the field at **r** from dq_1 and dq_2:

$$d\mathbf{E} = \frac{1}{4\pi\epsilon_0}\left(\frac{dq_1}{r_{01}{}^2} + \frac{dq_2}{r_{02}{}^2}\right)\hat{\mathbf{r}}_{01} \tag{48}$$

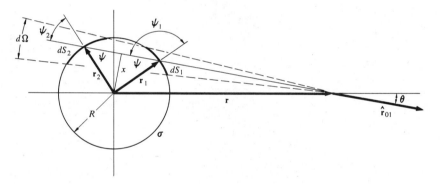

Figure 8 Section through a uniformly charged spherical shell. The electric field at the exterior point **r** is given by **(56)**.

The incremental charges dq_1 and dq_2 may be written as

$$dq_1 = \sigma\, dS_1 = \sigma\,\frac{r_{01}{}^2\, d\Omega_1}{\cos\psi_1} \qquad dq_2 = \sigma\, dS_2 = \sigma\,\frac{r_{02}{}^2\, d\Omega_2}{\cos\psi_2} \tag{49}$$

Substituting into **(48)** we obtain, with **(33)**,

$$d\Omega_2 = -d\Omega_1 = \sin\theta\, d\theta\, d\phi \tag{50}$$

and

$$\cos\psi_2 = -\cos\psi_1 = \cos\psi \tag{51}$$

$$d\mathbf{E} = \frac{\sigma}{2\pi\epsilon_0}\,\frac{\sin\theta\, d\theta\, d\phi}{\cos\psi}\,\hat{\mathbf{r}}_{01} \tag{52}$$

Integrating $d\phi$ through 2π about the **r** axis, we obtain the projection of $\hat{\mathbf{r}}_{01}$ onto the axis:

$$\int_0^{2\pi} \hat{\mathbf{r}}_{01}\, d\phi = 2\pi \cos \theta \hat{\mathbf{r}} \qquad (53)$$

We finally may write for the total field at **r**:

$$\mathbf{E} = \frac{\sigma}{\epsilon_0} \hat{\mathbf{r}} \int \frac{\sin \theta\, d \sin \theta}{\cos \psi} \qquad (54)$$

From Figure 8 we may reexpress the angle θ in terms of ψ:

$$x = R \sin \psi = r \sin \theta \qquad (55)$$

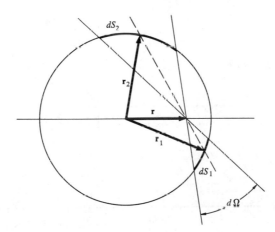

Figure 9 Section through a uniformly charged spherical shell. The electric field at an interior point is zero since the contributions from dS_1 and dS_2 cancel in pairs.

Integrating over ψ between 0 and $\pi/2$ we obtain:

$$\mathbf{E} = \frac{\sigma}{\epsilon_0} \frac{R^2}{r^2} \hat{\mathbf{r}} \int_0^{\pi/2} \sin \psi\, d\psi = \frac{1}{4\pi\epsilon_0} \frac{q}{r^2} \hat{\mathbf{r}} \qquad (56)$$

where we have used the relation $q = 4\pi R^2 \sigma$ to express the field in terms of the total charge on the sphere. We observe that **(56)** is exactly the same as if all the charge on the sphere were concentrated at the origin. To obtain the field inside the sphere we move the field position to the interior as shown in Figure 9. Now we have, in place of **(48)**,

$$d\mathbf{E} = \frac{1}{4\pi\epsilon_0} \left(\frac{dq_1}{r_{01}{}^2} - \frac{dq_2}{r_{02}{}^2} \right) \hat{\mathbf{r}}_{01} = 0 \qquad (57)$$

where dq_1 and dq_2 are still given by **(49)**. For the field point outside the sphere, the contributions from the two segments were equal and added. Within the spherical shell the two contributions cancel exactly. Since all segments cancel in pairs, we have the result that the field within the sphere is everywhere zero.

As an application of Gauss's law **(39)**, we recompute the field for a uniformly charged spherical shell. We sketch in Figure 10 a shell of radius R and surface charge σ. We wish to compute the electric field at position **r**. By substituting $\sigma \, dS_1$ for $\rho(\mathbf{r}_j) \, dV_j$ in **(6)**, we obtain for the field:

$$\mathbf{E} = \frac{1}{4\pi\epsilon_0} \int_{S_1} \frac{\sigma}{r_{01}^2} \, \hat{\mathbf{r}}_{01} \, dS_1 \tag{58}$$

Although the integration of **(58)** is cumbersome it can be done, as we have seen.

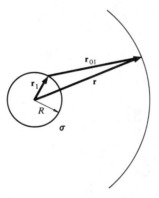

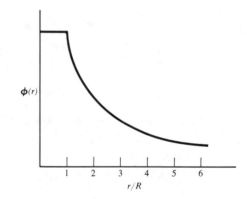

Figure 10   A shell of radius R and uniform surface charge density σ. The field is obtained through the use of a spherical Gaussian surface of radius r.

Figure 11   Potential of a uniformly charged spherical shell. The potential is constant within the shell.

A more elegant way of finding **E** is to observe that **E** must be normal to the surface of a concentric sphere of radius r and must be uniform over the surface of the sphere. Substituting into **(39)** we obtain for $r > R$:

$$E \times 4\pi r^2 = \frac{1}{\epsilon_0} 4\pi R^2 \sigma \quad \text{and} \quad \mathbf{E} = \frac{\sigma R^2 \hat{\mathbf{r}}}{\epsilon_0 \, r^2} = \frac{q\hat{\mathbf{r}}}{4\pi\epsilon_0 \, r^2} \tag{59}$$

For $r < R$:

$$E \times 4\pi r^2 = 0 \quad \text{and} \quad \mathbf{E} = 0 \tag{60}$$

We note that the vanishing of the field within a spherical charge distribution is strictly a consequence of the inverse square law for the electric field and thus provides a sensitive test of this law.[5] This was the approach taken by Cavendish some 10 years before Coulomb in largely unpublished research. In his treatise Maxwell recognized the elegance and precision of the Cavendish experiment.

The potential may be computed from **(20)** since $\mathbf{E}$ is everywhere radial. The potential may be written from **(19)** by substituting $\sigma \, dS_1$ for $\rho \, dV_1$:

$$\phi(\mathbf{r}) = \frac{1}{4\pi\epsilon_0} \int_{S_1} \frac{\sigma}{r_{01}} \, dS_1 \tag{61}$$

but may be obtained more simply from **(20)**. For $r \geq R$ we obtain:

$$\phi(r) = \frac{\sigma R^2}{\epsilon_0 \, r} = \frac{q}{4\pi\epsilon_0 \, r} \tag{62}$$

For $r \leq R$ the potential is constant since the interior field is zero:

$$\phi(r) = \phi(R) = \frac{\sigma R}{\epsilon_0} \tag{63}$$

We sketch $\phi(r)$ in Figure 11.

We next consider the field of a uniform spherical *volume* distribution of charge as shown in Figure 12. Again we may use **(6)** with the understanding that $\rho(\mathbf{r}_1) = \rho$ for dV_1 within the spherical volume and $\rho(\mathbf{r}_1) = 0$ otherwise. And as for the spherical shell we do better to use Gauss's law **(39)** and to write for $r \geq R$:

$$E \times 4\pi r^2 = \frac{4\pi R^3 \rho}{3\epsilon_0} \qquad \text{or} \qquad \mathbf{E} = \frac{\rho R^3}{3\epsilon_0 \, r^2} \hat{\mathbf{r}} = \frac{q}{4\pi\epsilon_0 \, r^2} \hat{\mathbf{r}} \tag{64}$$

For $r \leq R$ only a fraction of the charge is included within r:

$$E \times 4\pi r^2 = \frac{4\pi r^3 \rho}{3\epsilon_0} \qquad \text{or} \qquad \mathbf{E} = \frac{1}{3\epsilon_0} \frac{q\mathbf{r}}{V} = \frac{1}{3\epsilon_0} \rho\mathbf{r} \tag{65}$$

The potential is found from **(19)** and has the values:

$$\phi(r) = \frac{1}{4\pi\epsilon_0} \frac{q}{r} = \frac{\rho R^3}{3\epsilon_0 \, r} \qquad \text{for} \qquad r \geq R \tag{66}$$

[5] R. P. Feynman, R. B. Leighton, and M. Sands, *The Feynman Lectures on Physics*, Addison-Wesley, 1964, Section II:5-8. L. P. Fulcher and M. A. Telljohann, *Am. J. Phys.* **44**, 366 (1976). E. M. Purcell, *Electricity and Magnetism*, Berkeley physics course, Volume 2, McGraw-Hill, 1965, Section 1.4. For a discussion of the relation between the inverse square law and photon mass, see J. D. Jackson, *Classical Electrodynamics*, second edition, Wiley, 1975, Section I.2. Also see A. S. Goldhaber and M. M. Nieto, "The Mass of the Photon," *Scientific American*, **234**(5), 86 (May 1976).

and

$$\phi(r) = \frac{\rho R^2}{2\epsilon_0} - \frac{\rho r^2}{6\epsilon_0} \qquad \text{for} \qquad r \le R \tag{67}$$

As shown in Figure 13, the potential drops parabolically out to $r = R$ and then as $1/r$ for $r > R$.

Exercise

---

20. Charge is distributed about a point with spherical symmetry:

$$\rho = \rho(r)$$

Show that the electric field is radial and given by:

$$E(r) = \frac{1}{\epsilon_0 r^2} \int_0^r \rho(r')r'^2 \, dr'$$

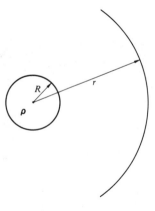

Figure 12 A sphere of radius R and uniform volume charge density ρ. The field is obtained through the use of a spherical Gaussian surface of radius r.

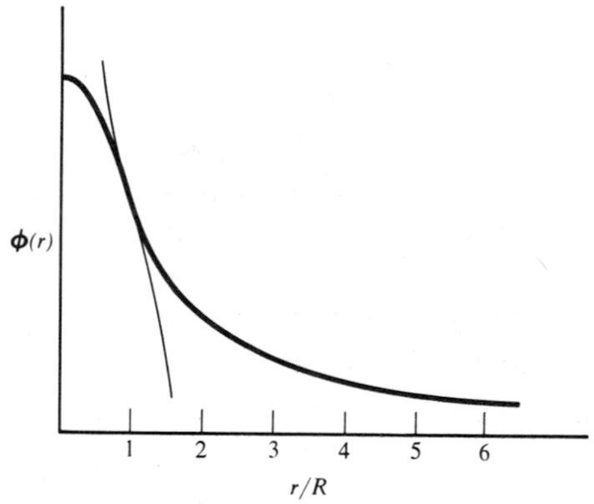

Figure 13 Potential of a uniformly charged sphere. The potential must be computed separately for $r \le R$ and for $r \ge R$. The two segments join with equal slope at $r = R$.

MOMENTS OF CHARGE DISTRIBUTIONS

Let us imagine that we wish to find the potential and field at some distance from a *limited* charge distribution $\rho(\mathbf{r}_1)$. Of course, **(19)** gives the potential for any charge distribution so that in principle we might think of repeatedly computing **(19)** at each position. As we will see, such a cumbersome procedure is not necessary if we are far enough away from the charges. Instead, we can characterize the distribution by its moments and write down in compact form the contribution that each moment makes to the potential and thus to the field. We emphasize that this procedure is useful only if we are well removed from the charges.

10   Expansion of the Potential. We show in Figure 14 the situation that we have in mind. We expand $1/r_{01}$ about an origin located within or close to the charge distribution. We write with the binomial expansion:

$$\frac{1}{r_{01}} = (r^2 - 2\mathbf{r} \cdot \mathbf{r}_1 + r_1^2)^{-1/2}$$

$$= (r^2)^{-1/2} - \tfrac{1}{2}(r^2)^{-3/2}(-2\mathbf{r} \cdot \mathbf{r}_1 + r_1^2) + \tfrac{3}{8}(r^2)^{-5/2}(-2\mathbf{r} \cdot \mathbf{r}_1 + r_1^2)^2 + \cdots \qquad \textbf{(68)}$$

Regrouping and retaining terms only to order $1/r^3$ we obtain, for **(68)**,

$$\frac{1}{r_{01}} = \frac{1}{r} + \frac{r_1}{r^2}(\hat{\mathbf{r}} \cdot \hat{\mathbf{r}}_1) + \frac{1}{2}\frac{r_1^2}{r^3}[3(\hat{\mathbf{r}} \cdot \hat{\mathbf{r}}_1)^2 - 1] + \cdots \qquad \textbf{(69)}$$

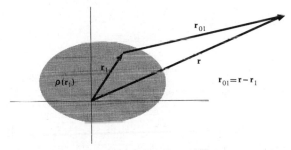

Figure 14   A limited charge distribution $\rho(\mathbf{r}_1)$. The potential at a distant position $\mathbf{r}$ may be expressed in terms of the multipole moments of the charge distribution.

The form of **(69)** suggests an expansion where a typical term is of the form r_1^l/r^{l+1} times an angular factor. Substituting into **(19)** we may write:

$$\phi(\mathbf{r}) = \phi_0(\mathbf{r}) + \phi_1(\mathbf{r}) + \phi_2(\mathbf{r}) + \cdots \qquad \textbf{(70)}$$

with

$$\phi_0(\mathbf{r}) = \frac{1}{4\pi\epsilon_0 r}\int_{V_1} \rho(\mathbf{r}_1)\, dV_1 \qquad \textbf{(71)}$$

$$\phi_1(\mathbf{r}) = \frac{\hat{\mathbf{r}}}{4\pi\epsilon_0 r^2} \cdot \int_{V_1} \mathbf{r}_1\rho(\mathbf{r}_1)\, dV_1 \qquad \textbf{(72)}$$

$$\phi_2(\mathbf{r}) = \frac{1}{8\pi\epsilon_0 r^3}\int_{V_1} r_1^2[3(\hat{\mathbf{r}} \cdot \hat{\mathbf{r}}_1)^2 - 1]\rho(\mathbf{r}_1)\, dV_1 \qquad \textbf{(73)}$$

Now ϕ_0 is just the potential that we would calculate if all the charge in the distribution were situated at the origin. The total charge:

$$q = \int_{V_1} \rho(\mathbf{r}_1)\, dV_1 \qquad \textbf{(74)}$$

is called the monopole moment of the distribution. The potential $\phi_1(\mathbf{r})$ arises from what is called the dipole moment of the distribution, defined as:

$$\mathbf{p} = \int_{V_1} \mathbf{r}_1 \rho(\mathbf{r}_1) \, dV_1 \tag{75}$$

The potential $\phi_2(\mathbf{r})$ arises from what is called the quadrupole moment of the distribution. The quadrupole moment is characterized by five independent components (rather than three for the dipole or one for the monopole). We may write it most compactly in terms of tensor products (Section A-10). The quadrupole tensor of the distribution is written as:

$$\mathbf{Q} = \int r_1^2 (3\hat{\mathbf{r}}_1 \hat{\mathbf{r}}_1 - \mathbf{I}) \rho(\mathbf{r}_1) \, dV_1 \tag{76}$$

Finally we may write (70) as a series expansion:

$$\phi(\mathbf{r}) = \frac{1}{4\pi\epsilon_0} \frac{q}{r} + \frac{1}{4\pi\epsilon_0} \frac{\hat{\mathbf{r}} \cdot \mathbf{p}}{r^2} + \frac{1}{8\pi\epsilon_0} \frac{\hat{\mathbf{r}} \cdot \mathbf{Q} \cdot \hat{\mathbf{r}}}{r^3} + \cdots \tag{77}$$

At large distance the dominant contribution to the potential will be from the monopole moment with a first-order correction from the dipole moment of the charge distribution. If the distribution is electrically neutral the monopole contribution vanishes and we have the dominant contribution from the dipole moment with a correction from the quadrupole moment. Following the quadrupole moment we may define an octupole moment and then a hexadecapole moment, and so on. In our discussion of dielectric materials in Chapter 3, we are particularly interested in the dipole moments of charge distributions. Quadrupole and higher moments are commonly used in characterizing the charge distributions of atomic nuclei.

Exercises

21. Develop (69) by Taylor expansion:

$$f(\mathbf{r},\mathbf{r}_1) = f(\mathbf{r},0) + \mathbf{r}_1 \cdot \mathbf{grad}_1 \, f(\mathbf{r},0) + \tfrac{1}{2}\mathbf{r}_1 \cdot \mathbf{grad}_1 \mathbf{grad}_1 \, f(\mathbf{r},0) \cdot \mathbf{r}_1 + \cdots$$

22. Show that the sum of the diagonal components of the quadrupole matrix is zero. Show that the quadrupole tensor is symmetric and that five independent coefficients are required to characterize the quadrupole moment.

23. Charge is distributed with spherical symmetry and bounded: $\rho(\mathbf{r}) = \rho(r)$. Show that *all* moments except the monopole vanish identically.

24. Show that the dipole moment of a distribution of charges is independent of origin as long as the total charge is zero.

11 **Discrete Charges.** We imagine a finite number of discrete charges q_1 to q_N at positions $\mathbf{r}_1$ to $\mathbf{r}_N$. Then from **(74)** the monopole moment of the distribution is simply the total charge:

$$q = \int \rho(\mathbf{r}_1)\, dV_1 = \sum_{i=1}^{N} q_i \tag{78}$$

The dipole moment of the distribution is, from **(75)**,

$$\mathbf{p} = \int_{V_1} \mathbf{r}_1 \rho(\mathbf{r}_1)\, dV_1 = \sum_{i=1}^{N} \mathbf{r}_i\, q_i \tag{79}$$

The quadrupole moment similarly may be written from **(76)**:

$$\mathbf{Q} = \sum_{i=1}^{N} r_i^2 (3\hat{\mathbf{r}}_i \hat{\mathbf{r}}_i - \mathbf{I}) q_i \tag{80}$$

Exercises

25. Write the potential of a pair of charges $+q$ and $-q$ at $+\frac{1}{2}a\hat{\mathbf{z}}$ and $-\frac{1}{2}a\hat{\mathbf{z}}$, respectively. Expand the potential for r/a large and identify the monopole, dipole, and quadrupole moments.

26. Charges $+q$ and $-q$ are placed alternately at the corners of a square. Compute the elements of the quadrupole tensor for axes parallel to the sides of the square. Repeat the calculation for axes parallel to the diagonals of the square.

27. Charges $+q$ are placed at the vertices of a regular tetrahedron and a charge $-4q$ at the center. Show that the monopole, dipole, and quadrupole moments are zero.

12 **Uniformly Charged Ellipsoid.** As a second example we find the moments of an ellipsoid of uniform charge density ρ and semiaxes, A, B, and C along x, y, and z as shown in Figure 15.

The monopole moment of the distribution is from **(74)**:

$$q = \int_V \rho\, dV = \frac{4\pi}{3}\rho ABC \tag{81}$$

The dipole moment taken with respect to the center of the ellipsoid is clearly zero since, for each dV at $\mathbf{r}$, there is a corresponding dV at $-\mathbf{r}$ and these two contributions to the dipole moment cancel in pairs. (We may generalize this statement to any distribution of charge that has a center of symmetry.)

The quadrupole moment, which is given by **(76)**, will have two kinds of terms. Terms that would be off the diagonal of the corresponding matrix:

$$Q_{ij} = \int (3x_i x_j - r^2\, \delta_{ij}) \rho(r)\, dV \tag{82}$$

involve an integral like

$$Q_{xy} = 3\rho \int xy \, dV \tag{83}$$

But since the ellipsoid has reflection symmetry, for each dV at (x,y) there is a corresponding dV at $(-x,y)$ and these contributions cancel in pairs. We are left with diagonal terms of the form:

$$Q_{xx} = \rho \int (3x^2 - r^2) \, dV \tag{84}$$

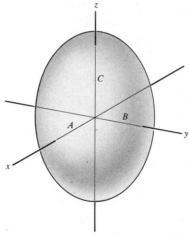

Figure 15   Ellipsoid of uniform charge density ρ. The semiaxes A,B, and C are along x,y, and z.

If we were integrating over a sphere of radius R **(84)** would be zero because, for a spherical volume, we have

$$\langle x^2 \rangle = \langle y^2 \rangle = \langle z^2 \rangle = \tfrac{1}{3}\langle r^2 \rangle = \frac{R^2}{5} \tag{85}$$

For an ellipse we do not have **(85)** and may expect to obtain a nonvanishing quadrupole moment. We may expand **(84)** into an integral of the form:

$$Q_{xx} = \rho \int (2x^2 - y^2 - z^2) \, dx \, dy \, dz \tag{86}$$

where the integral is over an ellipsoidal volume. We now make the change of variables:

$$x' = \frac{x}{A} \qquad y' = \frac{y}{B} \qquad z' = \frac{z}{C} \tag{87}$$

Then **(86)** becomes for the principal value along x:

$$Q_x = \rho ABC \int (2A^2 x'^2 - B^2 y'^2 - C^2 z'^2)\, dV \qquad (88)$$

But the integral of **(88)** is now over a sphere of unit radius and we may use **(81)** and **(85)** to write:

$$Q_x = \tfrac{1}{5} q(2A^2 - B^2 - C^2) \qquad (89)$$

Similarly, we must have for the other principal values:

$$Q_y = \tfrac{1}{5} q(2B^2 - C^2 - A^2) \qquad Q_z = \tfrac{1}{5} q(2C^2 - A^2 - B^2) \qquad (90)$$

Note that in the spherical limit $A = B = C = R$ the entire quadrupole moment vanishes. Note also that the trace of the quadrupole moment $Q_{xx} + Q_{yy} + Q_{zz}$ is always zero as it must be for *any* charge distribution.

Exercises

28. A charge q is distributed uniformly on a line of length L. Find the principal values of the quadrupole moment. Compare with **(89)** and **(90)**.

29. A charge q is distributed uniformly on a disc of radius R. Find the principal values of the quadrupole moment. Compare with **(89)** and **(90)**.

30. Charges q and $-q$ are uniformly distributed over coplanar concentric rings of radii a and b as shown. Show that the principal values of the quadrupole moment are:

$$Q_x = Q_y = -\tfrac{1}{2} Q_z = \tfrac{1}{2} q(b^2 - a^2)$$

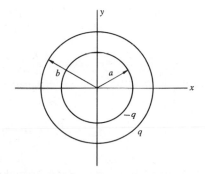

31. The quadrupole moment of the deuteron is $Q_z/e = 2.77 \times 10^{-31}$ m^2. Taking the density of nuclear matter to be 10^{18} kg/m^3 find the ratio C/A for the deuteron. Assume $A = B$.

13 **Electric Multipoles.** One may generate the potentials and fields corresponding to multipole moments using 2^l point charges. These simple structures usefully substitute for extended charge distributions.

1. The *electric monopole* $(l = 0)$ is a single charge q. Its potential, as we have seen, may be written for a charge at the origin as:

$$\phi_0(\mathbf{r}) = \frac{1}{4\pi\epsilon_0}\frac{q}{r} \tag{91}$$

and the electric field may be written as:

$$\mathbf{E}_0(\mathbf{r}) = -\nabla\phi_0(\mathbf{r}) = \frac{1}{4\pi\epsilon_0}\frac{q\hat{\mathbf{r}}}{r^2} \tag{92}$$

2. The *electric dipole* $(l = 1)$ is the limit of a pair of charges $+q$ and $-q$ separated by $\mathbf{a}$ as $\mathbf{a}$ goes to zero with the electric dipole moment $\mathbf{p} = q\mathbf{a}$ constant.

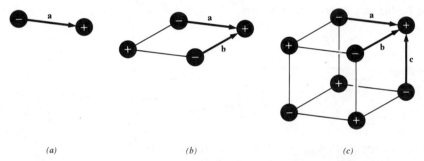

(a)   (b)   (c)

Figure 16 Electric multipoles. (*a*) A pair of charges $+q$ and $-q$ separated by a vector $\mathbf{a}$ form a dipole as $\mathbf{a}$ goes to zero with $\mathbf{p} = q\,\mathbf{a}$ constant. (*b*) A pair of dipoles $\mathbf{p}$ and $-\mathbf{p}$ form a quadrupole as $\mathbf{b}$ goes to zero with the electric quadrupole moment qab constant. (*c*) A pair of quadrupoles separated by $\mathbf{c}$ form an octupole as $\mathbf{c}$ goes to zero with $qabc$ constant.

This configuration is shown in Figure 16*a*. The potential from the two charges may be written as:

$$\phi_1(\mathbf{r}) = \phi_0(\mathbf{r}) - \phi_0(\mathbf{r} + \mathbf{a}) \tag{93}$$

In the limit that $\mathbf{a}$ goes to zero we have:

$$\phi_1(\mathbf{r}) = \phi_0(\mathbf{r}) - [\phi_0(\mathbf{r}) + \mathbf{a}\cdot\nabla\phi_0(\mathbf{r})] = -\mathbf{a}\cdot\nabla\phi_0(\mathbf{r}) = \mathbf{a}\cdot\mathbf{E}_0(\mathbf{r}) \tag{94}$$

Substituting from (**92**), we obtain the equivalent of (**72**):

$$\phi_1(\mathbf{r}) = \frac{q}{4\pi\epsilon_0}\frac{\mathbf{a}\cdot\hat{\mathbf{r}}}{r^2} = \frac{1}{4\pi\epsilon_0}\frac{\mathbf{p}\cdot\hat{\mathbf{r}}}{r^2} = -\frac{1}{4\pi\epsilon_0}\mathbf{p}\cdot\nabla\frac{1}{r} \tag{95}$$

The electric field is given by:

$$E_1(r) = -\nabla\phi_1(r) = \frac{1}{4\pi\epsilon_0} \frac{3\hat{r}(p \cdot \hat{r}) - p}{r^3} \tag{96}$$

3. The *electric quadrupole* ($l = 2$) shown in Figure 16b is the limit of a pair of dipoles $+p$ and $-p$ separated by b with b going to zero and the electric quadrupole moment qab constant. By analogy with the argument for the dipole potential we write:

$$\phi_2(r) = -b \cdot \nabla\phi_1(r) = b \cdot E_1(r) = \frac{q}{4\pi\epsilon_0} \frac{3\hat{r} \cdot (ab) \cdot \hat{r} - a \cdot b}{r^3} \tag{97}$$

which corresponds to **(76)** with:

$$Q = q(3ab + 3ba - 2I\,a \cdot b) \tag{98}$$

Note that **Q** must be symmetrical. The quadrupole field is given by:

$$E_2(r) = -\nabla\phi_2(r) \tag{99}$$

4. The *electric octupole* ($l = 3$) is shown in Figure 16c and is the limit of a pair of quadrupoles separated by c in the limit that c goes to zero with $qabc$ constant. The potential is given by:

$$\phi_3(r) = \quad c \cdot \nabla\phi_2(r) = c \cdot E_2(r) \tag{100}$$

and the electric field will be

$$E_3(r) = -\nabla\phi_3(r) \tag{101}$$

The hexadecapole ($l = 4$) is the limit of a pair of octupoles separated by a vector d in the limit that d goes to zero with $qabcd$ constant. In a similar way higher moments ($l = 5, 6, \ldots$) may be generated by means of vectors $e, f, \ldots$.

We note that there are no restrictions on the relative orientations of the vectors $a, b, c, d, e, f, \ldots$. Thus, for example, all the vectors may be parallel, developing a set of linear l-fold moments. While the dipole moment must be one-dimensional, the quadrupole moment will in general be two dimensional and the octupole moment three dimensional. Since the vectors $d, e, f, \ldots$ are in the space of a, b, c, the hexadecapole and higher moments are in general also three dimensional.

Exercises

32. Show that the principal directions of the quadrupole moment **(98)** are the bisectors of a and b and the normal to the plane containing a and b as shown in Figure 16b. Find the principal values of the quadrupole moment.

33. Show that the electric field of a dipole is normal to the direction of the dipole at an angle from the dipole:

$$\theta = \cos^{-1} 1/\sqrt{3} = 54.7356°$$

34. A hemispherical surface of radius R is bounded by the median plane of a dipole. Show that the flux through the spherical surface is given by:

$$\Psi = \frac{1}{2\epsilon_0} \frac{p}{R}$$

35. Show that in spherical coordinates the electric field of a dipole is of the form:

$$\mathbf{E} = \frac{p}{4\pi\epsilon_0} \frac{2\hat{\mathbf{r}} \cos\theta + \hat{\boldsymbol{\theta}} \sin\theta}{r^3}$$

36. A pair of dipoles $\mathbf{p} = p\hat{\mathbf{z}}$ are placed at $z = +a$ and $z = -a$. Sketch the field in a plane through the z axis. Locate the positions in the median plane at which the field reverses sign.

14   Forces and Torques on Electric Multipoles.

We want to have expressions for the forces on electric multipoles in terms of the exterior electric field and its derivatives.

1. The force *on* an electric monopole is proportional to the field:

$$\mathbf{F} = q\mathbf{E} \tag{102}$$

2. The force on a pair of charges $+q$ and $-q$ separated by $\mathbf{a}$ is

$$\mathbf{F} = q\mathbf{E}(\mathbf{r}) - q\mathbf{E}(\mathbf{r} - \mathbf{a}) \tag{103}$$

Now if we write for the field at $\mathbf{r} - \mathbf{a}$ (treating $\mathbf{a}$ as an infinitesimal):

$$\mathbf{E}(\mathbf{r} - \mathbf{a}) = \mathbf{E}(\mathbf{r}) - (\mathbf{a} \cdot \nabla)\mathbf{E}(\mathbf{r}) \tag{104}$$

then we have for the force:

$$\mathbf{F} = +q(\mathbf{a} \cdot \nabla)\mathbf{E}(\mathbf{r}) = (\mathbf{p} \cdot \nabla)\mathbf{E} \tag{105}$$

We may, if we wish, regard the gradient of $\mathbf{E}$ as a tensor and write (105) as:

$$\mathbf{F} = \mathbf{p} \cdot \nabla\mathbf{E} \tag{106}$$

From (106) the work to bring $\mathbf{p}$ into the field is given by:

$$W = -\int \mathbf{F} \cdot d\mathbf{r} = -\int \mathbf{p} \cdot \nabla\mathbf{E} \cdot d\mathbf{r} = -\int d\mathbf{r} \cdot \nabla\mathbf{E} \cdot \mathbf{p} = -\mathbf{p} \cdot \mathbf{E} \tag{107}$$

where we have used the property that $\mathbf{VE}$ is symmetric. In a similar way the forces on quadrupoles and higher moments may be written in terms of higher-order derivatives of the field.

In a quite similar way we may obtain an expression for the *torque* on electric multipoles:

1. The torque on an electric monopole is simply:

$$\tau = \mathbf{r} \times \mathbf{F} = q\mathbf{r} \times \mathbf{E} \tag{108}$$

2. The torque on a pair of charges $+q$ and $-q$ separated by $\mathbf{a}$ infinitesimal is, from **(108)**,

$$\tau = q\mathbf{r} \times \mathbf{E}(\mathbf{r}) - q(\mathbf{r} - \mathbf{a}) \times \mathbf{E}(\mathbf{r} - \mathbf{a}) = q\mathbf{a} \times \mathbf{E} + q\mathbf{r} \times (\mathbf{a} \cdot \mathbf{V})\mathbf{E}$$

$$= \mathbf{p} \times \mathbf{E} + \mathbf{r} \times \mathbf{p} \cdot \mathbf{VE} = \mathbf{p} \times \mathbf{E} + \mathbf{r} \times \mathbf{F} \tag{109}$$

where $\mathbf{F}$ is the force on the dipole as given by **(106)**.

Exercises

37. Let $\mathbf{E}(\mathbf{r})$, the field of a dipole $\mathbf{p}_1$ at the origin, make an angle θ with $\mathbf{r}$, where $\mathbf{r}$ makes an angle θ_1 with $\mathbf{p}_1$. Obtain the relation between the two angles:

$$\tan \theta_1 = 2 \tan \theta$$

38. A dipole $\mathbf{p}$ is rotated through an angle $d\theta$. Show that the incremental work done against the torque of the field may be written as:

$$dW = -\tau \cdot d\theta = -\mathbf{E} \cdot d\mathbf{p}$$

where $d\mathbf{p} = d\theta \times \mathbf{p}$ is the change in the dipole.

---

15 Examples of Electric Multipole Moments. As examples of electric moments we discuss the monopole, dipole, and quadrupole moments of nuclei, atoms, and molecules. In Table 1 we indicate some representative examples.

The dipole moment is generally given as a distance with the electron charge divided out. The quadrupole moment is similarly represented by an area with the electron charge divided out:

$$Q = Q_z/e = \frac{1}{e} \int_{V_1} (3z_1{}^2 - r_1{}^2)\rho(\mathbf{r}_1)\, dV_1 \tag{110}$$

A positive quadrupole moment indicates $3\langle z^2 \rangle > \langle r^2 \rangle$ and the nucleus is prolate. A negative quadrupole moment indicates $3\langle z^2 \rangle < \langle r^2 \rangle$ and the nucleus is oblate.

Table 1   Electric Moments

	Spin	Monopole Moment, e	Quadrupole Moment, 10^{-28} m$^2 e$
e	1/2	-1	0
n	1/2	0	0
p	1/2	1	0
d	1	1	$+0.00282$
He^4	0	2	0
Li^6	1	3	-0.008
I^{127}	5/2	53	-0.78

Nuclei are expected to exhibit only even moments (monopole, quadrupole, hexadecapole, etc.). This conclusion is based on the assumed symmetry of nuclear states.[6] As an experimental test of this assumption, a search for the electric dipole moment of the neutron indicates that if a moment exists, it is certainly less than the electron charge times 5×10^{-22} metres.[7]

The largest electric multipole that a nucleus can have is equal to twice the nuclear spin. Then nuclei of spin 0 and $\frac{1}{2}$ may have only monopole moments; nuclei of spin 1 and $\frac{3}{2}$ will have monopole and quadrupole moments, and so on. The requirement of no odd moments applies to atomic states as well. It does not apply to molecules because of the possibility of rotational states.

MACROSCOPIC ELECTRIC FIELD

In our discussion thus far we assumed that the charge density $\rho(\mathbf{r})$ is known precisely everywhere so that the potential and field may be known everywhere as well. For real physical problems it is not possible nor, fortunately, is it necessary to know the field with this kind of precision.

In condensed matter, for example, we are able to compute charge distributions at best approximately and thus we do not have more than an approximate idea of how the field and potential vary on what we will call a *microscopic* scale. Similarly, although we know some of the lower moments of nuclear charge distributions, the monopole and quadrupole moments, we do not know much about higher moments.

Fortunately, for physics, such precise knowledge of fields is rarely necessary. Let us imagine, for example, that we wish to study the effect of an electric field on a solid, a topic that will occupy our attention in Chapter 3. We approach the problem by asking for the average field over some extended region of the solid; we call this the *macroscopic* field. Next we ask for deviations from this average at the site of a

[6] See, for example, H. Frauenfelder and E. M. Henley, *Subatomic Particles*, Prentice-Hall, 1974.

[7] J. H. Smith, E. M. Purcell, and N. F. Ramsey, *Phys. Rev.* **108**, 120 (1957). An atomic beam magnetic resonance apparatus has been used in a test of parity and time-reversal invariance in atomic systems. From a study of the electric field-induced shift in radio frequency resonance of atomic cesium, an upper limit of 3×10^{-24} electron metres is obtained for the electric dipole moment of the electron. M. C. Weisskopf, J. P. Carrico, H. Gould, E. Lipworth, and T. S. Stein, *Phys. Rev. Letters* **21**, 1645 (1968).

particular atom or molecule. And finally we examine intraatomic or intramolecular fields. As we now see, the *macroscopic* field may be expressed in terms of moments of exterior and interior charge. For the interior charge, only the dipole moment of the distribution is required. If the region is large enough, the exterior charges are far away and the contribution of higher moments to the field is small, leaving the monopole and dipole moments of the exterior charge as the only moments that need be considered.

16 Spherical Average Field. We wish first to compute the average field over a spherical volume from *exterior charge*, a charge that is entirely outside the volume shown in Figure 17*a*.

   This result may be obtained most simply by first finding the force on a representative charge q_1 at position $\mathbf{r}_1$ with respect to the center of a uniformly charged sphere. The force on the sphere, which must be equal and opposite to the force on the charge, may be written in terms of the average field within the sphere.

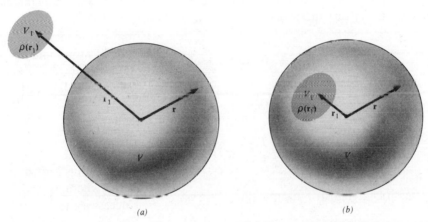

(a)   (b)

Figure 17   Average field over a spherical volume V. (*a*) The average field of charge outside the sphere is equal to the field at the center of the sphere. (*b*) The average field within the sphere is given by **(118)**.

From **(64)** the force on the charge q_1 is given by:

$$\mathbf{F}_1 = \frac{1}{4\pi\epsilon_0} \frac{q_1 \rho V}{r_1^2} \hat{\mathbf{r}}_1 \tag{111}$$

The force on the sphere must be equal and opposite to $\mathbf{F}_1$ and may be written as:

$$\mathbf{F} = -\mathbf{F}_1 = \int_V \rho \mathbf{E}(\mathbf{r}) \, dV \tag{112}$$

Comparing **(112)** and **(111)** we obtain:

$$\langle \mathbf{E} \rangle_V = \frac{1}{V} \int_V \mathbf{E}(\mathbf{r}) \, dV = -\frac{1}{4\pi\epsilon_0} \frac{q_1}{r_1^2} \hat{\mathbf{r}}_1 \tag{113}$$

This equation states that the average field from a point charge outside the sphere is equal to the field at the center of the sphere. We may now extend **(113)** to any distribution of charge exterior to V:

$$\langle \mathbf{E} \rangle_V = \mathbf{E}_{ext} = -\frac{1}{4\pi\epsilon_0} \int_{V_1} \frac{\rho(\mathbf{r}_1)}{r_1{}^2} \hat{\mathbf{r}}_1 \, dV_1 \tag{114}$$

This field is called the exterior field and will be used in the discussion of dielectrics in Section 3-7.

We next turn to the situation of Figure 17b where the charge is included within V. The force on a representative charge q_1 within a uniformly charged sphere is obtained from **(65)**:

$$\mathbf{F}_1 = \frac{\rho q_1}{3\epsilon_0} \mathbf{r}_1 \tag{115}$$

The force on the sphere must again be equal and opposite to $\mathbf{F}_1$:

$$\mathbf{F} = -\mathbf{F}_1 = \int_V \rho \mathbf{E}(\mathbf{r}) \, dV \tag{116}$$

Comparing **(115)** and **(116)** we obtain:

$$\langle \mathbf{E} \rangle_V = \frac{1}{V} \int_V \mathbf{E}(\mathbf{r}) \, dV = -\frac{1}{3\epsilon_0} \frac{q_1 \mathbf{r}_1}{V} \tag{117}$$

For a general charge distribution within V we have from **(117)**:

$$\langle \mathbf{E} \rangle_V = -\frac{1}{3\epsilon_0 V} \int_V \rho(\mathbf{r}_1) \mathbf{r}_1 \, dV_1 = -\frac{1}{3\epsilon_0} \frac{\mathbf{p}}{V} \tag{118}$$

where $\mathbf{p}$ is the dipole moment of the interior charge with respect to the origin as defined by **(79)**.

Finally, the average field from both exterior and interior charge may be written as:

$$\langle \mathbf{E} \rangle_V = \mathbf{E}_{ext} - \frac{1}{3\epsilon_0} \frac{\mathbf{p}}{V} \tag{119}$$

We show the situation schematically in Figure 18. We wish to find the average field over a spherical volume V of radius R as shown in Figure 18a. We find that the average field from charge exterior to V is equal to the field produced by the exterior charge at the center of the sphere shown in Figure 18b, while the average field from interior charge may be expressed in terms of the electric dipole moment of the sphere in Figure 18c.

We may obtain **(119)** in a different way by beginning with the expression for the electric field at **r** and integrating over a volume V to obtain the average field:

$$\langle \mathbf{E} \rangle_V = \frac{1}{V} \int_V \mathbf{E}(\mathbf{r}) \, dV = \frac{1}{V} \int_V dV \frac{1}{4\pi\epsilon_0} \int_{V_1} \frac{\rho(\mathbf{r}_1)\hat{\mathbf{r}}_{01}}{r_{01}{}^2} \, dV_1 \qquad (120)$$

Reversing the order of integration we may write **(120)** as:

$$\langle \mathbf{E} \rangle_V = \frac{1}{V} \int_{V_1} dV_1 \frac{1}{4\pi\epsilon_0} \int_V \frac{\rho(\mathbf{r}_1)\hat{\mathbf{r}}_{01}}{r_{01}{}^2} \, dV \qquad (121)$$

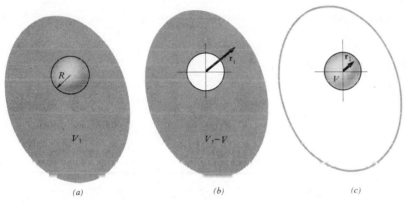

(a)   (b)   (c)

Figure 18   Average field from both exterior and interior charge. (*a*) We wish the average field within a sphere of radius R. (*b*) The average field of charge outside the sphere is equal to the field at the center of the sphere. (*c*) The average field of charge within the sphere is expressed in terms of the electric dipole moment of the sphere.

The integral over V is the *negative* of the field at $\mathbf{r}_1$ from a volume V uniformly charged to $\rho(\mathbf{r}_1)$. The form of this field for $r_1 < R$, the radius of the spherical volume V, is obtained from **(65)**:

$$\mathbf{E}(\mathbf{r}_1) = \frac{1}{3\epsilon_0} \rho(\mathbf{r}_1)\mathbf{r}_1 \qquad (122)$$

while the field for $r_1 > R$ is obtained from **(64)**

$$\mathbf{E}(\mathbf{r}_1) = \frac{1}{3\epsilon_0} \frac{\rho(\mathbf{r}_1)R^3}{r_1{}^2} \hat{\mathbf{r}}_1 = \frac{V}{4\pi\epsilon_0} \frac{\rho(\mathbf{r}_1)\hat{\mathbf{r}}_1}{r_1{}^2} \qquad (123)$$

We must therefore write **(121)** as the sum of two terms:

$$\langle \mathbf{E} \rangle_V = -\frac{1}{4\pi\epsilon_0} \int_{V_1-V} \frac{\rho(\mathbf{r}_1)\hat{\mathbf{r}}_1}{r_1{}^2} \, dV_1 - \frac{1}{3\epsilon_0} \frac{1}{V} \int_V \rho(\mathbf{r}_1)\mathbf{r}_1 \, dV_1 \qquad (124)$$

The first term is just the field at the center of the sphere from exterior charge and the second integral is the dipole moment of the interior charge in agreement with **(119)**.

Exercises

39. Prove that the force between two *uniformly* charged spheres of radius R_1 and R_2 is exactly the same as if all the charge were concentrated at their centers.
40. A spherical surface is constructed so as to enclose a distribution of charges with no charge outside. Show that the average electric field over the surface is zero. (Place a hypothetical surface charge σ on the shell and compute the forces.)
41. Show that the average of the electric field over a spherical *surface* which includes no charge is equal to the field at the center. Use the same method as in the preceding exercise.

POLARIZATION

An alternative way of writing (119) defines the mean polarization density **P** within V. We write for the *macroscopic field*:

$$\mathbf{E} = \mathbf{E}_{\text{ext}} - \frac{1}{3\epsilon_0} \mathbf{P} \tag{125}$$

with:

▶
$$\mathbf{P} = \frac{\mathbf{p}}{V} = \frac{1}{V} \left[\int_V \rho(\mathbf{r})\mathbf{r}\, dV + \int_S \sigma(\mathbf{r})\mathbf{r}\, dS \right] \tag{126}$$

where we have explicitly allowed for a surface charge density.

We emphasize that both **E** and **P** in (125) are *average quantities* while $\mathbf{E}_{\text{ext}}$ is determined at a *point*. The range R over which we average will depend on the problem. In a dielectric solid, R will usually be a distance large compared with interatomic distances but small compared with the size of the sample. For atomic problems, R should be small compared with an atomic radius but large compared with the scale on which either electron or nucleon structure is important.

17 **Polarization Charge.** Although it may seem surprising, we can describe a charge distribution entirely in terms of *polarization density* rather than in terms of *charge density*.[8] To develop such a description we begin with the potential of a dipole at the origin (95):

$$\phi(\mathbf{r}) = -\frac{1}{4\pi\epsilon_0}\, \mathbf{p} \cdot \nabla \frac{1}{r} \tag{127}$$

We may extend (127) to a distribution of dipoles at $\mathbf{r}_j$:

$$\phi(\mathbf{r}) = -\frac{1}{4\pi\epsilon_0} \sum_{j=1}^{n} \mathbf{p}_j \cdot \nabla \frac{1}{r_{0j}} \tag{128}$$

[8] E. Katz, *Am. J. Phys.* **33**, 306 (1965). See especially the appendix of this provocative paper.

where

$$\mathbf{r}_{0j} = \mathbf{r} - \mathbf{r}_j \qquad (129)$$

is the vector *from* the dipole *to* the field position $\mathbf{r}$.

In the limit that the dipoles become more numerous and smaller we may define a local polarization density:

$$\mathbf{P}(\mathbf{r}_j)\, dV_j = \mathbf{p}_j \qquad (130)$$

in terms of which the potential may be written:

$$\phi(\mathbf{r}) = -\frac{1}{4\pi\epsilon_0} \int_{V_1} \mathbf{P}(\mathbf{r}_1) \cdot \nabla \frac{1}{r_{01}}\, dV_1 \qquad (131)$$

We now show how **(131)** may be transformed into an expression of the form:

$$\phi(\mathbf{r}) = \frac{1}{4\pi\epsilon_0} \oint_{S_1} \frac{\sigma(\mathbf{r}_1)}{r_{01}}\, dS_1 + \frac{1}{4\pi\epsilon_0} \int_{V_1} \frac{\rho(\mathbf{r}_1)}{r_{01}}\, dV_1 \qquad (132)$$

where $\sigma(\mathbf{r}_1)$ is a *surface* charge density on S_1 and $\rho(\mathbf{r}_1)$ is a *volume* charge density within V_1. To do this, we recognize that the gradient of $1/r_{01}$ with respect to the field position $\mathbf{r}$ is just the negative of the gradient with respect to the source position:

$$\nabla \frac{1}{r_{01}} = -\frac{\hat{\mathbf{r}}_{01}}{r_{01}^{2}} = -\nabla_1 \frac{1}{r_{01}} \qquad (133)$$

Substituting into **(131)** and using the vector identity **(B-2)**:

$$\nabla_1 \cdot \frac{\mathbf{P}(\mathbf{r}_1)}{r_{01}} = \mathbf{P}(\mathbf{r}_1) \cdot \nabla_1 \frac{1}{r_{01}} + \frac{\nabla_1 \cdot \mathbf{P}(\mathbf{r}_1)}{r_{01}} \qquad (134)$$

we obtain:

$$\phi(\mathbf{r}) = \frac{1}{4\pi\epsilon_0} \int_{V_1} \nabla_1 \cdot \frac{\mathbf{P}(\mathbf{r}_1)}{r_{01}}\, dV_1 - \frac{1}{4\pi\epsilon_0} \int_{V_1} \frac{\nabla_1 \cdot \mathbf{P}(\mathbf{r}_1)}{r_{01}}\, dV_1 \qquad (135)$$

We transform the first term on the right by the divergence theorem **(B-11)** to obtain, finally,

$$\phi(\mathbf{r}) = \frac{1}{4\pi\epsilon_0} \oint_{S_1} \frac{\mathbf{P}(\mathbf{r}_1) \cdot d\mathbf{S}_1}{r_{01}} - \frac{1}{4\pi\epsilon_0} \int_{V_1} \frac{\nabla_1 \cdot \mathbf{P}(\mathbf{r}_1)}{r_{01}}\, dV_1 \qquad (136)$$

For $\mathbf{P}(\mathbf{r}_1) = 0$ on S_1 we have only the second term, which, on comparison with **(132)**, gives:

$$\blacktriangleright \qquad\qquad \rho(\mathbf{r}) = \rho_P(\mathbf{r}) = -\nabla \cdot \mathbf{P}(\mathbf{r}) \qquad\qquad\qquad \textbf{(137)}$$

That is, charge density is equal to the negative of the divergence of the polarization.

There are circumstances where it is useful to retain the first term of **(136)** and to regard $\mathbf{P}(\mathbf{r}_1) \cdot d\mathbf{S}_1$ as a surface charge density:

$$\sigma(\mathbf{r}_1) = \hat{\mathbf{n}} \cdot \mathbf{P}(\mathbf{r}_1) \qquad\qquad\qquad \textbf{(138)}$$

Let us imagine, for example, that we wish the contribution to the potential from only the polarization within V_1 as in Figure 19a. There may then be polarization on S_1 and we may expect a surface contribution to $\phi(\mathbf{r})$. The contribution from the exterior

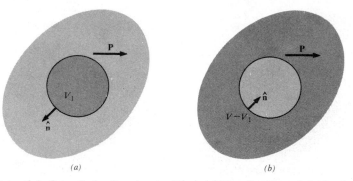

Figure 19   Potential of a polarization density $\mathbf{P}(\mathbf{r}_1)$. (*a*) For the potential of polarization within S_1 we must include the contribution of surface polarization charge as given by **(138)**. (*b*) The potential of the region outside the sphere includes a contribution from an equal and opposite surface charge density on S_1.

polarization will have a surface term just opposite to that from V_1 because the *outward* normal is reversed as shown in Figure 19b. When the contributions from V_1 and $V - V_1$ are added, the contribution from the interior surface clearly vanishes. Just such a fictitious interior surface is introduced in the computation of the local field as is discussed in Section 3-7. The surface charge associated with the exterior polarization leads to the *Lorentz field*.

An additional case of interest occurs when the outer surface S_1 just bounds a dielectric body, as shown in Figure 20a. Under these circumstances—where the polarization drops discontinuously to zero—we must include both volume and surface terms. What if we were to enlarge the bounding surface S_1 so that it is outside the dielectric body as shown in Figure 20b? Under these circumstances $\nabla \cdot \mathbf{P}$ has a singularity at the surface so that what was a surface contribution in Figure 20a becomes a volume contribution in Figure 20b.

Since we may make completely equivalent microscopic descriptions in terms of either polarization density or charge density, why do we bother to introduce $\mathbf{P}(\mathbf{r})$ at all?

As we have just seen, **(125)**, the macroscopic electric field is related to the polarization density. And as we will see in Chapter 3, when we discuss fields in dielectric media, the introduction of polarization density makes it possible to construct a local macroscopic description of a dielectric that is much simpler than a description in terms of microscopic charge density.

Finally, we may use Gauss's theorem **(B-11)** and **(137)** to express the volume polarization charge q_p in terms of a surface integral of the polarization:

$$\int_S \mathbf{P} \cdot d\mathbf{S} = \int_V \mathbf{\nabla} \cdot \mathbf{P} \, dV = -\int_V \rho_P \, dV \tag{139}$$

We have, then, for the volume polarization charge:

$$q_P = -\int_S \mathbf{P} \cdot d\mathbf{S} \tag{140}$$

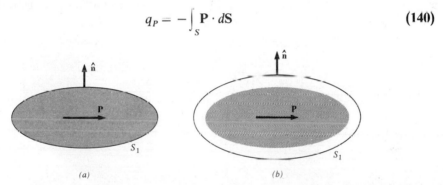

(a) (b)

Figure 20 Potential of a polarized dielectric. (*a*) We suppose that the polarization density drops discontinuously to zero at the surface. Under these circumstances we must include the contribution of surface polarization charge. (*b*) If we bound the dielectric by a surface S outside its physical surface, then the charge on the physical surface arises from the divergence of the polarization.

The surface polarization charge is the integral of **(138)**, which is just the negative of **(140)**, which is simply a statement of the electrical neutrality of a bounded distribution of electric dipoles.

Exercises

42. An *open* surface S carries a uniform normal surface polarization density:

$$\frac{d\mathbf{p}}{dS} = \sigma \, d \, \hat{\mathbf{n}}$$

Show that the potential at **r** is given by

$$\phi = -\frac{\sigma d}{4\pi\epsilon_0} \Omega$$

where Ω is the solid angle subtended at **r** by S.

43. A *closed* surface S carries a uniform normal surface polarization density as given in Exercise 42. Show that the potential within the closed surface is a constant and is given by:

$$\phi = -\frac{\sigma d}{\epsilon_0}$$

and that the potential outside is equal to zero. Such a charge distribution is called a *double layer*.

44. Show that the potential given in Exercise 42 is a solution of Laplace's equation **(43)**.

45. A cylinder of height L and radius R is uniformly polarized parallel to its axis. Show that the electric field at the center is given by:

$$\mathbf{E} = -\frac{1}{\epsilon_0}\mathbf{P}[1 - L/(4R^2 + L^2)^{1/2}]$$

46. Find the radial polarization density $P(r)$ equivalent to a uniformly charged sphere of radius R and charge density ρ.

47. By substituting **(137)** and **(138)** into **(126)** show that an identity is obtained. The proof will require the use of **(C-4)** and **(C-9)**.

48. Substitute **(137)** into:

$$\mathbf{F} = \int \rho(\mathbf{r})\mathbf{E}(\mathbf{r})\,dV$$

to obtain **(106)**. You will require **(C-4)** and **(C-9)**.

18 Current Density.

We write the charge flow of a single charge carrier as:

$$\mathbf{J} = q\mathbf{v} \tag{141}$$

The charge flow of a group of carriers is then:

$$\mathbf{J} = \sum_i q_i \mathbf{v}_i \tag{142}$$

By analogy with **(126)** we may define the current *density* as a limit:

$$\mathbf{j}(\mathbf{r}) = \lim \frac{\mathbf{J}}{V} = \lim \frac{1}{V}\int \rho(\mathbf{r})\mathbf{v}(\mathbf{r})\,dV = \rho(\mathbf{r})\mathbf{v}(\mathbf{r}) \tag{143}$$

19 Conservation of Charge.

The fact that electric charge is conserved may be expressed in terms of a differential relation between current density and charge density. We show in Figure 21 a volume V containing charge density $\rho(\mathbf{r})$ with current density $\mathbf{j}(\mathbf{r})$ on the surface S. Local conservation of electric charge requires that the rate of change of enclosed charge be expressed equivalently as either a surface or a volume integral:

$$\frac{\partial q}{\partial t} = \int_V \frac{\partial \rho(\mathbf{r})}{\partial t} \, dV = -\oint_S \mathbf{j}(\mathbf{r}) \cdot d\mathbf{S} \tag{144}$$

We may transform the right side of **(144)** by the divergence theorem **(B-11)**:

$$\oint_S \mathbf{j}(\mathbf{r},t) \cdot d\mathbf{S} = \int_V \nabla \cdot \mathbf{j}(\mathbf{r},t) \, dV \tag{145}$$

Substituting into **(144)** and rearranging, we obtain:

$$\int \left[\nabla \cdot \mathbf{j}(\mathbf{r},t) + \frac{\partial \rho(\mathbf{r},t)}{\partial t} \right] dV = 0 \tag{146}$$

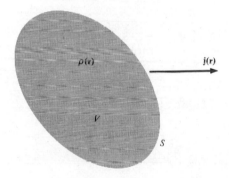

Figure 21   A volume V containing charge with current crossing the bounding surface S. As a consequence of local conservation the current and charge densities are related by **(147)**.

But since the volume V is completely arbitrary, the integrand of **(146)** must describe a *local* relation:

$$\nabla \cdot \mathbf{j}(\mathbf{r},t) = -\frac{\partial \rho(\mathbf{r},t)}{\partial t} \tag{147}$$

Exercises

49. For $\mathbf{j}(\mathbf{r},t)$ and $\rho(\mathbf{r},t)$ functions of the distance from the origin show that **(147)** may be written as:

$$\frac{1}{r^2} \frac{d}{dr} r^2 j(r,t) = -\frac{\partial}{\partial t} \rho(r,t)$$

50. A charge q is placed at the origin at $t = 0$ and allowed to diffuse freely. The expected charge distribution after a time t is:

$$\rho(r,t) = \left(\frac{6}{\pi}\right)^{1/2} \frac{q}{(4\pi/3)R^3} \, e^{-(3/2)(r^2/R^2)}$$

with $R^2 = 6D\,t$ where R is the root-mean-square displacement, and D is the diffusion constant. Show that the current

$$\mathbf{j}(\mathbf{r},t) = \frac{\mathbf{r}}{2t} \, \rho(r,t)$$

is a solution of **(147)**.

20 Polarization Current.

In terms of polarization current and charge **(147)** takes the form:

$$\nabla \cdot \mathbf{j}_P(\mathbf{r}) = -\frac{\partial}{\partial t} \rho_P(\mathbf{r}) = \frac{\partial}{\partial t} \nabla \cdot \mathbf{P}(\mathbf{r}) \tag{148}$$

where $\mathbf{j}_P$ is the density of flow of polarization charge q_P and we have substituted from **(137)**. The solution of **(148)** is:

▶
$$\mathbf{j}_P(\mathbf{r}) = \frac{\partial \mathbf{P}}{\partial t} + \mathbf{\Omega}(\mathbf{r}) \tag{149}$$

where $\mathbf{\Omega}(\mathbf{r})$ must be a solenoidal current, that is, it must have the property $\nabla \cdot \mathbf{\Omega}(\mathbf{r}) = 0$.

As an example we consider the uniform motion with velocity $\mathbf{v}$ of a polarized body. In this case we may write the current in terms of polarization charge. Substituting **(137)** into **(143)** we have:

$$\mathbf{j}(\mathbf{r}) = \mathbf{j}_P(\mathbf{r}) = -\mathbf{v}\nabla \cdot \mathbf{P}(\mathbf{r}) \tag{150}$$

For a uniformly moving body we may write:

$$\frac{\partial \mathbf{P}}{\partial t} = -(\mathbf{v} \cdot \nabla)\mathbf{P} \tag{151}$$

Substituting **(150)** and **(151)** into **(149)** we obtain:

$$\mathbf{\Omega}(\mathbf{r}) = -\mathbf{v}\nabla \cdot \mathbf{P} + (\mathbf{v} \cdot \nabla)\mathbf{P} = -\nabla \times (\mathbf{v} \times \mathbf{P}) \tag{152}$$

where the final equality follows from **(B-5)**. Substituting **(152)** into **(149)** we obtain, finally,

$$\mathbf{j}_P(\mathbf{r}) = \frac{\partial \mathbf{P}}{\partial t} - \nabla \times (\mathbf{v} \times \mathbf{P}) \tag{153}$$

which as expected has the form of **(149)**. We return to a discussion of polarization current in Chapters 5 and 11.

Exercises

51. Consider a cube polarized parallel to an edge. With **v** parallel to **P** identify **(150)** and **(151)**. Show that they are equivalent with, as expected, $\Omega = 0$.

52. Consider the situation described in Exercise 51 but with the velocity parallel to one of the other edges and thus normal to the polarization. Identify **(150)**, **(151)**, and **(152)**. Show that **(153)** is satisfied.

EXPANSION OF THE FIELD

The discussion of this part is in some sense the inverse of the earlier discussion of moments of a charge distribution. A charge distribution is usefully characterized in terms of its moments when the field position is at some distance from the charges. Here we imagine that we are in the midst of a distribution of charge and we wish to expand the field about some arbitrary point as origin. We first discuss the problem of a general distribution of charge. Then we consider a charge density that is uniform but bounded.

21 General Distribution of Charge. We imagine some general distribution of charge $\rho(\mathbf{r}_1)$ as shown in Figure 22a. The electric field at **r** is given from **(6)** by:

$$\mathbf{E}(\mathbf{r}) = \frac{1}{4\pi\epsilon_0} \int_{V_1} \frac{\rho(\mathbf{r}_1)}{r_{01}^2} \hat{\mathbf{r}}_{01} \, dV_1 \tag{154}$$

We wish to write the field at some position near the origin as an expansion of the field at the origin:

▶ $$\mathbf{E}(\mathbf{r}) = \mathbf{E}(0) + (\mathbf{r} \cdot \nabla)\mathbf{E}(0) + \cdots = \mathbf{E}(0) + \mathbf{r} \cdot \nabla\mathbf{E}(0) + \cdots \tag{155}$$

Taking the gradient of **(154)** we have, for the electric field gradient tensor,

$$\nabla\mathbf{E}(\mathbf{r}) = -\frac{1}{4\pi\epsilon_0} \int_{V_1} \frac{\rho(\mathbf{r}_1)}{r_{01}^3} (3\hat{\mathbf{r}}_{01}\hat{\mathbf{r}}_{01} - \mathbf{I}) \, dV_1 \tag{156}$$

At the origin we have, for **(154)**,

$$\mathbf{E}(0) = -\frac{1}{4\pi\epsilon_0} \int_{V_1} \frac{\rho(\mathbf{r}_1)}{r_1^{\ 2}} \hat{\mathbf{r}}_1 \, dV_1 \tag{157}$$

and for **(156)**:

$$\mathbf{V}\mathbf{E}(0) = -\frac{1}{4\pi\epsilon_0} \int_{V_1} \frac{\rho(\mathbf{r}_1)}{r_1^{\ 3}} (3\hat{r}_1\hat{r}_1 - \mathbf{I}) \, dV_1 \tag{158}$$

The difficulty with the theory that we have introduced lies in the evaluation of **(158)**. We see that the radial integration of **(158)** has a logarithmic divergence at $r_1 = 0$. This singularity must be taken out by the angular integration, but it is not always clear how one should proceed. For example, if $\rho(\mathbf{r}_1)$ is constant, the angular integration gives

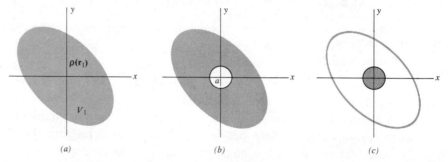

Figure 22 Expansion of the electric field. (*a*) A general distribution of charge is given by $\rho(\mathbf{r}_1)$. (*b*) The electric field gradient is obtained in two parts. The first part excludes the charge within a sphere of radius *a*. (*c*) The second part arises from a sphere of uniform density at the origin.

zero for **VE** and thus for the scalar of **VE**. But, as we have seen (Exercise 12), the scalar of **VE** gives the charge density:

$$\nabla E_S = \mathbf{V} \cdot \mathbf{E} = \rho/\epsilon_0 \tag{159}$$

To avoid problems of this sort we divide V_1 into two parts:

1. The integral given by **(158)** is evaluated over the region shown in Figure 22*b* that includes all of V_1 except for a sphere of radius *a* at the origin. This avoids the logarithmic singularity and any problem with the angular integration. In the limit that *a* goes to zero we obtain what is called the *principal value* of **(158)**.

2. The second region is a sphere of radius *a* centered at the origin, as shown in Figure 22*c*. For *a* small we may take the sphere to be uniformly charged. The electric field for small **r** is then, from **(65)**,

$$\mathbf{E}(\mathbf{r}) = \frac{1}{3\epsilon_0} \mathbf{r}\rho \tag{160}$$

And the field gradient is

$$\nabla \mathbf{E}(\mathbf{r}) = \frac{1}{3\epsilon_0} \rho(\mathbf{r}) \mathbf{I} \tag{161}$$

Letting a go to zero leaves **(161)** unaffected. Adding the contributions from the two regions we obtain in place of **(158)**:

$$\nabla \mathbf{E}(\mathbf{r}) = \frac{1}{3\epsilon_0} \rho(\mathbf{r}) \mathbf{I} - \frac{1}{4\pi\epsilon_0} \int_{V_1} \frac{\rho(\mathbf{r}_1)}{r_{01}^{3}} (3\hat{\mathbf{r}}_{01}\hat{\mathbf{r}}_{01} - \mathbf{I}) \, dV_1 \tag{162}$$

where it is understood that we are to take the principal value of the integral.

Exercise

53. Obtain the divergence of **E** from the scalar of **(162)** and verify that **(159)** is satisfied.

22 Uniform Charge Density. For a uniformly charged body of general shape we may write, for the electric field gradient,

$$\nabla \mathbf{E}(\mathbf{r}) = \frac{\rho}{\epsilon_0} \mathbf{N}(\mathbf{r}) \tag{163}$$

where the tensor **N** is given from **(162)** by:

$$\mathbf{N}(\mathbf{r}) = \frac{1}{3} \mathbf{I} - \frac{1}{4\pi} \int_{V} (3\hat{\mathbf{r}}_{01}\hat{\mathbf{r}}_{01} - \mathbf{I}) \frac{dV_1}{r_{01}^{3}} \tag{164}$$

We see from **(164)** that the trace of **N**, which is equal to the scalar of **N** (Section C-6), is equal to unity:

$$N_{xx} + N_{yy} + N_{zz} = 1 \tag{165}$$

For a uniformly charged body we may write, for the electric field near the origin, from **(155)**:

$$\mathbf{E}(\mathbf{r}) = \mathbf{E}(0) + \frac{\rho}{\epsilon_0} \mathbf{r} \cdot \mathbf{N}(0) + \cdots \tag{166}$$

Exercise

54. Integrate **(164)** over a uniformly charged sphere (using spherical coordinates) to obtain the result:

$$\mathbf{N} = \tfrac{1}{3}\mathbf{I}$$

23   Uniform Polarization Density. To generate a uniformly polarized body we superpose on a body of charge density ρ a second body of charge density $-\rho$ and displaced to $-\mathbf{a}$ as shown in Figure 23. The electric field resulting from the second body is then:

$$\mathbf{E}(\mathbf{r}) = -\mathbf{E}(0) - \frac{\rho}{\epsilon_0}(\mathbf{r} + \mathbf{a}) \cdot \mathbf{N} \tag{167}$$

Adding **(166)** and **(167)** we obtain:

$$\mathbf{E}(\mathbf{r}) = -\frac{1}{\epsilon_0}\mathbf{P} \cdot \mathbf{N} \tag{168}$$

where $\mathbf{P} = \rho\mathbf{a}$ is the polarization per unit volume. For a body of general shape, $\mathbf{N}$ may be regarded as a function of $\mathbf{r}$.

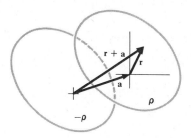

Figure 23   Generation of uniform polarization density. We take two identical but oppositely charged bodies to be displaced by a vector $\mathbf{a}$. The electric field is given by **(168)**.

Alternatively, the electric field may be regarded as arising from surface polarization charge **(138)**:

$$\sigma = \hat{\mathbf{n}} \cdot \mathbf{P} \tag{169}$$

to give:

$$\mathbf{E}(\mathbf{r}) = \frac{1}{4\pi\epsilon_0}\oint_{S_1} \frac{\sigma\hat{\mathbf{r}}_{01}}{r_{01}{}^2}\, dS_1 = \frac{1}{4\pi\epsilon_0}\int_{S_1} \frac{\hat{\mathbf{r}}_{01}}{r_{01}{}^2}\mathbf{P} \cdot d\mathbf{S}_1 \tag{170}$$

Exercises

55. A cylinder of height L and radius R is uniformly polarized transverse to its axis. By using **(165)** and the result of Exercise 45, show that the electric field at the center is given by:

$$\mathbf{E} = -\tfrac{1}{2}(\mathbf{P}/\epsilon_0)L/(4R^2 + L^2)^{1/2}$$

56. By writing the internal field in terms of surface polarization charge, obtain the relation:

$$\mathbf{N}(\mathbf{r}) = -\frac{1}{4\pi} \oint_{S_1} \frac{\hat{\mathbf{r}}_{01}\, dS_1}{r_{01}^{\,2}}$$

From this form show that the scalar of **N** is equal to unity as given by **(165)**.

57. Show that the field *outside* a uniformly polarized sphere of radius R and polarization density **P** is the same as the field of a dipole $\mathbf{p} = (4\pi/3)R^3\mathbf{P}$ located at the center of the sphere.

ELLIPSOIDAL CHARGE DISTRIBUTIONS

We have seen that charge distributions with spherical symmetry have a number of special properties. One of the simplest of these **(60)** is that the electric field is zero within a uniformly charged spherical shell. From this result we were able to show **(67)** that the potential within a uniformly charged sphere falls off quadratically and isotropically with distance from the center.

Newton showed for the gravitational field that if a region of uniform mass is bounded by two similar *ellipsoids* there is no interior field either. We first obtain this result and then show that the potential within a uniformly charged ellipsoid also falls off quadratically with distance from the center although the rate depends on direction.

24   Ellipsoidal Homoeoid. The region between similar curves is called a homoeoid. We show here, following Newton, that if such a region between ellipsoids is uniformly charged, the interior field is zero. We have already seen **(57)** that the field within a uniformly charged spherical shell may be written as an integral over segments that cancel in pairs. We redraw Figure 9 to emphasize in Figure 24a that the surface charge σ may be considered as a volume charge from a shell of thickness dR. The field at **r** from dq_1 and dq_2 is given **(57)** by:

$$d\mathbf{E} = \frac{1}{4\pi\epsilon_0}\left(\frac{dq_1}{r_{01}^{\,2}}\,\hat{\mathbf{r}}_{01} + \frac{dq_2}{r_{02}^{\,2}}\,\hat{\mathbf{r}}_{02}\right) = 0 \tag{171}$$

Now, an ellipsoidal homoeoid may be generated by a uniform expansion or contraction of the spherical shell along the axes of the ellipsoid. Figure 24b is generated, for example, by a contraction along x. This contraction takes **r** into **r'**. If we imagine that the charge is also contracted uniformly, then the charge originally within $d\Omega$ equals that within $d\Omega'$:

$$dq_1' = dq_1 \qquad dq_2' = dq_2 \tag{172}$$

In addition, parallel lengths remain parallel and in the same ratio:

$$\hat{\mathbf{r}}_{01}' = -\hat{\mathbf{r}}_{02}' \qquad \frac{r_{02}'}{r_{01}'} = \frac{r_{02}}{r_{01}} \tag{173}$$

so that we must have for the incremental field of the ellipsoidal homoeoid:

$$dE' = \frac{1}{4\pi\epsilon_0} \left(\frac{dq_1'}{r_{01}'^2} \hat{r}_{01}' + \frac{dq_2'}{r_{02}'^2} \hat{r}_{02}' \right) = 0 \tag{174}$$

We may now integrate **(174)** over the entire homoeoid to obtain the result that the interior field is identically zero, just as for the spherical shell!

25 **Uniformly Charged Ellipsoid.** We now wish to find the way in which the potential falls off with distance within a *uniformly* charged ellipsoid. We show in Figure 25a a position **r** at which we wish the field. We take **R** to be the vector from the origin to the bounding surface measured along **r**.

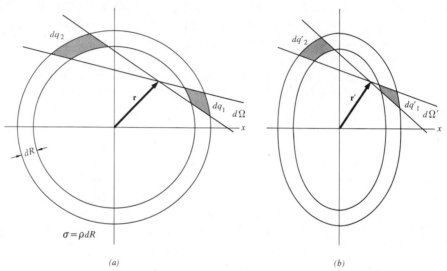

(a)   *(b)*

Figure 24 Ellipsoidal homoeoid. (*a*) The surface charge on a spherical shell may be regarded as the volume charge of a shell of thickness *dR*. (*b*) Uniform contraction or expansion of the shell along the *x* direction generates an ellipsoidal homoeoid.

Since there is no contribution to the field from the homoeoid beyond **r** we may compute the field from just that portion of ellipsoid shown in Figure 24*b*:

$$E(r) = E(uR) = \frac{\rho}{4\pi\epsilon_0} \int_{V_0 = u^3 V_1} \frac{\hat{r}_1}{r_1^2} \, dV_0 \tag{175}$$

Now the electric field as a function of **r** must simply be proportional, by scaling, to $u = r/R$. We have then:

$$E(r) = \frac{r}{R} E(R) \tag{176}$$

Comparing **(176)** (which is *exact* for a uniformly charged ellipsoid) with **(166)** (which is an expansion for a uniformly charged body of general shape), we must have:

$$\blacktriangleright \qquad \mathbf{E}(\mathbf{r}) = \frac{\rho}{\epsilon_0} \mathbf{r} \cdot \mathbf{N} \qquad (177)$$

with

$$\mathbf{E}(\mathbf{R}) = \frac{\rho}{\epsilon_0} \mathbf{R} \cdot \mathbf{N} \qquad (178)$$

where **N** is given by **(164)** and is *independent* of position.

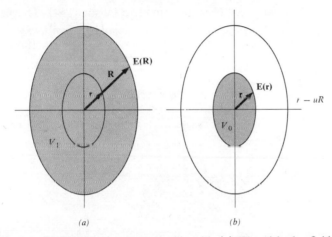

(a)   (b)

Figure 25 Potential of a uniformly charged ellipsoid. (*a*) We wish the field at a position **r** where **R** is the colinear vector the surface. (*b*) Since there is no contribution to the field from homoeoids beyond **r**, the field is given by **(176)**.

Exercises

58. Show that the potential within a uniformly charged ellipsoid is of the form:

$$\phi(\mathbf{r}) = \frac{\rho}{2\epsilon_0} \mathbf{r} \cdot \mathbf{N} \cdot \mathbf{r}$$

59. Show that the result of Exercise 58 is a solution of Poisson's equation.

26 Uniformly Polarized Ellipsoid. We first consider the field within the volume common to a pair of oppositely charged and similarly oriented ellipsoids displaced through a distance **a** as shown in Figure 26.

From **(178)** the field in the volume common to both ellipsoids must be given by:

$$\mathbf{E}(\mathbf{r}) = \frac{\rho}{\epsilon_0} \mathbf{r} \cdot \mathbf{N} - \frac{\rho}{\epsilon_0} (\mathbf{r} + \mathbf{a}) \cdot \mathbf{N} = -\frac{\rho}{\epsilon_0} \mathbf{a} \cdot \mathbf{N} \qquad (179)$$

and we see that the field in this region is perfectly uniform as a consequence of the fact that **(166)** is exact for an ellipsoid.

Now a uniformly *polarized* ellipsoid may be regarded as equivalent to the configuration shown in Figure 26 where the polarization density is given in the limit by:

$$\mathbf{P} = \rho \mathbf{a} \tag{180}$$

so that we may write for the internal electric field from **(161)**:

▶
$$\mathbf{E} = -\frac{1}{\epsilon_0} \mathbf{P} \cdot \mathbf{N} = -\frac{1}{\epsilon_0} \mathbf{N} \cdot \mathbf{P} \tag{181}$$

where we have used the fact that **N** is a symmetric tensor **(164)**. The field **E** in **(181)** is called the *depolarization field* and **N** is called the *depolarization tensor*. For $\hat{\mathbf{x}}$, $\hat{\mathbf{y}}$, and $\hat{\mathbf{z}}$

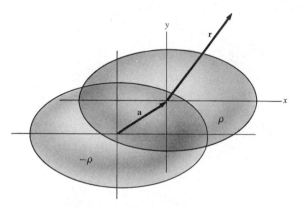

Figure 26 Uniformly polarized ellipsoid. Uniform polarization is generated by the displacement of oppositely charged and similarly oriented ellipsoids.

along principal axes of the ellipsoid, **N** will have only diagonal elements N_x, N_y, and N_z, which are called the *depolarization factors* of the ellipsoid. These factors have been computed from tabulated elliptic integrals.[9]

We finally wish to examine the sphere, the rod, and the disc as limiting forms of the ellipsoid:

1. For a sphere $A = B = C = R$. Then we expect, by symmetry, $N_x = N_y = N_z = \frac{1}{3}$. From **(181)** the depolarization field for a sphere is $\mathbf{E} = -\mathbf{P}/3\epsilon_0$.

2. For a long rod parallel to $\hat{\mathbf{z}}$ we take $A = B = R \ll C$. The field from a uniformly charged rod will be radial leading to $N_z = 0$. By symmetry $N_x = N_y = \frac{1}{2}$. From **(181)** there is no depolarization field for **P** parallel to the rod. If **P** is transverse, however, we obtain a depolarization field $\mathbf{E}_{xy} = -\mathbf{P}_{xy}/2\epsilon_0$.

---

[9] E. C. Stoner, *Phil. Mag.*, **36**, 803 (1945). See especially Professor Stoner's discussion of the history of this problem beginning with Dirichlet in 1839. A similar compilation has been prepared by J. A. Osborn, *Phys. Rev.*, **67**, 351 (1945).

3. For a thin disc normal to $\hat{x}$ we have $B = C = R \gg A$. For a uniformly charged disc the internal field will be normal to the plane of the disc so that $N_y = N_z = 0$ and $N_x = 1$. For a uniformly polarized disc there is no depolarization field for **P** normal to the axis. For the component of the polarization along x we have from (181) $E_x = -P_x/\epsilon_0$.

Exercise

60. A sphere is uniformly polarized along the $\hat{z}$ direction. The surface polarization charge density is then $\sigma_P = P \cos \theta$. Obtain an expression for the internal electric field at a general position arising from this charge. The integral may be carried out by an integration by parts along the $\hat{z}$ direction.

SUMMARY

1. The force on a stationary electric charge may be described in terms of an electric field:

$$\mathbf{F} = q\mathbf{E} \tag{3}$$

2. For a distribution of charge the field may be computed from

$$\mathbf{E}(\mathbf{r}) = \frac{1}{4\pi\epsilon_0} \int \frac{\rho(\mathbf{r}_1)}{r_{01}{}^2} \hat{\mathbf{r}}_{01} \, dV_1 \tag{6}$$

where $\mathbf{r}_{01} = \mathbf{r} - \mathbf{r}_1$ is the vector from the source charge $\rho(\mathbf{r}_1) \, dV_1$ to the field position **r**.

3. For a general charge distribution we may write the potential as:

$$\phi(\mathbf{r}) = \frac{1}{4\pi\epsilon_0} \int \frac{\rho(\mathbf{r}_1)}{r_{01}} \, dV_1 \tag{19}$$

4. The electric potential may be defined as the work required to move a unit charge to the position **r**:

$$\phi(\mathbf{r}) = -\int^{\mathbf{r}} \mathbf{E}(\mathbf{r}) \cdot d\mathbf{r} \tag{20}$$

Then the field may be written as the negative gradient of the potential:

$$\mathbf{E}(\mathbf{r}) = -\nabla\phi(\mathbf{r}) \tag{22}$$

5. The circulation of the field around a closed contour is zero leading to:

$$\mathscr{E} = \oint \mathbf{E}(\mathbf{r}) \cdot d\mathbf{r} = 0 \tag{25}$$

and

$$\mathbf{V} \times \mathbf{E} = 0 \tag{28}$$

from the *definition* of *curl* as circulation per unit area.

6. The flux through a closed surface is proportional to the enclosed charge

$$\Psi = \oint_S \mathbf{E}(\mathbf{r}) \cdot d\mathbf{S} = \frac{1}{\epsilon_0} \int_V \rho(\mathbf{r}) \, dV \tag{39}$$

which is called Gauss's law. Since the divergence of $\mathbf{E}$ is the flux per unit volume, we have:

$$\mathbf{V} \cdot \mathbf{E}(\mathbf{r}) = \frac{\rho(\mathbf{r})}{\epsilon_0} \tag{41}$$

7. The potential at a large distance from a bounded charge distribution may be expressed in terms of the moments of the distribution:

$$\phi(\mathbf{r}) = \frac{1}{4\pi\epsilon_0} \frac{q}{r} + \frac{1}{4\pi\epsilon_0} \frac{\hat{\mathbf{r}} \cdot \mathbf{p}}{r^2} + \frac{1}{8\pi\epsilon_0} \frac{\hat{\mathbf{r}} \cdot \mathbf{Q} \cdot \hat{\mathbf{r}}}{r^3} + \cdots \tag{77}$$

where the monopole moment is given by:

$$q = \int_{V_1} \rho(\mathbf{r}_1) \, dV_1 \tag{74}$$

the dipole moment is given by:

$$\mathbf{p} = \int_{V_1} \mathbf{r}_1 \rho(\mathbf{r}_1) \, dV_1 \tag{75}$$

and the quadrupole moment is given by:

$$\mathbf{Q} = \int_{V_1} r_1^2 (3\hat{\mathbf{r}}_1 \hat{\mathbf{r}}_1 - \mathbf{l}) \rho(\mathbf{r}_1) \, dV_1 \tag{76}$$

8. The same fields as produced by charge moments may be generated by multipoles. A single charge q gives a monopole field. Charges $+q$ and $-q$ separated by $\mathbf{a}$ give a dipole field in the limit that $\mathbf{a}$ goes to zero while the dipole moment $\mathbf{p} = q\mathbf{a}$ remains finite. Dipoles $+\mathbf{p}$ and $-\mathbf{p}$ separated by $\mathbf{b}$ give a quadrupole field in the limit that $\mathbf{b}$ goes to zero.

9. The average electrical field in a spherical volume V is given by $\langle \mathbf{E} \rangle_V = \mathbf{E}_{\text{ext}} - \mathbf{p}/3\epsilon_0 V$, where $\mathbf{p}$ is the dipole moment of the interior charge and $\mathbf{E}_{\text{ext}}$ is the field produced at the center of the sphere by *exterior* charge.

10. A local mean polarization density may be defined by the relation:

$$P = \frac{p}{V} \qquad (126)$$

where **p** is computed from the center of a spherical region. Charge density is associated with the divergence of the polarization:

$$\rho(\mathbf{r}) = \rho_P(\mathbf{r}) = -\nabla \cdot \mathbf{P}(\mathbf{r}) \qquad (137)$$

and current density with the rate of change of polarization density:

$$\mathbf{j}(\mathbf{r}) = \mathbf{j}_P(\mathbf{r}) = \frac{\partial}{\partial t}\mathbf{P}(\mathbf{r}) + \mathbf{\Omega}(\mathbf{r}) \qquad (149)$$

where $\mathbf{\Omega}(\mathbf{r})$ is a solenoidal contribution to the current.

11. The field may be expanded in the vicinity of a point in powers of **r**:

$$\mathbf{E}(\mathbf{r}) = \mathbf{E}(0) + \mathbf{r} \cdot \nabla\mathbf{E}(0) + \cdots \qquad (155)$$

The scalar of $\nabla\mathbf{E}(0)$ is equal to $\rho(0)/\epsilon_0$.

12. The field within a uniformly charged ellipsoid has the form:

$$\mathbf{E}(\mathbf{r}) = -\frac{\rho}{\epsilon_0}\mathbf{N} \cdot \mathbf{r} \qquad (177)$$

where N_x, N_y, and N_z are the principal values of a tensor characterizing the ellipsoid and sum to unity:

$$N_x + N_y + N_z = 1 \qquad (165)$$

13. The electric field within a uniformly polarized ellipsoid may be written in terms of N_x, N_y, and N_z, which are also called the depolarization factors of the ellipsoid:

$$E_x = -N_x P_x \qquad E_y = -N_y P_y \qquad E_z = -N_z P_z \qquad (181)$$

PROBLEMS

1. **Reciprocal vectors.** The potential is given at three nearby points $\mathbf{r}_1$, $\mathbf{r}_2$, and $\mathbf{r}_3$ not coplanar with the origin as ϕ_1, ϕ_2, and ϕ_3. The origin is at zero potential. Find the electric field at the origin to order r_i by writing $\mathbf{E}$ in vectors *reciprocal* to $\mathbf{r}_1$, $\mathbf{r}_2$, and $\mathbf{r}_3$. (See Section A-7.)

2. **Line charge.** A line of length L carries a lineal charge density λ. Show that the potential in the median plane may be written as

$$\phi(r) = \frac{\lambda}{4\pi\epsilon_0} \ln \frac{1 + \sin\theta}{1 - \sin\theta}$$

with $\tan\theta = L/2r$. Show for L much larger than r that the potential approaches

$$\phi(r) \simeq \frac{\lambda}{2\pi\epsilon_0} \ln \frac{L}{r}$$

What happens to the potential as L goes to infinity with r finite?

3. **Charged spherical cap.** A spherical cap of radius R and half angle θ carries a uniform surface charge σ. Show that the potential on the axis of the cap at a distance r from the center is given by

$$\phi(r) = \frac{\sigma R}{2\epsilon_0 r} [(r^2 - 2rR \cos\theta + R^2)^{1/2} - |r - R|]$$

Examine this expression in the limits $r \gg R$ and $r \ll R$. Show that for $r \sim R$ there is a field discontinuity given by σ/ϵ_0. Also examine the limits θ small and $\theta = \pi$.

4. **Charged rectangle.** A rectangle of sides $2a$ and $2b$ carries a uniform surface charge σ. Show that the field on the axis normal to the plane of the rectangle is given by:

$$E = \frac{\sigma}{\pi\epsilon_0} \tan^{-1} \frac{ab}{z(a^2 + b^2 + z^2)^{1/2}}$$

Assuming $a \leq b$, examine the limits $z \ll a$ and $z \gg b$.

5. **Charged ring.** A ring of radius R carries a charge λ per unit length. Obtain an expression for the electric field on the axis. Integrate this expression to obtain the potential. Obtain the potential directly from **(19)**. Interpret the limit $z/R \gg 1$ where z is the distance from the plane of the ring.

6. **Charged disc.** By integrating the field expression of Problem 5 over ring diameter, obtain the expression for the field on the axis of a disc of radius R and carrying a surface charge σ. Examine the limits $z/R \ll 1$ and $z/R \gg 1$.

7. **Charged cylinder.** Integrating the field of Problem 5 over length, obtain an expression for the field of a cylinder of length L and radius R and carrying a surface charge density σ on the cylindrical surface. Examine the limits $z \gg R$ and L, where z is the distance from the center of the cylinder, and $z \ll L$.

8. **Radial fields.** Find the radial field at a distance r from the axis of an infinitely long cylinder of radius R and carrying a surface charge density σ. Find the radial field for a solid cylinder of radius R and volume charge density ρ. Consider $r < R$ as well as $r > R$.

9. **Potential of an infinite cylinder.** By integrating the field of an infinite cylinder of radius R and charge density ρ, show that the potential outside the cylinder is given by:

$$\phi(r) = -\frac{\rho R^2}{2\epsilon_0} \ln \frac{r}{R}$$

where the zero of potential is taken arbitrarily at $r = R$. Show also that the potential inside the cylinder is given by:

$$\phi(r) = \frac{\rho}{4\epsilon_0} (R^2 - r^2)$$

10. **Uniform potential.** A closed equipotential surface contains no charge. Show that the interior electric field is zero and that the enclosed volume is therefore at the same potential as the bounding surface. Assume a local maximum or minimum and use **(41)** to show that the maximum or minimum must be flat. Show that if there were a saddle point as allowed by **(43)**, then **(29)** would require at least one local maximum or minimum as well. This problem may also be discussed in terms of the properties of harmonic functions.

11. **Intersecting spheres.** Two uniform spherical charge distributions of equal magnitude but opposite sign are superposed. Show that the field in the region common to both spheres is uniform. Find the magnitude and direction of the field. What does this result indicate for the field within a spherical cavity in an otherwise uniformly charged sphere?

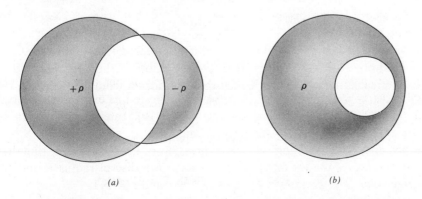

(a)   (b)

12. **Intersecting cylinders.** Find the electric field in the region common to a pair of oppositely charged cylinders of radius R and charge per unit length λ where the distance between the axes is a.

13. **Spherical charge distribution.** A particular charge distribution gives rise to the potential

$$\phi = \frac{1}{4\pi\epsilon_0} \frac{q}{r} e^{-r^2/a^2}$$

Find the electric field and from Gauss's law find the charge density. Show by direct integration that the total charge is zero. Why might you expect this result from the form of **E**?

14. **Hydrogen atom.** The *electronic* charge density in the hydrogen atom is given by:

$$\rho(\mathbf{r}) = \frac{-e}{\pi a_0{}^3} e^{-2r/a_0}$$

Show that the total electronic charge is $-e$. Obtain an expression for the electric field as a function of r including the field of the point proton at the nucleus:

$$\rho(\mathbf{r}) = +e\,\delta(\mathbf{r})$$

Make a plot of this field as a function of r/a_0.

15. **Lineal charge.** Charge is distributed along a line with density $\lambda(z)$. Show that sufficiently close to the line the *radial* component of the electric field is given by:

$$E_r = \frac{1}{2\pi\epsilon_0} \frac{\lambda}{r}$$

16. **Electron biprism.** An electron beam is deflected slightly by the field of a long wire carrying a lineal charge density λ. Show that the deflection is independent of the distance of the electron from the wire. Such an arrangement is called a biprism since electrons passing on opposite sides of the wire are deflected oppositely. See Problem 13-2 for the optical analog, the Fresnel biprism. [G. Möllenstedt and H. Düker, *Z. Phys.*, **145**, 377 (1956).]

17. **Dipole field.** An electric dipole of strength **p** is placed in a homogeneous electric field of strength **E** and is oriented in order to oppose the field. Show that there is a spherical surface of radius R such that the normal component of the resultant field vanishes everywhere on the surface. Find the value of R in terms of p and E. What is the value of E at the equator of the spherical surface?

18. **Quadrupole field.** Show that the field of a quadrupole is given by

$$\mathbf{E} = \frac{1}{8\pi\epsilon_0 r^4} (5\hat{\mathbf{r}}\hat{\mathbf{r}} \cdot \mathbf{Q} \cdot \hat{\mathbf{r}} - 2\hat{\mathbf{r}} \cdot \mathbf{Q})$$

19. **Scalar of quadrupole tensor.** Compute the total flux from a quadrupole by integrating over a sphere of radius R. Show that the electrical neutrality of a quadrupole requires that its scalar be zero.

20. **Multipole fields.** Plot the field along the x and y axes as a function of the distance from the origin for the configuration shown. Plot separately the difference between these fields and the monopole field. Obtain the quadrupole tensor for the distribution shown and plot the quadrupole fields along x and y. Compare the quadrupole field with the difference fields along x and y.

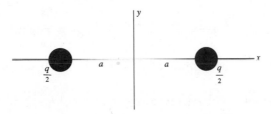

21. **Quadrupole moment.** A spherical shell carries a surface charge density:

$$\sigma(\theta) = (3 \cos^2 \theta - 1)\sigma$$

Find the monopole, dipole, and quadrupole moments of this charge distribution.

22. **Force on a quadrupole.** A quadrupole moment is placed in a nonuniform electric field $E(r)$. Show from **(106)** that the force may be written as:

$$\mathbf{F} = q(\mathbf{a} \cdot \nabla)(\mathbf{b} \cdot \nabla)\mathbf{E}(\mathbf{r})$$

23. **Force on a polarized body.** From **(106)** show that the force on an extended polarized body is given by:

$$\mathbf{F} = \int_V (\mathbf{P} \cdot \nabla)\mathbf{E} \, dV$$

By using **(C-4)** and **(C-9)**, show that the force may also be written as:

$$\mathbf{F} = -\int_V \mathbf{E}(\nabla \cdot \mathbf{P}) \, dV + \oint_S \mathbf{E}(\mathbf{P} \cdot d\mathbf{S})$$

Using **(137)** and **(138)** interpret this result.

24. **Lines of force.** The differential equation of a line of force is

$$\mathbf{E} \times \frac{d\mathbf{r}}{ds} = 0$$

where ds is an increment of length along the force line. Find the general equation for the lines of force of a dipole. Show that the line crossing a sphere of radius R centered on the dipole at $\theta = \pi/4$ from the direction of the dipole crosses the median plane at $r = 2R$.

25. **Uniformly polarized sphere.** Find the electric field and potential outside a uniformly polarized sphere. Find the internal electric field. Find the surface polarization charge. By constructing a small Gaussian cylinder through the surface, verify that the surface charge density **(138)** is consistent with Gauss's law.

26. **Ellipsoidal cavity.** A uniformly charged ellipsoid contains an ellipsoidal cavity whose center is displaced by **a** from the center of the charged body. If the ellipsoids are similar and similarly oriented, find the field within the cavity.

27. **Ellipsoidal homoeoid.** Show that the limiting surface charge density of an ellipsoidal homoeoid is given by:

$$\sigma = \rho_0\,\hat{\mathbf{n}}\cdot\mathbf{r}$$

where $\hat{\mathbf{n}}$ is normal to the ellipsoidal surface, and **r** is a vector from the center of the ellipsoid.

28. **Equipotential ellipsoid.** With reference to Problem 29, how must charge be distributed over the surface of an ellipsoid in order that the internal electric field be zero and the enclosed ellipsoid be at a uniform potential?

29. **Prolate ellipsoid.** The depolarizing factors of a prolate ellipsoid with semimajor axes $A \le B \ll C$ are given by:

$$N_x = \frac{B}{A+B} - \frac{1}{2}\frac{AB}{C^2}\ln\left(\frac{4C}{A+B}\right) + \frac{AB(A+3B)}{4(A+B)C^2}$$

$$N_y = \frac{A}{A+B} - \frac{1}{2}\frac{AB}{C^2}\ln\left(\frac{4C}{A+B}\right) + \frac{AB(3A+B)}{4(A+B)C^2}$$

$$N_z = \frac{AB}{C^2}\ln\left(\frac{4C}{A+B}\right) - \frac{AB}{C^2}$$

For the special case of a prolate *spheroid* $(A = B)$ compare with Exercises 45 and 55 in the limit $L \gg R$.

30. **Oblate ellipsoid.** The depolarizing factor of an oblate ellipsoid with semimajor axes $A \ge B \gg C$ are given by:

$$N_x = -\frac{BC}{C^2}\frac{1}{1-B^2/A^2}E(k) + \frac{BC}{A^2}\frac{1}{1-B^2/A^2}K(k)$$

$$N_y = \frac{C}{B}\frac{1}{1-B^2/A^2}E(k) - \frac{BC}{A^2}\frac{1}{1-B^2/A^2}K(k)$$

$$N_z = 1 - \frac{C}{B}E$$

where $E(k)$ and $K(k)$ are the complete elliptic integrals of argument $k^2 = 1 - B^2/A^2$. (See, for example, the *Mathematical tables of the Handbook of Chemistry and Physics*, Chemical Rubber Co.)

  Compare the depolarizing factors of an oblate *spheroid* $(A = B)$ with the results of Exercises 45 and 55 for the case $L \ll R$.

31. **Point dipole.** The volume of a uniformly polarized ellipsoid is allowed to shrink to zero while the polarization density $\mathbf{P}$ increases so that the product $\mathbf{p} = \mathbf{P}V$ remains constant. Show that in the limit of zero volume the electric field everywhere is given by:

$$\mathbf{E} = \frac{1}{4\pi\epsilon_0} \frac{3\hat{\mathbf{r}}(\hat{\mathbf{r}} \cdot \mathbf{p}) - \mathbf{p}}{r^3} - \frac{1}{\epsilon_0} \mathbf{N} \cdot \mathbf{p}\, \delta(\mathbf{r})$$

Compare with **(96)**. What is the limiting field of a polarized sphere?

32. **Electric stress tensor.** The force on a charge density $\rho(\mathbf{r})$ is from the integration of **(3)**:

$$\mathbf{F} = \int_V \rho(\mathbf{r})\mathbf{E}(\mathbf{r})\, dV$$

Show that the force may be written as an integral over the surface which bounds V:

$$\mathbf{F} = \oint_S d\mathbf{S} \cdot \mathbf{T}$$

where $\mathbf{T}$ is the electric stress tensor:

$$\mathbf{T} = \epsilon_0(\mathbf{EE} - \tfrac{1}{2}\mathbf{I}\, E^2)$$

Use **(41)**, **(C-4)**, **(C-3)** and Exercise 8, **(C-9)**, **(B-12)**, and **(C-28)**.

REFERENCES

Berkson, W., *Field of Force: The Development of a World View from Faraday to Einstein*, Halsted, 1974.

Tricker, R. A. R., *The Contributions of Faraday and Maxwell to Electrical Science*, Pergamon, 1966.

2 SOURCES OF THE ELECTRIC FIELD II

ELECTRIC ENERGY

In making the argument in the preceding chapter that the circulation of the electric field of static charges is everywhere zero we showed that the total amount of work done in carrying a test charge around a closed contour must be zero. This amounts to saying that energy is stored by virtue of the configuration of the charges and that this energy is equal to the amount of work which must be done to assemble the charge distribution.

1 **Charge Interaction.** Let us imagine a set of charges q_1, q_2, q_3, ..., q_N, which we wish to assemble into some configuration. We imagine that initially the charges are well separated. We also imagine that each charge is localized over a region small compared with the distance between charges in the final configuration. We begin by bringing charge q_1 to a position $\mathbf{r}_1$ as shown in Figure 1a. We next bring up q_2 to a position $\mathbf{r}_2$ as shown in Figure 1b. The amount of work done to bring up q_2 is given by:

$$W_2 = -\int \mathbf{F}_2 \cdot d\mathbf{r}_2 = -q_2 \int \mathbf{E}_2 \cdot d\mathbf{r}_2 = -\frac{q_1 q_2}{4\pi\epsilon_0} \int \frac{\hat{\mathbf{r}}_{21}}{r_{21}{}^2} \cdot d\mathbf{r}_2$$

$$= -\frac{q_1 q_2}{4\pi\epsilon_0} \int \frac{dr_{21}}{r_{21}{}^2} = \frac{1}{4\pi\epsilon_0} \frac{q_1 q_2}{r_{21}} \tag{1}$$

Next we bring up q_3 to a position $\mathbf{r}_3$ as shown in Figure 1c. The amount of work done on q_3 will be:

$$W_3 = -\int \mathbf{F}_3 \cdot d\mathbf{r}_3 = -q_3 \int \mathbf{E}_3 \cdot d\mathbf{r}_3 = -\frac{q_3}{4\pi\epsilon_0} \int \left(\frac{q_1 \hat{\mathbf{r}}_{31}}{r_{31}{}^2} + \frac{q_2 \hat{\mathbf{r}}_{32}}{r_{32}{}^2} \right) \cdot d\mathbf{r}_3$$

$$= -\frac{q_1 q_3}{4\pi\epsilon_0} \int \frac{dr_{31}}{r_{31}{}^2} - \frac{q_2 q_3}{4\pi\epsilon_0} \int \frac{dr_{32}}{r_{32}{}^2} = \frac{1}{4\pi\epsilon_0} \left(\frac{q_1 q_3}{r_{31}} + \frac{q_2 q_3}{r_{32}} \right) \tag{2}$$

We can see how this is going and should be able to write down the general term:

$$W_i = \frac{1}{4\pi\epsilon_0} \sum_{j=1}^{i-1} \frac{q_i q_j}{r_{ij}} \tag{3}$$

Finally, to obtain the total work done, we must sum over i from 1 to N:

$$W_{ext} = \sum_{i=1}^{N} W_i = \frac{1}{4\pi\epsilon_0} \sum_{i=1}^{N} \sum_{j=1}^{i-1} \frac{q_i q_j}{r_{ij}} \tag{4}$$

where the subscript *ext* reminds us that we compute the work against *external* forces only and do not consider *internal* forces associated with any possible structure of the q_i. Now instead of beginning with q_1 and then bringing up q_2, and so forth, we could have begun with q_N and then brought up q_{N-1}, and so on. The final configuration would

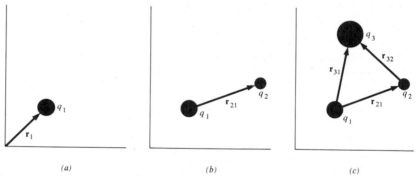

(a)   (b)   (c)

Figure 1 Charge interaction. (a) A charge q_1 is brought to a position r_1. (b) A second charge q_2 is then brought to r_2. (c) Next, a charge q_3 is brought to r_3.

be exactly the same with W containing exactly the same terms but summed in different order:

$$W_{ext} = \frac{1}{4\pi\epsilon_0} \sum_{i=1}^{N} \sum_{j=i+1}^{N} \frac{q_i q_j}{r_{ij}} \tag{5}$$

Now, if we add **(4)** and **(5)** and divide by 2 we obtain:

$$W_{ext} = \frac{1}{8\pi\epsilon_0} \sum_{i} \sum_{j}' \frac{q_i q_j}{r_{ij}} \tag{6}$$

where the prime on the second summation means that the term $i = j$ is excluded. Now the potential at the ith charge (treating q_i as if it were a test charge) is given from **(1-18)** by:

$$\phi_i = \frac{1}{4\pi\epsilon_0} \sum_{j}' \frac{q_i}{r_{ij}} \tag{7}$$

We emphasize that (7) excludes the $i = j$ term. Thus ϕ_i is *not* the potential that would be measured by an *independent* test charge q. It is the potential measured by q_i treated as a test charge and must therefore exclude any interaction of q_i with itself. Finally, substituting (7) into (6), we obtain for the work done:

$$\blacktriangleright \qquad W_{\text{ext}} = \tfrac{1}{2} \sum_i q_i \phi_i \qquad\qquad (8)$$

We emphasize that (8) is just the work done to assemble the configuration of charges. It does not include the self-energy of the charges, the work that must be done to build up the individual charges q_i in the first place. And W_{ext} may be positive or negative (or zero). For two charges only, for example, if the charges have the same sign then W_{ext} is positive; while if they have opposite signs, W_{ext} is negative.

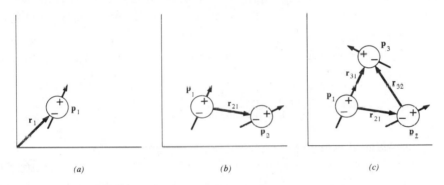

   (a)   (b)   (c)

Figure 2 Dipole interaction. (*a*) A dipole $\mathbf{p}_1$ is brought to $\mathbf{r}_1$. (*b*) A second dipole $\mathbf{p}_2$ is brought up to $\mathbf{r}_2$. (*c*) Next, a dipole $\mathbf{p}_3$ is brought to $\mathbf{r}_3$.

Exercises

1. Charges $-q$ are placed at the corners of a regular tetrahedron and a charge $+4q$ at the center. How much work is required to assemble this configuration of charges?
2. A uniformly charged sphere of radius R and charge density ρ is assembled by bringing successive layers of charge in from infinity. Show that the total amount of work that must be done is

$$W = \frac{3}{5} \frac{q^2}{4\pi\epsilon_0 R}$$

2 **Dipolar Interaction.** By analogy with the previous section, we wish to obtain expressions for the amount of work that must be done to bring together a group of electric dipoles. We first bring up dipole $\mathbf{p}_1$ as shown in Figure 2a. To bring up dipole $\mathbf{p}_2$ we must do an amount of work:

$$W_2 = -\int \mathbf{F}_2 \cdot d\mathbf{r}_2 \qquad\qquad (9)$$

The force on a dipole, from **(1-105)** is given by:

$$\mathbf{F}_2 = \mathbf{p}_2 \cdot \nabla \mathbf{E}_2 \tag{10}$$

where from **(1-96)** the field produced at $\mathbf{r}_2$ by $\mathbf{p}_1$ is:

$$\mathbf{E}_2 = \mathbf{E}(\mathbf{r}_2) = \frac{1}{4\pi\epsilon_0} \frac{3\hat{\mathbf{r}}_{21}(\mathbf{p}_1 \cdot \hat{\mathbf{r}}_{21}) - \mathbf{p}_1}{r_{21}{}^3} \tag{11}$$

Substituting **(10)** into **(9)** and integrating, we obtain:

$$W_2 = -\mathbf{p}_2 \cdot \mathbf{E}_2 = \frac{1}{4\pi\epsilon_0} \frac{\mathbf{p}_1 \cdot \mathbf{p}_2 - 3(\mathbf{p}_1 \cdot \hat{\mathbf{r}}_{21})(\mathbf{p}_2 \cdot \hat{\mathbf{r}}_{21})}{r_{21}{}^3} \tag{12}$$

Bringing up dipole $\mathbf{p}_3$ we obtain expressions similar to **(12)** for the interaction with $\mathbf{p}_1$ and $\mathbf{p}_2$. The argument follows in the same way as for charges and we obtain finally:

$$W_{\text{ext}} = \frac{1}{8\pi\epsilon_0} \sum_i \sum_j{}' \frac{\mathbf{p}_i \cdot \mathbf{p}_j - 3(\mathbf{p}_i \cdot \hat{\mathbf{r}}_{ij})(\mathbf{p}_j \cdot \hat{\mathbf{r}}_{ij})}{r_{ij}{}^3} \tag{13}$$

Now, the electric field at the jth dipole is given from **(11)** by:

$$\mathbf{E}_j = -\frac{1}{4\pi\epsilon_0} \sum_i{}' \frac{\mathbf{p}_i - 3\hat{\mathbf{r}}_{ij}(\mathbf{p}_i \cdot \hat{\mathbf{r}}_{ij})}{r_{ij}{}^3} \tag{14}$$

We may then write **(13)** as:

▶
$$W_{\text{ext}} = -\tfrac{1}{2} \sum_j \mathbf{p}_j \cdot \mathbf{E}_j \tag{15}$$

where $\mathbf{E}_j$ is the electric field at the site of the jth dipole produced by all the other dipoles. Again, we must emphasize that **(15)** is the work to bring fixed dipoles in from infinity and does not include the amount of work required to develop the dipole moments themselves.

Exercise

3. Demonstrate the identity:

$$\mathbf{p}_i \cdot \mathbf{E}_{ij} = \mathbf{p}_j \cdot \mathbf{E}_{ji}$$

where $\mathbf{E}_{ij}$ is the field of $\mathbf{p}_j$ at the position of $\mathbf{p}_i$ and $\mathbf{E}_{ji}$ is the field of $\mathbf{p}_i$ at the position of $\mathbf{p}_j$.

3 Mixed Interactions. We now want to consider the situation shown in Figure 3 where we have a mixture of charges and dipoles. We *could* develop expressions for the work required using the same procedures that we have already used for charges and dipoles considered separately. Instead, by considering an incremental change in charge and polarization we may obtain the required result directly.

The amount of work done on the distribution of charges and dipoles by an incremental change is given by:

$$dW_{\text{ext}} = \sum_i \phi_i \, dq_i - \sum_j \mathbf{E}_j \cdot d\mathbf{p}_j \tag{16}$$

where ϕ_i is the potential at $\mathbf{r}_i$ from all dipoles and charges except q_i. Although **(16)** appears to be limited to a situation where the values of the q_i and $\mathbf{p}_j$ are changed, but not their positions, it may actually be used to describe any small variation in the distribution of charges and dipoles. In particular, since we wish to exclude the energy required to establish the q_i and $\mathbf{p}_j$ we use **(16)** to describe the result of small

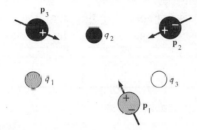

Figure 3 Mixed interactions. We consider the interaction between charges and dipoles.

displacements. To describe the displacement of q_i from $\mathbf{r}_i$ to $\mathbf{r}_i + d\mathbf{r}_i$, for example, we require two terms:

1. The first term is with $dq_i = -q_i$ and $\phi_i = \phi(\mathbf{r}_i)$.
2. The second term is with $dq_i = +q_i$ and $\phi_i = \phi(\mathbf{r}_i + d\mathbf{r}_i)$.

Even though the dq_i are not small, the change in potentials and fields at the other charges and dipoles will be small and dW_{ext} will be a small quantity.

Similarly, $\mathbf{E}_j$ is the field at $\mathbf{r}_j$ from all charges and dipoles *except* $\mathbf{p}_j$:

$$\phi_i = \sum_k{}' \frac{q_k}{r_{ik}} + \sum_l \frac{\mathbf{p}_l \cdot \hat{\mathbf{r}}_{il}}{r_{il}^{\,2}} \tag{17}$$

$$\mathbf{E}_j = -\nabla \phi_j = -\nabla \left(\sum_k \frac{q_k}{r_{jk}} + \sum_l{}' \frac{\mathbf{p}_l \cdot \hat{\mathbf{r}}_{jl}}{r_{jl}^{\,2}} \right) \tag{18}$$

Summing **(16)** over the displacements of q_i and $\mathbf{p}_j$ we obtain:

$$W_{\text{ext}} = \tfrac{1}{2} \sum_i q_i \phi_i - \tfrac{1}{2} \sum_j \mathbf{p}_j \cdot \mathbf{E}_j \tag{19}$$

The factors of one-half simply correct for the fact that the summations count all pair-interactions twice, as we have seen in obtaining **(6)** and **(13)**.

Exercises

4. Explain why the summations of **(19)** do not include the self-energy of the dipoles or charges.
5. Obtain the expression for the energy of interaction of a charge q and a dipole **p**. Write the energy alternatively in terms of the potential at q and the field at **p**. Show that the two expressions are equivalent.

4   Continuous Charge. We now wish to investigate the limit of **(19)** as we bring up more and more charges and dipoles while letting the magnitudes of the individual charges and dipoles go to zero. In this limit we wish to write the external work in terms of the macroscopic potential $\phi(\mathbf{r})$ and the macroscopic field $\mathbf{E}(\mathbf{r})$ rather than in terms of the local potential ϕ_i and field $\mathbf{E}_j$ as in **(19)**.

We treat the charge q_i as a sphere of radius R_i and charge density ρ_i. From **(1-67)** the potential at the center of the sphere is given by:

$$\phi(\mathbf{r}_i) = \phi_i + \frac{1}{2\epsilon_0} \rho_i R_i{}^2 \tag{20}$$

In the continuum limit we allow R_i to go to zero with ρ_i remaining finite. In this limit we have from **(20)**

$$\phi(\mathbf{r}_i) = \phi_i \tag{21}$$

so that the macroscopic potential is just equal to the potential from all other charge in the limit that the individual charges go to zero.

Similarly, we wish to treat the dipoles in the continuum limit. We have from **(1-125)** for the macroscopic field:

$$\mathbf{E}(\mathbf{r}_j) = \mathbf{E}_j - \frac{1}{3\epsilon_0} \mathbf{P}(\mathbf{r}_j) \tag{22}$$

where $\mathbf{E}_j$ as given by **(19)** is what in the context of Section 1-16 we called the external field.

The total charge density is the sum of the density of charges q_i and the divergence of the polarization density associated with the $\mathbf{p}_j$ from **(1-137)**:

$$\rho(\mathbf{r}) = \rho_i - \nabla \cdot \mathbf{P}(\mathbf{r}) \tag{23}$$

Finally, substituting **(21)**, **(22)** and **(23)** into **(19)** we obtain for the external work:

$$\blacktriangleright \qquad W_{\text{ext}} = \tfrac{1}{2} \int_V \rho(\mathbf{r})\phi(\mathbf{r})\, dV - \frac{1}{6\epsilon_0} \int_V P^2(\mathbf{r})\, dV \qquad (24)$$

In obtaining **(24)** we have used **(B-2)**, converting $\nabla \cdot [\phi(\mathbf{r})\mathbf{P}(\mathbf{r})]$ to an integral over S, which we have allowed to go to zero on the assumption that the polarization density on S is zero.

We note that **(24)** is not uniquely determined by the charge density $\rho(\mathbf{r})$ and the potential $\phi(\mathbf{r})$ but depends on what we take to be the local polarization density $\mathbf{P}(\mathbf{r})$. The reason for this ambiguity is that **(24)** is not the *total* work required to build up an arbitrary charge distribution, but excludes the internal energy of the individual dipoles $\mathbf{p}_j$, which does not go to zero, even in the continuum limit!

In obtaining **(24)** we considered only mixed interactions between charges and dipoles. Had we also included quadrupoles and possibly higher multipoles we would have had additional correction terms to **(24)** for the internal energy of these moments.[1]

Exercises

6. A sphere of radius R carries a charge q uniformly distributed through its volume. Using **(24)**, find the work required to build up the charge.

7. A sphere of radius R is uniformly polarized with polarization density $\mathbf{P}$. Using **(24)** find the work required to build up the polarization. Use the surface equivalent of **(24)**:

$$W_{\text{ext}} = \tfrac{1}{2} \oint_S \sigma(\mathbf{r})\phi(\mathbf{r})\, dS - \frac{1}{6\epsilon_0} \int_V P^2(\mathbf{r})\, dV$$

where $\sigma = \hat{\mathbf{n}} \cdot \mathbf{P}$ is the surface polarization charge density.

5   Average Potential on a Spherical Shell.

It is a consequence of Laplace's equation **(1-43)**:

$$\nabla^2 \phi(\mathbf{r}) = 0 \qquad (25)$$

that the average of ϕ over a spherical surface for which **(25)** is satisfied in the interior volume is simply equal to the value of ϕ at the center of the sphere. We prove this *theorem* by using the reciprocity of the energy in a way similar to the use of force reciprocity in Chapter 1 for finding the average electric field on a sphere.

We wish to find the average potential on a spherical surface S from a charge $\rho(\mathbf{r}_1)$ contained within V_1, which is wholly outside the sphere as shown in Figure 4a. In order

---

[1] See G. Russakoff, "A Derivation of the Macroscopic Maxwell Equations," *Am. J. Phys.* **38**, 1188 (1970) for a discussion that includes microscopic quadrupole moments. See also A. N. Kaufman, "Definition of Macroscopic Electrostatic Field," *Am. J. Phys.* **29**, 626 (1961).

to do this problem we first treat the simpler problem of the energy of interaction between a point charge q_1 at $\mathbf{r}_1$ and a uniformly charged spherical shell of radius R and surface charge density σ as shown in Figure 4b.

As we have seen **(1-62)**, the potential at $\mathbf{r}$ is given by

$$\phi(\mathbf{r}) = \frac{\sigma R^2}{\epsilon_0 \, r} \tag{26}$$

Then the energy of interaction between q_1 and the sphere is given from **(1)** by:

$$U = q_1 \phi(r_1) = \frac{q_1}{\epsilon_0} \frac{\sigma R^2}{r_1} \tag{27}$$

Now, another way of looking at the problem is to consider the energy of the surface charge σ in the potential developed around q_1:

$$U = \oint_S \phi(\mathbf{r}) \, \sigma \, dS \tag{28}$$

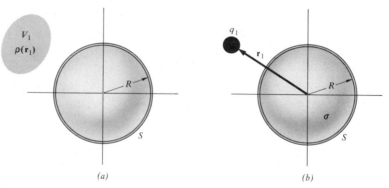

(a)   (b)

Figure 4   Average potential on a spherical surface. (*a*) The charge density $\rho(\mathbf{r}_1)$ is taken to be wholly outside the spherical surface S. (*b*) The potential is obtained from a calculation of the interaction between a point charge q_1 and a uniformly charged spherical shell.

Comparing **(28)** with **(27)** we have the result:

$$\langle \phi(\mathbf{r}) \rangle_S = \frac{1}{4\pi R^2} \oint_S \phi(\mathbf{r}) \, dS = \frac{q}{4\pi \epsilon_0 \, r_1} \tag{29}$$

which is just the potential at the center of the sphere! We may now extend q_1 to a charge distribution $\rho(\mathbf{r}_1)$. For each contribution to the potential we must have a relation of the form **(29)** from which it follows that for any source of potential outside S we have for the average over the spherical surface:

$$\langle \phi(\mathbf{r}) \rangle_S = \frac{1}{4\pi R^2} \oint_S \phi(\mathbf{r}) \, dS = \phi(0) \tag{30}$$

Exercises

8. By considering the energy of interaction between a point charge and a *hypothetical* uniformly charged sphere, show that the average potential within any spherical *volume* is equal to the potential at the center as long as there is no interior charge.

9. A spherical shell of radius R contains total charge q. Show that the average potential on the surface is $q/4\pi\epsilon_0 R$.

6 Energy Density of an Electric Field. We now look at **(24)** from an entirely different point of view in which the work W is regarded as going into building up the electric field rather than into assembling a charge configuration. For a static charge distribution, the two views of W are equivalent. If charges are moving (and fields are changing) the two ways of looking at W are different and, as we see much later, only the field view is correct.

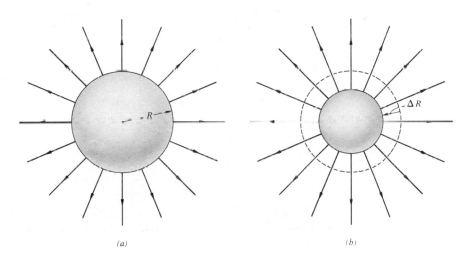

(a)   (b)

Figure 5   Energy density of an electric field. (*a*) A spherical shell of radius R carries a surface charge density σ. (*b*) The sphere is compressed from R to $R - \Delta R$. The amount of work is given by **(34)**.

Before discussing the general case, we consider the specific example shown in Figure 5. We imagine a spherical shell of radius R and surface charge density σ. As we may see, from **(1-59)**, the field just outside the surface is

$$E = \frac{1}{4\pi\epsilon_0} \frac{q}{R^2} = \frac{\sigma}{\epsilon_0} \tag{31}$$

and from **(1-63)** the potential is:

$$\phi = \frac{1}{4\pi\epsilon_0} \frac{q}{R} \tag{32}$$

The amount of work required to assemble such a distribution is, from **(24)** and **(32)**,

$$W = \tfrac{1}{2} \int \sigma\phi \, dS = \tfrac{1}{2}q\phi = \frac{q^2}{8\pi\epsilon_0 R} \tag{33}$$

Let us imagine that we perform work on the charge distribution and compress the sphere from radius R to $R - \Delta R$. The amount of work done from **(33)** is:

$$\Delta W = \frac{1}{8\pi\epsilon_0} \frac{q^2}{R^2} \Delta R = \frac{\sigma^2}{\epsilon_0} 2\pi R^2 \, \Delta R \tag{34}$$

We now wish to interpret **(34)** in terms of field energy. As we have seen **(31)** the field E just outside the surface is σ/ϵ_0 so that **(34)** may be written as

$$\Delta W = \tfrac{1}{2}\epsilon_0 E^2 \, \Delta V \tag{35}$$

where $\Delta V = 4\pi R^2 \, \Delta R$ is the additional volume made available to the *field* when the sphere is compressed. Since the field everywhere else is unaffected, we may think of the work ΔW going into the establishment of the field E in ΔV. Finally, we may write the total amount of work done to bring the charge q in from infinity (uniformly spread over a spherical surface of decreasing radius) as

$$W = \tfrac{1}{2}\epsilon_0 \int E^2 \, dV \tag{36}$$

where the integral is over *all space*. By conservation of energy we say that the work W goes to establish a field energy U given by **(36)**.

We now show that **(36)** is really much more general and in fact applies to any *charge* distribution. We do this by transforming **(24)**, which we now call U, the stored energy. Since we have, from **(1-42)** $\rho(\mathbf{r}) = -\epsilon_0 \nabla^2\phi(\mathbf{r})$, we may write

$$U = -\tfrac{1}{2}\epsilon_0 \int \phi(\mathbf{r}) \nabla^2\phi(\mathbf{r}) \, dV \tag{37}$$

where we have assumed that the polarization density $\mathbf{P}(\mathbf{r})$ is zero.

We transform the integrand of **(37)** by making use of the vector identity **(B-2)**:

$$\nabla \cdot (\phi\nabla\phi) = \phi\nabla^2\phi + \nabla\phi \cdot \nabla\phi = \phi\nabla^2\phi + E^2 \tag{38}$$

Substituting into **(37)** we obtain:

$$U = \tfrac{1}{2}\epsilon_0 \int_V E^2 \, dV - \tfrac{1}{2}\epsilon_0 \int_V \nabla \cdot (\phi\mathbf{E}) \, dV \tag{39}$$

Transforming the second integral by Gauss's theorem **(B-14)** we may write:

$$U = \tfrac{1}{2}\epsilon_0 \int_V E^2 \, dV - \tfrac{1}{2}\epsilon_0 \oint_S \phi \mathbf{E} \cdot d\mathbf{S} \tag{40}$$

Let us imagine that the total amount of charge is bounded and finite. Then as the surface S is extended to infinity the second integral must go to zero. This may be argued from our discussion in Section 1-10 of the moments of a charge distribution. The potential at large distances r must drop off as $1/r$ or faster while the field falls off as $1/r^2$ or faster. The surface S increases as r^2 so that the second integral goes to zero as $1/r$. In other words, the second integral goes to $\tfrac{1}{2}q\phi(\infty)$, where q is the total charge and is finite and $\phi(\infty)$ is the potential at infinity, which must go to zero for a bounded charge distribution. We are left, then, with the same result as **(36)**:

$$\blacktriangleright \qquad U = \tfrac{1}{2}\epsilon_0 \int E^2 \, dV \tag{41}$$

where, in order that we may neglect the surface term, the integral must be over *all space*.

As is apparent from the form of **(41)**, the energy in the electric field is positive. Thus the first term of **(24)**, which is equivalent to **(41)**, must be positive. The total external work, as given by the difference of the two terms in **(24)** may be positive or negative.

Exercises

10. Obtain **(34)** from the field acting on incremental charges $\sigma \, dS$ at the surface. Explain why the field used must be just half that given by **(31)**.
11. Beginning with **(41)**, obtain the first term of **(24)**.
12. Using **(41)** find the energy of a uniformly charged sphere of radius R and carrying a total charge q. What fraction of the energy is associated with the interior volume of the sphere?
13. A distribution of charge $q_1 = \int_{V_1} \rho \, dV$ is bounded by a surface S_1 of potential ϕ_1. Show that energy stored *outside* V_1 may be written as

$$U = \tfrac{1}{2}\epsilon_0 \int_{V-V_1} E^2 \, dV = \tfrac{1}{2}q_1\phi_1$$

7 Localization of Electric Energy. Finally, we discuss the sense in which

$$u = \frac{\Delta U}{\Delta V} = \tfrac{1}{2}\epsilon_0 E^2 \tag{42}$$

may be considered to be a *local* energy density. Although the argument leading to **(36)** strongly suggests that ΔW goes into the local buildup of field energy, this is not really clear in the transformation of **(24)** into **(40)**. At the same time, there is nothing in the present discussion that precludes our thinking of **(42)** as a local relation. We return to this question in Section 11-13, where we will see that momentum as well as energy may be associated with fields and that *local* expressions must be developed for the flow of both momentum and energy through fields if these quantities are to be conserved.

FORCE ON A CHARGE SHEET

We show that the force on a charge sheet is simply given by the product of the total charge in the sheet and the *mean* of the fields on opposite sides of the sheet. This result is exact and may be taken to the limit of an infinitesimally thin layer, which may be treated in terms of a surface charge density σ.

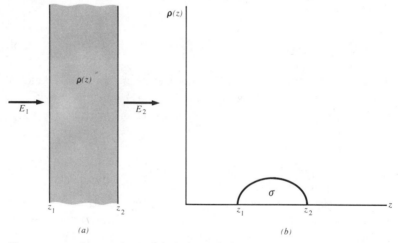

Figure 6 Force on a charge sheet. (*a*) A bounded one-dimensional charge distribution is described by $\rho(z)$. (*b*) The integrated volume charge density is given by σ.

8 **[Planar Sheet].** We show in Figure 6*a* a bounded one-dimensional charge distribution $\rho(z)$. We take E_1 and E_2 to be the fields on the two sides of the distribution.

From **(1-41)** we have

$$\mathbf{V} \cdot \mathbf{E} = \frac{\rho}{\epsilon_0} \tag{43}$$

Assuming that the charge density is a function only of z, we need consider just the field along z, which we write as $E(z)$. Then **(43)** becomes:

$$\frac{dE(z)}{dz} = \frac{\rho(z)}{\epsilon_0} \tag{44}$$

We may integrate **(44)** to obtain:

$$E(z) = E_1 + \frac{1}{\epsilon_0} \int_{z_1}^{z} \rho(z) \, dz \tag{45}$$

For $z \geq z_2$ the field becomes:

$$E(z) = E_2 = E_1 + \frac{1}{\epsilon_0} \int_{z_1}^{z_2} \rho(z) \, dz = E_1 + \frac{\sigma}{\epsilon_0} \tag{46}$$

where σ is the *total* charge per unit area.

Now the force on a section of width Δz and area S may be written as:

$$\Delta F = \rho(z) E(z) S \, \Delta z \tag{47}$$

Substituting from **(44)** for $\rho(z)$ we may write:

$$\Delta F = \epsilon_0 \, E(z) \frac{dE(z)}{dz} S \, \Delta z \tag{48}$$

The total force on the section of area S is the integral of **(48)** over z:

$$F = \epsilon_0 \int_{z_1}^{z_2} E(z) \frac{dE(z)}{dz} S \, dz = \epsilon_0 S \int_{E_1}^{E_2} E \, dE$$

$$= \tfrac{1}{2}\epsilon_0 \, S(E_2{}^2 - E_1{}^2) = \tfrac{1}{2}\epsilon_0 \, S(E_1 + E_2)(E_2 - E_1) \tag{49}$$

From **(46)** we have $\epsilon_0(E_2 - E_1) = \sigma$ so that we may finally write **(49)** in vector form as:

$$\mathbf{F} = \tfrac{1}{2}(\mathbf{E}_1 + \mathbf{E}_2)\sigma S \tag{50}$$

which is the product of the mean of the exterior fields with the total charge.

9 [**Mechanical Pressure**]. The force acting on the charge sheet must be equilibrated by a difference in mechanical pressure on the two sides. If we call the pressures P_1 and P_2, we may write, for the system in equilibrium,

$$F + (P_1 - P_2)S = 0 \tag{51}$$

Comparing **(51)** with **(49)** we write, for the mechanical pressure,

$$P_{\text{mech}} = P_0 + \tfrac{1}{2}\epsilon_0 \, E^2 \tag{52}$$

which, apart from the ambient pressure P_0, is just the expression given by **(42)** for the local energy density. We return to a consideration of stresses on a medium supporting electric charge in our discussions of dielectrics and again when we examine the behavior of changing electric and magnetic fields.

STABILITY OF ELECTRIC CHARGE

The discussion of the preceding section of equilibrating mechanical and electrical forces suggests that a system of distributed charge under electrical forces alone can not be stable. The argument was somewhat restricted, though, and it may be desirable to discuss the stability of charge from a microscopic point of view.

10   Earnshaw's Theorem. We imagine a static distribution of point charges q_i as shown in Figure 7. We select a charge q, which we regard as a test charge, and construct a sphere of radius R about this charge. The sphere is taken to be small enough that q is the

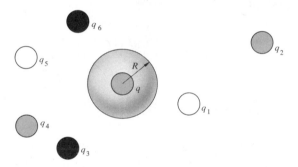

Figure 7   Stability of electric charge. We imagine a distribution of point charges q_i, selecting a particular charge q, which we regard as a test charge.

only charge within the sphere. We have seen **(30)** that the average potential on a spherical surface that contains no *source* charge is the same as the potential at the center. But for q to be in stable equilibrium, all points on the sphere would have to be at higher potential than the center, which cannot be the case. Then there must be at least one direction along which the charge can lower its energy through a displacement.

Since there is no stable position for any charge, the entire distribution will coalesce into a reduced number of point charges, which must then disperse to infinity.

The argument may be made directly from Laplace's equation in a way that may be instructive. The work required to *displace* q is given by:

$$\Delta W = q\Delta\phi = q(\Delta\mathbf{r} \cdot \nabla)\phi + \tfrac{1}{2}q(\Delta\mathbf{r} \cdot \nabla)^2\phi + \cdots \tag{53}$$

For the charge to be stable, ΔW must be positive for all displacements $\Delta\mathbf{r}$. We expect that the first term on the right side of **(53)**, which is just the field, will vanish.

Now, the second term, which is the curvature in ϕ along r, must be positive if the system is to be stable. But if the curvature is to be positive along all directions, then the *total* curvature:

$$\left(\frac{d^2\phi}{dx^2} + \frac{d^2\phi}{dy^2} + \frac{d^2\phi}{dz^2}\right)\phi = \nabla^2\phi \tag{54}$$

must be positive. But the potential must satisfy Laplace's equation **(1-43)**:

$$\nabla^2\phi = 0 \tag{55}$$

so that the total curvature is zero. In general, there will be some *directions* for which ϕ is a minimum, and others for which the curvature is negative and ϕ is a maximum.[2] And q will be unstable with respect to small displacements along such directions.

For the case of a continuous distribution of charge, we must make the argument differently since it is not possible to isolate continuous charge. The easiest argument to make is that the first term on the right side of **(53)** can not be zero everywhere within a continuous charge distribution and thus there must be forces acting on the charges.

Let us assume, however, that E *is* zero everywhere within a distribution of charge. Then the field may have no divergence. But if the field has no divergence within the region, the charge density must be zero, which is contrary to our assumption. We conclude, then, that any continuous distribution of charge must be accompanied by internal electric fields and in the absence of other than electrical forces can not be in static equilibrium.

That the potential cannot be a local minimum was first shown in 1842 by S. Earnshaw, who was treating an analogous elastic problem. The theorem was applied by Maxwell to the electrostatic case. As stated by Maxwell,[3] Earnshaw's theorem is:

A charged body placed in a field of electric force cannot be in stable equilibrium.

Thus, for example, the electron and proton will either require other than electrical forces for stabilization or cannot be viewed as static structures. Even assuming stable electrons and protons, we cannot build static atoms but must have kinetic energy present. And when we look at the interaction between ions we cannot expect stable structures if only Coulomb (i.e., central and $1/r^2$) forces are acting. We discuss real dielectric materials in the next chapter and examine the interactions required for the stabilization of these structures.

---

[2] Should all second derivatives vanish identically, higher derivatives of ϕ must be examined. See R. Weinstock, *Am. J. Phys.*, **44**, 392 (1976).

[3] J. C. Maxwell, *A Treatise on Electricity and Magnetism*, Third Edition, Dover, 1954, Section 116. See also W. T. Scott, "Who was Earnshaw?", *Am. J. Phys.*, **27**, 418 (1959).

SUMMARY

1. The energy of a distribution of discrete charges may be written as the work to bring the charges together:

$$W_{ext} = \tfrac{1}{2} \sum_i q_i \, \phi_i \tag{8}$$

where ϕ_i is the potential at $\mathbf{r}_i$ from all the other charges.

2. The energy of a distribution of discrete dipoles may be written as the work to bring the dipoles together:

$$W_{ext} = -\tfrac{1}{2} \sum_j \mathbf{p}_j \cdot \mathbf{E}_j \tag{15}$$

where $\mathbf{E}_j$ is the electric field at $\mathbf{r}_j$ from all the other dipoles.

3. For a bounded continuous distribution of charge and polarization, the work can always be written in terms of the *total* charge and polarization density:

$$W_{ext} = \tfrac{1}{2} \int \rho(\mathbf{r}) \phi(\mathbf{r}) \, dV - \frac{1}{6\epsilon_0} \int_V P^2(\mathbf{r}) \, dV \tag{24}$$

4. The average potential on the surface of a charge-free sphere is equal to the potential at the center of the sphere.

5. The work to establish a *charge* distribution may be written in terms of the electric field

$$W = \tfrac{1}{2}\epsilon_0 \int E^2(\mathbf{r}) \, dV \tag{41}$$

and $\Delta W / \Delta V = \tfrac{1}{2}\epsilon_0 E(\mathbf{r})^2$ may be interpreted as a *local* energy density.

6. The electric force on a charge distribution must be equilibrated locally by mechanical pressure with the pressure difference equal to the difference in the local energy density.

7. A static distribution of charge cannot be stabilized by Coulomb forces alone. This statement applies both to distributed and to discrete charge and is called Earnshaw's theorem.

PROBLEMS

1. **Interaction between charges.** By superposition, the field of charges q_1 and q_2 may be written as $\mathbf{E}_1 + \mathbf{E}_2$. Substituting into (36), the cross term should represent the interaction energy $q_1 q_2 / 4\pi\epsilon_0 r_{12}$. Show this explicitly.

2. **Interacting charge distributions.** A charge distribution $\rho_1(\mathbf{r})$ gives a potential $\phi_1(\mathbf{r})$; a charge distribution $\rho_2(\mathbf{r})$ gives a potential $\phi_2(\mathbf{r})$. What is the potential of

$\rho_1(\mathbf{r}) + \rho_2(\mathbf{r})$? From **(24)** obtain expressions for the work required to assemble $\rho_1(\mathbf{r})$, $\rho_2(\mathbf{r})$, and $\rho_1(\mathbf{r}) + \rho_2(\mathbf{r})$. Obtain an expression for the interaction energy between $\rho_1(\mathbf{r})$ and $\rho_2(\mathbf{r})$. Show that the interaction energy is equivalent to the work required to bring the distributions together:

$$W = \int \rho_1(\mathbf{r})\phi_2(\mathbf{r})\, dV = \int \rho_2(\mathbf{r})\phi_1(\mathbf{r})\, dV$$

3. **Division of charge.** Two spherical shells of radius R_1 and R_2 are separated by a distance r between centers. Charges q_1 and q_2 are distributed uniformly over the surfaces of the respective spheres with the total charge $q_1 + q_2 = Q$ a constant. How should Q be divided in order to minimize the energy?

4. **Concentric shells.** Two concentric shells of radius R_1 and R_2 have charge q_1 and q_2, respectively. Obtain the fields in the regions:

$$r < R_1 \qquad R_1 < r < R_2 \qquad R_2 < r$$

Obtain the potentials on the two spheres. Using **(36)** compute the total energy.

5. **Charged shell.** Compare Problem 4 with a second distribution in which the charge q_1 has been transferred from the inner to the outer shell. What is the energy of the system now? By how much has the energy changed? Is the energy higher or lower? Does the answer depend on the relative signs of q_1 and q_2? Using **(36)** compute the field energy between the two spheres when q_1 is on the inner shell and compare with the answer obtained from the difference in potential.

6. **Interpenetrating spheres.** Two spheres of equal radius R are uniformly charged with ρ_1 and ρ_2 and allowed to interpenetrate. Obtain an expression for the energy of interaction in terms of the distance between centers. Compare with the expression for point charges.

7. **Energy of a polarized sphere.** A sphere of radius R is uniformly polarized with polarization density **P**. Using **(15)** obtain the expression for the amount of work that must be done against external forces to bring the polarization together. Using **(36)** find the energy stored in the electric field within the spherical volume and the amount outside this volume. Compare the sum of the terms contributing to **(36)** with **(15)**.

8. **Stability of the Bohr atom.** The Bohr model of the hydrogen atom spreads the electron over a circular orbit of radius a_0 with the nucleus at the center. Can the nucleus be in stable equilibrium with respect to the electron distribution? For what displacement directions is the nucleus unstable?

9. **Electrostatic equilibrium.** Charges $-q_1$ are at the corners of an equilateral triangle. A charge q is placed at the center. Find the value of q such that the force on all four charges is zero. From **(6)** compute the amount of work required to bring the ensemble to this configuration. Obtain this answer by a physical argument. Would you expect this charge configuration to be stable, unstable, or neutral? Explain.

10. **Electromechanical equilibrium.** Three particles of mass m and charge q are suspended from the same point by strings of length l. Find the equilibrium positions of particles. Is this configuration stable? Explain why this does not violate Earnshaw's theorem.

11. **Charge oscillation.** Four charges, each of magnitude $-q_1$ are fixed at the corners of a square of side a. A particle of mass m and charge q is placed at the center of the square. Show that the charge q is in stable equilibrium for displacements normal to the plane of the square and find the frequency of oscillation for small displacements. Show that *any* displacement with a component in the plane of the square is unstable.

3 DIELECTRIC MEDIA I

We are now ready to discuss real materials. Our considerations of Chapters 1 and 2 were not quite realistic in several respects. First, we considered arbitrary configurations of continuous and point charges without much concern for how charge is actually distributed in nature. Second, we paid no attention to mechanisms for stabilizing charge distributions nor did we consider how charges respond to electric fields. And, as we saw finally in our discussion of Earnshaw's theorem, electrostatics provides only a partial description of the interactions among charges. Something else is required if charge distributions are to be stable. Yet, the results of Chapters 1 and 2 are essential to our analysis of dielectric media in that we now have techniques for finding the fields and what we have called the "external" energy associated with the presence of electric charge.

ATOMIC AND MOLECULAR STRUCTURE

We begin by considering how the simplest of atoms—hydrogen—is stabilized. Following this we discuss the response of atoms and ions to an electric field. Finally we examine the simplest of molecules, the diatomic alkali halides, their response to an electric field, and their stabilization.

1 **Stability of Atomic Structure.** The key to the stability of atoms and molecules is that the charges associated with these structures are not static but are in rapid motion. Thus we must consider not only the potential energy associated with the relative positions of the charges but also the kinetic energy associated with their motion. As an example we consider the hydrogen atom, a single electron moving in the Coulomb field of a single proton. For the moment we treat this atom as Niels Bohr first imagined it: an electron moving with angular momentum L on a circle of radius R with the proton at the center as shown in Figure 1.

The energy of the hydrogen atom U is written as the sum of the kinetic energy U_K and the potential energy U_P:

$$U = U_K + U_P \qquad (1) \quad 75$$

where the potential energy is given by:

$$\triangleright \qquad\qquad U_P = -\frac{1}{4\pi\epsilon_0}\frac{e^2}{R} \qquad\qquad (2)$$

and the kinetic energy is given by:

$$\triangleright \qquad\qquad U_K = \tfrac{1}{2}mv^2 = \frac{L^2}{2mR^2} \qquad\qquad (3)$$

where $L = mvR$ is the angular momentum.

The energy expressions given by (2) and (3) are restricted by the highly simplified, even nonphysical, model that we have introduced for the hydrogen atom. To indicate the restricted nature of such expressions we use an open flag, which contrasts with the solid flag used for results that are not so restricted.

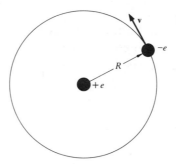

Figure 1   Bohr model of the hydrogen atom. An electron moves in the Coulomb field of a proton.

Rather than treat the problem classically with the radial force equated to the radial acceleration times the mass, we find the minimum energy for a given angular momentum. This approach is closer to the spirit of modern quantum mechanics as well as to our previous discussion of charge stability.[1] For the problem considered here, we obtain the classical result.

To find the minimum energy we substitute (2) and (3) into (1) and differentiate with respect to R. At the radius of minimum energy we obtain:

$$\frac{\partial U}{\partial R} = \frac{1}{4\pi\epsilon_0}\frac{e^2}{R^2} - \frac{L^2}{mR^3} = 0 \qquad\qquad (4)$$

---

[1] This approach is called the *variation method* and was first used by Lord Rayleigh in computing the vibration frequencies of mechanical systems. See J. W. S. Rayleigh, *The Theory of Sound*, Second Edition, Macmillan, 1894; Dover edition, 1945. Volume 1, Section 88. For a discussion of the stability of point charges, see F. J. Dyson and A. Lenard, "Stability of Matter," *J. Math. Phys.* **8**, 423 (1967); **9**, 698 (1968). For a discussion of lattice stability see F. J. Dyson, "Chemical Binding in Classical Coulomb Lattices," *Ann. Phys.* **63**, 1 (1971).

Multiplying the right side of **(4)** by R and comparing with **(2)** and **(3)** we find, at the minimum energy,

$$U_K = -\tfrac{1}{2}U_P \tag{5}$$

Substituting into **(1)** we obtain:

$$U = -U_K = \tfrac{1}{2}U_P \tag{6}$$

That is, the minimum energy is just the negative of the kinetic energy or half the

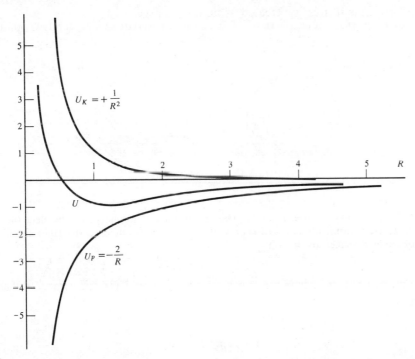

Figure 2   Normalized energies as a function of normalized radius for the Bohr atom. At $R = 1$ the total energy $U = -1$ is just the negative of the kinetic energy $U_K = +1$.

potential energy. In Figure 2 we plot *normalized* energies and radii in order to illustrate the dependence of energy on distance. Note that at $R = 1$ the total energy $U = -1$, which is just the negative of the kinetic energy $U_K = +1$.

With our present knowledge of atomic structure we must modify the simple Bohr picture in three ways:

1. The electron is not to be thought of as localized at a point, but is spread over a spherical shell with a density that is a function of the spherical angles θ and ϕ.

2. In addition to the spread in angle, the electron is also spread in radius, with the density always falling off exponentially at large radius. Associated with the radial spread is a radial kinetic energy.

3. The potential energy of the electron must now be taken as an average over the charge distribution:

$$U_P = -\frac{1}{4\pi\epsilon_0}\left\langle\frac{e^2}{r}\right\rangle = \frac{e}{4\pi\epsilon_0}\int\frac{\rho(\mathbf{r})}{r}\,dV \tag{7}$$

where $\rho(\mathbf{r})$ is the electron density. We emphasize that (7) is *not* the classical expression (2-24) for the energy of a charge distribution. The difference is that (2-24) contains the total potential rather than solely the nuclear contribution to the potential as discussed in Section D-2. In analyzing real materials in classical terms we will have to be careful that for some calculations the electron may be regarded as a distributed charge while for other problems it must be treated as a point charge.

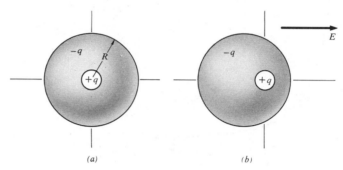

Figure 3   Polarizability of a free atom. (*a*) The nucleus is a point charge *q*. The electronic charge is represented by a uniformly charged sphere. (*b*) In an electric field to the right the electronic charge sphere is displaced to the left.

For many-electron atoms we may think of a total electronic charge density $\rho(\mathbf{r})$ such that we have

$$\int\rho(\mathbf{r})\,dV = -Ne \tag{8}$$

where N is the number of electrons. The energy of such an atom will contain terms corresponding to (a) the kinetic energies of the individual electrons, (b) the potential energy associated with the interaction of each electron with the nucleus, and (c) the potential energy of interaction among electrons.

2 **Electronic Polarizability.** We now discuss the polarizability of free atoms. For this purpose we make a very simple model of an atom. We represent the nucleus by a localized charge $+q$ and the electrons by a charge $-q$ spread *uniformly* throughout a sphere of radius R, as shown in Figure 3a. If we now apply an electric field **E** to the right as shown in Figure 3b there will be a force on the nucleus to the right and on the electrons to the left, resulting in a relative displacement **r** of the electronic charge to the *left*.

As we have seen **(1-117)** the average electric field within a sphere as the result of an interior charge $+q$ at $-\mathbf{r}$ is given by:

$$\langle \mathbf{E} \rangle = -\frac{q}{3\epsilon_0}\frac{-\mathbf{r}}{V} = \frac{q}{4\pi\epsilon_0}\frac{\mathbf{r}}{R^3} \tag{9}$$

At equilibrium we may then write for the force on the electron sphere:

$$F = -q\left(\mathbf{E} + \frac{q}{4\pi\epsilon_0}\frac{\mathbf{r}}{R^3}\right) = 0 \tag{10}$$

where $\mathbf{E}$ is the external field. The induced electric dipole moment is given by:

$$\mathbf{p} = -q\mathbf{r} = 4\pi\epsilon_0 R^3\mathbf{E} \tag{11}$$

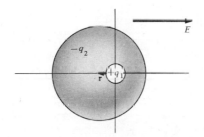

Figure 4   Polarizability of an ion. The nuclear charge q_1 need not be equal in magnitude to the electronic charge $-q_2$.

where we have substituted from **(10)**. We define the atomic polarizability:

$$\alpha = \frac{p}{E} = 4\pi\epsilon_0 R^3 \tag{12}$$

where we have substituted from **(11)**. For a uniformly charged sphere, α is a constant as long as the field is below the critical value at which the negative charge is stripped away from the positive charge. (In real atoms, the polarizability shows nonlinear behavior at fields well below this value.)

We now consider the electronic polarization of an *ion*, as shown in Figure 4, for which the *magnitudes* of the nuclear charge q_1 and the electron charge q_2 are not the same. Here we again find the force on the electrons with the nucleus held fixed. At equilibrium the force on the electron sphere may be written as:

$$F = -q_2\left(\mathbf{E} + \frac{q_1}{4\pi\epsilon_0}\frac{\mathbf{r}}{R^3}\right) = 0 \tag{13}$$

which leads to an electric dipole moment as measured with respect to the *nucleus*:

$$\mathbf{p} = -q_2\,\mathbf{r} = 4\pi\epsilon_0\,R^3\,\frac{q_2}{q_1}\,\mathbf{E} \tag{14}$$

The electronic polarizability of the ion is then given as:

▷
$$\alpha = \frac{p}{E} = 4\pi\epsilon_0\,R^3\,\frac{q_2}{q_1} \tag{15}$$

Table 1   Atomic Radii and Electronic Polarizabilities of Neutral Atoms.
Radii are in 10^{-10} m. Polarizabilities are expressed as $\alpha/4\pi\epsilon_0$ in 10^{-30} m^3

H	He	Li	Be
$\alpha/4\pi\epsilon_0 = 0.66$	$\alpha/4\pi\epsilon_0 = 0.201$	$\alpha/4\pi\epsilon_0 = 12$	$R = 1.06$
O	F	Ne	Na	Mg
$R = 0.66$	$R = 0.64$	$R = 1.58$		$R = 1.40$
		$\alpha/4\pi\epsilon_0 = 0.390$	$\alpha/4\pi\epsilon_0 = 27$	
S	Cl	Ar	K	
$R = 1.04$	$R = 0.99$	$R = 1.88$		
		$\alpha/4\pi\epsilon_0 = 1.62$	$\alpha/4\pi\epsilon_0 = 34$	
Se	Br	Kr		
$R = 1.14$	$R = 1.11$	$R = 2.00$		
		$\alpha/4\pi\epsilon_0 = 2.46$		
Te	I	Xe		
$R = 1.32$	$R = 1.28$	$R = 2.17$		
		$\alpha/4\pi\epsilon_0 = 3.99$		

We list in Table 1 atomic radii and/or polarizabilities of a number of neutral atoms as determined experimentally from the deflection of an atomic beam in an electric field gradient. In Table 2 we give values for the radii R and electronic polarizabilities of various ions in inert gas configurations.[2] As in the above discussion, the dipole moment is with respect to the nucleus.

Note that the electronic polarizabilities of the alkali ions are much smaller than for the neutral atom. This suggests that most of the polarizability arises from the outer electron that is stripped away to form the positive ion. Comparing isoelectronic

[2] The electronic polarizabilities of the alkali and halogen ions have been obtained empirically by minimizing the standard deviation between prediction and experiment. See J. Pirenne and E. Kartheuser, "On the Refractivity of Ionic Crystals," *Physica*, **30**, 2005 (1964). The ionic radii are averages for the NaCl-type alkali halides, M. P. Tosi and F. G. Fumi, *J. Phys. Chem. Solids (GB)* **25**, 31, 45 (1964).

systems in the same column of Table 2, we see that the polarizability increases as the nuclear charge becomes less positive. This is because the charge cloud is not pulled in so tightly and the radius R is larger.

3 **Polarization Energy.** An alternative way of discussing the polarization of an atom or ion is in terms of the internal energy associated with its polarization. For the ion shown in Figure 4, for example, we may distinguish two contributions to the energy:

1. The first contribution is the self energy of the polarized ion and is simply the work which must be done to accomplish the polarization:

$$U_{int} = -\int_0^r F(r)\,dr = \frac{1}{8\pi\epsilon_0}\frac{q_1 q_2 r^2}{R^3} \tag{16}$$

Table 2 Ionic Radii and Electronic Polarizabilities of Inert Gas Configurations. Radii are in 10^{-10} m. Polarizabilities are expressed as $\alpha/4\pi\epsilon_0$ in 10^{-30} m^3.

H $R = 1.54$ $\alpha/4\pi\epsilon_0 = 10.0$	He $\alpha/4\pi\epsilon_0 = 0.201$	Li$^+$ $R = 0.90$ $\alpha/4\pi\epsilon_0 = 0.029$	Be^{2+} $R = 0.30$ $\alpha/4\pi\epsilon_0 = 0.008$	
O^{2-} $R = 1.46$ $\alpha/4\pi\epsilon_0 = 3.88$	F$^-$ $R = 1.19$ $\alpha/4\pi\epsilon_0 = 0.867$	Ne $R = 1.58$ $\alpha/4\pi\epsilon_0 = 0.390$	Na$^+$ $R = 1.21$ $\alpha/4\pi\epsilon_0 = 0.312$	Mg^{2+} $R = 0.65$ $\alpha/4\pi\epsilon_0 = 0.094$
S^{2-} $R = 1.90$ $\alpha/4\pi\epsilon_0 = 10.2$	Cl$^-$ $R = 1.65$ $\alpha/4\pi\epsilon_0 = 3.063$	Ar $R = 1.88$ $\alpha/4\pi\epsilon_0 = 1.62$	K$^+$ $R = 1.51$ $\alpha/4\pi\epsilon_0 = 1.136$	Ca^{2+} $R = 0.94$ $\alpha/4\pi\epsilon_0 = 0.47$
Se^{2-} $R = 1.91$ $\alpha/4\pi\epsilon_0 = 10.5$	Br$^-$ $R = 1.80$ $\alpha/4\pi\epsilon_0 = 4.276$	Kr $R = 2.00$ $\alpha/4\pi\epsilon_0 = 2.46$	Rb$^+$ $R = 1.65$ $\alpha/4\pi\epsilon_0 = 1.758$	Sr^{2+} $R = 1.10$ $\alpha/4\pi\epsilon_0 = 0.86$
Te^{2-} $R = 2.11$ $\alpha/4\pi\epsilon_0 = 14.0$	I$^-$ $R = 2.01$ $\alpha/4\pi\epsilon_0 = 6.517$	Xe $R = 2.17$ $\alpha/4\pi\epsilon_0 = 3.99$	Cs$^+$ $R = 1.80$ $\alpha/4\pi\epsilon_0 = 3.015$	Ba^{2+} $R = 1.29$ $\alpha/4\pi\epsilon_0 = 1.55$

Writing $p = -q_2 r$ and taking the expression for α from **(15)** we may write this energy as:

$$\blacktriangleright \qquad U_{int} = \frac{1}{2}\frac{p^2}{\alpha} \tag{17}$$

Although **(17)** has been obtained from the highly simplified model which we have used, it is actually quite a general result, as indicated by the solid flag.

2. The second contribution to the energy is from the interaction with the external field **(1-107)**:

$$U_{ext} = -pE \tag{18}$$

To find the equilibrium value of the polarization we simply take the total energy:

$$U = U_{int} + U_{ext} = \frac{1}{2}\frac{p^2}{\alpha} - pE \tag{19}$$

and set the derivative with respect to the polarization equal to zero:

$$\frac{\partial U}{\partial p} = \frac{p}{\alpha} - E = 0 \tag{20}$$

which gives $p = \alpha E$ as before. There is nothing new in this procedure. Its advantage in dealing with an interacting system is that we can write an expression for the energy and then by minimizing the total energy find the various polarizations.

Exercises

1. Starting with **(1-105)** for the force on a dipole and assuming $\mathbf{p} = \alpha\mathbf{E}$, show that the work done to bring the dipole into the field is $-\frac{1}{2}\mathbf{p} \cdot \mathbf{E}$.
2. Charges $+q$ and $-q$ are held at equilibrium separation R by a spring of force constant k. The internal energy of the system is given by $\frac{1}{2}k(r-R)^2$ where r is the actual separation between charges. The system is brought into a region of uniform electric field $\mathbf{E}$. How much work must be done to carry the system into the field? What is the change in internal energy? What is the change in external energy?

4 Diatomic Alkali Halide Molecules. We now examine molecules formed from the combination of an alkali metal and a halogen. To a good approximation we may regard this molecule as being composed of the two ions that are formed when an electron is fully transferred from the alkali metal to the halogen. We show such an ion in Figure 5. In addition to the mutual attraction of the ions, two other processes must be considered:

1. The field between the ions exerts a force to the right on the electronic charge. This force produces a displacement of the electrons tending to reduce the dipole moment of the molecule.

2. If there is to be an equilibrium separation, then below some interionic distance the energy of interaction must increase again. If we imagine classically that the spheres are allowed to interpenetrate, we will obtain an equilibrium separation from Coulomb forces only. The minimum will be very "soft" however. A more realistic treatment of the problem requires that the kinetic energy of the electrons increase

rapidly with charge interpenetration. This has the effect of preventing much penetration and produces well-defined ionic radii.

We wish to compute the resultant dipole moment, considering for the moment the internuclear distance R to be a parameter. The energy may be written *approximately* for fixed R as:

$$U = -\frac{1}{4\pi\epsilon_0}\frac{e^2}{R} + \frac{1}{4\pi\epsilon_0}\frac{e(p_1 + p_2)}{R^2} - \frac{1}{4\pi\epsilon_0}\frac{2p_1p_2}{R^3} + \left(\frac{p_1^2}{2\alpha_1} + \frac{p_2^2}{2\alpha_2}\right) \tag{21}$$

where p_1 and p_2 are positive for polarization to the right in Figure 5.

We observe that the total energy as given by **(21)** is composed of four terms:

1. The first term is simply the Coulomb interaction between the two ions. At short distances we should also include the repulsive energy between ion cores.

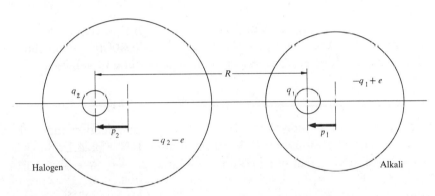

Figure 5 Alkali halide molecule. Each ion is polarized by the field of the other ion. The internuclear distance R is treated as a parameter.

2. The second term is the energy of interaction between each of the dipoles and the field produced by the other ion.

3. The third term is the energy of interaction between two dipoles and follows from **(2-12)**. In writing this term as a dipolar interaction we are assuming that the electronic displacements are small compared with the internuclear distance.

4. The last term is the internal energy of polarization of the two ions as given by **(17)**.

We minimize **(21)** with respect to p_1 and p_2:

$$\frac{4\pi\epsilon_0 R^3}{\alpha_1}p_1 - 2p_2 = -eR \tag{22}$$

$$-2p_1 + \frac{4\pi\epsilon_0 R^3}{\alpha_2}p_2 = -eR \tag{23}$$

Solving for p_1 and p_2 we obtain:

$$p_1 = -\frac{\beta\alpha_1(1 + 2\beta\alpha_2)}{1 - 4\beta^2\alpha_1\alpha_2} eR \tag{24}$$

$$p_2 = -\frac{\beta\alpha_2(1 + 2\beta\alpha_1)}{1 - 4\beta^2\alpha_1\alpha_2} eR \tag{25}$$

where $\beta = 1/4\pi\epsilon_0 R^3$ is a parameter. The total polarization of the molecule may be written as:

$$p = eR + p_1 + p_2 \tag{26}$$

Substituting (24) and (25) into (26) we obtain:

$$p = eR\left[1 - \frac{\beta(\alpha_1 + \alpha_2) + 4\beta^2\alpha_1\alpha_2}{1 - 4\beta^2\alpha_1\alpha_2}\right] \tag{27}$$

We now consider two limiting cases of (27):

1. We first discuss the case in which the polarizability of the alkali ion may be neglected with respect to the halogen. Setting $\alpha_1 = 0$ we obtain for (27):

$$p = eR(1 - \beta\alpha_2) \tag{28}$$

If R is sufficiently small we may have $\beta\alpha_2 \simeq 1$ and the polarization induced on the halogen may nearly cancel the internuclear contribution to the polarization.

2. The second limit is that of equal polarizabilities $\alpha_1 \simeq \alpha_2$. We obtain from (27) for this case:

$$p = eR\left(1 - \frac{2\alpha\beta}{1 - 2\alpha\beta}\right) \tag{29}$$

For this case $\alpha\beta$ is expected to be small compared with unity and depolarization is not so important.

The minimum energy of the molecule shown in Figure 5 is obtained by substituting (24) and (25) into (21). From (22) and (23) we may see that the energy is of the form:

$$U = U_0\left(1 - \frac{p_1 + p_2}{2eR} + \frac{2p_1p_2}{e^2R^2}\right) \tag{30}$$

where $U_0 = -e^2/4\pi\epsilon_0 R$ is the Coulomb energy. As long as at least one of the induced moments is small we may neglect the term in the product of the dipole moments and write approximately:

$$U \simeq U_0\left(\frac{3}{2} - \frac{p}{2eR}\right) \tag{31}$$

where p is the *total* moment as given by (26).

In Table 3 we give the observed values of the polarization p and the equilibrium internuclear distance R_{eq} for the diatomic alkali halide molecules. The dipole moments have been determined by molecular beam electron resonance and the internuclear distances by millimeter wave molecular beam absorption spectroscopy.[3] The ratio p/eR_{eq} is to be compared with (27). Note that the smallest ratio is for HI, which is the case approximated by (28). The ratio appears to be closest to one for generally similar ions.

Although we find it convenient here to express the polarization in electron-metres, such a unit is considered to be outside the International System (SI) because its magnitude depends on the experimentally determined charge of the electron. Thus,

Table 3   **Internuclear Distances and Electric Dipole Moments of Diatomic Alkali Halide Molecules.** Distances are in 10^{-14} m. Dipole moments are in 10^{-14} electron metres.

	Fluoride (F)	Chloride (Cl)	Bromide (Br)	Iodide (I)
Hydrogen (H)	$p = 0.3803e$	$p = 0.2275e$	$p = 0.1724e$	$p = 0.0932e$
	$R_{eq} = 0.9168$	$R_{eq} = 1.2745$	$R_{eq} = 1.4146$	$R_{eq} = 1.6091$
Lithium (Li)	$p = 1.3084e$	$p = 1.4752e$	$p = 1.5045e$	$p = 1.5379e$
	$R_{eq} = 1.5639$	$R_{eq} = 2.0207$	$R_{eq} = 2.7104$	$R_{eq} = 2.3919$
Sodium (Na)	$p = 1.6913e$	$p = 1.8679e$	$p = 1.8929e$	$p = 1.9176e$
	$R_{eq} = 1.9260$	$R_{eq} = 2.3609$	$R_{eq} = 2.5020$	$R_{eq} = 2.7114$
Potassium (K)	$p = 1.7819e$	$p = 2.1318e$	$p = 2.2075e$	$p = 2.25e$
	$R_{eq} = 2.1716$	$R_{eq} = 2.6667$	$R_{eq} = 2.8208$	$R_{eq} = 3.0478$
Rubidium (Rb)	$p = 1.7725e$	$p = 2.1826e$		$p = 2.39e$
	$R_{eq} = 2.2704$	$R_{eq} = 2.7869$	$R_{eq} = 2.9447$	$R_{eq} = 3.1768$
Cesium (Cs)	$p = 1.6403e$	$p = 2.1566e$	$p = 2.25e$	$p = 2.43e$
	$R_{eq} = 2.3455$	$R_{eq} = 2.9064$	$R_{eq} = 3.0722$	$R_{eq} = 3.3152$

the standard SI unit of polarization is the coulomb metre. A commonly used unit of polarization is named after Debye. In the International System we have approximately:

$$1 \text{ debye} = \frac{10^{-21}}{c} = 3.3356409 \times 10^{-30} \text{ coulomb metre}$$

In Table 4 are given the experimental binding energies U_{exp} of the diatomic alkali halides, the Coulomb energy U_0 at the *observed* internuclear distance, and U_{calc} the energy calculated when account is taken of depolarization as given by (31). U_{calc} is expected to be 10 to 20 percent more negative than U_{exp}, the difference being the short range repulsive energy that arises from increased electronic kinetic energy.[4] The present classical theory appears to work well except possibly for HI, where the binding is even greater than calculated. This is probably because our assumption of a rigid electron sphere is too restricted. Allowing the charge on the iodine to deform should yield an increased bond energy.

[3] F. J. Lovas and E. Tiemann, *J. Phys. Chem. Ref. Data*, **3**, 609 (1974).

[4] Experimental binding energies are from F. R. Bichowsky and F. D. Rossini, *The Thermochemistry of the Chemical Substances*, Reinhold, 1936. See also E. S. Rittner, *J. Chem. Phys.*, **19**, 1030 (1951).

Table 4   Binding Energies of the Diatomic Alkali Halide Molecules. Energies are in electron volts.

	Fluoride (F)	Chloride (Cl)	Bromide (Br)	Iodide (I)
Hydrogen (H)	$U_{exp} = -16.55$	$U_{exp} = -14.44$	$U_{exp} = -14.02$	$U_{exp} = -13.64$
	$U_0 = -15.705$	$U_0 = -11.298$	$U_0 = -10.179$	$U_0 = -8.948$
	$U_{calc} = -20.300$	$U_{calc} = -15.939$	$U_{calc} = -14.648$	$U_{calc} = -13.163$
Lithium (Li)	$U_{exp} = -8.57$	$U_{exp} = -6.93$	$U_{exp} = -6.58$	$U_{exp} = -6.13$
	$U_0 = -9.207$	$U_0 = -7.126$	$U_0 = -6.634$	$U_0 = -6.020$
	$U_{calc} = -9.960$	$U_{calc} = -8.088$	$U_{calc} = -7.652$	$U_{calc} = -7.095$
Solium (Na)	$U_{exp} = -6.97$	$U_{exp} = -5.69$	$U_{exp} = -5.49$	$U_{exp} = -5.13$
	$U_0 = -7.476$	$U_0 = -6.099$	$U_0 = -5.755$	$U_0 = -5.310$
	$U_{calc} = -7.932$	$U_{calc} = -6.736$	$U_{calc} = -6.456$	$U_{calc} = -6.087$
Potassium (K)	$U_{exp} = -6.78$	$U_{exp} = -5.11$	$U_{exp} = -4.91$	$U_{exp} = -4.55$
	$U_0 = -6.630$	$U_0 = -5.340$	$U_0 = -5.105$	$U_0 = -4.724$
	$U_{calc} = -7.225$	$U_{calc} = -5.876$	$U_{calc} = -5.660$	$U_{calc} = -5.342$
Rubidium (Rb)	$U_{exp} = -6.39$	$U_{exp} = -4.96$	$U_{exp} = -4.76$	$U_{exp} = -4.40$
	$U_0 = -6.342$	$U_0 = -5.167$	$U_0 = -4.890$	$U_0 = -4.533$
	$U_{calc} = -7.037$	$U_{calc} = -5.727$		$U_{calc} = -5.094$
Cesium (Cs)	$U_{exp} = -6.15$	$U_{exp} = -4.75$	$U_{exp} = -4.58$	$U_{exp} = -4.38$
	$U_0 = -6.139$	$U_0 = -4.954$	$U_0 = -4.687$	$U_0 = -4.343$
	$U_{calc} = -7.062$	$U_{calc} = -5.593$	$U_{calc} = -5.31$	$U_{calc} = -4.923$

5   The Hydrogen Bond. As we have seen for the hydrogen halides, the bond energy will be substantially increased if the negative ion is highly polarizable and the positive ion is small.

This effect is sufficiently strong that hydrogen can form a stable bond with two negative ions.[5] We show in Figure 6 a hydrogen ion between two negative ions. We may write the energy of the bond in a form similar to **(21)**:

$$U = -\frac{2e^2}{4\pi\epsilon_0\,R} + \frac{e^2}{8\pi\epsilon_0\,R} - 2\beta eRp + \frac{\beta p^2}{4} + \frac{p^2}{\alpha} \tag{32}$$

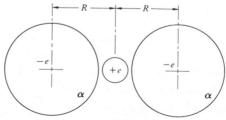

Figure 6   Hydrogen bond. A positive hydrogen ion can form a stable bond with two negative ions.

[5] G. C. Pimentel and A. L. McClellan, *Hydrogen Bond*, Freeman, 1960; M. D. Joesten and L. J. Schaad, *Hydrogen Bonding*, Dekker, 1974.

where $\beta = 1/4\pi\epsilon_0 R^3$ and p is the dipole moment induced on *each* of the ions. To find the polarization we minimize (32) with respect to p to obtain:

$$-2eR + \frac{p}{2} + \frac{2p}{\alpha\beta} = 0 \tag{33}$$

Solving for p we obtain:

$$p = \frac{4\alpha\beta}{4 + \alpha\beta} eR \tag{34}$$

To find the minimum energy we first multiply (33) by $p/2$ and subtract from (32):

$$U = -\frac{3e^2}{8\pi\epsilon_0 R} - \beta eRp \tag{35}$$

Now substituting from (34) we obtain, finally, for the bond energy:

$$U = -\frac{3e^2}{8\pi\epsilon_0 R} \left(1 + \frac{8}{3}\frac{\alpha\beta}{4 + \alpha\beta}\right) \tag{36}$$

With the model that we have used for electronic polarizability we expect that the minimum value of (36), which will occur when the ions are in contact with the hydrogen ion, will correspond to $\alpha\beta \sim 1$. Then the bond energy will be

$$U \simeq -\frac{3e^2}{8\pi\epsilon_0 R} \left(1 + \tfrac{8}{15}\right) \tag{37}$$

and we obtain about a 50 percent increase in the bond energy from the polarizability of the negative ions.

STRUCTURE OF IONIC SOLIDS

All the halides of Li, Na, K, and Rb (and CsF) crystallize in the rocksalt or NaCl structure shown in Figure 7. Note that for the cell shown the halogens are at the corners and face centers of the cube. The alkalis are displaced from each halogen by just half a lattice constant along one cube edge. The internuclear distances as given in Table 5 are seen to be 10 to 20 percent larger than for the diatomic molecules. The values given are determined by X-ray diffraction.

The other Cs halides and (at high temperature) RbCl crystallize in the structure shown in Figure 8 and called the CsCl structure. Here, with halogens at the corners of a cube there is single alkali ion at the body center. The internuclear distances are given in Table 6.

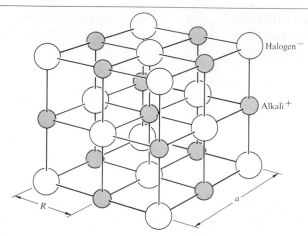

Figure 7   Rocksalt structure. The halogens are at the corners and face centers of a cube. The alkalis are displaced along a cube edge from each halogen by half a lattice constant.

Table 5   Internuclear Distances, NaCl Structure

	F	Cl	Br	I
Li	0.2014 nm	0.2570 nm	0.2751 nm	0.3000 nm
Na	0.2317 nm	0.2820 nm	0.2989 nm	0.3237 nm
K	0.2674 nm	0.3147 nm	0.3298 nm	0.3533 nm
Rb	0.2815 nm	0.3291 nm	0.3445 nm	0.3671 nm
Cs	0.3004 nm			

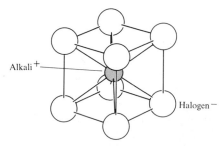

Figure 8   CsCl structure. The halogens are at the corners of a cube with a single alkali ion at the body center.

Table 6   Internuclear Distances, CsCl Structure

	Cl	Br	I
Rb	0.324 nm		
Cs	0.356 nm	0.3712 nm	0.3966 nm

The hydrogen halides form more complex structures with lower symmetry. That even the molecules are different is evident from Table 3, where it may be seen that the dipole moments are quite small.

6 Electrostatic Energy. An ionic crystal is bound by classical electrostatic interaction of the form discussed in Chapter 2. We obtained there (2-6) for a distribution of point charges:

$$W_{ext} = U = \frac{1}{8\pi\epsilon_0} \sum_{i=1}^{2N} \sum_{j=1}^{2N}{}' \frac{q_i q_j}{r_{ij}} \tag{38}$$

for a crystal of N molecules or $2N$ ions. An alternative way of writing (38) is in terms of the potential at each ion site (2-8):

$$U = \tfrac{1}{2} \sum_{i=1}^{2N} q_i \phi_i \tag{39}$$

where

$$\phi_i = \frac{1}{4\pi\epsilon_0} \sum_{j=1}^{2N}{}' \frac{q_j}{r_{ij}} \tag{40}$$

is the potential. If we fix our attention on a single ion-pair or molecule, the energy will be

$$\frac{U}{N} = \tfrac{1}{2}(q_1 \phi_1 + q_2 \phi_2) \tag{41}$$

where q_1 and q_2 represent the alkali and halogen ions respectively. Now the charges are given by $q_1 = -q_2 = e$ and the potentials are also negatives of each other. Then the energy per molecule will be:

$$\frac{U}{N} = e\phi_1 = \frac{e^2}{4\pi\epsilon_0} \sum_{j=2}^{2N} \frac{\pm 1}{r_{1j}} \tag{42}$$

where the choice of plus or minus sign in the summation depends on whether the jth ion is positive or negative. The sum indicated in (42) is written as:

$$\sum_{j=2}^{2N} \frac{\pm 1}{r_{ij}} = -\frac{\alpha}{R} \tag{43}$$

where R is the nearest neighbor internuclear distance. The quantity α (not to be confused with the polarizability) is called the *Madelung constant* and has the values:

NaCl structure $\alpha = 1.747565$

CsCl structure $\alpha = 1.762675$

Note that the electrostatic energy of the diatomic molecule is $e^2/4\pi\epsilon_0 R_{eq}$. Notwithstanding the fact that R is about 20 percent larger than R_{eq} the electrostatic binding of the crystal is about 50 percent larger than for the molecule because of the interaction with other ions.

In contrast with the molecule there is no electric field acting on the ions in the crystal and the ions are thus unpolarized.

DIELECTRIC SUSCEPTIBILITY

We have seen that an atom or molecule will be polarized by the action of an electric field and we write the polarization as:

$$\mathbf{p} = \alpha \mathbf{E}_{loc} \tag{44}$$

where by $\mathbf{E}_{loc}$ we designate the field acting *on* the atom or molecule. This would be the field seen by a test charge if we could remove the atom or molecule in question while leaving all other charges unaffected. As we have discussed, the field that is most readily measured is the macroscopic field, which we have defined for preciseness as the average over a spherical volume **(1-119)**. We wish now to relate the polarization density:

$$\mathbf{P} = n\mathbf{p} \tag{45}$$

where n is the number of molecules per unit volume, to the macroscopic field. To do this we must obtain a relation between the macroscopic field and the local field.

7 **Local Field.** We wish to obtain the field at the site of a molecule in what we assume to be a uniformly polarized medium, which is shown in Figure 9. In order to obtain the local electric field we divide the medium into two regions:

1. A near-zone region within a sphere of radius R is defined for R large compared with the separation between molecules but sufficiently small that the macroscopic field may be considered as uniform over the sphere.

2. The region outside the central sphere is regarded as being sufficiently removed that its contribution to the local field is the same as that of a uniformly polarized medium. The two regions are sketched in Figure 10a and b. The contribution to the field from the dipoles shown in Figure 10a is called the dipolar field $\mathbf{E}_{dip}$. The contribution to the field from the far zone shown in Figure 10b when taken at the center of the sphere is the exterior field $\mathbf{E}_{ext}$. We have for the sum of the two fields:

$$\mathbf{E}_{loc} = \mathbf{E}_{ext} + \mathbf{E}_{dip} \tag{46}$$

We first find the contribution to the field at the central site from the exterior region. We have already obtained **(1-119)** an expression for the macroscopic field,

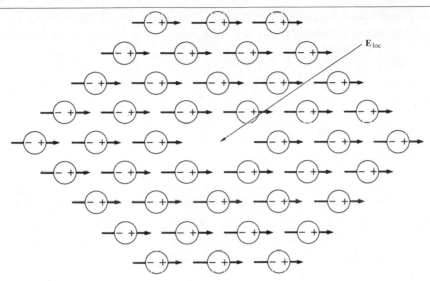

Figure 9 Local field. We wish the field at the site of a molecule in what we take to be a uniformly polarized medium.

sometimes called the Maxwell field, which we may use here without further calculation:

$$\mathbf{E} = \mathbf{E}_{ext} - \frac{1}{3\epsilon_0}\mathbf{P} \tag{47}$$

In **(47)** $\mathbf{E}$ is the macroscopic electric field in the medium, and $\mathbf{E}_{ext}$ is the field at the center of a sphere with no included charge. The vector $\mathbf{P}$ is the polarization density

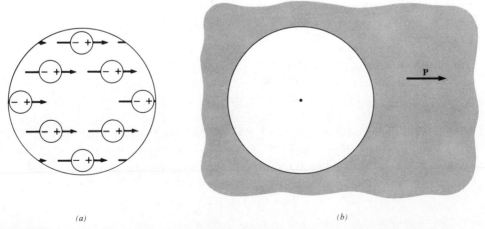

(a)   *(b)*

Figure 10 Calculation of local field. (*a*) A near-zone region within a sphere of radius R is defined for R large. (*b*) The region outside the central sphere is treated as if it were uniformly polarized.

in the medium. Since $\mathbf{E}_{ext}$ is *exactly* the field that we need to describe the situation in Figure 10b, we may invert (47) to obtain the exterior field:

$$\mathbf{E}_{ext} = \mathbf{E} + \frac{1}{3\epsilon_0}\,\mathbf{P} \tag{48}$$

Substituting (48) into (46) we obtain:

$$\mathbf{E}_{loc} = \mathbf{E} + \frac{1}{3\epsilon_0}\,\mathbf{P} + \mathbf{E}_{dip} \tag{49}$$

The field $\mathbf{P}/3\epsilon_0$ was first obtained by H. A. Lorentz[6] and is called the Lorentz field. This field was computed from the polarization charge $\sigma_P = \hat{\mathbf{n}} \cdot \mathbf{P}$ on the spherical surface shown in Figure 10b. By using (1-119) we have been able to avoid this calculation.

Finally, we must calculate the field at the central molecule or atom from similar polarized molecules within the near-zone. Dipole fields fall off with distance as $1/r^3$ and we might expect rapid convergence of the field. But the number of dipoles in dr increases as r^2 so that in fact we get rather slow convergence of the dipole field, and special techniques must be employed to compute such fields.

One special case, where the central molecule is at a site of cubic symmetry, yields a very simple result—that the dipole field is zero. We may demonstrate this result explicitly by writing the dipole sum over lattice sites and showing that it vanishes by symmetry.

An alternative way of showing that the dipole field is zero makes use of the properties of symmetric tensors. To obtain the field of the dipoles in Figure 10a we sum (1-96):

$$\mathbf{E}_{dip}(0) = \frac{1}{4\pi\epsilon_0}\sum_i \frac{3\hat{\mathbf{r}}_i(\mathbf{p}\cdot\hat{\mathbf{r}}_i) - \mathbf{p}}{r_i^{\,3}} = \frac{1}{4\pi\epsilon_0}\frac{\mathbf{p}\cdot\mathbf{T}}{R^3} \tag{50}$$

with

$$\mathbf{T} = R^3\sum_i \frac{3\hat{\mathbf{r}}_i\hat{\mathbf{r}}_i - \mathbf{I}}{r_i^{\,3}} \tag{51}$$

where R is a lattice distance. Now as is apparent from (51) $\mathbf{T}$ is a symmetric tensor with zero scalar T_S. Since $\mathbf{T}$ is symmetric we must be able to find axes for which $\mathbf{T}$ is diagonal with principal values $T_x = T_y = T_z$. But we also have $T_x + T_y + T_z = 0$ from which we conclude that $\mathbf{T}$ is identically zero and the dipolar field is zero:

$$\mathbf{E}_{dip}(0) = 0 \tag{52}$$

[6] H. A. Lorentz, *The Theory of Electrons*, Teubner, 1909. Dover reprint volume, 1952.

From (49) and (52) we obtain for sites of cubic symmetry:

$$\mathbf{E}_{loc} = \mathbf{E} + \frac{1}{3\epsilon_0}\mathbf{P} \tag{55}$$

Exercises

3. The Lorentz field $\mathbf{P}/3\epsilon_0$ may be considered to arise from polarization charge on the surface of the sphere shown in Figure 10b and as given by (1-138). Use this approach to compute the field at the center of the spherical cavity.

4. Beginning with the expression for the potential within a uniformly charged sphere, obtain the expression for the *field* within a uniformly polarized sphere.

5. Dipoles **p** are at the vertices of a regular octahedron and are normal to the basal plane. Show that the electric field at the center is zero. Repeat the calculation for **p** in the basal plane. Show from this result that the field must be zero for any general direction of the dipoles.

6. Dipoles are at the corners of a cube and oriented along a cube edge. Show that the electric field at the cube center is zero for this orientation (as well as for any orientation of parallel dipoles).

7. Dipoles are at the vertices of a regular tetrahedron and oriented in a general direction. Show that the electric field at the center of the tetrahedron is zero.

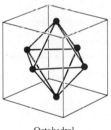
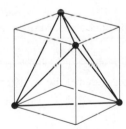
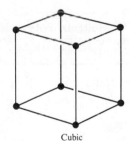

Octahedral   Tetrahedral   Cubic

8. A medium of uniform polarization **P** is bounded by concentric spheres of radius R_1 and R_2. Show that the dipolar field within the inner sphere is everywhere zero.

9. From the result of Exercise 8 explain why $\mathbf{E}_{dip}$ converges rather than diverging logarithmically.

10. A dielectric sphere is *uniformly* polarized in an external field $\mathbf{E}_0$. What is the macroscopic field $\mathbf{E}$? Show that the polarization density is given by:

$$\mathbf{P} = n\alpha\mathbf{E}_0$$

Use the result of Exercise 8 that the dipolar field is zero for a uniformly polarized dielectric.

8 [**Molecular Polarizability**]. We have already discussed polarization resulting from the displacement of the electron distribution with respect to the nucleus. In the case of an ionic solid we have in addition a contribution from the relative displacement of the ions. If X_1 and X_2 are the nuclear coordinates of the positive and negative ions, as shown in Figure 11, then we write for the *ionic* contribution to the polarization:

$$p_{12} = q_{12}(X_1 - X_2) = \alpha_{12} E_{\text{loc}} \tag{56}$$

where α_{12} is called the ionic polarizability and X_1 and X_2 are measured with respect to the nuclear positions in zero electric field.

Within the model that we have been discussing, **(56)** represents a rigid relative displacement of the two ions. That is, the electron distribution is assumed to move together with the nuclei and without distortion. Now, as we have discussed, the repulsive interaction between ions is associated with increased electronic kinetic energy and at least some small electronic distortion as well. Thus, the *internal* energy associated

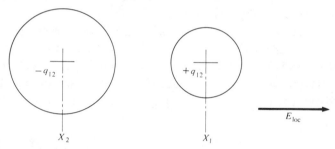

Figure 11   Molecular polarizability. The nuclear coordinates of the positive and negative ions are X_1 and X_2.

with the ionic polarization must in general be written in terms of the electronic displacements x_1 and x_2:

$$U_{12} = \frac{1}{2} \frac{q_{12}^2}{\alpha_{12}} (x_1 - x_2)^2 \tag{57}$$

The equation **(57)** is of the form of **(17)** except that it involves electronic rather than nuclear coordinates.

Applying an electric field to an ionic diatomic molecule produces a relative displacement of two ions, as we have just discussed. But, it also produces an electronic polarization of each of the ions, which from **(14)** and **(15)** may be written as:

$$p_1 = -q_1(x_1 - X_1) = \alpha_1 E_{\text{loc}} \tag{58}$$

$$p_2 = -q_2(x_2 - X_2) = \alpha_2 E_{\text{loc}} \tag{59}$$

where $-q_1$ and $-q_2$ are the electronic charges and we have assumed the same local field at both ions. The total moment produced by a field is the sum of **(56)**, **(58)**, and **(59)**:

$$p = p_1 + p_2 + p_{12} = -q_1 x_1 - q_2 x_2 + (q_1 + q_{12})X_1 + (q_2 - q_{12})X_2 \qquad (60)$$

where we have rearranged the terms so as to exhibit separately the contributions from the displacement of the electronic and nuclear charge. The moment per molecule may be written as

$$p = p_1 + p_2 + p_{12} = (\alpha_1 + \alpha_2 + \alpha_{12})E_{\text{loc}} \qquad (61)$$

and the molecular polarizability is just the sum of the polarizabilities:

$$\alpha = p/E_{\text{loc}} = \alpha_1 + \alpha_2 + \alpha_{12} \qquad (62)$$

The polarization per unit volume will be given by:

$$P = \frac{Np}{V} = \frac{N(\alpha_1 + \alpha_2 + \alpha_{12})}{V} E_{\text{loc}} = \frac{N\alpha}{V} E_{\text{loc}} \qquad (63)$$

9 Clausius-Mossotti Relation. To find the relation between polarization density and macroscopic field, it remains only to substitute the expression for the local field **(55)** into **(63)** and solve for P:

$$P = \frac{N\alpha}{V}\left(E + \frac{1}{3\epsilon_0}P\right) \qquad (64)$$

We define the *susceptibility* by the relation:

$$P = \chi\epsilon_0 E \qquad (65)$$

Substituting for P into **(64)** we obtain on dividing through by E:

$$\chi\epsilon_0 = \frac{N\alpha}{V}\left(1 + \frac{\chi}{3}\right) \qquad (66)$$

On rearranging terms **(66)** yields the Clausius-Mossotti relation, which expresses the connection between the susceptibility and the molecular polarizability:

$$\frac{\chi}{3 + \chi} = \frac{1}{3\epsilon_0}\frac{N\alpha}{V} \qquad (67)$$

Solving **(67)** for χ we obtain:

$$\chi = \frac{N\alpha/\epsilon_0 V}{1 - N\alpha/3\epsilon_0 V} \qquad (68)$$

How does the ionic polarizability α_{12} compare with the sum of the electronic polarizabilities $\alpha_1 + \alpha_2$? In Table 7 we list the observed susceptibilities of the crystalline alkali halides and $\alpha/4\pi\epsilon_0$ as computed from (67). For comparison we list $(\alpha_1 + \alpha_2)/4\pi\epsilon_0$ from Table 2 and the difference, $\alpha_{12}/4\pi\epsilon_0$. As we can see, the ionic and electronic polarizabilities are comparable for the light alkali halides. For the heavier alkali halides the electronic polarizability dominates.

How large is the correction term in the denominator of (68)? From (67) this term is just $\chi/(3 + \chi)$. The correction is largest for LiBr and is 0.8, which means that the observed susceptibility is five times what it would be if there were no Lorentz field!

As our discussion of polarizability has developed, α and χ describe the response to a static electric field. The Clausius-Mossotti relation is commonly written as a relation between the refractive index for the propagation of light and the molecular polarizability, as will be discussed in Chapter 12. Under these conditions, α includes

Table 7   Electronic and Ionic Polarizabilities in 10^{-30} m^3

$\chi \quad \dfrac{\alpha_1 + \alpha_2}{4\pi\epsilon_0}$ $\dfrac{\alpha}{4\pi\epsilon_0} \quad \dfrac{\alpha_{12}}{4\pi\epsilon_0}$	Fluoride (F)		Chloride (Cl)		Bromide (Br)		Iodide (I)	
Lithium (Li)	7.9	.90	11.0	3.09	12.2	4.30		6.55
	2.8	1.9	6.4	3.3	8.0	3.7		
Sodium (Na)	4.1	1.18	4.9	3.38	5.4	4.59		6.83
	3.4	2.2	6.6	3.2	8.2	3.6		
Potassium (K)	4.5	2.00	3.85	4.20		5.41	4.1	7.65
	5.5	3.5	8.36	4.16			12.2	4.6
Rubidium (Rb)	5.5	2.62		4.82		6.03	4.4	8.28
	6.9	4.3					14.0	5.7
Cesium (Cs)		3.88	6.2	6.08		7.29	4.65	9.53
			14.5	8.4			14.36	4.83

only the electronic terms α_1 and α_2 since the ions are unable to respond to the rapidly oscillating electric field. We list in Table 8 for the alkali halides $n^2 - 1$ where n is the refractive index for the propagation of sodium light, $\lambda = 589.3$ nm. We also tabulate $\alpha/4\pi\epsilon_0$ calculated from (67) with $\chi = n^2 - 1$. We observe that the computed polarizabilities are in good agreement with the electronic polarizabilities $(\alpha_1 + \alpha_2)/4\pi$ as given in Table 7. This agreement is not accidental since the empirical electronic polarizabilities as given in Table 2 are based on optical data.

10   Frequency Dependent Polarizability. In this section we work through the response of a single atom to an alternating electric field. As we will see, the atomic polarizability begins from (12) at low frequencies and rises with frequency to a resonance in which the electrons oscillate with respect to the nuclear masses. At still higher frequencies the motion of the electrons is finally limited by their inertia. This approach may be extended to diatomic molecules yielding three resonances, one from the vibration of the ion masses and two from electronic vibration.

We write for the motion of the electrons:

$$m \frac{d^2\mathbf{r}}{dt^2} = -q\mathbf{E}_{\text{eln}} = -\frac{q^2}{\alpha}(\mathbf{r} - \mathbf{R}) - q\mathbf{E}_{\text{loc}} \tag{69}$$

and for the motion of the nuclei:

$$M \frac{d^2\mathbf{R}}{dt^2} = q\mathbf{E}_{\text{nuc}} = -\frac{q^2}{\alpha}(\mathbf{R} - \mathbf{r}) + q\mathbf{E}_{\text{loc}} \tag{70}$$

Table 8 Susceptibility and Polarizability in 10^{-30} m^3 at Optical Frequency

$n^2 - 1$	$\alpha/4\pi\epsilon_0$	Fluoride (F)		Chloride (Cl)		Bromide (Br)		Iodide (I)	
Lithium (Li)		0.9363	0.928	1.762	2.998	2.183	4.187	2.822	6.249
Sodium (Na)		0.785	1.232	1.1385	2.944	1.6935	4.601	2.1489	6.759
Potassium (K)		0.858	2.030	1.220	4.302	1.430	5.528	1.812	7.929
Rubidium (Rb)		0.954	2.570	1.229	4.946	1.4118	6.247	1.7139	8.588
Cesium (Cs)		1.184	3.663	1.6955	5.988	1.8846	7.254	2.1955	9.689

where the interaction term is obtained from **(9)** and **(12)**. From **(70)** and **(69)** we obtain an equation for the polarization:

$$\frac{d^2\mathbf{p}}{dt^2} + \frac{q^2}{\alpha\mu}\mathbf{p} = \frac{q^2}{\mu}\mathbf{E}_{\text{loc}} \tag{71}$$

with the polarization given by:

$$\mathbf{p} = q(\mathbf{R} - \mathbf{r}) \tag{72}$$

and the reduced mass defined by:

$$\frac{1}{\mu} = \frac{1}{m} + \frac{1}{M} \tag{73}$$

We take for the electric field and the polarization:

$$\mathbf{E}_{\text{loc}}(t) = \mathbf{E}_{\text{loc}} \cos \omega t \qquad \mathbf{p}(t) = \alpha(\omega)\mathbf{E}_{\text{loc}} \cos \omega t \tag{74}$$

Substituting **(74)** into **(71)** we obtain for the frequency dependent polarizability:

$$\alpha(\omega) = \frac{\alpha}{1 - \omega^2 \alpha \mu / q^2} \tag{75}$$

This expression is plotted in Figure 12. We note the presence of a resonance at

$$\omega_{\text{res}} = \frac{q}{(\alpha\mu)^{1/2}} \tag{76}$$

At frequencies high compared with ω_{res} the polarizability falls off as:

$$\alpha(\omega) \simeq -\frac{q^2}{\omega^2 \mu} = -\frac{\alpha\omega_{\text{res}}^2}{\omega^2} \tag{77}$$

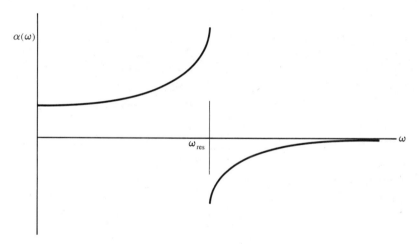

Figure 12   Frequency-dependent polarizability. The polarizability is dominated by a resonance at a frequency given by **(76)**.

11   Oscillator Strength.
The oscillator strength f of a resonance is defined by the relation

$$\frac{e^2}{m} f = \alpha\omega_{\text{res}}^2 \tag{78}$$

where e and m are the electronic charge and mass. The oscillator strength is a measure of the amount of charge associated with the resonance as may be seen by substituting **(76)** into **(78)**:

$$f = \frac{m}{\mu}\left(\frac{q}{e}\right)^2 \tag{79}$$

Our model of rigid electronic distributions places all the oscillator strength at a single frequency ω_{res}. Actually there is a whole spectrum of atomic excitations, and we must write in general:

$$\alpha(\omega) = \frac{e^2}{m} \sum_i \frac{f_i}{\omega_i^2 - \omega^2} \tag{80}$$

Many solids have electronic excitations at low frequencies as well as in the ultraviolet and into the X-ray region. Because their contribution to the static polarizability varies as $1/\omega_i^2$, they may be significant even though they have very little oscillator strength. We have, at zero frequency,

$$\alpha(0) = \frac{e^2}{m} \sum_i \frac{f_i}{\omega_i^2} \tag{81}$$

At very high frequency we have for the polarizability:

$$\alpha(\omega) \simeq -\frac{e^2}{m\omega^2} \sum_i f_i \tag{82}$$

But at very high frequency the motion of the electrons is limited only by their inertia:

$$Nm \frac{d^2\mathbf{r}}{dt^2} = -\omega^2 Nm\mathbf{r} = -Ne\mathbf{E} \tag{83}$$

and the polarizability is then given by:

$$\alpha(\omega) = -\frac{Ne}{E} r = -\frac{Ne^2}{m\omega^2} \tag{84}$$

Comparing **(84)** and **(82)** we have the important *sum rule*:

$$\sum_i f_i = N \tag{85}$$

where N is the number of electrons per atom.

FERROELECTRICITY

In the materials that we have thus far considered, polarization is developed only in the presence of an external field. There is a coupling between molecules through the Lorentz field but this coupling is not sufficiently strong to do more than enhance the susceptibility.

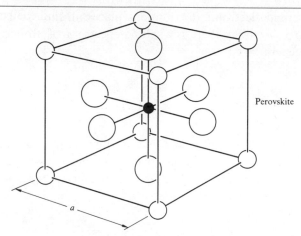

Figure 13   Perovskite structure. Divalent oxygen ions are at the face centers with a small highly charged ion at the body center.

12 Perovskite Structure. We now discuss a class of ionic compounds having a polarized state which forms spontaneously from an unpolarized state. These materials show very high dielectric susceptibilities, which makes them of very great utility in a number of electronic applications. The crystals that we discuss all form in the perovskite structure, shown in Figure 13.

   In all of the crystals that we consider the ions at the face center positions are divalent oxygen O^{2-}. The ion at the body center is tetravalent titanium Ti^{4+} or pentavalent niobium Nb^{5+} or tantalum Ta^{5+}. The ions at the corners of the cube are monovalent or divalent as required for the neutrality of the molecule. What makes this material so responsive to an electric field is the high electronic polarizability of the divalent oxygens. Although the central ions are in sites of cubic symmetry, the surrounding oxygens are not. The oxygens become polarized by the displaced central ions as shown in Figure 14a with the top and bottom oxygens polarized up and the side oxygens polarized down. The polarization in turn produces a field that enhances the central ion displacement as shown in Figure 14b.

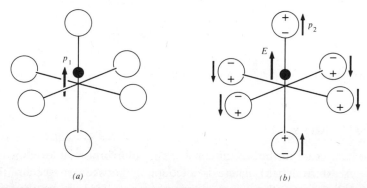

(a)   (b)

Figure 14   Polarization of perovskite. (a) The oxygens are polarized by the displaced central ion. (b) The field of the polarized oxygen ions enhances the displacement of the central ion.

We designate the polarization of the central ion as p_1 and the polarization of the oxygens by p_2.

The energy may be written as the sum of three terms: the interaction with the field, the polarization energy of the dipoles, and the interaction energy between the polarization of the central ion and of the oxygens. The contribution of the Lorentz field is absorbed into α_1 and α_2:

$$\frac{U}{N} = -(p_1 + p_2)E + \frac{1}{2}\frac{p_1{}^2}{\alpha_1} + \frac{1}{2}\frac{p_2{}^2}{\alpha_2} - \beta p_1 p_2 \tag{86}$$

where β is a measure of the strength of the coupling.

To find the polarizations we minimize (86) with respect to p_1 and p_2 to obtain:

$$\frac{p_1}{\alpha_1} - \beta p_2 = E \tag{87}$$

$$-\beta p_1 + \frac{p_2}{\alpha_2} = E \tag{88}$$

Solving for p_1 and p_2 we obtain:

$$p_1 = \frac{1 + \beta\alpha_2}{1 - \beta^2\alpha_1\alpha_2}\,\alpha_1 E \tag{89}$$

$$p_2 = \frac{1 + \beta\alpha_1}{1 - \beta^2\alpha_1\alpha_2}\,\alpha_2 E \tag{90}$$

The polarization density may be written as:

$$P = \frac{N}{V}(p_1 + p_2) = \frac{\alpha_1 + \alpha_2 + 2\beta\alpha_1\alpha_2}{1 - \beta^2\alpha_1\alpha_2}\frac{N}{V}E \tag{91}$$

Solving for P we obtain,

$$P = \epsilon_0 \chi E \tag{92}$$

with

$$\epsilon_0 \chi = \frac{\alpha_1 + \alpha_2 + 2\beta\alpha_1\alpha_2}{1 - \alpha_1\alpha_2 \beta^2}\frac{N}{V} \tag{93}$$

As long as we have $\beta^2\alpha_1\alpha_2 < 1$ there exists a situation in which the susceptibility may be large but where in the absence of a field there is no spontaneous polarization. This is called the paraelectric region. Should the denominator go to zero, (93) indicates an infinite polarization in the presence of a finite field. This is clearly a nonphysical result and arises from our limiting (86) to terms in $p_1{}^2$ for the polarization energy.

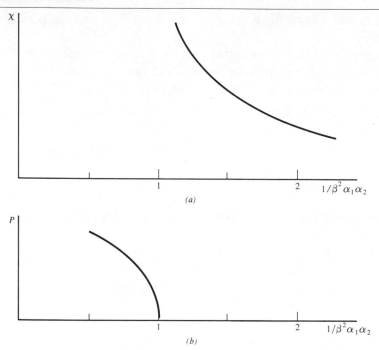

Figure 15 Polarization of perovskite. (*a*) The susceptibility χ is given as a function of an interaction parameter. (*b*) The polarization P is given as a function of the same parameter.

To treat the case of spontaneous polarization we add to **(86)** an additional term that represents a nonlinear limiting of the displacement of the central ion:

$$\frac{U}{N} = \frac{1}{2}\frac{p_1{}^2}{\alpha_1}\left(1 + \frac{1}{2}\frac{p_1{}^2}{p_0{}^2}\right) + \frac{1}{2}\frac{p_2{}^2}{\alpha_2} - \beta p_1 p_2 \qquad (94)$$

To determine the polarization we now minimize **(94)** with respect to p_1 and with respect to p_2 to obtain:

$$\frac{p_1}{\alpha_1} + \frac{p_1{}^3}{\alpha_1 p_0{}^2} - \beta p_2 = 0 \qquad (95)$$

$$\frac{p_2}{\alpha_2} - \beta p_1 = 0 \qquad (96)$$

Table 9   Perovskite Ferroelectrics

	$P, C\,m^{-2}$	a
$BaTiO_3$	0.26	0.40118 nm
$KNbO_3$	0.3	0.4007 nm
$PbTiO_3$	>0.5	0.3960 nm

Eliminating p_2 and solving for p_1 we obtain:

$$p_1 = 0 \tag{97}$$

$$p_1 = (\beta^2 \alpha_1 \alpha_2 - 1)^{1/2} p_0 \tag{98}$$

If the interactions are weak corresponding to $\beta^2 \alpha_1 \alpha_2 < 1$, then the first solution is the only possibility in the absence of a field. For sufficiently strong interactions that $\beta^2 \alpha_1 \alpha_2 > 1$ both **(97)** and **(98)** are possible solutions. However **(98)** has the lower energy as may be seen by substituting into **(94)**.

In Figure 15 we plot the susceptibility and the polarization P as a function of $1/\beta^2 \alpha_1 \alpha_2$.

At room temperature $SrTiO_3$ is paraelectric with a susceptibility of approximately 300. In Table 9 we list a number of perovskite ferroelectrics with their polarization and lattice constants.

13 **Ferroelectric Domains.** As we have seen **(2-20)** the external energy of a polarized material may be written as:

$$U_{ext} = -\tfrac{1}{2} \int \mathbf{P} \cdot \mathbf{E} \, dV \tag{99}$$

If **E** is a depolarization field we have **(1-181)**:

$$\mathbf{E} = -\frac{1}{\epsilon_0} \mathbf{N} \cdot \mathbf{P} \tag{100}$$

and the external energy is positive:

$$U_{ext} = \frac{1}{2\epsilon_0} \int \mathbf{P} \cdot \mathbf{N} \cdot \mathbf{P} \, dV \tag{101}$$

In order to minimize this energy, the ferroelectric material forms domains as shown in Figure 16. Now if a field is applied, the domain walls will move to set up an

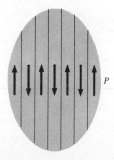

Figure 16 Ferroelectric domains. Regions of opposing polarization alternate through the sample.

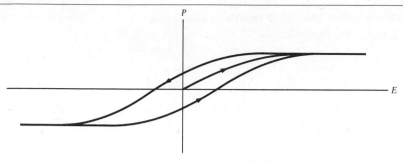

Figure 17   Volume polarization. In an applied field a polarization develops. The polarization curve is similar to the magnetization curve of a ferromagnetic material.

average volume polarization of the form shown in Figure 17. Ferroelectrics show many of the features that have long been familiar from the study of ferromagnetic materials.

Exercise

11. An ellipsoidal ferroelectric is polarized in an external electric field applied along a principal axis. Find the remanent polarization when the external field is reduced to zero. Use a graphical solution between **(100)** and Figure 17.

SUMMARY

1. Atoms are stabilized by electronic kinetic energy. The total energy U of a hydrogen atom may be regarded as the sum of the potential energy:

$$U_P = -\frac{1}{4\pi\epsilon_0}\frac{e^2}{r} \tag{2}$$

and a kinetic energy:

$$U_K = \tfrac{1}{2}mv^2 \tag{3}$$

For a classical atom with angular momentum L the minimum energy is just half the potential energy.

2. Atoms and ions may be polarized by an electric field. For a model in which the electrons are uniformly distributed through a sphere of radius R the polarizability is given by:

$$\alpha = \frac{p}{E} = 4\pi\epsilon_0 R^3 \tag{12}$$

as long as the atom is neutral. If the nuclear charge q_1 and the electronic charge $-q_2$ do not balance, then the polarizability, taken with respect to the *nucleus* is:

$$\alpha = \frac{p}{E} = 4\pi\epsilon_0 \frac{q_2}{q_1} R^3 \qquad (15)$$

3. A polarizable atom or ion is stabilized by an energy:

$$U_{\text{int}} = \frac{1}{2}\frac{p^2}{\alpha} \qquad (17)$$

This energy is in addition to usual external energy $U_{\text{ext}} = -\mathbf{p} \cdot \mathbf{E}$.

4. A molecule formed from polarizable positive and negative ions will have an electric dipole moment reduced from the rigid ion value by the depolarizing effect of the intraionic field. The effect of this reduction is to increase the binding.

5. The hydrogen bond is an extreme example of the bonding of negative and highly polarizable ions by a positive hydrogen ion.

6. The alkali halides form highly symmetrical crystal structures with a field energy per molecule about 75 percent greater than for an isolated molecule with the same internuclear separation.

7. The field $\mathbf{E}_{\text{loc}}$ acting on a molecule in a solid is the sum of the macroscopic electric field $\mathbf{E}$, the Lorentz field $\mathbf{P}/3\epsilon_0$, and a dipolar field $\mathbf{E}_{\text{dip}}$, which vanishes at sites of cubic symmetry.

8. The molecular polarizability may be written as the sum of ionic and electronic polarizabilities. The ionic polarizability is associated with a change in the internuclear distance.

9. The polarization of a ferroelectric material is induced by the large and negative energy associated with the coupling between ionic and electronic dipoles. The polarization is stabilized by higher order terms in the polarization energy. To minimize the external energy, ferroelectrics usually order in domains. A mean polarization is then induced by displacement of domain walls.

PROBLEMS

1. **Molecular beam electric deflection.** Polarizability of neutral atoms is measured by the deflection of a beam moving at thermal velocity through a transverse electric field and field gradient. Take the transverse field to be 15×10^6 V/m and the field gradient to be 3×10^8 V/m^2. The effective temperature of the beam is 300 K and the deflecting region is 0.1 m long. Compute the expected deflections of Li, Na, and K.

2. **Forces.** A *small* polarizable sphere and a charge q are separated by a distance r. Find the force on the sphere. Find the force on the charge.

3. **Torques.** A small polarizable sphere is located at a position $\mathbf{r}$ with respect to a dipole $\mathbf{p}$. Find the forces on the sphere and on the dipole. Find the torque acting on the dipole. Show that the sum of all torques about an arbitrary origin vanishes.

4. **Nonlinear polarizability.** An improved (but still classical) treatment of the polarizability of the hydrogen atom would consider that the electronic charge is not uniform but has the form:

$$\rho(r) = \frac{-e}{\pi a_0{}^3} e^{-2r/a_0}$$

where a_0 is the Bohr radius. Writing the polarizability as:

$$\alpha(E) = \alpha_0 + \alpha_1 E + \ldots\ldots$$

find α_0 and α_1. What happens if the electric field is reversed?

5. **Dipole moments of ammonialike molecules.** The following molecules have the structure shown in the figure with the parameters given in the table.

	$R, 10^{-10}$ m	θ	$p, 10^{-10}$ m
NH_3	1.008	107.3	0.306 e
PH_3	1.142	93.3	0.11 e
PF_3	1.535	100	0.213 e
PCl_3	2.043	100.1	0.16 e
AsH_3	1.519	91.8	0.33 e
AsF_3	1.712	102.0	0.586 e
$AsCl_3$	2.161	98.4	0.33 e

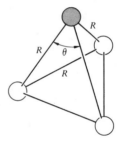

Compute the expected electric dipole moment if the binding were completely ionic (i.e., the halogens each had charge $-e$). Express the observed dipole moment as a fraction of the moment with ionic bonding. For which molecules is the discrepancy largest? Why?

6. **Polarizability tensor.** By an argument similar to that given in Appendix C for the elastic compliance tensor, show that the polarizability tensor must be symmetric. As a consequence any molecule, no matter how unsymmetric, must have three orthogonal principal axes of the polarizability tensor.

7. **Principal axes.** The molecular polarizability is in general a tensor quantity. Find the principal axes for the following molecules. Indicate which, if any, of the principal values are equal. Disregard the permanent moment and consider only the induced moment.

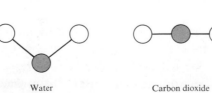

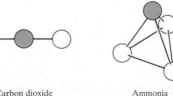

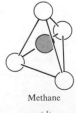

Water	Carbon dioxide	Ammonia	Methane
(a)	(b)	(c)	(d)

8. Electrostatic energy. Find the energy *per molecule* of the following structures:

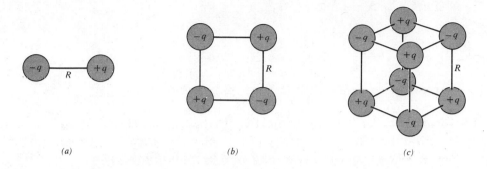

(a)   (b)   (c)

9. Madelung constant. Obtain the Madelung constant for a one-dimensional lattice of atoms of alternating charge:

10. Square lattice. Obtain the Madelung constant for a square lattice of atoms of alternating charge. To improve the convergence of your series it may be helpful to take neutral charge groupings.

11. Electric dipole moment of RbBr. The electric dipole moment of RbBr has not been measured. Using the ionic polarizabilities from Table 2 and the known equilibrium separation R_{eq}, compute a theoretical dipole moment. Compare your result with experimentally observed trends as reflected in Table 3.

12. Electron affinity of the halogens. An electron is bound to a neutral atom in part as a result of its polarization of the atom. Taking electronic polarizabilities and atomic radii from Table 1, compute the electron affinities of the negative halogen ions and compare with the values given below:

Halogen	Electron Affinity
F	3.448 eV
Cl	3.613 eV
Br	3.363 eV
I	3.063 eV

13. Electron affinity of the alkali metals. The energy with which an electron is bound to a neutral atom is called the electron affinity. These energies are tabulated below for the alkalis [T. A. Patterson et al., *Phys. Rev. Lett.*, **32**, 189 (1974)]. Using the model developed in the chapter, compute this energy using the atomic radii and polarizabilities given in Table 1.

Alkali	Electron Affinity
Li	0.62 eV
Na	0.548 eV
K	0.501 eV
Rb	0.486 eV
Cs	0.471 eV

14. **Dipolar field.** An alternative proof that the dipolar field at a site of cubic symmetry is zero starts with the symmetry of the potential. Show that the potential in the vicinity of a site of cubic symmetry must be of the form:

$$\phi(x,y,z) = A(x^2 + y^2 + z^2) + B(x^4 + y^4 + z^4) + C(x^2y^2 + y^2z^2 + z^2x^2) + \ldots.$$

Obtain $A = 0$ from the requirement that the charge density of the central site is excluded. If each charge q is now replaced by a dipole $\mathbf{p}$, obtain the form of the potential. Finally take the gradient to obtain the electric field and evaluate the field at $\mathbf{r} = 0$.

15. **Cubic near-zone.** Since $\mathbf{E}_{loc}$ must be independent of the exact shape of the surface dividing the near-zone and far-zone regions and since $\mathbf{E}_{dip}$ must vanish as long as the near-zone region has cubic symmetry then the local field must be given by **(55)** for *any* site that has cubic symmetry. Show this for a cubic near-zone. That is, compute the electric field at the center of an empty cube in a uniformly polarized medium by using **(1-124)**. Use the result of Problem 1-5 for the field of a charged rectangle.

16. **Ferroelectricity of $BaTiO_3$.** Assuming only nearest neighbor dipole interactions, estimate β for the perovskite structure (Figure 14). Taking the electronic polarizability of O^{2-} from Table 2, estimate the required ionic polarizability of Ti^{+4} for ferroelectric ordering. The actual value for $\alpha_1/4\pi\epsilon_0$ is 9.2×10^{-30} m^3.

REFERENCES

Anderson, J. C., *Dielectrics*, Chapman and Hall, 1967.

Bottcher, C. J. F., *Theory of electric polarization*, Second Edition, Elsevier, 1973.

Frisch, O. R., "Molecular Beams," *Scientific American*, **212** (5), 58 (May, 1965).

Kittel, C., *Introduction to Solid State Physics*, Fifth Edition, Wiley, 1976.

Ramsey, N. F., *Molecular Beams*, Oxford, 1963.

von Hippel, A. R., *Dielectrics and Waves*, MIT Press, 1954.

4 DIELECTRIC MEDIA II

This chapter brings together the discussions of Chapter 1, 2, and 3. In Chapters 1 and 2 we *assumed* a charge distribution and calculated fields and potentials and the energy associated with the distribution. We never really considered *how* the distribution was assembled or maintained, and we saw, in fact, that an *electrostatic* distribution of charge can never be stable of itself.

In Chapter 3 we discussed real media. We have seen that material charge distributions like atoms, which involve only Coulomb interactions, are stabilized by kinetic energy. Given stable matter we next considered the response to an external field, which we *assumed* in the sense of Chapter 1. We found that atoms and molecules and solids are polarized by such fields. And as we have just seen, there are some special cases where the internal interactions are sufficiently strong that the lowest energy state even in the absence of an external field may be one in which the ions are polarized.

DIELECTRIC MEDIA AND EXTERNAL CHARGE

We now wish to consider problems in which the field acting on a dielectric arises from charges that are introduced from outside into the dielectric medium. Most authors designate such charges as "free" and distinguish them from the "bound" charge of the medium. Charge in conducting media is similarly designated as "free." Because we will wish later to describe conducting media in terms of a frequency-dependent susceptibility, we make a distinction instead between "external" charge and "induced" charge.[1] We may regard "external" charge as postulated in that we *assume* some arbitrary charge density to which the medium responds. And the "induced" charge is the response of the medium. In this sense, we are doing a partial problem in that we need not be concerned with the response of the "external" charge to the medium. When we go on to discuss conducting media in the next chapter we will find it useful to associate all internal charge with a polarization density even though the

---

[1] This use of "external" was introduced by L. Onsager in his classic paper "Electric Moments of Molecules in Liquids," *J. Am. Chem. Soc.*, **58**, 1486 (1936).

dielectric susceptibility of such media may be very different from that of insulating dielectrics.

1 **Electric Displacement.** As a way of keeping track of the distinction between the external charge density $\rho_{ext}(\mathbf{r})$ and induced polarization charge density, we define a new vector $\mathbf{D}$ which is called the electric displacement:[2]

$$\blacktriangleright \qquad\qquad \mathbf{D} = \epsilon_0 \mathbf{E} + \mathbf{P} \qquad\qquad (1)$$

The SI units of $\mathbf{D}$ are the same as those of $\mathbf{P}$, coulombs per square metre (C/m^2). Taking the divergence of **(1)** we have:

$$\nabla \cdot \mathbf{D} = \epsilon_0 \nabla \cdot \mathbf{E} + \nabla \cdot \mathbf{P} \qquad\qquad (2)$$

Now as we have seen **(1-41)** the divergence of $\mathbf{E}$ is proportional to the *total* charge density:

$$\nabla \cdot \mathbf{E} = \frac{1}{\epsilon_0} \rho(\mathbf{r}) \qquad\qquad (3)$$

We have also seen **(1-137)** that the divergence of the polarization density generates a charge density, which we take to be the induced charge:

$$-\nabla \cdot \mathbf{P}(\mathbf{r}) = \rho_{ind}(\mathbf{r}) \qquad\qquad (4)$$

Substituting **(3)** and **(4)** into **(2)** we obtain:

$$\nabla \cdot \mathbf{D} = \rho(\mathbf{r}) - \rho_{ind}(\mathbf{r}) \qquad\qquad (5)$$

As we have also seen **(2-23)** that the total charge density may be decomposed into isolated (or external) charge density and polarization (or induced) charge density:

$$\rho(\mathbf{r}) = \rho_{ext}(\mathbf{r}) + \rho_{ind}(\mathbf{r}) \qquad\qquad (6)$$

so that finally we have:

$$\nabla \cdot \mathbf{D} = \rho_{ext}(\mathbf{r}) \qquad\qquad (7)$$

The special property of the displacement $\mathbf{D}$ is that its sources of divergence are *external* charges only. Now, what one regards as external charge will depend on the

---

[2] The vector $\mathbf{D}$ was originally given the name displacement because of the analogy between $\mathbf{F} \cdot d\mathbf{r}$ and $\mathbf{E} \cdot d\mathbf{D}$, which we will discuss later. Some authors call $\mathbf{D}$ the electric flux density because of its connection with external charge, as discussed in this section. We retain the name displacement for $\mathbf{D}$ since the name is in common use and is unlikely to cause confusion.

problem. Since the total charge $\rho(\mathbf{r})$ is determined we *must* develop a consistent definition of what we call polarization. As we have seen (1) the definition of **D** automatically takes care of this for us.

Exercises

1. The field in the vicinity of an external charge q in a homogeneous isotropic medium is given by:

$$\mathbf{E} = \frac{q}{4\pi\epsilon_0\, r^2}\, e^{-r/\lambda}\, \hat{\mathbf{r}}$$

Show that the medium surrounding the charge *must* have a polarization density:

$$\mathbf{P} = \frac{q}{4\pi r^2}\left(1 - e^{-r/\lambda}\right)\hat{\mathbf{r}}$$

2 Circulation of the Displacement. In our discussions of the electric field **E** we used only two properties in ultimately finding an expression for the potential. The first was the divergence of **E** (3) and the second was the statement of zero circulation:

$$\mathscr{E} = \oint \mathbf{E} \cdot d\mathbf{r} = 0 \tag{8}$$

which made it possible to define a potential.

We have seen (7) that the sources of the divergence of **D** are external charges. What are its sources of circulation? From (1) we write:

$$\mathscr{D} = \oint_C \mathbf{D} \cdot d\mathbf{r} = \oint_C \epsilon_0\, \mathbf{E} \cdot d\mathbf{r} + \oint_C \mathbf{P} \cdot d\mathbf{r} \tag{9}$$

Since the first term on the right vanishes (8) we have the circulation of **D**:

$$\mathscr{D} = \oint_C \mathbf{P} \cdot d\mathbf{r} \tag{10}$$

This means that the line integral of **D** depends on the path and we cannot expect to be able to write **D** as the gradient of a scalar as we could for **E**.

Exercises

2. Obtain for a homogeneous, isotropic, and linear dielectric the relation

$$\nabla \times \mathbf{D} = 0$$

3. A sphere of radius R contains only a radial polarization density:

$$\mathbf{P}(\mathbf{r}) = P\hat{\mathbf{r}}/r^2$$

where P is constant. Show that $\mathbf{D}$ is zero both inside and outside the sphere.

---

3. Permittivity. For the problems that we will be doing, we assume that the polarization density is everywhere proportional to the macroscopic electric field **(3-65)**:

$$\mathbf{D} = \epsilon_0 \mathbf{E} + \mathbf{P} = \epsilon_0(1 + \chi)\mathbf{E} = \epsilon \mathbf{E} \tag{11}$$

Table 1   Permittivity of Solids

	ϵ/ϵ_0
Aluminum oxide (corundum)	13.27 and 11.28
Calcium carbonate (calcite)	8.5 and 8.0
Calcium fluoride (fluorite)	7.36
Silicon dioxide (quartz)	4.34 and 4.27
Sodium chloride (rocksalt)	6.12
Titanium dioxide (rutile)	86 and 170
Pyrex glass	4.0 to 6.0
Silica glass	3.81
Vycor glass	3.8 to 3.9
Acrylics (plexiglas, lucite)	3.5 to 5.5
Fluoroplastics (teflon)	2.1
Nylons	3.7 to 5.5
Polycarbonates (lexan, merlon)	3.1
Polyesters (mylar)	3.6
Polyethylenes	1.6 to 2.4
Polypropylenes	2.2 to 2.6
Polystyrenes (dylene, styron)	2.4 to 4.8
Vinyl polymers	3.2 to 3.6

Thus the permittivity of the medium is written as:

$$\epsilon = \epsilon_0(1 + \chi) \tag{12}$$

We may still be able to describe media that are inhomogeneous by letting the permittivity be a function of position. For media that are anisotropic the permittivity becomes a tensor and we write:

▶

$$\mathbf{D} = \bar{\bar{\epsilon}} \cdot \mathbf{E} \tag{13}$$

It is left as Exercise 15 to show that the permittivity tensor must be symmetric. Thus, it is always possible to find axes for which the permittivity matrix is diagonal. For a crystal with threefold or higher rotational symmetry, the permittivity is independent of direction in the transverse plane. As will be discussed in Section 12-15, a crystal without rotational symmetry has two directions of propagation for which the *transverse* permittivity is isotropic. These directions are called the optic axes. For a crystal with threefold or higher rotational symmetry, the two optic axes are merged with the rotation axis.

We give in Table 1 the permittivities at room temperature of a number of minerals, glasses, and plastics. Where two values are given for a crystalline solid, the first is taken perpendicular to the optic axis and the second is parallel to this axis. A range of values indicates variable properties.

DIELECTRIC SCREENING

We now consider the response of a dielectric to divergent fields such as might arise from the presence of external charge within the dielectric medium. As we will see, the dielectric polarization produces an effective screening charge, which reduces the forces between external charges in the dielectric.

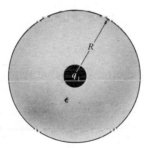

Figure 1 External charge at the center of a dielectric sphere. The medium is assumed to be homogeneous and isotropic.

4 Charge at the Center of a Dielectric Sphere. As a first example we show in Figure 1 a sphere of radius R and permittivity ϵ with an external charge q_1 at the center. We assume that the medium is homogeneous and isotropic.

From Gauss's theorem **(B-11)** we have for the flux of **D**:

$$\Psi_D = \oint_S \mathbf{D} \cdot d\mathbf{S} = 4\pi r^2 D = \int_V \mathbf{\nabla} \cdot \mathbf{D} \, dV = q_1 \tag{14}$$

which gives:

$$D = \frac{1}{4\pi} \frac{q_1}{r^2} \tag{15}$$

Now **(14)** has the same form whether it is written within the sphere or outside and this is one reason for using the displacement. We may now find the field from **(11)**. For $r < R$ we have:

$$E = \frac{D}{\epsilon} = \frac{1}{4\pi\epsilon}\frac{q_1}{r^2} \tag{16}$$

For $r > R$ we have:

$$E = \frac{D}{\epsilon_0} = \frac{1}{4\pi\epsilon_0}\frac{q_1}{r^2} \tag{17}$$

Let us now see how the same problem would be solved without the introduction of **D**. We write Gauss's law for the electric field but we must include both induced and external charge:

$$\oint_S \mathbf{E} \cdot d\mathbf{S} = \frac{1}{\epsilon_0}\int_V [\rho_{ext}(\mathbf{r}) + \rho_{ind}(\mathbf{r})]\, dV = \frac{q_1}{\epsilon_0} - \frac{1}{\epsilon_0}\int_V \nabla \cdot \mathbf{P}\, dV \tag{18}$$

By spherical symmetry and using the divergence theorem we obtain:

$$4\pi r^2 E = \frac{q_1}{\epsilon_0} - \frac{1}{\epsilon_0}\oint_S \mathbf{P} \cdot d\mathbf{S} = \frac{q_1}{\epsilon_0} - 4\pi r^2 \frac{P}{\epsilon_0} \tag{19}$$

For $r < R$ the polarization is given by $P = \epsilon_0 \chi E$ so that we may write:

$$E = \frac{1}{4\pi\epsilon_0(1+\chi)}\frac{q_1}{r^2} = \frac{1}{4\pi\epsilon}\frac{q_1}{r^2} \tag{20}$$

For $r > R$ the polarization is zero and we write:

$$E = \frac{1}{4\pi\epsilon_0}\frac{q_1}{r^2} \tag{21}$$

We may now look at this problem in terms of induced charge. From **(20)** the charge at the center of the sphere is:

$$\epsilon_0 \oint_S \mathbf{E} \cdot d\mathbf{S} = \frac{q_1}{1+\chi} = q - \frac{\chi}{1+\chi}q_1 \tag{22}$$

and there must be surrounding the external charge q an induced charge given by:

$$q_{ind} = -\frac{\chi}{1+\chi}q_1 \tag{23}$$

Because of the discontinuity in the normal component of **E** there must also be a charge layer at the surface. This charge will be given by ϵ_0 times the change in the flux of **E**:

$$\Delta\Psi = \oint_{S^+} \mathbf{E}\cdot d\mathbf{S} - \int_{S^-} \mathbf{E}\cdot d\mathbf{S} = \frac{q_1}{\epsilon_0} - \frac{q_1}{\epsilon} = \frac{\chi}{1+\chi}\frac{q_1}{\epsilon_0} \tag{24}$$

Thus there is an induced charge on the surface:

$$q_{ind} = \epsilon_0\,\Delta\Psi = \frac{\chi}{1+\chi}\,q_1 \tag{25}$$

or a surface charge density:

$$\sigma_{ind} = \frac{q_{ind}}{4\pi R^2} = \frac{\chi}{1+\chi}\frac{q_1}{4\pi R^2} = \epsilon_0\,\chi E = P \tag{26}$$

The induced charge at the center **(23)** is just balanced by the induced charge at the surface **(25)** as it must be if the dielectric is to remain electrically neutral.

Of the three approaches to the problem the first is fastest but hides the physics. The third probably gives more emphasis to polarization charge than is warranted given the different physical processes operating. This approach may be extremely useful, however, in solving problems.

Exercises

4. A uniformly charged sphere of radius R and total charge q is introduced into a medium of permittivity ϵ. Find the induced charge density σ_{ind} on the surface of the sphere. Show that the total induced charge is independent of the radius of the sphere and is given by **(25)**.
5. A uniform surface charge density σ_{ext} is placed on one surface of an infinite dielectric slab of permittivity ϵ and thickness d. What is the total surface charge density on each surface? What are the electric fields inside and outside the medium? Find **D** in the three regions.
6. A uniform surface charge density σ_{ext} covers a dielectric sphere. Show that the induced charge density is zero.
7. A uniformly charged sphere of radius R is surrounded by a dielectric whose permittivity is a function of r, the distance from the center of the sphere. Show that if the permittivity is proportional to $(R/r)^2$, then the electric field in the surrounding region will be constant in magnitude. Obtain an expression for the charge density in the dielectric.

5 Charge Interactions.

5 Charge Interactions. We consider a general distribution of charge in a *homogeneous*, isotropic, and *infinite* dielectric. From **(7)** and **(11)** we may write:

$$\mathbf{V} \cdot \mathbf{E} = \frac{1}{\epsilon} \rho_{\text{ext}}(\mathbf{r}) \tag{27}$$

where $\rho_{\text{ext}}(\mathbf{r})$ is the external charge only. Since the field is *always* the negative gradient of the potential we have Poisson's equation in a dielectric medium:

$$\nabla^2 \phi = -\frac{1}{\epsilon} \rho_{\text{ext}}(\mathbf{r}) \tag{28}$$

Since the divergence of $\mathbf{E}$ may be written in terms of the total microscopic charge density **(3)**:

$$\mathbf{V} \cdot \mathbf{E} = \frac{1}{\epsilon_0} \rho(\mathbf{r}) \tag{29}$$

we have from **(27)** and **(29)** the relation between total charge density and external charge density:

$$\rho(\mathbf{r}) = \frac{\epsilon_0}{\epsilon} \rho_{\text{ext}}(\mathbf{r}) \tag{30}$$

We may, if we wish, characterize both the induced charge and the external charge by a polarization density:

$$\mathbf{P}_{\text{ext}}(\mathbf{r}) = \frac{1}{V} \int_V \rho_{\text{ext}}(\mathbf{r}) \, \mathbf{r} \, dV \tag{31}$$

$$\mathbf{P}(\mathbf{r}) = \frac{1}{V} \int_V \rho_{\text{ind}}(\mathbf{r}) \, \mathbf{r} \, dV \tag{32}$$

Using **(30)** we obtain for the relation between *total* polarization density and the polarization density of the external charge alone:

$$\mathbf{P}_{\text{ext}}(\mathbf{r}) + \mathbf{P}(\mathbf{r}) = \frac{\epsilon_0}{\epsilon} \mathbf{P}_{\text{ext}}(\mathbf{r}) \tag{33}$$

Exercise

8. For a medium that is isotropic but not homogeneous show that the equivalent of **(27)** is:

$$\mathbf{V} \cdot \mathbf{E} = \frac{1}{\epsilon} \rho_{\text{ext}} - \frac{1}{\epsilon} \mathbf{E} \cdot \mathbf{V}\epsilon$$

Note that even though the density of external charge is zero, the density of induced charge need not be zero.

BOUNDARY CONDITIONS

We now modify our model to one of a polarizable dielectric with well-defined boundaries and interacting with a distribution of external charge. We assume that we know the electric field and the displacement on one side of an interface between two dielectrics as in Figure 2a and wish to find the electric field and displacement on the other side.

6 Continuity of Field and Displacement. From **(8)** the circulation of **E** around any contour, such as that shown in Figure 2b, must be zero. But if the contour is chosen close to the surface, the only contributions are from parallel components of the field and we have

$$\hat{z} \times \hat{n} \cdot E_1 = \hat{z} \times \hat{n} \cdot E_2 \tag{34}$$

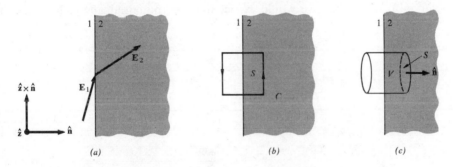

(a)   *(b)*   *(c)*

Figure 2 Boundary conditions. (*a*) The electric field on one side of the interface between two dielectrics is given. (*b*) From the circulation of **E** around a contour, the tangential component of **E** must be continuous. (*c*) From the flux of **D** through a closed surface *S*, the normal component of **D** must be continuous.

or

$$\hat{n} \times E_1 = \hat{n} \times E_2 \tag{35}$$

since $\hat{z}$ is an arbitrary unit vector in the surface.

From **(7)** and the divergence theorem we have that the flux of **D** from any closed surface *S* is equal to the external charge within the surface. If we compute the flux of **D** from the surface *S* shown in Figure 2c under the condition that there is *no* external charge at the interface, we obtain:

$$\hat{n} \cdot D_1 = \hat{n} \cdot D_2 \tag{36}$$

The above equations **(35)** and **(36)** give in simplest form the two continuity conditions and are all that is required.

Exercises

9. The angle between $\mathbf{E}_1$ and $\hat{\mathbf{n}}$ is θ_1. Find the angle θ_2 between $\mathbf{E}_2$ and $\hat{\mathbf{n}}$ for linear dielectrics.

10. Show that the surface polarization charge density at the interface is given by:

$$\sigma_P = \hat{\mathbf{n}} \cdot (\mathbf{P}_1 - \mathbf{P}_2) = \epsilon_0 \frac{\epsilon_1 - \epsilon_2}{\epsilon_2} \hat{\mathbf{n}} \cdot \mathbf{E}_1$$

11. Obtain **(35)** directly by applying **(B-13)** to the volume shown in Figure 2c.

7   Uniqueness Theorem. As we have seen **(1-19)** if $\rho(\mathbf{r})$ is known over all space we may write the potential everywhere as an integral over the charges. We may then compute $\mathbf{E}$ from the gradient of the potential. Similarly, given a distribution of external charge and polarizable dielectrics, we would expect that the potential and field everywhere

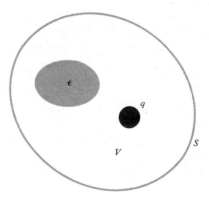

Figure 3   Uniqueness theorem. A volume V contains external charge and dielectrics.

would be determined even though actually finding them might be something of a problem.

We now consider the more limited problem of finding the potential and field not over all space but only within a volume V bounded by a surface S. We imagine that within the volume V are external charges and dielectrics as indicated in Figure 3. We might expect that to determine ϕ within V we would need to know both ϕ and $\mathbf{E}$ over the surface. As we will show, all we need over S is ϕ or the normal component of $\mathbf{D}$ but not both.

To demonstrate this remarkable result we assume two solutions ϕ_1 and ϕ_2 and show that if they both have the same postulated sources of $\mathbf{D}$ within V and satisfy the appropriate boundary conditions, then the two solutions must be identical. In this sense ϕ is determined by the charges and dielectrics within V together with only partial boundary conditions on S.

Associated with ϕ_1 and ϕ_2 there will be fields $\mathbf{E}_1$ and $\mathbf{E}_2$ and displacements $\mathbf{D}_1$ and $\mathbf{D}_2$. We form the function:

$$\Delta(\mathbf{r}) = (\phi_1 - \phi_2)(\mathbf{D}_1 - \mathbf{D}_2) \tag{37}$$

From the divergence theorem **(B-11)** we have:

$$\int_V \mathbf{\nabla} \cdot \Delta(\mathbf{r}) \, dV = \oint_S \Delta(\mathbf{r}) \cdot d\mathbf{S} \tag{38}$$

Now, the divergence of $\Delta(\mathbf{r})$ may be written from **(B-2)** as

$$\mathbf{\nabla} \cdot \Delta(\mathbf{r}) = (\phi_1 - \phi_2)(\mathbf{\nabla} \cdot \mathbf{D}_1 - \mathbf{\nabla} \cdot \mathbf{D}_2) - (\mathbf{E}_1 - \mathbf{E}_2) \cdot (\mathbf{D}_1 - \mathbf{D}_2) \tag{39}$$

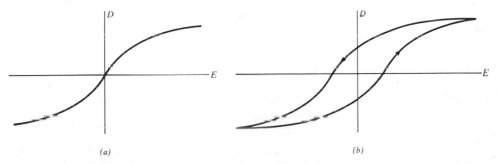

(a)   (b)

Figure 4 Variation of D with E. (a) The displacement is taken to be a single-valued mono-tonically increasing function of E. (b) The displacement is a double-valued function of E exhibiting hysteresis.

But the first term on the right side of **(39)** is zero since we require that $\mathbf{D}_1$ and $\mathbf{D}_2$ have the same sources within V. Then substituting **(37)** and **(39)** into **(38)** we obtain:

$$\int_V (\mathbf{E}_1 - \mathbf{E}_2) \cdot (\mathbf{D}_1 - \mathbf{D}_2) \, dV = -\oint_S (\phi_1 - \phi_2)(\mathbf{D}_1 - \mathbf{D}_2) \cdot d\mathbf{S} \tag{40}$$

Now, if *either* ϕ or $\hat{\mathbf{n}} \cdot \mathbf{D}$ is established everywhere on the surface, then the right-hand side of **(40)** will be zero since we must require that ϕ or $\mathbf{D}$ meet these minimal boundary conditions. This leaves us with:

$$\int_V (\mathbf{E}_1 - \mathbf{E}_2) \cdot (\mathbf{D}_1 - \mathbf{D}_2) \, dV = 0 \tag{41}$$

Now, if $\mathbf{D}$ is a single-valued monotonically increasing function of $\mathbf{E}$ as in Figure 4a then the integrand of **(41)** will be everywhere positive. Then the only way in which **(41)** can be satisfied is for $\mathbf{E}_1 = \mathbf{E}_2$ and thus $\mathbf{D}_1 = \mathbf{D}_2$ everywhere within V. We have shown then that there may be no difference between solutions which have the correct divergence of $\mathbf{D}$ and have the proper value of ϕ or $\hat{\mathbf{n}} \cdot \mathbf{D}$ on the boundary.

What is the significance of a variation of D with E as shown in Figure 4b? This is a hysteresis curve of the form discussed for ferroelectrics where the presence of domain walls makes $\mathbf{D}$ a double-valued function of $\mathbf{E}$ depending on whether the field is increasing or decreasing. Under such circumstances our theorem fails and the two solutions *may* be different since the left side of **(40)** may have negative as well as positive contributions. Under such conditions knowing only ϕ or the normal component of $\mathbf{D}$ is not sufficient to establish the state of the system. We must as well know the "history" of the dielectric.

Exercise

12. External charges $+q$ and $-q$ are distributed on equipotential surfaces S_1 and S_2 with S_2 entirely enclosing S_1. Show that $\hat{\mathbf{n}} \cdot \mathbf{D}$ must be identically zero on the outside of S_2. By the uniqueness theorem show that the electric field must be zero everywhere outside S_2 and that S_2 is at zero potential.

INTERACTION ENERGY

In analogy with the discussion at the end of Chapter 2 we now wish to consider the energy of interaction of a system of external charges and dielectrics. We imagine that we have, as shown in Figure 5, a dielectric body, which we polarize by bringing it into a region where there are fixed external charges q_i. Because we wish to consider the possibility that the dielectric may be nonlinear or may even exhibit hysteresis, we must describe in a very precise way how we mean to assemble the charges and dielectrics:

1. We first bring up the external charges alone. This is a problem that we have already considered in Chapter 2, and we may take over the results of that analysis.

2. We imagine that the dielectric body or bodies are at infinity and that they are initially depolarized. We bring up all the dielectric bodies at once with their relative positions fixed. In this way, the mechanical work that must be performed depends only on the fields of the fixed charges, which we call the *external* electric field $\mathbf{E}_{ext}(\mathbf{r})$.

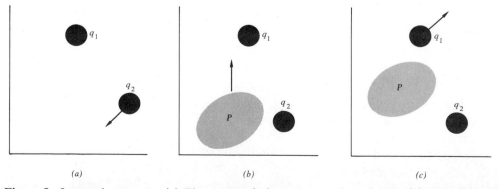

(a)   (b)   (c)

Figure 5 Interaction energy. (*a*) First external charges q_i are brought up. (*b*) Next initially depolarized dielectric bodies are brought up. (*c*) With the charges and dielectric in place we allow for a modification associated with displacement of the q_i.

3. With the charges and dielectrics in place, we may consider modifications associated with displacing the q_i.

The work required to bring the charges together is from **(2-6)**:

$$W = \tfrac{1}{2} \sum_{i,j}' q_i \phi_{ij} \tag{42}$$

where ϕ_{ij} is the ordinary Coulomb potential between pairs of charges. To bring up the dielectrics we must perform mechanical work **(2-10)**:

$$W_{\text{mech}} = - \int_V dV \int (\mathbf{P} \cdot \mathbf{V}) \mathbf{E}_{\text{ext}} \cdot d\mathbf{r} = - \int_V dV \int \mathbf{P} \cdot d\mathbf{E}_{\text{ext}} \tag{43}$$

Adding **(42)** and **(43)** we have for the total work:

$$W = \tfrac{1}{2} \sum_{ij}' q_i \phi_{ij} - \int_V dV \int \mathbf{P} \cdot d\mathbf{E}_{\text{ext}} \tag{44}$$

Now, since the interaction between q_i and $\mathbf{P(r)}$ may be written either in terms of the potentials at $\mathbf{r}_i$ from $\mathbf{P(r)}$ or in terms of the external fields at $\mathbf{P(r)}$ from the q_i, we have the identity:

$$\sum_i q_i \left(\phi_i - \sum_j' \phi_{ij} \right) = - \int_V dV \, \mathbf{P} \cdot \mathbf{E}_{\text{ext}} \tag{45}$$

where the quantity in parentheses is the contribution of $\mathbf{P}$ to ϕ_i. Eliminating ϕ_{ij} between **(44)** and **(45)** we obtain, for W, the symmetric form:

$$W = \tfrac{1}{2} \sum_i q_i \phi_i + \tfrac{1}{2} \int_V dV \, \mathbf{P} \cdot \mathbf{E}_{\text{ext}} - \int_V dV \int \mathbf{P} \cdot d\mathbf{E}_{\text{ext}} \tag{46}$$

If we now make a variation in the q_i as shown in Figure 5c the amount of work done may simply be written in terms of the change in charge:

$$\delta W = \sum \phi_i \, \delta q_i \tag{47}$$

In writing **(47)** we assume that the dielectrics are in place so that there is no *mechanical* work associated with displacement of the polarization.

8 Superposition. For linear dielectrics we may write by superposition that the potential at the *site* of q_i must be linearly dependent on the other external charges q_j:

$$\tag{48}$$

where the π_{ij} are called the coefficients of potential. We emphasize that even though in the presence of a dielectric it may not be possible to obtain the π_{ij} in any simple way, yet we know that as long as the dielectrics are linear we must be able to superpose the solutions.

With reference to Figure 5, bringing up q_i alone yields a potential $\phi_i(\mathbf{r})$. The gradient of this potential gives the fields $\mathbf{E}_i$ and $\mathbf{D}_i$ everywhere. If instead we bring up q_j alone, we have a potential $\phi_j(\mathbf{r})$ and fields $\mathbf{E}_j$ and $\mathbf{D}_j$. We argue that if charges q_i and q_j are brought up simultaneously, then the potential will be $\phi_i(\mathbf{r}) + \phi_j(\mathbf{r})$ with fields $\mathbf{E}_i + \mathbf{E}_j$ and $\mathbf{D}_i$ and $\mathbf{D}_j$.

We use the uniqueness theorem with the surface S at infinity so that $\phi(\mathbf{r})$ is zero on S. Since $\mathbf{D}_i + \mathbf{D}_j$ gives the correct distribution of external charge with both q_i and q_j brought up, then $\phi_i(\mathbf{r}) + \phi_j(\mathbf{r})$ must be the unique solution to the combined problem.

9 Electric Work.

By substituting **(48)** into **(47)** we obtain for the incremental work:

$$\delta W = \sum_{ij}{}' \pi_{ij}\, q_j\, \delta q_i \tag{49}$$

We argue next that the total amount of work required to assemble a distribution of charges in the presence of linear dielectrics cannot depend on the order in which the charges are assembled. This result was easily established for external charge alone **(2-6)** by an explicit expression for the external work. In other words, if we start with some distribution of charges that we alter in such a way that we ultimately return to the same distribution, the total work done must be zero. If the work were not zero we would have dissipated energy in the dielectric. But the fact that $\mathbf{P}$ is a single-valued function of $\mathbf{E}$ **(3-65)** ensures that no energy is dissipated. It follows then that **(49)** must be a perfect differential, and we have

$$\pi_{ij} = \pi_{ji} \tag{50}$$

With this condition we may integrate **(49)** to obtain

▶
$$W = \tfrac{1}{2} \sum_{ij}{}' \pi_{ij}\, q_i\, q_j = \tfrac{1}{2} \sum_i q_i\, \phi_i \tag{51}$$

where we have substituted from **(41)** to obtain the second equality.

Although the form of **(51)** is quite simple, all the complications of the problem are absorbed in the ϕ_i, which we have not attempted to write explicitly. We may get further insight into where the work goes by comparing **(46)** with the work required to bring together a distribution of external charges and polarized bodies as given by **(2-19)**:

$$W_{\text{ext}} = \tfrac{1}{2} \sum_i q_i\, \phi_i - \tfrac{1}{2} \sum_j \mathbf{p}_j \cdot \mathbf{E}_{\text{ext}} \tag{52}$$

Comparing (46) and (52) we obtain:

$$W = W_{\text{ext}} + \int dV \int \mathbf{E}_{\text{ext}} \cdot d\mathbf{P} \tag{53}$$

What is the physical significance of (53)? The second term must be the *internal* energy of the dipoles, which are here considered as part of a dielectric medium. If we call this energy W_{int} we then have:

▶
$$W_{\text{int}} = W - W_{\text{ext}} = \int dV \int \mathbf{E}_{\text{ext}} \cdot d\mathbf{P} \tag{54}$$

That is, the total work done on the system may be regarded as the sum of the energy required to assemble fixed moments and the internal energy required to establish their polarization.

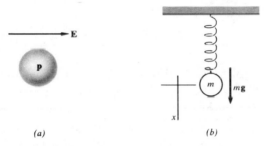

(a)   *(b)*

Figure 6 Polarizable atom. (*a*) The energy is given by (**55**). (*b*) A mass on a spring in a gravitational field behaves in an analogous way.

To make the discussion more explicit we consider the simple example of a polarizable atom in a uniform electric field as shown in Figure 6a. For the energy of such a system, *excluding the energy to establish* **E** we have had (**3-19**)

$$W = -pE + \frac{1}{2}\frac{p^2}{\alpha} \tag{55}$$

where the first term is the interaction between the polarization and the *external* field and is what we have called external energy. The second term is the interaction energy between the charges comprising the dipole and is what we have called internal energy. To find the equilibrium polarization we minimize (55) with respect to p to obtain:

$$p = \alpha E \tag{56}$$

Using (56) we may now rewrite (55) as:

$$W = W_{\text{ext}} + W_{\text{int}} = -pE + \tfrac{1}{2}pE = -\tfrac{1}{2}pE \tag{57}$$

From (53) we are led to this same expression if we take $\mathbf{E}_{ext} = \mathbf{E}$ and assume a linear relation between $\mathbf{P}$ and $\mathbf{E}$ as in (56).

What we are discussing here is a characteristic of the energy of any linear system. We may write, for example, for the mass on a spring shown in Figure 6b:

$$W = -mgx + \tfrac{1}{2}kx^2 \tag{58}$$

where x is the displacement from equilibrium. Minimizing the energy with respect to x we obtain:

$$x = -\frac{mg}{k} \tag{59}$$

and the work may be written as:

$$W = W_{ext} + U_{int} = -mgx + \tfrac{1}{2}mgx \tag{60}$$

where the internal energy, which is stored in the spring, is analogous to the internal energy of the polarizable atom, which is stored in the relative displacement of the electronic charge and the nuclear charge.

Exercises

13. A charge q is brought from infinity to a position $\mathbf{r}$ with respect to a small dielectric sphere of polarizability α. Show that the amount of work done is:

$$W = -\frac{q^2\alpha}{32\pi^2\epsilon_0^2 r^4}$$

14. A dipole $\mathbf{p}$ is brought from infinity to a position $\mathbf{r}$ with respect to a small dielectric sphere of polarizability α. Show that the work done is:

$$W = -\frac{\alpha}{32\pi^2\epsilon_0^2}\frac{3(\mathbf{p}\cdot\hat{\mathbf{r}})^2 - p^2}{r^6}$$

15. Let q_j be a distribution of charges that produce ϕ_i by (48). Let us now replace at the same positions the charges q_j by q_j' to obtain potentials ϕ_i'. By (50) prove *Green's reciprocation theorem*:

$$\sum_i \phi_i q_i' = \sum_i \phi_i' q_i$$

10 Field Energy. Just as we developed **(2-6)** as a field energy **(2-36)**, we may write **(46)** in terms of fields. To accomplish this we first eliminate the singular energy of point charges by writing **(47)** for distributed charge:

$$\delta W = \int_V \phi \, \delta \rho_{ext} \, dV \tag{61}$$

where $\rho_{ext}(\mathbf{r})$ is the external charge only. We use the vector identity **(B-2)**:

$$\nabla \cdot [\phi \, \delta \mathbf{D}] = \phi \, \delta \rho_{ext} - \mathbf{E} \cdot \delta \mathbf{D} \tag{62}$$

Substituting into **(61)** and using the divergence theorem we obtain:

$$\delta W = \int_V \mathbf{E} \cdot \delta \mathbf{D} \, dV + \oint_S \phi \, \delta \mathbf{D} \cdot d\mathbf{S} \tag{63}$$

By letting the surface S go to infinity the second term vanishes since $\oint \delta \mathbf{D} \cdot d\mathbf{S}$, which is the incremental flux of $\mathbf{D}$, is constant at large distances, and $\phi(\mathbf{r})$ drops off at least as fast as $1/r$. We have, then, for the *energy* of a configuration of charge and polarizable dielectrics:

$$U = \int dV \int \mathbf{E} \cdot d\mathbf{D} \tag{64}$$

For a *linear* medium described by **(11)** we may integrate **(64)** to obtain:

▶
$$U = \tfrac{1}{2} \int dV \int d(\mathbf{E} \cdot \mathbf{D}) = \tfrac{1}{2} \int \mathbf{E} \cdot \mathbf{D} \, dV \tag{65}$$

How is **(64)** related to our earlier expressions for the external field energy **(2-24)** and **(2-41)**:

▶
$$U_{ext} = \tfrac{1}{2} \epsilon_0 \int E^2 \, dV - \frac{1}{6\epsilon_0} \int P^2 \, dV \tag{66}$$

Subtracting **(66)** from **(64)** and using the definition of the displacement **(1)** we obtain:

▶
$$U_{int} = U - U_{ext} = \int dV \int \left(\mathbf{E} + \frac{1}{3\epsilon_0} \mathbf{P} \right) \cdot d\mathbf{P} \tag{67}$$

Using **(1-125)** to relate the external field and the macroscopic field:

$$\mathbf{E}_{ext} = \mathbf{E} + \frac{1}{3\epsilon_0} \mathbf{P} \tag{68}$$

we obtain for (67):

▶
$$U_{int} = \int dV \int \mathbf{E}_{ext} \cdot d\mathbf{P} \qquad (69)$$

which is identical with (54).

The expression given by (64) for the total energy is given in terms of macroscopic fields and is unambiguous. On the other hand, the expression for the internal energy (69) depends on what we take to be the external field. Although the total energy does not depend on our choice for $\mathbf{E}_{ext}$, the division between what we call internal and external energy does depend on this choice. For example, if we imagine that we are assembling microscopic dipoles, as in the discussion of Sections 2-3 and 2-4, then the interaction between dipoles, which is given by the Lorentz field $\mathbf{P}/3\epsilon_0$, is treated as external energy.

On the other hand, the discussion in this chapter of the interaction energy assumed macroscopic polarizable bodies in fixed relative position. Within such a discussion not only the interaction between induced dipoles but also the interaction between polarizable bodies is treated as internal energy.

Exercises

16. By an argument similar to that given in Section C-9 for the elastic compliance tensor and using (69) show that the permittivity tensor must be symmetric.
17. For a homogeneous but not necessarily isotropic dielectric obtain the relation:

$$\mathbf{\nabla} \times \mathbf{D} = \bar{\bar{\epsilon}} \cdot \mathbf{\nabla} \times \mathbf{E} = 0$$

Use the result of Exercise 16.
18. For a homogeneous but not necessarily linear medium plot D versus E and indicate the area represented by the integral of (64).

11 [**Forces on Dielectric Media**]. From (1-106) we have, for the force on a dipole in an electric field gradient,

$$\mathbf{F} = \mathbf{p} \cdot \mathbf{\nabla E} \qquad (70)$$

The force on the dipoles within some limited volume V_1 may then be written as:

$$\mathbf{F} = \int_{V_1} \mathbf{P} \cdot \mathbf{\nabla E} \, dV_1 = \int_{V_1} (\epsilon - \epsilon_0)\mathbf{E} \cdot \mathbf{\nabla E} \, dV_1 = \tfrac{1}{2} \int_{V_1} (\epsilon - \epsilon_0)\nabla E^2 \, dV_1 \qquad (71)$$

We may now write (71) as:

$$\mathbf{F} = \tfrac{1}{2} \int_{V_1} \mathbf{\nabla}[(\epsilon - \epsilon_0)E^2] \, dV_1 - \tfrac{1}{2} \int_{V_1} E^2 \mathbf{\nabla}\epsilon \, dV_1 \qquad (72)$$

By the gradient theorem **(B-12)** we may transform the first term to an integral over the surface:

$$\mathbf{F} = -\tfrac{1}{2}\int_{V_1} E^2 \nabla\epsilon \, dV_1 + \tfrac{1}{2}\int_{S_1} (\epsilon - \epsilon_0)E^2 \, d\mathbf{S} \tag{73}$$

The integrand of the second term, being normally directed acts like a pressure. We may write, then, for the force per unit volume:

$$\mathbf{f} = -\tfrac{1}{2}E^2\nabla\epsilon - \nabla P \tag{74}$$

where

$$P = -\tfrac{1}{2}(\epsilon - \epsilon_0)E^2 \tag{75}$$

is an effective electric pressure.

To describe the motion of a dielectric fluid or the strains within a dielectric solid, we must consider, in addition to **(74)**, body forces that may not be represented by macroscopic fields.[3]

CAPACITANCE

A practical electrical capacitor is a device for storing electric charge at a difference in electric potential. Most capacitors have conducting elements (metals or semiconductors) and insulating dielectric elements although, as we will see in Chapter 5, the distinction is not always so clear. Most introductory treatments of this subject assume ideal conductors, which are taken to be at a uniform potential. Because we wish to avoid this unrealistic assumption, we define capacitance in a more abstract way.

12 Coefficients of Capacitance. We invert **(48)** to write for the external charges:

$$q_i = \sum_j C_{ij}\phi_j \tag{76}$$

where the C_{ij} are called the coefficients of capacitance and ϕ_j does *not* include the potential of q_j. The symmetry of the coefficients of potential **(50)** ensures the symmetry of the coefficients of capacitance:

$$C_{ij} = C_{ji} \tag{77}$$

---

[3] For a treatment of the motion of dielectric fluids, see J. R. Melcher and G. I. Taylor, "Electrohydrodynamics: A Review of the Role of Interfacial Shear Stresses," *Annual Review of Fluid Mechanics*, Vol. 1, Annual Reviews, Palo Alto, 1969, Sears and Van Dyke, editors. See also the film "Electric Fields and Moving Media" written by J. R. Melcher and produced by Educational Development Center, Newton, MA.

Alternatively, we may write δW from **(47)** and **(76)**:

$$\delta W = \sum_{ij} C_{ij} \phi_i \, \delta \phi_j \tag{78}$$

and the requirement that δW be a perfect differential leads to **(77)**.

In SI units with q_i in coulombs and ϕ_j in volts, C_{ij} is given in farads, which, from **(76)**, is equivalent to coulombs per volt. The farad is defined as "the capacitance of a capacitor between the plates of which there appears a difference of potential of one volt when it is charged by a quantity of electricity equal to one coulomb."

We now imagine a set of charges q_i located at positions $\mathbf{r}_i$. The coefficient of capacitance of $\mathbf{r}_i$ with respect to $\mathbf{r}_j$ is given by:

$$C_{ij} = \frac{q_i}{\phi_j} \tag{79}$$

with the stipulation that all sites except the jth site are at zero potential. For i not equal to j the C_{ij} are called coefficients of induction and a potential ϕ_j is described as inducing a charge q_i at the ith site under the condition that all the potentials (including ϕ_i) and except for ϕ_j are zero. The coefficient of capacitance for $i = j$ is usually called simply the capacitance of the ith site and written with a single subscript:

$$C_i = C_{ii} \tag{80}$$

13   Extended Charge. Although **(48)** and **(76)** were written for point charges, they may be generalized to distributed charge. We may continue to use **(48)** by including the term for $i = j$. And the potential ϕ_j in both **(48)** and **(76)** will now include the potential of q_j. With practical capacitors we are concerned with a limited number of charges q_i distributed over conductors in such a way that the potentials ϕ_i on each conductor are constant or nearly constant. We consider here as an example a spherical distribution of charge.

We have, as shown in Figure 7, two concentric spherical shells separated by a medium of permittivity ϵ. The shells are of radius R_1 and R_2 and carry charges q_1 and q_2.

We obtain for the potential on the outer sphere:

$$\phi_2 = \frac{q_1 + q_2}{4\pi\epsilon_0 R_2} \tag{81}$$

and for the potential on the inner sphere:

$$\phi_1 = \phi_2 + \frac{q_1}{4\pi\epsilon} \left(\frac{1}{R_1} - \frac{1}{R_2} \right) \tag{82}$$

By comparison with **(76)** we obtain, for the coefficients of capacitance,

$$C_{11} = -C_{12} = -C_{21} = 4\pi\epsilon \frac{R_1 R_2}{R_2 - R_1} \tag{83}$$

$$C_{22} = C_{11} + 4\pi\epsilon_0 R_2 \tag{84}$$

From **(80)** we have, for the capacitance of the inner shell (with the outer shell at zero potential),

$$C_1 = C_{11} = 4\pi\epsilon \frac{R_1 R_2}{R_2 - R_1} \tag{85}$$

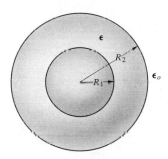

Figure 7 Extended charge. Two concentric spherical shells are separated by a medium of permittivity ϵ.

and for the outer shell (with the inner shell at zero potential),

$$C_2 = C_{22} = 4\pi\epsilon_0 R_2 + 4\pi\epsilon \frac{R_1 R_2}{R_2 - R_1} \tag{86}$$

Exercises

19. Show that the SI unit of permittivity is the farad per metre.
20. Two coaxial cylindrical distributions of charge with radii R_1 and R_2 are separated by a dielectric of permittivity ϵ. Find the capacitance per metre of the inner and outer conductors.
21. Two infinite sheets of charge are separated by a dielectric of permittivity ϵ and thickness d. Find the capacitance per square metre for either of the sheets.
22. Charges $+q$ and $-q$ are distributed on equipotential surfaces S_1 and S_2. Show that the difference in potential between the two surfaces is given by:

$$V = \phi_1 - \phi_2 = \frac{C_{22} - C_{11}}{C_{11}C_{22} - C_{12}C_{21}} q$$

For concentric spheres obtain the simple result:

$$V = q/C_{11}$$

23. Charges $+q$ and $-q$ are distributed on equipotential surfaces S_1 and S_2 with S_2 entirely enclosing S_1. Use the result of Exercise 12 to obtain for the difference in potential between the surfaces:

$$V = q/C_{11}$$

and the relation

$$C_{12} = C_{21} = -C_{11}$$

CHARGES NEAR A DIELECTRIC

We wish to find the field and energies when an external charge is brought near the interface between two dielectrics as in Figure 8.

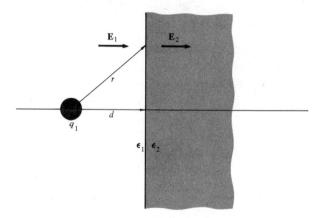

Figure 8   Charge near a dielectric interface. An external charge q_1 is brought to within a distance d of the interface.

We first write the electric field for this problem in terms of both the external charge q_1 and the induced polarization charge:

1. Surrounding q_1 will be **(23)** an induced charge:

$$q_{ind} = -\frac{\chi_1}{1 + \chi_1} q_1 \tag{87}$$

so that the total charge will be

$$q_1 + q_{ind} = \frac{q_1}{1 + \chi_1} = \frac{\epsilon_0}{\epsilon_1} q_1 \tag{88}$$

2. At the interface between the two dielectrics there will be an induced *surface* charge σ_{ind}. This surface charge will produce fields with normal components $\sigma_{ind}/2\epsilon_0$ on either side and directed away from the surface for positive σ_{ind}.

3. There is no other induced charge in the problem since within either of the dielectrics except at the position of q we have $\mathbf{V} \cdot \mathbf{D} = 0$. With the relation $\mathbf{E} = \mathbf{D}/\epsilon$, we have $\mathbf{V} \cdot \mathbf{E} = 0$ and thus no other charge.

The *normal* component of the field to the left of the interface may be written as:

$$\hat{\mathbf{n}} \cdot \mathbf{E}_1 = \frac{1}{4\pi\epsilon_1} \frac{q_1}{r^2} \frac{d}{r} - \frac{\sigma_{ind}}{2\epsilon_0} \tag{89}$$

The *normal* component of the field to the right of the interface may be written as:

$$\hat{\mathbf{n}} \cdot \mathbf{E}_2 = \frac{1}{4\pi\epsilon_1} \frac{q_1}{r^2} \frac{d}{r} + \frac{\sigma_{ind}}{2\epsilon_0} \tag{90}$$

From (**36**) we must have:

$$\hat{\mathbf{n}} \cdot \mathbf{D}_1 = \epsilon_1 \hat{\mathbf{n}} \cdot \mathbf{E}_1 = \hat{\mathbf{n}} \cdot \mathbf{D}_2 = \epsilon_2 \hat{\mathbf{n}} \cdot \mathbf{E}_2 \tag{91}$$

Substituting (**89**) and (**90**) into (**91**) and solving for σ_{ind} we obtain:

$$\frac{\sigma_{ind}}{2\epsilon_0} = \frac{\epsilon_1 - \epsilon_2}{\epsilon_1 + \epsilon_2} \frac{1}{4\pi\epsilon_1} \frac{q_1}{r^2} \frac{d}{r} \tag{92}$$

Substituting (**92**) into (**89**) and (**90**) we have for the normal components of the fields:

$$\hat{\mathbf{n}} \cdot \mathbf{E}_1 = \frac{1}{4\pi\epsilon_1} \frac{q_1}{r^2} \frac{d}{r} \left(1 - \frac{\epsilon_1 - \epsilon_2}{\epsilon_1 + \epsilon_2}\right) \tag{93}$$

$$\hat{\mathbf{n}} \cdot \mathbf{E}_2 = \frac{1}{4\pi\epsilon_1} \frac{q_1}{r^2} \frac{d}{r} \left(1 + \frac{\epsilon_1 + \epsilon_2}{\epsilon_1 - \epsilon_2}\right) = \frac{1}{4\pi\epsilon_2} \frac{2\epsilon_2}{\epsilon_1 + \epsilon_2} \frac{q_1}{r^2} \frac{d}{r} \tag{94}$$

Now that we have all the charges from (**88**) and (**92**) we may, in principle, at least, write the potential as:

$$\phi(\mathbf{r}) = \frac{1}{4\pi\epsilon_1} \frac{q_1}{r_{01}} + \frac{1}{4\pi\epsilon_0} \int_{S_1} \frac{\sigma_{ind}(\mathbf{r}_2)}{r_{02}} dS_2 \tag{95}$$

and from this expression compute the total field.

14 Method of Images. For problems of this kind there exists another technique that permits us to write the solution for $\phi(\mathbf{r})$ at sight. This is called the *method of images*.[4] We recall that we know the divergence of **D** within the medium to the left of the interface and we also know $\hat{\mathbf{n}} \cdot \mathbf{E}_1$ and thus $\hat{\mathbf{n}} \cdot \mathbf{D}_1$ on the interface. As we have seen from our uniqueness theorem, any solution that satisfies these conditions must in fact be the unique solution of the problem. We show in Figure 9 a charge configuration that may be seen to give **(93)** with:

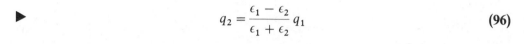

$$q_2 = \frac{\epsilon_1 - \epsilon_2}{\epsilon_1 + \epsilon_2} q_1 \tag{96}$$

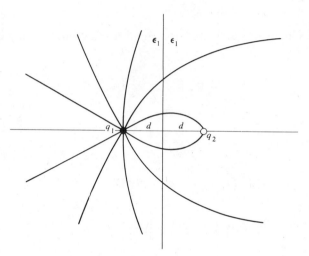

Figure 9   Image charge. The field within medium 1 is the same as that arising from q_1 and an image charge q_2.

The potential to the left of the interface is then

$$\phi(\mathbf{r}) = \frac{1}{4\pi\epsilon_1} \frac{q_1}{r_{01}} + \frac{1}{4\pi\epsilon_1} \frac{q_2}{r_{02}} \tag{97}$$

with $\mathbf{r}_{0i} = \mathbf{r} - \mathbf{r}_i$. We emphasize that **(97)** is the solution *only* to the left of the interface, that is, only in the limited region for which it is intended.

We may similarly find a solution in region 2 involving image charges in region 1. The field given by **(94)** would arise from a charge:

$$q_3 = \frac{2\epsilon_2}{\epsilon_1 + \epsilon_2} q_1$$

[4] For a more general application of the method of images, see B. G. Dick, *Am. J. Phys.* **41**, 1289 (1973) and **42**, 428 (1974) and M. Zahn, *Am. J. Phys.* **44**, 1132 (1976).

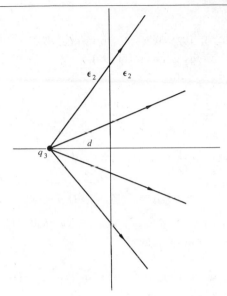

Figure 10   Image charge. The field in medium 2 is the same as that arising from a single image charge q_3

at the position of q_1 and in a medium of permittivity ϵ_2 as shown in Figure 10. Again, we may invoke our uniqueness theorem to conclude that the field in region 2 must be that of the image charge shown in Figure 10.

Finally we show in Figure 11 the field in both media arising from the presence of q_1. This field, which is a composite of Figures 9 and 11 is directly obtainable from **(95)** without the use of images.

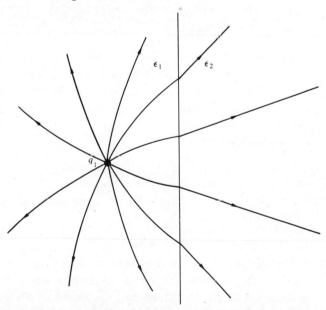

Figure 11   Composite field. The actual field is the composite of fields arising from the images shown in Figures 9 and 10.

Exercises

---

24. Show by integration of **(92)** over the interface that the total surface charge is equal to $(\epsilon_0/\epsilon_1)q_2$ where q_2 is given by **(96)**.

25. Beginning with the expression

$$W_{\text{ext}} = \frac{q}{4\pi\epsilon_1}\int_S \frac{\sigma_{\text{ind}}}{r}\,dS + \tfrac{1}{2}\int_S \sigma_{\text{ind}}\,\phi(\mathbf{r})\,dS$$

show that the *external* work may also be written as:

$$W_{\text{ext}} = \frac{\epsilon_0}{\epsilon_1}(q_1 + \tfrac{1}{2}q_2)\phi(\mathbf{r})$$

where $\phi(\mathbf{r})$ is the potential of q, at $\mathbf{r}$.

---

15 **Interaction Energy.** The work required to bring the external charge q_1 to within a distance d of the interface is given by:

$$W = -\int^d \mathbf{F}\cdot d\mathbf{r} \tag{98}$$

with

$$\mathbf{F} = \frac{q_1 q_2}{4\pi\epsilon_0}\frac{\hat{\mathbf{r}}_{12}}{r_{12}{}^2} = \frac{q_1 q_2}{16\pi\epsilon_0}\frac{\hat{\mathbf{n}}}{d^2} \tag{99}$$

We carry out the integration to obtain:

$$W = \frac{q_1 q_2}{16\pi\epsilon_0 d} = \frac{1}{2}\frac{1}{4\pi\epsilon_0}\frac{q_1 q_2}{r_{12}} = \tfrac{1}{2}q_1\phi \tag{100}$$

where ϕ is the potential of the image charge q_2 evaluated at the position of the external charge. We note as expected that **(100)** is in agreement with **(51)**. Although the arguments leading to the factor of one-half are different, they must involve the same physics since the amount of work done depends only on the final state and not on how it is reached.

Using **(96)** we may write **(100)** as:

$$W = \frac{1}{8\pi\epsilon_0}\frac{\epsilon_1 - \epsilon_2}{\epsilon_1 + \epsilon_2}\frac{q^2}{2d} \tag{101}$$

For $\epsilon_1 > \epsilon_2$ the work is positive and the charge is repelled from its image of the same sign. On the other hand for $\epsilon_1 < \epsilon_2$ as shown in Figures 9 to 11, the work is negative, the image is opposite in sign, and q_1 is attracted to the interface.

By superposition we find that if a distribution of charges is brought to the vicinity of an interface, the resultant field is that from the superposition of all the charges and their images. For $\epsilon_1 < \epsilon_2$ the effect of the image charges is to reduce the interaction energy between charges that are close to the interface.

Exercise

26. Show that the electric field acting on q_1 as obtained from **(99)** is equal to minus the gradient of ϕ as obtained from **(100)**. This relation is valid even though the image charge moves with q_1. Explain why the motion of the image may be neglected.

DIELECTRIC ELLIPSOIDS

We conclude this chapter by considering a class of problems that involve an ellipsoidal interface between dielectrics in the presence of a uniform *applied* field. The simplest case to consider is that of a dielectric ellipsoid in vacuum as shown in Figure 12a. We generalize this in Figure 12b to an ellipsoid of permittivity ϵ_1 in a medium of permittivity ϵ_2. Finally, as a specific example of this second case, we consider an ellipsoidal cavity in a medium of permittivity ϵ as shown in Figure 12c.

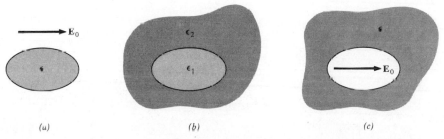

(a)   *(b)*   *(c)*

Figure 12 Dielectric ellipsoid. (*a*) The simplest case is that of a dielectric ellipsoid in vacuum. (*b*) We generalize the vacuum problem to an ellipsoid of permittivity ϵ_1 in a medium of permittivity ϵ_2. (*c*) As a special case we may then consider an ellipsoidal cavity in a medium of permittivity ϵ.

16 **Uniform Field.** To obtain the fields for the configuration of Figure 12a, we simply look for a self-consistent solution. As we may see from **(1-181)** the macroscopic field within a uniformly polarized dielectric ellipsoid is uniform and is given by:

$$\mathbf{E} = \mathbf{E}_0 - \frac{1}{\epsilon_0}\, \mathbf{N} \cdot \mathbf{P} \tag{102}$$

Now, if the polarization is related to the internal field by **(3-65)**

$$\mathbf{P} = \epsilon_0\, \chi \mathbf{E} = (\epsilon - \epsilon_0)\mathbf{E} \tag{103}$$

we may eliminate the polarization between **(102)** and **(103)** to obtain

$$\mathbf{E} = \mathbf{E}_0 - \frac{\epsilon - \epsilon_0}{\epsilon_0}\, \mathbf{N} \cdot \mathbf{E} \tag{104}$$

which, given the depolarizing tensor **N**, may be solved for **E**. The polarization may then be determined from **(103)**.

The solution represented by **(104)** is not only self-consistent but is unique as well. As we saw earlier, the only requirement for uniqueness is that the boundary conditions be satisfied: the field well away from the ellipsoid must approach **E**. Now the field outside the ellipsoid may be regarded as the sum of $\mathbf{E}_0$ and the field produced by surface polarization charge. Since the field from polarization charge falls as $1/r^3$ it follows that at sufficient distance the boundary condition is satisfied.

We may obtain an explicit expression for the polarization charge from **(36)** that takes the form:

$$\epsilon \,\hat{\mathbf{n}} \cdot \mathbf{E} = \epsilon_0 \,\hat{\mathbf{n}} \cdot \mathbf{E}_0 \tag{105}$$

where $\hat{\mathbf{n}}$ is the outward normal from the ellipsoidal surface and $\mathbf{E}_0$ is the field outside the ellipsoid as shown in Figure 13a. From the application of Gauss's law at the surface we have:

$$\hat{\mathbf{n}} \cdot \mathbf{E} + \frac{\sigma}{\epsilon_0} = \hat{\mathbf{n}} \cdot \mathbf{E}_0 \tag{106}$$

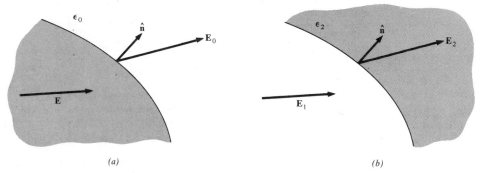

(a)   *(b)*

Figure 13   Ellipsoidal surface. (*a*) The field $\mathbf{E}_0$ is just outside the ellipsoidal surface in vacuum. (*b*) In a surrounding medium of permittivity ϵ_2 we take the field to be $\mathbf{E}_2$.

where σ is the surface polarization charge. Eliminating $\mathbf{E}_0$ between **(105)** and **(106)** we have for the surface polarization charge density:

$$\frac{\sigma}{\epsilon_0} = \frac{\epsilon - \epsilon_0}{\epsilon_0} \,\hat{\mathbf{n}} \cdot \mathbf{E} \tag{107}$$

As remarked above, the surface charge given by **(107)** generates a uniform *internal* field given by the second term on the right of **(104)**. The external field may be written in terms of a moment expansion of the surface charge distribution. For the special case of a dielectric sphere we have only an electric dipole moment and the field outside is the sum of the external field and the field of a dipole at the center of the sphere.

Exercises

27. A dielectric sphere of permittivity ϵ and radius R is placed in a uniform electric field. What is the magnitude of the induced dipole moment? Obtain expressions for the electric field inside and outside the dielectric sphere. Show that these solutions satisfy the boundary conditions **(34)** and **(35)**.

28. A long thin dielectric rod is placed at an angle θ with respect to a uniform applied field $\mathbf{E}_0$. Find the magnitude and direction of the induced polarization. Find the magnitude and direction of the torque on the rod.

---

17 Dielectric Medium. We now go on to the situation shown in Figure 12*b* where we have a dielectric ellipsoid of permittivity ϵ_1 in a medium of permittivity ϵ_2. In place of **(105)** we now write, for Figure 13*b*,

$$\epsilon_1 \hat{\mathbf{n}} \cdot \mathbf{E}_1 = \epsilon_2 \hat{\mathbf{n}} \cdot \mathbf{E}_2 \tag{108}$$

The application of Gauss's law at the surface gives in place of **(106)**:

$$\hat{\mathbf{n}} \cdot \mathbf{E}_1 + \frac{\sigma}{\epsilon_0} = \hat{\mathbf{n}} \cdot \mathbf{E}_2 \tag{109}$$

Here the surface polarization charge σ has contributions from the polarization of both media. Eliminating $\mathbf{E}_2$ between **(108)** and **(109)** we obtain for the polarization charge:

$$\frac{\sigma}{\epsilon_0} = \frac{\epsilon_1 - \epsilon_2}{\epsilon_2} \hat{\mathbf{n}} \cdot \mathbf{E}_1 \tag{110}$$

By comparing **(110)** with **(107)** and **(107)** with **(104)** we write in place of **(104)**:

$$\blacktriangleright \qquad \mathbf{E}_1 = \mathbf{E}_2 - \frac{\epsilon_1 - \epsilon_2}{\epsilon_2} \mathbf{N} \cdot \mathbf{E}_1 \tag{111}$$

Exercise

29. A sphere of permittivity ϵ_1 is surrounded by a medium of permittivity ϵ_2 in a uniform field $\mathbf{E}$. Show that the field in the sphere is given by:

$$\mathbf{E}_1 = \frac{3\epsilon_2}{\epsilon_1 + 2\epsilon_2} \mathbf{E}$$

---

18 Ellipsoidal Cavity. Finally, we consider the situation shown in Figure 12*c* with an ellipsoidal cavity in a medium of permittivity ϵ. Replacing ϵ_1 by ϵ_0 and ϵ_2 by ϵ we write, in place of **(111)**,

$$\mathbf{E}_0 = \mathbf{E} + \frac{\epsilon - \epsilon_0}{\epsilon} \mathbf{N} \cdot \mathbf{E}_0 \tag{112}$$

We consider three limiting cases: a sphere, a rod, and a disc as shown in Figure 14. For the case of a spherical cavity we have $\mathbf{N} = 1/3$ and we may write the internal field, which is called the *cavity* field, as

$$\mathbf{E}_{cav} = \frac{3\epsilon}{2\epsilon + \epsilon_0} \mathbf{E} \tag{113}$$

For a long rod whose axis is parallel to the external field we have $\mathbf{N} \cdot \mathbf{E} = 0$ with the result:

$$\mathbf{E}_0 = \mathbf{E} \tag{114}$$

For a disc whose axis is along the external field we have $\mathbf{N} \cdot \mathbf{E} = \mathbf{E}$, giving:

$$\epsilon_0 \mathbf{E}_0 = \epsilon \mathbf{E} \tag{115}$$

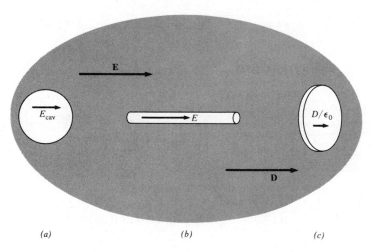

(a)   (b)   (c)

Figure 14   Ellipsoidal cavities. (*a*) A spherical cavity is the limit for which all three axes are equal. (*b*) A rod-shaped cavity is the limit of an extremely prolate spheroid. (*c*) A disc-shaped cavity is the limit of an extremely oblate spheroid.

Alternatively, **(115)** may be written as:

$$\mathbf{D}_0 = \mathbf{D} \tag{116}$$

We may also write **(114)** and **(116)** directly from the boundary conditions **(34)** and **(35)**. These expressions are occasionally used to define the macroscopic fields **E** and **D** in terms of cavity fields.

Exercise

30. A dielectric sphere of permittivity ϵ is placed in a uniform external field **E**. What is the field within a small spherical cavity in the dielectric?

SUMMARY

1. In discussing the interaction between polarizable media and external charge we introduce a new vector, the electric displacement:

$$\mathbf{D} = \epsilon_0 \mathbf{E} + \mathbf{P} \tag{1}$$

This vector has the property that its divergence is equal to the external charge density. The circulation of $\mathbf{D}$ is equal to the circulation defined by the relation:

$$\mathbf{D} = \bar{\bar{\epsilon}} \cdot \mathbf{E} \tag{13}$$

2. At the interface between two dielectric media the tangential components of $\mathbf{E}$ and the normal components of $\mathbf{D}$ must be continuous for $\sigma_{\text{ext}} = 0$.

3. In general, a unique potential is determined within a region by specifying the enclosed charge and either the potential or the normal component of the displacement on the bounding surface. The exception is where the displacement is not a single-valued function of the electric field.

4. The total energy associated with the interaction between external charges and dielectric media may be written as a sum over the external charges only:

$$W = \tfrac{1}{2} \sum_{i=1}^{N} q_i \phi_i \tag{51}$$

The total energy exceeds the external energy (which may be written as the sum over *all* charges) by an internal energy:

$$W_{\text{int}} = \int dV \int \mathbf{E}_{\text{ext}} \cdot d\mathbf{P} \tag{54}$$

5. The energy may also be written as an integral over all space:

$$U = \tfrac{1}{2} \int \mathbf{E} \cdot \mathbf{D} \, dV \tag{65}$$

This energy differs from the external energy:

$$W_{\text{ext}} - \tfrac{1}{2}\epsilon_0 \int E^2 \, dV - \frac{1}{6\epsilon_0} \int P^2 \, dV \tag{66}$$

by a term that may also be described as the internal energy:

$$U_{\text{int}} = \int dV \int \mathbf{E}_{\text{ext}} \cdot d\mathbf{P} \tag{69}$$

6. If a charge is brought near the interface between two dielectrics, it is acted on by a field equivalent to that from an image charge equal to

$$\frac{\epsilon_1 - \epsilon_2}{\epsilon_1 + \epsilon_2} q_1 \tag{96}$$

where ϵ_1 and ϵ_2 are the permittivities of the two media.

7. For a dielectric ellipsoid in a dielectric medium the internal field $\mathbf{E}_1$ may be related to the external field $\mathbf{E}_2$ by the relation:

$$\mathbf{E}_1 = \mathbf{E}_2 - \frac{\epsilon_1 - \epsilon_2}{\epsilon_2} \mathbf{N} \cdot \mathbf{E}_1 \tag{111}$$

Special conditions such as a dielectric ellipsoid in vacuum or a cavity in a dielectric may be treated as special cases. The limiting cases of a sphere, a rod, or a disc may also be obtained simply from the given relation.

PROBLEMS

1. **Charged dielectric sphere.** A dielectric sphere of radius R and permittivity ϵ also carries a uniform volume charge ρ_{ext}. Find the electric field both inside and outside the sphere. Find the field energy **(69)** within the volume of the sphere and outside the sphere.

2. **Polarization charge.** An external charge q in a medium of permittivity ϵ_1 is near the planar interface with a medium of permittivity ϵ_2. Show that the polarization charges at the interface in the two media are:

$$q_{P1} = \frac{\epsilon_2}{\epsilon_1} \frac{\epsilon_1 - \epsilon_0}{\epsilon_1 + \epsilon_2} q \qquad q_{P2} = -\frac{\epsilon_2 - \epsilon_0}{\epsilon_1 + \epsilon_2} q$$

3. **Interaction near a dielectric surface.** A pair of charges of magnitude q approach each other at a distance a from the surface of a dielectric as shown in the figure. Find the force as a function of r, the distance between charges. How much work must be done to bring the charges to within r? Discuss the limits $a = 0$ and $a = \infty$.

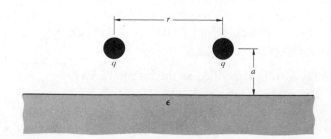

4. **Reaction field.** An electric dipole $\mathbf{p}$ is placed at the center of a spherical cavity of volume V in a dielectric of permittivity ϵ. In addition to the direct field of the dipole the induced polarization of the medium generates a uniform reaction field within the cavity. This field may be obtained from (113) by writing the cavity field as the sum of the macroscopic field, the Lorentz field, and the reaction to the *removal* of the moment $\mathbf{p} = \mathbf{P}V$ from the uniformly polarized dielectric. Use this argument to obtain

$$\mathbf{E}_{\text{react}} = -\frac{2}{3}\frac{\epsilon - \epsilon_0}{2\epsilon + \epsilon_0}\frac{\mathbf{p}}{\epsilon_0 V}$$

5. **Polarized sphere in a dielectric.** A ferroelectric sphere of radius R and polarization P is introduced into a medium of permittivity ϵ. Find the electric field in the surrounding medium. Let us imagine that the permittivity is increased gradually from ϵ_0 to ϵ. Use (64) to obtain an expression for the incremental work. Finally, integrate this expression up to ϵ to find the total change in energy. Check limiting cases.

6. **Hysteresis.** For a dielectric for which $\mathbf{P}$ is not a single-valued function of $\mathbf{E}$, show that work may be performed even though a full cycle is completed. Show that the work per unit volume is given by:

$$w = \oint \mathbf{E} \cdot d\mathbf{D}$$

and that, with reference to Figure 4b, the integral is equal to the enclosed area.

7. **van der Waals Interaction.** Show that the interaction energy between a pair of molecules of polarizability α and fluctuating root-mean-square moments

$$\langle p^2 \rangle^{1/2} \simeq eR$$

is given by:

$$U = -\frac{\alpha_1 \alpha_2}{(4\pi\epsilon_0 r_{12}{}^3)^2} \cdot \frac{e^2}{4\pi\epsilon_0}\left(\frac{1}{R_1} + \frac{1}{R_2}\right)$$

where an average has been taken over fluctuation directions. This expression assumes that the fluctuation period is long compared with the time for a light signal to pass back and forth between the molecules [F. London, *Z. f. Physik*, **63**, 245 (1930)].

8. **[Permittivity of composites].**[5] A material is compared of layers of permittivity $\bar{\bar{\epsilon}}_1$ and $\bar{\bar{\epsilon}}_2$ in planes with normal $\hat{\mathbf{n}}$. If the respective fractional volumes are f_1 and f_2, show that the mean permittivity is given by:

$$\bar{\bar{\epsilon}} = f_1 \bar{\bar{\epsilon}}_1 + f_2 \bar{\bar{\epsilon}}_2 - f_{12}\,\mathbf{aa}$$

---

[5] This problem was suggested by J. R. Hauser.

with

$$f_{12} = \frac{f_1 f_2}{\hat{\mathbf{n}} \cdot (f_2 \bar{\bar{\epsilon}}_1 + f_1 \bar{\bar{\epsilon}}_2) \cdot \hat{\mathbf{n}}}$$

and

$$\mathbf{a} = \hat{\mathbf{n}} \cdot (\bar{\bar{\epsilon}}_1 - \bar{\bar{\epsilon}}_2)$$

9. **Capacitance change.** A small dielectric ellipsoid of permittivity ϵ_1 is inserted into a dielectric medium of permittivity ϵ_2 between a pair of capacitor plates. The ellipsoid is oriented so that the electric field is along a principal axis. Show that the capacitance between the plates is given by:

$$C \left[1 - \frac{\epsilon_1 - \epsilon_2}{\epsilon_2 + N(\epsilon_1 - \epsilon_2)} \frac{v}{V} \right]^{-1}$$

Here C is the original capacitance, v is the volume of the dielectric ellipsoid, V is the volume of the dielectric medium, and N is the depolarization factor of the ellipsoid. This expression is most easily obtained from the expression for the energy of the system, which may be written as the sum of three terms: (1) the energy of the charged capacitor without the ellipsoid, (2) the self energy of the polarized ellipsoid within an infinite dielectric medium, and (3) the coupling between the capacitor and the ellipsoid.

REFERENCES

Moore, A. D., "Electrostatics," Scientific American **226** (3), 46 (March 1972).
Moore, A. D., Editor, *Electrostatics and its Applications*, Wiley, 1973.

5 CONDUCTING MEDIA

In this chapter we discuss the interaction between electric fields and those media in which charge may be freely, or nearly freely, transported. This is in contrast with our discussion of dielectrics, where we saw that the application of a static field produced a polarization but no steady current. In conducting media, as we will see, there is a continuing flow of charge in the presence of a static field. This is the important distinction between dielectrics and conductors—their response to a *static* field. Although the response of a dielectric and of a conductor to an alternating field may be different in detail we can usually treat them in a very similar way.

We begin our discussion with the consideration of a two-component plasma, a gas of positive ions and electrons, and develop an expression for the relation between current density and applied field. This same theory may be applied to electrical conduction in solids as long as the electron concentration is not too high. In the case of high electron concentrations, as in metals, we develop a related but essentially phenomenological theory.

We then turn to the screening of electric fields by plasmas and by metals. We consider first the screening of a static field and then the screening of an alternating field. As part of this discussion we demonstrate the existence of freely propagating longitudinally polarized modes, called plasma waves.

TRANSPORT OF ELECTRICAL CHARGE

We imagine, as shown in Figure 1, a plasma of positive ions of mass M and charge e and electrons of mass m and charge $-e$. The concentrations of the two charge carriers are taken to be equal to $n = N/V$.

1 Charge Flow. Now if an electric field is applied to an *ion*, the ion accelerates according to the equation:

$$M \frac{d\mathbf{V}}{dt} = e\mathbf{E} \tag{1}$$

and the velocity is then given by:

$$\mathbf{V} = \mathbf{V}_0 + \frac{e}{M}\,\mathbf{E}t \tag{2}$$

The charge flow *per ion* may be written as:

$$\mathbf{J}_p = e\mathbf{V} = e\mathbf{V}_0 + \frac{e^2}{M}\,\mathbf{E}t \tag{3}$$

where the subscript p identifies the positive charge carrier. For the *electrons* we have:

$$m\frac{d\mathbf{v}}{dt} = -e\mathbf{E} \tag{4}$$

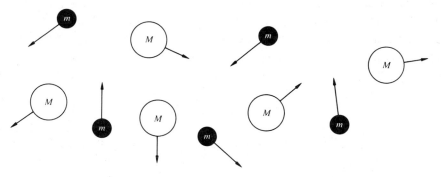

Figure 1   Charge plasma. The positive ions have mass M and charge e. The negative ions have mass m and charge $-e$.

and a velocity given by:

$$\mathbf{v} = \mathbf{v}_0 - \frac{e}{m}\,\mathbf{E}t \tag{5}$$

The electron charge flow will then be:

$$\mathbf{J}_n = -e\mathbf{v} = -e\mathbf{v}_0 + \frac{e^2}{m}\,\mathbf{E}t \tag{6}$$

Comparing **(6)** with **(3)** we see that the component of the current *induced by the field* has the same sign for the two carriers.

We now consider the effect of collisions between the two carriers. We imagine that an electron and ion collide as shown in Figure 2. How much current is carried after the collision? A simple argument suggests that the answer is zero on the average.

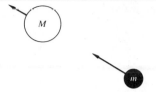

Figure 2   Electron-ion collision. After a collision the current is nearly zero.

It is convenient to think of two-body collisions such as that shown in Figure 2 in a frame in which the center of mass is at rest. We show in Figure 3a the motion depicted in Figure 2, but in a frame moving with the center of mass of ion and electron. Now, if the impact parameter between electron and ion is regarded as variable, then, to a good approximation, all outgoing directions are equally probable so that, on the average, there is no current carried in the center of mass frame after the collision. But what about current carried *by* the center of mass? Since the net charge associated with the center of mass is zero the center of mass carries no current. We conclude then that on the average the current in the laboratory frame following a collision is nearly zero.

The mechanism by which current is produced is then the following; after a collision there is on the average no current carried by the colliding electron and ion. Following the collision both carriers are accelerated by the field to produce a charge flow per pair:

$$\mathbf{J} = \mathbf{J}_p + \mathbf{J}_n = \left(\frac{e^2}{M} + \frac{e^2}{m}\right)\mathbf{E}t \tag{7}$$

If the mean time between collisions is τ, then, on the average, a carrier will at any given time have been accelerating for a time τ since its last collision and we may write, for the charge flow,

$$\mathbf{J} = \left(\frac{e^2}{m}\tau + \frac{e^2}{M}\tau\right)\mathbf{E} \tag{8}$$

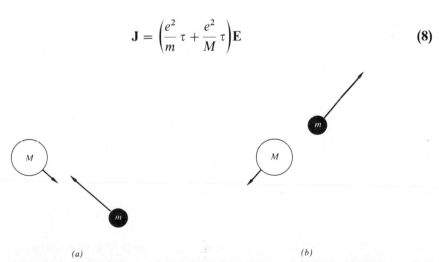

(a)   (b)

Figure 3   Center of mass frame. (a) The electron and ion approach with fixed impact parameter. (b) After the collision the outgoing direction depends on the value of the impact parameter.

The current *density* will be given by:

$$\mathbf{j} = n\mathbf{J} = \left(\frac{ne^2}{m}\,\tau + \frac{ne^2}{M}\,\tau\right)\mathbf{E} \tag{9}$$

The electrical conductivity is defined by the relation

$$\mathbf{j} = \sigma\mathbf{E} \tag{10}$$

which gives for the conductivity:

▶
$$\sigma = \frac{ne^2}{m}\,\tau + \frac{ne^2}{M}\,\tau \tag{11}$$

Since the mass of the electrons is less than 10^{-3} the mass of the ions, most of the current is carried by the electrons and we can ignore the contribution of the ions to **(11)**.

Exercises

1. If for any carrier the mean time since the last collision is τ and the mean time to the next collision is also τ, explain why the mean time *between* collisions may be τ and *not* 2τ [see D. E. Tilley, *Am. J. Phys.*, **44**, 597 (1976)].
2. Why in discussing collisions is it not necessary to consider collisions between positive carriers or collisions between negative carriers, but only collisions between positive and negative carriers?
3. What is the justification for using the same collision time τ for positive and negative carriers in **(8)**?

2 **Carrier Mobility.** From **(8)** the charge carriers drift through the field with velocities:

$$\langle\mathbf{V}\rangle = \frac{\mathbf{J}_p}{e} = \frac{e}{M}\,\tau\mathbf{E} \qquad \langle\mathbf{v}\rangle = -\frac{\mathbf{J}_n}{e} = -\frac{e}{m}\,\tau\mathbf{E} \tag{12}$$

The mobilities of the carriers are defined by the relations:

$$\langle\mathbf{V}\rangle = \mu_p\mathbf{E} \qquad \langle\mathbf{v}\rangle = \mu_n\mathbf{E} \tag{13}$$

Combining **(12)** and **(13)** we obtain for the mobilities:

$$\mu_p = \frac{e}{M}\,\tau \qquad \mu_n = -\frac{e}{m}\,\tau \tag{14}$$

3 **Transport in Solids.** For a solid, as long as the concentration of electrons is not too high, we may regard the carriers as scattering independently off the *lattice* of ions or

atoms. Because the ions are fixed in a lattice, current is carried only by the electrons and we have:

▶
$$\sigma = \frac{ne^2}{m}\tau \tag{15}$$

We assume that the collisions, which occur with mean time τ, completely relax the current but this is only a simplifying assumption.

For a metal we may not regard the carriers as scattering *independently* but must instead pay proper attention to their quantum behavior, which prevents an electron from scattering into a state already occupied by another electron. (This condition also exists for low density electron gases, but because the likelihood of such scattering is small we may ignore it.)

If we imagine that all the electrons are accelerated together, the rate of change of **j** from *fields* has the form:

$$\left[\frac{d\mathbf{j}}{dt}\right]_{\text{fields}} = ne\frac{d\mathbf{v}}{dt} = \frac{ne^2}{m}\mathbf{E} \tag{16}$$

We further assume that the rate of change of **j** as the result of *collisions* is a simple exponential:

$$\left[\frac{d\mathbf{j}}{dt}\right]_{\text{collisions}} = -\frac{\mathbf{j}}{\tau} \tag{17}$$

Adding (16) and (17) we have, for the total rate of change of current,

$$\frac{d\mathbf{j}}{dt} = \left[\frac{d\mathbf{j}}{dt}\right]_{\text{fields}} + \left[\frac{d\mathbf{j}}{dt}\right]_{\text{collisions}} = \frac{ne^2}{m}\mathbf{E} - \frac{\mathbf{j}}{\tau} \tag{18}$$

In steady state $d\mathbf{j}/dt = 0$ and we have:

$$\mathbf{j} = \frac{ne^2}{m}\tau\mathbf{E} = \sigma\mathbf{E} \tag{19}$$

which gives the same form as (15) for σ. Thus we can use (15) for either a dilute or a concentrated electron gas although the mechanism by which the current is relaxed is somewhat different in the two cases.

The reciprocal conductivity is called the resistivity and is denoted by the symbol ρ. (It is unfortunate that ρ and σ are used both for volume and surface charge density and for resistivity and conductivity. It is almost always possible to tell from the context which is which, but one must be careful.) The SI unit of resistivity is the ohm-metre

(Ωm). Conductivity is given is siemens per metre (S/m). We give in Table 1 the room temperature conductivities and carrier densities of a number of metals. The carrier velocities are calculated assuming that the electrons behave like free charges. In Table 2 we list the room temperature mobilities of the positive and negative carriers for a number of semiconductors.

Table 1   Metals

Metal	Conductivity	Carrier Density	Carrier Velocity
Aluminum	3.65×10^7 S/m	18.06×10^{28}/m^3	2.02×10^6 m/s
Copper	5.88	8.45	1.57
Silver	6.21	5.85	1.39
Lead	0.48	13.20	1.82
Tin	0.91	14.48	1.88
Zinc	1.69	13.10	1.82
Gold	4.55	5.90	1.39

Table 2   Semiconductor Mobilities

Semiconductor	Negative Carrier	Positive Carrier
Diamond	-0.18 m^2/V $\cdot$ s	$+0.12$ m^2/V $\cdot$ s
Germanium	-0.45	$+0.35$
Indium antimonide	-7.7	$+0.075$
Silicon	-0.13	$+0.05$

Exercise

4. From the information given in Table 1, compute the collision time τ for aluminum, copper, and silver.

4 Convective Derivative. The *total* derivative with respect to time, which is called the convective derivative, is taken in the frame which moves with each carrier leading to the relation for the charge density:

$$\frac{d\rho}{dt} = \frac{\partial\rho}{\partial t} + \langle \mathbf{v} \cdot \nabla\rho \rangle \tag{20}$$

where $\mathbf{v}$ is a carrier velocity. The effect of carrier relaxation is to reduce the correlation between the velocity of a carrier and its contribution to the charge density so that the second term may be ignored. An important problem where this term must be considered is given by the anomalous skin effect, which will be discussed in Section 9-16.

RESISTANCE

We assume a steady and solenoidal distribution of current. From the condition that charge does not accumulate we have, from **(1-147)**,

$$\mathbf{V} \cdot \mathbf{j}(\mathbf{r}) = -\frac{\partial}{\partial t} \rho(\mathbf{r}) = 0 \tag{21}$$

With reference to the situation shown in Figure 4, we imagine that potentials ϕ_1 and ϕ_2 are established and that current flows between these potentials with **(21)** holding in the intermediate region. The incremental current crossing an increment of area is written as:

$$dI = \mathbf{j} \cdot d\mathbf{S} \tag{22}$$

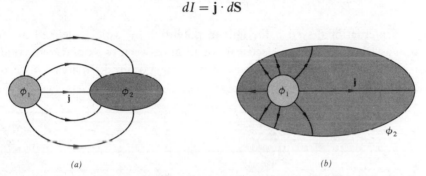

(a)   (b)

Figure 4   Resistance. (a) Current flows between an equipotential surface ϕ_1 and an equipotential surface ϕ_2. (b) The equipotential surface ϕ_2 may enclose the equipotential surface ϕ_1.

We have, then, for the *current* flowing from ϕ_1 to ϕ_2:

$$I = \oint_S \mathbf{j} \cdot d\mathbf{S} \tag{23}$$

The international unit of current is the ampere, which is defined in Section K-1.

We write, for the potential difference between ϕ_1 and ϕ_2,

$$V = \phi_1 - \phi_2 = -\int_{\phi_1}^{\phi_2} d\phi = \int_1^2 \mathbf{E} \cdot d\mathbf{r} \tag{24}$$

The resistance R between ϕ_1 and ϕ_2 is defined by the relation:

$$V = IR \tag{25}$$

The SI unit of resistance is the ohm, which is defined as "the electrical resistance between two points of a conductor when a constant difference of potential of one volt, applied between these two points, produces in this conductor a current of one ampere."

5 Dissipation. The rate at which work is done on the current $\mathbf{j(r)}$ is written as:

$$\frac{dW}{dt} = \int \mathbf{j} \cdot \mathbf{E}\, dV = \int \mathbf{j} \cdot \mathbf{E}(d\mathbf{S} \cdot d\mathbf{r}) = \int \mathbf{j} \cdot d\mathbf{S} \int \mathbf{E} \cdot d\mathbf{r} \tag{26}$$

where we have taken $d\mathbf{S}$ parallel to $\mathbf{E}$ and $d\mathbf{r}$ parallel to $\mathbf{j}$. To allow for the possibility that the medium may be anisotropic, the current density $\mathbf{j}$ need not be parallel to the electric field $\mathbf{E}$. Performing the integrations of **(26)** and using the definition of resistance we have:

$$\frac{dW}{dt} = IV = I^2R = V^2/R \tag{27}$$

In the international system the unit of potential, the volt, is defined in terms of dissipation as "the difference of electrical potential between two points of a conducting wire carrying a current of one ampere, when the power dissipated between these points is equal to one watt."

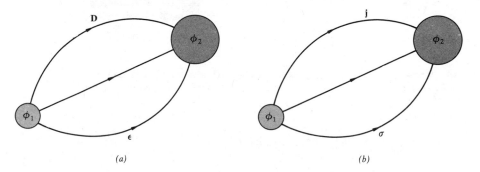

(a)   *(b)*

Figure 5 Resistance and capacitance. (*a*) The capacitance between a pair of equipotentials is proportional to the flux $\mathbf{D}$. (*b*) The resistance between a pair of equipotentials is inversely proportional to the total current.

6 Resistance and Capacitance. We demonstrate here a very simple relationship of the capacitance between a pair of equipotentials as shown in Figure 5*a* to the resistance between the same pair of equipotentials as shown in Figure 5*b*.

First we show that the electric field distribution is the same in both cases for uniform media. For Figure 5*a* we have, in the absence of external charge,

$$\nabla^2 \phi = -\nabla \cdot \mathbf{E} = -\frac{1}{\epsilon} \nabla \cdot \mathbf{D} = 0 \tag{28}$$

As we have discussed in Section 4-7, the potential and thus $\mathbf{E}$ and $\mathbf{D}$ are uniquely determined by the boundary conditions establishing ϕ_1 and ϕ_2.

For Figure 5b we have, as long as the current is solenoidal,

$$\nabla^2 \phi = -\nabla \cdot \mathbf{E} = -\frac{1}{\sigma} \nabla \cdot \mathbf{j} = 0 \tag{29}$$

Thus, the potential is a solution of Laplace's equation for both cases.
We define the capacitance

$$C = \frac{Q}{V} = \frac{\oint \mathbf{D} \cdot d\mathbf{S}}{\int \mathbf{E} \cdot d\mathbf{r}} = \epsilon \frac{\oint \mathbf{E} \cdot d\mathbf{S}}{\int \mathbf{E} \cdot d\mathbf{r}} \tag{30}$$

We have for the resistance from **(25)**,

$$R - \frac{V}{I} - \frac{\int \mathbf{E} \cdot d\mathbf{r}}{\oint \mathbf{j} \cdot d\mathbf{S}} - \frac{1}{\sigma} \frac{\int \mathbf{E} \cdot d\mathbf{r}}{\oint \mathbf{E} \cdot d\mathbf{S}} \tag{31}$$

Multiplying **(30)** and **(31)** we have:

$$RC = \frac{\epsilon}{\sigma} \tag{32}$$

which states that the *RC product* of a pair of equipotential surfaces is independent of the geometry of the surfaces and depends only on the ratio of permittivity to conductivity in the intervening medium.

FREQUENCY DEPENDENT CONDUCTIVITY

In this part we examine the solution of **(18)** when **E** is a sinusoidal function of the time. We write in scalar notation:

$$E(t) = E \cos \omega t \tag{33}$$

$$j(t) = j \cos (\omega t - \phi) \tag{34}$$

where we have allowed for the possibility that $j(t)$ and $E(t)$ may not be in phase. Substituting into **(18)** we obtain:

$$-\omega j \sin(\omega t - \phi) + \frac{j}{\tau} \cos(\omega t - \phi) = \frac{\sigma}{\tau} E \cos \omega t \tag{35}$$

We expand the trigonometric functions in (35) and equate the coefficients of sin ωt and of cos ωt to obtain:

$$j = \frac{\sigma E}{\cos \phi + \omega \tau \sin \phi} \tag{36}$$

$$\tan \phi = \omega \tau \tag{37}$$

The denominator of (36) may be written on substitution from (37) as:

$$\cos \phi + \tan \phi \sin \phi = \cos \phi + \sin^2 \phi/\cos \phi = 1/\cos \phi \tag{38}$$

The magnitude of the current is then given by:

$$j = \sigma E \cos \phi = \frac{\sigma E}{[1 + (\omega \tau)^2]^{1/2}} \tag{39}$$

In Figure 6a we plot $E(t)$ followed by $j(t)$ in 6b, c, and d for $\omega \tau = 0$, $\omega \tau = 1$ and $\omega \tau \gg 1$. At $\omega \tau = 0$ the current is in phase with the voltage. As the frequency is increased the magnitude of the current drops and the maximum of the current occurs at a later time. We describe this delay in the current by saying that the current *lags* the voltage.

Next we consider the rate at which work is done by the electric field E on the medium. The rate at which work is done on a single charged particle by a field is:

$$\frac{dW}{dt} = Fv = Eqv = E(t)J(t) \tag{40}$$

Then the work done per unit volume is:

$$\frac{dw}{dt} = n\frac{dW}{dt} = EnJ = E(t)j(t) \tag{41}$$

Substituting from (33) and (34):

$$\frac{dw}{dt} = Ej \cos \phi \cos^2 \omega t + \tfrac{1}{2}Ej \sin \phi \sin 2\omega t \tag{42}$$

The second term is the work that goes into kinetic energy, which is a periodic function of the time at twice the frequency of the field and averages to zero. The first term has a nonzero average. Since the average of $\cos^2 \omega t$ is 1/2, we obtain for the average rate at which work is done:

$$\left\langle \frac{dw}{dt} \right\rangle = \tfrac{1}{2}Ej \cos \phi \tag{43}$$

Substituting from **(39)** we obtain:

$$\left\langle \frac{dw}{dt} \right\rangle = \tfrac{1}{2}\sigma E^2 \cos^2 \phi = \frac{1}{2} \frac{\sigma E^2}{1 + (\omega\tau)^2} \tag{44}$$

7 Complex Conductivity. By using complex notation it is possible to write the frequency dependent conductivity in quite compact form. We first briefly develop the use of

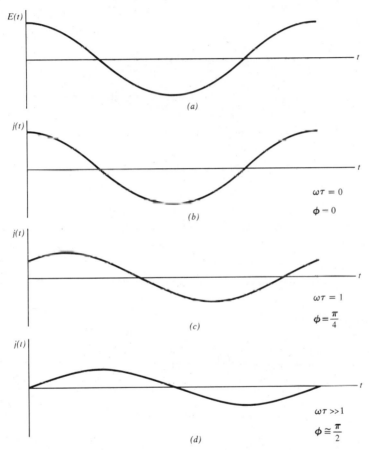

Figure 6 Transport in a sinusoidal field. (*a*) The electric field is a sinusoidal function of time. (*b*) At zero frequency the current is in phase with the electric field. (*c*) When the driving frequency is equal to the reciprocal of the relaxation time, the current lags the electric field by $\pi/4$. (*d*) When the driving frequency is much higher than the relaxation frequency the current lags by nearly $\pi/2$.

complex notation in the solution of a differential equation such as **(18)**. As we have seen, the solution of the equation:

$$\frac{dj}{dt} + \frac{j}{\tau} = \frac{\sigma}{\tau} E \cos \omega t \tag{45}$$

is:

$$j(t) = j \cos(\omega t - \phi) \tag{46}$$

where j and ϕ are given by **(37)** and **(39)**.

Now if we shift the phase of the driving field by $\pi/2$ we expect that the phase of the current will also be shifted by $\pi/2$. (This is just equivalent to changing the zero of time.) Then, for an equation:

$$\frac{dj}{dt} + \frac{j}{\tau} = \frac{\sigma}{\tau} E \sin \omega t \tag{47}$$

we expect a solution:

$$j(t) = j \sin(\omega t - \phi) \tag{48}$$

Now, since **(18)** is a linear equation we may apply the principle of superposition. If we sum two driving fields, the resultant current will be the sum of the individual currents. The difference here is that we generate a *complex* field:

$$\underline{E}(t) = E \cos \omega t - iE \sin \omega t = Ee^{-i\omega t} \tag{49}$$

As in **(49)** we will designate a complex quantity by the use of a rule below the symbol.

This is not of course a physically realizable field. The linearity of the equations still provides for this kind of superposition, however, and we expect a complex current:

$$\underline{j}(t) = j \cos(\omega t - \phi) - ij \sin(\omega t - \phi) = je^{-i(\omega t - \phi)} \tag{50}$$

The complex conductivity is defined as:

$$\underline{\sigma}(\omega) = \frac{\underline{j}(t)}{\underline{E}(t)} = \frac{j}{E} e^{i\phi} = \sigma \cos \phi e^{i\phi} \tag{51}$$

where the third equality follows from **(39)**. If we write the complex conductivity as

$$\underline{\sigma}(\omega) = \sigma'(\omega) + i\sigma''(\omega) \tag{52}$$

we have:

$$\sigma'(\omega) = \frac{j}{E} \cos \phi = \sigma \cos^2 \phi = \frac{1}{1 + (\omega\tau)^2} \sigma \tag{53}$$

$$\sigma''(\omega) = \frac{j}{E} \sin \phi = \sigma \sin \phi \cos \phi = \frac{\omega\tau}{1 + (\omega\tau)^2} \sigma \tag{54}$$

In Figure 7 we plot **(53)** and **(54)**, which follow from the equation of motion **(18)**. We observe that, by expanding **(51)**, we may write for the *real current*:

$$j(t) = \sigma'(\omega)E \cos \omega t + \sigma''(\omega)E \sin \omega t \tag{55}$$

Thus $\sigma'(\omega)$ gives the in-phase component of the current and $\sigma''(\omega)$ gives the component of the current that is $\pi/2$ out of phase. We note that $\sigma''(\omega)$ positive corresponds to a *lagging* current. This is the opposite of conventional circuit analysis where a *negative* imaginary component corresponds to a phase lag. The reason for the difference is that conventional circuit theory takes a time dependence $e^{+i\omega t}$ or $e^{+j\omega t}$ whereas we have taken $e^{-i\omega t}$. Our choice has been dictated by conventional treatments of complex propagating waves, which are written as $e^{i(kx-\omega t)}$ rather than $e^{i(\omega t-kx)}$. It is unfortunate that there are these two conventions and we must be careful.

From **(43)** for the rate at which work is done and **(53)** for the real part of the complex conductivity we may write

$$\left\langle \frac{dw}{dt} \right\rangle = \tfrac{1}{2}\sigma'(\omega)E^2 = \tfrac{1}{2}\,\mathrm{Re}\,j^*(t)E(t) \tag{56}$$

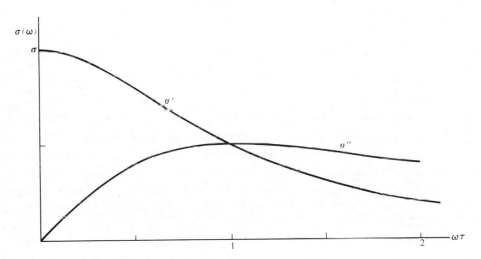

Figure 7 Complex conductivity. The real part of the conductivity is represented by σ' and the imaginary part by σ''.

The power factor (PF) is the ratio of the rate at which work *is* done to the rate at which work *would* be done if the current were in phase with the field:

$$\mathrm{PF} = \frac{\sigma'(\omega)}{[\sigma'(\omega)^2 + \sigma''(\omega)^2]^{1/2}} = \cos\phi \tag{57}$$

We list in Table 3 the power factor at various frequencies for some of the plastics listed in Table 4-1.

8 [**Equivalent Network for Conduction**]. From Section E-3 the equivalent network representation of **(18)** is shown in Figure 8. The voltage generator represents the field $E(t)$. The inertial term (τ/σ) is represented by an inductance and the relaxation

Table 3   Power Factor of Plastics

	Frequency		
Plastic	60 Hz	1 kHz	1 MHz
Lexan	0.0009	0.0021	0.010
Mylar	0.0055	0.0187	0.0208
Plexiglas	0.05 − 0.06	0.04 − 0.06	0.02 − 0.03
Teflon	< 0.0003	< 0.0003	< 0.0007

term by a resistance. We may define a complex resistivity

$$\underline{\rho}(\omega) = \frac{E(t)}{\underline{j}(t)} = \frac{1}{\underline{\sigma}(\omega)} = \frac{1}{\sigma} - i\omega\frac{\tau}{\sigma} \tag{59}$$

Again, we note that in conventional network analysis one writes $Z = R + iX$ and $X = \omega L - 1/\omega C$. Here we have the reverse sign for the definition of X because of our original choice for the complex time dependence of $\underline{E}(t)$.

9 [Dielectric Relaxation]. Here we discuss polarization processes substantially different from those considered in Section 3-10. The processes considered there were best characterized by their resonance frequencies and oscillator strengths with typical resonance frequencies occurring at optical frequencies. The processes we consider here are dominated by relaxation rather than by resonance. Typical relaxation times may be as long as seconds or as short as 10^{-10} s but seldom shorter. Such processes might well have been considered in Section 3-10 together with other polarization processes. The reason for discussing them here in a chapter on conduction is that at intermediate frequencies these processes are often indistinguishable from normal conduction.

As examples of processes that are dominated by relaxation we may mention the orientation in a gas or liquid of molecules carrying permanent electric dipole moments. Thus, the diatomic alkali halides would be expected to show dielectric relaxation.

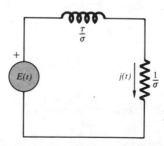

Figure 8 Equivalent network. The equivalent network for relaxation is an inductor in series with a resistor.

A much more common example is that of water, the molecules of which carry a permanent electric dipole moment. In addition there are other processes, particularly in disordered solids like the glasses, which are not at present very well understood.

We write for the rate of change of polarization of our system:

$$\frac{dP}{dt} = \left[\frac{dP}{dt}\right]_{\text{fields}} + \left[\frac{dP}{dt}\right]_{\text{collisions}} \tag{60}$$

which expresses the total rate of change of polarization as the sum of a rate induced by fields and a rate induced by collisions. For the processes that we consider here the first term is of little importance except at the highest frequencies and we ignore it for the moment. We write for the second term the phenomenological expression:

$$\frac{dP}{dt} = \frac{\epsilon_0 \chi E - P}{\tau} \tag{61}$$

where τ is the characteristic relaxation time. To solve **(61)** we use the just developed method of complex time dependence. We write:

$$\underline{E}(t) = E e^{-i\omega t} \qquad \underline{P}(t) = P e^{-i(\omega t - \phi)} \tag{62}$$

Substituting into **(61)**, we obtain, for the complex susceptibility,

$$\underline{\chi}(\omega) = \frac{P e^{i\phi}}{\epsilon_0 E} = \frac{\chi}{1 - i\omega\tau} \tag{63}$$

Writing the complex susceptibility $\underline{\chi}(\omega)$ as the sum of real and imaginary parts:

$$\underline{\chi}(\omega) = \chi'(\omega) + i\chi''(\omega) \tag{64}$$

we have:

$$\chi'(\omega) = \frac{P}{\epsilon_0 E} \cos \phi = \frac{1}{1 + (\omega\tau)^2} \chi \tag{65}$$

$$\chi''(\omega) = \frac{P}{\epsilon_0 E} \sin \phi = \frac{\omega\tau}{1 + (\omega\tau)^2} \chi \tag{66}$$

The rate at which work is done by the field may be written from **(41)** with the polarization current $j = j_P = dP/dt$,

$$\frac{dw}{dt} = E(t) \frac{dP(t)}{dt} \tag{67}$$

Since **(67)** is quadratic in the field we cannot directly apply superposition. The safest procedure is to take, for $P(t)$ and $E(t)$, the real parts of **(62)**:

$$E(t) = E \cos \omega t \qquad P(t) = P \cos(\omega t - \phi) \tag{68}$$

Substituting into **(67)** we obtain:

$$\frac{dw}{dt} = -\omega E P \cos \omega t \sin(\omega t - \phi)$$

$$= \omega E P \sin \phi \cos^2 \omega t - \tfrac{1}{2}\omega P E \cos \phi \sin 2\omega t \tag{69}$$

Averaging **(69)** we obtain with the use of **(66)**:

$$\left\langle \frac{dw}{dt} \right\rangle = \tfrac{1}{2}\omega P E \sin \phi = \tfrac{1}{2}\omega \epsilon_0 E^2 \chi''(\omega) \tag{70}$$

For a dielectric, ϕ is called the *loss angle* and the *dissipation factor* (DF) is defined as:

$$DF = \tan \phi = \frac{\chi''(\omega)}{\chi'(\omega)} \tag{71}$$

The quality factor (QF or Q) is the inverse of the dissipation factor. A *power factor* (PF) may also be defined for a dielectric and takes the form:

$$PF = \sin \phi = \chi''(\omega)/|\chi(\omega)| \tag{72}$$

Exercises

5. Plot **(65)** and **(66)** on semilog paper and note the symmetry of the curves:

$$\chi'(\omega) + \chi'(1/\omega\tau^2) = 1 \qquad \chi''(\omega) = \chi''(1/\omega\tau^2)$$

6. Show by eliminating $\omega\tau$ between **(65)** and **(66)** that the equation

$$(\chi')^2 + (\chi'')^2 = \chi\chi'$$

is obtained. Plot this curve, which is called the Debye semicircle.

7. Show that **(70)** may also be obtained from the expression:

$$\left\langle \frac{dw}{dt} \right\rangle = \tfrac{1}{2} \operatorname{Re}\left\{ E(t) \frac{d}{dt} P^*(t) \right\}$$

where Re{ } means the real part of the expression in braces and $P^*(t)$ is the complex conjugate of $P(t)$.

10   [**Equivalent Network for Polarization**]. The network equivalent to **(61)** is best suggested if dP/dt is written as the polarization current j_P:

$$E(t) = \frac{\tau}{\epsilon_0 \chi} j_P + \frac{1}{\epsilon_0 \chi} p \tag{73}$$

The equivalent network is shown in Figure 9. On comparing Figure 6 with Figure 8 we see that the behavior may be quite similar in an intermediate frequency range where both circuits are resistive. Does the absence of an inductive component in Figure 9 mean that polarization current may be induced to arbitrarily high frequencies? The answer is no. In neglecting the field term in **(60)** we have dropped the inertial terms that must ultimately limit the polarization current in just the way that the conduction current is limited.

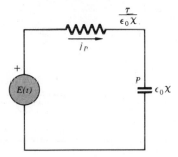

Figure 9   Equivalent network. The equivalent network for polarization is a capacitor in series with a resistor.

SCREENING OF STATIC FIELDS

In the preceding discussion we assumed a uniform field applied to a two-component plasma, which, on the average, was electrically neutral. It is not hard to create a situation in which the plasma is *forced* to develop a local charge density as we see in Figure 10. Here an electric field is simply applied normal to a slab of plasma. Let us imagine that at the moment the field is turned on the plasma is everywhere electrically neutral. But as we have seen from **(2)** and **(5)** the electric field will cause a drift of positive ions to the right and electrons to the left. These ions and electrons are confined by the boundaries of the plasma and a surface charge is built up on S. Note that if we have a surface charge σ there will be established within the plasma a macroscopic field:

$$E = E_0 + \frac{\sigma}{\epsilon_0} \tag{74}$$

An *approximately* correct description of the situation is to say that charge builds up until the interior is entirely screened from the field and we have

$$\sigma = -\epsilon_0 E_0 \tag{75}$$

This description is somewhat oversimplified in that it suggests that all the screening charge lies in a thin layer near the surface. Actually, if the carrier density is sufficiently low, the distance over which the field is screened may be appreciable. Second, the approximate explanation does not give us any idea of how much of the screening is done by electrons and how much by ions. In order to discuss these questions we must consider the problem of the diffusion of electrons and ions in regions where their concentration is not uniform.

11   Diffusion of Carriers. We now imagine small variations in electron and ion concentration throughout the volume of the plasma. (We assume though that these variations are not great enough to affect τ.) For the moment we assume that there are no electric fields; we will take up the effects of electric fields separately.

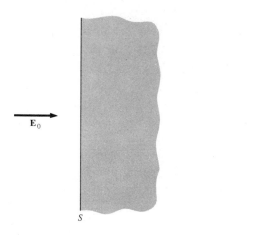

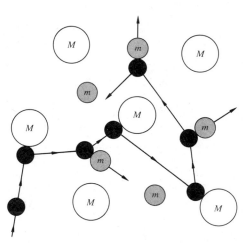

Figure 10   Field screening. An electric field $\mathbf{E}_0$ is applied normal to the surface of a bounded plasma.

Figure 11   Carrier diffusion. An electron represented by a solid sphere is followed as it moves through a plasma, scattering off both positive and negative carriers.

Let us take a second look at our picture of a plasma of colliding electrons and ions and follow an electron or an ion as it moves through the plasma. We indicate in Figure 11 what we would see as the motion of an electron. The electron moves through the plasma, colliding with ions and with other electrons, changing its direction with each collision. If we let $\mathbf{d}_j$ be the displacement of the electron between the $j-1$ and j collisions, we have, for the total displacement,

$$\mathbf{R} = \mathbf{d}_1 + \mathbf{d}_2 + \mathbf{d}_3 + \cdots + \mathbf{d}_N = \sum_{i=1}^{N} \mathbf{d}_i \qquad (76)$$

To find the mean square displacement in N collisions we write:

$$R^2 = \sum \mathbf{d}_i \cdot \mathbf{d}_j \qquad (77)$$

Now, if the $\mathbf{d}_i$ are uncorrelated, we are left in (77) with only the terms for $i = j$:

$$\langle R^2 \rangle = \sum_{i=1}^{N} d_i^2 = N \langle d^2 \rangle \tag{78}$$

How far does the carrier move along a particular direction, say $\hat{x}$? The total displacement may be decomposed along rectangular axes:

$$\mathbf{R} = X\hat{x} + Y\hat{y} + Z\hat{z} \tag{79}$$

and we have, for the mean square displacement,

$$\langle R^2 \rangle = \langle X^2 \rangle + \langle Y^2 \rangle + \langle Z^2 \rangle \tag{80}$$

Now, since the plasma is assumed isotropic we expect $\langle X^2 \rangle = \langle Y^2 \rangle = \langle Z^2 \rangle$ and

$$\langle X^2 \rangle = \tfrac{1}{3} N \langle d^2 \rangle = \frac{T}{3\tau} \langle d^2 \rangle \tag{81}$$

where $T = N\tau$ is the time for N collisions, and τ is the mean collision time between electrons and ions. As we have seen, collisions between electrons do not relax the current. Neither do they inhibit diffusion so that the relaxation time τ in (81) will be equal to the current relaxation time (17).

We now look at (81) from a slightly different point of view. Instead of regarding T as a definite time in which there is a spread in the displacement along $\hat{x}$, we take X to be a definite quantity with T to be the *mean* time for a displacement X. With this understanding we may write (81) as

$$X^2 = \tfrac{1}{3} \langle N \rangle \langle d^2 \rangle = \frac{1}{3} \frac{T}{\tau} \langle d^2 \rangle \tag{82}$$

To obtain the mean drift velocity at X we differentiate (82) with respect to T to obtain:

$$\frac{dX}{dT} = \frac{X}{2T} = \frac{\langle d^2 \rangle}{6X\tau} = \frac{\langle v^2 \rangle \langle t^2 \rangle}{6X\tau} = \frac{1}{3X} \langle v^2 \rangle \tau \tag{83}$$

where $\langle v^2 \rangle$ is the mean-square carrier speed and we use (Problem 1) $\langle t^2 \rangle = 2\tau^2$.

Now let us imagine a plasma density $\rho(x)$ as shown in Figure 12. What will be the mean current density at x? Carriers diffusing through position x have diffused through a distance X during a time T where the relationship between X and T is given by (83). We write the density of these carriers at $x - X$ and $t - T$ as $\rho(x - X, t - T)$. We have, then, for the current at position x and time t:

$$j(x,t) = \frac{1}{2} \frac{D}{X} \left[\rho(x - X, t - T) - \rho(x + X, t - T) \right] \tag{84}$$

where

$$D = \tfrac{1}{3}\langle v^2 \rangle \tau \tag{85}$$

is called the diffusion constant. Assuming that the charge density is a *slowly varying function of position* and time, we write:

$$\rho(x - X, t - T) = \rho(x,t) - X\frac{\partial}{\partial x}\rho(x,t) - T\frac{\partial}{\partial t}\rho(x,t) + \cdots \tag{86}$$

$$\rho(x + X, t - T) = \rho(x,t) + X\frac{\partial}{\partial x}\rho(x,t) - T\frac{\partial}{\partial t}\rho(x,t) + \cdots \tag{87}$$

Substituting **(86)** and **(87)** into **(84)** we obtain:

$$j(x,t) = -D\frac{\partial}{\partial x}\rho(x,t) \tag{88}$$

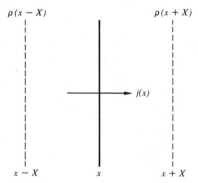

Figure 12   Plasma diffusion. We assume a charge density $\rho(x)$ and find the mean current density at x.

which is the diffusion equation in one dimension.

In three dimensions the diffusion current is given by *Fick's law*:

$$[\mathbf{j}(\mathbf{r})]_{\text{diffusion}} = -D\nabla\rho(\mathbf{r}) \tag{89}$$

In addition we may have a component of the current driven by *fields* **(19)**:

$$[\mathbf{j}(\mathbf{r})]_{\text{fields}} = \sigma(\mathbf{r})\mathbf{E}(\mathbf{r}) \tag{90}$$

To obtain the total current we add **(89)** and **(90)**:

$$\mathbf{j}(\mathbf{r}) = \sigma(\mathbf{r})\mathbf{E}(\mathbf{r}) - D\nabla\rho(\mathbf{r}) \tag{91}$$

In writing **(91)** we assume that the interaction between carriers may be described entirely in terms of the macroscopic electric field $\mathbf{E}(\mathbf{r})$ and that we need not consider collisions between carriers as a driving force. In Section 16 we consider situations where carrier densities change very gradually, and we include pressure gradients as well as macroscopic fields. Under such conditions it would be redundant to consider carrier diffusion as well.

Limiting ourselves to steady-state problems for which the total current is zero we obtain, from **(91)**,

$$\nabla \rho(\mathbf{r}) = \frac{\sigma}{D} \mathbf{E}(\mathbf{r}) = -\frac{\sigma}{D} \nabla \phi(\mathbf{r}) \tag{92}$$

In addition we have Gauss's law modified for a dielectric medium **(4-26)**:

$$\nabla \cdot \mathbf{E}(\mathbf{r}) = \frac{1}{\epsilon} \rho(\mathbf{r}) \tag{93}$$

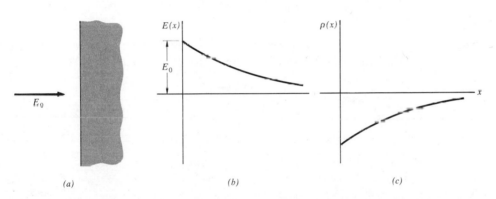

(a)   (b)   (c)

Figure 13   Field screening. (*a*) A uniform static field $\mathbf{E}_0$ is normal to the surface of a conducting medium. (*b*) The electric field decays exponentially into the medium. (*c*) A volume charge density is developed and decays exponentially into the medium.

where ρ is the total volume charge density and may include a uniform background charge density provided by the positive ions. Eliminating $\rho(\mathbf{r})$ between **(92)** and **(93)** we have **(B-10)** for $\mathbf{E}$ irrotational:

$$\blacktriangleright \qquad \nabla[\nabla \cdot \mathbf{E}(\mathbf{r})] = \nabla^2 \mathbf{E}(\mathbf{r}) = k_D \mathbf{E}(\mathbf{r}) \tag{94}$$

where k_D is called the *Debye wavenumber* and is given by the *Debye-Hückel* formula:

$$k_D{}^2 = \frac{\sigma}{\epsilon D} = \frac{1}{\lambda_D{}^2} \tag{95}$$

The distance λ_D is called the *Debye length*.

We now wish to obtain a solution to **(94)** for the situation shown in Figure 13*a* where a uniform static field $\mathbf{E}_0$ is normal to the surface of a conducting medium.

A particular solution of **(94)** is of the form:

$$\mathbf{E}(\mathbf{r}) = \mathbf{E}e^{-\mathbf{k}_D \cdot \mathbf{r}} \tag{96}$$

where $\mathbf{k}_D$ is a fixed vector of magnitude $1/\lambda_D$ and $\mathbf{r}$ is a general position vector. For the special case considered here we take $\mathbf{k}_D$ along $\hat{\mathbf{x}}$ and write:

$$E(x) = E_0\, e^{-x/\lambda_D} \tag{97}$$

The charge density plotted in Figure 13c may be written from **(81)**:

$$\rho(x) = \epsilon \mathbf{V} \cdot \mathbf{E}(x) = -\frac{\epsilon}{\lambda_D}\, E_0\, e^{-x/\lambda_D} \tag{98}$$

Exercise

---

8. Show that the ratio D/μ is numerically equal to $\frac{2}{3}$ the mean carrier energy, expressed in electron-volts where μ is the mobility **(14)**.

Table 4   Metals

	n	$m\langle v^2 \rangle/2e$	λ_D
Aluminum	18.06×10^{28} m^{-3}	11.63 V	0.52×10^{-10} m
Copper	8.45	7.00	0.59
Silver	5.85	5.48	0.63

12   Screening by Electrons. For a solid, the ions or atoms are fixed in the lattice and the screening is entirely by mobile electrons. The Debye length is then given by:

$$\lambda_D{}^2 = \frac{\epsilon D}{\sigma} = \frac{\epsilon m \langle v^2 \rangle}{3ne^2} \tag{99}$$

For a low density plasma as in a semiconductor, $\frac{1}{2}m\langle v^2 \rangle = \frac{3}{2}kT$ is the mean thermal kinetic energy. For a metal on the other hand $\frac{1}{2}m\langle v^2 \rangle = \frac{3}{2}kT_F$, where T_F is called the Fermi temperature, is very much larger because of the quantum nature of the gas. In Table 4 we list carrier concentrations, mean carrier energies, and screening lengths λ_D for several metals at room temperature.

It is instructive to compare the Debye length λ_D with the mean distance between particles r_0. Writing r_0 in terms of the relation

$$\frac{4\pi}{3} r_0{}^3 n = 1 \tag{100}$$

we obtain, from **(99)**, the relation

$$\frac{\lambda_D{}^2}{r_0{}^2} = \frac{2}{9} \frac{\frac{1}{2}m\langle v^2\rangle}{e^2/4\pi\epsilon r_0} \tag{101}$$

The numerator of the right side of **(101)** is the mean kinetic energy per particle, and the denominator is the potential energy of two charges at the mean separation r_0. Because in writing **(89)** we have assumed that the charge density is slowly varying, the present treatment is valid only in the limit that λ_D is large compared with r_0. Through **(101)** we require that the coulomb interaction between charges be small compared with the kinetic energy of the carriers.

Because of charge screening the *mean* energy of interaction between charges is of order $e^2/4\pi\epsilon_0\lambda_D$. The *plasma parameter* is given by the ratio of the mean kinetic energy to the mean screened interaction energy:

$$\frac{\frac{1}{2}m\langle v^2\rangle}{e^2/4\pi\epsilon\lambda_D} = 6\pi n\lambda_D{}^3 \tag{102}$$

13 Metal-Insulator-Metal Capacitors. For a wide variety of applications in electrical power transmission and in electronics it is desirable to store electrical energy through charge held at some potential. A device that performs this function is commonly called a capacitor. We show in Figure 14a a simple planar configuration. In Figure 14b we sketch the charge density, which has two components, the screening charge in the metal and the polarization charge in the dielectric insulator. If σ is the integrated surface charge density on the electrodes, then the displacement D within the dielectric is given by

$$D = \sigma \tag{103}$$

and the electric field is given by:

$$E = \frac{D}{\epsilon} = \frac{\sigma}{\epsilon} \tag{104}$$

Within the metal electrode on the left the electric field is given by **(103)**:

$$E = \frac{\sigma}{\epsilon_0} e^{x/\lambda_D} \tag{105}$$

Within the metal electrode on the right E is given by:

$$E = \frac{\sigma}{\epsilon_0} e^{-(x-d)/\lambda_D} \tag{106}$$

Integrating the field across the metal-insulator-metal capacitor we obtain as shown in Figure 11*c*:

$$V = \frac{\sigma}{\epsilon} d + 2 \frac{\sigma}{\epsilon_0} \lambda_D \qquad (107)$$

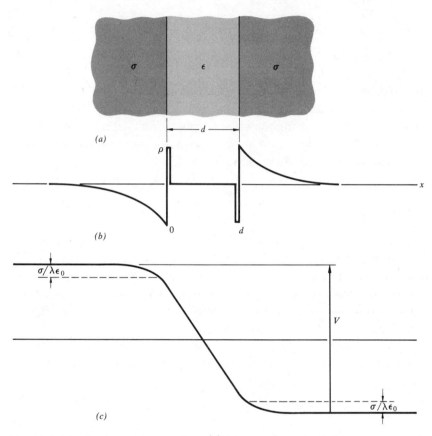

(a)

(b)

(c)

Figure 14 Metal-insulator-metal capacitor. (*a*) Media of conductivity σ are separated by a medium of permittivity ϵ. (*b*) The charge density has two components, the screening charge in the metal and the polarization charge in the dielectric insulator. (*c*) The potential is obtained by integrating the field and is given by (**107**).

The capacitance C may be defined in terms of the work required to charge the capacitor:

$$W = \tfrac{1}{2}QV = \tfrac{1}{2}CV^2 = \frac{1}{2}\frac{Q^2}{C} \qquad (108)$$

Taking the potential from (**92**) we obtain for the capacitance:

$$C = \frac{Q}{V} = \frac{\epsilon A}{d + 2\epsilon\lambda_D/\epsilon_0} \qquad (109)$$

where A is the cross-sectional area. For most capacitors d is a macroscopic dimension large compared with $1/\lambda_D$ and we obtain approximately:

$$C \simeq \frac{\epsilon A}{d} \tag{110}$$

In capacitors for which the dielectric is formed from a thin evaporated or oxidized layer the screening distance may be significant.

The derived SI unit of capacitance is the farad, which is equivalent to a coulomb per volt. From **(110)** we see the origin of farads per metre as the unit of permittivity.

Exercises

9. Show that the work given by **(108)** is larger than the electrostatic energy:

$$W = \tfrac{1}{2} \int \rho\phi \, dV = \tfrac{1}{2} \int ED \, dV$$

10. Thin film aluminum-aluminum oxide-aluminum capacitors are prepared by anodic oxidation of an aluminum substrate. The oxide formed (Al_2O_3) has a relative permittivity ϵ/ϵ_0 of 10. Capacitors designed to work up to 50 V have an insulating layer about 0.1 μm thick. What is the expected capacitance per square centimetre?

11. Thin film tantalum-tantalum oxide-gold capacitors have been prepared with a capacitance up to 5 μF/cm^2. The relative permittivity of anodic tantalum oxide (Ta_2O_5) is 27. What is the thickness of the oxide layer?

14 Schottky Barriers. A capacitor for which the potential is a quadratic function of the stored charge may be formed from a semiconductor and a metal. Such capacitors, which are called Schottky barriers, operate by depleting majority carrier charge from the semiconductor and are thus polarized. We show in Figure 15a a semiconductor-metal barrier and the associated charge density. In Figure 15b we show the charge density for small applied voltages. Since the maximum charge density is limited by the majority carrier concentration, once the stored charge exceeds a critical value

$$\sigma_c = \frac{Q_c}{A} = ne\lambda_D \tag{111}$$

the charge density profile changes to that shown in Figure 15c. This is because the maximum possible charge density in the semiconductor is limited to $-ne$, which occurs when the region is depleted of positive carriers.

As long as the charge is less than the critical charge the potential across the barrier is given by:

$$V = \sigma\left(\frac{\lambda}{\epsilon} + \frac{\lambda_0}{\epsilon_0}\right) \tag{112}$$

where ϵ and λ are the permittivity and screening length for the semiconductor and ϵ_0 and λ_0 are the corresponding quantities for the metal. The capacitance then takes the form:

$$C = \frac{Q}{V} = \frac{A}{\lambda/\epsilon + \lambda_0/\epsilon_0} \tag{113}$$

At higher charge densities the potential across the barrier may be written as:

$$V = \left(\frac{\lambda}{\epsilon} + \frac{\lambda_0}{\epsilon_0}\right) + \frac{\lambda}{2\epsilon}\,\sigma_c\left(\frac{Q}{Q_c} - 1\right)^2 \tag{114}$$

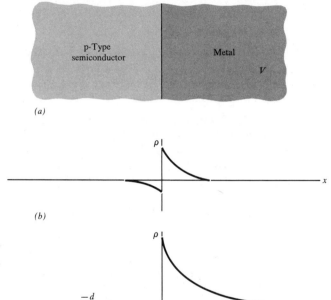

(a)

(b)

(c)

Figure 15 Schottky barrier. (*a*) A Schottky barrier is formed from a metal and a doped semiconductor. (*b*) The charge density through the barrier is shown for small applied voltages. (*c*) At large voltages the charge in the semiconductor is fully depleted near the interface.

In the limit that Q is very large we have

$$V \simeq \frac{\lambda}{2\epsilon\sigma_c}\frac{Q^2}{A^2} \tag{115}$$

We may in this region define a differential capacitance:

$$C' = \frac{dQ}{dV} \simeq \left(\frac{2\epsilon\sigma_c}{\lambda V}\right)^{1/2} A \tag{116}$$

and we see that for large potential across the barrier the capacitance decreases as one over the square root of the potential. Devices of this kind are used in applications where it is desired to control the value of a capacitance with a biasing potential.

15 **Charge Screening.** We discussed the screening of normal fields at the interface between different media. As a second example we discuss the screening of the field of an external charge q by a plasma. Here we begin to see the point of distinguishing external charge: that is, that the charge q is not a component of the plasma and is thus precluded from diffusing. We show in Figure 16 a charge q distributed over a sphere of radius a and surrounded by a screening charge $\rho(r)$ in a medium of permittivity ϵ.

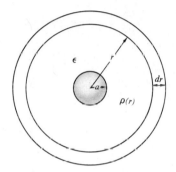

Figure 16   Screening a charge. An external charge q is embedded in a plasma of permittivity ϵ. A screening charge density $\rho(r)$ is formed.

By symmetry the electric field will be radial and a function only of r. The charge density similarly will be a function only of the radial distance. Under these conditions we may write **(92)** as:

$$\frac{d\rho}{dr} = \epsilon k_D{}^2 E = -\epsilon k_D{}^2 \frac{d\phi}{dr} \tag{117}$$

Applying Gauss's law to the spherical shell shown in Figure 16 we obtain

$$d\Phi = d(4\pi r^2 E) = \frac{1}{\epsilon} 4\pi r^2 \rho \, dr \tag{118}$$

Eliminating the electric field between **(117)** and **(118)** we obtain, as an equation for the charge density,

$$\frac{1}{r^2} \frac{d\rho}{dr} \left(r^2 \frac{d\rho}{dr} \right) = k_D{}^2 \rho \tag{119}$$

To solve **(119)** we introduce a quantity that has the units of surface charge density:

$$s(r) = r\rho(r) \tag{120}$$

Substituting **(120)** into **(119)** we obtain, for $s(r)$, the equation

$$\frac{d^2 s}{dr^2} = k_D{}^2 s \tag{121}$$

which has a solution of the form:

$$s = s_0\, e^{-r/\lambda_D} \tag{122}$$

To evaluate s_0 we substitute into **(117)** at the surface of the charge sphere and obtain:

$$\frac{d}{dr}\,\rho(a) = -\frac{s_0}{a^2}\left(1 + \frac{a}{\lambda_D}\right)e^{-a/\lambda_D} = \frac{k_D{}^2}{4\pi}\frac{q}{a^2} \tag{123}$$

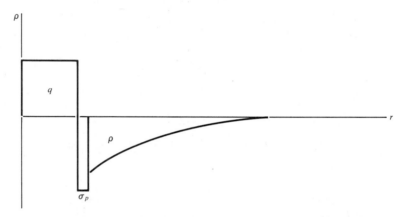

Figure 17   Charge components. Three components—the external charge, the polarization charge, and the plasma charge—characterize the charge distribution.

We assume for simplicity that the radius of the charge sphere is small compared with the screening length so that we may neglect a/λ_D. In this limit we have for the charge density:

$$\rho = -\frac{k_D{}^2}{4\pi}\frac{q}{r}e^{-r/\lambda_D} \tag{124}$$

We show in Figure 17 a plot of the three components of charge:

1. The external charge is a sphere of uniform density.

2. Surrounding the external charge is a surface polarization charge such that the total polarization charge is given by $q_P = -(1 - \epsilon_0/\epsilon)q$.

3. Finally outside is an exponentially decreasing plasma charge density.

We may obtain the potential by integrating **(117)**:

$$\phi(r) = -\frac{\lambda_D{}^2}{\epsilon}\,\rho(r) = \frac{1}{4\pi\epsilon}\frac{q}{r}e^{-r/\lambda_D} \tag{125}$$

The potential is the same as in a dielectric insulator except for the additional exponential attentuation resulting from the mobile charge carriers.

Exercises

12. Show that the sum of the polarization charge and plasma charge just balance the external charge.
13. A sphere of radius R bounds a medium of permittivity ϵ and screening length λ_D. An external charge q is placed at the center. For $k_D R$ large describe the charge density near the outside surface of the sphere.

PLASMA WAVES

We have discussed the response of a plasma to a uniform field that is a sinusoidal function of the time. We have also analyzed the screening of static fields by a plasma. These treatments may in principle be extended to the discussion of any uniform time-varying field and to any space-varying static field. We do not as yet, however, have a description of the response of a plasma to a field that is varying both in time and space.

The most satisfactory way of discussing the combined problem is to consider the excitations of a plasma. We make the simplifying assumption of a one-component electron plasma in a background medium that may be characterized by a permittivity ϵ. As we will see, the excitations of the plasma are waves of charge density, which propagate through the medium. Our theory ultimately will yield a relation between frequency ω and wavevector $k = 2\pi/\lambda$, where λ is the wavelength. This is called the *dispersion relation* for the medium. Although our discussion is greatly simplified and phenomenological, it does establish a result with which the dispersion relations of real plasmas may be compared.

16 Generalized Ohm's Law. We write an equation for the rate of change of current, which is the sum of three terms:

$$\frac{d\mathbf{j}}{dt} = \left[\frac{d\mathbf{j}}{dt}\right]_{\text{fields}} + \left[\frac{d\mathbf{j}}{dt}\right]_{\text{collisions}} + \left[\frac{d\mathbf{j}}{dt}\right]_{\text{pressure}} \tag{126}$$

1. The first term is the rate of change of current from *fields*, which we have written **(16)** as:

$$\left[\frac{d\mathbf{j}}{dt}\right]_{\text{fields}} = \frac{ne^2}{m}\mathbf{E} = \frac{\sigma}{\tau}\mathbf{E} \tag{127}$$

2. The second term is the rate of change of current from *collisions*. We have written this term for a *uniform* plasma (17) as:

$$\left[\frac{d\mathbf{j}}{dt}\right]_{\text{collisions}} = -\frac{\mathbf{j}}{\tau} \tag{128}$$

which yields exponential relaxation to zero current.

3. The final term results from the gas kinetic pressure of the electrons:

$$p = \tfrac{2}{3}nU \tag{129}$$

where U is the mean kinetic energy per electron. The force per unit volume is the gradient of the pressure:

$$\frac{d\mathbf{F}}{dV} = -\tfrac{2}{3}\nabla(nU) = -\tfrac{2}{3}\gamma U\nabla n \tag{130}$$

where γ is a numerical factor that allows for the possibility that the compression may be neither isothermal nor hydrostatic. In describing forces within the plasma by a pressure, we must assume that the pressure changes slowly over some characteristic distance within the plasma. For a neutral gas this distance is r_0, the mean separation between atoms or molecules. For gradual variations we may use the fluidlike concept of pressure. For more rapid variations we must treat the particlelike character of the gas explicitly. For a charged plasma, as we have seen, the interactions are much longer range, up to the Debye length λ_D. Thus the use of (130) assumes wavenumbers k less than the Debye wavenumber $k_D = 1/\lambda_D$. We write, for the pressure term,

$$\left[\frac{d\mathbf{j}}{dt}\right]_{\text{pressure}} = nq\left[\frac{d\mathbf{v}}{dt}\right]_{\text{pressure}} = \frac{q}{m}\frac{d\mathbf{F}}{dV} = -\frac{2}{3}\frac{\gamma U}{m}q\nabla n = -s^2\,\nabla\rho \tag{131}$$

where s is of the order of the electron thermal velocity.

Substituting (127), (128), and (131) into (126) we obtain, for the rate of change of current,

$$\frac{d\mathbf{j}}{dt} + \frac{\mathbf{j}}{\tau} = \frac{\sigma}{\tau}\mathbf{E} - s^2\,\nabla\rho \tag{132}$$

Taking the divergence of (132) we obtain an equation in the charge density:

$$\frac{\partial}{\partial t}\frac{d\rho}{dt} + \frac{1}{\tau}\frac{\partial\rho}{\partial t} + \frac{\sigma}{\epsilon\tau}\rho = s^2\,\nabla^2\rho \tag{133}$$

where we have written the divergence of the electric field as arising (4-27) from screened plasma charge:

$$\nabla\cdot\mathbf{E} = \frac{\rho}{\epsilon} \tag{134}$$

In order to investigate the form of possible propagating charge density waves, we assume a solution for the charge density of the form

$$\rho(\mathbf{r},t) = e^{-\alpha \mathbf{k} \cdot \mathbf{r}} \cos(\mathbf{k} \cdot \mathbf{r} - \omega t) \tag{135}$$

where α is an attenuation constant. Substituting into (133) we obtain two equations—one by equating the coefficients of the in-phase component of $\rho(\mathbf{r},t)$, and the other from the out-of-phase component:

$$\omega^2 = \frac{\sigma}{\epsilon \tau} + (1 - \alpha^2)k^2 s^2 \tag{136}$$

$$\omega = 2\alpha\tau k^2 s^2 \tag{137}$$

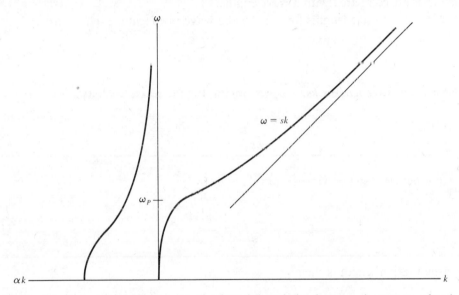

Figure 18 Dispersion curve. To the right is plotted the relation between the wavenumber k and the frequency ω. To the left is plotted the relation between the attenuation constant αk and the frequency ω.

We introduce the plasma frequency ω_p through the relation:

$$\omega_p{}^2 = \frac{\sigma}{\epsilon \tau} = \frac{ne^2}{\epsilon m} \tag{138}$$

Eliminating α between (136) and (137) and using (138) we write:

$$\omega^2 = \frac{\omega_p{}^2 + k^2 s^2}{1 + 1/(4k^2 s^2 \tau^2)} \tag{139}$$

We plot (139) to the right in Figure 18, which is called the *dispersion curve*. To the left we plot the attentuation constant αk, which may be obtained from **(137)** and the phase velocity ω/k. The equation **(139)** may conveniently be discussed in two limits:

1. At long wavelengths we have, from **(139)**, a wave with phase velocity:

$$u = \frac{\omega}{k} \simeq 2\omega_p \tau s \tag{140}$$

The wave attentuates in a distance that may be obtained from **(137)**:

$$\frac{1}{\alpha k} = \frac{2s^2\tau}{\omega/k} \simeq \frac{s}{\omega_p} \tag{141}$$

which is short compared with a wavelength.

2. At short wave lengths for $ks\tau > 1$ the wave propagates as:

$$\omega^2 = \omega_p{}^2 + k^2 s^2 \tag{142}$$

At *very* short wave lengths $ks > \omega_p$ we obtain, for the phase velocity,

$$v_p = \frac{\omega}{k} \simeq s \tag{143}$$

The attenuation constant is given in this limit by

$$\alpha = \frac{1}{2k^2 s^2 \tau} \simeq \frac{1}{2\omega\tau} \tag{144}$$

The attentuation length is then

$$\frac{1}{\alpha k} \simeq \frac{2\omega\tau}{k} \simeq 2s\tau \tag{145}$$

and the attenuation is characterized by a relaxation time rather than by a distance.

We may combine **(136)** and **(137)** into a single complex relation through the introduction of a complex propagation constant. We write in addition to **(135)** the solution:

$$\rho(\mathbf{r},t) = e^{-\alpha \mathbf{k} \cdot \mathbf{r}} \sin(\mathbf{k} \cdot \mathbf{r} - \omega t) \tag{146}$$

We introduce the complex combination of **(135)** and **(146)**:

$$\underline{\rho}(\mathbf{r},t) = \rho e^{-\alpha \mathbf{k} \cdot \mathbf{r}}[\cos(\mathbf{k} \cdot \mathbf{r} - \omega t) + i\sin(\mathbf{k} \cdot \mathbf{r} - \omega t)]$$
$$= \rho e^{i[(1 + i\alpha)\mathbf{k} \cdot \mathbf{r} - \omega t]} \tag{147}$$

Introducing the complex propagation constant:

$$\underline{\mathbf{k}} = \mathbf{k}' + i\mathbf{k}'' = \mathbf{k}(1 + i\alpha) \tag{148}$$

we have for the complex charge density,

$$\underline{\rho}(\mathbf{r},t) = \rho e^{i(\underline{\mathbf{k}} \cdot \mathbf{r} - \omega t)} \tag{149}$$

Substituting (149) into (133) we obtain the complex dispersion relation:

$$\blacktriangleright \qquad \omega\left(\omega + \frac{i}{\tau}\right) = \omega_p^2 + \underline{k}^2 s^2 \tag{150}$$

Exercises

14. Show that the complex combination of (136) and (137) leads to (150).
15. The free electron concentration in the earth's ionosphere is about 10^{14} per cubic metre. Find the plasma frequency. Find the plasma frequency of copper. See Table 4.
16. At zero frequency, (133) leads to the equation for the charge density:

$$\nabla\rho^2 - \frac{\sigma/\tau}{\epsilon s^2}\,\rho$$

Compare this equation with (94).

17 **Dielectric Function.** We obtain here the form of the *dielectric function* $\epsilon(k,\omega)$ of a one-component plasma. As we have seen [(1-137) and (1-149)], we may describe the behavior of charges in terms of polarization density rather than in terms of charge density and current. We have then for a plasma with $\mathbf{\Omega} = 0$:

$$\frac{\partial \mathbf{P}}{\partial t} = \mathbf{j} \qquad \mathbf{\nabla} \cdot \mathbf{P} = -\rho \tag{151}$$

where $\mathbf{P}$ is the polarization density of the charge carriers. [Note that (151) automatically gives the law of conservation of charge (1-147).]

Substituting (151) into (132) and integrating once with respect to the time, we obtain:

$$\frac{d}{dt}\frac{\partial \mathbf{P}}{\partial t} + \frac{1}{\tau}\frac{\partial \mathbf{P}}{\partial t} - s^2\,\mathbf{\nabla}(\mathbf{\nabla} \cdot \mathbf{P}) = \frac{\sigma}{\tau}\mathbf{E} \tag{152}$$

We rewrite (152) in terms of the plasma frequency as given by (138):

$$\frac{d}{dt}\frac{\partial \mathbf{P}}{\partial t} + \frac{1}{\tau}\frac{\partial \mathbf{P}}{\partial t} - s^2\,\mathbf{\nabla}(\mathbf{\nabla} \cdot \mathbf{P}) = \epsilon_{\text{ion}}\,\omega_p^2\mathbf{E} \tag{153}$$

where in place of ϵ we now write ϵ_{ion} to emphasize that only the ionic charge contributes to ϵ. In most of our discussion we are able to assume that ϵ_{ion} is very slowly varying with ω so that the obtained frequency dependence or *dispersion* is associated with the response of the electrons. Such will be the case in an electron-ion plasma at frequencies above the electron plasma frequency. This is because the motion of the much more massive ions is negligible at these frequencies. At very low frequencies, on the other hand, the electrons and ions move together and ϵ_{ion} is dispersive as well. In this regime an additional mode of propagation, called a positive ion wave, is obtained.

We assume a complex propagating electric field as in **(149)**:

$$\underline{\mathbf{E}}(\mathbf{r},t) = \underline{\mathbf{E}}e^{i(\underline{\mathbf{k}} \cdot \mathbf{r} - \omega t)} \tag{154}$$

and a polarization density:

$$\underline{\mathbf{P}}(\mathbf{r},t) = \underline{\mathbf{P}}e^{i(\underline{\mathbf{k}} \cdot \mathbf{r} - \omega t)} \tag{155}$$

Substituting **(154)** and **(155)** into **(153)** we obtain:

$$-\omega\left(\omega + \frac{i}{\tau}\right)\underline{\mathbf{P}} + s^2\underline{\mathbf{k}}(\underline{\mathbf{k}} \cdot \underline{\mathbf{P}}) = \epsilon_{ion}\,\omega_p{}^2\underline{\mathbf{E}} \tag{156}$$

We introduce the complex *dielectric function*:

$$\underline{\epsilon}(\underline{k},\omega) = \epsilon_{ion} + \underline{P}/\underline{E} \tag{157}$$

where ϵ_{ion} represents contributions to the polarization from the background medium. For a *longitudinal wave*, $\mathbf{P}$ and $\mathbf{k}$ are parallel and we have, from **(156)** for the dielectric function,

$$\blacktriangleright \qquad \underline{\epsilon}_L(\underline{k},\omega) = \epsilon_{ion}\left[1 - \frac{\omega_p{}^2}{\omega(\omega + i/\tau) - \underline{k}^2 s^2}\right] \tag{158}$$

For a transverse wave $\mathbf{P}$ and $\mathbf{k}$ will be perpendicular and the dielectric function is given from **(157)** by:

$$\blacktriangleright \qquad \epsilon_T(\underline{k},\omega) = \epsilon_{ion}\left(1 - \frac{\omega_p{}^2}{\omega(\omega + i/\tau)}\right) \tag{159}$$

A plasma wave is purely longitudinal and **(158)** applies. Later, when we discuss electromagnetic waves, which are transverse in an isotropic medium, we will use **(159)**.

We note that substituting the dispersion relation for a plasma wave **(150)** into **(158)** leads to:

$$\epsilon_L(\underline{k},\omega) = 0 \tag{160}$$

as the condition for the free propagation of a plasma wave. Alternatively, we may exhibit **(160)** as a resonance denominator by writing the polarization $\mathbf{P}(\mathbf{r},t)$ in terms of the displacement $\mathbf{D}(\mathbf{r},t)$. From **(157)** we have:

$$\mathbf{P}(\mathbf{r},t) = \epsilon_L(\underline{k},\omega)\mathbf{E}(\mathbf{r},t) - \epsilon_{\text{ion}}\,\mathbf{E}(\mathbf{r},t) \tag{161}$$

Writing for the displacement:

$$\mathbf{D}(\mathbf{r},t) = \epsilon_{\text{ion}}\,\mathbf{E}(\mathbf{r},t) + \mathbf{P}(\mathbf{r},t) \tag{162}$$

we obtain, by eliminating $\mathbf{E}(\mathbf{r},t)$ between **(161)** and **(162)**,

$$\mathbf{P}(\mathbf{r},t) = \left(1 - \frac{\epsilon_{\text{ion}}}{\epsilon_L(k,\omega)}\right)\mathbf{D}(\mathbf{r},t) \tag{163}$$

From **(158)** we have for the coefficient of $\mathbf{D}(\mathbf{r},t)$:

$$1 - \frac{\epsilon_{\text{ion}}}{\epsilon_L(\underline{k},\omega)} = \frac{\omega_p^{\,2}}{\omega_p^{\,2} + \underline{k}^2 s^2 - \omega(\omega + i/\tau)} \tag{164}$$

At resonance **(164)** goes to infinity, indicating that a finite plasma wave may propagate even though $\mathbf{D}(\mathbf{r},t)$ is everywhere zero.

18 Perfect Conductivity. For most electrostatic problems involving fields and conductors we may take over the results for the corresponding dielectric configuration by making a suitable choice of the dielectric function. Ideally, we let the screening length:

$$\lambda_D \quad \left(\frac{\epsilon D}{\sigma}\right)^{1/2} = \frac{1}{\omega_p}\left(\frac{D}{\tau}\right)^{1/2} \tag{165}$$

go to zero. This requires by **(99)** that the carrier concentration and thus by **(138)** the plasma frequency go to infinity. For most problems we can neglect carrier pressure leading to an isotropic dielectric function:

$$\epsilon(\omega) = \epsilon_{\text{ion}}\left(1 - \frac{\omega_p^{\,2}}{\omega(\omega + i/\tau)}\right) \tag{166}$$

In the limit that ω_p goes to infinity we see from **(166)** that the dielectric function goes to *negative* infinity.

Exercise

17. For low frequencies and long wavelengths the dielectric function $\epsilon(k,\omega)$ goes to negative infinity. Use this property for the image obtained for a charge near a dielectric to obtain the result for a conductor.

DISPLACEMENT CURRENT

We have seen **(4-63)** that the incremental amount of work required to alter a configuration of charges and dielectrics may be written as:

$$\delta W = \int \mathbf{E(r)} \cdot \delta \mathbf{D(r)} \, dV \tag{167}$$

where $\mathbf{E(r)}$ is the electric field and $\mathbf{D(r)}$ is the electric displacement, whose divergence is equal to the density of external charge. If we regard $\mathbf{D}$ and $\mathbf{E}$ as functions of the time we may write from **(167)** for the rate at which work is done:

$$\frac{dW}{dt} = \int \mathbf{E(r},t) \cdot \frac{d\mathbf{D(r},t)}{dt} \, dV \tag{168}$$

We introduce the *displacement current* density:

$$\mathbf{j}_D(\mathbf{r},t) = \frac{d\mathbf{D(r},t)}{dt} \tag{169}$$

and write for the work done per unit volume:

$$\frac{dw}{dt} = \frac{1}{V}\frac{dW}{dt} = \mathbf{E(r},t) \cdot \mathbf{j}_D(\mathbf{r},t) \tag{170}$$

which now has the form of **(41)**, suggesting the origin of the term displacement current.

19 **[Conducting Dielectric].** As an example of the use of the idea of displacement current, we discuss the problem of a conducting dielectric, or "leaky capacitor" in a uniform field. We consider a medium of permittivity ϵ and conductivity σ. We write for the displacement:

$$D = \epsilon_0 E + P + \int j \, dt \tag{171}$$

where P is the uniform polarization density of the dielectric and $\int j \, dt$ is the uniform polarization density associated with the charge carriers. The displacement current is given by:

$$j_D = \frac{dD}{dt} = \epsilon_0 \frac{dE}{dt} + \frac{dP}{dt} + j \tag{172}$$

To establish the field E and polarization P and to maintain the current j, work must be done per unit volume at a rate:

$$\frac{dw}{dt} = Ej_D = \epsilon_0 E \frac{dE}{dt} + E \frac{dP}{dt} + Ej \qquad (173)$$

Under steady-state conditions E and P will be constant and uniform and work is required only to maintain the current:

$$\frac{dw}{dt} = Ej \qquad (174)$$

Let us now imagine that we stop doing work to maintain the current. We do not initially alter the distribution of charge and thus P and E are initially unchanged. We can do this only by interrupting the displacement current. Setting (172) to zero we have:

$$\epsilon_0 \frac{dE}{dt} + \frac{dP}{dt} + j = 0 \qquad (175)$$

We write, for the polarization and current density,

$$P = \epsilon_0 \chi E = (\epsilon - \epsilon_0)E \qquad j = \sigma E \qquad (176)$$

Substituting into (175) we obtain

$$\epsilon \frac{dE}{dt} + \sigma E = 0 \qquad (177)$$

which has the solution:

$$E(t) = E e^{-t/\tau} \qquad (178)$$

where the relaxation time is given by:

$$\tau = \frac{\epsilon}{\sigma} \qquad (179)$$

and is the time for the decay of the field within a conducting dielectric.

20 [Equivalent Network]. The network equivalent to a conducting dielectric is shown in Figure 19. We show the conduction current j flowing through a resistance $1/\sigma$ and the polarization density P appearing as charge on a capacitance $(\epsilon - \epsilon_0)$. The displacement current j_D charges in addition a capacitance ϵ_0 with charge $\epsilon_0 E$.

When the source of energy, represented by a cell, is applied to the system, a displacement current given by **(160)** flows until steady-state conditions are reached, following which a steady current $j = \sigma E$ flows. Work on the system is terminated without the redistribution of charge simply by removing the cell. The charges on the capacitors and the current through the resistor are initially unchanged. But energy is required to maintain the current and this energy must come from the capacitors, resulting in a relaxation of the field **E**, described by **(178)** and **(179)**.

We may now apply this discussion to our earlier consideration of plasma waves and the dielectric function. The condition for the free propagation of plasma waves is zero displacement *current* and thus zero *displacement* at any nonvanishing frequency ω. Zero displacement together with a nonzero field **E** requires that the dielectric function $\epsilon(k,\omega)$ must be zero.

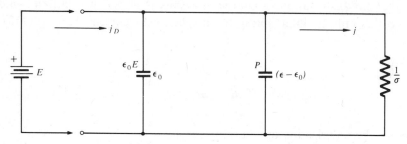

Figure 19   Equivalent network. The equivalent network of a conducting dielectric is a pair of capacitors in parallel with a conductor.

SUMMARY

1. In a two-component plasma the current is relaxed by scattering between the two charge carriers and the conductivity is given by:

$$\sigma = Ne^2\tau\left(\frac{1}{m} + \frac{1}{M}\right) \tag{11}$$

where m and M are the masses of the two carriers and τ is the mean collision time. In a metal, where there is only one charge carrier and collisions are with the lattice, we write

$$\sigma = \frac{Ne^2}{m}\tau \tag{15}$$

Here τ is a phenomenological relaxation time for decay of the current.
2. Currents may be induced in a plasma either by an electric field or by a gradient in concentration of the components of the plasma. These currents act to screen electric fields according to the law:

$$\nabla^2\mathbf{E}(\mathbf{r}) = k_D^2\mathbf{E}(\mathbf{r}) \tag{94}$$

as long as the fields are not sufficiently strong that either of the components of the plasma is depleted. For a two-component plasma the Debye wavenumber is given by:

$$k_D^2 = \frac{3ne^2}{\epsilon} \left(\frac{1}{M \langle V^2 \rangle} + \frac{1}{m \langle V^2 \rangle} \right)$$

3. For a metal-insulator-metal capacitor the capacitance is given by:

$$C = \frac{\epsilon A}{d + 2\epsilon \lambda_D / \epsilon_0} \tag{109}$$

where A is the area of the plates, d is the thickness of the dielectric insulator and ϵ is the permittivity; λ_D is the screening length in the metal. For most capacitors, the screening length in the metal may be neglected although it does limit the capacitance of very thin dielectric layers.

4. The capacitance of a Schottky barrier at low potentials is given by

$$C = \frac{A}{\lambda/\epsilon + \lambda_0/\epsilon_0} \tag{113}$$

where ϵ and λ refer to the semiconductor and ϵ_0 and λ_0 to the metal. As the potential is increased, charge is depleted from the semi-conductor. At high potentials, the potential difference is a quadratic function of stored charge:

$$V = \frac{\lambda}{2\epsilon A} \frac{Q^2}{Q_c} \tag{115}$$

where $Q_c = NeA\lambda_D$ is the critical charge for depletion. Such a device may be used with its differential capacitance dQ/dV controlled by the mean potential bias.

5. Charge density waves propagate through a plasma with a dispersion relation:

$$\omega(\omega + i/\tau) = \omega_p^2 + \underline{k}^2 s^2 \tag{150}$$

6. A plasma may be characterized by a dielectric function, which, for longitudinal waves, has the form

$$\epsilon_L(\underline{k}, \omega) = \epsilon_{\text{ion}} \left(1 - \frac{\omega_p^2}{\omega(\omega + i/\tau) - k^2 s^2} \right) \tag{158}$$

For transverse waves we obtain:

$$\epsilon_T(\underline{k}, \omega) = \epsilon_{\text{ion}} \left(1 - \frac{\omega_p^2}{\omega(\omega + i/\tau)} \right) \tag{159}$$

PROBLEMS

1. **Random collisions.** If dt/τ is the probability that a particle will collide in dt, show that the probability that a particle that last collided at $t = 0$ will suffer its next collision between t and $t + dt$ is $e^{-t/\tau} dt/\tau$. For exponentially distributed collision times show that the mean collision time $\langle t \rangle$ equals τ. Show that the mean-square collision time $\langle t^2 \rangle$ equals $2\tau^2$.

2. **Collision length.** For d, the distance between collisions, $\langle d^2 \rangle$ is the mean-square collision length. By using the final result of Problem 1, obtain the relation:

$$\langle d^2 \rangle = \langle v^2 \rangle \langle t^2 \rangle = 2\langle v^2 \rangle \tau^2$$

where it must be assumed that v and t are uncorrelated.

3. **Resistivity ratio.** The resistivity ratio of a metal is the resistivity at room temperature (20°C) divided by the resistivity at the temperature of liquid helium. For very high purity aluminum the resistivity ratio is 45,000. Find the collision time τ and mean free path $v\tau$ for electrons in high purity aluminum at 4.2 K. The velocity v is given in Table 1.

4. **Impurity scattering.** At sufficiently low temperature metallic conductivity is limited by scattering from impurities. Assuming a scattering cross section $\sigma \sim 10^{-20}$ square metre from impurities in aluminum, estimate the impurity content in parts per million (ppm) for aluminum with a resistivity ratio of 45,000 as described in Problem 3.

5. **Conductivity of sea water.** The principal ions present in sea water are given below with their concentrations and mobilities at 0°C. Find the contribution of each ion to the conductivity. Find the total conductivity from these ions at 0°C and compare with the standard value of 4.8 siemens per metre.
(This data was developed from the Smithsonian Physical Tables, 1956.)

Ion	Concentration, m^{-3}	Mobility, m^2/V-s
Na^+	276 $\times 10^{24}$	2.7×10^{-8}
K^+	5.8	4.2
Mg^{2+}	31.5	5.2
Cl^-	332	4.3
SO_4^{2-}	6.1	8.6

6. **Cole-Cole plot.** K. S. and R. H. Cole [*J. Chem. Phys.*, **9**, 341 (1941)] have been able to fit the frequency dependent susceptibility of a wide range of dielectrics with the empirical formula

$$\chi(\omega) = \frac{\chi}{1 + (-i\omega\tau)^{1-\alpha}}$$

Show by eliminating $\omega\tau$ that one obtains the equation

$$(\chi')^2 + (\chi'')^2 = \chi(\chi' - \chi'' \tan \tfrac{1}{2}\alpha\pi)$$

Show further that this equation is the arc of a circle which passes through the origin and is centered at $(\tfrac{1}{2}\chi, -\tfrac{1}{2}\chi \tan \tfrac{1}{2}\alpha\pi)$.

7. **Logarithmic symmetry.** Show that the Cole-Cole function indicated in Problem 6 preserves the logarithmic symmetry of the complex susceptibility:

$$\chi(\omega) + \chi^*(1/\omega\tau^2) = \chi$$

8. **Capacitance and resistance.** The capacitance of a pair of electrodes in a medium of permittivity ϵ is $C = Q/V$. If the dielectric medium is replaced by a conducting medium, show that the resistance $R = V/I$ measured between the electrodes satisfies the relation:

$$RC - \frac{\epsilon}{\sigma}$$

Show that this relation may also be obtained from the complex permittivity and the admittance $\underline{Y} = I/V = 1/R - I\omega C$.

9. **Resistance between hemispheres.** A pair of metallic spheres are half immersed in a medium of conductivity σ. If the separation b between spheres is large compared with their radius a, show that the electrical resistance between the spheres is:

$$R = \frac{1}{\pi\sigma}\left(\frac{1}{a} - \frac{1}{b}\right)$$

10. **Diffusion equation.** It is possible to obtain (89) by using the probability distribution of a diffusing charge. We use the result of Exercise 1-50 that if a charge is released from the origin at $t = 0$ and allowed to diffuse, the current at $\mathbf{r}$ at time t is given by:

$$\mathbf{j}(\mathbf{r},t) = \frac{\mathbf{r}}{2t}\rho(\mathbf{r},t) = \frac{\mathbf{r}}{2t}\left(\frac{6}{\pi}\right)^{1/2}\frac{q}{(4\pi/3)R^3}e^{-(3/2)(r^2/R^2)}$$

with $R^2 = 6Dt$. We now imagine that at $t = 0$ we have a charge distribution:

$$\rho(\mathbf{r}_1) = \rho(\mathbf{r}) + \mathbf{r}_{10} \cdot \nabla\rho(\mathbf{r}) + \cdots$$

(a) Show that the diffusion current $\mathbf{j}(\mathbf{r},t)$ is given by

$$\mathbf{j}(\mathbf{r},t) = -\frac{1}{2t}\left(\frac{6}{\pi}\right)^{1/2}\frac{\nabla\rho(\mathbf{r})}{(4\pi/3)R^3} \cdot \int_{V_1}\mathbf{r}_{01}\mathbf{r}_{01}e^{-(3/2)(r_{01}^2/R^2)}\,dV_1$$

(b) Carry out the integration to obtain:

$$\mathbf{j}(\mathbf{r}) = -D\nabla\rho(\mathbf{r})$$

Note that as long as the expansion of $\rho(\mathbf{r})$ is cut off with the gradient term, the current is independent of the time.

11. **Einstein relation.** A low density plasma with charge density $\rho(\mathbf{r})$ is in equilibrium in an electrostatic potential $\phi(\mathbf{r})$. The charge density and potential are related by:

$$\rho(\mathbf{r}) = \rho_0\, e^{-q\phi/kT}$$

Obtain the expression for the diffusion current $\mathbf{j}_{\text{diff}} = -D\nabla\rho$. Writing the conduction current as:

$$\mathbf{j}_{\text{cond}} = \sigma(\mathbf{r})E(\mathbf{r}) = \rho(\mathbf{r})\mu E(\mathbf{r})$$

and setting the total current to zero, obtain the Einstein relation

$$D = \frac{\mu}{q}kT$$

Here $\sigma(\mathbf{r})$ is the electrical conductivity, and μ is the carrier mobility.

12. **Ambipolar diffusion.** In a high density two-component plasma an electric field is established that inhibits the more rapidly diffusing carriers so that the charge density $\rho(\mathbf{r}) = \rho_{\text{eln}}(\mathbf{r}) + \rho_{\text{ion}}(\mathbf{r})$ and the current density $\mathbf{j}(\mathbf{r}) = \mathbf{j}_{\text{eln}}(\mathbf{r}) + \mathbf{j}_{\text{ion}}(\mathbf{r})$ both remain essentially zero. By writing an equation of the form of **(91)** for each carrier, obtain the relation:

$$\mathbf{j}_{\text{eln}}(\mathbf{r}) = -D_a\, \nabla\rho_{\text{eln}}(\mathbf{r}) \qquad \mathbf{j}_{\text{ion}}(\mathbf{r}) = -D_a\, \nabla\rho_{\text{ion}}(\mathbf{r})$$

where

$$D_a = \frac{\sigma_{\text{ion}} D_{\text{eln}} + \sigma_{\text{eln}} D_{\text{ion}}}{\sigma_{\text{eln}} + \sigma_{\text{ion}}} = \frac{(D/\mu)_{\text{eln}} + (D/\mu)_{\text{ion}}}{(1/\mu)_{\text{eln}} + (1/\mu)_{\text{ion}}}$$

is called the ambipolar diffusion coefficient.

13. **Dielectric relaxation of water.** The dielectric susceptibility of water at very low frequencies is 80. The susceptibility drops through the microwave range to a value in the infrared of 3.5. Assuming a single relaxation time $\tau = 5 \times 10^{-12}$ s, plot the real and imaginary part of the relative permittivity ϵ'/ϵ_0 as a function of frequency $\omega/2\pi$.

14. **Contact resistance.** Contact is made to a resistive layer of thickness d and resistivity ρ by a conductor that makes a small depression displacing a volume V of material bounded by a surface S. Wide area contact is made to the opposite side of the layer.

(a) Show that the measured resistance is

$$R = \epsilon_0 \rho / C - \rho / 6\pi d$$

where C is the capacitance of a conductor of volume $2V$ bounded by S and its reflection.

(b) Obtain the contact resistance for S a hemisphere.

(c) Taking the capacitance of a thin disc of radius a to be $C = 8\epsilon_0 a$, obtain the contact resistance of a circular contact of radius a.

15. **Field effect transistor.** The field effect transistor is a device in which transverse current is carried by the screening charge. The normal field is applied as a potential across an insulating layer as shown in the figure. Find the sensitivity dR/dV in ohms per volt in terms of the carrier concentration, the carrier mobility, and the dimensions.

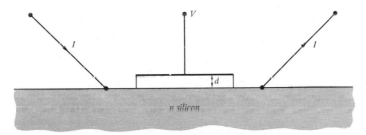

16. **Charge near a conducting surface I.** Find the surface charge σ for a charge q a distance d from a conducting surface using the requirement that $\mathbf{E} = 0$ within the conductor. Show that an image charge $-q$ produces the same normal component of field *outside* the conductor.

17. **Charge near a conducting surface II.** A charge q is placed near a conducting surface at zero potential. Find the charge that together with q ensures that the *transverse* electric field is zero at the surface. Compare your answer with Problem 16.

18. **Charge near a conducting surface III.** Obtain an expression for the amount of work required to bring a charge to within d of a conducting surface. Find the electrostatic potential at $\mathbf{r}$ from the surface charges. Does the expression

$$W = \tfrac{1}{2} q_0 \, \phi_0$$

give the correct answer? Show that the internal energy

$$W_{\text{int}} = W - W_{\text{ext}}$$

is zero with

$$W_{\text{ext}} = \frac{1}{2} \sum_i q_i \phi_i$$

and the sum is over *all* charges. Explain how the surface charge configuration is stable even with $W_{int} = 0$.

19. **Charged conducting ellipsoid.** Charge is deposited on a conducting ellipsoid in an initially field-free region. Show that the charge distributes over the surface of the ellipsoid so that the charge density σ is proportional to $\hat{n} \cdot r$, the normal component of the vector from the center of the ellipsoid. See Problem 1-29 on ellipsoidal homoeoids.

20. **Conducting ellipsoid in an electric field.** A neutral conducting ellipsoid is placed in a uniform electric field **E**. Show that the interior is screened by a surface charge density:

$$\sigma = \hat{n} \cdot \mathbf{N}^{-1} \cdot \mathbf{E}$$

where $\hat{n}$ is the surface normal and $\mathbf{N}^{-1}$ is the inverse depolarization tensor and is defined by $\mathbf{N} \cdot \mathbf{N}^{-1} = \mathbf{I}$.

21. **Charge near a conducting sphere.** An external charge q_1 is at a distance r_1 from the center of a perfectly conducting sphere of radius R. Show that q_1 together with an image q_2 at r_2 from the center make the spherical surface an equipotential. The geometry and values of q_2 and r_2 are shown below:

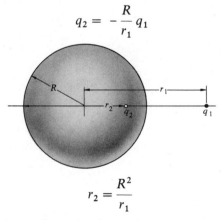

$$q_2 = -\frac{R}{r_1} q_1$$

$$r_2 = \frac{R^2}{r_1}$$

22. **Capacitance increase.** A small conducting ellipsoid is inserted into the dielectric medium between a pair of capacitor plates so that a principal axis is parallel to the electric field. Show that the capacitance is given by:

$$\frac{C}{1 - v/NV}$$

where v is the volume of the ellipsoid, V is the volume of the dielectric, N is the depolarizing factor of the ellipsoid, and C is the original capacitance. Use the results of Problem 4-9.

23. **Electrostatic plasma modes.** Show that a lossless plasma contained by an insulating ellipsoid will have a uniform mode ($k = 0$) given by

$$\omega_p^2 = \frac{N\sigma}{\epsilon\tau}$$

where N is a depolarization factor of the ellipsoid. Explain why this frequency differs from the long wavelength extrapolation of **(136)**.

24. **Positive ion waves.** This chapter's discussion of wave propagation in an electron-ion plasma considered only motion of the electrons, leading to a single longitudinal wave, the electron plasma wave. By allowing also for ion motion, a second mode, the positive ion wave is obtained. To obtain the dispersion relation for this wave assume that the frequency is sufficiently high that the displacement of the ions is limited by their inertia while at the same time the frequency is sufficiently low that the electrons screen the ions as if their displacement were static.

    (a) Obtain in this limit the expressions

$$\chi_{eln} = \frac{ne^2}{\epsilon_0 k_B T k^2} \qquad \chi_{ion} = -\frac{ne^2}{\epsilon_0 M \omega^2}$$

where T is the temperature and k_B is the Boltzmann constant.

    (b) Finally from the condition

$$\epsilon(\omega,k) = \epsilon_0(1 + \chi_{eln} + \chi_{ion}) = 0$$

obtain the dispersion relation for the positive ion waves.

    (c) Discuss the behavior in the limits of small and large k.

25. **Acoustic ion waves.** The analysis of the preceding problem neglects pressure gradients. Show that if the force on the positive ions from pressure gradients is taken into consideration, the ion susceptibility takes the form

$$\chi_{ion} = \frac{ne^2/\epsilon_0}{\gamma k^2 k_B T - M \omega^2}$$

where γ is the ratio of specific heats. Find the dispersion relation for acoustic ion waves.

26. **Metallic plasma waves.** In a metal the mean square electron velocity is given by

$$\langle v^2 \rangle = \tfrac{3}{5} v_F^2$$

where v_F is called the Fermi velocity and is proportional to the one-third power of the electron concentration.

    (a) From the dependence of mean electron concentration on energy show that γ in **(130)** is equal to $\tfrac{5}{3}$.

    (b) Show that the dispersion relation for electron plasma waves in a metal is given by

$$\omega^2 = \omega_p^2 + \tfrac{1}{3} k^2 v_F^2$$

27. **Low frequency plasma wave.** A low frequency, long wavelength plasma wave is propagated through a two-component plasma. For

$$\omega \ll (\sigma_{eln} + \sigma_{ion})/\epsilon_0 \qquad \text{and} \qquad k^2 \ll (2/\epsilon_0)(\sigma_{eln} + \sigma_{ion})/(D_{eln} + D_{ion})$$

show that the complex dispersion relation takes the form

$$i\omega \simeq D_a k^2$$

where D_a is the ambipolar diffusion coefficient defined in Problem 12. This is of the same form as the dispersion relation for heat conduction and, as we see later in Section 9-15, for the attenuation of an alternating magnetic field into a conductor.

28. **Charged conducting sphere.** Charge q is placed on a conducting sphere of permittivity ϵ, screening parameter k_D, and radius R. Assume a charge density of the form:

$$\rho = \frac{\sigma_0}{r}\left(e^{k_D r} - e^{-k_D r}\right)$$

(a) Use the fact that the total charge is q to obtain σ_0.
(b) By **(117)** obtain $E(r)$. Is the value of $E(R)$ consistent with Gauss's law?
(c) Use **(117)** to obtain $\phi(r)$ such that $\phi(R) = q/4\pi\epsilon_0 R$.
(d) Write ρ, E, and ϕ in the limits $k_D \to 0$ and $k_D \to \infty$. What are these limits, and are the results obtained in this way consistent with the results obtained directly?

29. **Conducting sphere in a uniform field.** A conducting sphere of permittivity ϵ, screening parameter k_D, and radius R is placed in a uniform electric field directed along z. Assume a screening charge density of the form:

$$\rho = \frac{d}{dz}\left[\frac{\sigma_0}{r}\left(e^{k_D r} - e^{-k_D r}\right)\right]$$

$$= \frac{\sigma_0 z}{r^3}\left[(k_D r - 1)e^{k_D r} + (k_D r + 1)e^{-k_D r}\right]$$

(a) Evaluate σ_0 by **(117)** where E includes the external field.
(b) Compute the dipole moment p and obtain an expression for the polarizability. Does this charge distribution have any higher moments?
(c) Find the internal electric field. Examine the expression for the field and for the dipole moment in the limit $k_D \to \infty$. Interpret the limit $k_D \to 0$. Compare with Problem 1-25.

30. **Dielectric tensor.** From **(158)** and **(159)** show that the dielectric tensor must be of the form:

$$\bar{\bar{\epsilon}} = \epsilon_{\text{ion}}\mathbf{I} - \frac{\epsilon_{\text{ion}}\,\omega_p^2}{\omega(\omega + i/\tau)}\left[\mathbf{I} + \frac{\mathbf{k}\mathbf{k}s^2}{\omega(\omega + i/\tau) - k^2 s^2}\right]$$

31. **Resistance increase.** A small insulating ellipsoid is inserted into the conducting medium between a pair of electrodes with a principal axis parallel to the field. Show that the resistance between the electrodes is given by:

$$R\left(1 + \frac{1}{1 - N}\frac{v}{V}\right)$$

where R is the original resistance, v is the volume of the ellipsoid, V is the volume of the medium, and N is the depolarization factor of the ellipsoid. See Problem 22 and Problem 4-9.

REFERENCES

Chen, F. F., *Introduction to Plasma Physics*, Plenum, 1974.

Ichimaru, S., *Basic Principles of Plasma Physics*, Benjamin, 1973.

Holt, E. H. and R. E. Haskell, *Foundations of Plasma Dynamics*, Macmillan, 1965.

Keefe, D., "Collective-Effect Accelerators," *Scientific American*, **226**(4), 22 (April, 1972).

Kittel, C., *Introduction to Solid State Physics*, Fifth Edition, Wiley, 1976.

Spitzer, L., *Physics of Fully Ionized Gases*, Second Edition, Interscience, 1962.

6 SOURCES OF THE ELECTROMAGNETIC FIELD I

The first five chapters of this text have paralleled Parts I and II of James Clerk Maxwell's *Treatise on Electricity and Magnetism*, first published in 1873.[1] Part I treated the electrostatics of charges in vacuum and dielectric media. Part II was concerned with electric current in conductors and dielectrics. Although we now know a great deal more about the microscopic constitution of conductors and dielectrics, there is little that one may do to improve on Maxwell's organization of the first two parts of his treatise.

In Part III Maxwell discusses the interaction between magnetic bodies and finally in Part IV he develops the theory of electromagnetism, which begins with the observation that electric currents interact with magnetic bodies and culminates in the electromagnetic theory of light.

Maxwell's theory was extended by Lorentz[2] to the discussion of electromagnetic phenomena in moving systems. Lorentz was able to show that "many electromagnetic actions are entirely independent of the motion of the system." Thus, an observer in a uniformly moving medium would have no way of knowing that the medium was in motion. Lorentz believed though that in the *absence of matter* Maxwell's equations held only for a particular coordinate system distinguished from all other coordinate systems by its state of rest.

As Einstein observed: "Maxwell's electrodynamics ... when applied to moving bodies, leads to asymmetries which do not appear to be inherent in the phenomena."[3] The fact that the then prevalent interpretation of electromagnetism "seemed to restrict the inertial system more strongly than did classical mechanics" was not lost on Einstein. As he has remarked, "this circumstance, which from the empirical point of view

---

[1] J. C. Maxwell, *A Treatise on Electricity and Magnetism*, Third Edition, Clarendon Press, 1891. Reprinted in two volumes by Dover Publications, 1954.

[2] H. A. Lorentz, *Electromagnetic Phenomena in a System Moving with Any Velocity Less than That of Light*. Reprinted in *The Principle of Relativity, a Collection of Original Memoirs on the Special and General Theory of Relativity*, Methuen and Company, Ltd., 1923. Reprinted by Dover Publications, 1952.

[3] A. Einstein, *On the Electrodynamics of Moving Bodies*. Reprinted in *The Principle of Relativity*, ibid. For a new translation of this paper, see H. M. Schwartz, *Am. J. Phys.* **45**, 18, 512 (1977). See also A. I. Miller, *Am. J. Phys.* **44**, 912 (1976); **45**, 1040 (1977).

appeared completely unmotivated, was bound to lead to the theory of special relativity."[4] Einstein formulated his theory in terms of two postulates. The first he called the principle of relativity, "that the same laws of electrodynamics and optics will be valid for all frames of reference for which the equations of mechanics hold good." The second postulate was "that light is always propagated in empty space with a definite velocity c which is independent of the state of motion of the emitting body."[3]

With this chapter we depart from the historical development just outlined. Instead, we use the relativistic transformation of forces as developed in Appendix F to show that the discoveries of Oersted and Ampère of the interaction between currents and magnets and between currents themselves are a consequence of Coulomb's law. Having established this connection, we return to a discussion of the interaction between currents, followed in the next chapter by an analysis of the magnetic properties of material media.

INTERACTION BETWEEN MOVING CHARGES

We begin our discussion with a consideration of the force on a test charge in the field of a source charge. We discuss four cases:

1. The source charge and the test charge are both at rest. This, as we have seen, is the interaction on which all of electrostatics is based.

2. The test charge is in motion in the field of a source charge at rest.

3. The source charge is in uniform motion and the test charge is at rest.

4. Both the source charge and the test charge are in motion.

First, what do we mean when we describe the source charge and the test charge as being at rest? Maxwell and Lorentz believed in a unique rest frame in which the laws of physics had special meaning. With Einstein we must give up any such notion. Thus, all we can mean when we say that the charges are at rest is that they are at rest with respect to the *observer*. Under these conditions we have the Coulomb force **(1-2)**:

$$\mathbf{F}(\mathbf{r}) = \frac{1}{4\pi\epsilon_0} \frac{q_1 q}{r_{01}{}^2} \hat{\mathbf{r}}_{01} \tag{1}$$

where q_1 is the magnitude of the source charge, q is the magnitude of the test charge, and $\mathbf{r}_{01} = \mathbf{r} - \mathbf{r}_1$ is a vector from the source charge at $\mathbf{r}_1$ to the test charge at the field position $\mathbf{r}$.

1   Charge Invariance. What force acts on the test charge when it moves with velocity $\mathbf{v}$ in the field of the source charge? Here we rely on an experiment that shows, as discussed in Chapter 1, that the force on the test charge is independent of its velocity.

---

[4] A. Einstein, *Mein Weltbild*, Querido Verlag, 1934. See *Ideas and Opinions: Albert Einstein*, Crown Publishers, Inc., 1954. Reprinted by Dell Publishing Co., 1973.

In other words, the magnitude of the test charge q is invariant and we may write the force on a charge moving in the vicinity of static charges as

$$F(\mathbf{r}) = q\mathbf{E}(\mathbf{r}) \tag{2}$$

which requires as well that the electric field $\mathbf{E}(\mathbf{r})$ be the force per unit charge independent of the velocity $\mathbf{v}$ of the test charge.

The third case that we consider is one in which the source charge q_1 is in motion with velocity $\mathbf{v}_1$ and the test charge q is at rest with respect to the observer as shown in Figure 1a. To write the force on the test charge we make a transformation to a frame moving with velocity:

$$\mathbf{V} = \mathbf{v}_1 \tag{3}$$

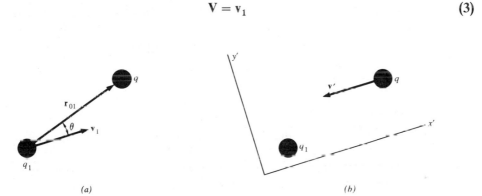

(a)   (b)

Figure 1 Force on a test charge at rest in the field of a moving source charge. (a) Both the source charge q_1 and the test charge q are invariant. (b) We transform to a frame in which the source charge is at rest.

as shown in Figure 1b. In this frame the source charge is at rest and the test charge moves with velocity:

$$\mathbf{v}' = -\mathbf{V} = -\mathbf{v}_1 \tag{4}$$

Postulating q_1 to be the source charge in the moving frame as well, the force on q is given by:

$$\mathbf{F}'(\mathbf{r}') = \frac{1}{4\pi\epsilon_0} \frac{q_1 q}{r_{01}'^3} \mathbf{r}_{01}' \tag{5}$$

which must, as we have argued, be independent of the velocity of the test charge.

To find the force acting on the test charge in the frame of the observer we must transform back to the rest frame using the vector form of the transformation developed in Section F-10:

$$\mathbf{F} = \gamma\mathbf{F}' - (\gamma - 1)\hat{\mathbf{V}}(\hat{\mathbf{V}} \cdot \mathbf{F}') + (\gamma/c^2)\mathbf{v} \times (\mathbf{V} \times \mathbf{F}') \tag{6}$$

with

$$\gamma = \frac{1}{(1 - V^2/c^2)^{1/2}} \tag{7}$$

and where the Lorentz transformation, which connects $\mathbf{r}, t$ with $\mathbf{r}', t'$ is given in Section F-4 and may be written in vector form as

$$\mathbf{r}' = \mathbf{r} + (\gamma - 1)\hat{\mathbf{V}}(\hat{\mathbf{V}} \cdot \mathbf{r}) - \gamma \mathbf{V}t \tag{8}$$

The separation between q_1 and q in the primed frame at time t' may be written as:

$$\mathbf{r}_{01} = \mathbf{r}' - \mathbf{r}_1' \tag{9}$$

and in terms of the unprimed coordinates at the corresponding time t as:

$$\mathbf{r}_{01}' = \mathbf{r}_{01} + (\gamma - 1)\hat{\mathbf{V}}(\hat{\mathbf{V}} \cdot \mathbf{r}_{01}) \tag{10}$$

provided that (1) the source charge q_1 is at rest in the primed frame and (2) the two vectors $\mathbf{r}$ and $\mathbf{r}'$, which give the position of the test charge in the two frames, are determined at times t and t' connected by a Lorentz transformation. The determination of $\mathbf{r}_1$ is simultaneous with that of $\mathbf{r}$ in the unprimed frame, and the determination of $\mathbf{r}_1'$ is simultaneous with that of $\mathbf{r}'$ in the primed frame. This means that $\mathbf{r}_1$ and $\mathbf{r}_1'$ are *not* connected by a Lorentz transformation. The square of the distance between charges $r_{01}'^2$ then is from **(F-45)**:

$$r_{01}'^2 = \gamma^2 r_{01}^2(1 - \beta^2 \sin^2 \theta) \tag{11}$$

where θ is the angle between $\mathbf{r}_{01}$ and $\mathbf{V}$ and we write $\beta = V/c$.

Since the test charge is at rest in the unprimed frame, we may set $\mathbf{v} = 0$ in **(6)** to obtain the expression for the force on q:

$$\mathbf{F} = \gamma \mathbf{F}' - (\gamma - 1)\hat{\mathbf{V}}(\hat{\mathbf{V}} \cdot \mathbf{F}') \tag{12}$$

where $\mathbf{F}'$ is given by **(5)**. Substituting **(5)** into **(12)** with $\mathbf{r}_{01}'$ given by **(10)** and r_{01}' given by **(11)** we obtain:

▶ $$\mathbf{E}(\mathbf{r}) = \frac{\mathbf{F}(\mathbf{r})}{q} = \frac{1}{4\pi\epsilon_0} \frac{q_1}{\gamma^2 r_{01}^2(1 - \beta^2 \sin^2 \theta)^{3/2}} \hat{\mathbf{r}}_{01} \tag{13}$$

We note that although the field intensity is no longer isotropic the field is still directed away from the *present position* of the source charge.[5] We must recognize though that

---

[5] For a description of computer-generated films of the electric field of moving charges, see J. C. Hamilton and J. L. Schwartz, *Amer. J. Phys.*, **39**, 1540 (1971).

what is meant by the "present position" depends on the frame. If in the primed frame as shown in Figure 2a we make a measurement of the force on q when q_1 is at the origin, then the position of q_1 in the unprimed frame as shown in Figure 2b is to the right of the origin.

The total flux of the electric field is invariant as may be seen by integrating (13) over a sphere of radius r_{01}:

$$\Psi = \oint \mathbf{E(r)} \cdot d\mathbf{S} = \frac{q_1}{2\gamma^2 \epsilon_0} \int_0^{\pi} (1 - \beta^2 \sin^2 \theta)^{-3/2} \sin \theta \, d\theta \tag{14}$$

where we have carried out the integral over ϕ. To perform the integral over θ we make the substitution:

$$\tan \psi = -\beta\gamma \cos \theta \tag{15}$$

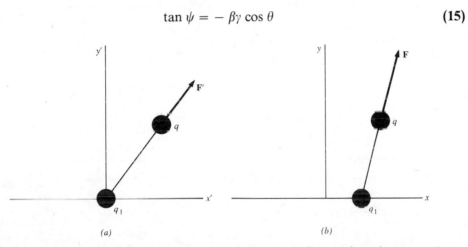

(a)   (b)

Figure 2 Present position of the source charge. (a) In the primed frame, where the source charge is at rest, the force F" acts along the line connecting the source and test charges. (b) In the unprimed frame the force is along the line connecting the test charge and the transformed position of the source charge.

Substituting into (14) we have:

$$\Psi = \frac{q_1}{2\beta\epsilon_0} \int_{-\sin^{-1}\beta}^{\sin^{-1}\beta} \cos \psi \, d\psi = \frac{q_1}{\epsilon_0} \tag{16}$$

which is exactly the same result as with q_1 at rest. We may regard (16) as a statement of the *invariance* of source charge, which is a consequence of the invariance of test charge and the Lorentz transformation.

A direct experimental test of (16) has been performed by John King[6] with the

---

[6] J. G. King, *Phys. Rev. Lett.*, **5**, 562 (1960). For a report of measurements on an isolated macroscopic body, see R. W. Stover, T. I. Moran, and J. W. Trischka, *Physical Review* **164**, 1599 (1967). See also J. G. King, "The Neutrality of Molecules," *Scientific American* (in press).

apparatus shown in Figure 3. Gas is allowed to escape from an inner container while its potential is measured with an electrometer. A deionizer prevents the escape of ions. With this apparatus an upper limit of 3×10^{-20} e is obtained per atom of hydrogen. Given that the motion of nuclear and electronic charge within the atom are expected to be very different, charge must be essentially independent of its state of motion.

Thus, the King experiment is really more general than **(16)**, which we have deduced for a uniformly moving charge. We may then write Gauss's law for any closed surface:

$$\Psi = \oint_S \mathbf{E} \cdot d\mathbf{S} = \frac{1}{\epsilon_0} \int_V \rho(\mathbf{r},t)\, dV \tag{17}$$

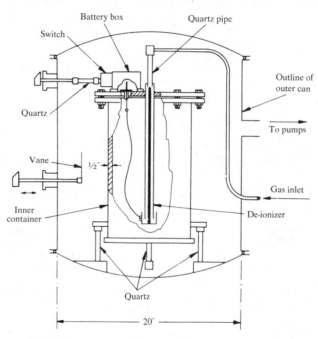

Figure 3   Apparatus for establishing the electrical neutrality of molecules. Gas molecules are allowed to escape from a container whose potential is measured with an electrometer.

independent of the motion of the enclosed charges. By letting the volume V shrink to zero we have the result

$$\blacktriangleright \qquad\qquad \mathbf{div}\, \mathbf{E}(\mathbf{r},t) = \frac{1}{\epsilon_0} \rho(\mathbf{r},t) \tag{18}$$

again regardless of the motion of the charges.

Exercises

1. Take the transformed charges as shown in Figure 1*b* to be q_1' and q'. Show that the source charge as determined in the unprimed frame is then $q_1'q'/q$. What does the King experiment indicate with respect to this charge? Finally, what can be concluded about q/q' and q_1'/q_1?

2. Let dS be an increment of area that moves with q_1. Show that the flux of $\mathbf{E}$ through dS, $d\Psi = \mathbf{E} \cdot d\mathbf{S}$ is invariant under a Lorentz transformation.

---

2   The Magnetic Field. The final case that we consider is one in which the source charge moves with velocity $\mathbf{v}_1$ and the test charge with velocity $\mathbf{v}$ as shown in Figure 4a. As for the previous case, we examine the interaction between q and q_1 in a frame moving with the source charge as given by **(3)** and shown in Figure 4b. Since the source charge is at rest in the primed frame, the force continues to be given by **(5)**. Transforming back into the original frame we obtain **(6)** for $\mathbf{F}'$. But since the test charge is in motion we must include the third term in **(6)**. Substituting **(5)** into **(6)** we obtain for the force on the test charge:

$$\mathbf{F}(\mathbf{r}) = \frac{1}{4\pi\epsilon_0} \frac{q_1 q}{\gamma^2 r_{01}{}^2 (1 - \beta^2 \sin^2 \theta)^{3/2}} \left[\hat{\mathbf{r}}_{01} + \frac{1}{c^2}\mathbf{v} \times (\mathbf{v}_1 \times \hat{\mathbf{r}}_{01}) \right] \qquad (19)$$

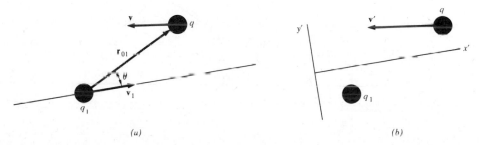

Figure 4   Force on a moving test charge in the field of a moving source charge. (a) The source charge moves with velocity $\mathbf{v}_1$ and the test charge with velocity $\mathbf{v}$. (b) In the primed frame the source charge is at rest and the test charge moves with the transformed velocity $\mathbf{v}'$.

The expression given by **(19)** differs from anything that we have considered thus far in that it contains a component of force depending on the velocity of the test charge. In order to describe the velocity-dependent part of the interaction we introduce the magnetic field $\mathbf{B}$ through the expression for the force on a moving test charge:

▶ $$\mathbf{F}(\mathbf{r}) = q(\mathbf{E} + \mathbf{v} \times \mathbf{B}) \qquad (20)$$

This expression is called the electromagnetic or Lorentz force. (As we will see, the expression is much more general than we have been able to show here, where we have been limited to uniformly moving source charges.) We have, then, for the electric and magnetic fields originating from a uniformly moving source charge:

$$\mathbf{E}(\mathbf{r},t) - \frac{1}{4\pi\epsilon_0} \frac{q_1}{\gamma^2 r_{01}{}^2 (1 - \beta^2 \sin^2 \theta)^{3/2}} \hat{\mathbf{r}}_{01} \qquad (21)$$

$$\mathbf{B}(\mathbf{r},t) = \frac{1}{c^2} \mathbf{v}_1 \times \mathbf{E}(\mathbf{r},t) \qquad (22)$$

We emphasize that $\mathbf{v}_1$ as written in (22) is the velocity of the *uniformly* moving sources of the field $\mathbf{E}(\mathbf{r},t)$. By superposition, however, we may use (21) and (22) to obtain the electric and magnetic fields for any distribution of uniformly moving charges, even though the velocities are not all the same.

Exercise

3. A particle of charge q and rest mass M moves in a uniform magnetic field $\mathbf{B}$. Show that the component of the velocity transverse to the field is constant in magnitude and rotates uniformly at a frequency

$$\omega = \frac{qB}{\gamma M}$$

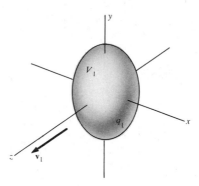

Figure 5   Field of a moving sphere of charge. A charge q_1 uniformly distributed through a spherical volume V_1 moves with velocity $\mathbf{v}_1$.

3 Field of a Moving Sphere of Charge. We are now in a position to write the electric and magnetic fields for any distribution of charge moving with uniform velocity. One case that is reasonably simple and that we will find useful later is that of a moving uniformly charged sphere. If we wish to write the fields for velocities approaching that of light we must specify in which frame the distribution is spherical. The situation that is physically meaningful (besides being much easier to calculate) is for the distribution to be spherical in a frame moving with the charge. We emphasize that the charge distribution will not be spherical in the laboratory frame, but will be contracted along the direction of motion into an oblate spheroid.

For simplicity of notation we write the fields at a time at which the center of charge coincides with the origin as shown in Figure 5. Our program will be to write the electric fields both interior and exterior to the sphere in the moving frame, transform back to the laboratory frame, and then by (22) find the magnetic fields.

The electric field within the sphere is given in the moving frame by (1-65):

$$\mathbf{E}'(\mathbf{r}') = \frac{q_1}{3\epsilon_0 V_1'}\mathbf{r}' \tag{23}$$

In the laboratory frame we obtain the simple result from (10) and (12):

$$E(r) = \frac{\rho_1}{3\epsilon_0} r_{01} \tag{24}$$

and a magnetic field from (24) and (22):

$$B(r) = \frac{1}{3\epsilon_0 c^2} j_1 \times r_{01} \tag{25}$$

where ρ_1 is the charge density q_1/V_1 in the laboratory frame and $j_1 = \rho_1 v_1$ is the laboratory current density.

Exterior to the sphere the field in the moving frame is the same as for a point charge and thus the electric and magnetic fields in the laboratory frame will simply be given by (21) and (22).

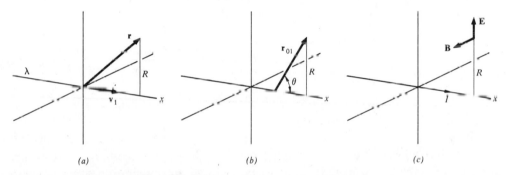

(a)   (b)   (c)

Figure 6 Field of a charged beam. (a) The beam has linear charge density λ and moves along the $\hat{x}$ axis with velocity v_1. (b) The electric field is obtained by integrating over the angle θ. (c) The magnetic field **B** is normal both to **E** and to the current direction.

4 Field of a Charged Beam. We next consider a charged beam moving with velocity v_1 along the x axis as shown in Figure 6a. We assume a linear charge density λ such that

$$dq_1 = \lambda \, dx \tag{26}$$

is the charge increment. The current is given by:

$$I = v_1 \frac{dq_1}{dx} = \lambda v_1 \tag{27}$$

Our procedure will be to first obtain the electric field and then use (22) to obtain the magnetic field. By direct integration of (21) we write:

$$E(r) = \frac{1}{4\pi\epsilon_0} \int_{-\infty}^{\infty} \frac{\hat{r}_{01} \lambda \, dx}{\gamma^2 r_{01}{}^2 (1 - \beta^2 \sin^2 \theta)^{3/2}} \tag{28}$$

Rewriting **(28)** as an integration over θ as shown in Figure 6b, we obtain:

$$\mathbf{E}(R) = \frac{\lambda \hat{\mathbf{R}}}{4\pi\epsilon_0 R} \int_0^\pi \frac{\sin\theta\, d\theta}{\gamma^2(1 - \beta^2 \sin^2\theta)^{3/2}} \tag{29}$$

where we have used the fact that the field component parallel to the wire cancels. Making the substitution **(15)** we may integrate **(29)**:

$$\mathbf{E}(R) = \frac{\lambda \hat{\mathbf{R}}}{4\pi\epsilon_0 R} \int_0^\pi \frac{\gamma \sin\theta\, d\theta}{\sec^3\psi} = \frac{\lambda \hat{\mathbf{R}}}{4\pi\epsilon_0 R} \int_{-\sin^{-1}\beta}^{\sin^{-1}\beta} \frac{1}{\beta} \cos\psi\, d\psi = \frac{\lambda \hat{\mathbf{R}}}{2\pi\epsilon_0 R} \tag{30}$$

Now **(30)** for a given λ is exactly the same as with the charges at rest. This may also be seen from a Gauss's law argument. Since **(13)** is symmetrical front and back, the electric field must be normal to the beam. But the total flux from a length dx is by **(16)** the same as if the charges were at rest and thus the electric field must be the same also. Now there is a magnetic field given by **(22)**:

$$\mathbf{B}(R) = \frac{1}{c^2} \mathbf{v}_1 \times \mathbf{E}(R) = \frac{I}{2\pi\epsilon_0 c^2} \frac{\hat{\mathbf{x}} \times \hat{\mathbf{R}}}{R} \tag{31}$$

where we have used **(27)**. As shown in Figure 6c the magnetic field **B** is normal to both **E** and to the current direction.

By using the fact that **(30)** depends on the velocity of the charges in the beam only through λ we can write **B(r)** in an alternate form, which we will need in the next section. We set $\gamma = 1$ and $\beta = 0$ in **(28)**. Then as long as we are integrating over the *entire* beam we may write for *any* velocity:

$$\mathbf{E}(\mathbf{r}) = \frac{1}{4\pi\epsilon_0} \int_{-\infty}^\infty \frac{\hat{\mathbf{r}}_{01}\lambda\, dx}{r_{01}^{\,2}} \tag{32}$$

and the magnetic field **(31)** may be written as:

$$\mathbf{B}(\mathbf{r}) = \frac{1}{c^2} \mathbf{v}_1 \times \mathbf{E}(\mathbf{r}) = \frac{\mu_0 I_1}{4\pi} \int_{C_1} \frac{d\mathbf{r}_1 \times \mathbf{r}_{01}}{r_{01}^{\,2}} \tag{33}$$

where

$$\mu_0 = \frac{1}{\epsilon_0 c^2} \tag{34}$$

is called the permeability of vacuum. In SI units (Section K-2) we have $\mu_0/4\pi = 10^{-7}$ henry per metre. The magnetic field **B** is expressed in tesla. And the ampere (Section K-1) is defined in terms of the force between current filaments as "that constant current which, if maintained in two straight parallel conductors of infinite length, of negligible cross-section, and placed one metre apart in vacuum, would produce between these conductors a force equal to 2×10^{-7} newton per metre of length."

Exercise

4. A particle of charge q_1 moves with constant velocity $\mathbf{v}_1$. If its closest approach to an observer is $-\mathbf{R}_1$ show that the time integral of the electric field is given by

$$\int_{-\infty}^{\infty} \mathbf{E}(t)\, dt = \frac{1}{2\pi\epsilon_0} \frac{q_1 \hat{\mathbf{R}}_1}{R_1}$$

CURRENT FILAMENTS AND DISTRIBUTIONS

We introduce the idea of a current filament as the superposition of two beams whose charge densities cancel but whose currents do not. The simplest example is to superpose on the charged beam of Figure 6 a static line charge of density $-\lambda$. The resulting electric field will then be zero and the magnetic field will be given by:

$$\mathbf{B}_1(\mathbf{r}) = \frac{\mu_0 I_1}{2\pi} \frac{\hat{\mathbf{x}}_1 \times \hat{\mathbf{R}}_1}{R_1} \tag{35}$$

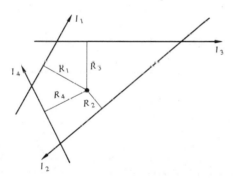

Figure 7 Distribution of linear current filaments. The magnetic field $\mathbf{B}(\mathbf{r})$ is a sum of the fields of the individual filaments.

5 Filament Synthesis. For some distribution of linear filaments as shown in Figure 7 we write the magnetic field:

$$\mathbf{B}(\mathbf{r}) = \sum_i \mathbf{B}_i(\mathbf{r}) = \frac{\mu_0}{2\pi} \sum_i \frac{I_i \hat{\mathbf{x}}_i \times \hat{\mathbf{R}}_i}{R_i} \tag{36}$$

In order to simplify the form of (36) we obtain, from (31) and (33),

$$\frac{\hat{\mathbf{x}}_i \times \hat{\mathbf{R}}_1}{R_i} = \frac{1}{2} \int_{C_i} \frac{d\mathbf{r}_i \times \hat{\mathbf{r}}_{0i}}{r_{0i}^2} \tag{37}$$

where C_i is the contour of the filament. We obtain, then,

$$\mathbf{B}(\mathbf{r}) = \sum_i \mathbf{B}_i(\mathbf{r}) = \frac{\mu_0}{4\pi} \sum_i \int_{C_i} \frac{I_i\, d\mathbf{r}_i \times \hat{\mathbf{r}}_{0i}}{r_{0i}^2} \tag{38}$$

6 Biot-Savart Law. As the current filaments become more and more numerous and the currents become smaller, we make the identification shown in Figure 8:

$$\mathbf{j}(\mathbf{r}_1)\,dV_1 = \sum_i I_i\,d\mathbf{r}_i \tag{39}$$

permitting us to write **(38)** as:

▶
$$\mathbf{B}(\mathbf{r}) = \frac{\mu_0}{4\pi}\int_{V_1}\frac{\mathbf{j}(\mathbf{r}_1)\times\hat{\mathbf{r}}_{01}}{r_{01}{}^2}\,dV_1 \tag{40}$$

where $\mathbf{j}(\mathbf{r}_1)$ is the current density associated with a superposition of filaments.

It has unfortunately not been shown in a simple and convincing way that *any* steady current distribution may be written as a superposition of current filaments and such a result is not obvious or even well known! We regard **(40)**, which is called the Biot-Savart law, as applying to any steady distribution of *solenoidal* currents. A solenoidal vector (Section B-15) is a vector whose divergence is everywhere equal to zero.

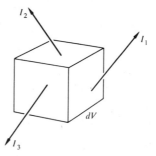

Figure 8   Limit of infinitesimal current filaments. The volume current density may be expressed as a sum over current filaments.

In some treatments **(40)** is used in differential form to describe the field of an increment of current $\mathbf{j}(\mathbf{r})\,dV$. This is not strictly speaking correct since the fields of charges within a limited region depend on the charge velocities and not the current alone. Only when the integral is carried over a complete contour do the fields become independent of velocity.

Exercise

5. Show from **(40)** that the magnetic field $\mathbf{B}(\mathbf{r})$ may be written in the form:

$$\mathbf{B}(\mathbf{r}) = \frac{\mu_0}{4\pi}\int_{V_1}\frac{\mathbf{\nabla}_1\times\mathbf{j}(\mathbf{r}_1)}{r_{01}}\,dV_1$$

where $\mathbf{\nabla}_1$ is the gradient operator with respect to $\mathbf{r}_1$.

7 Magnetic Circulation and Ampère's Law. The magnetic circulation is defined by the integral over a closed contour:

$$\mathcal{B} = \oint_C \mathbf{B} \cdot d\mathbf{r} = \int_S d\mathbf{S} \cdot \nabla \times \mathbf{B} \tag{41}$$

where the second equality follows from Stokes's law. For a single current filament we obtain the circulation of $\mathbf{B}$ by integrating (35):

$$\mathcal{B}_1 = \frac{\mu_0 I_1}{2\pi} \oint_C \frac{\hat{\mathbf{x}}_1 \times \hat{\mathbf{R}}_1 \cdot d\mathbf{r}}{R_1} \tag{42}$$

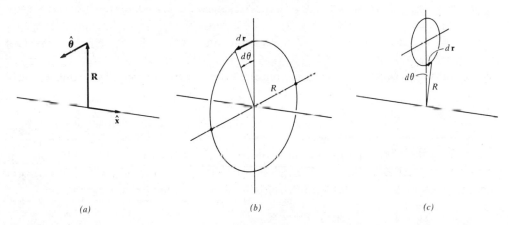

(a)   (b)   (c)

Figure 9 Magnetic circulation. (a) The unit vector $\hat{x}$ is along the current direction and $\hat{\theta}$ is in the azimuthal direction. (b) When the contour encloses the filament there is a circulation. (c) When the contour does not enclose the filament there is no circulation.

We write as shown in Figure 9a:

$$\hat{\mathbf{x}} \times \hat{\mathbf{R}} = \hat{\mathbf{\theta}} \quad \text{and} \quad \hat{\mathbf{\theta}} \cdot d\mathbf{r} = R \, d\theta \tag{43}$$

Substituting into (42) we obtain:

$$\mathcal{B}_1 = \frac{\mu_0 I_1}{2\pi} \oint_C d\theta \tag{44}$$

Now if the contour encloses the current filament as shown in Figure 9b we have:

$$\mathcal{B}_1 = \mu_0 I_1 \tag{45}$$

while if the contour does not enclose the filament as shown in Figure 9c we have zero for (44). We now consider a general distribution of currents, which we assume may be written as a superposition of current filaments. Then the magnetic field may be written in the form of (38) and we have for the circulation:

$$\mathscr{B} = \sum_i \mathscr{B}_i = \mu_0 \sum I_i \tag{46}$$

where the summation is only over those filaments enclosed by the contour. By using (39) we obtain:

$$\mathscr{B} = \oint_C \mathbf{B} \cdot d\mathbf{r} = \mu_0 \int_S \mathbf{j}(\mathbf{r}) \cdot d\mathbf{S} \tag{47}$$

where the integral is over the surface enclosed by the contour. Comparing (41) and (47) we finally obtain Ampère's law:

▶
$$\nabla \times \mathbf{B}(\mathbf{r}) = \mu_0 \mathbf{j}(\mathbf{r}) \tag{48}$$

as a consequence of the fact that the contour is arbitrary and may be shrunk to zero.

Exercise

6. By taking the curl of (40) show that Ampère's law (48) may be obtained. You will want to express $\hat{\mathbf{r}}_{01}/r_{01}^2$ as $-\nabla \, 1/r_{01}$ and use (B-5). The use of (1-47) will give $\mathbf{j}(\mathbf{r})$ while (C-9) makes it possible to drop the second term for $\mathbf{j}(\mathbf{r})$ solenoidal.

VECTOR POTENTIAL AND MAGNETIC FIELD

In order to simplify the calculation of magnetic fields it is convenient to relate the current distribution to a potential and then calculate the magnetic field from the potential. For this purpose we define a *vector* potential in what is called the *Coulomb gauge*:

▶
$$\mathbf{A}(\mathbf{r}) = \frac{\mu_0}{4\pi} \int_{V_1} \frac{\mathbf{j}(\mathbf{r}_1)}{r_{01}} dV_1 \tag{49}$$

8 **Vector Poisson Equation.** We now examine the properties of the vector potential. First we write for the divergence of $\mathbf{A}(\mathbf{r})$:

$$\nabla \cdot \mathbf{A}(\mathbf{r}) = \frac{\mu_0}{4\pi} \int_{V_1} \mathbf{j}(\mathbf{r}_1) \cdot \nabla \frac{1}{r_{01}} dV_1 = -\frac{\mu_0}{4\pi} \int_{V_1} \mathbf{j}(\mathbf{r}_1) \cdot \nabla_1 \frac{1}{r_{01}} dV_1 \tag{50}$$

where we have used (B-2). We may now rewrite the right-hand side of (50):

$$\nabla \cdot \mathbf{A}(\mathbf{r}) = -\frac{\mu_0}{4\pi} \int_{V_1} \nabla_1 \cdot \frac{\mathbf{j}(\mathbf{r}_1)}{r_{01}} dV_1 + \frac{\mu_0}{4\pi} \int_{V_1} \frac{\nabla_1 \cdot \mathbf{j}(\mathbf{r}_1)}{r_{01}} dV_1 \tag{51}$$

The first term on the right may be converted to a surface integral, and, as long as V_1 includes all currents, this term is zero. The second term is zero assuming that there is no accumulation of charge and thus, by **(1-147)**, no divergence of the current density. We obtain, then, for the divergence of the vector potential as long as there is no accumulation of charge:

$$\nabla \cdot \mathbf{A}(\mathbf{r}) = 0 \tag{52}$$

Next we look at the curl of the vector potential using **(B-3)**:

$$\nabla \times \mathbf{A}(\mathbf{r}) = \frac{\mu_0}{4\pi} \int_{V_1} \nabla \times \frac{\mathbf{j}(\mathbf{r}_1)}{r_{01}} \, dV_1 = \frac{\mu_0}{4\pi} \int_{V_1} \nabla \frac{1}{r_{01}} \times \mathbf{j}(\mathbf{r}_1) \, dV_1 \tag{53}$$

where the gradient operates only on r_{01}. Taking the gradient of $1/r_{01}$ we obtain **(40)** so that we have the result

$$\mathbf{B}(\mathbf{r}) = \nabla \times \mathbf{A}(\mathbf{r}) \tag{54}$$

which establishes the sense in which $\mathbf{A}(\mathbf{r})$ is a potential.

Taking the curl of **(54)** we obtain:

$$\nabla \times \mathbf{B} = \nabla \times (\nabla \times \mathbf{A}) = \nabla(\nabla \cdot \mathbf{A}) - \nabla^2 \mathbf{A} \tag{55}$$

Using **(52)** and **(48)** we may write:

$$\nabla^2 \mathbf{A}(\mathbf{r}) = -\mu_0 \mathbf{j}(\mathbf{r}) \tag{56}$$

which is the vector equivalent of Poisson's equation.

9 **Magnetic Flux.** The flux of **B** through a surface is defined **(B-37)** by the expression:

$$\Phi = \int_S \mathbf{B} \cdot d\mathbf{S} = \int_S (\nabla \times \mathbf{A}) \cdot d\mathbf{S} = \oint_C \mathbf{A} \cdot d\mathbf{r} = \mathscr{A} \tag{57}$$

where $\mathscr{A}$ is the circulation of **A**.

It may readily be seen from Figure 10 that the total flux through a *closed* surface (CS) must be zero. Let us divide the surface into two sections and apply **(57)** to each section. But the circulation of **A** around the two segments must be equal and opposite so that the total flux

$$\Phi_{CS} = \Phi_1 + \Phi_2 = 0 \tag{58}$$

It follows that the divergence of **B** must be everywhere zero from the definition of the divergence of a vector **(B-39)**:

$$\nabla \cdot \mathbf{B} = 0 \tag{59}$$

A more direct way of obtaining **(59)** is to take the divergence of **(54)**; since the divergence of the curl of a vector is everywhere zero, we obtain **(59)**.

Exercise

7. Obtain $\mathbf{V} \cdot \mathbf{B} = 0$ by taking the divergence of **(40)**.

10   Summary. We summarize this part of the chapter by exhibiting the analogous relations among static charge, potential and electric field as compared with steady current, vector potential, and magnetic field. The scalar and vector potentials are related to charge and current densities by:

$$\phi(\mathbf{r}) = \frac{1}{4\pi\epsilon_0} \int_{V_1} \frac{\rho(\mathbf{r}_1)}{r_{01}} \, dV_1 \qquad \mathbf{A}(\mathbf{r}) = \frac{\mu_0}{4\pi} \int_{V_1} \frac{\mathbf{j}(\mathbf{r}_1)}{r_{01}} \, dV_1 \tag{60}$$

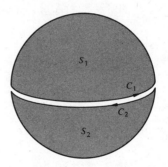

Figure 10   Magnetic flux through a closed surface. By dividing the surface by a contour it may be seen that the total flux must be zero.

The electric field is given by:

$$\mathbf{E}(\mathbf{r}) = -\mathbf{V}\phi(\mathbf{r}) = \frac{1}{4\pi\epsilon_0} \int \frac{\rho(\mathbf{r}_1)\hat{\mathbf{r}}_{01}}{r_{01}^2} \, dV_1 \tag{61}$$

and the magnetic field by:

$$\mathbf{B}(\mathbf{r}) = \mathbf{V} \times \mathbf{A}(\mathbf{r}) = \frac{\mu_0}{4\pi} \int_{V_1} \frac{\mathbf{j}(\mathbf{r}_1) \times \hat{\mathbf{r}}_{01}}{r_{01}^2} \, dV_1 \tag{62}$$

The divergences of the two fields are:

$$\mathbf{V} \cdot \mathbf{E}(\mathbf{r}) = -\mathbf{V}^2\phi(\mathbf{r}) = \frac{\rho(\mathbf{r})}{\epsilon_0} \qquad \mathbf{V} \cdot \mathbf{B}(\mathbf{r}) = 0 \tag{63}$$

The curls of the fields are:

$$\mathbf{V} \times \mathbf{E}(\mathbf{r}) = 0 \qquad \mathbf{V} \times \mathbf{B}(\mathbf{r}) = -\mathbf{V}^2\mathbf{A}(\mathbf{r}) = \mu_0 \mathbf{j}(\mathbf{r}) \tag{64}$$

CURRENT CONFIGURATIONS

In this section we obtain expressions for the magnetic field **B** for a number of specific current configurations. We have chosen examples that illustrate some of the techniques for obtaining fields and the limiting forms of the field at large distances.

11 **Field of a Uniform Current.** We first consider a cylindrical conductor of radius R carrying a current of uniform density **j** parallel to the $\hat{\mathbf{z}}$ direction as shown in Figure 11. We have already obtained **(35)** the field of a single current filament along $\hat{\mathbf{z}}$:

$$\mathbf{B}(\mathbf{r}) = \frac{\mu_0 I}{2\pi} \frac{\hat{\mathbf{z}} \times \hat{\boldsymbol{\rho}}}{\rho} \tag{65}$$

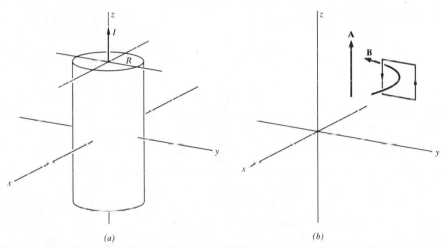

(a)   (b)

Figure 11 Field of a uniform current. (*a*) A current I is uniformly distributed over a cylinder of radius R (*b*) The magnetic field is in the azimuthal direction and the vector potential is in the direction of the current.

Although we might integrate **(65)** over the cross section of the conductor, we may obtain the result much more simply by using the cylindrical symmetry of the problem and Ampère's law **(48)**. Integrating around a circle centered on the z axis we obtain:

$$2\pi\rho B(\rho) = \mu_0 \int_0^\rho j(\rho) 2\pi\rho \, d\rho \tag{66}$$

where **B** is in the azimuthal direction and ρ is the transverse distance.

For radius ρ less than R we obtain from **(66)**:

$$\mathbf{B}(\rho) = \tfrac{1}{2}\mu_0 j\rho \hat{\boldsymbol{\phi}} \tag{67}$$

For radius ρ greater than R we obtain:

$$\mathbf{B}(\rho) = \tfrac{1}{2}\mu_0 j \frac{R^2}{\rho} \hat{\boldsymbol{\phi}} = \frac{\mu_0 I}{2\pi\rho} \hat{\boldsymbol{\phi}} \tag{68}$$

The vector potential for this problem is obtained by substitution into **(49)**:

$$\mathbf{A}(\rho) = \frac{\mu_0}{4\pi} \int_{V_1} \frac{dV_1}{r_{01}} \frac{I}{\pi R^2} \hat{\mathbf{z}} \tag{69}$$

where we have written $\mathbf{j}(\mathbf{r}_1) = I\hat{\mathbf{z}}/\pi R^2$ within the volume V_1. Now the integral over V_1 is just $1/\rho \, c^2$ times the electrostatic potential of an infinite cylinder of charge density ρ as given in Problem 1-9. We have then for the vector potential outside the cylinder:

▶
$$\mathbf{A}(\rho) = -\frac{\mu_0}{2\pi} I \ln \frac{\rho}{R} \hat{\mathbf{z}} \tag{70}$$

and inside the cylinder:

▶
$$\mathbf{A}(\rho) = \frac{\mu_0}{4\pi} I\left(1 - \frac{\rho^2}{R^2}\right) \hat{\mathbf{z}} \tag{71}$$

The zero of potential of **(70)** and **(71)** is at $\rho = R$ as for the corresponding electrostatic problem. The reason for this is that $\mathbf{A}(\rho)$ as given by **(49)** diverges for an infinite cylinder.

Exercises

8. From **(70)** and **(71)** obtain **(68)** and **(67)** by using **(54)**. Note from **(B-93)** that for $\mathbf{A} = A(\rho)\hat{\mathbf{z}}$ we have the simple relation:

$$\nabla \times \mathbf{A} = -\hat{\boldsymbol{\phi}}\partial A_z/\partial\rho$$

9. A current I flows longitudinally on the surface of an infinite cylinder of radius R. Find the magnetic field $\mathbf{B}$ and vector potential $\mathbf{A}$ both outside and inside the cylinder.

12 **Field of a Current Ring.** The second example that we consider is the field of a current I flowing on a ring of radius R as shown in Figure 12. The vector potential may be obtained from **(49)** and has the form:

$$\mathbf{A}(\mathbf{r}) = \frac{\mu_0}{4\pi} I \oint_c \frac{d\mathbf{r}_1}{r_{01}} \tag{72}$$

where we have used the limiting form of a current contour:

$$\mathbf{j}(\mathbf{r}_1) \, dV_1 = I \, d\mathbf{r}_1 \tag{73}$$

and where r_{01} is given by:

$$r_{01} = (r^2 + R^2 - 2rR \sin\theta \cos\phi_1)^{1/2} \tag{74}$$

We now write $d\mathbf{r}_1$ in the form:

$$d\mathbf{r}_1 = -\hat{\boldsymbol{\rho}} R \sin \phi_1 \, d\phi_1 + \hat{\boldsymbol{\phi}} R \cos \phi_1 \, d\phi_1 \tag{75}$$

where $\hat{\boldsymbol{\rho}}$ and $\hat{\boldsymbol{\phi}}$ are in the radial and azimuthal directions at the *field position*. Since r_{01} is an even function of ϕ the radial component of **A** integrates to zero and we are left with:

$$\mathbf{A}(\mathbf{r}) = \frac{\mu_0}{4\pi} I \, \hat{\boldsymbol{\phi}} \int_0^{2\pi} \frac{R \cos \phi_1 \, d\phi_1}{r_{01}} \tag{76}$$

We may obtain the vector potential *either* near the z axis and for all values of r *or* at large values of r and all orientations by expanding $1/r_{01}$:

$$\frac{1}{r_{01}} \simeq \frac{1}{(r^2 + R^2)^{1/2}} + \frac{rR \sin \theta \cos \phi}{(r^2 + R^2)^{3/2}} + \cdots \tag{77}$$

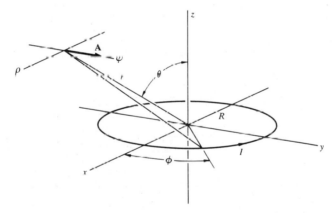

Figure 12 Field of a current ring. A current I flows on a ring of radius R lying in the xy plane.

Substituting **(77)** into **(76)** we obtain for the vector potential:

$$\blacktriangleright \qquad \mathbf{A}(\mathbf{r}) = \tfrac{1}{4} \mu_0 I \frac{R^2 r \sin \theta}{(r^2 + R^2)^{3/2}} \, \hat{\boldsymbol{\phi}} \tag{78}$$

1. *Near the axis* we have approximately:

$$\mathbf{A}(\mathbf{r}) \simeq \tfrac{1}{4}\mu_0 I \frac{R^2 r \sin \theta}{(z^2 + R^2)^{3/2}} \, \hat{\boldsymbol{\phi}} \tag{79}$$

The simplest way of obtaining the magnetic field is from the integral form of **(54)**:

$$\int_S \mathbf{B}(\mathbf{r}) \cdot d\mathbf{S} = \oint_C \mathbf{A}(\mathbf{r}) \cdot d\mathbf{r} \tag{80}$$

Integrating **(79)** over a circle centered on the $\hat{z}$ axis we obtain:

$$\blacktriangleright \qquad \mathbf{B}(z) = \tfrac{1}{2}\mu_0 I \, \frac{R^2}{(z^2 + R^2)^{3/2}} \, \hat{z} \qquad\qquad (81)$$

2. *At large distances* **(78)** takes the form:

$$\mathbf{A}(\mathbf{r}) = \tfrac{1}{4}\mu_0 I \, \frac{R^2}{r^2} \, \sin \theta \, \hat{\phi} \qquad\qquad (82)$$

From **(B-98)** we obtain for the magnetic field **B**:

$$\mathbf{B} = \nabla \times \mathbf{A} = \frac{\hat{\mathbf{r}}}{r \sin \theta} \frac{\partial}{\partial \theta} (\sin \theta A_\phi) - \frac{\hat{\boldsymbol{\theta}}}{r} \frac{\partial}{\partial r} (r A_\phi) \qquad (83)$$

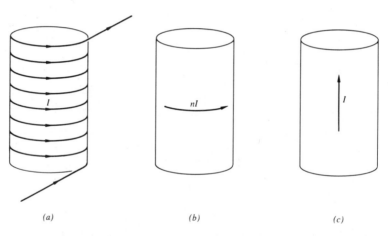

(a)   (b)   (c)

Figure 13   Field of a long solenoid. (*a*) A current *I* advances on a cylindrical surface with *n* turns per metre. (*b*) The azimuthal surface current density is *nI*. (*c*) There is in addition an axial current *I* arising from the advance of the circulating current.

Substituting from **(82)** for **A** we obtain:

$$\blacktriangleright \qquad \mathbf{B} = \tfrac{1}{4}\mu_0 IR^2 \frac{2\hat{\mathbf{r}} \cos \theta + \hat{\boldsymbol{\theta}} \sin \theta}{r^3} \qquad\qquad (84)$$

which has the same form as the field of an electric dipole, written in spherical coordinates. See Exercise 1-35.

13   Field of a Long Solenoid. We show in Figure 13*a* a long solenoid of *n* turns per metre, radius *R*, and current *I*. We may find the field within the solenoid by summing the fields of the individual turns. For a very long solenoid we may let the limits of integration go to infinity and obtain from **(81)** for the field on the axis:

$$\blacktriangleright \qquad B = \int_{-\infty}^{\infty} B(z)n \, dz = \tfrac{1}{2}\mu_0 I \int_{-\infty}^{\infty} \frac{nR^2 \, dz}{(R^2 + z^2)^{3/2}} = \mu_0 nI \qquad (85)$$

In fact, for an infinitely long solenoid, **(85)** must express the value of the magnetic field *everywhere inside* as may be shown from symmetry together with Ampère's law.

Another way of regarding a tightly wound solenoid is as the superposition of Figure 13*b* and *c*. The surface current density *nI* gives for the exterior field as may be shown by integrating over the contour shown in Figure 14*a* and using **(48)**:

$$B = 0 \qquad (86)$$

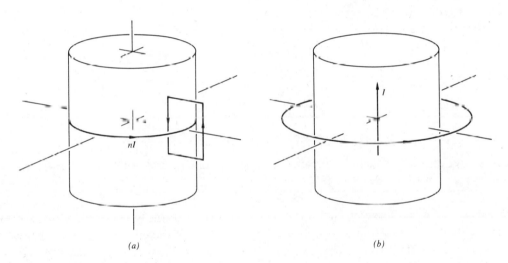

(*a*)

(*b*)

Figure 14 Contours for the evaluation of the magnetic field. (*a*) The axial field is evaluated on a contour that encloses the circulating current. (*b*) The azimuthal field is evaluated on a contour that encloses the axial current.

The current shown in Figure 13*c* arises because the current must advance along the solenoid. The field of this component is given by **(68)** and may be obtained by integrating **(48)** over the contour shown in Figure 14*b*.

The vector potential of a long solenoid (where we neglect the current advance) may be obtained from the integration of **(78)**. It is simpler though to obtain the vector potential inside and outside the solenoid by the integration of **(79)** and **(82)**. We obtain inside from **(79)**:

$$\blacktriangleright \qquad \mathbf{A}(\rho) = \tfrac{1}{4}\mu_0 nI \hat{\boldsymbol{\phi}} R^2 \rho \int_{-\infty}^{+\infty} \frac{dz}{(R^2 + z^2)^{3/2}} = \tfrac{1}{2}\mu_0 nI \rho \, \hat{\boldsymbol{\phi}} \qquad (87)$$

The vector potential outside is obtained from **(82)**:

$$\blacktriangleright \quad \mathbf{A}(\rho) = \tfrac{1}{4}\mu_0\, nIR^2 \hat{\boldsymbol{\phi}} \int_{-\infty}^{+\infty} \frac{\rho\, dz}{(\rho^2 + z^2)^{3/2}} = \tfrac{1}{2}\mu_0\, nI\, \frac{R^2}{\rho}\, \hat{\boldsymbol{\phi}} \qquad (88)$$

where ρ is now the distance from the axis.

Exercises

10. Show by symmetry that for an infinitely long solenoid the internal magnetic field must everywhere be longitudinal.

11. Show as a result of Exercise 10 that the field must also be uniform inside and therefore equal everywhere inside to **(85)**.

12. Using the result of Exercise 11 obtain **(86)**.

13. From **(87)** and **(88)** obtain **(85)** and **(86)** by using **(54)**. Note from **(B-93)** that for $\mathbf{A}(\mathbf{r}) = \hat{\boldsymbol{\phi}} A_\phi(\rho)$ we may write:

$$\nabla \times \mathbf{A} = \frac{\hat{\mathbf{z}}}{\rho}\, \frac{\partial}{\partial \rho}\, \rho A_\phi$$

14. By integrating **(85)** between $-\tfrac{1}{2}L$ and $+\tfrac{1}{2}L$ obtain the magnetic field **B** at the center of a *finite* solenoid. Expanding in powers of R/L find the leading term in the correction to the result of **(85)**.

---

14   Field of a Torus. We show in Figure 15a a coil of N turns wound on a torus of some general cross section S. If we neglect the advance of the current, or equivalently assume N separate current-carrying coils, we may write from symmetry that **(49)** must have the form

$$\mathbf{A} = \hat{\boldsymbol{\rho}} A_\rho(\rho,z) + \hat{\mathbf{z}} A_z(\rho,z) \qquad (89)$$

Also, by symmetry, the magnetic field must be of the form

$$\mathbf{B} = \hat{\boldsymbol{\rho}} B_\rho(\rho,z) + \hat{\mathbf{z}} B_z(\rho,z) + \hat{\boldsymbol{\phi}} B_\phi(\rho,z) \qquad (90)$$

That is, **B** must be independent of ϕ. We now show given the form of **(89)** and **(90)** that **(54)** allows only an azimuthal component of **B**.

To obtain the radial component of **B** we use **(57)** over the contour shown in Figure 15b. Now with $A_\phi = 0$ and since A_z may not be a function of ϕ we must have:

$$B_\rho(\rho,z) = 0 \qquad (91)$$

To obtain the $\hat{z}$ component of **B** we integrate over the contour shown in Figure 15c. Since again we have $A_\phi = 0$ and since A_ρ may not be a function of ϕ we have:

$$B_z(\rho,z) = 0 \qquad (92)$$

leaving only $B_\phi(\rho,z)$. To determine the azimuthal component of **B** we use **(47)**. If the contour is within S, the right side of **(47)** is $\mu_0 NI$ and we have for the internal field:

▶
$$\mathbf{B} = \frac{\mu_0}{2\pi} \frac{NI}{\rho} \hat{\phi} \qquad (93)$$

Outside the torus the right side of **(47)** is zero and we obtain zero for the azimuthal field.

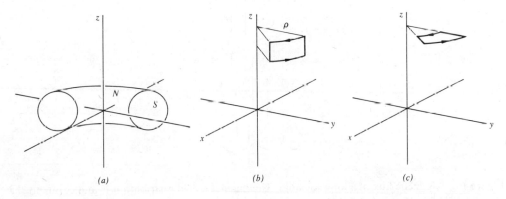

Figure 15 Field of a torus. (*a*) A coil of N turns is wound on a torus of general cross-section S. (*b*) The flux through a contour normal to $\hat{\rho}$ must be zero. (*c*) The flux through a contour normal to $\hat{z}$ must similarly be zero.

Exercise

15. If the torus is wound from a single long wire, the external field will not be zero. How must the above argument be modified?

What symmetry argument is used to justify the form of **(89)**, that is, that there is no azimuthal component of **A** nor may **A** be a function of ϕ?

MAGNETIC FLUX CONFIGURATIONS

Although it is always possible, at least in principle, to obtain the vector potential from the distribution of currents by **(49)**, there are many situations where it is simpler and more useful to obtain the vector potential from **B**(**r**). To show how this may be done we argue from the analogous relation between $\mathbf{B}(\mathbf{r})$ and $\mathbf{j}(\mathbf{r})$. From (59), (48), and (40) we write:

$$\nabla \cdot \mathbf{B} = 0 \qquad \nabla \times \mathbf{B} = \mu_0 \mathbf{j}(\mathbf{r}) \qquad \mathbf{B}(\mathbf{r}) = \frac{\mu_0}{4\pi} \int_{V_1} \frac{\mathbf{j}(\mathbf{r}_1) \times \hat{\mathbf{r}}_{01}}{r_{01}^{2}} \, dV_1 \qquad (94)$$

We have from (52) and (54) for the vector potential:

$$\nabla \cdot \mathbf{A} = 0 \qquad \nabla \times \mathbf{A} = \mathbf{B}(\mathbf{r}) \qquad (95)$$

By analogy we must then have for the vector potential:

$$\blacktriangleright \qquad \mathbf{A}(\mathbf{r}) = \frac{1}{4\pi} \int_{V_1} \frac{\mathbf{B}(\mathbf{r}_1) \times \hat{\mathbf{r}}_{01}}{r_{01}^{2}} \, dV_1 \qquad (96)$$

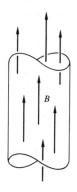

Figure 16   Magnetic flux of a long solenoid. The magnetic field distribution is analogous to the current distribution in a long wire.

As examples of the use of (96) we shall obtain the vector potential of a long solenoid and of a torus.

Exercises

---

16. Obtain (96) from (49) and (48). Use (B-3) and then (B-13).
17. Take the curl of (96) to obtain (54). Use (B-75) or (1-47) with (C-4), (C-9), and (59).

---

15   **Potential of a Long Solenoid.** We begin with the magnetic flux of a long solenoid as given by (85) and (86) and as shown in Figure 16. As may be seen, the magnetic flux is analogous to the uniform current shown in Figure 11, permitting us to write within the solenoid by analogy with (67):

$$\mathbf{A}(\rho) = \tfrac{1}{2} B \rho \hat{\boldsymbol{\phi}} = \tfrac{1}{2} \mu_0 n I \rho \, \hat{\boldsymbol{\phi}} \qquad (97)$$

which is the result of **(87)**. Outside the solenoid we have by analogy with **(68)**:

$$A(\rho) = \tfrac{1}{2} B \frac{R^2}{\rho} \hat{\phi} = \frac{\Phi}{2\pi\rho} \hat{\phi} \tag{98}$$

which is equivalent to **(88)**.

16 Potential of a Torus. The magnetic field **B** of a torus has only an azimuthal component as given by **(93)**. The field falls off as $1/\rho$ within the torus and is zero outside. Assuming that the cross section of the torus is small, we may write the vector potential on the axis by analogy with **(83)**:

▶
$$A(z) = \tfrac{1}{2} \Phi \frac{R^2}{(z^2 + R^2)^{3/2}} \hat{z} \tag{99}$$

where the total flux is given by:

$$\Phi = \int_S \mathbf{B} \cdot d\mathbf{S} = \frac{1}{2\pi} \mu_0 NI \int_S \frac{dS}{\rho} \tag{100}$$

and **B** has been obtained from **(93)**.

The vector potential at large distance and in a general direction may be written by comparing **(83)** and **(84)** as:

▶
$$A(r,\theta) = \tfrac{1}{4} \Phi R^2 \frac{2\hat{r} \cos\theta + \hat{\theta} \sin\theta}{r^3} \tag{101}$$

MOMENTS OF CURRENT DISTRIBUTIONS

Our discussion here parallels that of Chapter 1 of moments of charge distributions. We first write an expansion for the vector potential well away from a limited distribution of steady currents. We interpret the terms in the expansion as arising from moments of the distribution. We finally discuss forces and torques on magnetic dipoles and examples of magnetic multipoles.

We have from **(49)** the expression for the vector potential associated with a distribution of steady currents $\mathbf{j}(\mathbf{r})$. If we are well removed from the currents we may expand the denominator of **(49)**:

$$\frac{1}{r_{01}} = [r^2 - 2\mathbf{r} \cdot \mathbf{r}_1 + r_1^2]^{-1/2} = \frac{1}{r} + \frac{r_1}{r^2}(\hat{r} \cdot \hat{r}_1) + \frac{r_1^2}{2r^3}[3(\hat{r} \cdot \hat{r}_1)^2 - 1] + \cdots \tag{102}$$

We have written **(102)** only to order $1/r^3$ but may extend the series to as high order as is required with a typical term of the form $(r_1/r)^l(1/r)$ times an angular factor.

Corresponding to the expansion of $1/r_{01}$ we may write the vector potential in a series expansion:

$$\mathbf{A}(\mathbf{r}) = \mathbf{A}_0(\mathbf{r}) + \mathbf{A}_1(\mathbf{r}) + \mathbf{A}_2(\mathbf{r}) + \dots \tag{103}$$

where the individual terms are given by:

$$\mathbf{A}_0(\mathbf{r}) = \frac{\mu_0}{4\pi r} \int_{V_1} \mathbf{j}(\mathbf{r}_1)\, dV_1 \tag{104}$$

$$\mathbf{A}_1(\mathbf{r}) = \frac{\mu_0 \hat{\mathbf{r}}}{4\pi r^2} \cdot \int_{V_1} \mathbf{r}_1 \mathbf{j}(\mathbf{r}_1)\, dV_1 \tag{105}$$

$$\mathbf{A}_2(\mathbf{r}) = \frac{\mu_0}{8\pi r^3} \int_{V_1} r_1^2 [3(\hat{\mathbf{r}} \cdot \hat{\mathbf{r}}_1)^2 - 1]\mathbf{j}(\mathbf{r}_1)\, dV_1 \tag{106}$$

17   Monopole Moment. We first show that $\mathbf{A}_0(\mathbf{r})$, which is proportional to the monopole moment of the current distribution, is zero for solenoidal currents. From **(C-4)** we write the tensor identity:

$$\mathbf{V}_1 \cdot (\mathbf{j}\mathbf{r}_1) = \mathbf{r}_1 \mathbf{V}_1 \cdot \mathbf{j}(\mathbf{r}_1) + \mathbf{j} \cdot \mathbf{V}_1 \mathbf{r}_1 \tag{107}$$

For solenoidal currents the first term on the right of **(107)** vanishes. The second term is simply the current density $\mathbf{j}(\mathbf{r}_1)$. Substituting into **(104)** we have

$$\mathbf{A}_0(\mathbf{r}) = \frac{\mu_0}{4\pi r} \int_{V_1} \mathbf{V}_1 \cdot (\mathbf{j}\mathbf{r}_1)\, dV_1 \tag{108}$$

By **(C-9)** we write:

$$\mathbf{A}_0(\mathbf{r}) = \frac{\mu_0}{4\pi r} \oint_{S_1} d\mathbf{S}_1 \cdot \mathbf{j}(\mathbf{r}_1)\mathbf{r}_1 \tag{109}$$

Now, if S_1 bounds the current distribution then $\mathbf{j}(\mathbf{r}_1)$ must be zero on the surface, and $\mathbf{A}_0(\mathbf{r})$ vanishes.

18   Magnetic Dipole Moment. The magnetic dipole moment associated with a distribution of currents is defined as:

▶
$$\boldsymbol{\mu} = \tfrac{1}{2} \int_{V_1} \mathbf{r}_1 \times \mathbf{j}(\mathbf{r}_1)\, dV_1 \tag{110}$$

For the special case of current I flowing on a contour C_1 we may write **(110)** as:

$$\boldsymbol{\mu} = \tfrac{1}{2} I \oint_{C_1} \mathbf{r}_1 \times d\mathbf{r}_1 = I \int_{S_1} d\mathbf{S}_1 = I S_1 \tag{111}$$

where dS_1 is the increment of area swept out by $d\mathbf{r}_1$. The current I is the rate at which charge passes any point on the contour. Then if the total moving charge distributed over the contour is q and the time for the charge to make one circuit of the contour is τ, we may alternatively write the magnetic moment as

$$\boldsymbol{\mu} = \frac{q}{\tau}\,\mathbf{S}_1 \tag{112}$$

We now show that (105) may be expressed in terms of (110). We write for the vector potential of a current I flowing on a contour C_1:

$$\mathbf{A}_1(\mathbf{r}) = \frac{\mu_0}{4\pi}\,\frac{\hat{\mathbf{r}}\cdot\mathbf{T}}{r^2} \tag{113}$$

with

$$\mathbf{T} = \int_{V_1} \mathbf{r}_1 \mathbf{j}(\mathbf{r}_1)\,dV_1 = I \oint_{C_1} \mathbf{r}_1 d\mathbf{r}_1 \tag{114}$$

Integrating (114) by parts we have:

$$\mathbf{T} = I \oint_{C_1} d(\mathbf{r}_1\mathbf{r}_1) - I \oint_{C_1} d\mathbf{r}_1\mathbf{r}_1 \tag{115}$$

Since the integrals are taken over a complete contour the first term is zero. Adding (114) and (115) we write for $\mathbf{T}$:

$$\mathbf{T} = \tfrac{1}{2}I \oint_{C_1} (\mathbf{r}_1 d\mathbf{r}_1 - d\mathbf{r}_1\mathbf{r}_1) \tag{116}$$

Taking the scalar product of (116) with $\hat{\mathbf{r}}$ we have:

$$\hat{\mathbf{r}}\cdot\mathbf{T} = \tfrac{1}{2}I \oint_{C_1} (\hat{\mathbf{r}}\cdot\mathbf{r}_1)\,d\mathbf{r}_1 - (\hat{\mathbf{r}}\cdot d\mathbf{r}_1)\mathbf{r}_1 = \tfrac{1}{2}I \oint_{C_1} (\mathbf{r}_1 \times d\mathbf{r}_1) \times \hat{\mathbf{r}} \tag{117}$$

By comparison with (111) we write, for (113),

$$\blacktriangleright \qquad \mathbf{A}_1(\mathbf{r}) = \frac{\mu_0}{4\pi}\,\frac{\hat{\mathbf{r}}\cdot\mathbf{T}}{r^2} = \frac{\mu_0}{4\pi}\,\frac{\boldsymbol{\mu}\times\hat{\mathbf{r}}}{r^2} \tag{118}$$

For a steady solenoidal current distribution we write for the local current density:

$$\mathbf{j}(\mathbf{r})\,dV = \sum_i I_i\,d\mathbf{r}_i \tag{119}$$

which allows us to write:

$$\mathbf{T} = \int_{V_1} \mathbf{r}_1 \mathbf{j}(\mathbf{r}_1)\, dV_1 = \tfrac{1}{2} \int_{V_1} [\mathbf{r}_1 \mathbf{j}(\mathbf{r}_1) - \mathbf{j}(\mathbf{r}_1)\mathbf{r}_1]\, dV_1 \tag{120}$$

In this way we may write **(118)** for the general case with the magnetic moment given by **(110)**.

Exercises

18. Show that **(82)** and **(118)** are equivalent.

19. Use **(C-8)** with $\mathbf{A} = \mathbf{j}(\mathbf{r}_1)$ and $\mathbf{B} = \mathbf{C} = \mathbf{r}_1$ to demonstrate that the tensor:

$$\mathbf{T} = \int_{V_1} \mathbf{r}_1 \mathbf{j}(\mathbf{r}_1)\, dV_1 = -\int_{V_1} \mathbf{j}(\mathbf{r}_1)\mathbf{r}_1\, dV_1$$

is antisymmetric.

19 Potential and Field of a Dipole. We introduce the magnetic dipole as the limit of a current ring with R going to zero and I to infinity such that the magnetic moment **(111)** remains finite. In this limit **(118)** is valid to $r = 0$:

$$\mathbf{A}(\mathbf{r}) = \frac{\mu_0}{4\pi}\, \boldsymbol{\mu} \times \frac{\hat{\mathbf{r}}}{r^2} = -\frac{\mu_0}{4\pi}\, \boldsymbol{\mu} \times \boldsymbol{\nabla}\frac{1}{r} \tag{121}$$

To obtain the magnetic field $\mathbf{B}(\mathbf{r})$ we take the curl of **(121)**:

$$\mathbf{B}(\mathbf{r}) = \boldsymbol{\nabla} \times \mathbf{A}(\mathbf{r}) = -\frac{\mu_0}{4\pi}\, \boldsymbol{\nabla} \times \left(\boldsymbol{\mu} \times \boldsymbol{\nabla}\frac{1}{r} \right) = \frac{\mu_0}{4\pi}\, (\boldsymbol{\mu} \times \boldsymbol{\nabla}) \times \boldsymbol{\nabla}\frac{1}{r}$$

$$= -\frac{\mu_0}{4\pi}\left[\boldsymbol{\mu}\nabla^2 \frac{1}{r} + (\boldsymbol{\mu} \cdot \boldsymbol{\nabla})\frac{\mathbf{r}}{r^3} \right] \tag{122}$$

Using **(1-47)** and expanding the second term we write for the field of a magnetic current moment:

$$\blacktriangleright \qquad \mathbf{B}(\mathbf{r}) = \frac{\mu_0}{4\pi} \frac{3\hat{\mathbf{r}}(\boldsymbol{\mu} \cdot \hat{\mathbf{r}}) - \boldsymbol{\mu}}{r^3} + \mu_0 \boldsymbol{\mu}\, \delta(\mathbf{r}) \tag{123}$$

The first term of **(123)** is exactly analogous to the field of the electric dipole **(1-96)**. What, however, is the origin of the second term? The difference between the electric and magnetic dipole fields is illustrated in Figure 17. In Figure 17a we show the field of the electric dipole, which diverges from the charges. The field of a magnetic dipole, however, has no divergence and is positive at the origin. The two fields are identical except at the origin with the difference between the fields expressed by the Dirac delta function. Because the first term in **(123)** also diverges at $r = 0$ with an integrable singularity we must be extremely careful in evaluating the field at $r = 0$ or in integrating over the origin. This problem was discussed for the limiting case of a uniformly polarized ellipsoid in Problem 1-33. It is discussed for the corresponding magnetic case in Problem 29.

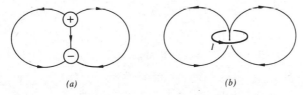

(a)   *(b)*

Figure 17   Magnetic fields of charge and current dipoles. (*a*) The field of a charge dipole diverges from the charges. (*b*) The field of a current dipole has no divergence.

Exercises

20. Show from **(122)** that the flux through a hemisphere of radius R and bounded by the plane normal to the dipole is:

$$\Phi = \tfrac{1}{2}\mu_0 \frac{\mu}{R}$$

21. Show that the integral of **(123)** for $r > 0$ gives zero. Thus, any contribution to the integral over all space must come from the origin.

22. Show from **(122)** or **(123)** that the line integral of **B(r)** between two positions at infinite distance from a current loop is equal to zero as long as the contour does not pass through the loop:

$$\int_C \mathbf{B(r)} \cdot d\mathbf{r} = 0$$

20   Forces and Torques on Magnetic Moments. In general, we have for the force on a distribution of currents:

$$\mathbf{F} = \int_V \mathbf{j(r)} \times \mathbf{B(r)}\, dV \qquad (124)$$

We wish to express **(124)** in terms of the moments of the current distribution and derivatives of the external magnetic field **B**. To do this we first write **(124)** for a current I flowing on a single contour:

$$\mathbf{F} = I \oint_C d\mathbf{r} \times \mathbf{B} \qquad (125)$$

From **(125)** we have by Stokes's theorem **(B-16)**:

$$F = I \int_S (d\mathbf{S} \times \mathbf{V}) \times \mathbf{B} \tag{126}$$

If **B(r)** is changing slowly we may take **B** out of the integral to obtain:

▶
$$F = (\boldsymbol{\mu} \times \mathbf{V}) \times \mathbf{B} \tag{127}$$

where we have used **(111)** for the magnetic moment.

For a volume distribution of currents we use **(119)** to go from **(125)** to **(124)**. We write then for **(126)**:

$$F = \sum_i I_i \int_{S_i} (d\mathbf{S}_i \times \mathbf{V}) \times \mathbf{B} \tag{128}$$

Again, if **B** is changing slowly throughout the distribution of currents we have **(127)** with:

$$\boldsymbol{\mu} = \sum_i \boldsymbol{\mu}_i = \sum_i I_i \mathbf{S}_i \tag{129}$$

By using the vector identities **(B-6)** and **(B-7)** we may write **(127)** as

▶
$$F = (\boldsymbol{\mu} \cdot \mathbf{V})\mathbf{B} - \boldsymbol{\mu}(\mathbf{V} \cdot \mathbf{B}) + \boldsymbol{\mu} \times (\mathbf{V} \times \mathbf{B}) \tag{130}$$

We had in Chapter 1 for the analogous polarization problem:

$$F = q\mathbf{E} + (\mathbf{p} \cdot \mathbf{V})\mathbf{E} \tag{131}$$

In the present case we have a current distribution with no monopole component. We expect by comparison with the electric field problem only the first term of **(130)**. The two additional terms, which arise in the presence of magnetic charge or external currents, are shown in Figure 18a and 18b.

If there is a magnetic charge g within a current loop as shown in Figure 18a, then there will be a net force downward on the loop. If an external current I flows out from from the paper as shown in Figure 18b, there will be a force to the right. In the absence of magnetic charge or external currents, we have only the first term, which is the same as for the electric dipole.

Our present considerations are limited to steady solenoidal currents as sources of the magnetic field **B**. As we will see in Section 11-8, where the currents are not solenoidal and there is accumulation of charge we may obtain locally a *curl* of **B** even in the absence of currents. Under such circumstances the *form* of the force on a current ring will be different from that on a magnetic charge dipole even in the absence of external currents.

The torque on a current distribution may be written as:

$$\boldsymbol{\tau} = \int_V \mathbf{r} \times [\mathbf{j}(\mathbf{r}) \times \mathbf{B}] \, dV \tag{132}$$

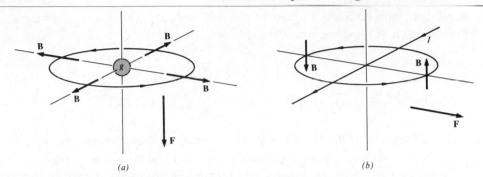

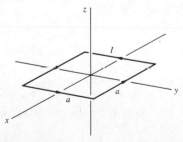

Figure 18   Force on a magnetic moment. (*a*) The presence of magnetic charge at the position of the moment results in a downward force. (*b*) The presence of current results in a force in the plane of the moment.

Writing **(132)** for a current I flowing on a contour C_1 we have:

$$\tau = I \oint_{C_1} \mathbf{r}_1 \times (d\mathbf{r}_1 \times \mathbf{B}) = I\mathbf{B} \cdot \oint_{C_1} \mathbf{r}_1 \, d\mathbf{r}_1 - I\mathbf{B} \oint_{C_1} \mathbf{r}_1 \cdot d\mathbf{r}_1 \tag{133}$$

as long as $\mathbf{B}(\mathbf{r})$ is slowly varying. The second integral is a perfect differential and vanishes over a closed path. The first integral is **(114)** so that we may write:

$$\tau = \mathbf{B} \cdot \mathbf{T} = \boldsymbol{\mu} \times \mathbf{B} \tag{134}$$

where the argument is the same as that leading to **(118)**. The argument for a general current distribution proceeds in the same way with the moment given by **(110)**.

Exercises

23. Show that when the magnetic moment $\boldsymbol{\mu}$ is parallel to $\mathbf{B}$ the force is given by $\mathbf{F} = \mu \nabla B$.

24. Show that for $\mathbf{j}(\mathbf{r}) = 0$ the tensor $\nabla \mathbf{B}$ is symmetric.

25. A current I flows around a square of side a lying in the xy plane. If the magnetic field is $\mathbf{B}(\mathbf{r})$ show that for $\mathbf{B}$ slowly varying in space, the force is

$$\mathbf{F} = \mu\left(\hat{\mathbf{y}} \times \frac{\partial \mathbf{B}}{\partial x} - \hat{\mathbf{x}} \times \frac{\partial \mathbf{B}}{\partial y}\right)$$

Expand this expression and show that it is equivalent to **(127)**.

26. A current I flows around a square of side a lying in the xy plane. If the magnetic field is slowly varying show that the torque is given by:

$$\boldsymbol{\tau} = \mu[\hat{\mathbf{x}} \times (\hat{\mathbf{y}} \times \mathbf{B}) - \hat{\mathbf{y}} \times (\hat{\mathbf{x}} \times \mathbf{B})]$$

Expand this expression and show that it is equivalent to **(134)**.

21   Examples of Magnetic Multiple Moments. As examples of magnetic multiple moments we discuss the monopole, dipole, quadrupole, and octupole moments of fundamental particles, nuclei, and atoms. In Table 1 we indicate some representative examples.

Moments are given in nuclear magnetons $\mu_N = \mu_B/1836.15 = 5.05082 \times 10^{-27}$ joule per tesla times the appropriate power of the metre (-1 for the monopole, 0 for the dipole, 2 for the octupole, etc.). A track etched from lexan sheets carried by a balloon in the upper atmosphere has been interpreted as possibly arising from a magnetic monopole.[7]

Table 1   Magnetic Multipole Moments in Nuclear Magnetons

	Spin	Monopole	Dipole	Octupole
g_0		6.51595×10^{18} m^{-1}	0	0
e	1/2	0	-1832.2	0
n	1/2	0	-1.91315	0
p	1/2	0	2.792777	0
d	1	0	0.857409	0
He^4	0	0	0	0
Li^6	1	0	0.822014	0
I^{127}	5/2	0	2.8094	0.3×10^{-28} m^2

Nuclei are expected to exhibit only odd magnetic multipole moments (dipole, octupole, etc.). The largest magnetic multipole that a nucleus can have is equal to twice the nuclear spin. Thus nuclei of spin 0 may have no moments; nuclei of spin $\frac{1}{2}$ and 1 may have only magnetic dipole moments; nuclei of spin $\frac{3}{2}$ and 2 may have both dipole and octupole moments, and so on.[8]

The determination of the nuclear magnetic octupole moment of I^{127} has been performed by the atomic beam magnetic resonance method.[9] The only other reported study is for In^{115}, which has a spin of $\frac{9}{2}$.[10]

---

[7] P. B. Price, E. B. Shirk, W. Z. Osborne, and L. S. Pinsky, *Phys. Rev. Lett.*, **35**, 487 (1975). For a discussion of ionization by magnetic monopoles, see R. Snyder, *Am. J. Phys.* **44**, 1181 (1976). For the alternate proposal of a heavy antimatter nucleus, see R. Hagstrom, *Phys. Rev. Lett.* **38**, 729 (1977).

[8] See, for example, H. Frauenfelder and E. M. Henley, *Subatomic Particles*, Prentice-Hall, 1974.

[9] V. Jaccarino, J. G. King, R. A. Satten, and H. H. Stroke, *Phys. Rev.*, **94**, 1798 (1954).

[10] P. Kusch and T. G. Eck, *Phys. Rev.*, **94**, 1799 (1954).

MACROSCOPIC MAGNETIC FIELD

The present discussion parallels the earlier discussion of the macroscopic electric field. Our expressions for the vector potential **(49)** and for the magnetic field **(40)** assume a precise knowledge of the distribution of currents. If we are well removed from these currents our discussion of moments shows that the magnetic field arises primarily from the dipole moment of the current distribution. Another situation for which we need not know the currents precisely is where we wish the average value of $\mathbf{B}(\mathbf{r})$ over some region of space.

22 Average Magnetic Field. To make the discussion explicit, we ask for the average of the magnetic field over a spherical volume

$$\langle \mathbf{B} \rangle_V = \frac{1}{V} \int_V \mathbf{B}(\mathbf{r}) \, dV \tag{135}$$

with $V - (4\pi/3)R^3$. Using **(40)** for $\mathbf{B}(\mathbf{r})$ we obtain:

$$\langle \mathbf{B} \rangle_V = \frac{1}{V} \int_V dV \frac{\mu_0}{4\pi} \int_{V_1} \frac{\mathbf{j}(\mathbf{r}_1) \times \hat{\mathbf{r}}_{01}}{r_{01}{}^2} \, dV_1 \tag{136}$$

Reversing the order of integration yields:

$$\langle \mathbf{B} \rangle_V = \int_{V_1} dV_1 \frac{\mu_0}{4\pi V} \int_V \frac{\mathbf{j}(\mathbf{r}_1) \times \hat{\mathbf{r}}_{01}}{r_{01}{}^2} \, dV \tag{137}$$

We evaluate **(137)** by an argument very similar to that used in Chapter 1 to find the average electric field over a sphere, where we were able to use previously obtained results for the electric field of a uniformly charged sphere. We notice in a similar way that the second integral in **(137)** is just the negative of the field at $\mathbf{r}_1$ from a uniform current $\mathbf{j}(\mathbf{r}_1)$ distributed over a sphere:

$$\mathbf{B}(\mathbf{r}_1) = \frac{\mu_0}{4\pi} \int_V \frac{\mathbf{j}(\mathbf{r}_1) \times \hat{\mathbf{r}}_{10}}{r_{10}{}^2} \, dV = \frac{\mu_0}{4\pi} \mathbf{j}(\mathbf{r}_1) \times \int_V \frac{\hat{\mathbf{r}}_{10}}{r_{10}{}^2} \, dV \tag{138}$$

As we have seen, the value of $\mathbf{B}(\mathbf{r}_1)$ depends on whether r_1 is less than or greater than R, the radius of the current sphere. For r_1 less than R we have, from **(25)**,

$$\mathbf{B}(\mathbf{r}_1) = \frac{\mu_0}{3} \mathbf{j}(\mathbf{r}_1) \times \mathbf{r}_1 \tag{139}$$

For r_1 greater than R the magnetic field is the same, as we have seen, as the field of a current $\mathbf{J} = \mathbf{j}(\mathbf{r}_1)V$ at the center of the sphere. We may now write **(137)** as the sum of two terms, the contribution to the average field from interior currents plus the contribution from exterior currents:

$$\langle \mathbf{B} \rangle_V = -\frac{\mu_0}{3V} \int_V \mathbf{j}(\mathbf{r}_1) \times \mathbf{r}_1 \, dV - \frac{\mu_0}{4\pi} \int_{V_1 - V} \frac{\mathbf{j}(\mathbf{r}_1) \times \mathbf{r}_1}{r_1{}^2} \, dV_1 \tag{140}$$

From **(110)** we may write the first integral in terms of the magnetic moment of those currents contained within the sphere. The second term, which we call $\mathbf{B}_{\text{ext}}$ is the field at the center of the spherical cavity from exterior currents. Finally, we may write for the average magnetic field within a spherical volume:

$$\blacktriangleright \qquad \langle \mathbf{B} \rangle_V = \mathbf{B}_{\text{ext}} + \tfrac{2}{3}\mu_0 \, \frac{\mu}{V} \tag{141}$$

23 **Magnetization Density.** We define a macroscopic field quantity, the magnetization density, from the relation:

$$\mathbf{M}(\mathbf{r}) = \frac{\mu}{V} \tag{142}$$

where μ is the magnetic moment of a sphere of radius R centered at $\mathbf{r}$. From **(110)** we have then:

$$\blacktriangleright \qquad \mathbf{M}(\mathbf{r}) = \frac{1}{2V_1} \left[\int_{V_1} \mathbf{r}_1 \times \mathbf{j}(\mathbf{r}_1) \, dV_1 + \oint_{S_1} \mathbf{r}_1 \times \mathbf{i}(\mathbf{r}_1) \, dS_1 \right] \tag{143}$$

where $\mathbf{i}(\mathbf{r}_1)$ is a surface current. The size of the region over which we average to obtain $\mathbf{M}(\mathbf{r})$ depends on the problem. In a magnetic solid the appropriate volume is usually an atomic cell. We may, however, let V_1 in **(143)** go to zero so that $\mathbf{M}(\mathbf{r})$ becomes a microscopic field quantity. By regarding $\mathbf{M}(\mathbf{r})$ in this way we find it possible to treat magnetic problems using scalar potential theory.

For example, we may use $\mathbf{M}(\mathbf{r})$ to write **(127)** and **(134)**, the expressions for the force and torque on a magnetic dipole, for an extended current or magnetization distribution. We replace the magnetic moment μ by $\mathbf{M}dV$. Then **(127)** becomes:

$$d\mathbf{F} = dV(\mathbf{M} \times \nabla) \times \mathbf{B} \tag{144}$$

and the force on an extended magnetization distribution becomes

$$\mathbf{F} = \int_V dV(\mathbf{M} \times \nabla) \times \mathbf{B} \tag{145}$$

For the torque we write, from **(134)**,

$$d\boldsymbol{\tau} = \mathbf{M}(\mathbf{r}) \times \mathbf{B}(\mathbf{r}) \, dV \tag{146}$$

Integrating, we obtain:

$$\boldsymbol{\tau} = \int_V \mathbf{M}(\mathbf{r}) \times \mathbf{B}(\mathbf{r}) \, dV \tag{147}$$

24 **Magnetization Current.** We generalize **(118)**, which is the vector potential of a magnetic dipole:

$$\blacktriangleright \qquad A(r) = \frac{\mu_0}{4\pi} \int_{V_1} \frac{M(r_1) \times \hat{r}_{01}}{r_{01}{}^2} \, dV_1 \qquad (148)$$

We transform the integrand of **(148)** by two vector identities:

$$M(r_1) \times \frac{\hat{r}_{01}}{r_{01}{}^2} = M(r_1) \times \nabla_1 \frac{1}{r_{01}} = \frac{\nabla_1 \times M(r_1)}{r_{01}} - \nabla_1 \times \frac{M(r_1)}{r_{01}} \qquad (149)$$

Substituting **(149)** into **(148)** and using **(B-13)** we have:

$$A(r) = \frac{\mu_0}{4\pi} \int_{V_1} \frac{\nabla_1 \times M(r_1)}{r_{01}} \, dV_1 + \frac{\mu_0}{4\pi} \int_{S_1} \frac{M(r_1) \times dS_1}{r_{01}} \qquad (150)$$

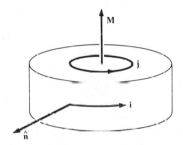

Figure 19 Magnetization current. The volume current flows azimuthally within the magnetized medium and the surface current flows on those surfaces that contain a component of the magnetization.

Comparing **(150)** with **(49)** we identify the curl of the magnetization with the current:

$$\blacktriangleright \qquad j_M(r) = \nabla \times M(r) \qquad (151)$$

and the cross product of the magnetization and the surface normal with a surface current density:

$$\blacktriangleright \qquad i_M(r) = M(r) \times \hat{n} \qquad (152)$$

The surface and volume current are represented in Figure 19.

For the magnetization equal to zero on S_1 we are left with only the volume term. There are, however, two cases where we must include a surface term:

1. If $A(r)$ represents the potential from only part of the magnetization, then there may well be magnetization on S_1.

2. The more usual use of the surface term is for a body whose magnetization drops discontinuously to zero at the bounding surface, which is taken to be identical with S_1.

In Section 1-17 we discussed similar cases for the scalar potential of a volume polarization density. The arguments appropriate to the two kinds of situations are quite similar.

Any solenoidal current may be regarded as a magnetization current. For example, we obtained in Section 1-20 for the polarization current **(1-149)**:

$$\mathbf{j}_P(\mathbf{r}) = \frac{\partial \mathbf{P}}{\partial t} + \mathbf{\Omega}(\mathbf{r}) \tag{153}$$

where $\mathbf{\Omega}(\mathbf{r})$ is a solenoidal vector. It is usual to treat $\mathbf{\Omega}(\mathbf{r})$ as arising from a magnetization:

$$\mathbf{\Omega}(\mathbf{r}) = \nabla \times \mathbf{M} \tag{154}$$

which gives on substitution into **(153)** and using **(151)**:

$$\mathbf{j}_P(\mathbf{r}) = \frac{\partial \mathbf{P}}{\partial t} + \nabla \times \mathbf{M} = \frac{\partial \mathbf{P}}{\partial t} + \mathbf{j}_M(\mathbf{r}) \tag{155}$$

For the particular case of a uniformly moving dielectric we have from **(1-153)**:

$$\mathbf{j}_P(\mathbf{r}) = \frac{\partial \mathbf{P}}{\partial t} - \nabla \times (\mathbf{v} \times \mathbf{P}) \tag{156}$$

Comparing **(155)** and **(156)** we may write:

▶
$$\mathbf{M} = -\mathbf{v} \times \mathbf{P} \tag{157}$$

that is, a moving dielectric generates magnetization current and is therefore equivalent to a magnetization.

It is not essential that we regard $\mathbf{\Omega}(\mathbf{r})$ as the curl of a magnetization although this is conventionally done with the result that the field equations in moving media are considerably simplified.

Exercises

27. Substitute **(151)** into **(104)** and show that the monopole moment of steady solenoidal currents is zero. You will need **(B-13)**.
28. Substitute **(151)** and **(152)** into **(143)** to obtain an identity. The proof is long and will require the use of **(B-6)**, **(B-12)**, **(C-4)**, and **(C-9)**.
29. Substitute **(151)** into **(124)** to obtain **(130)**. The proof is long and you will require **(B-6)**, **(C-4)**, and **(C-9)**. Assume that **B** is slowly varying in space.

25 **Magnetic Scalar Potential.** We now show how, by use of the magnetization density, it is possible to find a *magnetic scalar potential*. We write the magnetic field as the sum of two field quantities:

$$\mathbf{B}(\mathbf{r}) = [\mathbf{B}(\mathbf{r}) - \mu_0 \mathbf{M}(\mathbf{r})] + \mu_0 \mathbf{M}(\mathbf{r}) \tag{158}$$

We now examine the properties of the first field quantity $\mathbf{B}(\mathbf{r}) - \mu_0 \mathbf{M}(\mathbf{r})$. Taking the curl of this quantity, we obtain from **(48)** and **(151)**:

$$\nabla \times [\mathbf{B}(\mathbf{r}) - \mu_0 \mathbf{M}(\mathbf{r})] = 0 \tag{159}$$

Taking the divergence of this quantity we have from **(59)**:

$$\nabla \cdot [\mathbf{B}(\mathbf{r}) - \mu_0 \mathbf{M}(\mathbf{r})] = - \mu_0 \nabla \cdot \mathbf{M}(\mathbf{r}) \tag{160}$$

Now, **(159)** and **(160)** are very much like the field equations of electrostatics. If we regard $\mathbf{M}(\mathbf{r})$ as the magnetic equivalent of $\mathbf{P}(\mathbf{r})$, then the right side of **(160)** is to be interpreted as a magnetic charge density. Just as the electrostatic field has zero circulation and thus may be written as the gradient of a scalar, we may write, from **(159)**,

$$\blacktriangleright \qquad \mathbf{B}(\mathbf{r}) = - \nabla \overline{\phi}(\mathbf{r}) + \mu_0 \mathbf{M}(\mathbf{r}) \tag{161}$$

where, from **(160)**, we have the magnetic equivalent of Poisson's equation:

$$\nabla^2 \overline{\phi}(\mathbf{r}) = \mu_0 \nabla \cdot \mathbf{M}(\mathbf{r}) \tag{162}$$

Just as for the electrostatic field, we may write the solution of **(162)** as an integral over the magnetic charge:

$$\blacktriangleright \qquad \overline{\phi}(\mathbf{r}) = \frac{\mu_0}{4\pi} \int_{V_1} \frac{\overline{\rho}_M(\mathbf{r}_1)}{r_{01}} \, dV_1 + \frac{\mu_0}{4\pi} \int_{S_1} \frac{\overline{\sigma}_M(\mathbf{r}_1)}{r_{01}} \, dS_1 \tag{163}$$

where the volume magnetic charge density is given by:

$$\blacktriangleright \qquad \overline{\rho}_M(\mathbf{r}_1) = - \nabla_1 \cdot \mathbf{M}(\mathbf{r}_1) \tag{164}$$

and the surface magnetic charge density by:

$$\overline{\sigma}_M(\mathbf{r}_1) = \hat{\mathbf{n}} \cdot \mathbf{M} \tag{165}$$

Exercises

30. Show that the magnetic field B within a torus is given by:

$$B = \mu_0 M = \mu_0 n I$$

31. Take the curl of **(148)** and use **(54)** and **(163)** with **(1-47)** or **(B-75)**, **(C-9)**, and **(C-4)** to obtain **(161)**.

32. A sphere of radius R bounds a radial magnetization:

$$\mathbf{M}(\mathbf{r}) = M\hat{\mathbf{r}}/r^2$$

where M is constant. Show that **B** is zero both inside and outside the sphere.

---

26 Demagnetization Field. We consider a uniformly magnetized ellipsoid as shown in Figure 20.

We wish to find the macroscopic magnetic field **B**. As we have seen, a scalar potential $\bar{\phi}$ may be found for the function $\mathbf{B} - \mu_0 \mathbf{M}$. The relation between $\bar{\phi}$ and the magnetization density **M** is analogous to that between ϕ and the polarization density **P** as discussed in Section 1-17. In this way we may take over the results of Section 1-26 by replacing $\mathbf{E}(\mathbf{r})$ by $\mathbf{B}(\mathbf{r}) - \mu_0 \mathbf{M}$ and $\mathbf{P}/\epsilon_0$ by $\mu_0 \mathbf{M}$. We obtain then from **(1-181)**:

▶
$$\mathbf{B}(\mathbf{r}) = \mathbf{B}_{\text{ext}} + \mu_0 \mathbf{M} - \mu_0 \, \mathbf{N} \cdot \mathbf{M} \tag{166}$$

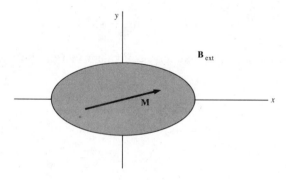

Figure 20   Uniformly magnetized ellipsoid. We wish to find the macroscopic magnetic field **B**.

where **N** is here called the *demagnetization* tensor. As for the electric problem, the magnetic field within a uniformly magnetized ellipsoid is itself uniform. As we saw in Section 1-22 the demagnetization tensor may be expressed in terms of principal values N_x, N_y, and N_z, which sum to unity:

$$N_x + N_y + N_z = 1 \tag{167}$$

The rod, the disc, and the sphere may be regarded as limiting cases of the general ellipsoid. The field $-\mu_0 \, \mathbf{N} \cdot \mathbf{M}$, which is called the demagnetization field, may be found for these structures from the arguments of Section 1-26.

27 Moments of a Magnetization Distribution. We may develop a discussion of the expansion of the scalar potential $\bar{\phi}(\mathbf{r})$, which closely follows our earlier discussion of the moments of a charge distribution. We treat here the same problem as discussed earlier where we wrote the vector potential in terms of moments of the current distribution. The present discussion is more useful in that it avoids the complications of curls and cross products.

By expanding $1/r_{01}$ in (163) for $\mathbf{r}$ well away from the magnetization distribution we obtain:

$$\bar{\phi}(\mathbf{r}) = \bar{\phi}_0(\mathbf{r}) + \bar{\phi}_1(\mathbf{r}) + \bar{\phi}_2(\mathbf{r}) + \cdots \tag{168}$$

with the individual terms given by:

$$\bar{\phi}_0(\mathbf{r}) = -\frac{\mu_0}{4\pi r}\int_{V_1} \mathbf{\nabla}_1 \cdot \mathbf{M}(\mathbf{r}_1)\, dV_1 \tag{169}$$

$$\bar{\phi}_1(\mathbf{r}) = -\frac{\mu_0\,\hat{\mathbf{r}}}{4\pi r^2}\cdot\int_{V_1} \mathbf{r}_1\mathbf{\nabla}_1 \cdot \mathbf{M}(\mathbf{r}_1)\, dV_1 \tag{170}$$

$$\bar{\phi}_2(\mathbf{r}) = -\frac{\mu_0}{8\pi r^3}\int_{V_1} r_1{}^{\boldsymbol{\cdot}}[3(\hat{\mathbf{r}}\cdot\hat{\mathbf{r}}_1)^{\boldsymbol{\cdot}} - 1]\mathbf{\nabla}_1 \cdot \mathbf{M}(\mathbf{r}_1)\, dV_1 \tag{171}$$

We may convert (169) to a surface integral over S_1. With no magnetization on the bounding surface, the integral is zero and we obtain zero for the monopole contribution to the scalar potential. For (170) we obtain the dipole moment of the distribution:

$$\boldsymbol{\mu} = \int_{V_1} \mathbf{r}_1\mathbf{\nabla}_1 \cdot \mathbf{M}(\mathbf{r}_1)\, dV_1 = \int_{V_1} \mathbf{M}(\mathbf{r}_1)\, dV_1 \tag{172}$$

where we have used (C-4) and (C-9). The next term in the expansion gives the quadrupole moment, and so on. Finally, we may write $\mathbf{B}(\mathbf{r})$ as a series expansion:

$$\mathbf{B}(\mathbf{r}) = -\mathbf{\nabla}\bar{\phi}_1(\mathbf{r}) - \mathbf{\nabla}\bar{\phi}_2(\mathbf{r}) - \cdots \tag{173}$$

when we have assumed that $\mathbf{M}(\mathbf{r}) = 0$ well away from the current distribution.

Exercise

33. From (171) obtain the expression for the magnetic quadrupole moment $\mathbf{Q}_M$ of a magnetization distribution with:

$$\bar{\phi}_2(\mathbf{r}) = \frac{\mu_0}{8\pi}\frac{\hat{\mathbf{r}}\cdot\mathbf{Q}_M\cdot\hat{\mathbf{r}}}{r^3}$$

SUMMARY

1. The electric field of a uniformly moving charge q_1 is given by:

$$\mathbf{E}(\mathbf{r},t) = \frac{1}{4\pi\epsilon_0} \frac{q_1}{\gamma^2 r_{01}{}^2 (1 - \beta^2 \sin^2 \theta)^{3/2}} \hat{\mathbf{r}}_{01} \qquad (13)$$

where θ is the angle between $\hat{\mathbf{r}}_{01}$ and $\mathbf{v}$.

2. The total flux as measured in the laboratory frame continues to be q_1/ϵ_0, which leads to a generalization of the differential form of Gauss's law:

$$\text{div } \mathbf{E}(\mathbf{r},t) = \frac{1}{\epsilon_0} \rho(\mathbf{r},t) \qquad (18)$$

3. A moving charge also produces a magnetic field:

$$\mathbf{B}(\mathbf{r},t) = \frac{1}{c^2} \mathbf{v}_1 \times \mathbf{E}(\mathbf{r},t) \qquad (22)$$

where the magnetic field is defined by the Lorentz force on a moving test charge:

$$\mathbf{F}(\mathbf{r},t) = q\mathbf{E}(\mathbf{r},t) + q\mathbf{v} \times \mathbf{B}(\mathbf{r},t) \qquad (20)$$

4. The magnetic field of a steady solenoidal current $\mathbf{j}(\mathbf{r}_1)$ is given by the law of Biot and Savart:

$$\mathbf{B}(\mathbf{r}) = \frac{\mu_0}{4\pi} \int_{V_1} \frac{\mathbf{j}(\mathbf{r}_1) \times \hat{\mathbf{r}}_{01}}{r_{01}{}^2} dV_1 \qquad (40)$$

where $\mu_0 = 1/\epsilon_0 c^2$ is the permeability of vacuum. This result is most easily obtained by decomposing $\mathbf{j}(\mathbf{r}_1)$ into straight current filaments.

5. From the filamentary decomposition of $\mathbf{j}(\mathbf{r}_1)$ we obtain Ampère's law:

$$\text{curl } \mathbf{B} = \nabla \times \mathbf{B} = \mu_0 \mathbf{j} \qquad (48)$$

6. The vector potential of a steady solenoidal current may be defined as

$$\mathbf{A}(\mathbf{r}) = \frac{\mu_0}{4\pi} \int_{V_1} \frac{\mathbf{j}(\mathbf{r}_1)}{r_{01}} dV_1 \qquad (49)$$

from which the magnetic field $\mathbf{B}$ may be obtained:

$$\mathbf{B}(\mathbf{r}) = \nabla \times \mathbf{A}(\mathbf{r}) \qquad (54)$$

and from which it follows that $\mathbf{B}(\mathbf{r})$ is solenoidal:

$$\nabla \cdot \mathbf{B}(\mathbf{r}) = 0 \tag{59}$$

7. The field of a long current-carrying cylinder of radius R and uniform current density j is azimuthal and outside the cylinder is given by

$$\mathbf{B}(\rho) = \frac{\mu_0 I}{2\pi\rho} \hat{\boldsymbol{\phi}} \tag{68}$$

while inside the field is given by:

$$\mathbf{B}(\rho) = \tfrac{1}{2}\mu_0 j\rho\hat{\boldsymbol{\phi}} \tag{67}$$

The vector potential is parallel to the axis of the cylinder and is given outside by.

$$\mathbf{A}(\rho) = -\frac{\mu_0}{2\pi} I \ln \frac{\rho}{R} \hat{\mathbf{z}} \tag{70}$$

where the zero of potential is taken arbitrarily at the surface of the cylinder. Inside the cylinder, the potential is given by

$$\mathbf{A}(\rho) = \frac{\mu_0}{4\pi} I \left(1 - \frac{\rho^2}{R^2}\right)\hat{\mathbf{z}} \tag{71}$$

8. The vector potential of a current ring of radius R and current I is azimuthal and is given well away from the ring by

$$\mathbf{A}(\mathbf{r}) = \tfrac{1}{4}\mu_0 IR^2 \frac{r \sin \theta}{(r^2 + R^2)^{3/2}} \hat{\boldsymbol{\phi}} \tag{78}$$

where θ is the angle between the normal to the ring and $\hat{\mathbf{r}}$.
9. The magnetic field on the axis of the ring is given by:

$$\mathbf{B}(z) \simeq \tfrac{1}{2}\mu_0 I \frac{R^2}{(z^2 + R^2)^{3/2}} \hat{\mathbf{z}} \tag{81}$$

At large distance from the ring and in a general direction the field is written in spherical coordinates as

$$\mathbf{B}(r,\theta) = \tfrac{1}{4}\mu_0 IR^2 \frac{2\hat{\mathbf{r}} \cos \theta + \hat{\boldsymbol{\theta}} \sin \theta}{r^3} \tag{84}$$

which is of the same form as the electric field of a dipole.

10. The magnetic field within a long solenoid of n turns per metre is uniform and is given by:

$$\mathbf{B} = \mu_0 \, nI\hat{\mathbf{z}} \tag{85}$$

The magnetic field outside is zero.

11. The vector potential outside a long solenoid is given by

$$\mathbf{A}(\rho) = \tfrac{1}{2}\mu_0 \, nI \, \frac{R^2}{\rho} \, \hat{\boldsymbol{\phi}} \tag{88}$$

and inside by

$$\mathbf{A}(\rho) = \tfrac{1}{2}\mu_0 \, nI\rho\hat{\boldsymbol{\phi}} \tag{87}$$

where the advance of the current has been neglected.

12. The magnetic field of a torus is given within the windings by:

$$\mathbf{B} = \frac{\mu_0}{2\pi} \frac{NI}{\rho} \, \hat{\boldsymbol{\phi}} \tag{93}$$

The field outside is zero.

13. The vector potential on the axis of a torus is given by:

$$\mathbf{A}(z) = \tfrac{1}{2}\Phi \, \frac{R^2}{(z^2 + R^2)^{3/2}} \, \hat{\mathbf{z}} \tag{99}$$

where $\Phi = \int \mathbf{B} \cdot d\mathbf{S}$ is the flux of the torus. The vector potential at large distance and in a general direction is given by:

$$\mathbf{A}(r,\theta) \simeq \tfrac{1}{4}\Phi R^2 \, \frac{2\hat{\mathbf{r}} \cos\theta + \hat{\boldsymbol{\theta}} \sin\theta}{r^3} \tag{101}$$

14. From the analogous relations between the vector potential $\mathbf{A}$, the magnetic field $\mathbf{B}$, and the current density $\mathbf{j}$, we may write the vector potential $\mathbf{A}$ as:

$$\mathbf{A}(\mathbf{r}) = \frac{1}{4\pi} \int_{V_1} \frac{\mathbf{B}(\mathbf{r}_1) \times \hat{\mathbf{r}}_{01}}{r_{01}{}^2} \, dV_1 \tag{96}$$

From this relation the vector potential may be obtained from the magnetic field $\mathbf{B}$ without the necessity of having to find the current density.

15. The vector potential of a bounded distribution of currents may be expanded in powers of $1/r$, which gives the moments of the distribution. The lowest moment, the monopole moment, is zero for solenoidal currents. The dipole moment of the distribution is written as:

$$\boldsymbol{\mu} = \tfrac{1}{2} \int_{V_1} \mathbf{r}_1 \times \mathbf{j}(\mathbf{r}_1) \, dV_1 \tag{110}$$

The vector potential arising from the dipole moment is written as:

$$\mathbf{A}_1(r) = \frac{\mu_0}{4\pi} \frac{\boldsymbol{\mu} \times \hat{\mathbf{r}}}{r^2} \tag{118}$$

16. Magnetic multipoles are limiting current distributions that produce the fields of multipole moments to $r = 0$. The magnetic dipole is the limit of a current ring of area $S = \pi R^2$ and current I as R goes to zero and I to infinity with the product $\mu = IS$ remaining constant. The field of a magnetic dipole is given by

$$\mathbf{B}(\mathbf{r}) = \frac{\mu_0}{4\pi} \frac{3\hat{r}(\boldsymbol{\mu} \cdot \hat{\mathbf{r}}) - \boldsymbol{\mu}}{r^3} + \mu_0 \boldsymbol{\mu} \delta(\mathbf{r}) \tag{123}$$

The first term is of the same form as that of an electric dipole. The second term is nonvanishing only at $r = 0$, establishing the difference between the field of a charge dipole and a current ring.

17. The force on a steady, solenoidal current distribution in a slowly varying external field is given by:

$$\mathbf{F} = (\boldsymbol{\mu} \cdot \nabla)\mathbf{B} - \boldsymbol{\mu}(\nabla \cdot \mathbf{B}) + \boldsymbol{\mu} \times (\nabla \times \mathbf{B}) \tag{130}$$

where $\boldsymbol{\mu}$ is the dipole moment of the distribution. The first term is analogous to the force on an electric dipole. The second term arises from a magnetic charge density within the current distribution while the third term arises from an external *current* density within the distribution.

18. The torque on a steady, solenoidal current distribution is given by:

$$\boldsymbol{\tau} = \boldsymbol{\mu} \times \mathbf{B} \tag{134}$$

where again $\boldsymbol{\mu}$ is the dipole moment of the distribution.

19. The macroscopic magnetic field is defined as the mean field $\mathbf{B}$ over a sphere of radius R. The macroscopic field may be expressed as

$$\langle \mathbf{B} \rangle_V = \mathbf{B}_{ext} + \tfrac{2}{3}\mu_0 \frac{\boldsymbol{\mu}}{V} \tag{141}$$

where $\mathbf{B}_{ext}$ is the field at the center of the sphere from currents outside the sphere, $\boldsymbol{\mu}$ is the dipole moment of currents *within* the sphere, and V is the volume of the sphere.

20. The mean magnetization density of a current distribution is given by

$$\mathbf{M}(\mathbf{r}) = \frac{\mu}{V} = \frac{1}{2V_1}\left[\int_{V_1} \mathbf{r}_1 \times \mathbf{j}(\mathbf{r}_1)\, dV_1 + \oint_{S_1} \mathbf{r}_1 \times \mathbf{i}(\mathbf{r}_1)\, dS_1\right] \qquad (143)$$

where $\mathbf{j}(\mathbf{r}_1)$ is a volume current density and $\mathbf{i}(\mathbf{r}_1)$ is a surface current density.

21. Writing the vector potential as an integral over a magnetization distribution

$$\mathbf{A}(\mathbf{r}) = \frac{\mu_0}{4\pi}\int_{V_1} \frac{\mathbf{M}(\mathbf{r}_1) \times \hat{\mathbf{r}}_{01}}{r_{01}{}^2}\, dV_1 \qquad (148)$$

leads to volume and surface magnetization currents:

$$\mathbf{j}_M = \nabla \times \mathbf{M}(\mathbf{r}) \quad (151) \qquad \mathbf{i}_M = \mathbf{M}(\mathbf{r}) \times \hat{\mathbf{n}} \qquad (152)$$

22. A uniformly moving polarization density generates a magnetization density:

$$\mathbf{M}(\mathbf{r}) = -\mathbf{v} \times \mathbf{P}(\mathbf{r}) \qquad (157)$$

23. A magnetic scalar potential may be defined by analogy with the electric scalar potential from the expression:

$$\bar{\phi}(r) = \frac{\mu_0}{4\pi}\int_{V_1} \frac{\bar{\rho}_M(\mathbf{r}_1)}{r_{01}}\, dV_1 + \frac{\mu_0}{4\pi}\int_{S_1} \frac{\bar{\sigma}_M(\mathbf{r}_1)}{r_{01}}\, dS_1 \qquad (163)$$

where $\bar{\rho}_M(\mathbf{r}_1)$ is a magnetic charge density given by:

$$\bar{\rho}_M(\mathbf{r}_1) = -\nabla_1 \cdot \mathbf{M}(\mathbf{r}_1) \qquad (164)$$

The magnetic field is given from the magnetic scalar potential by the relation

$$\mathbf{B}(\mathbf{r}) = -\nabla\bar{\phi}(\mathbf{r}) + \mu_0\mathbf{M}(\mathbf{r}) \qquad (161)$$

24. The internal field within a uniformly magnetized ellipsoid may be written as:

$$\mathbf{B} = \mathbf{B}_{\text{ext}} + \mu_0\mathbf{M} - \mu_0\,\mathbf{N}\cdot\mathbf{M} \qquad (166)$$

The first term is the external applied field. The second term is called the magnetic induction of the sample. The third term is the demagnetization field.

25. The scalar potential of a bounded magnetization distribution may be expanded in powers of $1/r$, giving the multipole moments of the magnetization. The moment expansion written in this form is much more tractable than written in terms of the current distribution.

PROBLEMS

1. **Moving disc.** A thin disc of radius R and charge density σ moves with velocity v parallel to its axis. Obtain the electric field on the axis.

2. **Moving shell.** A uniformly charged spherical shell moves in the x direction with velocity v_1. Show that the shell as observed in the laboratory is ellipsoidal. Show that the charge density on the moving shell is given by:

$$\sigma = \frac{\gamma q}{4\pi R^2}\left[1 + (\beta\gamma^2 x/R)^2\right]^{-1/2}$$

Compare this expression with the limiting surface charge density of an ellipsoidal homoeoid.

3. **Fields of a moving shell.** Find the electric and magnetic fields inside and outside the uniformly moving charge shell described in Problem 2.

4. **Moving charges.** A pair of charges q separated by $\mathbf{R}$ are in uniform motion with velocity $\mathbf{v}$ transverse to $\mathbf{R}$. Obtain an expression for the force of repulsion.

5. **Moving wires.** Two long parallel wires, separated by a distance D are each charged with λ coulombs per metre. Take the wires to be parallel to $\hat{\mathbf{x}}$.
   (a) Assuming the wires to lie in the xy plane, find the electric field at each wire resulting from the charge on the other wire. Find the force of repulsion in newtons per metre.
   (b) The wires are now set into motion with velocity $\mathbf{v}$ parallel to $\hat{\mathbf{x}}$. What is the charge density and current in the laboratory frame?
   (c) What electric field acts between the wires in the laboratory frame?
   (d) What magnetic field acts between the wires in the laboratory frame?
   (e) Obtain the force of interaction in the laboratory frame and compare with (a).

6. **Fields of a moving charge.** By taking the curl of **(22)** obtain the relation:

$$\text{curl } \mathbf{B}(\mathbf{r},t) = \frac{1}{c^2}\frac{\partial}{\partial t}\,\mathbf{E}(\mathbf{r},t)$$

By taking the curl of **(21)** obtain the relation:

$$\text{curl } \mathbf{E}(\mathbf{r},t) = -\frac{\partial}{\partial t}\,\mathbf{B}(\mathbf{r},t)$$

7. **Equation of motion.** A particle of relativistic mass m and charge q moves in an electric field $\mathbf{E}$ and magnetic field $\mathbf{B}$. Show that the equation of motion may be written as:

$$m\frac{d\mathbf{v}}{dt} = q\left(\mathbf{E} + \mathbf{v}\times\mathbf{B} - \frac{\mathbf{v}}{c^2}\,\mathbf{v}\cdot\mathbf{E}\right)$$

8. **Intersecting cylinders.** A pair of cylindrical conductors of radius R intersect as shown in the following diagram. If the currents are uniform and equal (but of

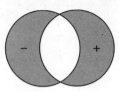

   opposite sign) and zero in the intersecting region, find the magnitude and direction of **B** in the intersecting region. Sketch the lines of flux.

9. **Saddle coils.** A uniform deflecting field in a cathode ray tube is produced by placing a pair of *cosine-wound* coils over the neck of the tube. Assuming a surface current density proportional to cosine θ on an infinitely long cylinder, show that the internal

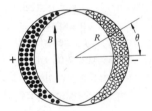

   field is uniform and find the value of the field in terms of the maximum surface current density and radius of the cylinder. Note that **i** is *parallel* to the axis of the cylinder. Compare with Problem 8. Such coils are called saddle coils because of the way in which the windings are brought back over the tube neck.

10. **Biot-Savart law.** Obtain the Biot-Savart law **(40)** by integrating Ampere's law **(48)** with the assumption **(59)** that the divergence of **B** is zero. Begin by writing **B** as the curl of the vector potential **(54)**. Write the integral form of **A** by analogy with the integral form of the scalar potential.

11. **Vector potential of a rotating shell.** A sphere of radius R carries a charge q, which is distributed uniformly over its surface. The sphere rotates with angular velocity $\boldsymbol{\omega}$. Show that the vector potential may be written in the form:

$$\mathbf{A}(\mathbf{r}) = -\frac{\mu_0}{4\pi}\frac{q}{4\pi R}\,\boldsymbol{\omega}\times\nabla\int_{V_1}\frac{dV_1}{r_{01}}$$

   From the potential of a uniformly charged sphere evaluate the integral both inside and outside the sphere. Show that outside the vector potential is that of a dipole. Show that within the shell the magnetic field is uniform and obtain its value.

12. **Short solenoid.** A solenoid of length L carries a current I and has n turns per metre. Show that the magnetic field on the axis is given by the expression:

$$B = \tfrac{1}{2}\mu_0\, nI(\cos\theta_1 + \cos\theta_2)$$

where θ_1 and θ_2 are the angles between the axis of the solenoid and lines from the axial position to the end windings.

13. **Helmholtz coils.** A pair of current rings of radius R are separated by a distance D as shown in the following diagram. Find the axial magnetic field near the midpoint

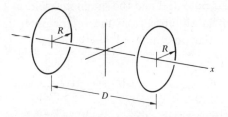

to order x^2. For what value of D is the field homogeneity greatest? Coils in this configuration are called Helmholtz coils.

14. **Vector potential of a uniform field.** Show that if the magnetic field **B** is uniform, the vector potential given by:

$$A(r) = \tfrac{1}{2}B \times r$$

is a solution of **(54)**.

15. **Symmetry of field and potential.** In general the vector potential and magnetic field may be written in cylindrical coordinates as

$$A = \hat{\rho}A_\rho(\rho,\phi,z) + \hat{\phi}A_\phi(\rho,\phi,z) + \hat{z}A_z(\rho,\phi,z)$$
$$B = \hat{\rho}B_\rho(\rho,\phi,z) + \hat{\phi}B_\phi(\rho,\phi,z) + \hat{z}B_z(\rho,\phi,z)$$

Because of the symmetry of certain distributions of currents, not all components of A and B will be present nor will those components that are present necessarily be a function of all three coordinates. Write the most general form allowed by symmetry for **A** and **B** for the following current configurations:

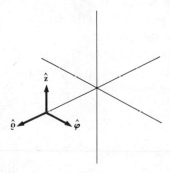

(a) Infinite straight wire.
(b) Finite solenoid—neglect axial current
(c) Infinite solenoid—neglect axial current
(d) Torus—neglect azimuthal current

16. **Magnetic shell.** The magnetic field of a current I flowing on a ring of radius R may be obtained by replacing the ring by a fictitious magnetic shell bounded by the ring. Replace the ring by a disc of thickness t and magnetization density $M = I/t$.
    (a) Show that **(159)** is satisfied.
    (b) Obtain the scalar potential and the magnetic field at large distances. Compare with the magnetic field obtained from the vector potential of the current loop.
    (c) Show that the scalar potential at $\mathbf{r}$ is given by:

$$\phi(\mathbf{r}) = \frac{\mu_0}{4\pi} I\Omega$$

where Ω is the solid angle subtended by the surface S at $\mathbf{r}$. Compare with Exercise 1-42.

17. **Magnetic quadrupole moment.** Obtain the magnetic quadrupole moment of a current I flowing around an ellipse of semiaxes A and B. Replace the ellipse by a magnetic shell of thickness t and magnetization density $M = I/t$ and use the result of Exercise 33.

18. **Magnetized cylinder.** A cylinder of radius R and length L carries a uniform axial magnetization $\mathbf{M}$. Find the magnetic field $\mathbf{B}$ on the axis both outside and inside the cylinder from the magnetic scalar potential. Compare with the field of a solenoid.

19. **Moving dielectric.** A dielectric disc of radius R moves parallel to its axis and normal to a magnetic field $\mathbf{B}$ with velocity $\mathbf{v}$. Show that a polarization density:

$$\mathbf{P} = \epsilon_0 \chi \mathbf{v} \times \mathbf{B}$$

is induced. Show in agreement with **(157)** that the polarization current may be associated with a magnetic moment:

$$\boldsymbol{\mu} = -\mathbf{v} \times \mathbf{p}$$

and a polarization current density $\partial \mathbf{P}/\partial t$.

20. **Rotating dielectric sphere.** A dielectric sphere of susceptibility χ and radius R rotates with angular frequency ω about an axis parallel to a magnetic field $\mathbf{B}$. Find the volume and surface polarization charge.

21. **Rotating spherical conductor.** A conducting sphere of volume V rotates at angular frequency ω about an axis parallel to a magnetic field $\mathbf{B}$. Show that within the sphere there is charge accumulation

$$q = 2\epsilon_0 \omega B$$

Find the density of surface charge.

22. **Work to rotate a current loop.** A current loop is rotated about the $\hat{\mathbf{z}}$ axis through 2π generating a closed volume V. Show that the work required to rotate the loop through 2π is given by

$$W = \oint \hat{\mathbf{z}} \cdot \boldsymbol{\tau} \, d\theta = -I \int_V \nabla \cdot \mathbf{B} \, dV$$

That is, the work is zero unless the volume includes net magnetic charge [C. H. Page, *Am. J. Phys.*, **42**, 490 (1974)].

23. **Magnetization of a rotating shell.** A sphere of radius R carries a charge q, which is distributed uniformly over its surface. The sphere rotates with angular velocity ω. Find the surface current $\mathbf{i}(\mathbf{r})$ and the equivalent volume magnetization. What is the volume magnetic charge density $\bar{\rho}(\mathbf{r})$ and the surface magnetic charge density $\bar{\sigma}(\mathbf{r})$. Show that the external magnetic field is the same as that of a dipole $\boldsymbol{\mu} = \mathbf{M}V$ at the origin. Show that the internal magnetic field $\mathbf{B}$ is uniform and obtain its value.

24. **Magnetized cylinder.** A cylinder of length L and radius R is uniformly magnetized along its axis. Find the form of the scalar potential on the axis. Show that the magnetic field at the center of a long cylinder is:

$$\mathbf{B} = \mu_0 \mathbf{M}(1 - 2R^2/L^2)$$

25. **Dipole field.** A magnetic dipole of strength μ is placed in a homogeneous magnetic field of strength B_{ext} and is oriented so as to *oppose* the field. Show that there is a spherical surface of radius R such that the normal component of the resultant field vanishes everywhere on the surface. Find the value of R in terms of μ and B_{ext}. What is the value of B at the equator of the spherical surface?

26. **Magnetic field of the earth.** Consider the magnetic field of the earth to arise from a uniformly magnetized sphere concentric with the earth and of radius less than the radius of the earth. Show that lines of $\mathbf{B}$ crossing the earth's surface at $45°$ longitude meet the equatorial plane at a distance from the center of the earth that is twice the earth's radius. Note that this result does not depend on the size of the magnetic region.

27. **Force on a magnetized body. I.**
    (a) From (130) show that the force on an extended magnetized body is given by:

$$\mathbf{F} = \int_V dV (\mathbf{M} \cdot \nabla)\mathbf{B}$$

in the absence of external currents.
    (b) Use (C-4) and (C-9) to obtain for the force the expression:

$$\mathbf{F} = -\int_V \mathbf{B}(\nabla \cdot \mathbf{M})\, dV + \oint_S (d\mathbf{S} \cdot \mathbf{M})\mathbf{B}$$

    (c) Use (164) and (165) to write this result in terms of magnetic charge.

28. **Force on a magnetized body. II.**
    (a) From (124) obtain for the force on a magnetized body the expression:

$$\mathbf{F} = \int_V (\nabla \times \mathbf{M}) \times \mathbf{B}\, dV - \oint_S \mathbf{B} \times (\mathbf{M} \times d\mathbf{S})$$

(b) Through the use of **(B-6)**, **(C-4)**, **(B-12)**, and **(C-9)** transform this expression to obtain:

$$\mathbf{F} = \int_V \bar{\rho}_M \mathbf{B} \, dV + \oint_S \bar{\sigma}_M \mathbf{B} \, d\mathbf{S})$$

It will be necessary to assume $\mathbf{V} \times \mathbf{B} = 0$ and $\mathbf{V} \cdot \mathbf{B} = 0$ as discussed in Section 20.

29. **Point moment.** The volume of a uniformly magnetized ellipsoid is allowed to shrink to zero while the magnetization density M increases so that the product $\mathbf{\mu} = \mathbf{M}V$ remains constant. Show that in the limit of zero volume the magnetic field is everywhere given by:

$$\mathbf{B} = \frac{\mu_0}{4\pi} \frac{3\hat{\mathbf{r}}(\hat{\mathbf{r}} \cdot \mathbf{\mu}) - \mathbf{\mu}}{r^3} - \mu_0(\mathbf{\mu} - \mathbf{N} \cdot \mathbf{\mu}) \, \delta(\mathbf{r})$$

Compare with **(123)** and with Problem 1-33. What is the limiting field of a magnetized sphere?

30. **Magnetic stress tensor.** Show that the expression for the force on a volume distribution of currents **(124)** may be written as an integral over the surface which bounds V:

$$\mathbf{F} = \oint_S d\mathbf{S} \cdot \mathbf{T}$$

where **T** is the magnetic stress tensor:

$$\mathbf{T} = (1/\mu_0)(\mathbf{BB} - \tfrac{1}{2}\mathbf{I}B^2)$$

Use **(48)**, **(C-1)**, **(C-3)**, **(C-4)** and **(59)**, **(C-9)**, **(B-12)**, and **(C-28)**.

REFERENCES

Berkson, W., *Fields of Force: The Development of a World View from Faraday to Einstein*, Wiley, 1974.

Furth, H. P., M. A. Levine, and R. W. Waniek, "Strong Magnetic Fields," *Scientific American*, **198**(2), 28 (February 1958).

Kolm, H. H. and A. J. Freeman, "Intense Magnetic Fields," *Scientific American*, **212**(4), 66 (April 1965).

Kolm, H. H., J. Oberteuffer, and D. Kellard, "High-Gradient Magnetic Separation," *Scientific American*, **233**(5), 46 (November 1975).

Purcell, E. M., *Electricity and Magnetism*, Berkeley physics course, Volume 2, McGraw-Hill, 1965.

7 MAGNETIC MEDIA

This chapter parallels Chapter 3, in which we discussed the properties of dielectric media. We begin in a similar way with the Bohr model of the hydrogen atom and the effect of a uniform axial magnetic field. We consider next the intrinsic moment of the electron followed by a discussion of transition elements. Finally, we discuss the properties of bulk magnetic media. We postpone to Chapter 10 a treatment of the interaction of magnetic media and external currents.

ATOMIC STRUCTURE

As we have seen in Section 6-2, a charge moving through an electric and magnetic field is acted on by the Lorentz force **(6-20)**:

$$\mathbf{F} = q(\mathbf{E} + \mathbf{v} \times \mathbf{B}) \tag{1}$$

We begin our discussion of magnetic media by examining the consequences of **(1)** for free atoms.

1 Hydrogen Atom. We first discuss the hydrogen atom on the basis of the Bohr model. As was pointed out in Section 3-1 this is a model in which the electron is localized and moves on a classical orbit. However, by taking the angular momentum constant and minimizing the energy we obtain a relationship between angular momentum and energy compatible with the results of quantum mechanics. The variational treatment that we used in that discussion was not essential. Its purpose was to introduce a method that *is* required in the present discussion of the effect of an applied axial magnetic field. We write the internal energy of the hydrogen atom as the sum of kinetic and potential energy **(3-1)**:

$$U = U_K + U_P \tag{2}$$

where the kinetic energy is given by

$$U_K = \tfrac{1}{2}mv^2 \tag{3}$$

and the potential energy is given **(3-2)** by:

$$U_P = - \frac{1}{4\pi\epsilon_0} \frac{q^2}{r} \tag{4}$$

as for the field-free atom with $q = e$, the elementary charge.

We now wish to find the conditions under which the total energy is a minimum. For the field-free problem we took the angular momentum mvr constant **(3-3)**. We could do this because we were able to imagine a virtual displacement δr of the radius without any change in angular momentum. In the presence of a magnetic field a virtual change in *radius* produces a virtual torque and a consequent change in angular momentum. For a circular orbit we obtain from **(1)** an angular impulse on the electronic charge $q = -e$:

$$\tau \delta t = Fr\,\delta t = -qB\frac{\delta r}{\delta t} r\,\delta t = -qBr\,\delta r = \delta(mvr) \tag{5}$$

Integrating **(5)** we obtain as a constant of the motion the *generalized angular momentum* (Section G-3):

$$L = mvr + \tfrac{1}{2}qBr^2 \tag{6}$$

By holding L constant as we minimize the energy we are able to obtain results that are in accord with quantum mechanics even though our model of the hydrogen atom is strictly classical. In terms of the angular momentum we obtain for the kinetic energy:

$$U_K = \frac{(L - \tfrac{1}{2}qBr^2)^2}{2mr^2} = \frac{L^2}{2mr^2} - \frac{LqB}{2m} + \frac{(qBr)^2}{8m} \tag{7}$$

Minimizing the total energy with respect to r we obtain:

$$\frac{\partial U}{\partial r} = \frac{1}{4\pi\epsilon_0}\frac{q^2}{r^2} - \frac{L^2}{mr^3} + \frac{(qB)^2}{4m} r = 0 \tag{8}$$

By rearranging **(8)** we may write:

$$\frac{1}{r} = \frac{q^2}{4\pi\epsilon_0}\frac{m}{L^2} + \frac{q^2B^2}{4L^2} r^3 \tag{9}$$

For B small we may approximate **(9)** by the expression:

$$\frac{1}{r} \simeq \frac{1}{R} + \frac{\pi\epsilon_0}{m} B^2 R^2 \tag{10}$$

where R is the stable radius **(3-4)** in the absence of a magnetic field. Substituting into the expression for the energy we obtain:

$$\triangleright \qquad U = U_K + U_P \simeq U_0 - \frac{q}{2m} BL + \frac{q^2}{8m} B^2 R^2 \qquad (11)$$

where

$$U_0 = \frac{L^2}{2mR^2} - \frac{q^2}{4\pi\epsilon_0 R} = -\frac{L^2}{2mR^2} \qquad (12)$$

is the energy in the absence of an applied magnetic field. Note that to second order in B, **(11)** is the same function of R that the sum of **(4)** and **(7)** is of r. This is because the energy in the absence of a magnetic field is a minimum at $r = R$. This result does not apply to **(4)** and **(7)** taken separately. From **(6-110)** and **(6)** we write for the magnetic moment:

$$\mu = \tfrac{1}{2}qvr = \frac{qL}{2m} - \frac{q^2 R^2}{4m} B \qquad (13)$$

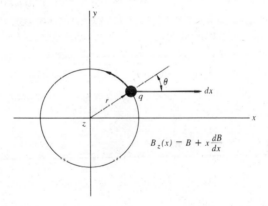

Figure 1 Atom moving into a magnetic field. The atom moves slowly compared with the orbital speed of the electron.

Eliminating the magnetic field B between **(11)** and **(13)** we have, for the internal energy,

$$U = 2U_0 + \frac{2m}{q^2 R^2} \mu^2 \qquad (14)$$

Taking the differential of **(11)** with respect to B we may write:

$$dU = -\left(\frac{qL}{2m} - \frac{q^2 R^2}{4m} B\right) dB = -\mu\, dB \qquad (15)$$

2 Orbital Diamagnetism. We now consider the problem of an atom moving into a magnetic field $B_z(x)$ as shown in Figure 1. For a displacement dx occurring in a time dt

we have, under the condition $dx/dt \ll v$,

$$d(m\mathbf{r} \times \mathbf{v}) = \langle \mathbf{r} \times \mathbf{F} \rangle \, dt = q\langle \mathbf{r} \times (\mathbf{v} \times \mathbf{B}) \rangle \, dt$$

$$= -q\langle \mathbf{B}r \cdot d\mathbf{r} \rangle = -q\mathbf{B}r \, dr - qr\langle B \cos \theta \rangle \, dx \qquad (16)$$

where the first of the final terms is from a change in radius and the second term is from a change in the center of the orbit. The field on the orbit will be given by

$$B(\theta) = B + r \cos \theta \, \frac{dB}{dx} \qquad (17)$$

where B is the field at the center. Substituting into (16) and performing the average over θ, we obtain:

$$d(mvr) = -qBr \, dr - \tfrac{1}{2}qr^2 \, dB = -d(\tfrac{1}{2}qBr^2) \qquad (18)$$

where $dB = (dB/dx) \, dx$. We see from (18) that the generalized angular momentum as given by (6) remains constant, and we have (15) for the change in internal energy.

From (6-127) the mechanical force that balances the force of the magnetic field is given by:

$$\mathbf{F}_{\text{mech}} = -(\boldsymbol{\mu} \times \mathbf{V}) \times \mathbf{B} = (\nabla \mathbf{B}) \cdot \boldsymbol{\mu} = \hat{\mathbf{x}}\mu \, \frac{dB}{dx} \qquad (19)$$

On comparison with (15) we obtain

$$dW = F_{\text{mech}} \, dx = -\mu \, dB = dU \qquad (20)$$

and we conclude that mechanical work goes into increasing *internal* energy.

For a gas with n atoms per unit volume the magnetization density will be given from (13) by:

$$\mathbf{M} = n\langle \boldsymbol{\mu} \rangle = \frac{nq}{2m} \langle \mathbf{L} \rangle - \frac{nq^2 \langle R^2 \rangle}{4m} \mathbf{B} \qquad (21)$$

If we assume that on entering the field all orientations of $\mathbf{L}$ are equally probable and that no redistribution takes place, then the first term on the right side of (21) averages to zero. For the second term we must average over all orientations of the orbit. This is most easily done by recognizing that only the component of $\mathbf{B}$ normal to the plane of the orbit produces a moment and that the induced moments must then be summed. A quick way of getting the correct result is to imagine the orbits oriented along $\hat{\mathbf{x}}$, $\hat{\mathbf{y}}$, and $\hat{\mathbf{z}}$. Only those orbits whose normals are along $\hat{\mathbf{z}}$ will produce a moment.

For a spherical charge distribution we take $\langle R^2 \rangle = (\frac{2}{3})\langle r^2 \rangle$ for all orientations to obtain finally:

$$\mathbf{M} = n\langle \boldsymbol{\mu} \rangle = -\frac{nq^2 \langle r^2 \rangle}{6m} \mathbf{B} \tag{22}$$

Defining the magnetic susceptibility by the relation:

$$\mu_0 M = \chi B \tag{23}$$

we obtain for the molar susceptibility the classical Langevin result:

$$\chi_{mol} = -\mu_0 \frac{N_A q^2 \langle r^2 \rangle}{6m} \tag{24}$$

We see from (22) that the induced magnetic moment is opposite to the field. Such a moment is called *diamagnetic* and the expression given by (24) is called the diamagnetic susceptibility.

For an atom with Z orbital electrons, each electron produces a moment of the form of (20) so that we have for the molar susceptibility

$$\blacktriangleright \qquad \chi_{mol} = -\mu_0 \frac{N_A Z e^2 \langle r^2 \rangle}{6m} \tag{25}$$

where r^2 is averaged over the electron orbits. In (24) and (25) N_A is the Avagadro number. In writing (24) we have used the relation between the volume susceptibility and the molar susceptibility $\chi = (N/V)\chi_{mol}$ where N/V is the number of *moles* per unit volume.

Exercise

1. A molecule contains N electrons whose orbits are confined to a *plane* with mean-square radius $\langle R^2 \rangle$. Show that the contribution to the diamagnetic susceptibility of the liquid from these electrons is:

$$\chi = -\frac{\mu_0 \, nNe^2 \langle R^2 \rangle}{12m}$$

3 Larmor Frequency. From (6) we may write for the angular frequency:

$$\omega = \frac{v}{R} = \frac{L}{mR^2} - \frac{qB}{2m} \tag{26}$$

The second term is called the Larmor frequency and is linear in B:

$$\blacktriangleright \qquad \omega_L = -\frac{qB}{2m} \tag{27}$$

A more general derivation of the Larmor frequency is given in Section G-3.

In the absence of a magnetic field a charge q rotates with angular velocity:

$$\omega_0 = \left(\frac{1}{4\pi\epsilon_0}\frac{q^2}{mR^3}\right)^{1/2} \tag{28}$$

as may be shown by equating the Coulomb force to the mass of the particle times its radial acceleration. Substituting (27) and (28) into (10) we have for the atomic radius:

$$\frac{1}{r} \simeq \frac{1}{R}\left(1 + \frac{\omega_L{}^2}{\omega_0{}^2}\right) \tag{29}$$

As may be observed from (29), as long as the Larmor frequency is small compared with the zero-field angular frequency, contraction of the orbit may be neglected.

4 **Orbital Paramagnetism.** If we allow an atomic gas to come into thermal equilibrium, there will be a redistribution of the orientations of $\mathbf{L}$ with a preponderance of moments oriented along the field. From (6-134) the atoms experience a torque:

$$\tau = \mu \times \mathbf{B} \tag{30}$$

which tends to turn them into the field. The work that must be done *against* the field is given by:

$$W = -\int_0^\theta \tau\, d\theta = \mu B \int_0^\theta \sin\theta\, d\theta = -\mu B \cos\theta \tag{31}$$

In thermal equilibrium the probability of angle θ varies as:

$$P(\theta)\, d\theta \propto e^{-W/kT}\, d\Omega \tag{32}$$

and the mean thermal moment is then given by

$$\langle\mu\rangle = \mu\langle\cos\theta\rangle = \mu\mathscr{L}(\mu B/kT) \tag{33}$$

where

$$\mathscr{L}(\mu B/kT) = \frac{\int_0^\pi \cos\theta\, e^{\mu B \cos\theta/kT} \sin\theta\, d\theta}{\int_0^\pi e^{\mu B \cos\theta/kT}\sin\theta\, d\theta} \tag{34}$$

is called the Langevin function. In the limit that μB is small compared with kT we may expand the exponential:

$$e^{+\mu B \cos \theta / kT} \sim 1 + \frac{\mu B \cos \theta}{kT} \tag{35}$$

Substituting into (34) and performing the integral we obtain:

$$\langle \mu \rangle \simeq \frac{\mu^2 B}{3kT} \tag{36}$$

Writing the susceptibility for this contribution to the magnetization as

$$\mu_0 \mathbf{M} = \mu_0 n \langle \boldsymbol{\mu} \rangle = \chi \mathbf{B} \tag{37}$$

we obtain from (36) the Curie law

▶
$$\chi = \frac{\mu_0 n \mu^2}{3kT} \tag{38}$$

for the paramagnetic susceptibility.

5 Spin Paramagnetism. The electron carries an intrinsic spin angular momentum:

$$S = \frac{h}{4\pi} = \frac{1}{4\pi} \times 6.6262 \times 10^{-34} \text{ joule seconds (J} \cdot \text{s)} \tag{39}$$

The quantity $k/2\pi$, which is usually written as $\hbar$, has the value 1.0546×10^{-34} J $\cdot$ s or equivalently kg $\cdot$ m^2/s. The electron has also an intrinsic magnetic moment:

$$\boldsymbol{\mu} = -(g\mu_B/\hbar)\mathbf{S} \tag{40}$$

where μ_B is called the Bohr magneton and has the value:

$$\mu_B = \frac{e\hbar}{2m} = 9.2741 \times 10^{-24} \text{ joule per tesla (J/T)} \tag{41}$$

The quantity $h/2\pi$ which is usually written as $\hbar$, has the value 1.0546×10^{-34} J $\cdot$ s or value 2.0023193 for a free electron. The ratio of (40) to (39) is called the magneto-mechanical ratio:

$$\gamma = \frac{\mu}{S} = -\frac{ge}{2m} \simeq -\frac{e}{m} \tag{42}$$

For an electron orbit, we have from **(13)**, disregarding the diamagnetic correction:

$$\gamma = \frac{\mu}{L} = -\frac{e}{2m} \tag{43}$$

and we see comparing **(42)** and **(43)** that the magnetomechanical ratio associated with the electron spin is approximately twice that for the orbital contribution.

For an electron with both spin and orbital angular momentum we write for the total angular momentum:

$$\mathbf{J} = \langle \mathbf{L} + \mathbf{S} \rangle \tag{44}$$

The magnetic moment is written as

▶ $$\boldsymbol{\mu} = -g_J \frac{e}{2m} \mathbf{J} = \gamma \mathbf{J} \tag{45}$$

where g_J depends on the relative orbital and spin contributions to the magnetic moment.

6 Transition Elements. There are four groups of elements in the periodic table whose magnetic properties are particularly pronounced:

1. The elements between calcium ($Z = 20$) and copper ($Z = 29$) have a net spin associated with partially shielded electrons in the inner 3d shell. This series is usually called the iron group. The elements are metals. Three, iron, cobalt, and nickel, are ordered ferromagnetically and two others, chromium and manganese, order antiferromagnetically.

2. The elements between strontium ($Z = 38$) and silver ($Z = 47$) have a net spin associated with partially shielded electrons in the inner 4d shell. The elements in this series are also metals but do not show magnetic ordering.

3. The elements following lanthanum ($Z = 57$) have a net spin associated with well shielded electrons in the inner 4f shell. These elements ending with lutetium are commonly called the rare earths but are also referred to as the lanthanides. The heavier rare earths show magnetic ordering. Gadolinium becomes ferromagnetic around room temperature while dysprosium orders at lower temperatures. The other heavy rare earths form ordered spiral structures.

4. The elements following actinium ($Z = 89$) have a net spin associated with well-shielded electrons in the inner 5f shell. These elements, which are called the actinides, appear to show magnetic ordering in some of their compounds but not in the pure metallic form.

MAGNETIC RESONANCE

One of the most effective ways of studying the magnetic properties of atoms and ions in beams and vapors and in solids is to excite the precession of the electronic magnetic moment by means of a radio frequency field. Let us imagine as shown in

Figure 2a that the magnetic moment of an atom is oriented at an angle θ to a uniform magnetic field **B**. There will then be a torque acting on the moment given by **(30)**.

7 **[Equation of Motion].** The rate of change of angular momentum is simply equal to the torque

$$\frac{d\mathbf{J}}{dt} = \tau \tag{46}$$

Combining **(30)** and **(46)** with **(45)** we have, for the equation of motion of the angular momentum,

▶ $$\frac{d\mathbf{J}}{dt} = \gamma(\mathbf{J} \times \mathbf{B}) \tag{47}$$

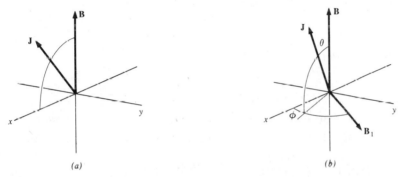

(a)   (b)

Figure 2 Excitation of magnetic resonance. (a) The angular momentum **J** precesses about the magnetic field **B**. (b) Resonance is excited by a transverse field **B**$_1$ oscillating at the frequency of precession of **J** about **B**.

For **B** constant the solution of **(47)** is uniform precession of **J** at a frequency:

▶ $$\omega = -\gamma\mathbf{B} \tag{48}$$

To excite nutation of the moment we apply a transverse field **B**$_1$ rotating at frequency ω given by **(48)** as shown in Figure 2b. The motion of **J** under the action of the constant field **B** and the rotating field **B**$_1$ is most easily found by writing $d\mathbf{J}/dt$ as the sum of a uniform precession plus motion in a frame rotating with **B**$_1$:

$$\frac{d\mathbf{J}}{dt} = \omega \times \mathbf{J} + \frac{\delta\mathbf{J}}{\delta t} = \gamma\mathbf{J} \times (\mathbf{B} + \mathbf{B}_1) \tag{49}$$

Under the conditions of **(48)** we obtain:

$$\frac{\delta\mathbf{J}}{\delta t} = \gamma\mathbf{J} \times \mathbf{B}_1 \tag{50}$$

The solution to **(50)** is precession of **J** about $\mathbf{B}_1$ in the rotating frame at a frequency:

▶
$$\omega_1 = \gamma B_1 \tag{51}$$

If **J** is initially parallel to the field, the angle with the field opens at a linear rate:

$$\theta = \gamma B_1 t \tag{52}$$

By applying the transverse field $\mathbf{B}_1$ for a time $t = \pi/(\gamma B_1)$ and then turning off the transverse field, we are able to reverse the direction of **J**. This is usually called a 180° pulse at frequency ω. Applying B_1 for a time $t = \pi/(2\gamma B_1)$ leaves **J** precessing in the plane transverse to **B** and is called a 90° pulse.

In the molecular beam magnetic resonance technique, **J** is reversed by a 180° pulse as seen by the moving atom. In resonance studies of vapors and solids the free precession of the magnetization is observed following a 90° pulse.

FERROMAGNETISM

We discuss next the properties of ordered magnetic materials. Although magnetic materials are tremendously varied, encompassing both metals and insulators, we will see that the general features of ferromagnetic ordering are given in terms of a mean field approximation. The formal theory is very much like the theory that we developed in Chapter 3 for ferroelectric materials. We divided the problem into two parts. First we developed a description of the development of polarization by an electric field and next we found a way in which the local field was modified by the development of polarization. Putting the two aspects of the problem together we found, first that at high temperatures, in what we called the paraelectric range, the susceptibility was enhanced by the interaction between dipoles. Second, we found a transition to an ordered "ferroelectric" phase in which a polarization is developed spontaneously in the absence of any external field.

There is an important difference, however, between electric and magnetic ordering. For the electric problem, polarization is developed by a real electric field produced by neighboring dipoles. For the magnetic problem the interaction field that produces the ordering of neighboring magnetic dipoles is actually associated with the interaction between charges rather than currents. Such an interaction, called an exchange interaction, is written for a pair of spins $\mathbf{S}_1$ and $\mathbf{S}_2$ as shown in Figure 3:

$$U = -2J\mathbf{S}_1 \cdot \mathbf{S}_2 \tag{53}$$

where U is the energy of interaction between the spins and J is called the exchange integral. From **(42)** we may rewrite **(53)** as:

$$U = -\frac{2J\mathbf{S}_1}{\gamma_1} \cdot \boldsymbol{\mu}_2 = -\boldsymbol{\mu}_2 \cdot \mathbf{B}_{\text{exch}} \tag{54}$$

where

$$\mathbf{B}_{exch} = \frac{2JS_1}{\gamma_1} \tag{55}$$

is called the exchange field.

8 Mean Field Approximation.

In the mean field approximation we generalize **(37)** to:

$$\mu_0 \mathbf{M} = \mu_0 n\langle\mathbf{\mu}\rangle = \chi(\mathbf{B} + \mathbf{B}_{exch}) \tag{56}$$

where the susceptibility is the same as in the absence of exchange interaction:

$$\chi = \frac{\mu_0 n\mu^2}{3kT} = \frac{C}{T} \tag{57}$$

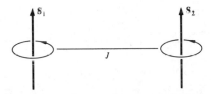

Figure 3 Exchange interaction. A pair of spins are coupled through an interaction given by **(53)**.

The quantity C is called the Curie constant. The exchange field is taken to be proportional to the magnetization:

$$\mathbf{B}_{exch} = \lambda\mu_0 \mathbf{M} \tag{58}$$

which is the essence of the mean field approximation. The quantity λ is called the exchange constant.

Substituting **(58)** into **(56)** we obtain:

$$\blacktriangleright \qquad \mu_0 \mathbf{M} = \frac{C}{T - \lambda C}B = \chi B \tag{59}$$

For $T > \lambda C$ we have the paramagnetic range in which M is proportional to B but is enhanced above the value for noninteracting moments. At a temperature $T_c = \lambda C$ a transition to an ordered phase occurs in which a spontaneous magnetization develops as the temperature is lowered. Because the magnetization M that develops may be an appreciable fraction of $n\mu$ we have to solve **(34)** exactly. The theory gives the curves shown in Figures 4a and 4b.

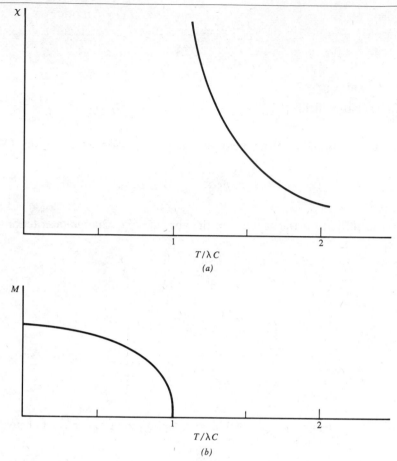

Figure 4 Temperature dependence of magnetic behavior. (*a*) The magnetic susceptibility for $T > \lambda C$ is enhanced above the value for noninteracting spins. (*b*) For $T < \lambda C$ a spontaneous magnetization develops. This magnetization may be an appreciable part of $n\mu$.

We list in Table 1 a number of ferromagnetic materials with their ferromagnetic transition temperature in kelvins and saturation induction $\mu_0 M$ in tesla.

Table 1   Ferromagnetic Materials

Material	$\mu_0 M$, T	T_c, K
Fe	2.186	1043
Co	1.817	1388
Ni	0.641	627
Gd	2.526	292
Cu_2MnAl	0.691	710
CrO_2	0.647	386
Fe_3O_4	0.603	858
$Y_3Fe_5O_{12}$	0.251	560

9 [**Ferromagnetic Resonance**]. We may study the precessional motion of an ordered magnetization by using (47) as long as **B** includes the demagnetization field (Section 6-26). We write for the macroscopic internal field:

$$\mathbf{B} = \mathbf{B}_{ext} + \mu_0 \mathbf{M} - \mu_0 \mathbf{N} \cdot \mathbf{M} \tag{62}$$

Taking $\mathbf{M} = n\gamma\mathbf{J}$, we obtain for the equation of motion:

$$\frac{d\mathbf{M}}{dt} = \gamma\mathbf{M} \times \mathbf{B} = \gamma\mathbf{M} \times (\mathbf{B}_{ext} - \mu_0 \mathbf{N} \cdot \mathbf{M}) \tag{63}$$

For a spherical sample $\mathbf{N} \cdot \mathbf{M}$ is parallel to $\mathbf{M}$ and there is no additional torque from the demagnetization field. In this case, the precession of the magnetization is the same as we have discussed for a single spin. For nonspherical shapes, however, the demagnetization field does exert a torque on the magnetization as first shown by Kittel[1] and the precessional frequencies will be correspondingly modified.

By using ferromagnetic resonance it has been possible to determine the magneto-mechanical ratios for ferromagnetic materials as a key to understanding their magnetic structure. Various devices in the microwave frequency range make use of the gyro-magnetic properties of ferromagnetic media.

10 Antiferromagnetism. There exist many materials for which the exchange field (58) is oppositely directed to **M** so that we write, in place of (58),

$$\mathbf{B}_{exch} = -(\theta/C)\mu_0 \mathbf{M} \tag{64}$$

where θ is called the *Curie-Weiss temperature*. This interaction leads to a magnetization:

$$\blacktriangleright \qquad \mu_0 \mathbf{M} = \frac{C}{T + \theta}\mathbf{B} = \chi\mathbf{B} \tag{65}$$

and we see that the magnetic susceptibility is reduced by the exchange interaction. Such materials generally undergo a transition to an ordered state in which strongly interacting moments are antiparallel. The temperature at which this transition takes place is called the *Neel temperature* and given by the symbol T_N. We list in Table 2 a number of iron group oxides which crystallize in the sodium chloride structure and are antiferromagnetic.

Table 2 Antiferromagnetic Materials

Material	Curie-Weiss Temperature, θ	Neel Temperature, T_N
NiO	2000 K	525 K
MnO	610	116
FeO	570	198
CoO	330	291

[1] C. Kittel, *Phys. Rev.*, **71**, 270 (1947); **73**, 155 (1948).

SUMMARY

1. The internal energy of an idealized hydrogen atom in the presence of a magnetic field may be written as:

$$U = U_K + U_P = U_0 - \frac{q}{2m} BL + \frac{q^2}{8m} B^2 R^2 + \cdots \tag{11}$$

The first term is the energy in the absence of a field. The second term is the interaction between the magnetic moment $qL/2m$ and the magnetic field **B**. The third term arises from the induced diamagnetic moment $-q^2\mathbf{B}R^2/4m$. It should be emphasized that these terms are all describable in terms of the internal potential and kinetic energy of the atom.
2. The mechanical work performed to bring an atom into a magnetic field goes to increase the internal energy of the atom.
3. The magnetic susceptibility of a diamagnetic atom is given by the classical Langevin result:

$$\chi = \frac{\mu_0\, nZe^2\langle r^2\rangle}{6m} \tag{25}$$

where n is the number of atoms per unit volume, Z is the number of electrons per atom, and $\langle r^2 \rangle$ is the mean-square atomic radius.
4. When an atom is placed in a magnetic field the electronic orbital frequencies are all shifted by the Larmor frequency:

$$\omega_L = -\frac{qB}{2m} \tag{27}$$

5. The paramagnetic susceptibility of noninteracting moments is given by the Curie law:

$$\chi = \frac{\mu_0\, n\mu^2}{3kT} \tag{38}$$

where μ is the effective magnetic moment per atom. In general we write:

$$\boldsymbol{\mu} = g_J \frac{e}{2m} \mathbf{J} = \gamma \mathbf{J} \tag{45}$$

where g_J is called the spectroscopic splitting factor and γ is the magnetomechanical ratio.
6. The precession of the angular momentum in a magnetic field is described by the gyromagnetic equation:

$$\frac{d\mathbf{J}}{dt} = \gamma(\mathbf{J} \times \mathbf{B}) \tag{47}$$

If **B** is a static field, the motion is one of precession about **B** at the angular frequency:

$$\omega = -\gamma \mathbf{B} \tag{48}$$

Adding a transverse radio frequency field $\mathbf{B}_1$ oscillating at ω produces nutational motion at a frequency

$$\omega_1 = \gamma B_1 \tag{51}$$

7. In the mean field approximation and above the transition temperature T_c, the magnetic susceptibility of a ferromagnetic material may be written as:

$$\chi = \frac{\mu_0 \, n\mu^2}{3k(T - T_c)} \tag{59}$$

In the ferromagnetic phase there is a spontaneous magnetization, which rises from zero at T_c to $n\mu$ at $T = 0$.

8. The excitation of resonance in a ferromagnetic sample is similar to that in a paramagnetic sample except that if the sample is nonspherical the demagnetization field will have radio frequency components, which shift the resonance frequency from γB_{ext}.

9. For some materials the mean internal field is opposite rather than parallel to the induced moment. This has the effect of reducing the susceptibility to the value given by the Curie-Weiss law:

$$\chi = \frac{C}{T + \theta} \tag{65}$$

Such materials generally undergo a transition to an antiferromagnetic state at a Neel temperature T_N of the order of θ.

PROBLEMS

1. **Diamagnetic susceptibility of atomic hydrogen.** From the electronic charge density of atomic hydrogen:

$$\rho(r) = -\frac{e}{\pi a_0^{3}} e^{-2r/a_0}$$

obtain $\langle r^2 \rangle$, the mean-square radius, with $a_0 = 5.29177 \times 10^{-11}$ m. From (25) compute the molar susceptibility of atomic hydrogen and compare with the tabulated value:

$$\chi_{mol}/4\pi = -2.36 \times 10^{-12} \text{ m}^3/\text{mol}$$

2. **Additivity of diamagnetism.** To a good approximation it is possible to compute the diamagnetic susceptibility of acyclic hydrocarbons by summing the molar susceptibilities of individual radicals and groups. Use the tabulated susceptibilities of the following organic compounds to obtain the molar susceptibilities of the methyl radical (CH_3-), the methylene radical ($CH_2=$), the binary methine group ($=CH-$), the methylene group ($-CH_2-$), the tertiary methine group ($-CH-$), and the covalently bonded carbons: binary ($=C=$), tertiary ($-C=$), and quaternary ($-C-$).

Compound	Formula	$\chi_{mol}/4\pi$, 10^{-12} m^3/mole
Methane	CH_4	-12.27
Ethane	$(CH_3)_2$	-27.37
Propane	$(CH_3)_2CH_2$	-40.5
Ethylene	$CH_2:CH_2$	-15.3
Propene	$CH_3CH:CH_2$	-31.6
2 Methylpropene	$(CH_3)_2C:CH_2$	-44.4
2,3-Pentadiene	$CH_3CH:C:CHCH_3$	-49.1
2,2-Dimethylpropane	$(CH_3)_4C$	-63.1
2-Methylbutane	$(CH_3)_2CHCH_2CH_3$	-64.4

Using the susceptibilities of the above nine radicals and groups, compute the susceptibilities of the following compounds and compare with tabulated values:

Compound	Formula	$\chi_{mol}/4\pi$, 10^{-12} m^3/mol
Butane	$CH_3CH_2CH_2CH_3$	-57.4
2-Butene	$CH_3CH:CHCH_3$	-43.3
2-Methylpropene	$(CH_3)_2C:CH_2$	-44.4
2-Methyl-2-butene	$(CH_3)_2C:CHCH_3$	-54.14
Pentane	$CH_3CH_2CH_2CH_2CH_3$	-63.05

3. **Diamagnetism of cyclic hydrocarbons.** The benzene ring is formed of six methine ($=CH-$) groups at the vertices of a regular hexagon. Napthalene is formed of $N = 2$ fused benzene rings with two carbons ($=C-$) shared by both rings. By fusing additional benzene rings in various locations a system of fused polycyclic hydrocarbons may be constructed. We list a number of these compounds below with the measured susceptibility of the *liquid*.

From the susceptibility of diamond ($\chi_{mol} = -75 \times 10^{-12}$ m^3/mol) estimate the contribution to the diamagnetism of electrons distributed over the ring. Taking the radius of the ring to be the length of the carbon-carbon bond, 1.39×10^{-10} m, estimate the number of electrons per ring responsible for the diamagnetism. Repeat this calculation for the fused benzene rings. See M. Blumer, "Polycyclic Aromatic Compounds in Nature," *Scientific American*, **234**(3), 35 (March 1976).

N	Compound	Formula	$\chi_{mol}/4\pi$, 10^{-12} m^3/mole
1	Benzene	C_6H_6	-54.84
2	Napthalene	$C_{10}H_8$	-91.9
3	Anthracene	$C_{14}H_{10}$	-130
3	Phenanthrene	$C_{14}H_{10}$	-127.9
4	Pyrene	$C_{16}H_{12}$	-147.9
4	Chrysene	$C_{18}H_{12}$	-166.67
5	Pentacene	$C_{22}H_{14}$	-205.4
5	Perylene	$C_{20}H_{12}$	-166.8
5	Dibenzophenanthrene	$C_{22}H_{14}$	-200.5
7	Coronene	$C_{24}H_{12}$	-243.3
8	Pyranthrene	$C_{30}H_{16}$	-266.9
10	Ovalene	$C_{32}H_{14}$	-353.8

4. Diamagnetism of graphite. The molar susceptibility of graphite powder is

$$\chi_{mol} = -1090 \times 10^{-12} \text{ m}^3/\text{mol}$$

For reference the molar susceptibility of diamond is -75×10^{-12} m^3/mol and should be a measure of the atomic contribution to the susceptibility. To interpret the very considerable additional diamagnetism of graphite we observe that each carbon atom is on three hexagonal rings with a carbon-carbon bond distance $R = 1.418 \times 10^{-10}$ metres. Taking R to be the effective radius of a ring, compute the contribution to the diamagnetic susceptibility of one circulating electron per carbon atom and compare with the tabulated susceptibilities. If the mass of the circulating electrons were m^* rather than m, find the ratio m^*/m that would give agreement with the observed diamagnetism.

5. Gouy balance. A Gouy balance is used to determine the magnetic susceptibility from the force on a long rod-shaped sample in a nonuniform magnetic field. If B_1 and B_2 are the fields at the two ends of the sample, integrate **(6-130)** to obtain for the force

$$F = \mu_0 \chi \frac{V}{L} (B_2{}^2 - B_1{}^2)$$

where L is the length of the sample.

6. Faraday balance. A Faraday balance is used to determine the susceptibility from the force on a short sample in a magnetic field gradient. Show that the force may be written as:

$$\mathbf{F} = \tfrac{1}{2}\mu_0 \chi \, \nabla B^2$$

7. **[Langevin function]**. Show that the Langevin function **(34)** may be integrated to obtain:

$$\mathscr{L}(u) = \langle \cos \theta \rangle = \coth u - \frac{1}{u}$$

Obtain for small u

$$\mathscr{L}(u) = \frac{u}{3}\left(1 - \frac{u^2}{15} + \cdots \right)$$

For a tabulation of the Langevin function, see the *Handbook of Chemistry and Physics*, Chemical Rubber Co.

8. **Paramagnetism of oxygen.** The tabulated paramagnetic susceptibility of oxygen is given below:

T	phase	$\chi_{mol}/4\pi$, 10^{-9} m^3/mole
23.7	Solid, α	1.760
54.3	Solid, γ	10.200
70.8	Liquid	8.685
90.1	Liquid	7.699
293	Gas	3.449

From the room temperature susceptibility, use **(38)** to obtain the magnetic moment per molecule. Show that this moment is consistent with spin-only paramagnetism ($g = 2.0023$) from two parallel electron spins in a triplet state. From the data at the four lower temperatures determine the Curie-Weiss temperature in the various phases. How does the exchange interaction appear to vary with density? Solid oxygen becomes antiferromagnetic with a Neel temperature $T_N = 23.9$ K.

9. **Paramagnetism of nitric oxide.** The tabulated magnetic susceptibility of nitric oxide (NO) is given below:

T	Phase	$\chi_{mol}/4\pi$, 10^{-9} m^3/mol
90	Solid	0.0198
117.64	Liquid	0.1142
146.9	Gas	2.324
203.8	Gas	1.895
293	Gas	1.461

From the data in the gas phase and **(38)** obtain μ/μ_B at the three temperatures. The ground state of nitric oxide is weakly paramagnetic $S = \frac{1}{2}$, $L = 1$, $J = \frac{1}{2}$. Just above this state and separated by $\Delta U/k = 174.1$ K is a paramagnetic state with $S = \frac{1}{2}$, $L = 1$, $J = \frac{3}{2}$. Use this spectroscopic information to interpret the magnetic susceptibility in the gas. The very considerable reduction of the susceptibility in the liquid and solid results from the formation of the dimer N_2O_2, which is diamagnetic.

10. **Exchange field.** From the information given in Table 1, obtain the exchange field B_{exch} in iron, nickel, and cobalt.

11. **[Ferromagnetic resonance].** A magnetic field B_{ext} is applied along the $\hat{z}$ axis of an ellipsoidal ferromagnet with demagnetizing factors N_x, N_y, and N_z. Show from **(63)** that the frequency of free precession is given by:

$$\omega^2 = \gamma^2[B_{ext} + (N_y - N_z)\mu_0 M][B_{ext} + (N_x - N_z)\mu_0 M]$$

Obtain expressions for the frequency when the sample is a sphere, a disc, and a rod and with the field applied along a principal direction.

12. **Magnetic levitation electrometer.** The gravitational force on a small permeable particle is equilibrated by a magnetic field gradient. Show that the particle oscillates about its equilibrium position at a frequency:

$$\omega = (g/l)^{1/2}$$

where g is the acceleration of gravity and l is given below.

(a) For oscillations in the vertical direction obtain from the result of Problem 6 the relation:

$$\frac{1}{l} = \frac{1}{dB^2/dz}\frac{d^2B^2}{dz^2}$$

For the use of this technique in measuring the electrical charge on a small superconducting sphere, see G. S. LaRue, W. M. Fairbank, and A. F. Hebard, *Physical Review Letters* **38**, 1011 (1977).

(b) For oscillations in the transverse plane obtain the relation:

$$\frac{1}{l} = \frac{1}{dB^2/dz}\frac{d^2B^2}{dx\,dz}$$

For the use of this technique in measuring the electrical charge on a small iron cylinder, see G. Galinaro, M. Marinelli, and G. Morpurgo, *Physical Review Letters* **38**, 1255 (1977).

REFERENCES

Blumer, M., "Polycyclic Aromatic Compounds in Nature," *Scientific American*, **234**(3), 35 (March 1976).

Breslow, R., "The Nature of Aromatic Molecules," *Scientific American,* **227**(2), 32 (August 1972).

Kittel, C., *Introduction to Solid State Physics*, Fifth Edition, Wiley, 1976.

Selwood, P. W., *Magnetochemistry*, Second Edition, Interscience, 1956; *Chemisorption and Magnetization*, Academic, 1975.

Van Vleck, J. H., *Electric and Magnetic Susceptibilities*, Oxford, 1932.

8 MAGNETOCONDUCTING MEDIA I

In this chapter we return to conducting media, extending the discussion of Chapter 5 by considering the effect of static magnetic fields on the motion of charge carriers. We first examine the response of a one-component plasma to static magnetic and electric fields. Next we consider the response to an alternating electric field. Then we discuss the containment of a plasma by magnetic fields, followed by a discussion of superconductivity.

MAGNETOCONDUCTIVITY

In Chapter 5 we wrote for the rate of change of current (5-18):

$$\frac{d\mathbf{j}}{dt} = \left[\frac{d\mathbf{j}}{dt}\right]_{\text{fields}} + \left[\frac{d\mathbf{j}}{dt}\right]_{\text{collisions}} \tag{1}$$

which expresses the total rate of change of current as the sum of a term resulting from the action of applied fields plus a term resulting from collisions. In the presence of an electric field we wrote (5-16):

$$\left[\frac{d\mathbf{j}}{dt}\right]_{\text{fields}} = \frac{nq^2}{m}\mathbf{E} = \frac{\sigma}{\tau}\mathbf{E} \tag{2}$$

where we imagine n carriers per unit volume each carrying a charge q. For the collision term we made the simplifying approximation that all carriers relax with the same characteristic time (5-17):

$$\left[\frac{d\mathbf{j}}{dt}\right]_{\text{collisions}} = -\frac{1}{\tau}\mathbf{j} \tag{3}$$

In the presence of a magnetic field we replace $q\mathbf{E}$ in (2) by the Lorentz force (6-20) to obtain:

$$\left[\frac{d\mathbf{j}}{dt}\right]_{\text{fields}} = \frac{nq^2}{m}\,(\mathbf{E} + \mathbf{v} \times \mathbf{B}) = \frac{\sigma}{\tau}\,\mathbf{E} + \frac{q}{m}\,\mathbf{j} \times \mathbf{B} \tag{4}$$

1 Hall Effect. The Hall effect is the development of an electrostatic potential when current flows transverse to a magnetic field in a bounded medium.[1] We wish to obtain a solution to (1) with a static uniform magnetic field $\mathbf{B}$ parallel to the $\hat{\mathbf{z}}$ axis. In steady state, the total rate of change of current is zero and we have, from (3) and (4),

$$\mathbf{j} - \mu\,\mathbf{j} \times \mathbf{B} = \sigma\mathbf{E} \tag{5}$$

where

$$\mu = \frac{q}{m}\,\tau = \frac{\sigma}{nq} \tag{6}$$

is called the carrier mobility. Writing (5) in component form we obtain:

$$j_x - \mu B\,j_y = \sigma E_x \tag{7}$$

$$\mu B\,j_x + j_y = \sigma E_y \tag{8}$$

$$j_z = \sigma E_z \tag{9}$$

Solving for the currents we obtain:

$$j_x = \sigma\,\frac{E_x + \mu B E_y}{1 + (\mu B)^2} \tag{10}$$

$$j_y = \sigma\,\frac{-\mu B E_x + E_y}{1 + (\mu B)^2} \tag{11}$$

$$j_z = \sigma E_z \tag{12}$$

We see from (12) that the longitudinal current is unaffected by the presence of the magnetic field. A more realistic treatment of the scattering yields a reduction in current along the magnetic field as well, termed *longitudinal magnetoresistance.*

The configuration in which the Hall effect is observed is shown in Figure 1. Current flow is restricted to a direction parallel to the $\hat{\mathbf{x}}$ axis. Thus, by the constraints imposed we have $j_y = 0$. We have then from (11) and (10):

$$E_y = \mu B E_x = R B\,j_x \tag{13}$$

---

[1] Although the Hall effect may be treated in terms of electrostatic fields, there are also motional electromotive forces involved as discussed in Section 9-12.

where

$$R = \frac{1}{nq} \tag{14}$$

is called the Hall constant. From **(10)** and **(13)** the current is simply:

$$j_x = \sigma E_x \tag{15}$$

Any reduction in j_x resulting from a field along $\hat{z}$ is called *transverse magnetoresistance* and is absent under the assumption of **(3)**.

Note that **(13)** gives an electric field along $\hat{y}$. The physics of the development of this field is the following: let us imagine that initially current is flowing along $\hat{x}$ in the

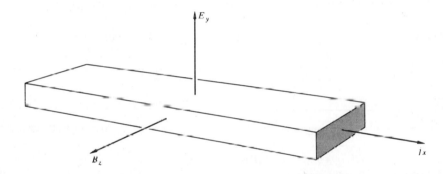

Figure 1 Hall effect configuration. Current flow is restricted to a direction parallel to the $\hat{x}$ axis.

absence of a magnetic field. When the magnetic field is turned on the Lorentz force will produce a transverse current which from **(10)** and **(11)** may be written as:

$$j_y = -\mu B j_x \tag{16}$$

Initially E_y is equal to zero. But since the sample is bounded transverse to $\hat{x}$, charge must accumulate on the surfaces and it is this charge that establishes the field E_y. In steady state the force produced by E_y must just balance the force on the carriers resulting from their motion through the magnetic field:

$$qE_y = qv_x B = q\mu BE_x \tag{17}$$

which is just the same as **(13)**.

The Hall constant R together with the conductivity permits a determination of the concentration and sign of the charge carriers as well as their mobility.

Exercises

1. From **(10)**, **(11)**, and **(12)** find the elements of the conductivity matrix.
2. Show that the magetoconductivity tensor has the form[2]

$$\bar{\bar{\sigma}} = \sigma_{\parallel} \,\hat{z}\hat{z} + \sigma_{\perp}(\mathbf{I} - \hat{z}\hat{z}) - \sigma_H \,\hat{z} \times \mathbf{I}$$

Identify $\sigma_{\parallel}$, $\sigma_{\perp}$, and the Hall conductivity σ_H.

---

2 Cyclotron Resonance. A charged particle injected into a magnetic field executes circular motion in a plane normal to the field. For velocities low compared with c, the frequency of rotation is independent of velocity. It was this principle that led E. O. Lawrence to develop the cyclotron in which charges were injected continuously into the magnetic field and accelerated by an electric field that alternated synchronously with the rotation of the charges. The same technique is used on a much small scale in the study of the electronic structure of conducting solids, in the investigation of plasmas, and in high-precision mass spectrometers.

The motion of a free charge q in a magnetic field $\mathbf{B}$ is given by

$$m\frac{d\mathbf{v}}{dt} = q\mathbf{v} \times \mathbf{B} \tag{18}$$

The solution to **(18)** can be regarded as a uniform rotation of the velocity vector:

$$\frac{d\mathbf{v}}{dt} = \boldsymbol{\omega}_c \times \mathbf{v} \tag{19}$$

where the angular frequency $\boldsymbol{\omega}_c$ is called the cyclotron frequency. Comparing **(18)** and **(19)** we have:

$$\boldsymbol{\omega}_c = -\frac{q}{m}\mathbf{B} \tag{20}$$

For a plasma characterized by a conductivity σ and relaxation time τ we have from **(1)**, **(3)**, and **(4)**:

$$\frac{d\mathbf{j}}{dt} = \frac{\sigma}{\tau}\mathbf{E} + \frac{q}{m}\mathbf{j} \times \mathbf{B} - \frac{1}{\tau}\mathbf{j} \tag{21}$$

We assume that a uniform magnetic field is applied along $\hat{z}$ and that a rotating electric field is applied transverse to $\mathbf{B}$:

$$E_x = E_1 \cos \omega t \qquad E_y = E_1 \sin \omega t \tag{22}$$

---

[2] Suggested by A. N. Kaufman.

Writing (21) in component form and using (20) and (22) we obtain:

$$\left(\tau\frac{dj_x}{dt} + j_x\right) + \omega_c\tau\, j_y = \sigma E_1 \cos\omega t \tag{23}$$

$$-\omega_c\tau\, j_x + \left(\tau\frac{dj_y}{dt} + j_y\right) = \sigma E_1 \sin\omega t \tag{24}$$

The simplest way of solving (23) and (24) for the currents is to multiply (24) by $-i$ and add to (23). We introduce the complex current density:

$$j(t) = j_x - i\, j_y \tag{25}$$

The equation for $j(t)$ takes the form:

$$\tau\frac{dj(t)}{dt} + (1 + i\omega_c\tau)\, j(t) = \sigma E_1 e^{-i\omega t} \tag{26}$$

We assume a solution to (26) of the form:

$$j(t) = \underline{j}\,e^{-i\omega t} \tag{27}$$

Substituting (27) into (26) we obtain:

$$\underline{j} = \frac{\sigma}{1 + i(\omega_c - \omega)\tau}\, E_1 \tag{28}$$

Substituting (28) back into (27) and using (25) to equate the real and imaginary parts we obtain, finally,

$$j_x = \sigma_1 E_1 \cos\omega t + \sigma_2 E_1 \sin\omega t \tag{29}$$
$$j_y = -\sigma_2 E_1 \cos\omega t + \sigma_1 E_1 \sin\omega t \tag{30}$$

where the components of the conductivity are given by:

$$\sigma_1(\omega) = \frac{1}{1 + (\omega_c - \omega)^2\tau^2}\,\sigma \tag{31}$$

$$\sigma_2(\omega) = -\frac{(\omega_c - \omega)\tau}{1 + (\omega_c - \omega)^2\tau^2}\,\sigma \tag{32}$$

Note that $\sigma_1(\omega)$ corresponds to the component of the current that rotates in phase with the electric field, while $\sigma_2(\omega)$ corresponds to the component of the current that lags $\pi/2$ behind the electric field.

In Figure 2 we plot the components of σ. The rate at which work is done on the plasma is given by:

$$\left\langle \frac{dw}{dt} \right\rangle = \langle j_x E_x \rangle + \langle j_y E_y \rangle = \sigma_1 E_1^2 \tag{33}$$

The absorption of energy is a maximum when the driving frequency ω is equal to the cyclotron frequency ω_c and drops off far from resonance inversely with the square of the frequency difference.

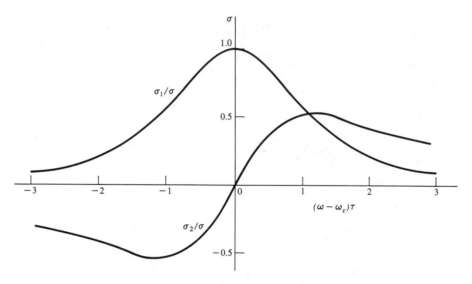

Figure 2   Conductivity near cyclotron resonance. The real part of the conductivity is symmetric about the resonance while the imaginary part is antisymmetric.

In order to discuss the energy stored it is convenient to view the problem in terms of polarization rather than current. Writing:

$$\mathbf{j} = \frac{d\mathbf{P}}{dt} \tag{34}$$

we have in place of (21):

$$\frac{d^2\mathbf{P}}{dt^2} + \boldsymbol{\omega}_c \times \frac{d\mathbf{P}}{dt} + \frac{1}{\tau}\frac{d\mathbf{P}}{dt} = \frac{\sigma}{\tau}\mathbf{E} \tag{35}$$

We may solve (35) to obtain:

$$P_x = \chi_1 E_1 \cos \omega t + \chi_2 E_1 \sin \omega t \tag{36}$$

$$P_y = -\chi_2 E_1 \cos \omega t + \chi_1 E_1 \sin \omega t \tag{37}$$

where the susceptibilities are given by:

$$\chi_1(\omega) = -\frac{(\omega_c - \omega)\tau}{1 + (\omega_c - \omega)^2\tau^2}\frac{\sigma}{\omega} = -\frac{\sigma_2}{\omega} \tag{38}$$

$$\chi_2(\omega) = \frac{1}{1 + (\omega_c - \omega)^2\tau^2}\frac{\sigma}{\omega} = \frac{\sigma_1}{\omega} \tag{39}$$

Now, the energy density may be written from **(4-57)** as:

$$\langle u \rangle = -\tfrac{1}{2}\langle \mathbf{P} \cdot \mathbf{E} \rangle = -\tfrac{1}{2}\langle P_x E_x \rangle - \tfrac{1}{2}\langle P_y E_y \rangle \tag{40}$$

Using **(36)** and **(37)** we obtain:

$$\langle u \rangle = -\tfrac{1}{2}\chi_1 E_1^2 = \frac{1}{2}\frac{\sigma_2}{\omega}E_1^2 \tag{41}$$

At low frequencies (below cyclotron resonance) **P** is in phase with **E**, and the energy is negative. Above cyclotron resonance **P** is out of phase with **E**, and the energy is positive as may be seen by comparing **(41)** with Figure 2.

PLASMA CONTAINMENT

We consider in this section two examples of the containment of a plasma by a magnetic field. The first involves the operation of the magnetron, a microwave oscillator with a radial electric field and an axial magnetic field. The second example is that of the magnetic mirror in which charge is reflected back along field lines from regions of high field intensity.

3 **Magnetron.** In the magnetron electrons are thermionically emitted from an axial cathode of radius a and accelerated toward a coaxial anode of radius R and at a potential ϕ as shown in Figure 3. A magnetic field **B** is imposed parallel to the common axis. In the absence of a magnetic field, carriers impinge on the anode and appear as current in the external circuit. The effect of a magnetic field is to curve the electrons away from the anode. If the anode contains a structure which is resonant at microwave frequencies, the electrons will excite resonance oscillations in the structure, generating microwave radiation. It is for this purpose that the magnetron is primarily used.

Because of the simple geometry we may completely neglect motion along the $\hat{\mathbf{z}}$ direction, which in any case will be very slight. In cylindrical coordinates we write the velocity and position vector as

$$\mathbf{v} = v_\rho \hat{\boldsymbol{\rho}} + v_\phi \hat{\boldsymbol{\phi}} \qquad \mathbf{r} = \rho \hat{\boldsymbol{\rho}} \tag{42}$$

where $\hat{\boldsymbol{\rho}}$ and $\hat{\boldsymbol{\phi}}$ are unit vectors along the radial and azimuthal directions.

The equation of motion for electrons is:

$$m \frac{d\mathbf{v}}{dt} = -e(\mathbf{E} + \mathbf{v} \times \mathbf{B}) \tag{43}$$

This problem has two constants of the motion:

1. As usual, the *energy* is a constant of the motion as long as there are no dissipative processes. The energy is given by:

$$U = \tfrac{1}{2}mv^2 - e\phi(\rho) \tag{44}$$

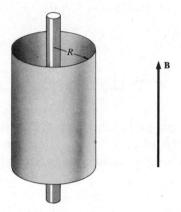

Figure 3 Magnetron configuration. The cathode and anode are coaxial. A magnetic field is parallel to the common axis.

where the potential is given by

$$\phi(\rho) = \frac{\ln(\rho/a)}{\ln(R/a)} \, \phi(R) \tag{45}$$

To show that U is constant we differentiate **(44)**:

$$\frac{dU}{dt} = m\mathbf{v} \cdot \frac{d\mathbf{v}}{dt} - e \frac{d\phi(\rho)}{dt} \tag{46}$$

We may write for the time rate of change of potential seen by the *moving* electron:

$$\frac{d\phi(\rho)}{dt} = \mathbf{v} \cdot \nabla\phi(\rho) = -\mathbf{v} \cdot \mathbf{E} \tag{47}$$

and we have, for the rate of change of energy,

$$\frac{dU}{dt} = \mathbf{v} \cdot \left(m \frac{d\mathbf{v}}{dt} + e\mathbf{E} \right) = -e\mathbf{v} \cdot \mathbf{v} \times \mathbf{B} = 0 \tag{48}$$

where the second equality is obtained from **(43)**.

2. The second constant of the motion is the *generalized angular momentum* (Section G-3):

$$\mathbf{L} = m\mathbf{r} \times \mathbf{v} - \tfrac{1}{2}eB\rho^2 \qquad (49)$$

which is the vector form of (7-6) with $q = -e$. To show that $\mathbf{L}$ is constant we differentiate (49) in a uniform magnetic field:

$$\frac{d\mathbf{L}}{dt} = m\mathbf{r} \times \frac{d\mathbf{v}}{dt} - eB\mathbf{r} \cdot \mathbf{v} \qquad (50)$$

Now from (43), we obtain

$$m\mathbf{r} \times \frac{d\mathbf{v}}{dt} = -e\mathbf{r} \times \mathbf{E} - e\mathbf{r} \times (\mathbf{v} \times \mathbf{B}) = eB\mathbf{r} \cdot \mathbf{v} \qquad (51)$$

where we have used the fact that $\mathbf{r}$ is parallel to $\mathbf{E}$ and perpendicular to $\mathbf{B}$ and have expanded the triple cross product. Substituting (51) into (50) we obtain the result that $\mathbf{L}$ is a constant of the motion

We now wish to evaluate the energy U and the angular momentum L. We have taken (45), the potential, to be zero at the cathode. If the electrons are emitted with negligible kinetic energy then the total energy U is zero and we have, from (44),

$$\tfrac{1}{2}mv^2 = \tfrac{1}{2}mv_\rho^2 + \tfrac{1}{2}mv_\phi^2 = e\phi(\rho) \qquad (52)$$

The angular momentum may similarly be evaluated at the cathode. Setting $v = 0$ and $r = a$ in (49) we obtain:

$$L = mv_\phi\rho - \tfrac{1}{2}eB\rho^2 = -\tfrac{1}{2}eBa^2 \qquad (53)$$

Eliminating v_ϕ between (52) and (53) we obtain:

$$\tfrac{1}{2}mv_\rho^2 = e[\phi(\rho) - \phi_c(\rho)] \qquad (54)$$

with

$$e\phi_c(\rho) = \tfrac{1}{8}m\omega_c^2 \left(\frac{\rho^2 - a^2}{\rho}\right)^2 \qquad (55)$$

We now have an equation for the radial velocity alone. The azimuthal kinetic energy appears as a potential, which is called the *centrifugal potential* $e\phi_c(\rho)$. In Figure 4 we plot the terms in (54) for a value of B sufficient to prevent the electrons from reaching the anode. In operation the voltage V is adjusted until the electrons just graze the anode.

4   Magnetic Mirror. We consider the motion of charge carriers in a static *inhomogeneous* magnetic field under two conditions:

1. The velocity of the charges parallel to the magnetic field is sufficiently low that in one cyclotron period the change in field is small:

$$\frac{1}{\omega_c}\frac{dB}{dt} = \frac{1}{\omega_c}v_z\frac{dB}{dz} \ll B \tag{56}$$

2. The second condition is that the transverse velocity is sufficiently low that the change in magnetic field over the orbit is small. Since the relevant distance is the radius of the cyclotron orbit:

$$\rho = \frac{v_\phi}{\omega_c} \tag{57}$$

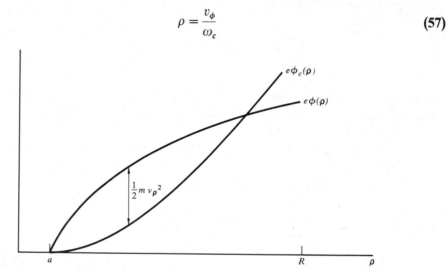

Figure 4   Kinetic and potential energy. The radial part of the kinetic energy is represented as the difference between the total potential energy and a centrifugal potential energy.

we have:

$$\rho\frac{dB}{d\rho} = \frac{1}{\omega_c}v_\phi\frac{dB}{dr} \ll B \tag{58}$$

We may combine **(57)** and **(58)** into a single vector equation:

$$\frac{1}{\omega_c}\mathbf{v}\cdot\nabla\mathbf{B} \ll \mathbf{B} \tag{59}$$

Under this condition, the orbit is a helix for which the magnetic field acts as a *guiding center* as shown in Figure 5.

By integrating the flux over the surface of a cylinder we may obtain a relation between the radial magnetic field and the rate of change of the intensity of the

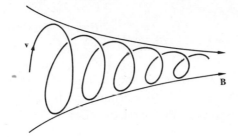

Figure 5   Orbit in an inhomogeneous field. The field acts as a guiding center for the helical orbit.

longitudinal field. We let $\hat{r}$, $\hat{\phi}$, and $\hat{z}$ be a local coordinate system for the field as shown in Figure 6. From the fact that the total outgoing flux must be zero we have:

$$\pi\rho^2 \frac{\partial B_z}{\partial z} + 2\pi\rho B_\rho = 0 \qquad (60)$$

which leads to:

$$B_\rho = -\tfrac{1}{2}\rho \frac{\partial B_z}{\partial z} \qquad (61)$$

Since the z direction is that of the magnetic field at $r = 0$ we are able to drop the subscript on B_z and write **(61)** in terms of the field magnitude B.

The radial field B_ρ exerts a Lorentz force parallel to $\hat{z}$, which acts in a direction to slow the charge:

$$m\frac{dv_z}{dt} = -qv_\phi B_\rho = \tfrac{1}{2}q\rho v_\phi \frac{\partial B}{\partial z} \qquad (62)$$

From the definition of the magnetic moment **(7-13)** we may rewrite **(62)** in more compact form. We have for the magnitude of μ, which is parallel to **B**:

$$\mu = \tfrac{1}{2}q\rho v_\phi \qquad (63)$$

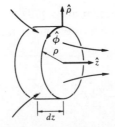

· **Figure 6**   Magnetic flux. By integrating the flux over the surface of a cylinder, a relation is obtained between the radial field and the rate of change of the longitudinal field.

In terms of the magnetic moment we then have, from (62),

$$m \frac{dv_z}{dt} = \mu \frac{\partial B}{\partial z} \tag{64}$$

We now show that as the particle moves into a more intense field its radius constricts so as to keep the enclosed flux constant and the azimuthal velocity increases so as to maintain the constancy of the magnetic moment. Taking the scalar product of **v** with (18) we obtain:

$$\mathbf{v} \cdot \frac{d\mathbf{v}}{dt} = v_z \frac{dv_z}{dt} + v_\phi \frac{dv_\phi}{dt} = 0 \tag{65}$$

We have the relation (57) for the azimuthal velocity:

$$v_\phi = \omega_c \rho = -\frac{q}{m} B \rho \tag{66}$$

Substituting (66) into (65) we may write:

$$m \frac{dv_z}{dt} = q \frac{v_\phi}{v_z} \frac{d}{dt} (\rho B) = q v_\phi \frac{\partial}{\partial z} (\rho B) \tag{67}$$

Comparing (67) with (62) we have:

$$\tfrac{1}{2} \rho \frac{\partial B}{\partial z} = \frac{\partial}{\partial z} (\rho B) \tag{68}$$

Expanding the right side of (68) we obtain:

$$\tfrac{1}{2} \rho \frac{\partial B}{\partial z} + B \frac{\partial \rho}{\partial z} = 0 \tag{69}$$

This equation has the simple solution that the flux through the orbit:

$$\Phi = \pi \rho^2 B \tag{70}$$

is constant as may be seen by differentiating (70).

We now show that the magnetic moment μ and the angular momentum L are also constant. From (63) we have:

$$\mu = \tfrac{1}{2} q \rho v_\phi = \tfrac{1}{2} q \omega_c \rho^2 = -\frac{q^2}{2m} B \rho^2 = -\frac{q^2}{2\pi m} \Phi \tag{71}$$

and we see that the magnetic moment is constant as well. We may regard (71) as the condition for *complete diamagnetism*.

From (7-6) the generalized angular momentum is given approximately by:

$$L = mv_\phi \rho + \tfrac{1}{2}q\rho^2 B \tag{72}$$

From (63) we may write:

$$mv_\phi \rho = \frac{2m}{q}\mu \tag{73}$$

We may now express (72) in terms of the magnetic moment and flux from (73) and (69):

$$L = \frac{2m}{q}\mu + \frac{1}{2\pi}q\Phi \tag{74}$$

Finally, using (71) for the relation between the induced moment and the enclosed flux, we have, for complete diamagnetism,

$$L = -\frac{1}{2\pi}q\Phi = \frac{m}{q}\mu \tag{75}$$

Having established the constancy of the angular momentum, we now return to an examination of (64). Since μ is constant and opposite to **B**, the force on q is toward decreasing magnetic field. We may integrate (64) to obtain

$$\tfrac{1}{2}mv_z^2 - \mu B = U \tag{76}$$

where U is a constant. From (70) and (66) we have:

$$-\mu B = \tfrac{1}{2}m\left(\frac{qB_\rho}{m}\right)^2 = \tfrac{1}{2}mv_\phi^2 \tag{77}$$

By substituting (77) into (76) we find that U is the kinetic energy:

$$U = \tfrac{1}{2}mv_z^2 + \tfrac{1}{2}mv_\phi^2 = \tfrac{1}{2}mv^2 \tag{78}$$

That there is no radial kinetic energy is a consequence of (56), which gives a change in radius slow compared with the orbital motion.

We sketch in Figure 7 a situation in which charge is trapped between two regions of increased magnetic field intensity. The charge will spiral periodically back and forth between regions where all the kinetic energy is in the transverse motion. The reflecting regions are called *magnetic mirrors*.[3]

[3] For an interesting optical analog, see D. Stern, *Am. J. Phys.* **29**, 767 (1961).

We conclude the discussion of magnetic mirrors by obtaining an expression for the fraction of particle flux or current reflected by a magnetic mirror. We take as a reference point some position along a field line where the field is B and the charges are assumed to have an isotropic distribution of velocity directions. If θ is the angle which the velocity of a charge makes with the magnetic field direction, we may write the magnetic moment as

$$\mu = \tfrac{1}{2}q\rho v_\phi = \tfrac{1}{2}q\frac{v_\phi^2}{\omega_c} = -\frac{1}{2}\frac{mv^2 \sin^2 \theta}{B} = -\frac{U}{B}\sin^2 \theta \tag{79}$$

Now, if B_{max} is the maximum available field, a charge will be reflected back if it meets the condition:

$$-\mu B_{max} > U \tag{80}$$

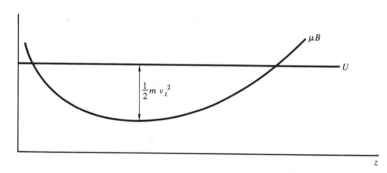

Figure 7   Magnetic mirrors. A charge may be trapped between two regions of increased magnetic field intensity.

By comparison of **(80)** with **(79)** we may express this condition in terms of the angle with the field:

$$\sin^2 \theta > \frac{B}{B_{max}} \tag{81}$$

We complete the argument by finding the fraction of *flux* associated with θ greater than the condition of **(81)**:

$$f = \frac{\int_\theta^{\pi/2} \cos\theta \sin\theta \, d\theta}{\int_0^{\pi/2} \cos\theta \sin\theta \, d\theta} = \cos^2\theta = 1 - \sin^2\theta = 1 - \frac{B}{B_{max}} \tag{82}$$

SUPERCONDUCTIVITY

We conclude this chapter by discussing some of the most common characteristics of superconducting media and their general interpretation. We first discuss the phenomenon of zero electrical resistivity which marked the discovery of super-conductivity in mercury at a temperature of 4.2 K. We next consider the screening of the magnetic field from a superconductor, which is called the Meissner effect and was first shown by F. London to be a macroscopic quantum phenomenon. Questions related to the stability of the superconducting state, the formation of domains, and the differences between what are called type I and type II superconductors are postponed to Chapter 10, where we discuss superconductors as magnetic media.

5 Perfect Conductivity. Superconductivity was discovered in 1911 in the laboratory of Kamerlingh Onnes in Leiden where helium gas had first been liquified some three

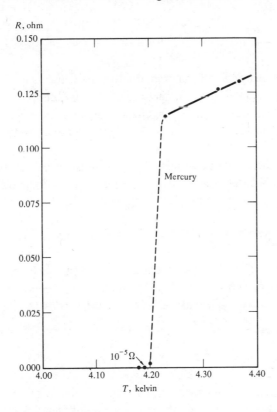

Figure 8 Resistance of mercury. The resistance was observed to drop by at least four orders of magnitude in a fraction of a kelvin.

years earlier. The availability of liquid helium made it possible to investigate the electrical properties of metals at temperatures down to about 1.5 K. We show in Figure 8 a plot of the resistance of a sample of mercury that was observed to drop by at least four orders of magnitude in a fraction of a kelvin. Since this early work, superconductivity has been discovered in nearly one-third of the pure elements. Considerable effort has been devoted to finding intermetallic compounds with high transition temperatures. Currently, the record is held by the alloy Nb_3Ge, which becomes superconducting at 23.2 K. The earliest studies of this material have been on films produced by a technique called sputtering. In Table 1 we list a number of superconducting materials and their transition temperatures.

Table 1   Superconductivity in Elements and Compounds

Selected Elements	T_c	Selected Compounds	T_c
Al	1.180	V_3Si	17.1
Cd	0.56	V_3Ga	16.5
In	3.4035	Nb_3Al	17.5
Sn	3.722	Nb_3Ge	23.2
Hg	4.153	Nb_3Sn	18.05
Tl	2.39		
Pb	7.193		

6   Flux Exclusion. The property of superconductors to which we wish to pay principal attention here is the screening of magnetic fields. This phenomenon, which was first observed in 1933, is called the Meissner effect. We show in Figure 9 the distribution of the magnetic field in the vicinity of a pair of parallel lead or tin wires.

The magnetic field is applied transverse to the wires and penetrates the wires as shown in Figure 9a at temperatures above the transition temperature. When the temperature is reduced below the transition **(96)** the magnetic field is expelled from the wires and is observed to concentrate in the region between the wires. This phenomenon clearly does not follow from perfect electrical conductivity alone since there are no electric fields produced that would induce currents in the process of lowering the temperature.

The explanation of the Meissner effect was given by F. London and may be understood in terms of our earlier discussion of the magnetic properties of atoms when applied to a macroscopic plasma. We introduced **(7-6)** the generalized angular momentum:

$$\mathbf{L} = m\mathbf{r} \times \mathbf{v} - \tfrac{1}{2}q\mathbf{B}r^2 \tag{83}$$

and have seen that $\mathbf{L}$ is rigorously a constant of the motion for a charge in a uniform magnetic field and a Coulomb potential of spherical symmetry. We also saw **(49)** that $\mathbf{L}$ was constant for a potential of cylindrical symmetry (the magnetron)

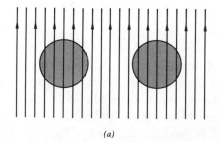

(a)

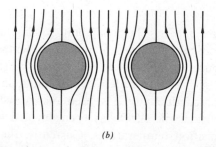

(b)

Figure 9   Flux exclusion, (a) A magnetic field is applied transverse to a pair of wires. At temperatures above the transition temperature, the flux penetrates the wires. (b) Below the transition the magnetic field is expelled and concentrates between the wires.

with a uniform magnetic field parallel to the axis. We have further seen (75) that L was an *approximate* constant of the motion for a charge moving in a nonuniform magnetic field as long as the field changed slowly over a cyclotron period and in the absence of electric fields.

For nonuniform magnetic fields the generalized angular momentum takes the form (G-27):

$$L = r \times (mv + qA) \qquad (84)$$

London proposed that flux exclusion was a consequence of the rigorous constancy of L in the superconducting state. We emphasize that this condition is *not* to be expected classically. One expects for a classical plasma in a magnetic field that the current everywhere:

$$j(r) = nqv(r) \qquad (85)$$

is zero given sufficient time for the plasma to equilibrate. Then for the classical gas L is determined solely by the vector potential A. London recognized that the constancy of L in the superconducting state was a quantum phenomenon although he was unable to offer a microscopic mechanism that would lead to the observed macroscopic quantum behavior. With the modern theory of superconductivity as advanced by Bardeen, Cooper, and Schreiffer we now understand that the coherency of the superconducting state arises from the formation of electron pairs.

We now show that the constancy over the sample of L leads for a simply connected sample (i.e., a sample for which any contour can be deformed into any other contour) to flux exclusion. We write (84) as:

$$L(r) = qr \times [\mu_0 \lambda_L^2 j(r) + A(r)] \qquad (86)$$

where $j(r)$ is given by (85) and λ_L the London screening length is given by

$$\mu_0 \lambda_L^2 = m/nq^2 = \omega_p^2/\epsilon \qquad (87)$$

Representative values of the London screening length, or penetration depth, are given in Table 2.

Table 2   London Penetration Depth

Element	λ_L, 10^{-10} m
Al	160
Sn	340
Cd	1100
Pb	370

Since $\mathbf{L}(\mathbf{r})$ is a constant vector its divergence is zero and we have from **(B-4)**:

$$\nabla \cdot \mathbf{L}(\mathbf{r}) = -q\mathbf{r} \cdot \nabla \times [\mu_0 \lambda_L^2 \mathbf{j}(\mathbf{r}) + \mathbf{A}(\mathbf{r})] = 0 \tag{88}$$

Since $\mathbf{r}$ is arbitrary we must have as a consequence of **(88)**:

$$\nabla \times [\mu_0 \lambda_L^2 \mathbf{j}(\mathbf{r}) + \mathbf{A}(\mathbf{r})] = 0 \tag{89}$$

In steady state we may have no charge accumulation and thus no divergence of the current:

$$\nabla \cdot \mathbf{j}(\mathbf{r}) = 0 \tag{90}$$

And we have seen for the vector potential in steady state **(6-52)**:

$$\nabla \cdot \mathbf{A}(\mathbf{r}) = 0 \tag{91}$$

Combining **(90)** and **(91)** we may write:

$$\nabla \cdot [\mu_0 \lambda_L^2 \mathbf{j}(\mathbf{r}) + \mathbf{A}(\mathbf{r})] = 0 \tag{92}$$

Next we look at the boundary conditions on $\mathbf{j}(\mathbf{r})$ and $\mathbf{A}(\mathbf{r})$. We assume an isolated sample so that no current may flow in or out, and we have on the bounding surfaces:

$$\hat{\mathbf{n}} \cdot \mathbf{j}(\mathbf{r}) = 0 \tag{93}$$

Finally, we must discuss the boundary conditions on $\mathbf{A}(\mathbf{r})$. From **(6-49)** we have:

$$\mathbf{A}(\mathbf{r}) = \frac{\mu_0}{4\pi} \int_{V_1} \frac{\mathbf{j}(\mathbf{r}_1)}{r_{01}} dV_1 \tag{94}$$

For a general current distribution $\mathbf{j}(\mathbf{r}_1)$ we are unable to say anything about the boundary conditions on $\mathbf{A}(\mathbf{r})$. We can, however, always find a new vector potential that we write as

$$\mathbf{A}_L(\mathbf{r}) = \mathbf{A}(\mathbf{r}) + \nabla \psi(\mathbf{r}) \tag{95}$$

which automatically gives

$$\mathbf{B}(\mathbf{r}) = \nabla \times \mathbf{A}(\mathbf{r}) = \nabla \times \mathbf{A}_L(\mathbf{r}) \tag{96}$$

We continue to require zero divergence

$$\nabla \cdot \mathbf{A}_L(\mathbf{r}) = \nabla \cdot \mathbf{A}(\mathbf{r}) = 0 \tag{97}$$

which requires that $\psi(\mathbf{r})$ be a solution of Laplace's equation

$$\nabla^2 \psi(\mathbf{r}) = 0 \tag{98}$$

We *impose* the boundary condition:

$$\hat{\mathbf{n}} \cdot \mathbf{A}_L(\mathbf{r}) = 0 \tag{99}$$

which has, as a consequence,

$$\hat{\mathbf{n}} \cdot \nabla \psi(\mathbf{r}) = -\hat{\mathbf{n}} \cdot \mathbf{A}(\mathbf{r}) \tag{100}$$

on the bounding surface. We have seen **(4-38)** that a solution of Laplace's equation is determined uniquely in a limited region by the normal component of its gradient on a bounding surface. We may conclude then that **(98)** and **(100)** determine $\psi(\mathbf{r})$ and we have on the boundary:

$$\hat{\mathbf{n}} \cdot [\mu_0 \lambda_L{}^2 \mathbf{j}(\mathbf{r}) + \mathbf{A}_L(\mathbf{r})] = 0 \tag{101}$$

As a consequence of **(89)** for V simply connected we may write by **(B-126)**:

$$\mu_0 \lambda_L{}^2 \mathbf{j}(\mathbf{r}) + \mathbf{A}_L(\mathbf{r}) = \nabla \eta(\mathbf{r}) \tag{102}$$

From **(92)** we have within V:

$$\nabla^2 \eta(\mathbf{r}) = 0 \tag{103}$$

and on S:

$$\hat{\mathbf{n}} \cdot \nabla \eta = 0 \tag{104}$$

As we have seen **(103)** and **(104)** determine η to within a constant. We may conclude then from **(103)** and **(104)** that η must be constant within V. Consequently, we have:

$$\mu_0 \lambda_L{}^2 \mathbf{j}(\mathbf{r}) + \mathbf{A}_L(\mathbf{r}) = 0 \tag{105}$$

The relation **(105)**, which we have shown to be a consequence of the constancy of $\mathbf{L}$ for a simply connected body, is called the *London equation*. In order to find the dependence of the supercurrent density $\mathbf{j}(\mathbf{r})$ we write Ampere's law as a second relation connecting $\mathbf{j}(\mathbf{r})$ and $\mathbf{A}(\mathbf{r})$:

$$\mu_0 \mathbf{j}(\mathbf{r}) = \nabla \times \mathbf{B}(\mathbf{r}) = \nabla \times [\nabla \times \mathbf{A}_L(\mathbf{r})] = -\nabla^2 \mathbf{A}_L(\mathbf{r}) \tag{106}$$

as a consequence of **(96)** and **(97)**. Combining **(105)** and **(106)** we have:

$$\nabla^2 \mathbf{A}_L(\mathbf{r}) = \frac{1}{\lambda_L{}^2} \mathbf{A}_L(\mathbf{r}) \tag{107}$$

Taking the curl of both sides of **(107)** we may write, alternatively,

$$\nabla^2 \mathbf{B}(\mathbf{r}) = \frac{1}{\lambda_L{}^2} \mathbf{B}(\mathbf{r}) \tag{108}$$

and finally, taking the curl of both sides of **(108)**, we have

$$\nabla^2 \mathbf{j}(\mathbf{r}) = \frac{1}{\lambda_L{}^2} \mathbf{j}(\mathbf{r}) \tag{109}$$

These three equations describe the way in which the vector potential, the magnetic field, and the supercurrent density vary within a superconductor.

As a simple example we imagine that the $x = 0$ plane marks the bounding surface of a superconductor. At the interface we assume a uniform field **B**:

$$\mathbf{B}(0, y, z) = \mathbf{B} \tag{110}$$

For x positive we have, as a solution of **(108)**,

$$\mathbf{B}(x) = \mathbf{B}(0)e^{-x/\lambda_L} \tag{111}$$

and the surface magnetic field is screened in a distance λ_L. The supercurrent density is obtained from **(106)**:

$$\mu_0 \mathbf{j}(x) = \nabla \times [\mathbf{B}(0)e^{-x/\lambda_L}] = -\frac{1}{\lambda_L} \hat{\mathbf{x}} \times \mathbf{B}(0)e^{-x/\lambda_L} \tag{112}$$

Similarly, the vector potential is given, from **(105)**,

$$\mathbf{A}_L(x) = \lambda_L \hat{\mathbf{x}} \times \mathbf{B}(0)e^{-x/\lambda_L} \tag{113}$$

7 **[Trapped Flux].** We now consider the behavior of a sample that is not simply connected, such as the ring shown in Figure 10. Two distinct kinds of contours are possible for such a body. We may construct C_1 as shown in Figure 10a for which we can always find a bounded surface S_1, which lies completely within the superconductor. Furthermore, we may shrink the contour C_1 and the surface S_1 to zero while always remaining within the superconductor. A second kind of contour is shown in Figure 10b. For this contour at least part of the bounded surface S_2 must lie outside the superconductor. And it is not possible to shrink S_2 to zero while keeping the contour within the superconductor. (See the discussion of connectivity in Section B-20.)

We now consider the following sequence of experiments using a *ring* as shown in Figure 11. First as shown in Figure 11a we place the ring in the normal state in a magnetic field B so that at least some flux links the ring. We now cool the ring so that it undergoes a transition to the superconducting state. As shown in Figure 11b,

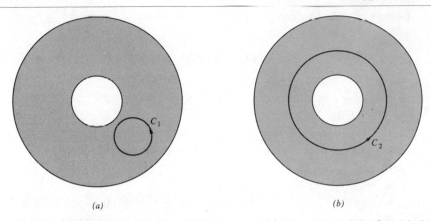

Figure 10 Contours in a non-simply connected body. (*a*) A contour of the first kind may be shrunk to a point without leaving the superconductor. (*b*) For a contour of the second kind, part of the bounded surface S_2 must lie outside the superconductor.

flux will now be excluded from the volume of the superconductor. We imagine that some flux Φ_0 now threads the ring.

We may still suppose that **(86)** is satisfied and that **(89)**, **(92)**, and **(101)** are satisfied as well. May we conclude that **(105)** is satisfied as for a simply connected body? No, we may not. All that we can conclude is that over contours of the first kind we have

$$\oint_{C_1} [\mu_0 \lambda_L{}^2 \mathbf{j}(\mathbf{r}) + \mathbf{A}_L(\mathbf{r})] \cdot d\mathbf{r} = 0 \qquad \textbf{(114)}$$

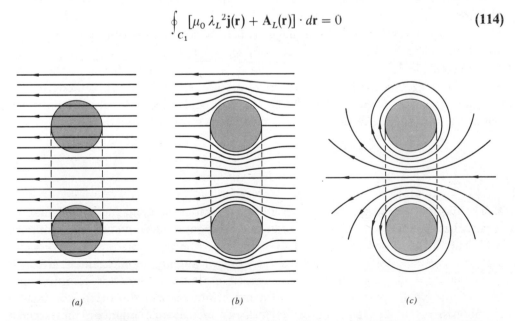

Figure 11 Trapped flux. (*a*) A magnetic field is applied to the ring in the normal state. (*b*) The ring is cooled so that it undergoes a transition to the superconducting state. Flux is excluded from the superconducting medium. (*c*) Finally, the external field is removed. It is observed that the flux trapped within the ring remains constant.

as a consequence of **(89)**. For contours of the second kind we have

$$\oint_{C_2} [\mu_0 \lambda_L{}^2 \mathbf{j}(\mathbf{r}) + \mathbf{A}_L(\mathbf{r})] \cdot d\mathbf{r} = \Phi_0 \qquad (115)$$

where Φ_0 is a constant flux. The expression **(115)** is a consequence of the fact that any two contours of the second kind differ by a contour of the first kind. We may readily show that Φ_0 is just the flux linking the contour. Let us take the contour C_2 well within the superconductor where $\mathbf{j}(\mathbf{r}) = 0$. We know that the contour integral of the vector potential is the enclosed flux from **(6-57)** and since **(115)** is constant it follows that Φ_0 must be the enclosed flux.

Finally, we turn off the external field as in Figure 11c. What is observed is that the trapped flux Φ_0 remains the same and is maintained by a persistent current in the superconductor. London's interpretation of flux trapping is that whatever value of **L** characterizes the situation of Figure 11b, this value is fixed and is not altered by removing the external field. This *rigidity* of the superconducting state that London supposed is now understood as a consequence of the quantization of the trapped flux Φ_0.

SUMMARY

1. If a magnetic field is imposed on a transversely bounded conductor, an electrostatic field will be established to balance the $\mathbf{v} \times \mathbf{B}$ Lorentz force. The development of this field is called the Hall effect. For a single carrier the field is given by $\mathbf{E} = R\mathbf{B} \times \mathbf{j}$ where $R = 1/nq$ is called the Hall constant.

2. A charged particle in a uniform magnetic field **B** executes uniform circular motion in a plane perpendicular to **B** at a frequency $\omega_c = -(q/m)B$.

3. A magnetron is a device for producing a large azimuthal current in the plane of a cylindrical anode. Electrons are thermionically emitted from a cathode on the axis and are accelerated toward the anode. An axial magnetic field produces the azimuthal motion. The critical magnetic field is given from the relation:

$$\omega_c = \frac{e}{m} B = \frac{R}{R^2 - a^2} \left(\frac{8eV}{m} \right)^{1/2} \qquad (55)$$

4. The motion of charged particles in a nearly homogeneous magnetic field may be regarded as the superposition of cyclotron motion about the field and motion along the field. The flux enclosed by the orbit $\Phi = \pi \rho^2 B$ and the magnetic moment $\mu = \frac{1}{2} q \rho v_\phi$ are both constants of the motion. The generalized angular momentum is given by $L = -q\Phi/2\pi = m\mu/q$, which may be regarded as the condition for complete diamagnetism. As the particles move into a more intense field the orbits contract to maintain constant flux and the orbital velocity increases to maintain the moment constant. Since the total kinetic energy is constant, the increase in orbital velocity will require a decrease in the longitudinal velocity. At a critical field given by $\mu B = -U$, where U is the total kinetic energy, the guiding center is reflected back along the field.

5. A large number of elements and compounds undergo a transition at low temperature to a superconducting state in which a dc current flows without loss. If a sample is cooled through the superconducting transition in a magnetic field, flux is largely excluded from the sample, penetrating only the London penetration depth λ_L, of the order of 10^{-8} m. If the sample is multiply connected, flux which is enclosed by the sample will be trapped. Both the exclusion of flux and the trapping of flux may be understood from the constancy of the generalized angular momentum $\mathbf{L} = \mathbf{r} \times (m\mathbf{v} + q\mathbf{A})$ throughout the sample.

PROBLEMS

1. **Hall effect in a two-component plasma.** Assuming carrier concentrations n for the negative carriers and p for the positive carriers and mobilities μ_n and μ_p, respectively, obtain the Hall constant at low magnetic fields:

$$R = \frac{E_y}{Bj_x} = \frac{1}{q} \frac{p\mu_p^2 - n\mu_n^2}{(p\mu_p + n\mu_n)^2}$$

Write an equation of the form of (5) for each component. Find the value of F_y for which the total transverse current vanishes.

2. **High field magnetoresistance.** A medium with two carriers of concentration n_1 and n_2, and charges q_1 and q_2, mobilities μ_1 and μ_2 carries a current I in the presence of a transverse magnetic field B.
   (a) If the sample is bounded in the other transverse direction show that the conductivity is given by

$$\sigma = \frac{\sigma_1}{1 + (\mu_1 B)^2} + \frac{\sigma_2}{1 + (\mu_2 B)^2} + \frac{\left[\dfrac{\sigma_1 \mu_1 B}{1 + (\mu_1 B)^2} + \dfrac{\sigma_2 \mu_2 B}{1 + (\mu_2 B)^2}\right]^2}{\dfrac{\sigma_1}{1 + (\mu_1 B)^2} + \dfrac{\sigma_2}{1 + (\mu_2 B)^2}}$$

   with $\sigma_1 = n_1 q_1 \mu_1$, and $\sigma_2 = n_2 q_2 \mu_2$.
   (b) Show that, in general, the high field resistivity approaches the saturation value

$$\rho = \frac{(n_1 q_1/\mu_1) + (n_2 q_2/\mu_2)}{(n_1 q_1 + n_2 q_2)^2}$$

   (c) For the particular case $n_2 = n_1$, $q_2 = -q_1$ and $\mu_2 = -\mu_1/\alpha$ (electrons and holes in equal concentration) show that the restivity increases quadratically with the field as:

$$\rho = \frac{\mu_1 B^2}{n_1 q_1 (1 + \alpha)}$$

3. **Hall probe.** The Hall effect is used for the measurement of magnetic fields by sending a current through a small semiconducting sample and measuring the transverse voltage developed. A typical probe uses a sample of indium antimonide 0.180 in. long, 0.080 in. wide, and 0.020 in. thick. The dominant carriers in indium antimonide are electrons with an intrinsic concentration of about $10^{22}/m^3$. Find the Hall voltage developed across the width of the sample when a current of 200 mA flows along the length in a field of one tesla.

4. **Clip-on current probe.** The Hall effect may be used to determine the dc current I flowing through a conductor without interrupting the circuit by the arrangement shown in the following diagram. Individual Hall probes are arranged to surround

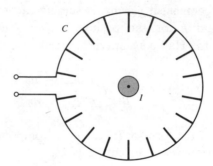

the conductor. The probe current flows through the individual detectors in series, and the Hall current is delivered in parallel. Show that the Hall current is proportional to the enclosed dc current I and is independent of where I is placed within the contour C or on the exact shape of C. The current I flows through the contour C.

5. **Hall effect in a flowing electrolyte.** An electrolyte of NaCl dissolved in water flows across a magnetic field at a velocity of 0.558 meters per second. Find the potential developed between a pair of probes separated by $d = 6.3$ mm in a magnetic field of 0.36 T. [See J. J. Wright and S. Van der Beken, "The Hall Effect in a Flowing Electrolyte," *Am. J. Phys.*, **40**, 245 (1972). See also the comment on this article by D. B. Geselowitz, *Am. J. Phys.*, **40**, 1183 (1972).]

6. **Relativistic cyclotron frequency.** A particle of rest mass M and charge q moves at high speed in a magnetic field **B**. Show that the equation of motion is

$$\frac{d\mathbf{v}}{dt} = \boldsymbol{\omega} \times \mathbf{v}$$

where

$$\boldsymbol{\omega} = -\frac{q\mathbf{B}}{\gamma M}$$

is the cyclotron frequency with $\gamma = (1 - v^2/c^2)^{-(1/2)}$.

7. **Planar magnetron.** In the planar magnetron the cathode and anode are flat parallel plates and the magnetic field is parallel to the plates. Electrons are emitted from the cathode at zero velocity and accelerated toward the anode. If the anode is at a distance d from the cathode and is at a relative potential V, show that the magnetic field that cuts off the anode current is given by:

$$B = (2mV/e)^{1/2}\, d$$

[E. Kay, J. Appl. Phys. **34**, 760 (1963)].

8. **Drift in crossed fields.** A particle of mass m and charge q moves in uniform electric and magnetic fields. Show that the motion is the sum of (a) acceleration along $\mathbf{B}$, (b) uniform drift along $\mathbf{E} \times \mathbf{B}$, and (c) cyclotron motion about $\mathbf{B}$. Find the rate of acceleration, the drift velocity, and the cyclotron frequency.

9. **Corbino disc.** A radial potential V is applied to a disc of inner radius a, outer radius R, and thickness d. Show that the current through the disc is given by:

$$I = \frac{2\pi\sigma d}{\ln R/a}\, V$$

A magnetic field B is imposed normal to the plane of the disc. Show that the current is now given by:

$$I = \frac{2\pi\sigma d}{\ln R/a}\, \frac{V}{1 + (\mu B)^2}$$

where μ is the carrier mobility.

10. **Magnetic field penetration.** A superconducting plate of thickness d is introduced into a uniform magnetic field $\mathbf{B}_{\text{ext}}$ so that $\mathbf{B}_{\text{ext}}$ is parallel to the plate. Show that the magnetic field within the superconductor is of the form:

$$\mathbf{B}(x) = \mathbf{B}_{\text{ext}}\, \frac{\cosh(x/\lambda_L)}{\cosh(d/2\lambda_L)}$$

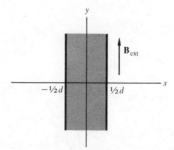

11. **Flux exclusion.** Flux is excluded from a superconducting sphere of radius R. Find the equivalent magnetization $\mathbf{M}(\mathbf{r})$ and actual surface current density $\mathbf{i}(\mathbf{r})$.

Find the moments of $\mathbf{M}(\mathbf{r})$ using the scalar potential and of $\mathbf{i}(\mathbf{r})$ using the vector potential. Obtain an expression for the field produced outside the sphere. What similarity do you see with Problem 6-25?

12. **Helmholtz vortex theorem.** The Helmholtz vortex theorem of hydrodynamics is

$$\nabla \times (\mathbf{v} \times \mathbf{w}) = \frac{\partial \mathbf{w}}{\partial t}$$

where

$$\mathbf{w} = \nabla \times \mathbf{v} + \frac{q}{m}\,\mathbf{B}$$

is a function of position. Show that as a consequence of the London equation $\mathbf{w}$ is zero within a superconductor.

REFERENCES

Cahill, L. J., "The Magnetosphere," *Scientific American*, **212**(3), 58 (March 1965).

Feynman, R. P., R. B. Leighton, and M. Sands, *The Feynman Lectures on Physics*, Addison-Wesley, 1965. See Volume III, Chapter 21 for an introduction to super-conductivity.

Geballe, T. H., "New Superconductors," *Scientific American*, **225**(5), 22 (November 1971).

Keefe, D., "Collective-Effect Accelerators," *Scientific American*, **226**(4), 22 (April 1972).

Kittel, C., *Introduction to Solid State Physics*, Fifth Edition, Wiley, 1976.

London, F., *Superfluids*, Wiley, 1950. Reprinted by Dover Publications.

Matthias, B. T., "Superconductivity," *Scientific American*, **197**(5), 92 (November 1957).

Spitzer, L., *Physics of Fully Ionized Gases*, Second Edition, Interscience, 1956.

9 SOURCES OF THE ELECTROMAGNETIC FIELD II

In this chapter we discuss the investigations of Michael Faraday of the magnetic generation of electric fields. To quote from Maxwell[1]:

> *Faraday, in his mind's eye, saw lines of force traversing all space where the mathematicians saw centres of force attracting at a distance: Faraday saw a medium where they saw nothing but distance: Faraday sought the seat of the phenomena in real actions going on in the medium, they were satisfied that they had found it in a power of action at a distance impressed on the electric fields.*
>
> *I would recommend the student, after he has learned, experimentally if possible, what are the phenomena to be observed, to read carefully Faraday's* Experimental Researches in Electricity. *He will find there a strictly contemporary historical account of some of the greatest electrical discoveries and investigations, carried on in an order and succession which could hardly have been improved if the results had been known from the first, and expressed in the language of a man who devoted much of his attention to the methods of accurately describing scientific operations and their results.*

Following the presentation of Faraday's discovery of electromagnetic induction, we discuss an important example—the screening of alternating magnetic fields by conductors through the induction of what are called eddy currents. Next, we use the law of induction to obtain an expression for the external work to bring together a distribution of magnetic moments. We are led, finally, to a discussion of magnetic energy density and magnetic pressure.

Our discussion of the magnetic generation of electric fields will begin with the Lorentz transformation as did our discussion of the electric generation of magnetic fields in Section 6-2. What we did in Section 6-2 was to show that the force on a test charge moving with velocity $\mathbf{v}$ in the vicinity of charge moving with velocity $\mathbf{v}_1$ is given by the Lorentz force **(6-20)**:

$$\mathbf{F} = q(\mathbf{E} + \mathbf{v} \times \mathbf{B}) \tag{1}$$

---

[1] From *A Treatise on Electricity and Magnetism* by James Clerk Maxwell, 3rd ed., 1891, Oxford University Press. See Michael Faraday, *Experimental Researches in Electricity*, Taylor and Francis, 1839, 1844, 1855. The original three volumes have been printed and bound as two volumes by Dover Publications.

where the magnetic field **B** is given by **(6-22)**:

$$\mathbf{B} = \frac{1}{c^2}\,\mathbf{v}_1 \times \mathbf{E} \tag{2}$$

This expression, which is called Maxwell's law, states that a moving electric field generates a magnetic field. We next discussed steady electric currents for which we obtained **(6-48)**:

$$\nabla \times \mathbf{B} = \mu_0\,\mathbf{j} \tag{3}$$

This equation, which is called Ampère's law, applies to any steady current, even though individual charges may be in accelerated motion. We relied for **(3)** on the argument that any distribution of steady currents may be written as a superposition of current filaments.

We begin this chapter with a discussion of the interaction between moving charges and magnetic moments. We will show that a moving magnetic moment generates an electric field, which can be related simply to the magnetic field that it produces. We may extend this discussion to any distribution of uniformly moving steady currents. Finally, we extend the discussion to nonuniformly moving currents through the research of Faraday.

INTERACTION BETWEEN MOVING CHARGES AND MOMENTS

We are already familiar with the problem of a test charge q moving in the vicinity of a magnetic moment $\boldsymbol{\mu}$. The force on the test charge is given by:

$$\mathbf{F} = q\mathbf{v} \times \mathbf{B} \tag{4}$$

where the magnetic field is given for $r > 0$ by **(6-123)**:

$$\mathbf{B(r)} = \frac{\mu_0}{4\pi}\frac{1}{r^3}\,[3\hat{\mathbf{r}}(\boldsymbol{\mu} \cdot \hat{\mathbf{r}}) - \boldsymbol{\mu}] = -\frac{\mu_0}{4\pi}\,(\boldsymbol{\mu} \cdot \nabla)\frac{\mathbf{r}}{r^3} \tag{5}$$

We show in Figure 1 the several quantities to which we refer in **(4)** and **(5)**.

1 **Force on a Test Charge.** What may we say about the force on a charge that is at rest in the vicinity of a moving magnetic moment as shown in Figure 2? This is not a problem that we have thus far considered. We should, however, by making a transformation to a frame moving with velocity:

$$\mathbf{V} = \mathbf{v}_1 \tag{6}$$

be able to find the force on a charge q moving with the transformed velocity $\mathbf{v}' = -\mathbf{v}_1$ as shown in Figure 3. By then transforming back into the laboratory frame we may

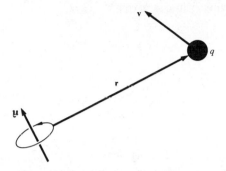

Figure 1 Force on a test charge. The test charge moves in the magnetic field of a moment, which is at rest.

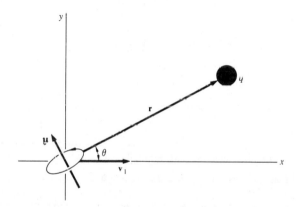

Figure 2 Force on a test charge. The test charge is at rest and the magnetic moment moves.

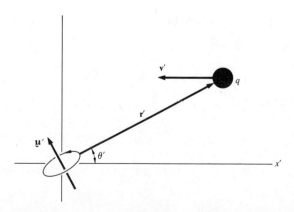

Figure 3 Force on a test charge. By transforming to a frame in which the moment is at rest, we may obtain the force on the test charge.

finally obtain an expression for the force on the charge in Figure 2. The force on the test charge q in the moving frame is given by:

$$\mathbf{F}'(\mathbf{r}') = q\mathbf{v}' \times \mathbf{B}'(\mathbf{r}') \tag{7}$$

with

$$\mathbf{B}'(\mathbf{r}') = -\frac{\mu_0}{4\pi} q(\boldsymbol{\mu}' \cdot \nabla') \frac{\mathbf{r}'}{r'^3} \tag{8}$$

We transform back into the laboratory frame by **(F-93)** and **(F-94)**. Since we have $F_x' = 0$, and the velocity of q is zero in the unprimed frame, we simply write

$$\mathbf{F} = \gamma \mathbf{F}' = \frac{\mu_0}{4\pi} \gamma q(\boldsymbol{\mu}' \cdot \nabla') \frac{\mathbf{v}_1 \times \mathbf{r}'}{r'^3} \tag{9}$$

where $\boldsymbol{\mu}'$ is the magnetic moment in the primed frame. Since q is at rest in the laboratory frame we are forced to interpret **(9)** in terms of an electric field:

$$\mathbf{E}(\mathbf{r},t) = \frac{\mu_0}{4\pi} \gamma(\boldsymbol{\mu}' \cdot \nabla') \frac{\mathbf{v}_1 \times \mathbf{r}'}{r'^3} \tag{10}$$

We will return to the interpretation of **(10)** after discussing the general case in which q is also in motion in the laboratory frame with velocity $\mathbf{v}$ as shown in Figure 1 and where the moment has velocity $\mathbf{v}_1$ as shown in Figure 2. The force on q in the primed frame is again given by **(7)** where $\mathbf{v}'$ is now the transformed velocity with components:

$$v_x' = \frac{v_x - v_1}{1 - v_x v_1/c^2} \qquad v_{yz}' = \frac{v_{yz}}{\gamma(1 - v_x v_1/c^2)} \tag{11}$$

Using **(F-90)** and **(F-91)** we obtain for the transverse components of force in the laboratory frame:

$$F_y = -\frac{\mu_0}{4\pi} \gamma q(\boldsymbol{\mu}' \cdot \nabla') \frac{(\mathbf{v} \times \mathbf{r})_y + v_1 z}{r'^3} \tag{12}$$

$$F_z = -\frac{\mu_0}{4\pi} \gamma q(\boldsymbol{\mu}' \cdot \nabla') \frac{(\mathbf{v} \times \mathbf{r})_z - v_1 y}{r'^3} \tag{13}$$

Substituting **(12)** and **(13)** into the expression for the longitudinal force **(F-89)** we obtain after some algebra:

$$F_x = -\frac{\mu_0}{4\pi} \gamma q(\boldsymbol{\mu}' \cdot \nabla') \frac{(\mathbf{v} \times \mathbf{r})_x}{r'^3} \tag{14}$$

where x, y, and z are the coordinates of q in the laboratory frame. Note that for the case $\mathbf{v} = \mathbf{v}_1$ the force $\mathbf{F}$ is zero as expected since q is at rest in the frame of the moment. For $\mathbf{v} = 0$ we obtain **(8)**, which as we remarked is to be interpreted as an electric field **(10)**. Comparing **(12)**, **(13)**, and **(14)** with **(1)** we write for the magnetic field:

$$\mathbf{B}(\mathbf{r},t) = -\frac{\mu_0}{4\pi}\,\gamma(\boldsymbol{\mu}' \cdot \nabla')\frac{\mathbf{r}}{r'^3} \tag{15}$$

Comparing **(10)** and **(15)** we have for the electric field the expression:

▶
$$\mathbf{E}(\mathbf{r},t) = -\mathbf{v}_1 \times \mathbf{B}(\mathbf{r},t) \tag{16}$$

We have written $\mathbf{B}(\mathbf{r},t)$ in terms of a combination of quantities in the primed and unprimed frames for simplicity. We should of course finally transform coordinates in **(15)** into the laboratory frame.

Comparing the expression for the magnetic field in the laboratory frame **(15)** with the expression for the field in the primed frame **(8)** we obtain the relations

▶
$$B_x(\mathbf{r},t) = B_x'(\mathbf{r}') \qquad B_{yz}(\mathbf{r},t) = \gamma B_{yz}'(\mathbf{r}') \tag{17}$$

where $\mathbf{r}$, t, and $\mathbf{r}'$ are connected by a Lorentz transformation.

Although we have obtained **(16)** and **(17)** for a moving magnetic moment, these relations apply to any uniformly moving distribution of steady solenoidal currents. This is because, as we have seen **(6-143)**, any distribution of volume and surface currents may be treated in terms of an equivalent volume magnetization density. In fact, as we will see in Section 11-2, both **(16)** and **(17)** may be obtained from the transformation properties of the Lorentz force alone without the need for specifying the connection between $\mathbf{E}$ and $\mathbf{B}$ and charges or currents.

Exercises

1. Verify **(12)** and **(13)**.
2. Verify **(14)**.
3. Show that uniformly moving currents may be synthesized from uniformly moving charges. Start from the filament synthesis of Section 6-6.

2 **Force on a Moving Magnetic Moment.** We now consider the force acting on a magnetic moment moving with velocity $\mathbf{v}$ in the field of a charge q_1 as shown in Figure 4 and then more generally in any stationary electric field. Transforming to the primed frame in which the moment is at rest, and q moves with velocity $\mathbf{v}_1' = -\mathbf{v}$, we obtain from **(6-22)** a magnetic field:

$$\mathbf{B}'(\mathbf{r}',t') = \frac{1}{c^2}\mathbf{v}_1' \times \mathbf{E}' = -\frac{1}{c^2}\mathbf{v} \times \mathbf{E}' = -\frac{\gamma}{c^2}\mathbf{v} \times \mathbf{E}(\mathbf{r}) \tag{18}$$

where $\mathbf{E}'$ is given from (6-21) by:

$$\mathbf{E}'(\mathbf{r}',t') = \frac{\gamma}{4\pi\epsilon_0} \frac{q\mathbf{r}_{01}'}{r_{01}'^3} \tag{19}$$

and $\mathbf{r}'$ is directed from the charge q to the moment $\boldsymbol{\mu}'$. The force on the moment in the primed frame is given by (6-127) as:

$$\mathbf{F}' = (\boldsymbol{\mu}' \times \nabla') \times \mathbf{B}' = -\frac{\gamma}{c^2}(\boldsymbol{\mu}' \times \nabla') \times (\mathbf{v} \times \mathbf{E}) \tag{20}$$

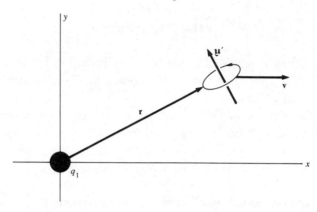

Figure 4 Force on a magnetic moment. A moment moves in the field of an electric source charge, which is at rest.

We transform (20) by writing the force components separately. We obtain, for the longitudinal component,

$$F_x' = -\frac{\gamma}{c^2}[(\boldsymbol{\mu}' \times \nabla')_y(\mathbf{v} \times \mathbf{E})_z - (\boldsymbol{\mu}' \times \nabla')_z(\mathbf{v} \times \mathbf{E})_y]$$

$$= -\gamma\frac{v}{c^2}\left[\left(\mu_z'\frac{\partial}{\partial x'} - \mu_x'\frac{\partial}{\partial z'}\right)E_y + \left(\mu_x'\frac{\partial}{\partial y'} - \mu_y'\frac{\partial}{\partial x'}\right)E_z\right] \tag{21}$$

In writing the expression for F_x, the longitudinal force in the laboratory frame, we we must be extremely careful. Because a magnetic moment is an extended current, it makes a difference whether the force is evaluated at fixed time in the primed frame or at fixed time in the laboratory frame. Since we wish the force at fixed time in the laboratory frame we write from (F-1), (F-2), and (F-13)

$$dx' = \gamma\, d(x - Vt) = \gamma\, dx \qquad dy' = dy \qquad dz' = dz \tag{22}$$

Using this transformation together with (F-93) we obtain

$$F_x = F_x' = -\gamma\frac{v}{c^2}\left[\left(\frac{\mu_z'}{\gamma}\frac{\partial}{\partial x} - \mu_x'\frac{\partial}{\partial z}\right)E_y + \left(\mu_x'\frac{\partial}{\partial y} - \frac{\mu_y'}{\gamma}\frac{\partial}{\partial x}\right)E_z\right] \tag{23}$$

Using $\nabla \times \mathbf{E} = 0$ in the unprimed frame we obtain for (23)

$$F_x = \frac{v}{c^2} \left(\mu_y' \frac{\partial}{\partial z} - \mu_z' \frac{\partial}{\partial y} \right) E_x \tag{24}$$

For the transverse components we obtain:

$$F_{yz} = F_{yz}'/\gamma = \pm \frac{1}{c^2} (\boldsymbol{\mu}' \times \mathbf{V}')_x (\mathbf{v} \times \mathbf{E})_{zy} = \frac{v}{c^2} \left(\mu_y' \frac{\partial}{\partial z} - \mu_z' \frac{\partial}{\partial y} \right) E_{yz} \tag{25}$$

where we have used the result of Exercise F-22 together with (22). Combining (24) and (25) we obtain:

$$\blacktriangleright \qquad \mathbf{F}(\mathbf{r}) = \frac{1}{c^2} (\mathbf{v} \times \boldsymbol{\mu}') \cdot \nabla\, \mathbf{E}(\mathbf{r}) \tag{26}$$

This equation is similar to that for the force on an electric dipole moment in an electric field gradient (1 105):

$$\mathbf{F}(\mathbf{r}) = \mathbf{p} \cdot \nabla\, \mathbf{E}(\mathbf{r}) \tag{27}$$

and suggests that a moving *magnetic* moment has an *electric* dipole moment:[2]

$$\blacktriangleright \qquad \mathbf{p} = \frac{1}{c^2} \mathbf{v} \times \boldsymbol{\mu}' \tag{28}$$

3 Spin-Orbit Coupling. As a result of the presence of a magnetic moment, the energy of an electron in an atom depends on the relative orientation of its spin and orbital angular momentum. This additional energy term is called the spin-orbit energy. Usual treatments of this problem transform to a frame moving instantaneously with the electron. In such a frame the nucleus is in motion, generating a magnetic field. The interaction of the magnetic moment of the electron with this magnetic field gives the coupling energy.

By using (28) we may look at this interaction from a somewhat different point of view. For an electron in uniform motion we write the coupling energy as:

$$U = -\mathbf{p} \cdot \mathbf{E} \tag{29}$$

where $\mathbf{p}$ is the electric dipole moment given by (28). Where the electron is in *rotational* motion the energy is only half as large as a result of Thomas precession[3]:

$$U = -\tfrac{1}{2}\mathbf{p} \cdot \mathbf{E} \tag{30}$$

---

[2] For an extended discussion of the electric dipole moment of a moving magnetic dipole, see: G. P. Fisher, *Am. J. Phys.*, **39**, 1528 (1971).

[3] See, for example, Robert Eisberg and Robert Resnick, *Quantum Physics*, Wiley, 1974, Appendix J. See also G. P. Fisher, "The Thomas Precession," *Am. J. Phys.*, **40**, 1772 (1972).

We have then by substituting **(28)** into **(30)**:

$$U = -\frac{1}{2c^2}\, \mathbf{E} \cdot \mathbf{v} \times \mathbf{\mu}' = -\frac{1}{2c^2}\, \mathbf{\mu}' \cdot \mathbf{v} \times \mathbf{E} \tag{31}$$

For a central electric field we have:

$$\mathbf{E} = -\nabla \phi = -\hat{\mathbf{r}}\,\frac{d\phi}{dr} \tag{32}$$

Writing the angular momentum as:

$$\mathbf{L} = m\mathbf{r} \times \mathbf{v} \tag{33}$$

we may write **(31)** as:

$$U = -\frac{1}{2mc^2}\, \mathbf{\mu}' \cdot \mathbf{L}\, \frac{1}{r}\frac{d\phi}{dr} \tag{34}$$

The magnetic moment of the electron may be written approximately (neglecting radiative corrections) as

$$\mathbf{\mu}' = -\frac{e}{m}\, \mathbf{S} \tag{35}$$

Finally, we write the spin-orbit energy as:

$$U = \frac{1}{2m^2 c^2}\, \mathbf{S} \cdot \mathbf{L}\, \frac{e}{r}\frac{d\phi}{dr} \tag{36}$$

Exercises

4. Estimate the spin-orbit energy splitting between the $^2p_{3/2}$ and $^2p_{1/2}$ states of atomic hydrogen.
5. Find the electric dipole moment of the moving electron in the $2p$ states of hydrogen. Express the moment in electron-metres and compare with the orbital radius.
6. In the frame of the moving electron the spin-orbit interaction may be regarded as the interaction between the magnetic moment of the electron and the magnetic field **B** generated by the motion of the nucleus. Determine the magnitude of **B** in the $2p$ states of hydrogen.

4 Magnetic Charge. Our discussion thus far has been limited to the magnetic fields of moving electric charges. The magnetic moments of current distributions, as we have seen in Section 6-17, are dipole and higher. Magnetic monopoles had been postulated in the theory of elementary particles[4] as described in Section D-5. They are expected to have a *magnetic* charge $\bar{q} = g$ an integral multiple of:

$$g_0/c = 68.518e = 1.097790 \times 10^{-17} \text{ C} \tag{37}$$

where e is the electron charge. The field of a magnetic charge $\bar{q}$ will be given by:

▶
$$\mathbf{B}(\mathbf{r}) = \frac{\mu_0}{4\pi} \frac{\bar{q}}{r_{01}^2} \hat{r}_{01} \tag{38}$$

To find the fields developed by moving magnetic charge we must repeat our earlier discussion. We show in Figure 5a magnetic charge $\bar{q}$ moving past a charge q at rest.

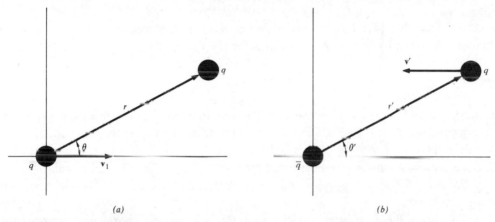

(a)   (b)

Figure 5 Force on a test charge. (a) A magnetic charge $\bar{q}$ moves with velocity $\mathbf{v}_1$ past an electric test charge q. (b) Transforming to a frame in which the magnetic charge is at rest, the test charge moves with velocity $\mathbf{v}'$.

In the frame in which $\bar{q}$ is at rest q moves with velocity $\mathbf{v}' = -\mathbf{v}_1$ as shown in Figure 5b. The force on the test charge q in the moving frame is given by:

$$\mathbf{F}' = q\mathbf{v}' \times \mathbf{B}' = \frac{\mu_0}{4\pi} \frac{q\bar{q}}{r'^2} \mathbf{v}' \times \hat{\mathbf{r}}' \tag{39}$$

Transforming to the laboratory frame we obtain for the electric field,

$$\mathbf{E}(\mathbf{r}) = \frac{\gamma}{q} \mathbf{F}' = -\frac{1}{4\pi\epsilon_0} \frac{\bar{q}/c}{\gamma^2 r^2 (1 - \beta^2 \sin^2 \theta)^{3/2}} \frac{1}{c} \mathbf{v}_1 \times \hat{\mathbf{r}} \tag{40}$$

To find the magnetic field we must consider both the magnetic and electric charges to be moving. In this way we find for the magnetic field:

$$\mathbf{B}(\mathbf{r}) = \frac{\mu_0}{4\pi} \frac{\bar{q}}{\gamma^2 r^2 (1 - \beta^2 \sin^2 \theta)^{3/2}} \hat{\mathbf{r}} \tag{41}$$

[4] K. W. Ford, "Magnetic Monopoles," *Scientific American,* **209**(6), 122 (December 1963).

Table 1 Fields of Electric and Magnetic Charge

Electric Charge	Magnetic Charge
$\mathbf{E} = \dfrac{1}{4\pi\epsilon_0}\,\dfrac{q\hat{\mathbf{r}}}{\gamma^2 r^2(1 - \beta^2\sin^2\theta)^{3/2}}$	$\mathbf{B} = \dfrac{\mu_0}{4\pi}\,\dfrac{\bar{q}\hat{\mathbf{r}}}{\gamma^2 r^2(1 - \beta^2\sin^2\theta)^{3/2}}$
$\mathbf{B} = \dfrac{1}{c^2}\,\mathbf{v}\times\mathbf{E}$	$\mathbf{E} = -\mathbf{v}\times\mathbf{B}$

Comparing **(40)** and **(41)** we see that **(16)** applies to magnetic charge as well as to steady currents as it must if the fields **E** and **B** are to be definable from the Lorentz force.

We note that **(41)** has exactly the same *form* as the electric field of a moving charge q as given by **(6-13)**. We have then for the flux of **B**:

$$\Phi = \mu_0\,\bar{q} \qquad (42)$$

which is analogous to **(6-16)**, and we have the result that the magnetic flux from a charge $\bar{q}$ is independent of its velocity with respect to the frame in which the flux is measured. We described this property of electric charge by characterizing q as a Lorentz scalar. We see that in a similar way postulated magnetic charge $\bar{q}$ is a Lorentz scalar.

We show for comparison in Table 1 the electric and magnetic fields of electric and magnetic charges.

To obtain the expression for the force on a magnetic charge we consider the situation shown in Figure 6 where a magnetic moment $\boldsymbol{\mu}$ is at rest in the field of a magnetic charge $\bar{q}$. The force on the magnetic moment is from **(6-127)** and **(38)**:

$$\mathbf{F} = (\boldsymbol{\mu}\times\nabla)\times\mathbf{B} = -\frac{\mu_0}{4\pi}\,\bar{q}\,(\boldsymbol{\mu}\times\nabla)\times\frac{\hat{\mathbf{r}}_{01}}{r_{01}^2} \qquad (43)$$

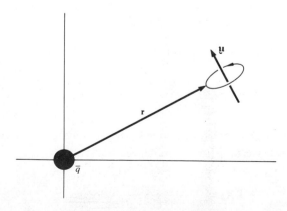

Figure 6 Force on a magnetic charge. The force must be equal and opposite to the force on the magnetic moment.

The force acting on the magnetic charge must be the negative of **(43)**. On comparison with **(6-122)** we are able to write then for the force on $\bar{q}$

$$\mathbf{F}(\mathbf{r}) = \bar{q}\mathbf{B}(\mathbf{r}) \tag{44}$$

Although we have obtained **(44)** from the interaction between $\bar{q}$ and a single magnetic moment, it is completely general since we can generate any steady solenoidal distribution of currents from a volume distribution of moments. If the sources of the magnetic field are in uniform motion, we transform to the moving frame in which we write **(44)**. Transforming back into the laboratory frame we obtain the Lorentz force for magnetic charge:

$$\blacktriangleright \qquad \mathbf{F} = \bar{q}\left(\mathbf{B} - \frac{1}{c^2}\mathbf{v} \times \mathbf{E}\right) \tag{45}$$

Since we cannot distinguish between an electric field that arises from a moving magnetic field and an electric field that arises from electric charge, **(45)** must, in fact, be independent of the sources of either **B** or **E**.[5]

Exercises
_____

7. From **(45)** show that the force on a moving magnetic dipole produced by magnetic charge is:

$$\mathbf{F} = \boldsymbol{\mu} \cdot \nabla\mathbf{B} - \frac{1}{c^2}\mathbf{v} \times (\boldsymbol{\mu} \cdot \nabla)\mathbf{E}$$

Compare the second term with the force on a moving magnetic dipole produced by electric current from **(26)**. See also **(6-130)**.

8. An electronic charge e and a magnetic monopole g are both in motion with the same velocity **v**. Use the Lorentz force expressions **(6-20)** and **(45)** to show that the force on each in the fields of the other is zero. Show how the same result may be obtained by transforming to a frame moving with $\mathbf{V} = \mathbf{v}$.

TRANSFORMATION OF CHARGE AND CURRENT

That charge in one frame should appear as current in another frame is perfectly reasonable. The converse, however, also exists as a consequence of special relativity. We imagine a frame in which there is solenoidal current and electrical neutrality. As we will see, as a consequence of the Lorentz transformation, a charge density will appear in a frame that moves with respect to this frame. The simplest way of discussing the transformation of charge and current is in terms of the invariance of local charge conservation.

_____

[5] For a discussion of **(45)** from the point of view of symmetry, see S. Frankel, *Am. J. Phys.*, **44**, 683 (1976).

5 Charge Conservation. We obtained in Section 1-19 the statement of local charge conservation:

$$\blacktriangleright \qquad \nabla \cdot \mathbf{j}(\mathbf{r},t) + \frac{\partial}{\partial t}\, \rho(\mathbf{r},t) = 0 \qquad\qquad (46)$$

Using the transformation of Section F-3 we may write (46) as:

$$\frac{\partial}{\partial x}\, \gamma(j_x - V\rho) + \frac{\partial}{\partial y}\, j_y + \frac{\partial}{\partial z}\, j_z + \frac{\partial}{\partial t}\, \gamma(\rho - j_x V/c^2) = 0 \qquad (47)$$

Now by the first postulate of relativity we must be able to write a statement of charge conservation in the primed frame:

$$\frac{\partial}{\partial x'}\, j_x' + \frac{\partial}{\partial y'}\, j_y' + \frac{\partial}{\partial z'}\, j_z' + \frac{\partial}{\partial t'}\, \rho' = 0 \qquad\qquad (48)$$

Comparing (47) and (48) we arrive at the transformation of charge and current:

$$\blacktriangleright \qquad\qquad j_x' = \gamma(j_x - V\rho) \qquad\qquad (49)$$

$$\blacktriangleright \qquad\qquad j_{yz}' = j_{yz} \qquad\qquad (50)$$

$$\blacktriangleright \qquad\qquad \rho' = \gamma(\rho - j_x V/c^2) \qquad\qquad (51)$$

Note the similarity between (49), (50), and (51) and the Lorentz transformation:

$$x' = \gamma(x - Vt) \qquad\qquad (52)$$

$$y' = y \qquad\qquad (53)$$

$$z' = z \qquad\qquad (54)$$

$$t' = \gamma(t - Vx/c^2) \qquad\qquad (55)$$

As may be seen, current density transforms like a displacement while charge density transforms like a time. As with the Lorentz transformation we may write the inverse transformation of current and charge by solving (49), (50), and (51) for the current and charge densities in the laboratory frame to obtain:

$$\blacktriangleright \qquad\qquad j_x = \gamma(j_x' + V\rho') \qquad\qquad (56)$$

$$\blacktriangleright \qquad\qquad j_{yz} = j_{yz}' \qquad\qquad (57)$$

$$\blacktriangleright \qquad\qquad \rho = \gamma(\rho' + j_x'V/c^2) \qquad\qquad (58)$$

Note that if we had simply interchanged and primed and unprimed components and the sign of V in (49), (50), and (51) we would also have obtained (56), (57), (58). Thus the transformation of current and charge is completely symmetric. This is to be expected since the transformation is a local one, connecting densities at positions and times related by the Lorentz transformation.

We may use the transformation of current and charge to find the way in which a uniformly moving electric or magnetic dipole moment transforms into the laboratory frame. We write the electric dipole moment in the moving frame as

$$\mathbf{p}' = \int \mathbf{r}' \rho(\mathbf{r}') \, dV' \tag{59}$$

Substituting from (51) and (52) and using (49) and (6-116) we obtain for the longitudinal component of the electric dipole moment in the laboratory frame:

$$p_x = p_x'/\gamma + qVt \tag{60}$$

where q is the total charge. The longitudinal component of $\mathbf{p}'$ is simply reduced by Lorentz contraction.

To obtain the transverse components of $\mathbf{p}$ we use (51), (49), (53), and (54) with (6-120) to obtain:

$$p_y = p_y' - (V/c^2)\mu_z' \tag{61}$$

$$p_z = p_z' + (V/c^2)\mu_y' \tag{62}$$

Note that (61) and (62) give the same result as (28) for a moving magnetic moment. The relations (60), (61), and (62) are not symmetric in that they require that both $\mathbf{p}'$ and $\boldsymbol{\mu}'$ be at rest in the primed frame. Since the electric and magnetic dipole moments represent an integration over volume, the obtained relations can not be local as with the current and charge densities.

SOURCES OF THE MAGNETIC FIELD

We next investigate the sources of flux and circulation of the magnetic field $\mathbf{B}$ produced by a *moving current* as given by (15). As we will see, we continue to have (6-59) even when the sources of $\mathbf{B}$ are in motion. On the other hand, Ampère's law (6-48) will be modified when the sources of $\mathbf{B}$ are in motion.

6 Magnetic Flux. We write for the divergence of the magnetic field:

$$\nabla \cdot \mathbf{B}(\mathbf{r},t) = \frac{\partial}{\partial x} B_x(\mathbf{r},t) + \frac{\partial}{\partial y} B_y(\mathbf{r},t) + \frac{\partial}{\partial z} B_z(\mathbf{r},t) \tag{63}$$

Substituting from (17) and the results of Section F-3 where $\mathbf{B}'(\mathbf{r}')$ is independent of t' we obtain:

$$\nabla \cdot \mathbf{B}(\mathbf{r},t) = \gamma \nabla' \cdot \mathbf{B}'(\mathbf{r}') \tag{64}$$

But $\mathbf{B}'(\mathbf{r}')$ is the field of a distribution of steady solenoidal currents for which we have, from **(6-59)**

$$\nabla' \cdot \mathbf{B}'(\mathbf{r}') = 0 \tag{65}$$

which gives then for currents in uniform motion

$$\nabla \cdot \mathbf{B}(\mathbf{r},t) = 0 \tag{66}$$

As we will see in Section 11-9, **(66)** applies to any motion of *electric* charge.

7 Magnetic Circulation. We write for the curl of the magnetic field:

$$\nabla \times \mathbf{B}(\mathbf{r},t) = \hat{\mathbf{x}} \left[\frac{\partial}{\partial y} B_z(\mathbf{r},t) - \frac{\partial}{\partial z} B_y(\mathbf{r},t) \right]$$

$$+ \hat{\mathbf{y}} \left[\frac{\partial}{\partial z} B_x(\mathbf{r},t) - \frac{\partial}{\partial x} B_z(\mathbf{r},t) \right]$$

$$+ \hat{\mathbf{z}} \left[\frac{\partial}{\partial x} B_y(\mathbf{r},t) - \frac{\partial}{\partial y} B_x(\mathbf{r},t) \right] \tag{67}$$

Substituting from **(17)** and the results of Section F-3 we write

$$(\nabla \times \mathbf{B})_x = \gamma \left(\frac{\partial}{\partial y'} B_z' - \frac{\partial}{\partial z'} B_y' \right) = \gamma (\nabla' \times \mathbf{B}')_x \tag{68}$$

$$(\nabla \times \mathbf{B})_y = \frac{\partial}{\partial z'} B_x' - \gamma^2 \frac{\partial}{\partial x'} B_z' = (\nabla' \times \mathbf{B}')_y - \beta^2 \gamma^2 \frac{\partial}{\partial x'} B_z' \tag{69}$$

$$(\nabla \times \mathbf{B})_z = \gamma^2 \frac{\partial}{\partial x'} B_y' - \frac{\partial}{\partial y'} B_x' = (\nabla' \times \mathbf{B}')_z + \beta^2 \gamma^2 \frac{\partial}{\partial x'} B_y' \tag{70}$$

Since $\mathbf{B}'(\mathbf{r}')$ is the magnetic field of a stationary distribution of currents we may write **(6-48)**

$$\nabla' \times \mathbf{B}'(\mathbf{r}') = \mu_0 \mathbf{j}'(\mathbf{r}') \tag{71}$$

Using **(56)** and **(57)** with $\rho' = 0$ we may combine **(68)**, **(69)**, and **(70)**:

$$\nabla \times \mathbf{B} = \mu_0 \mathbf{j} + \beta^2 \gamma^2 \frac{\partial}{\partial x'} (\hat{\mathbf{x}} \times \mathbf{B}') \tag{72}$$

Using **(17)** and **(F-27)** with $\mathbf{B}'(\mathbf{r}')$ independent of t' we may rewrite **(72)** as:

$$\mathbf{V} \times \mathbf{B} - \frac{1}{c^2}\,\mathbf{v}_1 \cdot \mathbf{V}(\mathbf{v}_1 \times \mathbf{B}) = \mu_0\,\mathbf{j} \tag{73}$$

In Section 11-9 we will obtain a more general form for **(73)**.

Exercise

---

9. Show that **(73)** may be written in the form:

$$\mathbf{V} \times \mathbf{B}(\mathbf{r},t) = \mu_0\left[\mathbf{j}(\mathbf{r},t) + \epsilon_0\,\frac{\partial}{\partial t}\,\mathbf{E}(\mathbf{r},t)\right]$$

where $\mathbf{E}(\mathbf{r},t)$ is given by **(16)**.

ELECTROMAGNETIC INDUCTION

The electric field given by **(16)** has properties very different from those of an electric field arising from charges. We first examine these properties for the special case that we have considered of a uniformly moving distribution of steady solenoidal currents. Following this introduction we will discuss the experimental investigations of Faraday and the generalization of these results.

8 Circulation of the Electric Field. The circulation of the electric field is given by:

$$\mathscr{E} = \oint_C \mathbf{E} \cdot d\mathbf{r} \tag{74}$$

where $\mathscr{E}$ is called the electromotive force.

We now show that **(74)** may be written in terms of the rate of change of magnetic flux through the contour C. Taking **(16)** for the electric field we obtain:

$$\mathscr{E} = -\oint_C \mathbf{v}_1 \times \mathbf{B}(\mathbf{r}) \cdot d\mathbf{r} = -\mathbf{v}_1 \cdot \oint_C \mathbf{B}(\mathbf{r}) \times d\mathbf{r} \tag{75}$$

We may reexpress **(75)** by finding the flux of **B** through the contour C:

$$\Phi = \oint_S \mathbf{B}(\mathbf{r}) \cdot d\mathbf{S} \tag{76}$$

as shown in Figure 7a. The change in flux through S in a time dt may be written as:

$$\frac{\partial \Phi}{\partial t} dt = \Phi(t + dt) - \Phi(t) = \oint_C \mathbf{B}(\mathbf{r}) \cdot d\mathbf{r} \times d\mathbf{s}$$

$$= d\mathbf{s} \cdot \oint_C \mathbf{B}(\mathbf{r}) \times d\mathbf{r} \tag{77}$$

where $(7b)$ $d\mathbf{s} = \mathbf{v}_1 \, dt$ is the displacement of the *flux* in dt. In writing (77) we have used the fact that the flux through S increases in dt by just the flux passing out the sides of the figure.

Comparing (75) and (77) we obtain, for the induced electromotive force,

$$\mathscr{E} = -\frac{\partial}{\partial t} \Phi \tag{78}$$

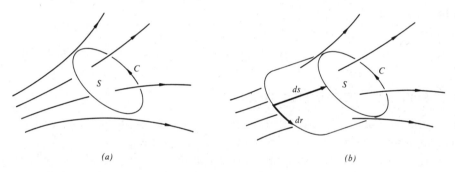

(a)   (b)

Figure 7   Circulation of the electric field. (*a*) The flux enclosed by C is obtained from (76). (*b*) The circulation of the electric field may be related to the rate of change of flux through C.

9 Flux of the Electric Field. The flux of the electric field is given by the relation:

$$\Psi = \oint_S \mathbf{E} \cdot d\mathbf{S} \tag{79}$$

Taking the electric field from (16) we have:

$$\Psi = -\oint_S \mathbf{v}_1 \times \mathbf{B}(\mathbf{r},t) \cdot d\mathbf{S} = -\mathbf{v}_1 \cdot \oint_S \mathbf{B}(\mathbf{r},t) \times d\mathbf{S} \tag{80}$$

From (B-13) we may write:

$$\Psi = \mathbf{v}_1 \cdot \int_V \nabla \times \mathbf{B}(\mathbf{r},t) \, dV = \mu_0 \mathbf{v}_1 \cdot \int_V \mathbf{j}(\mathbf{r},t) \, dV \tag{81}$$

where we have used (73).

Since Ψ may also be written as the integral of $\mathbf{V} \cdot \mathbf{E}$ over V and since V is arbitrary we obtain, from **(81)**,

$$\mathbf{V} \cdot \mathbf{E}(\mathbf{r},t) = \mu_0 \mathbf{v}_1 \cdot \mathbf{j}(\mathbf{r},t) \tag{82}$$

In discussing the electric field of moving charges, we obtained the generalization of Gauss's law:

$$\mathbf{V} \cdot \mathbf{E}(\mathbf{r},t) = \frac{1}{\epsilon_0} \rho(\mathbf{r},t) \tag{83}$$

Comparing **(82)** and **(83)** we *must* conclude that there is a charge density in the laboratory frame given by:

$$\blacktriangleright \qquad \rho(\mathbf{r},t) = \frac{1}{c^2} \mathbf{v}_1 \cdot \mathbf{j}(\mathbf{r},t) \tag{84}$$

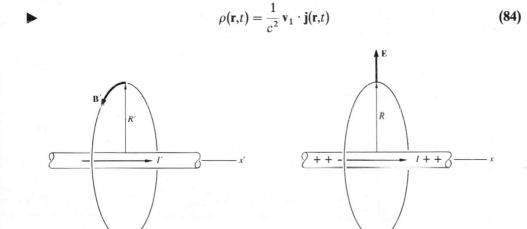

(a)   (b)

Figure 8   Generation of a charge by moving current. (*a*) In the primed frame the wire is electrically neutral and there is only a magnetic field given by **(85)**. (*b*) In the laboratory frame there is an electric field, which must arise from charge.

On comparison with the general transformation statement **(51)** we see that with $\rho(\mathbf{r}') = 0$ as for the present case we obtain **(84)**.

10   Charge Generation. The remarkable result **(84)** that a moving current generates an electric charge density deserves further discussion, and we will want to consider some specific examples. As a first example, we consider a wire carrying current along the $\hat{x}$ direction. We show in Figure 8*a* the situation in the primed frame in which the wire is electrically neutral and we have only a magnetic field:

$$B'(R') = \frac{\mu_0}{2\pi} \frac{I'}{R'} \tag{85}$$

In the laboratory frame $(8b)$, in which the wire moves with velocity $v_1\,\hat{\mathbf{x}}$, we have from (16) an electric field:

$$E(R) = v_1 B(R) \tag{86}$$

where the electric field is radial and the magnetic field is azimuthal with magnitudes:

$$E(R) = \frac{1}{2\pi\epsilon_0}\frac{\lambda}{R} \qquad B(R) = \frac{\mu_0}{2\pi}\frac{I}{R} \tag{87}$$

Substituting into (86) we obtain, for the charge per unit length,

$$\lambda = \epsilon_0\,\mu_0\,v_1 I = v_1 I/c^2 \tag{88}$$

which is consistent with (84).

The electric field $E(R)$ may be seen to arise from Lorentz contraction by the following argument.[6] We assume in the primed frame a current of positive carriers:

$$j_p{}' = \rho_p{}' v_p{}' \tag{89}$$

and a current of negative carriers:

$$j_n{}' = \rho_n{}' v_n{}' \tag{90}$$

We assume electrical neutrality in the primed frame:

$$\rho' = \rho_p{}' + \rho_n{}' = 0 \tag{91}$$

with a current:

$$j' = j_p{}' + j_n{}' \tag{92}$$

We transform to the unprimed frame, with respect to which the primed frame is moving with velocity $\mathbf{V} = \mathbf{v}_1$. As a result of Lorentz contraction we have for the charge density in the primed frame with respect to the unprimed frame:

$$\rho_p{}' = \rho_p\frac{(1 - v_p{}^2/c^2)^{1/2}}{(1 - v_p{}'^2/c^2)^{1/2}} = \gamma\rho_p(1 - v_p V/c^2) \tag{93}$$

where we have used (F-61) for the second equality. Similarly, we have for the negative carriers:

$$\rho_n{}' = \gamma\rho_n(1 - v_n V/c^2) \tag{94}$$

---

[6] A similar argument is given by E. M. Purcell, *Electricity and Magnetism*, Berkeley physics course volume 2, McGraw-Hill, 1965, Section 5.9. See also N. Arista and A. Lopez, *Am. J. Phys.* **43**, 525 (1975) and L. S. Lerner, *Am. J. Phys.* **41**, 724 (1973); **42**, 260 (1974).

with $\gamma = (1 - V^2/c^2)^{-1/2}$. Now the charge density in the unprimed frame is given by:

$$\rho = \rho_p + \rho_n \qquad (95)$$

and the current density is given by:

$$j = j_p + j_n = \rho_p v_p + \rho_n v_n \qquad (96)$$

Adding **(93)** and **(94)** and using **(91)** and **(96)** we obtain **(88)**.

Exercise

8. Obtain from **(93)** and **(94)** the transformation equations **(51)** and **(49)**:

$$\rho' = \gamma(\rho - jV/c^2)$$
$$j' = \gamma(j - \rho V)$$

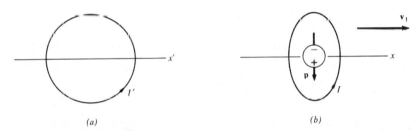

(a)   (b)

Figure 9 Moving current loop. (a) The loop is at rest in the primed frame and is electrically neutral. (b) In the laboratory frame, in which the loop is moving, it is contracted and has developed an electric dipole moment.

11   Electric Dipole Moment. As a second example of **(84)** we consider a moving current loop. We take the loop to be at rest in the primed frame as shown in Figure 9a. The loop is electrically neutral and carries in this frame a current I'. In the unprimed frame the loop moves to the right with velocity v_1. The loop is shown contracted along $\hat{x}$ as expected. But, in addition, an electric dipole moment is developed as shown in Figure 9b.

The origin of the electric dipole moment may be understood from the transformation of velocities. We write the Lorentz transformation for a charge q moving on the loop as:

$$x = \gamma(x' + v_1 t') \qquad t = \gamma(t' + x'v_1/c^2) \qquad (97)$$

Dividing the differentials of **(97)** we obtain for the velocity:

$$v_x = \frac{v_x' + v_1}{1 + v_x'v_1/c^2} \qquad (98)$$

We show in Figure 10a and 10b v_x' and v_x as functions of the time. The x component of velocity is sinusoidal in the primed frame but distorted as shown in the laboratory frame. Then a positive charge rotating counterclockwise will spend more time on the bottom half where $\mathbf{I}'$ and $\mathbf{v}_1$ are in the same direction. On the other hand, a negative charge rotating clockwise will spend more time on the top half where $\mathbf{I}'$ and $\mathbf{v}_1$ are antiparallel. Both charges contribute in the same way to the electric dipole moment.

We may use **(84)** to obtain a general expression for the induced electric dipole moment. The electric dipole moment is given by **(1-75)** as

$$\mathbf{p} = \int_{V_1} \mathbf{r}_1 \rho(\mathbf{r}_1)\, dV_1 \tag{99}$$

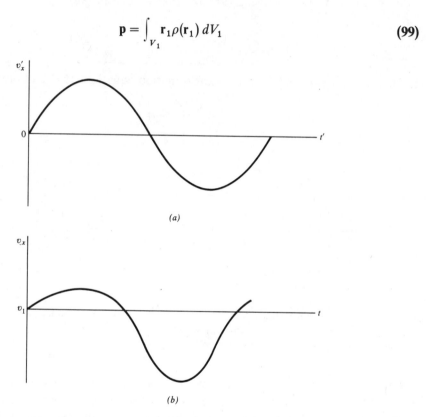

(a)

(b)

Figure 10   Longitudinal velocity. (a) In the primed frame the longitudinal velocity is sinusoidal. (b) In the laboratory frame the longitudinal velocity is distorted with the charge spending more time on the bottom half.

Substituting from **(84)** we obtain:

$$\mathbf{p} = \frac{1}{c^2} \int_{V_1} \mathbf{r}_1 [\mathbf{v}_1 \cdot \mathbf{j}(\mathbf{r}_1)]\, dV_1 \tag{100}$$

Now we write $\mathbf{j}(\mathbf{r}_1)$ as the sum of a polarization current and a magnetization current:

$$\mathbf{j}(\mathbf{r}_1) = \rho(\mathbf{r}_1)\mathbf{v}_1 + \mathbf{j}_M(\mathbf{r}_1) \tag{101}$$

Substituting **(101)** into **(100)** we obtain

$$\mathbf{p} = \frac{\gamma^2}{c^2} \int_{V_1} \mathbf{r}_1 [\mathbf{v}_1 \cdot \mathbf{j}_M(\mathbf{r}_1)] \, dV_1 \tag{102}$$

As shown in **(6-116)**, the tensor:

$$\mathbf{T} = \int_{V_1} \mathbf{r}_1 \, \mathbf{j}_M(\mathbf{r}) \, dV_1 \tag{103}$$

is antisymmetric so that we may write **p** as

$$\mathbf{p} = \frac{\gamma^2}{2c^2} \int_{V_1} (\mathbf{r}_1 \mathbf{j}_M - \mathbf{j}_M \mathbf{r}_1) \cdot \mathbf{v}_1 \, dV_1 = \frac{\gamma^2}{2c^2} \mathbf{v}_1 \times \int_{V_1} \mathbf{r}_1 \times \mathbf{j}_M \, dV_1 \tag{104}$$

Finally, from **(6-110)** we may write:

$$\mathbf{p} = \frac{\gamma^2}{c^2} \mathbf{v}_1 \times \boldsymbol{\mu} \tag{105}$$

For a body with magnetization $\mathbf{M}(\mathbf{r})$ moving with velocity $\mathbf{v}(\mathbf{r})$ we may write, for the polarization density from **(105)**,

$$\mathbf{P}(\mathbf{r}) = \frac{\gamma^2}{c^2} \mathbf{v}(\mathbf{r}) \times \mathbf{M}(\mathbf{r}) \tag{106}$$

As a check we take the divergence of **(106)** to obtain:

$$\rho(\mathbf{r}) = -\nabla \cdot \mathbf{P}(\mathbf{r}) = \frac{\gamma^2}{c^2} \nabla \cdot (\mathbf{M} \times \mathbf{v}) = \frac{\gamma^2}{c^2} \mathbf{v} \cdot \mathbf{j}_M(\mathbf{r}) \tag{107}$$

Substituting from (101) for $\mathbf{j}_M(\mathbf{r})$ we obtain:

$$\rho(\mathbf{r}) = \frac{1}{c^2} \mathbf{v} \cdot \mathbf{j}(\mathbf{r}) \tag{108}$$

which is just **(51)** with the primed frame electrically neutral.

Exercise

---

10. A charge q travels with speed s around a square of side a. In a frame in which the square moves with velocity **v** parallel to one of the edges, obtain an expression for **p** and compare with **(105)**.

FARADAY'S INVESTIGATIONS

We describe here very briefly the experimental investigations of Michael Faraday.[7] It is useful to consider a pair of circuits C_1 and C_2 as shown in Figure 11. We imagine that C_1 carries a current I_1 and we measure in C_2 an electromotive force $\mathscr{E}_2$

Faraday carried out three kinds of experiments:

1. He moved C_1 uniformly while it carried a constant current I_1 and measured $\mathscr{E}_2$ with a galvanometer. This is just the situation that we have been discussing and is described by **(78)**.

2. He moved C_2 uniformly while keeping C_1 fixed.

3. While holding C_1 and C_2 fixed, he varied the *magnitude* of I_1 and observed $\mathscr{E}_2$.

12 Motional Electromotive Force. We expect for Faraday's second experiment that a *unit charge* carried around C_2 will experience a *motional* electromotive force:

$$\mathscr{F} = \frac{1}{q}\oint_C \mathbf{F}\cdot d\mathbf{r}_2 = \oint_{C_2} \mathbf{v}_2 \times \mathbf{B}(\mathbf{r})\cdot d\mathbf{r}_2 \tag{109}$$

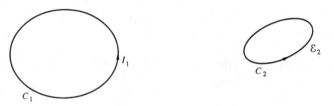

Figure 11   Interaction between a pair of circuits. The first circuit carries a current I_1. An electromotive force is measured in the second circuit.

Now by **(75)** we see that **(109)** may *also* be written as the negative of the rate of change of flux through the circuit:

$$\mathscr{F} = -\frac{d}{dt}\Phi_2 \tag{110}$$

And thus we expect, as Faraday observed, that the current induced in C_2 is the same whether C_1 or C_2 is moved—as long as the relative motion is the same.

Although motion of C_2 produces exactly the same effect as the equivalent relative motion of C_1, the charges are *not* driven by an electric field in the laboratory frame, but rather by the motion across the magnetic field as given by **(1)**. Einstein's 1905 paper[8] begins with the recognition of this asymmetry:

---

[7] A fairly complete discussion of Faraday's investigations is given by James Clerk Maxwell, *A Treatise on Electricity and Magnetism*, Part IV, Chapter III, Third Edition, Clarendon Press, 1891. Reprinted by Dover Publications, 1954.

[8] Einstein, A. "On the Electrodynamics of Moving Bodies," *Annalen der Physik*, **17**, 891 (1905). Translated in *The Principle of Relativity*, Methuen, 1923. Reprinted by Dover, 1952.

It is known that Maxwell's electrodynamics—as usually understood at the present time—when applied to moving bodies, leads to asymmetries which do not appear to be inherent in the phenomena. Take, for example, the reciprocal electrodynamic action of a magnet and a conductor. The observable phenomenon here depends only on the relative motion of the conductor and the magnet, whereas the customary view draws a sharp distinction between the two cases in which either the one or the other of these bodies is in motion. For if the magnet is in motion and the conductor at rest, there arises certain definite energy, producing a current at the places where parts of the conductor are situated. But if the magnet is stationary and the conductor in motion, no electric field arises in the neighborhood of the magnet. In the conductor, however, we find an electromotive force to which in itself there is no corresponding energy, but which gives rise— assuming equality of relative motion in the two cases discussed—to electric currents of the same path and intensity as those produced by the electric forces in the former case.

13   **Induced Electromotive Force.** Faraday's third experiment involved varying the *magnitude* of the current through C_1 while holding both C_1 and C_2 fixed. His remarkable discovery was that changing the flux through the second circuit in this way produced exactly the same response as did relative motion of the two circuits. Thus the circulation of the electric field is independent of the way in which the enclosed flux is changed:

$$\mathscr{E} = \oint_C \mathbf{E} \cdot d\mathbf{r} = -\frac{\partial}{\partial t}\Phi \tag{111}$$

From Stokes's theorem **(B-14)** we may also write the electromotive force as

$$\mathscr{E} = \int_S \mathbf{V} \times \mathbf{E(r)} \cdot d\mathbf{S} \tag{112}$$

Since the flux is the integral of the normal component of **B** over S, we may write for the electromotive force from **(111)**

$$\mathscr{E} = -\int_S \frac{\partial}{\partial t} \mathbf{B(t)} \cdot d\mathbf{S} \tag{113}$$

Since the contour C enclosing S is completely arbitrary we finally write from **(112)** and **(113)** the local relation

▶ $$\nabla \times \mathbf{E(r)} = -\frac{\partial}{\partial t}\mathbf{B(r)} \tag{114}$$

which is called *Faraday's law.*

    As we will show in Section 11-9, Faraday's law and Gauss's law applied to arbitrarily moving charges together with the Lorentz force law provide all that is required for a complete formulation of classical electromagnetic theory, given the postulates of relativity.

Exercises

11. A magnetic field changes everywhere *linearly* with the time. Obtain from **(114)** the expression for the resulting field:

$$\mathbf{E}(\mathbf{r}) = -\frac{1}{4\pi} \int \frac{\partial}{\partial t} \mathbf{B}(\mathbf{r}_1) \times \frac{\hat{\mathbf{r}}_{01}}{r_{01}^2} \, dV_1$$

Use the analogy between Ampère's law **(6-48)** and the Biot-Savart law **(6-40)**. Explain why the expression for **E** is valid only so long as the magnetic fields are changing linearly.

12. A magnetic field is a function of position and time $\mathbf{B}(\mathbf{r},t)$. Show that the convective derivative (Section 5-4) of the flux through a uniformly moving contour C is given by:

$$\frac{d}{dt} \Phi = \int_S \frac{\partial \mathbf{B}}{\partial t} \cdot d\mathbf{S} - \oint_C (\mathbf{v} \times \mathbf{B}) \cdot d\mathbf{r}$$

where **v** is the velocity of the contour.

14 **[Vector Potential].** We introduced the vector potential as **(6-49)**:

$$\mathbf{A}(\mathbf{r}) = \frac{\mu_0}{4\pi} \int_{V_1} \frac{\mathbf{j}(\mathbf{r}_1)}{r_{01}} \, dV_1 \tag{115}$$

where we assumed that the currents were both steady and solenoidal. We may relax the condition that the currents be steady as long as they are changing *slowly*. The condition basically is that the change in $\mathbf{j}(\mathbf{r}_1)$ be small during the time required for a light signal to cross the system of currents.

Now if we write for the magnetic field:

$$\mathbf{B}(\mathbf{r}) = \frac{\mu_0}{4\pi} \int \frac{\mathbf{j}(\mathbf{r}_1) \times \hat{\mathbf{r}}_{01}}{r_{01}^2} \, dV_1 \tag{116}$$

under the same conditions, then we may continue to write:

$$\mathbf{B}(\mathbf{r}) = \nabla \times \mathbf{A}(\mathbf{r}) \tag{117}$$

Combining **(117)** with Faraday's law **(114)** we have:

$$\nabla \times \mathbf{E} = -\frac{\partial}{\partial t} \nabla \times \mathbf{A} \tag{118}$$

Integrating **(118)** we write in the absence of charge (thus assuming solenoidal currents):

$$\mathbf{E} = -\frac{\partial \mathbf{A}}{\partial t} \tag{119}$$

Exercise

13. Show that the electric field in Exercise 11 may also be written as:

$$\mathbf{E}(\mathbf{r}) = -\frac{1}{4\pi} \int \frac{1}{r_{01}} \, \mathbf{V}_1 \times \frac{\partial}{\partial t} \, \mathbf{B}(\mathbf{r}_1) \, dV_1$$

$$= -\frac{\mu_0}{4\pi} \frac{\partial}{\partial t} \int \frac{\mathbf{j}(\mathbf{r}_1)}{r_{01}} \, dV_1 = -\frac{\partial}{\partial t} \, \mathbf{A}(\mathbf{r})$$

EDDY CURRENTS

We may combine Faraday's law **(114)**:

$$\mathbf{V} \times \mathbf{E} = -\frac{\partial \mathbf{B}}{\partial t} \tag{120}$$

with Ampère's law **(6-48)**:

$$\mathbf{V} \times \mathbf{B} = \mu_0 \mathbf{j} \tag{121}$$

and a phenomenological relation between current and field such as **(5-18)**:

$$\frac{d\mathbf{j}}{dt} + \frac{\mathbf{j}}{\tau} = \frac{\sigma}{\tau} \mathbf{E} \tag{122}$$

to obtain a description of the induction of electric current by a changing magnetic field. We must caution that **(121)** was obtained for steady solenoidal currents. We may extend **(121)** to changing currents as long as the rate of change is not too great. This point will be considered at length in Section 11-10. The condition on the rate of change of $\mathbf{j}$ is that in the time required for a light signal to propagate across the current distribution the change in $\mathbf{j}$ will have been negligible. We are still restricted however to solenoidal currents by the form of **(121)**:

$$\mathbf{V} \cdot \mathbf{j} = -\frac{\partial \rho}{\partial t} = 0 \tag{123}$$

The form of **(122)** is also a consequence of **(123)** in that diffusion current is absent. We have observed in Section 5-4 that **(122)** is not strictly a *local* relation between $\mathbf{j}$ and $\mathbf{E}$ in that the time derivative, which is called the *convective derivative* follows from Newton's second law and moves with the charges. The relation between this derivative and the local or partial derivative is:

$$\frac{d\mathbf{v}}{dt} = \frac{\partial \mathbf{v}}{\partial t} + \mathbf{v} \cdot \mathbf{V}\mathbf{v} \tag{124}$$

Multiplying **(124)** by the charge density ρ and substituting into **(122)** we have

$$\frac{\partial \mathbf{j}}{\partial t} + \rho\langle \mathbf{v} \cdot \nabla\mathbf{v}\rangle + \frac{\mathbf{j}}{\tau} = \frac{\sigma}{\tau}\mathbf{E} \tag{125}$$

where $\mathbf{v}$ is the velocity of an individual charge, and the average is taken over the charges contributing to ρ.

15 **Classical Skin Effect.** At low frequencies the electric field changes slowly with position and the second term in **(125)**, which is associated with the gradient in the current density, may be neglected. In this limit, then, we may ignore the distinction between the convective derivative and the local derivative. With this assumption we obtain from **(120)**, **(121)**, and **(122)** the equation for the magnetic field **B**:

$$\nabla^2\left(\mathbf{B} + \tau\frac{\partial \mathbf{B}}{\partial t}\right) = \mu_0\,\sigma\,\frac{\partial \mathbf{B}}{\partial t} \tag{126}$$

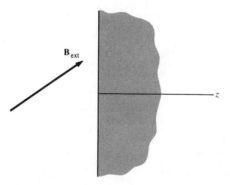

Figure 12   Skin effect. An oscillating magnetic field at the surface of a conductor penetrates into the conductor.

where τ is the relaxation time of the current carriers and σ is the conductivity of the medium. As an example of the solution of **(126)** we assume a magnetic field

$$\mathbf{B} = \mathbf{B}_{\text{ext}}\,e^{-i\omega t} \tag{127}$$

at the surface of a conductor, as shown in Figure 12. We assume within the conductor a field of the form:

$$\mathbf{B}(z,t) = \mathbf{B}_{\text{ext}}\,e^{i(kz - \omega t)} \tag{128}$$

where we have taken **B** to be continuous at the interface. Substituting into **(126)** we have for the propagation vector:

$$k^2 = \frac{i\omega\mu_0\,\sigma}{1 - i\omega\tau} \tag{129}$$

At low frequencies with $\omega\tau \ll 1$ we obtain

$$k = \frac{1+i}{\delta} \tag{130}$$

where

▶
$$\delta = \left(\frac{2}{\omega\mu_0\,\sigma}\right)^{1/2} \tag{131}$$

is called the classical skin depth. At high frequencies with $\omega\tau \gg 1$ (but where we may still neglect the gradient of the current) we obtain:

$$k = i(\mu_0\,\sigma/\tau)^{1/2} = i\omega_p/c \tag{132}$$

which corresponds to purely exponential attentuation into the conductor. The frequency ω_p **(5-138)** is the plasma frequency.

Exercises

14. Using Table 5-1 obtain the classical skin depth in copper at frequencies $\omega/2\pi = 10^8$, 10^{10}, and 10^{12}.
15. Show that the condition for neglecting the gradient term in **(125)** is that the carrier mean free path be short compared with the classical skin depth.
16. Using Table 5-1 obtain the mean free path $v\tau$ in copper at room temperature and compare with the classical skin depth as obtained in Exercise 14.

16 [**Anomalous Skin Effect**]. At higher frequencies it is not possible to neglect the gradient in the current. The exact treatment of transport in this range is a formidable mathematical problem, which is manageable only in what is called the extreme anomalous limit, where the current gradient makes the dominant contribution. The problem may be handled phenomenologically by writing:

$$\rho\langle\mathbf{v}\cdot\nabla\mathbf{v}\rangle = -\tfrac{1}{3}\alpha v\hat{\mathbf{n}}\cdot\nabla\mathbf{j} \tag{133}$$

where $\hat{\mathbf{n}}$ is normal to the surface of the conductor and α is a dimensionless parameter.[9] With the contribution of **(133)** we obtain in place of **(126)**

$$\nabla^2\left(\mathbf{B} - \tfrac{1}{3}\alpha v\tau\hat{\mathbf{n}}\cdot\nabla\mathbf{B} + \tau\frac{\partial\mathbf{B}}{\partial t}\right) = \mu_0\,\sigma\,\frac{\partial\mathbf{B}}{\partial t} \tag{134}$$

---

[9] G. E. H. Reuter and E. H. Sondheimer, *Proc. Roy. Soc.*, **A195**, 336 (1948). The parameter in **(133)** is found to equal $3\sqrt{3}/2\pi$. See also R. G. Chambers, *Proc. Roy. Soc.*, **A65**, 458 (1952) and R. B. Dingle, *Physica*, **19**, 311 (1953).

Again, assuming a magnetic field of the form of (127) at the surface of a conductor and of the form of (128) in the interior, we obtain

$$k^2(1 - \tfrac{1}{3}ik\alpha v\tau - i\omega\tau) = i\omega\mu_0\,\sigma \tag{135}$$

Under extreme anomalous conditions the gradient term is dominant and we have from (135):

$$\blacktriangleright \qquad k = \left(-\frac{3\omega\mu_0\,\sigma}{\alpha v\tau}\right)^{1/3} = (-\tfrac{2}{3}\alpha\delta^2\lambda_C)^{-1/3} \tag{136}$$

where δ is the classical skin depth computed from (131) and $\lambda_C = v\tau$ is the carrier mean free path.

17 **Surface Impedance.** The rate at which energy is delivered per unit area from the electric field to the medium may be written, from (5-56),

$$\frac{dW}{dt} = \tfrac{1}{2}\,\mathrm{Re}\int_0^\infty \mathbf{j}^* \cdot \mathbf{E}\,dz = \frac{1}{2\mu_0}\,\mathrm{Re}\int (\nabla \times \mathbf{B}^*)\cdot\mathbf{E}\,dz \tag{137}$$

where we have used (121). By (B-4) we may write (137) as

$$\frac{dW}{dt} = -\frac{1}{2\mu_0}\,\mathrm{Re}\int \nabla\cdot(\mathbf{E}\times\mathbf{B}^*)\,dz + \frac{1}{2\mu_0}\,\mathrm{Re}\int (\nabla\times\mathbf{E})\cdot B^*\,dz \tag{138}$$

From (120) we write:

$$(\nabla\times\mathbf{E})\cdot\mathbf{B}^* = i\omega\mathbf{B}\cdot\mathbf{B}^* \tag{139}$$

which is pure imaginary so that we may drop the second term in (138). The first term may be integrated to obtain

$$\frac{dW}{dt} = \frac{1}{2\mu_0}\,\mathrm{Re}[\hat{\mathbf{n}}\cdot\mathbf{E}\times\mathbf{B}^*] = \frac{1}{2\mu_0}\,\mathrm{Re}[\hat{\mathbf{n}}\times\mathbf{E}\cdot\mathbf{B}^*] \tag{140}$$

where the fields are evaluated at the surface. From (120) we write:

$$\nabla\times\mathbf{E} = ik\hat{\mathbf{n}}\times\mathbf{E} = i\omega\mathbf{B} \tag{141}$$

Substituting into (140) we obtain:

$$\blacktriangleright \qquad \frac{dW}{dt} = \frac{1}{2\mu_0^{\,2}}\,B^2\,\mathrm{Re}\,\frac{\mu_0\,\omega}{k} \tag{142}$$

We define the surface impedance as

▶
$$Z = R - iX = \frac{\mu_0 \, \omega}{k} \tag{143}$$

where R is the surface resistance and X the surface reactance. We may then write the energy transfer in terms of the surface resistance as:

$$\frac{dW}{dt} = \frac{1}{2\mu_0{}^2} B^2 R \tag{144}$$

Exercises

17. Show that the SI units of Z are ohms.

18. Show that in the low frequency classical skin depth limit the surface impedance is given by:

$$Z = -\tfrac{1}{2}(1 - i)\mu_0 \, \omega\delta$$

19. Show that exponential attentuation corresponds to a pure imaginary surface impedance. What can be said about the energy transfer into the medium when the surface impedance is pure imaginary?

18 Flux Decay. High conductivity metals are studied by observing the decay of magnetic flux when a uniform static field is suddenly switched off.[10] Although such measurements are normally performed on long thin rods about which a solenoid is wound, the analysis will be simpler if we assume that the magnetic field penetrates a thin slab as shown in Figure 13a. After the field is turned off, as shown in Figure 13b, the flux

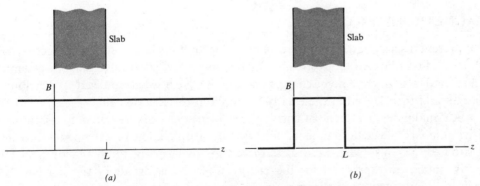

(a)   (b)

Figure 13 Flux decay. (a) A uniform magnetic field penetrates a thin slab. (b) After the field is turned off, the flux is maintained by eddy currents within the slab.

[10] C. P. Bean, R. W. Deblois, and L. B. Nesbit, *J. Appl. Phys.*, **30**, 1976 (1959). J. LePage, A. Bernalte, and D. A. Lindholm, *Rev. Sci. Instr.*, **39**, 1019 (1968). M. B. Kasen, *Phil. Mag. (GB)*, **21**, 599 (1970).

is maintained by eddy currents within the slab. To describe the decay of the flux we use **(126)**. Now, if we have a magnetic field of the form:

$$B(z) = B_n(t) \sin \frac{n\pi z}{L} \tag{145}$$

we obtain on substitution into **(126)**:

$$(\tau + \mu_0 \sigma L^2/n^2\pi^2)\frac{\partial B_n}{\partial t} + B_n = 0 \tag{146}$$

which gives:

$$B_n(t) = B_n(0)e^{-t/\tau_n} \tag{147}$$

with

▶
$$\tau_n = \tau + \mu_0 \sigma L^2/n^2\pi^2 \tag{148}$$

By writing the initial field as a superposition of sinusoidal fields of the form of **(145)** we find that each of the sinusoidal components decays with a characteristic time given by **(148)**. Experimentally, the long term decay will be a measure of the term corresponding to $n = 1$:

$$\tau_1 = \tau + \mu_0 \sigma L^2/\pi^2 \tag{149}$$

Exercise

20. High purity aluminum has been produced with a conductivity at 4.2 K, 45,000 times the room temperature value. Find the longest decay time at 4.2 K in a slab 1 cm thick.

MAGNETIC ENERGY

Our present discussion will to some extent parallel that of electric energy in Section 4-9, where we first obtained an expression for the amount of work necessary to assemble a distribution of discrete point charges and dipoles. Next we generalized to distributed charges and then looked at the energy as stored in the electric field.

An analogous treatment of currents considers the work required to assemble a distribution of magnetic moments, followed by a discussion of extended loops and distributed currents. Next, we discuss the energy in terms of the magnetic field and finally obtain an expression for the energy when some or all of the currents are regarded as magnetization currents.

We ask why this discussion could not have been presented earlier, as soon as we obtained an expression for the force on a test moment **(6-127)**:

$$\mathbf{F} = (\boldsymbol{\mu} \times \nabla) \times \mathbf{B} \tag{150}$$

where the magnetic field from the source moment is given by **(6-123)**:

$$\mathbf{B}(\mathbf{r}) = \frac{\mu_0}{4\pi} \frac{1}{r_{01}{}^3} [3\hat{\mathbf{r}}_{01}(\hat{\mathbf{r}}_{01} \cdot \boldsymbol{\mu}_1) - \boldsymbol{\mu}_1] \tag{151}$$

Although we might have thought of integrating **(150)** this would not have been correct, the reason being that we would have neglected the electromagnetic induction given by **(78)**.

This is the important distinction between calculating work against electric and magnetic forces. When a test charge is moved in the field of source charges, work need be done only on the test charge. This is because all that is required in addition is that the source charges be held fixed and this requires no work. The situation is different, as we shall see, when a test moment is moved in the field of source moments. It is necessary not only to hold the source moments fixed in position but also to maintain the *magnitude* of the moments in the presence of the electric fields generated by the moving test moment; and this requires work beside that done on the test moment.

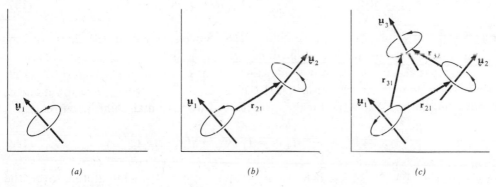

(a)	(b)	(c)

Figure 14 Interaction between magnetic moments. (a) A magnetic moment is at position $\mathbf{r}_1$. (b) A second moment is brought to position $\mathbf{r}_2$. (c) A third moment is brought to position $\mathbf{r}_3$.

19 Magnetic Moments. We imagine a magnetic moment $\boldsymbol{\mu}_1$ at position $\mathbf{r}_1$ as shown in Figure 14a and calculate the amount of work required to bring up a second moment $\boldsymbol{\mu}_2$ to a position $\mathbf{r}_2$ as shown in Figure 14b.[11]

The amount of work required to displace $\boldsymbol{\mu}_2$ through $d\mathbf{r}_2$ while keeping the magnitude of the moment constant is given by:

$$dW_2 = dW_{2\ \text{mech}} + dW_{2\ \text{elec}} \tag{152}$$

The mechanical work is performed against the force on $\boldsymbol{\mu}_2$ in the magnetic field of $\boldsymbol{\mu}_1$ **(6-123)**:

$$dW_{2\ \text{mech}} = -d\mathbf{r}_2 \cdot (\boldsymbol{\mu}_2 \times \nabla) \times \mathbf{B}_{21} = -d\mathbf{r}_2 \cdot (\nabla \mathbf{B}_{21}) \cdot \boldsymbol{\mu}_2 = -\boldsymbol{\mu}_2 \cdot d\mathbf{B}_{21} \tag{153}$$

---

[11] This section follows a closely related discussion by R. P. Feynman, R. B. Leighton, and M. Sands, *The Feynman Lectures on Physics*, Addison-Wesley, 1964, Section II-15-2.

where $\mathbf{B}_{21}$ is the field of $\boldsymbol{\mu}_1$ at $\boldsymbol{\mu}_2$. The electrical work is given by:

$$dW_{2 \text{ elect}} = -q_2 \mathscr{F}_2 = q_2 \frac{d}{dt} \Phi_{21} = I_2 \, d\Phi_{21} \tag{154}$$

where we think of $\boldsymbol{\mu}_2$ as arising from a circulating charge q_2 for which $\mathscr{F}_2$ **(109)** is the motional electromotive force induced around the path C_2. Now we may write the electrical work in the same form as the mechanical work:

$$dW_{2 \text{ elect}} = I_2 \mathbf{S}_2 \cdot d\mathbf{B}_{21} = \boldsymbol{\mu}_2 \cdot d\mathbf{B}_{21} \tag{155}$$

and we see that the two terms exactly cancel and we have:

$$dW_2 = 0 \tag{156}$$

There is another way in which we can demonstrate **(156)**. From **(1)** we have

$$dW_2 = -\mathbf{F} \cdot d\mathbf{r} = -q_2 \mathbf{E} \cdot d\mathbf{r} - q_2 \, d\mathbf{r} \cdot \mathbf{v} \times \mathbf{B} \tag{157}$$

For $\mathbf{E} = 0$ and writing $d\mathbf{r} = \mathbf{v} \, dt$ we obtain **(156)**. Since a magnetic field does no work on moving charges, no work need be done against the field.

We must also do work to maintain the current constant in $\boldsymbol{\mu}_1$. Here there is an electric field as we have seen from **(111)** but no mechanical work since the moment is not being moved. We write for the increment of work to maintain the current:

$$dW_1 = -\mathscr{E}_1 I_1 \, dt = I_1 \, d\Phi_1 = d(\boldsymbol{\mu}_1 \cdot \mathbf{B}_{12}) \tag{158}$$

where $\mathbf{B}_{12}$ is the magnetic field of $\boldsymbol{\mu}_2$ at the site of $\boldsymbol{\mu}_1$. Integrating over the displacement from infinity we obtain for the amount of work required to bring up the moment $\boldsymbol{\mu}_2$ to within a distance $\mathbf{r}_{21}$ of $\boldsymbol{\mu}_1$:

$$W_{12} = W_1 + W_2 = \boldsymbol{\mu}_1 \cdot \mathbf{B}_{12} \tag{159}$$

If we next bring up a third moment $\boldsymbol{\mu}_3$ we do no net work on $\boldsymbol{\mu}_3$. However, we must do additional work to maintain $\boldsymbol{\mu}_1$ constant:

$$W_{13} = \boldsymbol{\mu}_1 \cdot \mathbf{B}_{13} \tag{160}$$

and work to maintain $\boldsymbol{\mu}_2$ constant:

$$W_{23} = \boldsymbol{\mu}_2 \cdot \mathbf{B}_{23} \tag{161}$$

The total amount of work to establish the configuration shown in Figure 14c is given by:

$$W = W_{12} + W_{13} + W_{23} = \boldsymbol{\mu}_1 \cdot (\mathbf{B}_{12} + \mathbf{B}_{13}) + \boldsymbol{\mu}_2 \cdot \mathbf{B}_{23} \tag{162}$$

We may extend **(162)** to N moments and write at sight:

$$W = \sum_{i=1}^{N} \sum_{j=i+1}^{N} \boldsymbol{\mu}_i \cdot \mathbf{B}_{ij} \tag{163}$$

Now, we might have started by bringing up $\boldsymbol{\mu}_N$ first and then $\boldsymbol{\mu}_{N-1}$ and so on. For the first three moments the energy would then be:

$$W = \boldsymbol{\mu}_N \cdot (\mathbf{B}_{N, N-1} + \mathbf{B}_{N, N-2}) + \boldsymbol{\mu}_{N-1} \cdot \mathbf{B}_{N-1, N-2} \tag{164}$$

We may similarly extend **(164)** to all N moments to obtain

$$W = \sum_{i=1}^{N} \sum_{j=1}^{i-1} \boldsymbol{\mu}_i \cdot \mathbf{B}_{ij} \tag{165}$$

The energy of the final distribution of moments can not depend on the order in which the moments are assembled but only on the final distribution. Thus **(163)** and **(165)** must be identical and we may add them to obtain

$$W = \tfrac{1}{2} \sum_{i=1}^{N}{}' \sum_{j=1}^{N} \boldsymbol{\mu}_i \cdot \mathbf{B}_{ij} \tag{166}$$

where the prime means that the term corresponding to $i = j$ is to be excluded. Now, the magnetic field at $\mathbf{r}_i$ may be written as

$$\mathbf{B}_i = \sum_{j=1}^{N}{}' \mathbf{B}_{ij} \tag{167}$$

so that we may rewrite **(166)** as:

▶
$$W = \tfrac{1}{2} \sum_{i=1}^{N} \boldsymbol{\mu}_i \cdot \mathbf{B}_i \tag{168}$$

Exercise

21. Demonstrate the identity

$$\boldsymbol{\mu}_i \cdot \mathbf{B}_{ij} = \boldsymbol{\mu}_j \cdot \mathbf{B}_{ji}$$

20 **Magnetization and Current.** We next consider the work which must be done to bring together a distribution of magnetic moments $\boldsymbol{\mu}_i$ and currents I_j bounding surfaces S_j. From **(158)** we may write for the incremental work:

$$dW = \sum_i \boldsymbol{\mu}_i \cdot d\mathbf{B}_i + \sum_j I_j \, d\Phi_j \tag{169}$$

The arguments of Section 19 may be used to integrate **(169)** to obtain:

$$W_{\text{ext}} = \tfrac{1}{2} \sum_i \boldsymbol{\mu}_i \cdot \mathbf{B}_i + \tfrac{1}{2} \sum_j I_j \Phi_j \tag{170}$$

We now wish to take **(170)** to the continuum of a magnetization density $\mathbf{M}(\mathbf{r})$ and current density $\mathbf{j}(\mathbf{r})$. We must replace the magnetic field $\mathbf{B}_j$ at the site of the jth dipole by the macroscopic field **(6-141)**:

$$\mathbf{B}(\mathbf{r}) = \mathbf{B}_i + \tfrac{2}{3}\mu_0 \mathbf{M}(\mathbf{r}) \tag{171}$$

The flux Φ_j linking the jth current element may be written as:

$$\Phi_j = \int_{S_j} \mathbf{B}(\mathbf{r}) \cdot d\mathbf{S} = \int_{S_j} \nabla \times \mathbf{A}(\mathbf{r}) \cdot d\mathbf{S} = \oint_{C_j} \mathbf{A}(\mathbf{r}) \cdot d\mathbf{r} \tag{172}$$

where we have used **(6-54)** for the vector potential. The total current density in the continuum limit is the sum of two terms:

$$\mathbf{j}(\mathbf{r}) = \mathbf{j}_j + \nabla \times \mathbf{M}(\mathbf{r}) \tag{173}$$

where the second term is the magnetization current by **(6-151)**.

Substituting **(171)**, **(172)**, and **(173)** into the continuum limit of **(170)** we obtain:

$$W_{\text{ext}} = \tfrac{1}{2} \int_V \mathbf{M}(\mathbf{r}) \cdot \mathbf{B}(\mathbf{r}) \, dV - \tfrac{1}{2} \int_V \mathbf{A}(\mathbf{r}) \cdot \nabla \times \mathbf{M}(\mathbf{r}) \, dV$$

$$+ \tfrac{1}{2} \int_V \mathbf{j}(\mathbf{r}) \cdot \mathbf{A}(\mathbf{r}) \, dV - \tfrac{1}{3}\mu_0 \int_V M^2(\mathbf{r}) \, dV \tag{174}$$

Transforming the second term of **(174)** by **(B-4)** and assuming that the magnetization density is zero on the surface bounding V, we have finally for the external work:

▶
$$W_{\text{ext}} = \tfrac{1}{2} \int_V \mathbf{j}(\mathbf{r}) \cdot \mathbf{A}(\mathbf{r}) \, dV - \tfrac{1}{3}\mu_0 \int M^2(\mathbf{r}) \, dV \tag{175}$$

We note that **(175)** involves the magnetization density derived from moments $\boldsymbol{\mu}_i$. This is because the external work does *not* include the work required to establish these moments, but only the work necessary to bring them together with currents. On the other hand, **(175)** does include the work required to establish the currents.

Exercise

22. As we have discussed in Section 8-6, the vector potential **A** is determined by (172) only to within the gradient of a scalar. Show that (175) is unaffected by the choice of scalar. That is, establish the identity:

$$\int \mathbf{j}(\mathbf{r}) \cdot \nabla \psi(\mathbf{r}) \, dV = 0$$

where $\mathbf{j}(\mathbf{r})$ is a steady solenoidal distribution of currents and the volume of integration includes all currents.

---

21   **Magnetic Energy Density.** We may write (175) in terms of the magnetic field $\mathbf{B}(\mathbf{r})$ alone by using Ampère's law (3):

$$\mu_0 \, \mathbf{j}(\mathbf{r}) = \nabla \times \mathbf{B}(\mathbf{r}) \tag{176}$$

Substituting into (175) we obtain by (B-4):

$$W_{\text{ext}} = \frac{1}{2\mu_0} \int_V \mathbf{A}(\mathbf{r}) \cdot \nabla \times \mathbf{B}(\mathbf{r}) \, dV - \tfrac{1}{3}\mu_0 \int_V M^2(\mathbf{r}) \, dV = -\frac{1}{2\mu_0} \int_V \nabla \cdot [\mathbf{A}(\mathbf{r}) \times \mathbf{B}(\mathbf{r})] \, dV$$

$$+ \frac{1}{2\mu_0} \int_V \mathbf{B}(\mathbf{r}) \cdot \nabla \times \mathbf{A}(\mathbf{r}) \, dV - \tfrac{1}{3}\mu_0 \int_V M^2(\mathbf{r}) \, dV \tag{177}$$

By (B-11) and (171) we may write (177) as:

$$W_{\text{ext}} = -\frac{1}{2\mu_0} \int_S \mathbf{A}(\mathbf{r}) \times \mathbf{B}(\mathbf{r}) \cdot d\mathbf{S} + \frac{1}{2\mu_0} \int_V B^2(\mathbf{r}) \, dV - \tfrac{1}{3}\mu_0 \int_V M^2(\mathbf{r}) \, dV \tag{178}$$

Since $\mathbf{A}(\mathbf{r})$ falls off at least as $1/r$ from a limited distribution of currents and $\mathbf{B}(\mathbf{r})$ falls off at least as fast as $1/r^2$, the first term goes to zero as the bounding surface goes to infinity, and we are left with:

▶ $$W_{\text{ext}} = \frac{1}{2\mu_0} \int_V B^2(\mathbf{r}) \, dV - \tfrac{1}{3}\mu_0 \int_V M^2(\mathbf{r}) \, dV \tag{179}$$

22   **Localization of Magnetic Energy.** To get a feeling for the locatization of (179) we consider a current I of *surface* density i flowing uniformly along a cylinder of radius R as shown in Figure 15a. The magnetic field outside is given by:

$$B = \frac{\mu_0}{2\pi} \frac{I}{r} \tag{180}$$

The vector potential at $r = R$ is given from **(6-70)** as:

$$A = \frac{\mu_0}{2\pi} I \ln R_0/R \qquad \textbf{(181)}$$

where we have taken the zero of potential arbitrarily at *fixed* radius R_0. The amount of energy per unit length is then given from **(175)** to within an arbitrary constant as:

$$\frac{U}{L} = \frac{\mu_0}{4\pi} I^2 \ln R_0/R \qquad \textbf{(182)}$$

If we expand the radius of the cylinder by ΔR (15*b*) the energy is reduced by an amount:

$$\frac{\Delta U}{L} = -\frac{\mu_0}{4\pi} I^2 \frac{\Delta R}{R} \qquad \textbf{(183)}$$

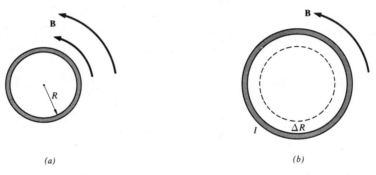

(a)   (b)

Figure 15 Localization of magnetic energy. (*a*) A current of surface density *i* flows uniformly along a cylindrical surface of radius R. (*b*) The radius of the cylinder is expanded by R, keeping the current constant.

Now, from **(179)** the amount of magnetic energy stored in the lost volume is given by:

$$\frac{\Delta U}{L} = -\frac{1}{2\mu_0} B^2(R) 2\pi R \Delta R = -\frac{\mu_0}{4\pi} \frac{I^2}{R} \Delta R \qquad \textbf{(184)}$$

which is just the answer given by **(179)**. Thus, the energy is reduced by just the energy stored in the magnetic field in ΔV if we regard:

▶
$$u = \frac{\Delta U}{\Delta V} = \frac{1}{2\mu_0} B^2(\mathbf{r}) \qquad \textbf{(185)}$$

as local energy density. The really compelling arguments for regarding **(185)** and **(2-42)** as local energy densities will be presented in Section 11-16 where we will see that an electromagnetic field carries momentum as well as energy and that we may describe the flow of both these quantities in terms of local fields.

Exercise

23. Obtain (184) from (179) and (180) without the use of the vector potential.

23 Magnetic Charge and Scalar Potential. An additional form for the work required to establish a magnetization distribution may be obtained by using the scalar potential (6-161):

$$-\nabla\overline{\phi}(\mathbf{r}) = \mathbf{B}(\mathbf{r}) - \mu_0\,\mathbf{M}(\mathbf{r}) \tag{186}$$

Substituting into (170) we obtain:

$$W_{\text{ext}} = \tfrac{1}{2}\int_V \mathbf{M}(\mathbf{r})\cdot[\mu_0\,\mathbf{M}(\mathbf{r}) - \nabla\phi(\mathbf{r})]\,dV - \tfrac{1}{3}\mu_0\int_V M^2(\mathbf{r})\,dV \tag{187}$$

Expanding and using (B-2), we write:

$$W_{\text{ext}} - \tfrac{1}{6}\mu_0\int_V M^2(\mathbf{r})\,dV \quad \tfrac{1}{2}\int_V \nabla\cdot[\overline{\phi}(\mathbf{r})\mathbf{M}(\mathbf{r})]\,dV \mid \tfrac{1}{2}\int_V \overline{\phi}(\mathbf{r})\,\nabla\cdot\mathbf{M}(\mathbf{r})\,dV \tag{188}$$

We may convert the second term to a surface integral by (B-11) As long as $\mathbf{M}(\mathbf{r})$ vanishes on S this term may be dropped and we have finally:

$$W_{\text{ext}} - \tfrac{1}{6}\mu_0\int_V M^2(\mathbf{r})\,dV + \tfrac{1}{2}\int_V \overline{\phi}(\mathbf{r})\,\nabla\cdot\mathbf{M}(\mathbf{r})\,dV \tag{189}$$

Using (6-164) we may rewrite (189) as:

▶
$$W_{\text{ext}} = \tfrac{1}{6}\mu_0\int_V M^2(\mathbf{r})\,dV - \tfrac{1}{2}\int_V \overline{\rho}_M(\mathbf{r})\overline{\phi}(\mathbf{r})\,dV \tag{190}$$

We note the difference between (190) and the corresponding expression for a polarization distribution (2-24):

$$W_{\text{ext}} = -\frac{1}{6\epsilon_0}\int_V P^2(\mathbf{r})\,dV + \frac{1}{2}\int_V \rho(\mathbf{r})\phi(\mathbf{r})\,dV \tag{191}$$

in that the terms are of reversed sign. Thus, we must be careful when treating magnetic problems in terms of the scalar potential and magnetic charge, particularly in obtaining the energy.

We note that all our energy calculations have been performed on the assumption that the magnetic currents remained constant in magnitude and were simply redistributed. Since whatever energies are *internal* to the magnetic moments are in this way kept constant, we may regard the energies which we have calculated as being *external* to the field sources. In the next chapter we consider the interaction between external currents and magnetoconducting material and will obtain expressions for the total energy stored in the interaction between such materials and external currents.

MAGNETIC FORCES

In this section we discuss the forces produced by magnetic fields on simple current configurations. We will be particularly interested in cases where the currents are induced by changes in the magnetic field or where the magnetic field arises from the current.

24 **Force on a Current Sheet.** To simplify the geometry we imagine as shown in Figure 16*a* a current flowing along the x direction and a magnetic field along the y direction.

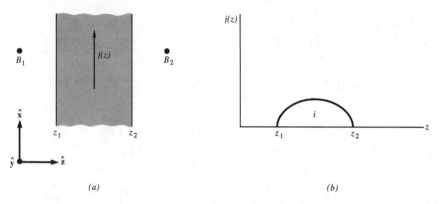

(a)   (b)

Figure 16   Force on a current sheet. (*a*) A current flows along the x direction with a magnetic field along the y direction. (*b*) The total current per unit area is the integral under the curve.

We imagine for simplicity that the currents are bounded along z and that the magnetic fields on the two sides are B_1 and B_2. From Ampère's law **(176)** we have:

$$\frac{d}{dz} B(z) = -\mu_0 j(z) \tag{192}$$

We integrate **(182)** to obtain:

$$B(z) = B_1 - \mu_0 \int_{z_1}^{z} j(z)\, dz \tag{193}$$

For $z > z_2$ the field becomes

$$B(z) = B_2 = B_1 - \mu_0 \int_{z_1}^{z_2} j(z)\, dz = B_1 - \mu_0 i \tag{194}$$

where i is the *total* current per unit area as shown in Figure 16*b*.

From **(1)** the force on a section of width Δz and area S may be written as

$$\Delta F = j(z)B(z)S\,\Delta z \tag{195}$$

Substituting from **(192)** for $j(z)$ we may write

$$\Delta F = -\frac{1}{\mu_0}B(z)\frac{dB(z)}{dz}S\,\Delta z \tag{196}$$

The total force on the section of area S is the integral of **(196)** over z:

$$\mathbf{F} = -\frac{1}{\mu_0}S\int_{B_1}^{B_2}B(z)\,dB(z) = \frac{1}{2\mu_0}S(B_1{}^2 - B_2{}^2)$$

$$= \frac{1}{2\mu_0}S(B_1 + B_2)(B_1 - B_2) \tag{197}$$

From **(194)** we have $B_1 - B_2 = \mu_0 i$ so that we may finally write, for **(197)**,

$$\mathbf{F} = \tfrac{1}{2}\mathbf{i} \times (\mathbf{B_1} + \mathbf{B_2})S \tag{198}$$

and we see that the force is equal to the current crossed into the *mean* of the external fields.

25 **Magnetic Pressure.** The force acting on the current sheet must be equilibrated in steady state by a difference in mechanical pressure on the two sides of the current sheet. If we call the pressures P_1 and P_2, we may write, for the system in equilibrium,

$$F + (P_1 - P_2)S = 0 \tag{199}$$

Comparing **(199)** with **(197)** we write for the mechanical pressure:

$$P_{mech} = P_0 - \frac{1}{2\mu_0}B^2 \tag{200}$$

Note that **(200)** is just the *negative* of the energy density **(141)**.

We have intentionally developed the present discussion in close parallel with the corresponding discussion of Section 2-8 of the force on a charge sheet and the equilibrating mechanical pressure. We found there that the pressure was equal to the *positive* of the energy density while here we find for magnetic fields the equilibrating pressure must be the *negative* of the energy density. This is to say that in the absence of equilibrating forces a charge sheet will disperse while a current sheet will concentrate.

The force per unit area **(197)** is usually written as the gradient of a *magnetic* pressure:

$$\mathbf{F} = -\nabla P_{mag} S \tag{201}$$

which gives:

$$\blacktriangleright \qquad P_{\text{mag}} = \frac{1}{2\mu_0} B^2 \qquad\qquad (202)$$

Comparing **(200)** and **(202)** we have in equilibrium:

$$P_{\text{mech}} + P_{\text{mag}} = P_0 \qquad\qquad (203)$$

Exercises

---

24. A solenoid with n turns per metre carries a current I. What is the magnetic pressure on the inside and outside of the solenoid? [C. H. Blanchard, *Am. J. Phys.*, **44**, 891 (1976).]

25. A solenoid with n turns per metre carries a current I. A uniform external field B_0 opposes the field within the solenoid? What are the internal and external magnetic pressures?

26. Show that the magnetic pressure on a solenoid is equal to $\frac{1}{2}iB$, where B is the internal field as expected from **(198)**.

27. Show that the magnetic pressure in atmospheres is about $4B^2$ with B expressed in tesla.

28. A thin conducting tube of radius R carries a current I. Show that a magnetic pressure

$$P_{\text{mag}} = \frac{\mu_0}{8\pi} \frac{I^2}{\pi R^2}$$

acts to compress the cylinder. For $I = 1000$ A and $R = 1$ cm find the pressure in atmospheres.

---

26 **Pinch Effect.** As an example we discuss very briefly an instability observed in ionized plasmas in which plasma current is contained by its own magnetic field. We may think of the instability as developing in the following way: at low applied electric fields the current density is generally uniform because of carrier diffusion. As the field is increased we may expect that, as the result of fluctuations, the current density will be larger in some places than in others. Where the current density is larger the magnetic fields bounding the current will be larger and the carriers will be prevented by the Lorentz force from diffusing out of the current-carrying region. Other carriers diffusing into the region will become trapped by the magnetic field until most of the plasma is contained within a single filament. As we will see, the Lorentz force is equilibrated by the increased gas pressure *within* the filament. Furthermore, although the plasma density falls off parabolically from the center of the filament, the carrier collision rate will fall off in the same way and we may assume that the conductivity σ and the current density j are constant up to a limiting radius R. The magnetic field within the filament is found from Ampère's law (176):

$$\oint_C \mathbf{B} \cdot d\mathbf{r} = \mu_0 \int_S \mathbf{j} \cdot d\mathbf{S} = \mu_0 \, \pi r^2 j \tag{204}$$

For an axially symmetric uniform current we obtain from (200) for the pressure gradient:

$$\frac{dP}{dr} = -jB = -\tfrac{1}{2}\mu_0 j^2 r \tag{205}$$

Integrating, we obtain for the pressure, which is zero for $r > R$:

$$P(r) = \int_R^r -\tfrac{1}{2}\mu_0 j^2 r \, dr = \tfrac{1}{2}\mu_0 j^2 (R^2 - r^2) \tag{206}$$

We assume for a gas that the carrier concentration is proportional to the pressure:

$$P(r) = kTn(r) \tag{207}$$

We obtain an expression for the number of carriers per unit length by integrating $n(r)$:

$$N = \int_0^R n(r) 2\pi r \, dr \tag{208}$$

Substituting from (206) and (207) into (208) we obtain:

$$NkT = \frac{\mu_0}{8\pi} I^2 \tag{209}$$

where

$$I = \pi R^2 j = \pi R^2 \sigma E \tag{210}$$

is the total current flowing in the filament. We may also regard (207) as determining the radius:

$$R = \left(\frac{I}{\pi \sigma E}\right)^{1/2} \tag{211}$$

where σ must itself be a function of I and R.

27 **Flux Compression.** We discuss in this section the compression of magnetic flux by mechanical forces. We demonstrate that if conduction losses can be neglected then the enclosed flux remains constant and all mechanical work goes into increasing the energy

in the magnetic field. As a first example, we consider the solenoid shown in Figure 17a with n turns per metre, radius R, length L, and an initial current I. The magnetic pressure from **(202)** is

$$P_{\text{mag}} = \frac{1}{2\mu_0} B^2 = \tfrac{1}{2}\mu_0 n^2 I^2 \tag{212}$$

The amount of mechanical work done to compress the solenoid through $\Delta V = 2\pi R L\,\Delta R$ is:

$$dW = -P_{\text{mag}}\,\Delta V \tag{213}$$

The increase in magnetic energy is just equal to the work done:

$$dU = \frac{1}{2\mu_0}\,d(B^2 V) = dW = -\frac{1}{2\mu_0} B^2\,\Delta V \tag{214}$$

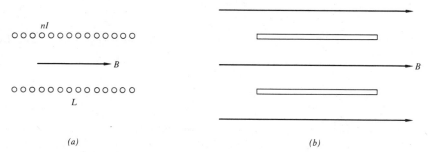

(a)   (b)

Figure 17   Flux compression. (a) A solenoid with n turns per metre carries a current I. (b) A metallic cylinder is compressed mechanically in a magnetic field B.

Integrating **(214)** we obtain:

$$BV = \Phi L = \text{const} \tag{215}$$

Since the magnetic flux is constant, the field and the current must increase inversely as the square of the radius:

$$B = \mu_0 nI = \frac{\Phi}{\pi R^2} \tag{216}$$

How are we to understand the increase in current when the enclosed flux remains constant? The increased current is produced by the Lorentz force acting on the carriers as they move across the field. The mean electric field along the wire will be given by

$$E = -\tfrac{1}{2}B\frac{dR}{dt} \tag{217}$$

The electromotive force around a single turn is:

$$\mathcal{E} = \oint \mathbf{E} \cdot d\mathbf{r} = -\pi RB\frac{dR}{dt} \tag{218}$$

The electromotive force results in an increased current limited by the mean increased magnetic field from **(111)**.

$$\mathcal{E} = \tfrac{1}{2}\pi R^2 \frac{dB}{dt} \tag{219}$$

Combining **(218)** and **(219)** we obtain:

$$\pi R^2 B = \Phi = \text{const} \tag{220}$$

The result for the compressed solenoid is very similar to the perfect diamagnetism of a plasma as discussed in Section 8-4. For both cases flux remains constant.

As a second example, we consider the metallic cylinder shown in Figure 17*b*. If the cylinder is compressed mechanically, currents will be induced in the cylinder to keep the enclosed flux constant. The argument is just the same as for the solenoidal winding. The mean Lorentz force must be just balanced by the mean electromotive force produced by the changing magnetic field which leads again to **(220)**.

By compressing a conducting cylinder with an explosive charge, fields up to 1000 T have been obtained, corresponding to magnetic pressures of ten million atmospheres or 10^{12} N/m².[12] These pressures may be communicated to a sample placed in an inner conducting tube as shown in Figure 18.[13]

Exercise
___

28. Show that the electric field within the contracting solenoid is given by:

$$E = -\tfrac{1}{2}r\frac{dB}{dt}$$

and that the electric field outside is zero. Note that the mean of these two values leads to **(219)**.
___

[12] F. Bitter, "Ultrastrong Magnetic Fields," *Scientific American*, **213**(1), 65 (July 1965).

[13] R. S. Hawke et al., "Method of Isentropically Compressing Materials to Several Megabars," *J. Appl. Phys.*, **43**, 2734 (1962).

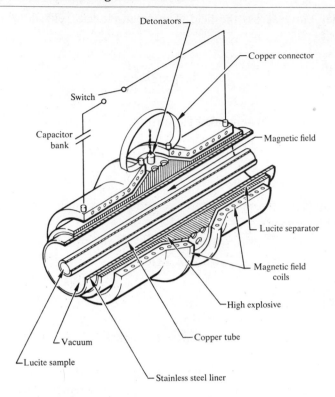

Figure 18 Ultrastrong magnetic fields. With an explosive charge fields up to 1000 tesla have been obtained.

SUMMARY

1. Moving sources of magnetic field produce an electric field:

$$\mathbf{E}(\mathbf{r},t) = -\mathbf{v}_1 \times \mathbf{B}(\mathbf{r},t) \tag{16}$$

where $\mathbf{v}_1$ is the velocity with which the magnetic field sources move.

2. The magnetic field in the laboratory is related to the magnetic field in the frame of the moving current source by:

$$B_x(\mathbf{r},t) = B_x{}'(\mathbf{r}') \qquad B_{yz}(\mathbf{r},t) = \gamma B_{yz}{}'(\mathbf{r}') \tag{17}$$

where the currents are moving along the x direction with $\gamma = (1 - v_1{}^2/c^2)^{-1/2}$.

3. A magnetic moment $\boldsymbol{\mu}'$ moving through an electric field experiences a force:

$$\mathbf{F} = \frac{1}{c^2} (\mathbf{v} \times \boldsymbol{\mu}' \cdot \nabla)\mathbf{E} \tag{26}$$

Comparison with the expression for the force on an electric dipole moment suggests that a moving magnetic moment has an electric dipole moment:

$$\mathbf{p} = \frac{1}{c^2} \mathbf{v} \times \boldsymbol{\mu}' \tag{28}$$

4. A hypothetical magnetic charge $\bar{q}$ is surrounded by a magnetic field

$$\mathbf{B}(\mathbf{r}) = \frac{\mu_0}{4\pi} \frac{\bar{q}}{r_{01}^2} \hat{\mathbf{f}}_{01} \tag{38}$$

which integrates to total flux

$$\Phi = \mu_0 \bar{q} \tag{42}$$

5. The equivalent of the Lorentz force for magnetic charge is

$$\mathbf{F} = \bar{q}\left(\mathbf{B} - \frac{1}{c^2} \mathbf{v} \times \mathbf{E}\right) \tag{45}$$

6. The invariance of local charge conservation

$$\nabla \cdot \mathbf{j}(\mathbf{r},t) + \frac{\partial}{\partial t} \rho(\mathbf{r},t) = 0 \tag{46}$$

leads to the transformation of charge and current:

$$j_x' = \gamma(j_x - V\rho) \qquad j_x = \gamma(j_x + V\rho') \tag{49} \tag{56}$$
$$j_{yz}' = j_{yz} \qquad j_{yz} = j_{yz}' \tag{50} \tag{57}$$
$$\rho' = \gamma(\rho - j_x V/c^2) \qquad \rho = \gamma(\rho' + j_x'V/c^2) \tag{51} \tag{58}$$

7. Uniformly moving sources of magnetic field act as sources of circulation for the electric field according to the local relation:

$$\nabla \times \mathbf{E} = -\frac{\partial \mathbf{B}}{\partial t} \tag{114}$$

This relationship, which is called Faraday's law, applies generally for any change in magnetic field.

8. Moving currents produce an electric charge density:

$$\rho(\mathbf{r}) = \frac{1}{c^2} \mathbf{v}_1 \cdot \mathbf{j}(\mathbf{r}) \tag{84}$$

where $\mathbf{v}_1$ is the velocity of the uniformly moving current distribution.

9. If a conductor is placed in an oscillatory magnetic field, the field will penetrate into the conductor by an amount

$$\delta = (2/\omega\mu_0\,\sigma)^{1/2} \tag{131}$$

where it is assumed that the frequency ω is much less than the collision rate $1/\tau$ and that the carrier mean free path λ_C is less than the skin depth δ.

10. If the carrier mean free path is much greater than the classical skin depth δ the field penetrates through a distance of order:

$$(\delta^2\lambda_C)^{1/3} \tag{136}$$

11. The rate at which energy is absorbed by a conducting medium may be written as:

$$\frac{dW}{dt} = \frac{1}{2\mu_0{}^2}\,B^2R \tag{142}$$

where B is the value of the magnetic field at the surface and R is the real part of the surface impedance:

$$Z = R - iX = \frac{\mu_0\,\omega}{k} \tag{143}$$

and where k is the wavevector at frequency ω for propagation in the medium.

12. Magnetic flux decays exponentially within a slab with a characteristic time:

$$\tau_n = \tau + \mu_0\,\sigma L^2/n^2\pi^2 \tag{148}$$

where τ is the collision time of the carriers and $n\pi/L$ is the wavevector of the magnetic field.

13. The work required to bring together a set of fixed magnetic moments $\boldsymbol{\mu}_i$ is given by:

$$W = \tfrac{1}{2}\sum_i \boldsymbol{\mu}_i \cdot \mathbf{B}_i \tag{168}$$

This work may be written for a continuous distribution of moments and currents as:

$$W_{\text{ext}} = \tfrac{1}{2}\int_V \mathbf{j}(\mathbf{r}) \cdot \mathbf{A}(\mathbf{r}) - \tfrac{1}{3}\mu_0\int M^2(\mathbf{r})\,dV \tag{175}$$

14. The energy stored in the magnetic field may also be written as:

$$W_{\text{ext}} = \frac{1}{2\mu_0}\int_V B^2(\mathbf{r})\,dV - \tfrac{1}{3}\mu_0\int M^2(\mathbf{r})\,dV \tag{179}$$

leading to a local energy density:

$$u = \frac{1}{2\mu_0} B^2(\mathbf{r}) \tag{185}$$

15. The energy stored in the magnetic field may be written in terms of the scalar potential as:

$$W_{ext} = \tfrac{1}{6}\mu_0 \int M^2 \, dV - \frac{1}{2} \int \bar{\rho}_M \, \bar{\phi} \, dV \tag{190}$$

Notice that this expression is the negative of the corresponding expression for the electrostatic field.

16. The force on a current sheet is equal to the current crossed into the mean of the magnetic fields on the two sides of the sheet.

17. Magnetic fields produce a pressure

$$P_{mag} = \frac{1}{2\mu_0} B^2(\mathbf{r}) \tag{202}$$

which for static conditions must be equilibrated by mechanical pressure.

18. If an ideal resistanceless solenoid is compressed the enclosed flux remains constant and additional energy in the magnetic field is produced by mechanical work.

PROBLEMS

1. **Transversely moving solenoid.** A long solenoid of circular cross-section has n' turns per metre and current I'. The equivalent volume magnetization is $M' = \mu_0 n'I'$. The magnetic field is $\mathbf{B}' = \mu_0 \mathbf{M}'$ inside and $\mathbf{B}' = 0$ outside. The solenoid is in motion with velocity $\mathbf{v}_1$ transverse to its axis.

   (a) Find the magnetic field $\mathbf{B}$ in the laboratory both inside and outside the solenoid.

   (b) Find the electric field both inside and outside the solenoid from (16).

   (c) From (106) find the polarization density of the solenoid and obtain the corresponding transverse depolarization field of a long cylinder. Describe the external electric field.

   (d) By comparing (b) and (c) find the contribution of (114) to $\mathbf{E}$ both inside and outside the solenoid.

2. **Longitudinally moving solenoid.** A long solenoid of circular cross-section has n' turns per metre and current I'. The solenoid is in motion with velocity $\mathbf{v}_1$ parallel to its length.

   (a) Find the magnetic field $\mathbf{B}$ in the laboratory both inside and outside the solenoid.

   (b) Find the electric field $\mathbf{E}$ both inside and outside the solenoid from (16).

3. **Hall effect.** Although the Hall effect is usually considered to be an electrostatic effect, this cannot be the whole story since the Hall voltage may drive a current in an external load and thus deliver power.

(a) Show that as a result of longitudinal drift there is a motional electromotive force appearing at the terminals shown in the following diagram and given by

$$\mathscr{F} = RjBw$$

where $R = 1/nq$ is the Hall constant as discussed in Section 8-1.

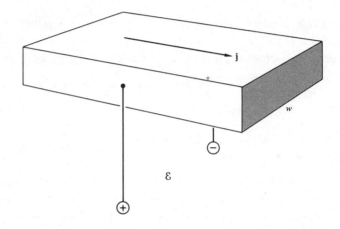

(b) Show that if a Hall current is permitted to flow, then there is a reverse longitudinal current, and the power delivered to the external load originates with the source driving the longitudinal current. [See M. Phillips, "Electromotive Force Again," *Am. J. Phys.*, **30**, 309 (1962).]

4. **Faraday disc.** A conducting disc rotates about its axis at frequency ω in a magnetic field **B** normal to the plane of the disc.

(a) Find the electric field developed in the disc so that the Lorentz force **(1)** on the mobile carriers is zero.

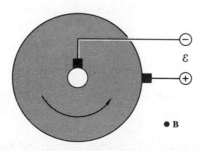

(b) Contacts are placed at the center of the disc and on the edge. Show that a motional electromotive force

$$\mathscr{F} = \tfrac{1}{2}\omega R^2 B$$

is developed between the terminals.

(c) If the disc were at rest and the sources of the magnetic field **B** were rotated instead, what would be the response of the disc?

5. **Rotating magnetic insulator.** An insulating magnetic sphere of radius R and magnetization density **M** rotates at angular frequency ω about an axis through its center and parallel to **M**.

(a) Obtain from **(28)** the approximate polarization density for $\omega R \ll c$.

(b) Show that the rotating sphere carries a uniform volume charge density and find its magnitude.

(c) Find the density of surface charge and establish explicitly that the sphere is electrically neutral.

(d) Find the electric field within the sphere.

6. **Rotating magnetic conductor.** A conducting magnetic sphere of radius R and magnetization density **M** rotates at angular frequency ω about an axis through its center and parallel to **M**.

(a) Find the magnetic field **B** within the sphere for $\omega R \ll c$.

(b) From the fact that the Lorentz force **(1)** on the mobile carriers is zero, find the electric field within the sphere.

7. **Unipolar induction.** A conducting magnetic sphere of radius R and magnetization density **M** rotates at angular frequency ω about an axis through its center and parallel to **M**. Contact is made to the sphere at one pole and on the equator as shown in the following diagram. Show that a motional electromotive force

$$\mathscr{F} = \tfrac{1}{3}\omega R^2 \mu_0 M$$

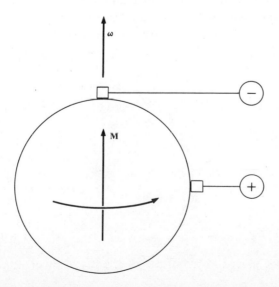

is developed at the terminals. [See J. W. Then, "Experimental Study of the Motional Electromotive Force," *Am. J. Phys.*, **30**, 411 (1962). See also E. P. Mosca, "Magnetic Forces Doing Work?," *Am. J. Phys.*, **42**, 295 (1974).]

8. **Rotating loop.** A conducting loop of radius R rotates about an axis in the plane of the loop through the center and normal to a magnetic field **B**. Obtain an expression for the electromotive force $\mathscr{F}$ induced in the loop.

9. **Betatron condition.** In the betatron, electrons are held in a circular orbit in a vacuum chamber by a magnetic field **B**. The electrons are accelerated by gradually increasing the magnitude of **B**. Show that the *average* magnetic field Φ/S over the plane of the orbit must equal *twice* the value of B at the radius of the orbit if the orbital radius is to remain fixed as the electron's energy is increased.

10. **Induced electric field.** A toroidal coil is fed by a voltage $V(t)$. If the total length of the wire is l, show that the electric field along the wire is given by:

$$E = V/l + \partial A/\partial t$$

where A is the vector potential associated with the enclosed magnetic field B.

11. **Lenz's law.** In 1834, Lenz pointed out that the induced field as given by **(114)** is in such a direction as to drive a current that maintains the enclosed flux. Verify this statement both for currents induced by the relative motion of a current loop and a field source and for varying source currents.

12. **Electromotive force.** Show that **(111)** is equivalent to the statement that the electromotive force is given by:

$$\mathscr{E} = -\frac{\partial}{\partial t} \oint_C \mathbf{A} \cdot d\mathbf{r}$$

13. **Rogowski coil.** A flexible helix with an axial return wire, as shown in the following diagram is called a Rogowski coil and is used as a clip-on ac current probe. Such a coil of n turns per metre and cross-section S surrounds a conductor carrying

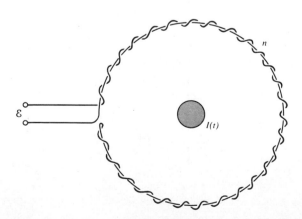

a current $I(t)$. Show that the voltage between the terminals is given by

$$\mathscr{E} = -\mu_0 nS \frac{dI}{dt}$$

independent of the path of the coil. Explain why the wire is returned along the axis of the helix.

14. **Spherical inductor.** An inductor is wound on a spherical surface such that the number of turns per unit length on the surface is proportional to $\cos \theta$, where θ is the angle with the equatorial plane. The inductor is placed in an inhomogeneous alternating field. Show that as long as there are no currents within the inductor the induced voltage is proportional to the rate of change of the axial field at the *center* of the inductor. See Section 6-22 and Problem 6-23.

15. **Eddy currents.** An alternating magnetic field $B \cos \omega t$ is imposed normal to a thin conducting disc of thickness t, radius R, and conductivity σ.
    (a) Find the induced current density as a function of radius.
    (b) Find the magnetization density in the disc and show that it is given by:

$$\mu_0 M = \frac{1}{2\pi} \left(\frac{r}{\delta}\right)^2 B \sin \omega t$$

where δ is the classical skin depth in the conductor.
    (c) Show that the induced currents produce a field at the center of the disc:

$$\Delta B = \frac{Rt}{2\pi\delta^2} B \sin \omega t$$

16. **Fields in conductors.** An electric field varies in a conductor as

$$E(r,t) = Ee^{i(\mathbf{k} \cdot \mathbf{r} - \omega t)}$$

where $\mathbf{k}$ is a complex propagation vector:

$$\mathbf{k} = \mathbf{k}' + i\mathbf{k}''$$

Obtain expressions for the complex magnetic field $\mathbf{B}(r,t)$ and the complex current density $\mathbf{j}(r,t)$. For a real electric field

$$E(r,t) = Ee^{-\mathbf{k}'' \cdot \mathbf{r}} \cos(\mathbf{k}' \cdot \mathbf{r} - \omega t)$$

obtain the corresponding magnetic field and current density.

17. **Electromagnetic levitation.** A metallic sphere of radius R large compared with a skin depth is situated in an alternating magnetic field whose intensity decreases with increasing height. Find the effective magnetization density of the sphere and

from (6-127) the force exerted by the field. Discuss this problem in terms of magnetic pressure.

18. **Eddy current heating.** Compute the rate at which energy is dissipated in the electromagnetically levitated sphere described in Problem 17.

19. **Flux velocity.** Magnetic flux penetrates a conductor to an effective depth:

$$\delta(t) = \frac{\Phi}{B_{\text{ext}}} = \frac{1}{B_{\text{ext}}} \int_0^\infty B(z,t)\, dz$$

Show that the electric field at the surface is given by:

$$E(t) = B_{\text{ext}} \frac{d}{dt} \delta(t)$$

20. **Flux diffusion.** A magnetic field $\mathbf{B}_{\text{ext}}$ is imposed on the surface of a conductor at $t = 0$. By assuming that the change in $\mathbf{B}$ within the conductor is slow we have from (126) the diffusion equation:

$$\nabla^2 \mathbf{B} = \mu_0\, \sigma\, \frac{\partial \mathbf{B}}{\partial t}$$

(a) Show that

$$\mathbf{B}(z,t) = \mathbf{B}_{\text{ext}}[1 - P(\tfrac{1}{2}z\sqrt{\mu_0\, \sigma/t})]$$

where $P(x)$ is the probability integral

$$P(x) = \frac{2}{\sqrt{\pi}} \int_0^x e^{-y^2}\, dy$$

is a solution of the differential equation and satisfies the condition $\mathbf{B}(0,t) = \mathbf{B}_{\text{ext}}$.

(b) Show that the electric field at the surface is given by

$$\mathbf{E}(0,t) = -\frac{1}{2\sqrt{\pi}} \frac{\hat{\mathbf{n}} \times \mathbf{B}_{\text{ext}}}{(\mu_0\, \sigma t)^{1/2}}$$

(c) Show that the effective penetration depth of the magnetic field (see Problem 19) is given by:

$$\delta(t) = \frac{1}{2\sqrt{\pi}} \left(\frac{t}{\mu_0\, \sigma}\right)^{1/2}$$

How long does it take for the flux to penetrate one centimetre into copper at room temperature?

21. **Flux diffusion into a perfect conductor.** We obtain **(I-29)** for the flux penetration depth:

$$\delta(t) \sim \frac{(\omega_0 t)^\beta}{k_0}$$

where the dispersion relation is given **(I-20)**:

$$k \simeq k_0 (\omega/\omega_0)^\beta$$

For a perfect conductor $(\tau = \infty)$ use **(136)** for the dispersion relation and estimate the time required for magnetic flux to diffuse through $d = 1$ cm in a metal of ordinary electron concentration. Note that to use **(136)** in this way, the electron mean free path would have to be at least equal to d.

22. **Pinch effect.** In the pinch effect the compression of the charge carriers increases the energy stored in the magnetic field. Work is done against mechanical forces as well as in compressing the enclosed gas. Demonstrate that the mechanical and magnetic energy are delivered from the generator which supplies the current. Assume for simplicity that a current I flows on a cylindrical surface.
(a) Obtain an expression for the additional magnetic energy.
(b) Obtain an expression for the work done against mechanical pressure when the radius of the cylinder is contracted by ΔR.
(c) What is the electromotive force induced per unit length in the circuit which drives the current?
(d) What is the work done by the generator against the induced electromotive force?
(e) Show that energy is conserved.

23. **Flux compression.** A current I flows on the surface of an inner conductor and back on an outer coaxial conductor.
(a) Find the energy stored per metre in the enclosed magnetic field.
(b) Find the magnetic pressure acting on the inner and outer conductors.
(c) The outer conductor is contracted by ΔR. How much work is done on the field?
(d) Show by energy balance that the enclosed flux remains constant.

REFERENCES

Kolm, H. H. and R. D. Thornton, "Electromagnetic Flight," *Scientific American*, **229**(4), 17 (October 1973).

Nasar, S. A. and I. Bodlea, *Linear Motion Electric Machines*, Wiley-Interscience, 1976.

10 MAGNETOCONDUCTING MEDIA II

This chapter is for magnetoconducting media what Chapter 4 was for dielectric media. We begin with a discussion of magnetic media and current configurations, where we introduce the magnetic field **H** for which the source of circulation is *external* currents only. We examine the properties of **H** and discuss some of the special properties of linear media. We examine the boundary conditions on **B** and **H** at an interface between different media. Next we discuss the energy of interaction between magnetic media and external currents. We discuss the screening of static magnetic fields by a magnetic medium. We conclude with a discussion of the energetics of ferromagnets and superconductors.

MAGNETIC MEDIA AND EXTERNAL CURRENTS

We now wish to discuss more complex problems in which the field acting on a magnetic medium arises explicitly from *external* currents.

1 Magnetic Field H. As a way of keeping track of the distinction between *external* currents and the *induced* currents whose properties are those of the medium, we define a new magnetic field[1] to which we give the symbol **H**:

$$\mathbf{B} = \mu_0\,\mathbf{H} + \mu_0\,\mathbf{M} \tag{1}$$

We write the total current as the sum of two components:

$$\mathbf{j} = \mathbf{j}_{ext} + \mathbf{j}_{ind} \tag{2}$$

---

[1] Some authors call **H** the magnetic field strength and **B** the magnetic flux density. Our preference would be to call **B** the magnetic *field* because of its place with **E** in the Lorentz force law. To avoid confusion we call **B** the magnetic field "**B**" and **H** the magnetic field "**H**".

where j_{ext} is the density of external currents and j_{ind} an induced *magnetization* current **(6-151)**:

$$j_{ind} = \nabla \times M \tag{3}$$

Taking the curl of **(1)** we obtain:

$$\nabla \times B = \mu_0 \nabla \times H + \mu_0 \nabla \times M \tag{4}$$

From Ampère's law **(6-48)** and **(2)**, **(3)**, and **(4)** we may write, for the curl of **H**

▶
$$\nabla \times H = j_{ext}(r) \tag{5}$$

From **(1)**, **(6-59)**, and **(6-164)** we may write the divergence of **H** as

▶
$$\nabla \cdot H = -\nabla \cdot M = \bar{\rho}_M(r) \tag{6}$$

Equations **5** and **6** determine **H** to within a constant field if the external currents and magnetization are given everywhere.

2 **Permeability.** For many problems we can assume that the magnetization density is everywhere proportional to the macroscopic magnetic field **B** **(7-23)**:

$$\mu_0 M(r) = \chi B(r) \tag{7}$$

We have then for the magnetic field **B**:

▶
$$B = \mu_0 H + \chi B = \mu H \tag{8}$$

where μ is the permeability of the medium and is related to the magnetic susceptibility by:

▶
$$\mu = \frac{1}{1 - \chi} \mu_0 \tag{9}$$

3 **Screening of Currents.** Multiplying **(5)** by the permeability μ we have:

$$\mu \nabla \times H = \nabla \times B = \mu j_{ext} = \mu_0 j \tag{10}$$

Thus, the total current is larger for χ positive than the external current by a factor:

$$\frac{\mu}{\mu_0} = \frac{1}{1 - \chi} \tag{11}$$

which is called the relative permeability. This behavior is to be contrasted with charge screening by dielectrics **(4-33)** where the total charge is *reduced* from the external charge by a factor:

$$\frac{\epsilon_0}{\epsilon} = \frac{1}{1+\chi}$$

(12)

where χ in **(12)** is the *dielectric* susceptibility.

4 Current on the Axis of a Magnetic Cylinder.

As an example of the use of the magnetic field **H** we consider a cylinder of radius R and permeability μ with a current I flowing on its axis as shown in Figure 1*a*. We regard the current I as external and assume that the medium is homogeneous and isotropic.

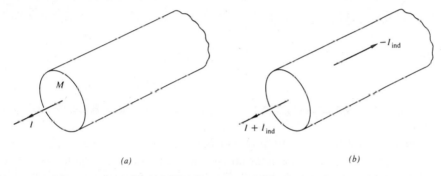

(a)   (b)

Figure 1   Current on the axis of a magnetic cylinder. (*a*) An external current I flowing on the axis magnetizes the cylinder. (*b*) The induced magnetization is equivalent to induced currents flowing on the axis and on the outside of the cylinder.

We have from **(5)** for the circulation of **H**:

$$\mathscr{H} = \oint_C \mathbf{H} \cdot d\mathbf{r} = \int_S \nabla \times \mathbf{H} \cdot d\mathbf{S} = \int_S \mathbf{j}_{\text{ext}} \cdot d\mathbf{S}$$

(13)

which leads to

$$H = \frac{I}{2\pi r}$$

(14)

Now **(13)** has the same form whether the contour is within the cylinder or outside. This is why it is convenient to introduce the magnetic field **H**, particularly for linear media. We may find **B** from **(8)**. For $r < R$:

$$B = \frac{\mu}{2\pi} \frac{I}{r}$$

(15)

For $r > R$:

$$B = \frac{\mu_0}{2\pi} \frac{I}{r}$$

(16)

Let us see how the same problem would be solved without explicitly introducing **H**. We write the circulation of the magnetic field **B** including both external and magnetization current:

$$\mathscr{B} = \oint_C \mathbf{B} \cdot d\mathbf{r} = \int_S \nabla \times \mathbf{B} \cdot d\mathbf{S} = \mu_0 I + \mu_0 \int_S \mathbf{j}_{ind} \cdot d\mathbf{S} \tag{17}$$

The total induced current within S is given by:

$$\int_S \mathbf{j}_{ind} \cdot d\mathbf{S} = \int_S \nabla \times \mathbf{M} \cdot d\mathbf{S} = \oint_C \mathbf{M} \cdot d\mathbf{r} = 2\pi r M(r) \tag{18}$$

Then from **(14)** and **(15)** we have

$$B(r) = \frac{\mu_0}{2\pi} \frac{I}{r} + \mu_0 M(r) \tag{19}$$

Using **(7)** and **(8)** we obtain **(15)** and **(16)** for the magnetic field **B** inside and outside the cylinder.

A third way of looking at this problem is in terms of an induced magnetization current I_{ind}, which flows with the external current I and on the outside surface of the conductor as shown in Figure 1*b*. From **(15)** the current flowing on the axis is

$$\frac{I}{\mu_0} \mathscr{B} = \frac{\mu}{\mu_0} I = I + \frac{\mu - \mu_0}{\mu_0} I \tag{20}$$

The second term in **(20)** is the induced current:

$$I_{ind} = \frac{\mu - \mu_0}{\mu_0} I = \frac{\chi}{1 - \chi} I \tag{21}$$

On the outside surface there must be an induced current $-I_{ind}$ or a magnetic *surface* current density:

$$i_{ind} = -\frac{\chi}{1 - \chi} \frac{I}{2\pi R} \tag{22}$$

BOUNDARY CONDITIONS

We imagine that we know the magnetic fields **B** and **H** on one side of an interface between two magnetic media and we wish to know **B** and **H** on the other side. We show this situation in Figure 2*a*.

5 Continuity of Magnetic Fields. From **(5)** we know that the circulation of **H** around any contour such as that shown in Figure 2b must be the enclosed external current. Under the usual condition that the external current at a magnetic interface is zero, we have:

$$\hat{z} \times \hat{n} \cdot \mathbf{H}_1 = \hat{z} \times \hat{n} \cdot \mathbf{H}_2 \tag{23}$$

or

▶ $$\hat{n} \times \mathbf{H}_1 = \hat{n} \times \mathbf{H}_2 \tag{24}$$

since $\hat{z}$ is an arbitrary vector in the plane.

We next compute the flux of **B** from the surface shown in Figure 2c. Since the flux of **B** through *any* closed surface must be zero, we have:

$$\hat{n} \cdot \mathbf{B}_1 = \hat{n} \cdot \mathbf{B}_2 \tag{25}$$

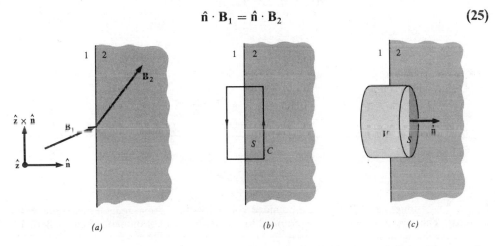

(a)   (b)   (c)

Figure 2 Boundary conditions on **B** and **H**. (*a*) Magnetic fields $\mathbf{B}_1$ and $\mathbf{B}_2$ exist near the interface between two media. (*b*) In the absence of external surface current, the circulation of **H** must be zero. (*c*) In the absence of magnetic charge, the flux of **B** must be zero.

Equations **24** and **25** together with the relative permeabilities of the two media determine the field $\mathbf{B}_2$ if $\mathbf{B}_1$ is known.

Exercises

1. Show that **B** is normal to the surface of a medium of infinite permeability.
2. Show that the surface of a medium of infinite permeability is a surface of constant magnetic scalar potential **(6-161)**.
3. Obtain **(24)** directly by applying **(B-13)** to the volume shown in Figure 2c.

6 Uniqueness Theorem. We imagine that within some bounded volume as shown in Figure 3 there are magnetic media and external currents. What must we know about magnetic media or external currents outside V in order to obtain the fields inside?

As we will see, all that is required is the *tangential* component of the vector potential **A** or of the magnetic field **H** on the surfaces but not both. And we do not require a knowledge of the normal components at all.

To show this, we imagine two solutions to the problem $\mathbf{B}_1$, $\mathbf{H}_1$, $\mathbf{A}_1$ and $\mathbf{B}_2$, $\mathbf{H}_2$, $\mathbf{A}_2$ where of course we must have

$$\nabla \times \mathbf{H}_1 = \nabla \times \mathbf{H}_2 = \mathbf{j}_{\text{ext}}(\mathbf{r}) \tag{25}$$

and

$$\mathbf{B}_1 = \mu(\mathbf{r})\mathbf{H}_1 \qquad \mathbf{B}_2 = \mu(\mathbf{r})\mathbf{H}_2 \tag{26}$$

We introduce a vector function:

$$\Delta(\mathbf{r}) = (\mathbf{A}_1 - \mathbf{A}_2) \times (\mathbf{H}_1 - \mathbf{H}_2) \tag{27}$$

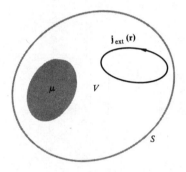

Figure 3   Uniqueness of **B**. Within a bounded volume V there are magnetic media and external currents. The magnetic field **B** is determined by the tangential component of either **A** or **H**.

From Gauss's theorem **(B-11)** we have:

$$\int_V \nabla \cdot \Delta(\mathbf{r})\, dV = \int_S (\mathbf{A}_1 - \mathbf{A}_2) \times (\mathbf{H}_1 - \mathbf{H}_2) \cdot d\mathbf{S} \tag{28}$$

By **(B-4)** we may expand the left side of **(28)**:

$$\int_V \nabla \cdot \Delta(\mathbf{r})\, dV = \int_V (\mathbf{H}_1 - \mathbf{H}_2) \cdot \nabla \times (\mathbf{A}_1 - \mathbf{A}_2)\, dV - \int_V (\mathbf{A}_1 - \mathbf{A}_2) \cdot \nabla \times (\mathbf{H}_1 - \mathbf{H}_2)\, dV \tag{29}$$

The second term is zero by **(25)**. Rewriting the first integral in terms of **B** and equating to **(28)** we have:

$$\int_V (\mathbf{B}_1 - \mathbf{B}_2) \cdot (\mathbf{H}_1 - \mathbf{H}_2)\, dV = \int_S (\mathbf{A}_1 - \mathbf{A}_2) \times (\mathbf{H}_1 - \mathbf{H}_2) \cdot d\mathbf{S} \tag{30}$$

Now, if either the *tangential* component of **H** or **A** is determined on the surface the right side of **(30)** will be zero. Further, if the field B is a monotonically increasing function of H, the left side will be positive definite and the only possible solution of **(27)** is for $\Delta(\mathbf{r}) = 0$, which gives

$$\mathbf{B}_1(\mathbf{r}) = \mathbf{B}_2(\mathbf{r}) \tag{31}$$

everywhere within V. Thus the solution within V is determined uniquely by the tangential component of either the vector potential **A** or the magnetic field **H** on the surface.

INTERACTION ENERGY

In analogy with the discussion at the end of Chapter 9 of the external energy of a system of magnetic moments, we now discuss the energy of interaction of a system of external magnetic moments and permeable media. We analyze the problem just as we did the corresponding dielectric problem. We consider, as shown in Figure 4, moments

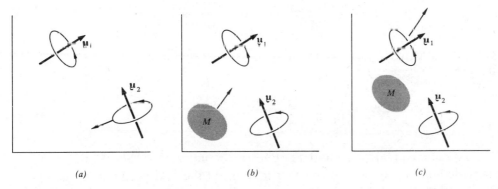

(a)   *(b)*   *(c)*

Figure 4 Interaction between magnetic moments and permeable media. (*a*) The magnetic moments are assembled in the absence of permeable media. (*b*) All magnetic media are brought up, keeping their relative positions fixed. (*c*) A variation is made in the system by displacing the external moments.

and permeable media, which must be brought up in a specified order inasmuch as we wish to allow for the possibility not only that the magnetic media may be nonlinear but also that they may exhibit hysteresis. Our procedure will be the following:

1. We first assemble the fixed moments μ_i in the absence of the permeable media. This is the problem which we discussed in Section 9-19, and we will be able to take over the result of that analysis.

2. We next bring up all magnetic media together keeping their relative positions fixed. At the same time we perform whatever electrical work is required to keep the μ_i fixed. In addition to the electrical work on the μ_i we will, of course, have to do mechanical work on the magnetic bodies.

3. Finally, we make a variation in the system by displacing the external moments μ_i.

The work required to initially assemble the fixed moments is from **(9-166)**:

$$W_{\text{initial}} = \tfrac{1}{2} \sum_{ij}{}' \mu_i \cdot \mathbf{B}_{ij} \tag{32}$$

where $\mathbf{B}_{ij}$ is the field of μ_j at the site of μ_i. Now, on bringing up the magnetic media we must perform mechanical work **(9-153)**:

$$W_{\text{mech}} = -\int_V dV \int \mathbf{M} \cdot d\mathbf{B}_{\text{ext}} \tag{33}$$

where $\mathbf{B}_{\text{ext}}$ is the field of the fixed moments only. This result follows from the requirement that we bring up the magnetic media rigidly so that it is not necessary to do mechanical work against the forces between magnetic bodies. In order to maintain the μ_i fixed we must perform, in addition, electrical work

$$W_{\text{elect}} = \sum_i \mu_i \cdot \left(\mathbf{B}_i - \sum_j{}' \mathbf{B}_{ij}\right) \tag{34}$$

where $\mathbf{B}_i$ is the final magnetic field at μ_i. To obtain the total work we add **(32)**, **(33)**, and **(34)**:

$$W = \sum_i \mu_i \cdot \mathbf{B}_i - \tfrac{1}{2} \sum_{ij}{}' \mu_i \cdot \mathbf{B}_{ij} - \int_V dV \int \mathbf{M} \cdot d\mathbf{B}_{\text{ext}} \tag{35}$$

We may simplify **(35)** by using the result that the interaction between the μ_i and the magnetic bodies may be written either in terms of the field produced at the μ_i or in terms of the external field at the magnetic bodies:

$$\int_V dV \, \mathbf{M} \cdot \mathbf{B}_{\text{ext}} = \sum_i \mu_i \cdot \left(\mathbf{B}_i - \sum_j{}' \mathbf{B}_{ij}\right) \tag{36}$$

Eliminating the sum over i and j between **(35)** and **(36)** we have, finally,

$$\blacktriangleright \qquad W = \tfrac{1}{2} \sum_i \mu_i \cdot \mathbf{B}_i + \tfrac{1}{2} \int_V dV \, \mathbf{M} \cdot \mathbf{B}_{\text{ext}} - \int_V dV \int \mathbf{M} \cdot d\mathbf{B}_{\text{ext}} \tag{37}$$

Finally we may *imagine* a small rearrangement of the fixed moments μ_i. The only work required is against the electromotive force induced in the *external* moments, permitting us to write:

$$\blacktriangleright \qquad \delta W = \sum_i \mu_i \cdot \delta \mathbf{B}_i \tag{38}$$

7 **Superposition.** To discuss *linear* media we write the magnetic field at the site of the ith moment:

$$\mathbf{B}_i = \sum_j{}' \bar{\bar{\lambda}}_{ij} \cdot \boldsymbol{\mu}_j \tag{39}$$

where the $\bar{\bar{\lambda}}_{ij}$ are tensor coefficients of induction and the $\boldsymbol{\mu}_j$ are external moments. In order to write (39) we must be able to establish a principle of superposition for magnetic fields. The argument closely parallels the discussion of Section 4-8 of the superposition of electric fields when external charge is brought up to a linear dielectric.

Bringing up an external moment $\boldsymbol{\mu}_i$ we have a vector potential $\mathbf{A}_i(\mathbf{r})$, which leads to magnetic fields $\mathbf{B}_i$ and $\mathbf{H}_i$. We imagine that S is at infinity so that the boundary condition is simply that $\mathbf{A}(\mathbf{r})$ is zero on S. Now if instead we were to bring up a moment $\boldsymbol{\mu}_j$ we would have a vector potential $\mathbf{A}_j(\mathbf{r})$ and magnetic fields $\mathbf{B}_j$ and $\mathbf{H}_j$. We argue that if $\boldsymbol{\mu}_i$ and $\boldsymbol{\mu}_j$ are brought up to positions $\mathbf{r}_i$ and $\mathbf{r}_j$, then the vector potential is $\mathbf{A}_i(\mathbf{r}) + \mathbf{A}_j(\mathbf{r})$. To establish this we use the uniqueness theorem, which requires only that $\mathbf{A}_i(\mathbf{r}) + \mathbf{A}_j(\mathbf{r})$ satisfy the boundary condition, which in this case is that the vector potential vanishes on S, and second that $\mathbf{H}(\mathbf{r})$ gives correctly the external currents.

8 **Magnetic Energy.** By substituting (39) into (38) we have the incremental work:

$$dW = \sum_{ij}{}' \boldsymbol{\mu}_i \cdot \bar{\bar{\lambda}}_{ij} \cdot d\boldsymbol{\mu}_i \tag{40}$$

As we have argued, the total amount of work can not depend on the order in which the moments are assembled as long as the magnetization density of the permeable medium is a single-valued function of $\mathbf{B}(\mathbf{r})$. This means that (40) must be a perfect differential and that we be able to write the work done in the form:

$$W = \tfrac{1}{2} \sum_{ij}{}' \boldsymbol{\mu}_i \cdot \bar{\bar{\lambda}}_{ij} \cdot \boldsymbol{\mu}_j = \tfrac{1}{2} \sum_i \boldsymbol{\mu}_i \cdot \mathbf{B}_i \tag{41}$$

Taking the differential of (41) and equating to (40) we obtain for the condition that (40) be a perfect differential:

$$\bar{\bar{\lambda}}_{ij} = \bar{\bar{\lambda}}_{ji} \tag{42}$$

where the tensor on the left has been transposed, that is the rows and columns of the corresponding matrix have been interchanged.

Exercises

4. Let $\boldsymbol{\mu}_j$ be a distribution of moments which produce $\mathbf{B}_i$ by (39). We now replace the moments $\boldsymbol{\mu}_j$ at positions $\mathbf{r}_j$ by $\boldsymbol{\mu}_j'$ and so obtain $\mathbf{B}_i'$. By (42) prove *Green's reciprocation theorem:*

$$\sum_i \mathbf{B}_i \cdot \boldsymbol{\mu}_i' = \sum_i \mathbf{B}_i' \cdot \boldsymbol{\mu}_i$$

5. Obtain an explicit expression for $\bar{\lambda}_{ij}$ in the absence of magnetic material and show that $\bar{\lambda}_{ij}$ is a symmetric tensor.

9 **Internal Energy.** If instead of a permeable medium, we had assembled a system of fixed magnetization and external moments, we would have had to perform from **(9-168)** an amount of work:

$$\blacktriangleright \qquad W_{\text{ext}} = \tfrac{1}{2} \sum_i \boldsymbol{\mu}_i \cdot \mathbf{B}_i + \tfrac{1}{2} \int_V dV\, \mathbf{M} \cdot \mathbf{B}_{\text{ext}} \qquad (43)$$

instead of the expression given by **(37)** for variable moments. The difference between **(37)** and **(43)** suggests an internal energy associated with the development of magnetization:

$$\blacktriangleright \qquad U_{\text{int}} = - \int_V dV \int \mathbf{M} \cdot d\mathbf{B}_{\text{ext}} \qquad (44)$$

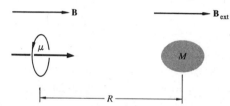

Figure 5   Internal energy. A magnetic moment and a small permeable body are separated by a distance R.

As a simple example of **(44)** we recall the diamagnetic atom. We saw in Section 7-1 that the internal energy of an idealized diamagnetic atom in a magnetic field **(7-11)** has the form:

$$U_{\text{int}} = U_0 - \frac{qL}{2m} B + \frac{q^2 R^2}{8m} B^2 \qquad (45)$$

Taking the differential of **(45)** we may write:

$$dU_{\text{int}} = -\left(\frac{qL}{2m} - \frac{q^2 R^2}{4m} B \right) dB = -\boldsymbol{\mu} \cdot d\mathbf{B} \qquad (46)$$

where $\boldsymbol{\mu}$ is the induced magnetic moment given by **(7-13)**. We see that **(46)** has the form of **(44)**.

As a second example we imagine a magnetic moment $\boldsymbol{\mu}$ and a small permeable body of magnetization $\mathbf{M}$ separated by a distance R as shown in Figure 5. The magnetic field at the magnetic body is given by

$$\mathbf{B}_{ext} = \frac{2\mu_0\,\mu}{R^3} \tag{47}$$

and the field at the fixed moment is given by

$$\mathbf{B} = \frac{2\mu_0\,MV}{R^3} \tag{48}$$

If we imagine that μ is fixed in position and the magnetic body is brought up, the amount of work performed is

$$W = -V\int \mathbf{M}\cdot d\mathbf{B}_{ext} + \mu\cdot\mathbf{B} \tag{49}$$

where the first term is the mechanical work performed on the magnetic body, and the second term is the electrical work required to keep μ constant. We have from **(47)** and **(48)**

$$\mu\cdot\mathbf{B} = \frac{2\mu_0\,V\mu\cdot\mathbf{M}}{R^3} = \mathbf{M}\cdot\mathbf{B}_{ext}\,V \tag{50}$$

so that we may write **(49)** as:

$$W = V\int \mathbf{B}_{ext}\cdot d\mathbf{M} \tag{51}$$

Alternatively, we may use **(50)** to write **(49)** in a more symmetrical form:

$$W = \tfrac{1}{2}\mu\cdot\mathbf{B} + \tfrac{1}{2}V\mathbf{M}\cdot\mathbf{B}_{ext} - V\int \mathbf{M}\cdot d\mathbf{B}_{ext} \tag{52}$$

The important result of **(51)** is that if the magnetization is induced parallel to $\mathbf{B}_{ext}$, then the work is positive. On the other hand, if $\mathbf{M}$ is opposite to $\mathbf{B}_{ext}$, then the work is negative.

If μ were a *real* magnet there would be no mechanism for maintaining μ constant, and we would have, from the first term of **(49)**,

$$W_{mech} = -2\mu_0\,V\int \mathbf{M}\cdot\mu\,d\,\frac{1}{R^3} \tag{53}$$

which is negative if the permeable medium is paramagnetic and positive if the medium is diamagnetic. Thus, we have established the already familiar result that magnets are attracted to paramagnetic bodies and repelled from diamagnetic bodies. Note that **(53)** is written so that only the gradient of $\mathbf{B}_{ext}$ contributes to the mechanical work and not the change in $\mathbf{B}_{ext}$ arising from a change in μ as the magnetic body is brought up.

10   Field Energy. We now wish to write **(37)** as an integral over fields. If we imagine a small rearrangement of the moments $\boldsymbol{\mu}_i$ we must do electrical work to maintain the moments constant **(38)**:

$$\delta W = \sum_i \boldsymbol{\mu}_i \cdot \delta \mathbf{B}_i \tag{54}$$

The change envisioned by **(54)** is *only* on the fixed moments and not on the magnetized bodies. To emphasize this we write the fixed moments in terms of external current through:

$$\boldsymbol{\mu}_i = \mathbf{M}_{\text{ext}}\, dV_i \tag{55}$$

where the curl of $\mathbf{M}_{\text{ext}}$ is the external current:

$$\nabla \times \mathbf{M}_{\text{ext}} = \mathbf{j}_{\text{ext}} = \nabla \times \mathbf{H} \tag{56}$$

and where the second equality follows from **(5)**:
To transform **(54)** we integrate the divergence of $\delta \mathbf{A} \times (\mathbf{H} - \mathbf{M}_{\text{ext}})$:

$$\int_V \nabla \cdot [\delta \mathbf{A} \times (\mathbf{H} - \mathbf{M}_{\text{ext}})]\, dV = \int_V (\mathbf{H} - \mathbf{M}_{\text{ext}}) \cdot \delta \mathbf{B}\, dV - \int_V \delta \mathbf{A} \cdot \nabla \times (\mathbf{H} - \mathbf{M}_{\text{ext}})\, dV \tag{57}$$

The left side may be converted into a surface integral, which goes to zero as the surface goes to infinity. The second term on the right is zero by **(56)**. In this way we may write, from **(54)**,

$$\delta U = \int_V dV\, \mathbf{M}_{\text{ext}}(\mathbf{r}) \cdot \delta \mathbf{B}(\mathbf{r}) = \int dV\, \mathbf{H}(\mathbf{r}) \cdot \delta \mathbf{B}(\mathbf{r}) \tag{58}$$

where the second integral is over all space.

Integrating **(58)** we obtain for the energy of a configuration of external currents and permeable media

$$\blacktriangleright \qquad U = \int dV \int \mathbf{H}(\mathbf{r}) \cdot d\mathbf{B}(\mathbf{r}) \tag{59}$$

Now if we regard **(9-179)** as leading to an external energy:

$$U_{\text{ext}} = \frac{1}{\mu_0} \int_V dV \int \mathbf{B}(\mathbf{r}) \cdot d\mathbf{B}(\mathbf{r}) - \tfrac{2}{3}\mu_0 \int_V dV \int \mathbf{M}(\mathbf{r}) \cdot d\mathbf{M}(\mathbf{r}) \tag{60}$$

Then the internal energy becomes from **(58)** and **(60)**:

$$U_{int} = U - U_{ext} = -\int_V dV \int \mathbf{M}(\mathbf{r}) \, d[\mathbf{B}(\mathbf{r}) - \tfrac{2}{3}\mu_0 \mathbf{M}(\mathbf{r})] = -\int_V dV \int \mathbf{M}(\mathbf{r}) \cdot d\mathbf{B}_{ext}$$

$$(61)$$

where we have used **(6-141)** for the external magnetic field. Comparing **(61)** with **(44)** we see that the two descriptions are identical.

The discussion of **(61)** closely follows that of **(4-69)**. What is meant by the external field depends on the problem. Although the total energy as given by **(59)** does not depend on what we take to be $\mathbf{B}_{ext}$, the separation between external and internal energy does depend on this choice. Thus in obtaining **(9-175)** we imagined a continuous distribution of magnetization and current with the interaction between moments given by the Lorentz field $-(2/3)\mu_0 \mathbf{M}(\mathbf{r})$ and treated as external energy. Similarly, the work required to establish the currents was treated as external energy.

On the other hand, the discussion of this chapter involves the interaction between external moments $\mathbf{\mu}_i$ and permeable bodies in fixed relative position. The external magnetic field is that of the external moments only and the interaction between magnetized bodies is included in the internal energy.

For linear media we are able to carry out the integrations. By assuming **(8)**, for example, we may integrate **(59)** to obtain for the total energy in terms of the fields:

$$\blacktriangleright \qquad\qquad U = \tfrac{1}{2}\int dV \, \mathbf{H}(\mathbf{r}) \cdot \mathbf{B}(\mathbf{r}) \qquad\qquad (62)$$

Exercises

6. Show that the internal energy of a uniformly magnetized linear ellipsoid is given by:

$$U_{int} = W_{int} = -\tfrac{1}{2}\mathbf{M} \cdot \mathbf{B}V + \tfrac{1}{2}(1 - N)\mu_0 M^2 V$$

7. Show from **(59)** that the energy in the magnetic field may be written as:

$$U = \int dV \int \mathbf{j}_{ext} \cdot d\mathbf{A}$$

Write $\mathbf{B}$ in terms of the vector potential and use **(B-4)**, **(B-11)**, and **(5)**. [C. H. Page, *Am. J. Phys.* **44**, 1104 (1976).]

INDUCTANCE

A magnetic inductor is a device for storing magnetic energy as the result of the field induced by an external current flowing through conducting windings. In order to reduce the volume of such a device and to minimize losses, a highly permeable core is normally used. In this section we discuss the interaction energy and electromotive force of inductors as an extension of our earlier theory of the interaction between external magnetic moments.

We write from **(6-111)** for an external magnetic moment:

$$\boldsymbol{\mu}_j = I_j \mathbf{S}_j \tag{63}$$

where we imagine a current I_j flowing around an area $\mathbf{S}_j$. For the flux that passes through the area $\mathbf{S}_i$ of a second moment we write from **(9-76)**:

$$\Phi_i = \mathbf{S}_i \cdot \mathbf{B}_i \tag{64}$$

Substituting **(63)** into **(38)** we write:

$$dW = \sum_i I_i \mathbf{S}_i \cdot d\mathbf{B}_i = \sum_i I_i \, d\Phi_i \tag{65}$$

where we have used **(64)**. Substituting **(39)** into **(64)** and including the term for $i = j$ we obtain:

▶
$$\Phi_i = \sum_j L_{ij} I_j \tag{66}$$

with

$$L_{ij} = \mathbf{S}_i \cdot \overline{\overline{\boldsymbol{\lambda}}}_{ij} \cdot \mathbf{S}_j \tag{67}$$

where L_{ij} is called the coefficient of inductance. For i equal to j, L_{ij} is called the self inductance. For i not equal to j, L_{ij} is called the mutual inductance and is usually given the symbol M_{ij}. From **(42)** we have the relation for the coefficients of inductance

$$L_{ij} = \mathbf{S}_i \cdot \overline{\overline{\boldsymbol{\lambda}}}_{ij} \cdot \mathbf{S}_j = \mathbf{S}_j \cdot \overline{\overline{\boldsymbol{\lambda}}}_{ij} \cdot \mathbf{S}_i = \mathbf{S}_j \cdot \overline{\overline{\boldsymbol{\lambda}}}_{ji} \cdot \mathbf{S}_i = L_{ji} \tag{68}$$

and we see that the L_{ij} are symmetric. By substituting **(66)** into **(65)** and using **(68)** we see that dW is a perfect differential, which we may integrate to obtain:

▶
$$W = \tfrac{1}{2} \sum_{ij} L_{ij} I_i I_j \tag{69}$$

11   Extended Currents. It is not possible to assume that the distance between real physical inductors is large compared with their size, and **(64)** must be written as an integral over $d\mathbf{S}_i$. Although we may think of an inductor as composed of small moments of the form of **(63)**, it is simpler to regard **(66)** as *defining* the coefficients of inductance with I_j the current of the jth inductor and Φ_i the flux through the ith inductor.

As an example of the computation of coefficients of inductance for extended currents we consider two *long* coaxial solenoids of radius R_1 and R_2, length l_1 and l_2, and with N_1 and N_2 turns as shown in Figure 6. We assume that l_1 is shorter than l_2.

With a current I_1 flowing through the inner inductor we have, from **(6-85)**,

$$B_1 = \mu H_1 = \mu \frac{N_1}{l_1} I_1 \tag{70}$$

The flux linking the inner inductor is, neglecting end effects,

$$\Phi_1 = \pi R_1^2 B_1 N_1 \tag{71}$$

where it is important to realize that the equivalent surface S of an inductor with N turns is N times the area of a single turn. From **(66)** we have for the self-inductance of the inner solenoid:

$$L_1 = L_{11} = \frac{\Phi_1}{I_1} = \mu_0 \pi R_1^2 N_1^2 / l_1 \tag{72}$$

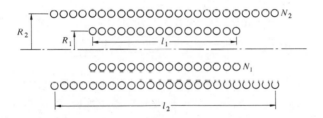

Figure 6 Coefficients of inductance. Two long coaxial solenoids are wound with N_1 and N_2 turns respectively.

The flux linking the outer inductor is

$$\Phi_2 = \pi R_1^2 B_1 N_2 \, l_1 / l_2 \tag{73}$$

since only a fraction of the turns l_1/l_2 are linked by B_1. This gives, for the mutual inductance,

$$L_{21} = \frac{\Phi_2}{I_1} = \mu_0 \pi R_1^2 N_1 N_2 / l_2 \tag{74}$$

We now perform a similar analysis with current flowing through the outer solenoid. With I_2 in the outer solenoid we obtain for the flux linking the outer solenoid:

$$\Phi_2 = \pi R_2^2 B_2 \, N_2 = \mu_0 \pi R_2^2 N_2^2 I_2 / l_2 \tag{75}$$

which leads to a coefficient of self-inductance:

$$L_2 = L_{22} = \mu_0 \pi R_2^2 N_2^2 / l_2 \tag{76}$$

The flux that links the inner solenoid is given by:

$$\Phi_1 = \pi R_1{}^2 B_2 N_1 = \mu_0 \pi R_1{}^2 N_1 N_2 I_2/l_2 \tag{77}$$

We obtain from **(66)** for the mutual inductance:

$$L_{12} = \frac{\Phi_1}{I_2} = \mu_0 \pi R_1{}^2 N_1 N_2/l_2 \tag{78}$$

We see comparing **(74)** and **(78)** that **(68)** is satisfied as it must be for **(65)** to be a perfect differential.

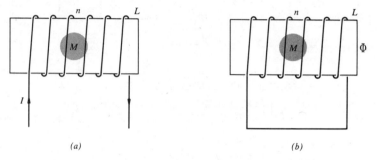

(a)   *(b)*

Figure 7   Interaction of an inductor with a magnetic sample. *(a)* A current I flows through a solenoid of inductance L containing a sample of magnetization M. *(b)* The solenoid is linked by a flux Φ given by **(85)**.

12   **Electromotive Force.** From **(9-111)** the electromotive force developed as a result of a change in flux is given by:

$$\blacktriangleright \qquad \mathscr{E}_i = -\frac{d\Phi_i}{dt} = -\sum_j L_{ij} \frac{dI_j}{dt} \tag{79}$$

which gives the coefficient of inductance in terms of the electromotive force and the rate of change of current. The SI unit of inductance is the henry, which is defined as the "inductance of a closed circuit in which an electromotive force of one volt is produced when the electric current in the circuit varies uniformly at a rate of one ampere per second."

As an example of the interaction of an inductor with a magnetic sample, we compute the amount of work that must be done to establish a current I in a solenoid of inductance L that contains a sample of magnetization M and volume V as shown in Figure 7. From **(79)** the amount of work required to establish the flux that links L is given by

$$W = -\int I \mathscr{E} \, dt = \int I \, d\Phi \tag{80}$$

We next need to determine the flux of **M** that links the inductor. The simplest way to obtain this result is to imagine the sample as an indicator of N_M turns and current I_M, which links the solenoid and use the symmetry of the mutual inductance **(68)**. The flux coupled to the sample is

$$\Phi_M = L_{12} I = \mu_0 n_M V I \tag{81}$$

which gives:

$$L_{12} = \mu_0 n_M V \tag{82}$$

The flux coupled from the sample into the solenoid is:

$$\Phi = L_{12} I_M = \mu_0 V n_N I_M = \mu_0 n M V \tag{83}$$

where we have used by **(6-152)**:

$$n_M I_M = M \tag{84}$$

Now the total flux that links the inductor is the sum of **(84)** and the flux of an empty solenoid:

$$\Phi = LI + \mu_0 n M V \tag{85}$$

Substituting **(85)** into **(80)** we have for the work:

$$W = \tfrac{1}{2} L I^2 + V \int \mathbf{B}_{\text{ext}} \cdot d\mathbf{M} \tag{86}$$

where by **(6-85)**:

$$B_{\text{ext}} = \mu_0 n I \tag{87}$$

is the field of a long solenoid. Alternatively, we may write **(86)** in terms of the flux that links L rather than in terms of the current to obtain:

$$W = \frac{1}{2} \frac{\Phi^2}{L} - \frac{\mu_0 n V}{L} \int M \, d\Phi \tag{88}$$

If we assume that the sample volume is small compared with that of the inductor we may approximate $d\Phi$ in the integral by $L \, dI$ permitting us to write approximately:

$$W \simeq \frac{1}{2} \frac{\Phi^2}{L} - V \int \mathbf{M} \cdot d\mathbf{B}_{\text{ext}} \tag{89}$$

In Figure 7 we have shown two configurations of the inductor. In Figure 7*a* we show a constant current. From **(86)** we see that the presence of the sample with **M** parallel to $\mathbf{B}_{\text{ext}}$ increases the amount of work that must be done and the energy stored in the inductor for the same current. On the other hand, if the sample is diamagnetic the energy stored is *less* than for an empty inductor. In Figure 7*b* we show an inductor that is shorted so that what remains constant is the *flux* rather than the current. With the flux constant we see from **(89)** that the energy is *reduced* by a ferromagnetic or paramagnetic sample. On the other hand, the energy is *increased* by a diamagnetic sample.

Exercises

---

8. Show that the mechanical work required to move a magnetic sample into a solenoid as shown in Figure 8 is given by:

$$W_{\text{mech}} = - V \int \mathbf{M} \cdot d\mathbf{B}_{\text{ext}}$$

Explain why this result is correct only if the current I is fixed.

9. Show that the SI unit of magnetic permeability is the henry per metre.

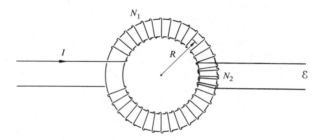

Figure 8 Magnetization measurement. The sample is formed in the shape of a ring, which is wound with both primary and secondary windings.

FERROMAGNETIC MATERIALS

In Chapter 7 we discussed very briefly the microscopic theory of ferromagnetism and saw that magnetization results from the ordering of electron moments as a result of the interaction between electrons. We saw that this interaction leads at temperatures above the transition temperature to an enhanced magnetic susceptibility. Our present discussion will be in terms of macroscopic quantities. We begin by discussing hysteresis in typical magnetic materials. We then discuss flux concentration. Finally we examine the shielding of magnetic fields by high permeability materials.

13 **Magnetization and Hysteresis.** A typical experimental arrangement for the study of magnetic materials is shown in Figure 8. The sample to be studied is formed in the shape of a ring, which is wound with both a primary magnetizing winding and a secondary winding that detects changes in magnetic flux. From (13) we have:

$$2\pi R H = N_1 I \tag{90}$$

Thus from a measurement of the current we have H in amperes per metre (A/m). From (79) we have for the voltage per turn in the secondary:

$$\mathscr{E} = -N_2 S \frac{dB}{dt} \tag{91}$$

where S is the cross-sectional area of the ring. The magnetic field B may then be obtained by integrating (91):

$$B = -\frac{1}{N_2 S} \int \mathscr{E} \, dt \tag{92}$$

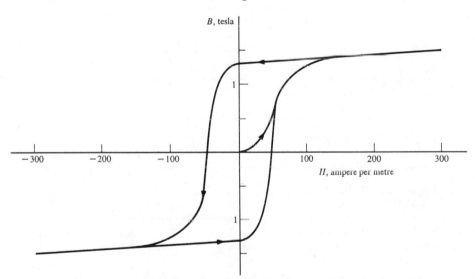

Figure 9 Plot of **B** versus **H** for iron. A curve of this form is taken by beginning with the material in the demagnetized state.

In the international system the unit of magnetic flux is the weber (Wb) and the unit of flux density is the tesla (T). An integration of the induced voltage gives directly the magnetic field B. In Figure 9 we show a typical plot of B versus H for iron. A curve of this form is taken by beginning with the material in the demagnetized state. This state may be achieved by heating the sample above the transition temperature and cooling in zero field. The same results may be achieved more easily by introducing into the primary windings a large alternating current which is gradually reduced. Then increasing the primary current until the magnetization:

$$M = \frac{B}{\mu_0} - H \tag{93}$$

has become saturated, we obtain what is called the initial magnetization curve. This curve typically has an inflection with the permeability

$$\mu(H) = \frac{B}{H} \tag{94}$$

increasing with current to a maximum value and then decreasing at higher currents. In Figure 10a we show the initial permeability as a function of H. It is generally more useful to show the permeability as a function of B as in Figure 10b.

As the magnetizing current is reduced the magnetization does not follow the initial curve, but is substantially higher. The value of B at zero current is called the remanent induction. The reverse demagnetizing field required to bring B to zero is called the coercive field. By increasing the reverse current still further, the sample may be taken

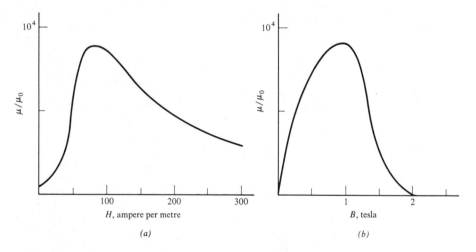

Figure 10 Permeability of iron. (*a*) The initial permeability is shown as a function of *H*. (*b*) The initial permeability as a function of *B* is generally more useful.

to saturation in the reverse direction. By reducing the reverse current to zero and increasing it in the positive direction, the cycle may be completed.

Exercise

10. A sphere of radius R has a uniform remanent magnetization M. Obtain an expression for for the internal energy, which is the energy of formation of the magnetization.

14 Ferromagnetic Domains. The fields **B** and **H** and the magnetization density **M** are all macroscopic quantities and represent an average over the volume of the sample. On a microscopic scale one finds, on the other hand, a local magnetization **M(r)** equal in magnitude to the saturation magnetization of the material. In the "demagnetized" state the magnetization averages to zero over an extended volume of the sample.

In highly permeable magnetic materials like iron or nickel the local magnetization changes from one direction to another quite gradually in several hundred atomic distances through what is called a domain wall. Normally, the magnetization rotates within the plane of the wall so that the condition

$$\bar{\rho}_M = -\nabla \cdot \mathbf{M}(\mathbf{r}) = 0 \tag{95}$$

is preserved within the ferromagnet.

As discussed in Chapter 7, the dominant interaction that leads to magnetic ordering is the local exchange interaction with an energy from **(7-54)**:

$$U_{\text{exch}} = -\tfrac{1}{2}\mathbf{M}(\mathbf{r}) \cdot \mathbf{B}_{\text{exch}}(\mathbf{r})\, V = -\tfrac{1}{2}\mu_0\, \lambda \langle M^2 \rangle\, V \tag{96}$$

where $\mathbf{M}(\mathbf{r})$ in **(96)** is regarded as a microscopic field quantity.

In addition to **(96)** we have from **(86)** for the work to magnetize the sample:

$$\blacktriangleright \qquad W = V \int \mathbf{B}_{\text{ext}} \cdot d\mathbf{M} = V \int \mu_0\, \mathbf{H} \cdot d\mathbf{M} + \tfrac{1}{2}\mu_0\, N \langle M \rangle^2 V \tag{97}$$

where the field quantities $\mathbf{H}$ and $\langle \mathbf{M} \rangle$ are macroscopic.

Domain walls are driven by $\mathbf{H}$ in such a way as to develop a magnetization in the direction of $\mathbf{H}$ and thus *increase* the total magnetic energy. Where the magnetic field $\mathbf{H}$ arises from external currents, the source of this energy is the electromotive force driving the external currents. For a ferromagnetic sample with $\mathbf{H} = 0$, which is to say no external current, we expect the stable state to be one in which the sum of **(96)** and **(97)** is a minimum. Since the magnetization in **(96)** is a local quantity the exchange energy is little affected by the formation of domains except for the energy necessary to form the domain wall itself. On the other hand, **(97)** is minimized by a macroscopic magnetization:

$$\langle \mathbf{M} \rangle = 0 \tag{98}$$

which characterizes the demagnetized state. Because domain walls are not always able to move freely the condition **(98)** is only an approximate one and there may be substantial remanence and hysteresis as shown in Figure 9.

15 [**Flux Concentration**]. We next consider the magnetization of a sample of ellipsoidal shape in a uniform external field as in Figure 11. We may write for the internal field:

$$B = B_{\text{ext}} + \mu_0(1 - N)M \tag{99}$$

where N is the demagnetization factor (Section 6-26) with the assumption that B_{ext} is applied along a principal axis of the ellipsoid.

Eliminating M between **(99)** and **(1)** we obtain:

$$B + \frac{1 - N}{N}\, \mu_0\, H = \frac{1}{N}\, B_{\text{ext}} \tag{100}$$

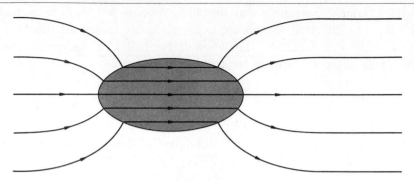

Figure 11   Flux concentration. An ellipsoidal sample is placed in a uniform external magnetic field.

To find B for some B_{ext} we plot the straight line given by **(100)** on the magnetization curve of the material as shown for cobalt in Figure 12 where we have taken $N = \frac{1}{2}$, which is the transverse demagnetizing factor of a long cylinder. In low external fields the permeability $\mu = B/H$ is high. Under these circumstances the second term on the left of **(100)** may be neglected and we have

$$B = \frac{1}{N} B_{ext} \tag{101}$$

That is, additional flux proportional to B_{ext} is drawn into the ellipsoid. At high fields the material is saturated and we may use **(99)**, which indicates that the additional flux through the ellipsoid is independent of B_{ext}.

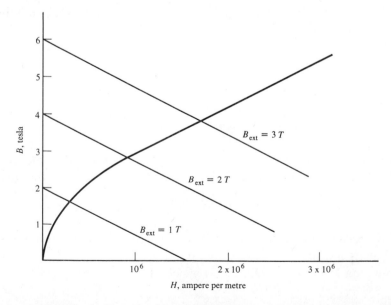

Figure 12   Internal magnetic field. The internal field B for some external field B_{ext} is obtained graphically.

Table 1   Magnetic Properties

	μ_{max}/μ_0	$\mu_0 M_s$	B_r	H_c
Cobalt	250	1,760 mT	490 mT	710 A/m
Iron	2,500	2,145	1,400	50
Nickel	2,500	610	400	60
Permalloy	350,000	750		0.3

In Table 1 we give room temperature values of the maximum permeability, μ_{max}, the saturation induction $\mu_0 M_s$, the remanent induction B_r, and the coercive field H_c. We give representative values for permalloy, which is the commercial name of a class of alloys containing primarily iron and nickel.

16 [Magnetic Shielding]. We now apply the previous discussion to the problem of a cylindrical magnetic shield in a transverse field as shown in Figure 13a. We assume that the external magnetic field is the same as for a highly permeable solid cylinder as shown in Figure 14b. Then the flux which penetrates the cylinder is:

$$2BRL = 4B_{ext}RL \tag{102}$$

where L is the length of the cylinder and R is the radius. But instead of the flux being distributed uniformly as in Figure 13b, it is concentrated in the magnetic shield. The flux density in the shield is given by:

$$B_0 = -2B_{ext}\frac{z}{d} = -\frac{2R}{d}B_{ext}\sin\theta \tag{103}$$

where d is the thickness of the shield. If the permeability of the shield material is μ, then the magnetic field H_θ within the shield is given by

$$H_\theta = -\frac{2R}{\mu d}B_{ext}\sin\theta \tag{104}$$

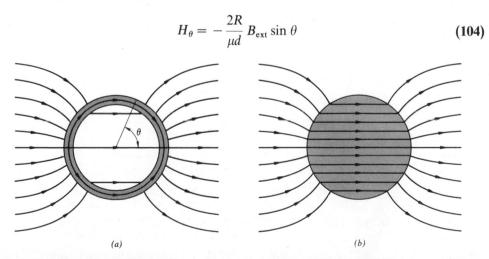

(a)   (b)

Figure 13 Magnetic shielding. (a) A cylindrical magnetic shield is placed across a magnetic field. (b) It is assumed that the field outside the cylinder is the same as for a solid cylinder of some equivalent permeability.

To obtain the magnetic field B_{ext} within the central volume we use **(24)** to obtain:

$$\frac{1}{\mu_0} B_{int} \sin \theta = \frac{2R}{\mu d} B_{ext} \sin \theta \tag{105}$$

From **(105)** we may write finally for the interior field in terms of the exterior field:

$$B_{int} = \frac{2R}{d} \frac{\mu_0}{\mu} B_{ext} \tag{106}$$

For high permeability materials and typical dimensions, the interior field may be as small as $\frac{1}{1000}$ of the exterior field.

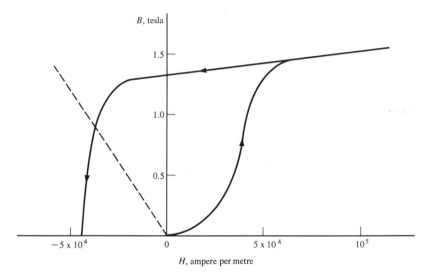

Figure 14   Permanent magnetization. The magnetization may be obtained graphically from the demagnetization curve.

By using a double shield the field within the central cylinder may be reduced considerably more than if the same total amount of material were used in the outer shield. In the limit that the inner radius is much smaller than the outer radius the two reduction factors multiply, giving an over all reduction by as much as 10^6.

17   Permanent Magnets. A useful permanently magnetized material should have a large induction B even in the presence of a moderate demagnetizing field. This induction should be relatively insensitive to shock, changes in temperature, or external magnetic fields. We show in Figure 14 the initial magnetization and demagnetization curves for Alnico 5, an alloy of aluminum, nickel, and cobalt that has been given special heat treatment. Also plotted in Figure 14 is **(100)** in zero external field and with a demagnetizing factor $N = 0.05$ as for a rod with a length to diameter ratio $L/D = 5$. The actual field of 0.9 T is at the intersection of the two curves. It is clear from Figure 14 that for a large induction both the remanence B_r and the coercive field H_c must be

Table 2 Magnetic Properties

	B_r	H_c	$(BH)_{max}$
Magnet steel	1,000 mT	18,000 A/m	7,000 A · T/m
Platinum 77% cobalt 23%	500	240,000	36,000
Alnico 5	1,250	46,000	36,000

large. Materials suitable for permanent magnets are characterized by either the product of B_r and H_c or by $(BH)_{max}$, which is the negative of the maximum product of B and H on the demagnetization curve. We give in Table 2 values of B_r, H_c, and $(BH)_{max}$ for several permanent magnetic materials

SUPERCONDUCTING MATERIALS

In this section we discuss briefly the characteristics of superconducting materials for the generation of intense magnetic fields and for the low-loss transmission of electric power. In both connections the magnetic properties of the superconductor are of crucial importance. In this discussion we characterize a superconductor by a magnetization density related to the superconducting currents by **(6-151)**:

$$\nabla \times \mathbf{M}(\mathbf{r}) = \mathbf{j}(\mathbf{r}) \tag{107}$$

Although we find it possible to discuss superconducting and ferromagnetic materials from very similar points of view, we must remember of course that while the currents responsible for ferromagnetism are strictly atomic, the currents responsible for superconductivity extend over macroscopic distances.

18 Type I Superconductivity. In our discussion of superconductivity in Section 8-6 we saw that as a consequence of the constancy of the generalized angular momentum, magnetic fields are screened from the interior of a superconductor in a distance λ_L of macroscopic dimensions. The assumption of coherence over macroscopic dimensions does not actually apply to all superconducting materials, but only to those materials which are highly ordered, such as the pure elements. These materials are called Type I and are characterized in the superconducting state by:

$$\mathbf{B}(\mathbf{r}) = 0 \tag{108}$$

From **(44)** the internal energy is given by:

$$dW_{int} = -V \int \mathbf{M} \cdot d\mathbf{B}_{ext} = -V \int \mu_0 \mathbf{M} \cdot d\mathbf{H} - V \int \mu_0 N \mathbf{M} \cdot d\mathbf{M} \tag{109}$$

where we have used:

$$d\mathbf{B}_{ext} = \mu_0 \, d\mathbf{H} + \mu_0 \, N \, d\mathbf{M} \tag{110}$$

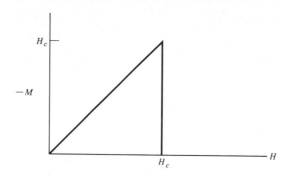

Figure 15   Type I superconductivity. The negative of the magnetization is plotted as a function of H.

Using **(108)** permits us to integrate **(109)** to obtain the internal energy:

$$W_{int} = \tfrac{1}{2}\mu_0 H^2 V - \tfrac{1}{2}\mu_0 NH^2 V \qquad\qquad (111)$$

As the external magnetic field is increased the internal energy as given by **(111)** increases until the superconducting state is no longer stable and a transition occurs to the normal state. The value of H at which the transition takes place for $N = 0$ is called the critical field H_c. We show in Figure 15 a plot of the negative of the magnetization as a function of H for a Type I superconductor. The critical field H_c is given at $T = 0$ in Table 3 for a number of Type I superconductors. By comparison with Table 8-1, it may be seen that the critical field at $T = 0$ is linearly proportional to the transition temperature T_c. The critical field falls smoothly with increasing temperature, going to zero at the critical temperature.

Table 3   Critical Field of Type I Superconductors

Superconductor	H_c
Cd	2.2 kA/m
In	22
Sn	24
Hg	32
Tl	14
Pb	64

Exercise

11. Show for a superconducting sphere that the magnetization is given by:

$$\mu_0 \mathbf{M} = -\tfrac{3}{2}\mathbf{B}_{ext}$$

Show that the magnetic field **B** just outside the sample and in the median plane is given by:

$$\mathbf{B} = \tfrac{3}{2}\mathbf{B}_{\text{ext}}$$

---

19 Type II Superconductivity. There exists a second class of superconducting materials for which **(108)** is satisfied only up to a field H_{c1} called the lower critical field. Above this field magnetic flux penetrates limited regions and the material continues to show superconducting properties. Finally, at an upper critical field H_{c2} some hundred times larger than H_{c1}, the field $\mathbf{B}_{\text{ext}}$ fully penetrates the sample, and there is a transition to the normal state. We show in Figure 16 a plot of magnetization as a function of H for a Type II superconductor.

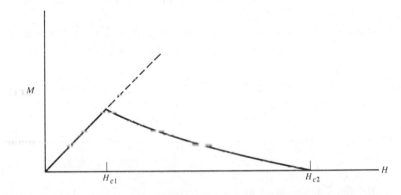

Figure 16 Type II superconductivity. The negative of the magnetization is plotted as a function of H. There exist two critical fields. At H_{c1} the magnetization deviates from that for perfect diamagnetism. At H_{c2} the magnetization finally goes to zero.

The internal energy is given by integrating **(109)** and is the area under the curve up to to the value of the internal field H.

SUMMARY

1. The magnetic field **H** is introduced through the relation:

$$\mathbf{B} = \mu_0\,\mathbf{H} + \mu_0\,\mathbf{M} \qquad\qquad (1)$$

which leads to the sources of **H**:

$$\mathbf{\nabla} \times \mathbf{H} = \mathbf{j}_{\text{ext}} \qquad \mathbf{\nabla} \cdot \mathbf{H} = \bar{\rho}_M \qquad\qquad (5) \quad (6)$$

2. For a linear medium we write

$$\mathbf{B} = \mu\mathbf{H} \qquad\qquad (8)$$

From $\mu_0\mathbf{M} = \chi\mathbf{B}$ we obtain for the relation between the permeability and the susceptibility:

$$\mu = \frac{\mu_0}{1 - \chi} \tag{9}$$

3. The boundary conditions on **B** and **H** at a planar interface between two media are that the normal component of **B** is continuous:

$$\hat{\mathbf{n}} \cdot \mathbf{B}_1 = \hat{\mathbf{n}} \cdot \mathbf{B}_2 \tag{25}$$

and the tangential component of **H** is continuous:

$$\hat{\mathbf{n}} \times \mathbf{H}_1 = \hat{\mathbf{n}} \times \mathbf{H}_2 \tag{24}$$

4. The magnetic field $\mathbf{B}(\mathbf{r})$ is uniquely determined within a bounded volume by the enclosed permeable media and external currents and by the tangential component of **H** or **A** on the boundary.

5. The work required to bring together fixed moments $\boldsymbol{\mu}_i$ in the presence of permeable media is given by:

$$W = \tfrac{1}{2}\sum_i \boldsymbol{\mu}_i \cdot \mathbf{B}_i + \tfrac{1}{2}\int_V dV\, \mathbf{M} \cdot \mathbf{B}_{\text{ext}} - \int_V dV \int \mathbf{M} \cdot d\mathbf{B}_{\text{ext}} \tag{37}$$

where $\mathbf{B}_{\text{ext}}$ is the field of the fixed moments only.

6. The work associated with a virtual rearrangement in the fixed moments is given by:

$$\delta W = \sum_i \boldsymbol{\mu}_i \cdot \delta\mathbf{B}_i \tag{38}$$

7. The assembly of fixed moments $\boldsymbol{\mu}_i$ and $\mathbf{M}(\mathbf{r})$ leads to the external work:

$$W_{\text{ext}} = \tfrac{1}{2}\sum_i \boldsymbol{\mu} \cdot \mathbf{B}_i + \tfrac{1}{2}\int_V dV\, \mathbf{M} \cdot \mathbf{B}_{\text{ext}} \tag{43}$$

8. The difference between the total work and the external work must be the internal energy associated with the development of the magnetization:

$$W_{\text{int}} = -\int_V dV\, \mathbf{M} \cdot d\mathbf{B}_{\text{ext}} \tag{44}$$

9. The energy of a configuration of external currents and permeable media may be written as:

$$U = \int dV \int \mathbf{H}(\mathbf{r}) \cdot d\mathbf{B}(\mathbf{r}) \tag{59}$$

For linear media the integration over the fields may be performed to give:

$$U = \tfrac{1}{2} \int dV \, \mathbf{H}(\mathbf{r}) \cdot \mathbf{B}(\mathbf{r}) \tag{62}$$

10. The interaction between current contours may be written in terms of currents and fluxes rather than in terms of moments and fields. We write for the flux linking the ith contour:

$$\Phi_i = \sum_j L_{ij} I_j \tag{66}$$

where the L_{ij} are called the coefficients of inductance. The work required to establish currents in the inductors may then be written as

$$W = \tfrac{1}{2} \sum_{ij} L_{ij} I_i I_j \tag{69}$$

11. The electromotive force developed in the ith circuit may be written as:

$$\mathscr{E}_i = -\frac{d}{dt}\Phi_i = -\sum_j L_{ij}\frac{d}{dt} I_j \tag{79}$$

12. Ferromagnetic materials are generally characterized by the dependence of the magnetic field $\mathbf{B}$ on the field $\mathbf{H}$. Normally, measurements are made on a magnetic ring for which $\mathbf{H}$ is determined by the magnetizing current, and $\mathbf{B}$ is measured by integrating the electromotive force developed in a secondary winding. Such materials are strongly nonlinear and may show substantial hysteresis. The amount of work done in completing a cycle of magnetization may be written as:

$$W = V \oint \mathbf{H} \cdot d\mathbf{B} \tag{97}$$

13. Although the magnetic properties of a superconductor are determined by surface screening currents, it is more satisfactory to develop a macroscopic description in terms of an effective magnetization whose curl is equal to the screening currents:

$$\mathbf{V} \times \mathbf{M}(\mathbf{r}) = \mathbf{j}(\mathbf{r})$$

which leads to the expression for the internal energy:

$$W_{\text{int}} = -\int dV \int \mathbf{M} \cdot d\mathbf{B}_{\text{ext}} \tag{109}$$

14. A type I superconductor fully excludes flux up to a critical external field $\mathbf{H} = \mathbf{H}_c$ at which a transition occurs to the normal state. This transition occurs when the internal magnetic energy exceeds the condensation energy associated with the superconducting state.

15. A type II superconductor allows some flux penetration above a lower critical field $\mathbf{H} = \mathbf{H}_{c1}$. At least some superconducting regions persist to an upper critical field $\mathbf{H} = \mathbf{H}_{c2}$ at which a transition to the normal state takes place.

PROBLEMS

1. **Magnetomotive force.** By analogy with the electromotive force, we define the magnetomotive force as the circulation of $\mathbf{H}$ over a contour C:

$$\mathscr{H} = \oint_C \mathbf{H} \cdot d\mathbf{r}$$

Show that the magnetomotive force is equal to the total *external* current included by the contour.

2. **Scalar potential.** By comparing **(6-161)** and **(1)** show that the magnetic field $\mathbf{H}$ may be written as the gradient of a scalar:

$$\mu_0 \mathbf{H} = -\nabla \bar{\phi}$$

Where the permeability μ is a function of position obtain the differential equation for $\bar{\phi}$:

$$\mu \nabla^2 \bar{\phi} + (\nabla \bar{\phi}) \cdot (\nabla \mu) = 0$$

3. **Boundary conditions.** Show that the boundary conditions between two permeable media may be written in terms of the scalar potentials on the two sides as:

$$\bar{\phi}_1 = \bar{\phi}_2 \qquad \mu_1 \hat{\mathbf{n}} \cdot \nabla \bar{\phi}_1 = \mu_2 \hat{\mathbf{n}} \cdot \nabla \bar{\phi}_2$$

4. **Permeable ellipsoid.** An ellipsoid of permeability μ_1 is placed in a medium of permeability μ_2, where there is a magnetic field $\mathbf{B}_{\text{ext}}$. Show that the magnetic field $\mathbf{B}$ within the ellipsoid is related to the external field by the expression

$$\mu_2 \mathbf{B} + (\mu_1 - \mu_2)\mathbf{N} \cdot \mathbf{B} = \mu_1 \mathbf{B}_{\text{ext}}$$

where $\mathbf{N}$ is the demagnetizing tensor of the ellipsoid. Use **(6-166)**, **(7-23)** and the arguments of Section 4-17.

5. **Boundary conditions on vector potential.** Show that the boundary conditions on $\mathbf{B}$ and $\mathbf{H}$ between two permeable media may be written as boundary conditions on

the vector potential:

$$\hat{n} \times A_1 = \hat{n} \times A_2 \qquad \frac{1}{\mu_1} \hat{n} \times (\nabla \times A_1) = \frac{1}{\mu_2} \hat{n} \times (\nabla \times A_2)$$

6. **Magnetic field analog.** The magnetic field generated by *very high* permeability magnetic materials may be shown to have the same spatial variation as the electric field produced by conductors of the same shape. Show by using the magnetic scalar potential that the two theories are analogous as long as there are no currents in the magnetic air gap. Use the result of Exercise 1.

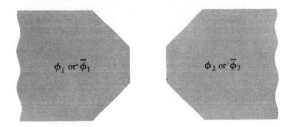

7. **Coaxial inductor.** A medium of permeability μ is enclosed by two infinite coaxial cylindrical conductors. The radius of the inner conductor is u and that of the outer conductor is b. If a current I flows through the inner conductor and back through the outer conductor, find the magnetic energy per metre and the inductance per metre.

8. **Parallel transmission line.** A parallel transmission line is constructed of two parallel wires of radius a and b whose centers are separated by c. A current I flows up one wire and down the other. Assuming that the currents are uniformly distributed over the wires, find the magnetic energy per metre from

$$W = \tfrac{1}{2} \int j_{ext}(r) \cdot A(r) \, dV$$

Show that the inductance per metre is given by:

$$L = \frac{\mu_0}{4\pi} \left[1 + 2 \ln (c^2/ab) \right]$$

Although this result is exact, the proof is rather complex. By assuming the wires well separated the expression for L is readily obtained.

9. **Coefficient of coupling.** A pair of inductors of inductance L_1 and L_2 have mutual inductance M. Show that the fraction of flux of 1 that links 2 is equal to the fraction of flux of 2 that links 1 and that this fraction, called the coefficient of coupling, is given by:

$$k = \frac{M}{(L_1 L_2)^{1/2}}$$

10. **Mutual inductance.** A rectangular circuit of sides $2a$ and $2b$ is placed so that its center is at a distance r from an infinite straight conductor as shown in the following diagram. Find the mutual inductance between the two circuits.

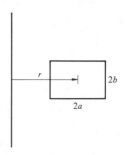

11. **Current ring.** A current I flows around a ring of radius R. Show that there is a tension in the ring:

$$T = \frac{1}{2\pi} I^2 \frac{dL}{dR}$$

where L is the inductance of the ring.

12. **Surface Reactance.** A solenoid is wound on a conducting rod. Show that the inductance of the solenoid is given by:

$$L = \frac{1}{\omega} X$$

where X is the surface reactance of the conductor **(9-143)**.

13. **Tapered iron whisker.** Very fine single crystal iron "whiskers" are grown with radii R of the order of 10 μm and lengths up to several millimetres. The whiskers carry a uniform magnetization density $\mathbf{M}$ along the length of the whisker and ordinarily are tapered. Assuming a very gradual taper $dR/dz \ll 1$, show that the transverse magnetic field at a distance r from the whisker is given by:

$$B_T = -\mu_0 M \frac{R}{r} \frac{dR}{dz}$$

14. **Ellipsoidal inductor.** An inductor is wound on an ellipsoid of permeability μ with the plane of the windings normal to a principal axis. The turns are evenly spaced axially with n turns per metre.

(a) Show that the magnetic field B is given by:

$$B = \mu H = \frac{\mu N I}{1 + \mu N/\mu_0 (N - 1)}$$

(b) Show that the inductance is:

$$L = \frac{\mu n^2 V}{1 + \mu N/\mu_0 (N - 1)}$$

(c) Show that the fraction f of the integrated field energy **(62)** that is stored within the ellipsoidal volume is:

$$f = \frac{1}{1 + \mu N/\mu_0 (N - 1)}$$

15. **Remanent magnetization.** A magnetized medium is reduced to its remanent state with $j_{ext}(\mathbf{r}) = 0$ everywhere. Show that the integral

$$\int \mathbf{B} \cdot \mathbf{H} \, dV$$

over all space is identically zero.

16. **Ferromagnetic resonance.** Using the results of Section 7-7 obtain an expression for the permeability tensor in the vicinity of ferromagnetic resonance. Show that below the resonance $\mu_0 M_x$ may exceed B_x, and the transverse permeability will be negative.

17. **Magnetic circuit.** A ring of Alnico contains a small air gap of width d. Show that in the remanent state the magnetic energy within the gap is just half the field energy within the magnetic material. Explain why the maximum available field in the gap is determined by the maximum BH product in the material.

18. **Domain rotation.** Where domain walls are absent or are present but pinned, a macroscopic magnetization is developed through the rotation of $\mathbf{M}$ in the macroscopic field $\mathbf{B}$. Assume an internal energy of the form

$$U_{int} = K \sin^2 \phi$$

where ϕ is the angle between $\mathbf{M}$ and the preferred magnetization direction. A magnetization field $\mathbf{B}$ is applied at an angle θ as shown in the following diagram.

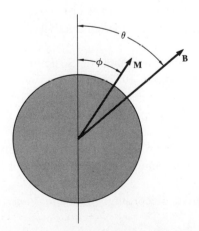

(a) For $\theta = \pi/2$ plot the component of the magnetization along **B** as a function of the magnitude of **B**.

(b) For $\theta = 0$ obtain the magnetization curve for a complete cycle. Show that for this orientation a square hysteresis loop is obtained.

19. **Eddy currents.** The current through a long solenoid containing an iron core of diameter D, is suddenly changed. Show that the characteristic time for magnetic flux to diffuse into the core is of order

$$\tau = \pi D^2 \sigma \mu$$

Evaluate τ for a rod one inch in diameter. Take for the conductivity of soft iron 1.02×10^7 S/m. Assume that the relative permeability is near the peak value of 10^4. Compare with Problem 9-14 and explain very carefully why the time is proportional to the relative permeability.

20. **Inductance change.** A small ellipsoidal inductor of permeability μ_1 is inserted into a medium of permeability μ_2, which forms the core of an inductor L. The ellipsoid is oriented so that a principal axis is along the magnetic field of the inductor. Show that the inductance is given by

$$L\left[1 + \frac{\mu_1 - \mu_2}{\mu_2 + N(\mu_1 - \mu_2)}\frac{v}{V}\right]$$

where L is the original inductance, v is the volume of the ellipsoid, V is the volume of the core, and N is the demagnetizing factor of the ellipsoid. Use **(86)** and the result of Problem 4.

REFERENCES

Becker, J. J., "Permanent Magnets," *Scientific American*, **223** (6), 92 (December 1970).

Bobeck, A. H. and H. E. D. Scovil, "Magnetic Bubbles," *Scientific American*, **224** (6), 78 (June 1971).

Bozorth, R. M., "Magnetic Materials," *Scientific American*, **192** (1), 68 (January 1955).

Brown, W. F., Jr., *Magnetostatic Principles in Ferromagnetism*, North Holland, 1962.

Buchhold, T. A., "Applications of Superconductivity," *Scientific American*, **202** (3), 74 (March 1960).

Essman, U. and H. Träuble, "The Magnetic Structure of Superconductors," *Scientific American*, **224** (3), 74 (March 1971).

Keffer, F., "The Magnetic Properties of Materials," *Scientific American*, **217** (3), 222 (September 1967).

Jensen, H., "The Airborne Magnetometer," *Scientific American*, **204** (6), 151 (June 1961).

Kittel, C., *Introduction to Solid State Physics*, Fifth Edition, Wiley, 1976.

Kunzler, J. E. and M. Tannenbaum, "Superconducting Magnets," *Scientific American*, **206** (6), 60 (June 1962).

Ragosine, V. E., "Magnetic Recording," *Scientific American*, **221** (5), 58 (November 1969).

Sampson, W. B., P. P. Craig, and M. Strongin, "Advances in Superconducting Magnets," *Scientific American*, **216** (3), 114 (March 1967).

Snowden, D. P., "Superconductors for Power Transmission," *Scientific American*, **226** (1), 84 (January 1972).

11 SOURCES OF THE ELECTROMAGNETIC FIELD III

In this chapter we complete the theory of electromagnetic fields and obtain a set of four equations for the sources of flux and circulation of electric and magnetic fields. These four equations were fully formulated by James Clerk Maxwell in 1865. The history of their development is available in his *Treatise on Electricity and Magnetism.*[1] Maxwell's work stands as a landmark in theoretical physics. His careful scholarship produced an integration of the insights of Faraday with the results of potential theory. He recognized that existing theory was incomplete, and his completion of the theory to give an electromagnetic description of light was a rare act of theoretical creativity.

Although we have no wish to minimize Maxwell's contribution, the line of development that we have been following is somewhat different from his. We began with a discussion following Maxwell of the interaction between charges. For the interaction between steady currents, we adopted the approach of Einstein[2] and made use of the transformation properties of forces between moving frames. Our discussion of electromagnetic induction was a synthesis of the Einstein and Maxwell approaches. In this chapter we complete the theory by finding expressions for the sources of circulation and flux of the magnetic field in the presence of nonsolenoidal currents. We first present the theory as it may be developed from the Lorentz force law, special relativity, and the sources of circulation and flux of the electric field. Then we examine Maxwell's solution to the problem—a correct solution without the benefit of our contemporary insight into relativity. Following this we discuss energy and momentum in the electromagnetic field and then radiation from oscillating dipoles and moving charges and currents.

---

[1] *A Treatise in Electricity and Magnetism* by James Clerk Maxwell, 3rd ed., 1891, Oxford University Press. Reprinted in two volumes by Dover Publications, 1954.

[2] A. Einstein, "On the Electrodynamics of Moving Bodies." Reprinted in *The Principle of Relativity*, Metheun and Company, Ltd., 1923. Reprinted by Dover Publications, 1952.

TRANSFORMATION OF FIELDS

The transformation of electromagnetic fields follows from the transformation properties of forces. Having obtained general expressions for the way in which electric and magnetic fields transform, we look at two special cases, uniformly moving charges as discussed in Section 6-2, and uniformly moving currents as discussed in Section 9-1.

1   **Lorentz Force.** We introduced in Section 6-2 **(6-20)** the expression for the force on a moving charge:

$$\mathbf{F} = q(\mathbf{E} + \mathbf{v} \times \mathbf{B}) \tag{1}$$

This expression is called the electromagnetic or Lorentz force.

Evidence for its correctness comes from many parts of physics. The acceleration of electrons and protons to velocities approaching that of light and their subsequent deflection by magnetic fields provides a verification of **(1)** to high precision. The absence of any measurable deflection of a beam of neutral atoms by either an electric or magnetic field again provides a verification of **(1)** since the motions of atomic electrons and nuclear particles are quite different. We know only that on the average their velocities must be the same and thus from experiment that the force is *linear* in the velocity. The experiments of Coulomb and of Ampère are a verification of the Lorentz force law as well. Although we like to think of the charges on a conductor as residing motionless on the surface, in fact these charges are in random motion at high velocities. And the motion of charge through a conductor is only a slow drift superimposed on the rapid diffusion of the carriers. Where we find for accelerated charges that the force does not *appear* to follow **(1)** we will conclude that **E** and **B** are affected by the motion of the charge. In this sense we may regard **(1)** as a *defining* equation for the electric and magnetic fields. Thus, we may use the transformation properties of forces to give us the transformation properties of fields.

2   **Transformation of Fields.** We now describe the motion of the same charge q as in **(1)** but from a frame moving with velocity **V** with respect to the original frame. From the principle of relativity we expect to be able to write an equation:

$$\mathbf{F}' = q'(\mathbf{E}' + \mathbf{v}' \times \mathbf{B}') \tag{2}$$

We know from experiments like the King experiment, described in Section 6-1, that charge does not depend on velocity and so we may write for the electric flux:

$$\Psi' = \frac{1}{\epsilon_0} q' = \Psi = \frac{1}{\epsilon_0} q \tag{3}$$

since as we have seen **(6-16)** Ψ is independent of the velocity of q. If Ψ were to change with the frame in which the observation is made, the principle of relativity would be violated.

From **(F-85)** and **(F-86)** we have for the relation between $\mathbf{F'}$ and $\mathbf{F}$:

$$F_x' = \frac{F_x - (V/c^2)\mathbf{F} \cdot \mathbf{v}}{1 - v_x V/c^2} \tag{4}$$

$$F_{yz}' = \frac{F_{yz}}{\gamma(1 - v_x V/c^2)} \tag{5}$$

where the x axis is taken as the direction of $\mathbf{V}$. We also require the transformation of velocities **(F-24)** and **(F-25)**:

$$v_x' = \frac{v_x - V}{1 - v_x V/c^2} \tag{6}$$

$$v_{yz}' = \frac{v_{yz}}{\gamma(1 - v_x V/c^2)} \tag{7}$$

Substituting **(1)** and **(2)** into **(4)** and **(5)** we obtain

$$E_x' + v_y' B_z' - v_z' B_y' = \frac{E_x + v_y B_z - v_z B_y - (V/c^2)(E_x v_x + E_y v_y + E_z v_z)}{1 - v_x V/c^2} \tag{8}$$

$$E_y' + v_z' B_x' - V_x' B_z' = \frac{E_y + v_z B_x - v_x B_z}{\gamma(1 - v_x V/c^2)} \tag{9}$$

$$E_z' + v_x' B_y' - v_y' B_x' = \frac{E_z + v_x B_y - v_y B_x}{\gamma(1 - v_x V/c^2)} \tag{10}$$

If we substitute from **(7)** for the velocities in the primed frame and equate coefficients of v_y and v_z on the two sides of **(8)**, we obtain:

$$E_x' = E_x \tag{11}$$

$$B_y' = \gamma\left(B_y + \frac{V}{c^2} E_z\right) = \gamma\hat{y} \cdot \left(\mathbf{B} - \frac{1}{c^2} \mathbf{V} \times \mathbf{E}\right) \tag{12}$$

$$B_z' = \gamma\left(B_z - \frac{V}{c^2} E_y\right) = \gamma\hat{z} \cdot \left(\mathbf{B} - \frac{1}{c^2} \mathbf{V} \times \mathbf{E}\right) \tag{13}$$

Substituting into **(9)** and **(10)** we obtain the remaining field relations:

$$B_x' = B_x \tag{14}$$

$$E_y' = \gamma(E_y - V B_z) = \gamma\hat{y} \cdot (\mathbf{E} + \mathbf{V} \times \mathbf{B}) \tag{15}$$

$$E_z' = \gamma(E_z + V B_y) = \gamma\hat{z} \cdot (\mathbf{E} + \mathbf{V} \times \mathbf{B}) \tag{16}$$

Exercises

1. Show that if the magnetic field **B** lies in the yz plane and is everywhere orthogonal to **E** such that the ratio

$$cB_y/E_z = -cB_z/E_y$$

is constant and less than one, then it is possible to find a frame in which there is only an electric field.

2. Show that if the electric field **E** lies in the yz plane and is everywhere orthogonal to **B** such that the ratio:

$$E_z/cB_y = -E_y/cB_y$$

is constant and less than one, then it is possible to find a frame in which there is only a magnetic field.

3 **Lorentz Invariants.** Taking the product of (**11**) and (**14**), of (**12**) and (**15**), and of (**13**) and (**16**) we have:

$$E_x{}'B_x{}' = E_x B_x \tag{17}$$

$$E_y{}'B_y{}' = \gamma^2 E_y B_y + \frac{\gamma^2 V}{c^2} E_y E_z - \gamma^2 V B_y B_z - \frac{\gamma^2 V^2}{c^2} E_z B_z \tag{18}$$

$$E_z{}'B_z{}' = \gamma^2 E_z B_z - \frac{\gamma^2 V}{c^2} E_y E_z + \gamma^2 V B_y B_z - \frac{\gamma^2 V^2}{c^2} E_y B_y \tag{19}$$

Adding (**17**), (**18**), and (**19**) we have that the scalar product of **E**′ and **B**′:

$$\mathbf{E}' \cdot \mathbf{B}' = \mathbf{E} \cdot \mathbf{B} \tag{20}$$

is Lorentz invariant, which is to say that the value of (**20**) is independent of the frame in which **E** and **B** are determined.

We obtain a second invariant from the squares of the fields. From (**11**), (**15**), and (**16**) we have

$$E'^2 = E_x{}'^2 + E_y{}'^2 + E_z{}'^2$$

$$= E_x{}^2 + \gamma^2(E_y{}^2 + E_z{}^2) + \gamma^2 V^2(B_y{}^2 + B_z{}^2) - 2\gamma^2 \mathbf{V} \cdot \mathbf{E} \times \mathbf{B} \tag{21}$$

From (**12**), (**13**), and (**14**) we have:

$$B'^2 = B_x{}'^2 + B_y{}'^2 + B_z{}'^2$$

$$= B_x{}^2 + \gamma^2(B_y{}^2 + B_z{}^2) + \frac{\gamma^2 V^2}{c^4}(E_y{}^2 + E_z{}^2) - \frac{2\gamma^2}{c^2} \mathbf{V} \cdot \mathbf{E} \times \mathbf{B} \tag{22}$$

By taking the indicated combination of **(21)** and **(22)** we obtain:

$$E'^2 - c^2 B'^2 = E^2 - c^2 B^2 \tag{23}$$

as a second Lorentz invariant.

Exercises

3. Show that if the magnitude of **E** is greater than that of c**B** in one frame, then it must be greater at the corresponding position in all frames.
4. Show that if **E** and **B** are orthogonal at **r** in one frame, they are orthogonal at the corresponding position in all frames.
5. Show that if **E** and **B** make an acute angle in one frame at position **r**, they make an acute angle at the corresponding position in all frames.

4 Uniformly Moving Charge. We consider the special case in which we have some general distribution of charge moving uniformly with velocity **V**. Then the charge will be at rest in the primed frame and we have **B**′ = 0. This is the case which we introduced in Section 6-2 for finding the forces between steady currents. From **(12)** and **(13)** we have for the magnetic field in the *laboratory* frame:

$$\mathbf{B} = \frac{1}{c^2} \mathbf{V} \times \mathbf{E} \tag{24}$$

which is just **(6-22)**. For the electric fields we obtain:

$$E_x = E_x' \qquad E_{yz} = \gamma E_{yz}' \tag{25}$$

5 Uniformly Moving Current. As a second special case we imagine a general distribution of steady currents moving uniformly with velocity **V**. The currents will be steady in the primed frame and we will have **E**′ = 0, which is the case discussed in Section 9-1. From **(15)** and **(16)** we have for the electric field in the laboratory frame **(9-16)**;

$$\mathbf{E} = -\mathbf{V} \times \mathbf{B} \tag{26}$$

and for the magnetic fields:

$$B_x = B_x' \qquad B_{yz} = \gamma B_{yz}' \tag{27}$$

THE MAXWELL EQUATIONS

We start from a description of the sources of the electric field and obtain, from the transformation of fields, expressions for the sources of the magnetic field. Following this we discuss the contribution of Maxwell.

6   Sources of the Electric Field. We write expressions for the sources of flux and circulation of the electric field which we take as the first two Maxwell equations. The source of flux of the electric field is from *experiment* Gauss's law:

$$\Psi = \frac{1}{\epsilon_0} q \tag{28}$$

with no restrictions placed on the motion of the enclosed charge. This is equivalent to **(3)** which we discussed in Section 6-1. As we have seen, **(28)** may readily be written in differential form:

▶ ME I
$$\nabla \cdot \mathbf{E} = \frac{1}{\epsilon_0} \rho \tag{29}$$

We write *Faraday's law* for the circulation of the electric field:

$$\mathcal{E} = -\frac{\partial \Phi}{\partial t} \tag{30}$$

where Φ is the magnetic flux through a surface S, and $\mathcal{E}$ is the electromotive force around the bounding contour. We introduced **(30)** as an *experimental* law with essentially no restrictions on the frame in which the circulation of $\mathbf{E}$ is measured or on the mechanisms through which $\mathbf{B}$ is changed. By writing **(30)** in differential form we have the second Maxwell equation:

▶ ME II
$$\nabla \times \mathbf{E} = -\frac{\partial \mathbf{B}}{\partial t} \tag{31}$$

7   Conservation of Charge. We take as an *experimental* fact that electrical charge is conserved. As a consequence of local charge conservation we have the equation of continuity, written in the unprimed frame **(1-147)**, as

$$\nabla \cdot \mathbf{j} + \frac{\partial \rho}{\partial t} = 0 \tag{32}$$

The principle of relativity requires that we must have an equation of the form of **(32)** in any frame. Taking **(9-56)**, **(9-57)**, and **(9-58)** for the transformation of current and charge and **(F-27)**, **(F-28)**, and **(F-29)** for the transformation of derivatives, we should be able to obtain from **(32)** the same expression in the primed frame. We write for the components of the left side of (32):

$$\frac{\partial j_x}{\partial x} = \gamma^2 \frac{\partial j_x'}{\partial x'} + \gamma^2 V \frac{\partial \rho'}{\partial x'} - \gamma^2 \frac{V}{c^2} \frac{\partial j_x'}{\partial t'} - \gamma^2 \frac{V^2}{c^2} \frac{\partial \rho'}{\partial t'} \tag{33}$$

$$\frac{\partial j_{yz}}{\partial yz} = \frac{\partial j_{yz}'}{\partial yz'} \tag{34}$$

$$\frac{\partial \rho}{\partial t} = -\gamma^2 \frac{V^2}{c^2} \frac{\partial j_x'}{\partial x'} - \gamma^2 V \frac{\partial \rho'}{\partial x'} + \gamma^2 \frac{V}{c^2} \frac{\partial j_x'}{\partial t'} + \gamma^2 \frac{\partial \rho'}{\partial t'} \tag{35}$$

Adding (33), (34), and (35) we obtain:

$$\mathbf{V} \cdot \mathbf{j} + \frac{\partial \rho}{\partial t} = \mathbf{V}' \cdot \mathbf{j}' + \frac{\partial \rho'}{\partial t'} \tag{36}$$

This is the result that we expected since the transformation of charge and current was obtained by requiring (36). Although the procedure may appear circular it is based firmly on the principal of relativity. Equation 36 states that the left side of (32) is invariant under a Lorentz transformation. Thus, *if* charge is conserved in one frame it is conserved in all frames. We will now use a similar argument to obtain the two remaining Maxwell equations.

8 Sources of the Magnetic Field. Given the transformation equations for fields, charges, and currents, and for coordinates and time, we may transform ME I (29) in much the way that we transformed (32). The transformation is quite straightforward and yields

$$\mathbf{V}' \cdot \mathbf{E}' - \frac{1}{\epsilon_0} \rho' = \gamma \left(\mathbf{V} \cdot \mathbf{E} - \frac{1}{\epsilon_0} \rho \right) - \gamma \mathbf{V} \cdot \left[\mathbf{V} \times \mathbf{B} - \mu_0 \left(\mathbf{j} + \epsilon_0 \frac{\partial \mathbf{E}}{\partial t} \right) \right] \tag{37}$$

What we see is that there is a subsidiary condition for (29) to be invariant under a Lorentz transformation, namely that the bracketed term on the right of (37) vanish, and this becomes the third Maxwell equation:

▶ ME III $$\qquad\qquad \mathbf{V} \times \mathbf{B} = \mu_0 \left(\mathbf{j} + \epsilon_0 \frac{\partial \mathbf{E}}{\partial t} \right) \tag{38}$$

Taking the divergence of (38) we obtain

$$\mathbf{V} \cdot (\mathbf{V} \times \mathbf{B}) = \mu_0 \left(\mathbf{V} \cdot \mathbf{j} + \epsilon_0 \frac{\partial}{\partial t} \mathbf{V} \cdot \mathbf{E} \right) = \mu_0 \left(\mathbf{V} \cdot \mathbf{j} + \frac{\partial \rho}{\partial t} \right) \tag{39}$$

The left side of (39) is zero by (B-9) and the right side is zero by (29) and (32). As we will see, it was this property of (38) that led Maxwell to induce its form. What about the invariance of ME III under a Lorentz transformation? We may rewrite **(37)** interchanging the primed and unprimed variables and the sign of **V** and in this way obtain the transformation relation for ME III. What we find of course is that ME I is the condition for the invariance of ME III. Thus ME I and ME III form a coupled pair of equations.

We next examine the transformation properties of ME II **(31)**. Using the same procedure as for **(36)** and **(37)** we obtain:

$$\left(\mathbf{\nabla}' \times \mathbf{E}' + \frac{\partial \mathbf{B}'}{\partial t'}\right)_x = \gamma\left(\mathbf{\nabla} \times \mathbf{E} + \frac{\partial \mathbf{B}}{\partial t}\right)_x - \gamma V\, \mathbf{\nabla} \cdot \mathbf{B} \tag{40}$$

$$\left(\mathbf{\nabla}' \times \mathbf{E}' + \frac{\partial \mathbf{B}'}{\partial t'}\right)_{yz} = \left(\mathbf{\nabla} \times \mathbf{E} + \frac{\partial \mathbf{B}}{\partial t}\right)_{yz} \tag{41}$$

The condition for the invariance of the x component of ME II becomes ME IV:

▶ ME IV $\mathbf{\nabla} \cdot \mathbf{B} = 0$ \hfill (42)

Exercises

6. Beginning with **(42)** and performing a Lorentz transformation obtain Faraday's law **(31)**. Thus ME II and ME IV constitute a coupled pair of equations.
7. Beginning with **(40)** interchange the primed and unprimed fields and reverse the sign of **V** to obtain ME II. You will need to regard the direction of **V** as arbitrary.
8. By interchanging primed and unprimed quantities and the sign of **V** in **(37)** obtain the relation

$$\mathbf{\nabla}' \times \mathbf{B}' - \mu_0\left(\mathbf{j}' + \epsilon_0 \frac{\partial}{\partial t'}\,\mathbf{E}'\right) = \gamma\left[\mathbf{\nabla} \times \mathbf{B} - \mu_0\left(\mathbf{j} + \epsilon_0 \frac{\partial}{\partial t}\,\mathbf{E}\right)\right] + \frac{\gamma}{c^2}\,\mathbf{V}\left(\mathbf{\nabla} \cdot \mathbf{E} - \frac{1}{\epsilon_0}\,\rho\right)$$

which shows that ME I is required for the invariance of ME III.
9. Show that ME III together with charge conservation requires that the time rate of change of ME I is identically zero. Thus, if ME I hold in any frame, it must hold in all frames.
10. Show that ME II requires that the time rate of change of ME IV is zero. Thus, if ME IV holds in any frame it must hold in all frames.

9   Maxwell's Contribution. Maxwell is justifiably regarded as the father of modern electromagnetic theory. The four equations that carry his name are a synthesis of Faraday's field concept and the results of potential theory, which had developed as a branch of mathematics. As important as was this scholarly contribution, Maxwell is primarily credited for recognizing that Ampère's law, which we have written as

$$\mathbf{\nabla} \times \mathbf{B} = \mu_0 \mathbf{j} \tag{43}$$

could not be correct for other than solenoidal currents.[3] Maxwell defined a current by the relation

$$\mathbf{\nabla} \times \mathbf{H} = \mathbf{j}_H \tag{44}$$

Taking the divergence of **(44)** he obtained:

$$\mathbf{\nabla} \cdot \mathbf{j}_H = 0 \tag{45}$$

which, as Maxwell observed, is the condition for the motion of an incompressible fluid. He recognized that **(45)** requires that $\mathbf{j}_H$ include the variation of the electric displacement $\mathbf{D}$ as well as the current density that we have called external current $\mathbf{j}_{ext}$. Maxwell wrote in effect:

$$\mathbf{j}_H = \mathbf{j}_{ext} + \frac{\partial \mathbf{D}}{\partial t} = \mathbf{\nabla} \times \mathbf{H} \tag{46}$$

where the divergence of the displacement is the density of external charge **(4-7)**

$$\mathbf{\nabla} \cdot \mathbf{D} = \rho_{ext} \tag{47}$$

Taking the divergence of **(46)** we obtain

$$\mathbf{\nabla} \cdot \mathbf{j}_H = \mathbf{\nabla} \cdot \mathbf{j}_{ext} + \frac{\partial \rho_{ext}}{\partial t} \tag{48}$$

The right side is the continuity equation for external charge and current and must be zero, leading to **(45)**.

Finally, we show that **(46)** and **(38)** are equivalent. Writing

$$\mathbf{B} = \mu_0 (\mathbf{H} + \mathbf{M}) \tag{49}$$

$$\mathbf{D} = \epsilon_0 \mathbf{E} + \mathbf{P} \tag{50}$$

we have, taking the curl of **(49)** and using **(46)** and **(50)**,

$$\mathbf{\nabla} \times \mathbf{B} = \mu_0 (\mathbf{\nabla} \times \mathbf{H} + \mathbf{\nabla} \times \mathbf{M}) = \mu_0 \left(\mathbf{j}_{ext} + \frac{\partial \mathbf{P}}{\partial t} + \mathbf{\nabla} \times \mathbf{M} + \epsilon_0 \frac{\partial \mathbf{E}}{\partial t} \right) \tag{51}$$

But the first three terms on the right side of **(51)** are the three contributions to the current **(6-155)**:

$$\mathbf{j} = \mathbf{j}_{ext} + \frac{\partial \mathbf{P}}{\partial t} + \mathbf{\nabla} \times \mathbf{M} \tag{52}$$

[3] A. M. Bork, "Maxwell, Displacement Current, and Symmetry," *Am. J. Phys.*, **31**, 854 (1963).

which is just the statement of **(38)**. In interpreting **(52)** we must realize that a moving polarized body generates a magnetization current and a moving magnetized body generates a polarization current **(9-101)**.

10 **Magnetic Charge.** The fourth Maxwell equation **(42)** is a statement that there is no magnetic equivalent of electric charge and as such must be regarded as a tentative experimental statement. Monopoles have been postulated in the theory of elementary particles and as discussed in Section D-5 are expected to have a magnetic charge some integral multiple of

$$g_0/c = 68.5180e = 1.097788 \times 10^{-17} \, C \tag{53}$$

where e is the electron charge. If $\bar{q}$ is the total magnetic charge within a volume V, the flux of **B** will be given **(9-42)** by

$$\Phi = \mu_0 \bar{q} \tag{54}$$

The differential relation corresponding to **(54)** is:

$$\nabla \cdot \mathbf{B} = \mu_0 \bar{\rho} \tag{55}$$

where $\bar{\rho}$ is the magnetic charge density. Now it is clear from **(40)** that we cannot modify ME IV without also modifying Faraday's law, ME II. The appropriate form for **(31)** is suggested by Maxwell's argument. We write:

$$\nabla \times \mathbf{E} = -\mu_0 \bar{\mathbf{j}} - \frac{\partial \mathbf{B}}{\partial t} \tag{56}$$

where $\bar{\mathbf{j}}$ is the magnetic current density and is *presumably* related to the charge density $\bar{\rho}$ by an equation of continuity:

$$\nabla \cdot \bar{\mathbf{j}} + \frac{\partial}{\partial t} \bar{\rho} = 0 \tag{57}$$

Taking the divergence of **(56)** we obtain **(57)**. Then ME II and ME IV would have to be replaced by **(56)** and **(55)**. Note the very interesting result that a magnetic current has around it a circulation of the electric field. Thus one might hope to detect the presence of magnetic monopoles by an electric current induced in a coil through which a monopole current flows.[4]

---

[4] Luis W. Alvarez et al., "A Magnetic Monopole Detector Utilizing Superconducting Elements," *Rev. Sci. Instr.*, **42**, 326 (1971). For a brief discussion of the generation of fluxoids by magnetic monopoles, see Section D-5.

Exercises

11. Show that an orbiting magnetic monopole produces the same external electric field as does an electric dipole. Obtain an expression for the apparent electric dipole moment in terms of g, ω, and R.

12. Show that a pair of rotating magnetic monopoles separated by a vector **b**

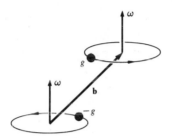

    constitutes an apparent electric quadrupole. Obtain an expression for the quadrupole moment in terms of q, ω, R, and **b**.

13. Obtain the relation for magnetic charge density:

$$\bar{\rho}_H = \bar{\rho} + \bar{\rho}_M$$

where $\bar{\rho}_H = \mathbf{V} \cdot \mathbf{H}$ is an effective magnetic charge density. We may call $\bar{\rho}$ the *true* charge density and $\bar{\rho}_M$ a pseudo-charge density since we know that magnetization is produced by currents of electric charge.

14. Show that the duality transformation:

$$\mathbf{E}' = -c\mathbf{B}$$

$$\mathbf{B}' = \frac{1}{c}\mathbf{E}$$

$$(q',\rho',\mathbf{j}') = -\frac{1}{c}(\bar{q},\bar{\rho},\bar{\mathbf{j}})$$

$$(\bar{q}',\bar{\rho}',\bar{\mathbf{j}}) = c(q,\rho,\mathbf{j})$$

leaves the Maxwell equations **(29)**, **(38)**, **(55)**, and **(56)** invariant. What happens to the Lorentz force:

$$\mathbf{F} = q(\mathbf{E} + \mathbf{v} \times \mathbf{B})$$

Compare with **(9-50)**. [See E. Katz, "Concerning the Number of Independent Variables of the Classical Electromagnetic Field," *Am. J. Phys.*, **33**, 306 (1965).]

11 **Electromagnetic Theory of Light.** By operating on the Maxwell equations we obtain wave equations for **E** and **B** with current and charge density as wave sources. Taking the curl of ME II and substituting from ME III we obtain:

$$\mathbf{\nabla} \times (\mathbf{\nabla} \times \mathbf{E}) = -\frac{\partial}{\partial t} \mu_0 \left(\mathbf{j} + \epsilon_0 \frac{\partial \mathbf{E}}{\partial t} \right) \tag{58}$$

Expanding the double curl and simplifying we obtain:

$$\nabla^2 \mathbf{E} - \frac{1}{c^2} \frac{\partial^2 \mathbf{E}}{\partial t^2} = \mu_0 \frac{\partial \mathbf{j}}{\partial t} + \frac{1}{\epsilon_0} \mathbf{\nabla}\rho \tag{59}$$

We obtain a similar equation for **B** by taking the curl of ME III:

$$\mathbf{\nabla} \times (\mathbf{\nabla} \times \mathbf{B}) = \mu_0 \mathbf{\nabla} \times \mathbf{j} + \frac{1}{c^2} \frac{\partial}{\partial t} (\mathbf{\nabla} \times \mathbf{E}) \tag{60}$$

Using ME II and ME IV we obtain:

$$\nabla^2 \mathbf{B} - \frac{1}{c^2} \frac{\partial^2 \mathbf{B}}{\partial t^2} = -\mu_0 \mathbf{\nabla} \times \mathbf{j} \tag{61}$$

We have discussed in Section 5-16 the propagation of longitudinal plasma waves with $\mathbf{\nabla} \times \mathbf{j} = 0$ for which an electric field but *no* magnetic field is excited. Here we have propagation of a *coupled* electromagnetic wave. In the absence of currents and charge we have as a possible solution of **(59)**:

$$\mathbf{E}(\mathbf{r},t) = f(\hat{\mathbf{n}} \cdot \mathbf{r} - ct)\mathbf{E} \tag{62}$$

where **E** is a constant vector. From ME I we obtain

$$\mathbf{\nabla} \cdot \mathbf{E}(\mathbf{r},t) = \mathbf{E} \cdot \mathbf{\nabla}f = (\hat{\mathbf{n}} \cdot \mathbf{E})f' = 0 \tag{63}$$

where f' is the derivative of f with respect to its argument. We have then the result that **E** must be perpendicular to $\hat{\mathbf{n}}$. From ME II we have

$$-\frac{\partial \mathbf{B}(\mathbf{r},t)}{\partial t} = \mathbf{\nabla} \times \mathbf{E}(\mathbf{r},t) = f'\hat{\mathbf{n}} \times \mathbf{E} \tag{64}$$

Integrating over the time we have:

$$c\mathbf{B}(\mathbf{r},t) = \hat{\mathbf{n}} \times \mathbf{E}(\mathbf{r},t) \tag{65}$$

and we see that $\mathbf{E}(\mathbf{r},t)$, $\mathbf{B}(\mathbf{r},t)$, and $\hat{\mathbf{n}}$ are mutually perpendicular for a wave of the form of **(62)**. The ratio of the magnitude of $\mathbf{E}$ to the magnitude of $\mathbf{B}$ for such a wave is c.

A closely related quantity is the *intrinsic impedance* of vacuum:

$$Z_0 = \frac{E}{H} = \mu_0 \frac{E}{B} = \mu_0 c = 1/\epsilon_0 c \tag{66}$$

Exercises

15. Show that in an isotropic medium characterized by permitivity ϵ and permeability μ the electric and magnetic fields may be given by

$$\mathbf{E}(\mathbf{r},t) = f(\hat{\mathbf{n}} \cdot \mathbf{r} - vt)\, \mathbf{E}$$
$$v\mathbf{B}(\mathbf{r},t) = f(\hat{\mathbf{n}} \cdot \mathbf{r} - vt)\, \hat{\mathbf{n}} \times \mathbf{E}$$

where $v = (\epsilon\mu)^{-1/2}$ is the velocity of propagation in the medium.

16. Show that the impedance of vacuum is given by:

$$Z_0 = 376.730\ \Omega$$

Verify the SI unit of impedance.

12 Speed of Light.

The present best value for the speed of light

$$c = 299\,792\,457.4 \pm 1.1\ \text{m/s}$$

has been obtained by an elaborate scheme in which the frequency of a methane-stabilized He–Ne infrared laser is measured against the frequency standard of the National Bureau of Standards. At the same time the wavelength is measured against the krypton 605.7 nm line, which defines the metre.[5] The estimated error in c is less than 1 percent of that in the previously accepted value, obtained from the simultaneous measurement of frequency and wavelength of a stabilized submillimeter wave klystron oscillator.

ENERGY AND MOMENTUM

We have obtained an expression for the energy stored in a static configuration of *external* charges and dielectric media **(4-64)**:

$$U_{\text{elec}} = \int_V dV \int \mathbf{E} \cdot d\mathbf{D} \tag{67}$$

[5] K. M. Evenson, J. S. Wells, F. R. Petersen, B. L. Danielson, G. W. Day, R. L. Barger, and J. L. Hall, *Phys. Rev. Lett.*, **29**, 1346 (1972). For a recent review, see J. F. Mulligan, *Am. J. Phys.*, **44**, 960 (1976).

By considering the work on external charges we indicated the plausibility of a *local* energy density:

$$u_{\text{elec}} = \int \mathbf{E} \cdot d\mathbf{D} \tag{68}$$

We have obtained similar expressions for the magnetic energy of a system of *external* currents and magnetic media **(10-59)**:

$$U_{\text{mag}} = \int_V dV \int \mathbf{H} \cdot d\mathbf{B} \tag{69}$$

and have similarly obtained as a plausible expression for the local energy density:

$$u_{\text{mag}} = \int \mathbf{H} \cdot d\mathbf{B} \tag{70}$$

We begin by discussing the mechanical energy and momentum of interacting charges and currents and show that neither mechanical energy nor momentum is conserved. We argue that the most natural way of describing the situation is to retain conservation of energy and momentum and to identify energy and momentum with the fields as well as with mass motion. The fact that fields may act as agents for the transfer of mechanical energy and momentum from one place and time to another place and time makes such a point of view particularly compelling.

We go on to obtain expressions for the flow of field energy, the Poynting vector, and for the momentum density, Maxwell stress tensor, and the angular momentum density. As applications of the theory we discuss the Aharanov-Bohm effect and the angular momentum of an electron and magnetic monopole.

13 Mechanical Energy and Momentum. We show by the simple example of the interaction between a *slowly* moving charge and a magnetic moment that neither *mechanical* energy nor momentum is conserved. The absence of conservation of kinetic energy is familiar from mechanics. By introducing potential energy we preserve overall conservation of energy. As we have seen, we may readily describe the potential energy of charges at rest or of steady currents. For charges in motion it is possible to write the equivalent of a potential energy in terms of the previous history of the particles. But if we wish to describe the energy in terms of *present* local quantities we must use the electric and magnetic fields.

When we examine the conservation of momentum the field concept becomes even more compelling. In classical mechanics we have Newton's third law and mechanical momentum is conserved. As soon as we allow for the propagation of an interaction at finite velocity we lose mechanical momentum conservation and are forced, if we wish to preserve the conservation of momentum as a law of physics, to identify momentum with the fields themselves. In describing the interaction between charges and currents, mechanical momentum is not conserved even to first order in the velocity of the moving charge. We may understand the absence of momentum conservation as a consequence of the fact that magnetic interactions are fundamentally relativistic.

We show in Figure 1 a magnetic moment $\boldsymbol{\mu}$ at the origin and a charge q moving with velocity of magnitude small compared with c. From **(6-122)** we have for the magnetic field of the dipole:

$$\mathbf{B(r)} = -\frac{\mu_0}{4\pi} \, \boldsymbol{\nabla} \times \left(\boldsymbol{\mu} \times \boldsymbol{\nabla} \frac{1}{r} \right) \tag{71}$$

The force on the moving charge is from the Lorentz force **(1)**

$$\mathbf{F}_q = -\frac{\mu_0}{4\pi} \, q\mathbf{v} \times \left[\boldsymbol{\nabla} \times \left(\boldsymbol{\mu} \times \boldsymbol{\nabla} \frac{1}{r} \right) \right] \tag{72}$$

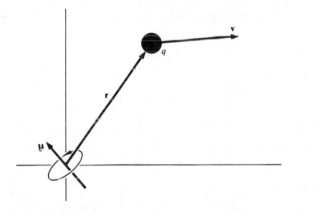

Figure 1 Interaction between a moving test charge and a magnetic moment. The test charge moves with velocity $\mathbf{v}$ in the field of the moment.

We now transform and rewrite **(72)** through a series of identities. Expanding the triple product we write:

$$\mathbf{F}_q = -\frac{\mu_0}{4\pi} \, q \left\{ \left[\boldsymbol{\nabla}\left(\boldsymbol{\mu} \times \boldsymbol{\nabla} \frac{1}{r} \right) \right] \cdot \mathbf{v} - \mathbf{v} \cdot \boldsymbol{\nabla}\left(\boldsymbol{\mu} \times \boldsymbol{\nabla} \frac{1}{r} \right) \right\} \tag{73}$$

By using **(C-6)** we rewrite the force on q as:

$$\mathbf{F}_q = \frac{\mu_0}{4\pi} \, q \left(\boldsymbol{\nabla}\boldsymbol{\nabla} \frac{1}{r} \right) \times \boldsymbol{\mu} \cdot \mathbf{v} + \frac{\mu_0}{4\pi} \, q\mathbf{v} \cdot \left(\boldsymbol{\nabla}\boldsymbol{\nabla} \frac{1}{r} \right) \times \boldsymbol{\mu} \tag{74}$$

To obtain the force on the magnetic moment we write **(6-22)** for the magnetic field produced by the moving charge:

$$\mathbf{B(r)} = \frac{1}{c^2} \, \mathbf{v} \times \mathbf{E(r)} \simeq -\frac{\mu_0}{4\pi} \, q\mathbf{v} \times \boldsymbol{\nabla} \frac{1}{r} \tag{75}$$

where the approximate equality is for nonrelativistic velocities. From **(6-127)** the force on the moment is given by:

$$\mathbf{F}_\mu = (\boldsymbol{\mu} \times \nabla) \times \mathbf{B} = -\frac{\mu_0}{4\pi} q(\boldsymbol{\mu} \times \nabla) \times \left(\mathbf{v} \times \nabla \frac{1}{r}\right) \tag{76}$$

We now transform and rewrite the force on the moment so that we may make a comparison with the force on the charge. Expanding **(76)** we write

$$\mathbf{F}_\mu = -\frac{\mu_0}{4\pi} q \left\{ \left[\nabla \left(\mathbf{v} \times \nabla \frac{1}{r} \right) \right] \cdot \boldsymbol{\mu} - \boldsymbol{\mu} \nabla \cdot \left(\mathbf{v} \times \nabla \frac{1}{r} \right) \right\} \tag{77}$$

By **(C-6)** and **(B-4)** we write the force as:

$$\mathbf{F}_\mu = \frac{\mu_0}{4\pi} q \left\{ \left[\left(\nabla \nabla \frac{1}{r} \right) \times \mathbf{v} \right] \cdot \boldsymbol{\mu} + \boldsymbol{\mu} \left(\nabla \times \nabla \frac{1}{r} \right) \cdot \mathbf{v} \right\} \tag{78}$$

The second term is identically zero by **(B-8)**. Interchanging $\boldsymbol{\mu}$ and $\mathbf{v}$ reverses the sign of the first term:

$$\mathbf{F}_\mu = -\frac{\mu_0}{4\pi} q \left[\left(\nabla \nabla \frac{1}{r} \right) \times \boldsymbol{\mu} \right] \cdot \mathbf{v} \tag{79}$$

Adding **(74)** and **(79)** we have for the sum of the forces:

$$\mathbf{F}_q + \mathbf{F}_\mu = \frac{\mu_0}{4\pi} q\mathbf{v} \cdot \left(\nabla \nabla \frac{1}{r} \right) \times \boldsymbol{\mu}$$

$$= \frac{\mu_0}{4\pi} q \frac{3(\hat{\mathbf{r}} \cdot \mathbf{v})\hat{\mathbf{r}} - \mathbf{v}}{r^3} \times \boldsymbol{\mu} \tag{81}$$

which is nonvanishing to first order in the velocity of the charge. Since there are no other forces acting we must conclude that mechanical momentum is *not* conserved.

Exercise

17. An electric charge q moves slowly past a *magnetic charge* dipole. Write the force on q and on the dipole. See Exercise 9-7. Show that the force on q is equal and opposite to the force on the dipole. Compare with the interaction between q and an electric *current* moment.

14 Interaction Between Charges and Currents. In this section we generalize the discussion of the interaction between a charge and a magnetic moment to the interaction between a charge and any distribution of steady solenoidal currents.[6]

---

[6] G. T. Trammel, *Phys. Rev.*, **134**, B1183 (1964).

A slowly moving charge generates a magnetic field given by (75). Then the force on a general current distribution may be written as:

$$F_j \simeq -\frac{\mu_0}{4\pi} q \int_{V_1} j(r_1) \times \left(v \times \nabla_1 \frac{1}{r_{01}}\right) dV_1 \tag{82}$$

By writing

$$\nabla \frac{1}{r_{01}} = -\nabla_1 \frac{1}{r_{01}} \tag{83}$$

and changing the order of the cross product we may write:

$$F_j \simeq -\frac{\mu_0}{4\pi} q(v \times \nabla) \times \int_{V_1} \frac{j(r_1)}{r_{01}} dV_1 \tag{84}$$

Now we recognize the integral as the vector potential in the Coulomb gauge (6-49) so that we may write in general for the force on the currents

$$F_j \simeq -q(v \times \nabla) \times A(r) \tag{85}$$

where we emphasize that $A(r)$ is the vector potential at the site of the moving charge.
For the force on the *charge* we have simply:

$$F_q = qv \times B(r) = qv \times [\nabla \times A(r)] \tag{86}$$

Adding (85) and (86) we obtain for the sum of the forces:

$$F_j + F_q \simeq -q(v \cdot \nabla)A = -q\frac{d}{dt} A(r) \tag{87}$$

where $A(r)$ again is the vector potential of the currents at the site of the charge.

Exercise

18. Show that (81) is of the form of (87).

15 Poynting Vector. We consider a volume V, shown in Figure 2, surrounded by a surface S across which there is energy flow characterized by a vector Π called the Poynting vector after J. Poynting, who in 1884 developed the present theory. From energy conservation we may write:

$$\oint_S \Pi \cdot dS + \frac{\partial U}{dt} = \frac{dW}{dt} \tag{88}$$

where U is the sum of **(67)** and **(69)** and dW/dt is the work done on the field by external charges and currents. By the divergence theorem **(B-11)** we have:

$$\oint_S \boldsymbol{\Pi} \cdot d\mathbf{S} = \int_V \boldsymbol{\nabla} \cdot \boldsymbol{\Pi} \, dV \tag{89}$$

Since the volume V is arbitrary we may shrink V to obtain the local relation:

$$\boldsymbol{\nabla} \cdot \boldsymbol{\Pi} + \frac{\partial u}{\partial t} = \frac{dw}{dt} \tag{90}$$

where w is the work done per unit volume. As long as ϵ and μ are constant we have, from **(68)** and **(70)**,

$$\frac{\partial u}{\partial t} = \frac{\partial}{\partial t}\left(u_{\text{elec}} + u_{\text{mag}}\right) = \mathbf{E} \cdot \frac{\partial \mathbf{D}}{\partial t} + \mathbf{H} \cdot \frac{\partial \mathbf{B}}{\partial t} \tag{91}$$

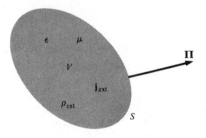

Figure 2   Poynting vector. A volume V is surrounded by a surface S across which there is an energy flow.

Using **(31)** and **(46)** we may rewrite this expression as:

$$\frac{\partial u}{\partial t} = \mathbf{E} \cdot \boldsymbol{\nabla} \times \mathbf{H} - \mathbf{E} \cdot \mathbf{j}_{\text{ext}} - \mathbf{H} \cdot \boldsymbol{\nabla} \times \mathbf{E} = -\boldsymbol{\nabla} \cdot (\mathbf{E} \times \mathbf{H}) - \mathbf{E} \cdot \mathbf{j}_{\text{ext}} \tag{92}$$

where the final equality follows from **(B-4)**. Transposing terms we write:

$$\boldsymbol{\nabla} \cdot (\mathbf{E} \times \mathbf{H}) + \frac{\partial u}{\partial t} = -\mathbf{E} \cdot \mathbf{j}_{\text{ext}} \tag{93}$$

Comparing **(90)** and **(93)** we have for the work done via the external current:

$$\frac{dw}{dt} = -\mathbf{E} \cdot \mathbf{j}_{\text{ext}} \tag{94}$$

Including a solenoidal term we have for the Poynting vector:

▶
$$\Pi = E \times H + \Omega \tag{95}$$

with $\mathbf{V} \cdot \mathbf{\Omega} = 0$.

We now consider two simple examples to show how the Poynting vector may be used. We show in Figure 3 a capacitor with circular plates of radius R and separation d in a medium characterized by ϵ and μ. If the charge on the plates is increased, we have, from (44) and (46),

$$\mathbf{V} \times \mathbf{H} = \frac{\partial \mathbf{D}}{\partial t} \tag{96}$$

since there are no external currents in the plates and $D = \sigma_{\text{ext}}$ equals the external charge density on the plates. From (96) we obtain for the magnetic field H at radius r:

$$H = \tfrac{1}{2}r \frac{\partial D}{\partial t} = \tfrac{1}{2}\epsilon r \frac{\partial E}{\partial t} \tag{97}$$

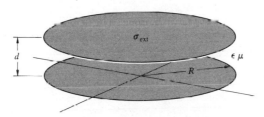

Figure 3 Energy flow. Energy flows into the region between a pair of circular plates of radius R and separation d.

If the charge is increased only very slowly we may neglect the energy in the magnetic field to obtain for a cylinder of radius r and height d:

$$U = U_{\text{elec}} = \tfrac{1}{2}\epsilon E^2 \pi r^2 d \tag{98}$$

The flow of energy *into* this region may be written as:

$$-2\pi r \, d \, \Pi = 2\pi r \, d \, EH = \pi r^2 \, d \, \epsilon E \frac{\partial E}{\partial t} \tag{99}$$

and we see that (98) and (99) are in agreement with (88). The curious aspect of this result is that the flow of energy into the intervening region is not normal to the plates but rather is parallel to their plane. Although this discussion is certainly not the usual point of view of circuit theory it is perfectly consistent internally and provides an alternative way of viewing lumped circuit problems.

To take an example involving magnetic energy we imagine a cylinder of length l and radius R as shown in Figure 4 on which we have a surface current I surrounded by a medium characterized by μ and ϵ. From (44) and (46) we have for this problem:

$$\nabla \times \mathbf{H} = \mathbf{j}_{\text{ext}} \qquad H = \frac{I}{2\pi r} \tag{100}$$

The current is increased slowly to produce an electric field that satisfies the relation

$$\hat{\boldsymbol{\phi}} \cdot \nabla \times \mathbf{E} = -\frac{\partial}{\partial r} E_z = -\frac{\partial B}{\partial t} = -\mu \frac{\partial H}{\partial t} = -\frac{\mu}{2\pi r} \frac{\partial I}{\partial t} \tag{101}$$

Integrating from the surface of the cylinder out to r, we obtain:

$$E_z(r) = E(R) + \frac{\mu}{2\pi} \frac{dI}{dt} \ln \frac{r}{R} \tag{102}$$

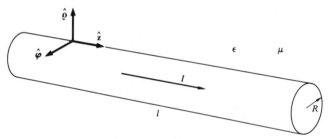

Figure 4 Magnetic energy flow. Energy flows into the magnetic field along the radial direction.

If the current is increased slowly most of the energy is in the magnetic field and we may write:

$$U = U_{\text{mag}} = \tfrac{1}{2}\mu l \int_R^r H^2 2\pi r \, dr = \frac{1}{4\pi} \mu l I^2 \ln \frac{r}{R} \tag{103}$$

We obtain, using the Poynting vector,

$$-2\pi r l \, \Pi = 2\pi r l \, EH = \frac{1}{2\pi} \mu l \, I \frac{dI}{dt} \ln \frac{r}{R} + l \, E(R) I \tag{104}$$

Again (103) and (104) are in agreement with (88) but this time there is work done on the field by the external current I.

Exercises

19. The energy stored in an inductor is written as

$$U = \tfrac{1}{2} L I^2$$

Imagine a long solenoid in which the current is changing. Integrate the Poynting vector just inside the windings, and show that the flux of the Poynting vector is equal to dU/dt.

20. Using Table 9-1, obtain the expressions for the Poynting vector of uniformly moving electric and magnetic charge. Show that the Poynting vectors have the same dependence on $\mathbf{r}$ and are related by a factor $(\bar{q}/cq)^2$.

---

16 Field Momentum. We obtain an expression for the momentum carried by the electromagnetic field by an argument similar to that employed for the energy flux. We show in Figure 5 a volume V containing a material medium characterized by ϵ and μ. If $\mathbf{p}$ is the *total* momentum in V (exclusive of any momentum associated with external charges and currents), we write, for the rate of change of momentum,

$$\frac{d\mathbf{p}}{dt} = \mathbf{F} + \oint d\mathbf{S} \cdot \mathbf{T} \tag{105}$$

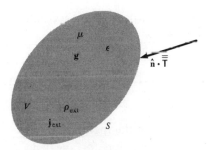

Figure 5   Field momentum. The momentum flow into V is given by the normal component of the Maxwell stress tensor.

where $\mathbf{F}$ is the volume force exerted by *external* charges and currents within V:

$$\mathbf{F} = -\int_V \rho_{ext}\mathbf{E}\, dV - \int_V \mathbf{j}_{ext} \times \mathbf{B}\, dV \tag{106}$$

and $\mathbf{T}$ is a tensor that represents the momentum transferred through the surface and is called the Maxwell stress tensor. We may convert (105) to a local relation:

$$\frac{\partial \mathbf{g}}{\partial t} = -\rho_{ext}\mathbf{E} - \mathbf{j}_{ext} \times \mathbf{B} + \nabla \cdot \mathbf{T} \tag{107}$$

where $\mathbf{g}$ is the local momentum density. By using (46) and (47) we may rewrite (107) as:

$$\frac{\partial \mathbf{g}}{\partial t} = \frac{\partial}{\partial t}(\mathbf{D} \times \mathbf{B}) - (\nabla \cdot \mathbf{D})\mathbf{E} + \mathbf{B} \times (\nabla \times \mathbf{H}) + \mathbf{D} \times (\nabla \times \mathbf{E}) + \nabla \cdot \mathbf{T} \tag{108}$$

From **(108)** we make the identification to within a constant:

$$\mathbf{g(r)} = \mathbf{D(r)} \times \mathbf{B(r)} \tag{109}$$

$$\nabla \cdot \mathbf{T} = (\nabla \cdot \mathbf{D})\mathbf{E} - \mathbf{B} \times (\nabla \times \mathbf{H}) - \mathbf{D} \times (\nabla \times \mathbf{E}) \tag{110}$$

In order to obtain the form of **T** it is necessary to make certain assumptions about the spatial variation in ϵ and μ. We must also be careful about the way in which we handle "mechanical" forces. If **(105)** is to be a complete statement **T** must include all the forces transmitted through S and not just those normally associated with macroscopic fields.[7] As a simple example we assume a homogeneous isotropic and linear medium. That is, both ϵ and μ are taken to be independent of position and the orientation and magnitude of the fields. Under these circumstances **T** is a symmetric tensor and may be written as:

$$\mathbf{T} = \mathbf{DE} + \mathbf{BH} - \tfrac{1}{2}\mathbf{I}(\mathbf{D} \cdot \mathbf{E} + \mathbf{B} \cdot \mathbf{H}) \tag{111}$$

By taking the divergence of **(111)** and using **(B-6)** we obtain **(110)**.

Comparing **(109)** with **(95)** we write:

$$\mathbf{\Pi(r)} = \mathbf{E} \times \mathbf{H} = \frac{1}{\epsilon\mu}\, \mathbf{D} \times \mathbf{B} = v^2\mathbf{g(r)} \tag{112}$$

where v is the speed of propagation in the medium.

As an *example* we find the expression for the field momentum of a magnetic field **B** in the presence of an external charge q:

$$\mathbf{p}_{\text{field}} = \int \mathbf{D} \times \mathbf{B}\, dV = \int \mathbf{D} \times (\nabla \times \mathbf{A})\, dV \tag{113}$$

Expanding the integrand of **(113)** by **(C-1)** we write in tensor notation:

$$\mathbf{D} \times (\nabla \times \mathbf{A}) = (\nabla\mathbf{A}) \cdot \mathbf{D} - (\mathbf{D} \cdot \nabla)\mathbf{A} \tag{114}$$

From **(C-2)** we have in the absence of polarization or changing magnetic flux, which is equivalent to the assumption of **(75)**:

$$\mathbf{A} \times (\nabla \times \mathbf{D}) = (\nabla\mathbf{D}) \cdot \mathbf{A} - (\mathbf{A} \cdot \nabla)\mathbf{D} = 0 \tag{115}$$

---

[7] In discussing the momentum carried by an electromagnetic wave in a refractive medium one must be particularly careful. See R. Peierls, *Proc. Roy. Soc. (Lond.)* **A347**, 475 (1976) and H.-K. Wong and K. Young, *Am. J. Phys.*, **45**, 195 (1977). See also W. H. Furry, *Am. J. Phys.*, **37**, 621 (1969) and J. A. Arnaud, *Am. J. Phys.*, **42**, 71 (1974). For a discussion of electrostrictive effects, see H. J. Juretschke, *Am. J. Phys.*, **45**, 277 (1977).

From **(C-3)** we write

$$\nabla(\mathbf{A} \cdot \mathbf{D}) = (\nabla\mathbf{A}) \cdot \mathbf{D} + (\nabla\mathbf{D}) \cdot \mathbf{A} \tag{116}$$

and from **(C-4)**

$$\nabla \cdot (\mathbf{AD}) = (\mathbf{A} \cdot \nabla)\mathbf{D} \qquad \nabla \cdot (\mathbf{DA}) = \mathbf{A}(\nabla \cdot \mathbf{D}) + (\mathbf{D} \cdot \nabla)\mathbf{A} \tag{117}$$

Adding **(114)**, **(115)**, **(116)**, and **(117)** we obtain the identity

$$\mathbf{D} \times (\nabla \times \mathbf{A}) = \nabla(\mathbf{A} \cdot \mathbf{D}) - \nabla \cdot (\mathbf{AD} + \mathbf{DA}) + \mathbf{A}(\nabla \cdot \mathbf{D}) \tag{118}$$

By **(C-6)** and **(B-12)** we write from **(113)** for a finite volume

$$\mathbf{p}_{\text{field}} = \int_V \mathbf{A}(\nabla \cdot \mathbf{D}) \, dV + \oint_S \mathbf{A} \cdot \mathbf{D} \, dS - \oint_S dS \cdot (\mathbf{AD} + \mathbf{DA}) \tag{119}$$

For a *uniformly* moving charge the displacement **D** falls off as $1/r^2$ as we have seen. Since the vector potential **A** falls off as $1/r$ or faster, the two surface integrals must go to zero as S goes to infinity and we have:

$$\mathbf{p}_{\text{field}} = \int \mathbf{A}(\nabla \cdot \mathbf{D}) \, dV = q\mathbf{A} \tag{120}$$

where we have used $\nabla \cdot \mathbf{D} = \rho_{\text{ext}}$ and integrated over the charge q. Associating a mass m with q we have for the total momentum:

$$\mathbf{p} = \mathbf{p}_{\text{ext}} + \mathbf{p}_{\text{field}} = m\mathbf{v} + q\mathbf{A} \tag{121}$$

which is the usual expression **(G-9)** for the momentum conjugate to **v**. We see from this example that the term $q\mathbf{A}$ is just the field momentum as long as the charge q is moving uniformly.

Exercises

21. Show that if momentum density **g** flows with velocity **v**, then the Maxwell stress tensor is of the form:

$$\mathbf{T} = \mathbf{vg}$$

22. Develop the connection between **(120)** and **(87)**.

23. A magnetic charge $\bar{q}$ is at the origin, and an electric charge q is at **r**. Use **(109)** and symmetry to show that the total field momentum is zero. What may be concluded about the field momentum of a charge q and a magnetic *charge* dipole $\boldsymbol{\mu}$?

17 **[Aharonov-Bohm Effect].** The presence of the vector potential in the expression for the momentum of a particle-field system **(121)** follows from our analysis of the effect of external charges and currents. As we have seen, however, the force on the charge q is the Lorentz force **(1)** and depends on the fields **E** and **B**.

As discussed in Sections D-1 and G-1 the wave function of a system is characterized by a phase:

$$\phi = \int \mathbf{k} \cdot d\mathbf{r} \tag{122}$$

where $\hbar\mathbf{k}$ is the *total* momentum:

$$\hbar\mathbf{k} = m\mathbf{v} + q\mathbf{A} \tag{123}$$

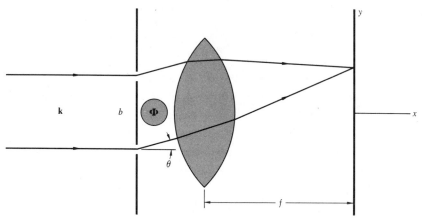

Figure 6   Aharonov-Bohm effect. A beam of electrons is directed at a pair of parallel slits behind which is a long solenoid. After passing through the slits the electrons are focused onto a screen.

and we see that even though the *classical* behavior of a system evidently depends only on the magnetic field **B**, the quantum mechanical *phase* at least depends directly on **A**. In 1956, Aharonov and Bohm[8] suggested a possible interference experiment to demonstrate the role of the vector potential in the behavior of a quantum system.[9] We show in Figure 6 their proposal.

A beam of electrons is to be directed at a pair of parallel slits behind which is a long solenoid. The electrons after passing through the slits are focused onto a screen by a lens of focal length f. Without current through the solenoid we expect destructive interference between electrons passing through the two slits for a phase difference

$$\phi_1 - \phi_2 = kb \sin \theta = (2n + 1)\pi \tag{124}$$

[8] Y. Aharonov and D. Bohm, *Phys. Rev.*, **115**, 485 (1956); **123**, 1511 (1961); **125**, 2192 (1962); **130**, 1625 (1963).

[9] R. P. Feynman, R. B. Leighton, and M. Sands, *The Feynman Lectures on Physics*, Addison-Wesley, 1964, Section II-15-5.

where b is the separation between slits. Then at a position on the screen given by

$$y = f \tan \theta \cong \frac{(2n + 1)\pi}{k} \frac{f}{b} \qquad (125)$$

there should be a null in the intensity pattern.

Now if a current is sent through the solenoid behind the slits, producing a flux Φ, we expect from **(122)** and **(123)** an additional shift in relative phase for the electrons passing through the two slits:

$$\Delta(\phi_1 - \phi_2) = \oint \mathbf{k} \cdot d\mathbf{r} = \frac{q}{\hbar} \oint \mathbf{A} \cdot d\mathbf{r} = \frac{q}{\hbar} \Phi \qquad (126)$$

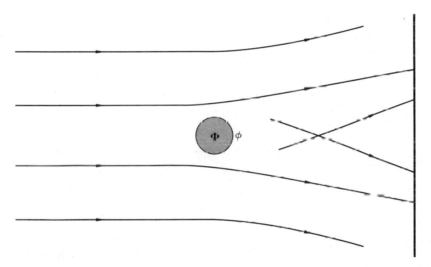

Figure 7 Experimental realization. By placing an electrostatic potential on the solenoid, it acts like a Fresnel biprism.

and a shift in the interference pattern by an amount:

$$\Delta y = \frac{q\Phi}{\hbar k} \frac{f}{b} = \frac{q\Phi}{mv} \frac{f}{b} \qquad (127)$$

The experiment has been performed in a number of laboratories[10] with an arrangement generally of the form shown in Figure 7.

The magnetic flux Φ is enclosed in a solenoid of microscopic dimensions. By placing a negative electrostatic potential ϕ on the solenoid, a barrier is established that prevents electrons from interacting with the magnetic field of the solenoid. The

[10] R. G. Chambers, *Phys. Rev. Lett.*, **5**, 3 (1960). H. Boersch et al., *Z. Phys.*, **167**, 72 (1962); **169**, 263 (1962). G. Möllenstedt et al., *Physik Bl.*, **18**, 299 (1962); *Naturwissenschaften*, **49**, 81 (1962); H. A. Fowler et al., *J. Appl. Phys.*, **32**, 1153 (1961).

charged solenoid also acts as a Fresnel biprism (see Problem 1-16) in producing interference fringes on a fluorescent screen. All observers have obtained agreement with the predictions of quantum mechanics. This observation of electron interference as predicted by quantum mechanics has stimulated a great deal of interest in the effect and in the relation between classical and quantum theory.[11] The principal concern has been with the observation of a phenomenon that appears to have no classical analog even though the expression for the fringe shift **(127)** involves purely classical quantities. And the fact that the vector potential is a generally unfamiliar field quantity has not made the elucidation of the problem easier. Yet there are some aspects of the classical problem that have not been extensively discussed and that we consider here.[6] Following this discussion we suggest very briefly a quantum interpretation that regards the electrons and the solenoid as a coherently interacting system.

Our concern is with a charge q passing a long solenoid through which there is flux Φ as shown in Figure 8. From **(6-98)** the vector potential is given by:

$$A(r) = \frac{\Phi}{2\pi\rho} \hat{\phi} \tag{128}$$

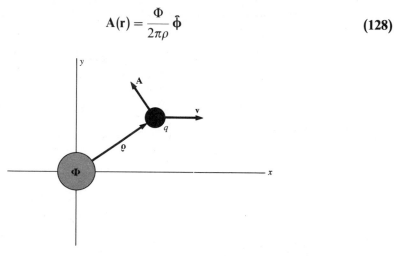

Figure 8   Interaction between a charge and a solenoid. A charge q moves with velocity v past a long solenoid containing a flux Φ.

As we have seen in Section 6-13, the magnetic field outside a long solenoid is equal to zero, and thus there is *no* force on the moving charge q. How then are we to understand that as the position of q changes, and consequently A at the position of q changes, there is a change in the momentum of the system **(121)**?

To examine this question we look not at the charge q, but at the *solenoid*. Even though there may be no force on q, there is the possibility of a force on the solenoid. We write the force on the solenoid as:

$$F = \int_V j_{ext} \times B \, dV \tag{129}$$

[11] H. Ehrlichson, *Am. J. Phys.*, **38**, 162 (1970); **40**, 1707 (1972). J. Woodilla and H. Schwarz, *Am. J. Phys.*, **39**, 111 (1971). T. H. Boyer, *Am. J. Phys.*, **40**, 56, 1708 (1972).

where we regard the flux Φ as established by an external current, and $\mathbf{B}$ is the field produced by the moving charge q. Writing the external current in terms of the external magnetic field by Ampère's law (6-48):

$$\nabla \times \mathbf{B}_{ext} = \mu_0 \mathbf{j}_{ext} \tag{130}$$

and writing the magnetic field of $\mathbf{q}$ in terms of its electric field (6-22)

$$\mathbf{B} = \frac{1}{c^2} \mathbf{v} \times \mathbf{E} \tag{131}$$

we may write the force on the solenoid as:

$$\mathbf{F} = -\epsilon_0 \int_V (\mathbf{v} \times \mathbf{E}) \times (\nabla \times \mathbf{B}_{ext}) \, dV \tag{132}$$

We are now faced with a considerable amount of vector algebra in order to transform (132) into familiar form. We first expand the integrand of (132) by (C-1):

$$(\mathbf{v} \times \mathbf{E}) \times (\nabla \times \mathbf{B}_{ext}) = (\nabla \mathbf{B}_{ext}) \cdot (\mathbf{v} \times \mathbf{E}) \quad (\mathbf{v} \times \mathbf{E}) \quad \nabla \mathbf{B}_{ext} \tag{133}$$

We have from (C-3)

$$\nabla(\mathbf{B}_{ext} \cdot \mathbf{v} \times \mathbf{E}) = (\nabla \mathbf{B}_{ext}) \cdot \mathbf{v} \times \mathbf{E} + [\nabla(\mathbf{v} \times \mathbf{E})] \cdot \mathbf{B}_{ext} \tag{134}$$

and from (C-4)

$$\nabla \cdot [\mathbf{B}_{ext}(\mathbf{v} \times \mathbf{E})] = \mathbf{B}_{ext} \cdot \nabla(\mathbf{v} \times \mathbf{E}) \tag{135}$$

where we have used $\nabla \cdot \mathbf{B}_{ext} = 0$. We also write from (C-4)

$$\nabla \cdot [(\mathbf{v} \times \mathbf{E})\mathbf{B}_{ext}] = (\mathbf{v} \times \mathbf{E}) \cdot \nabla \mathbf{B}_{ext} \tag{136}$$

where we have used

$$(\nabla \times \mathbf{E}) \cdot \mathbf{v} = -\frac{\partial \mathbf{B}}{\partial t} \cdot \mathbf{v} = (\mathbf{v} \cdot \nabla)\mathbf{B} \cdot \mathbf{v}$$

$$= \frac{1}{c^2}(\mathbf{v} \cdot \nabla)(\mathbf{v} \times \mathbf{E}) \cdot \mathbf{v} = 0 \tag{137}$$

Subtracting (134) and adding (135) and (136) to (133) we obtain:

$$(\mathbf{v} \times \mathbf{E}) \times (\nabla \times \mathbf{B}_{ext}) = [\nabla \times (\mathbf{v} \times \mathbf{E})] \times \mathbf{B}_{ext} + \nabla(\mathbf{B}_{ext} \cdot \mathbf{v} \times \mathbf{E})$$

$$- \nabla \cdot [\mathbf{B}_{ext}(\mathbf{v} \times \mathbf{E}) + (\mathbf{v} \times \mathbf{E})\mathbf{B}_{ext}] \tag{138}$$

Substituting **(138)** into **(132)** and letting the volume go to infinity we drop the gradient and divergence terms by Gauss's and Stokes's theorems to obtain:

$$\mathbf{F} = -\epsilon_0 \int [\mathbf{\nabla} \times (\mathbf{v} \times \mathbf{E})] \times \mathbf{B}_{\text{ext}} \, dV \tag{139}$$

Expanding the triple product we write:

$$\mathbf{F} = -\epsilon_0 \int (\mathbf{v}\mathbf{\nabla} \cdot \mathbf{E} - \mathbf{v} \cdot \mathbf{\nabla}\mathbf{E}) \times \mathbf{B}_{\text{ext}} \, dV \tag{140}$$

Now the divergence of $\mathbf{E}$ is zero everywhere but at q, where $\mathbf{B}_{\text{ext}}$ is zero so that the first term does not contribute to the integral. We are left with the second term, which may be transformed by the identity

$$\frac{\partial \mathbf{E}}{\partial t} = -\mathbf{v} \cdot \mathbf{\nabla}\mathbf{E} \tag{141}$$

to yield:

$$\mathbf{F} = -\epsilon_0 \int \frac{\partial \mathbf{E}}{\partial t} \times \mathbf{B}_{\text{ext}} \, dV = -\epsilon_0 \frac{d}{dt} \int \mathbf{E} \times \mathbf{B}_{\text{ext}} \, dV \tag{142}$$

But the right side of **(142)** is just the negative of the rate of change of field momentum **(113)**:

$$\mathbf{F} = -\frac{d}{dt}\,\mathbf{p}_{\text{field}} = -q\,\frac{d\mathbf{A}}{dt} \tag{143}$$

If the solenoid were restrained by the mechanical force:

$$\mathbf{F}_{\text{mech}} = -\mathbf{F} = \frac{d}{dt}\,\mathbf{p}_{\text{field}} \tag{144}$$

then we would obtain the result that the field momentum arises from the mechanical force on the solenoid.

   Our interpretation of the Aharonov-Bohm experiment is then in terms of an external mechanical force acting on the system. The wave vector of the *entire system* changes according to the equation

$$\hbar\,\frac{d\mathbf{k}}{dt} = \mathbf{F}_{\text{mech}} \tag{145}$$

Integrating **(145)** we have

$$\mathbf{k} = \frac{1}{\hbar} \int \mathbf{F}_{\text{mech}} \, dt = \frac{q}{\hbar}\,\mathbf{A}(\mathbf{r}) \tag{146}$$

where $A(r)$ is the vector potential at the site of q. We interpret the phase in (122) as that of the coupled charge-solenoid *system* under the action of an external force.

It may be objected that although our discussion points to the importance of external forces, it does not indicate why the charge q and the solenoid should be regarded as a coherent system when there is no force on q and therefore presumably no possibility of transfer of *mechanical* momentum from the solenoid to the charge. We may understand this aspect of the problem by assuming for the moment that there are no external mechanical forces, and the solenoid has finite mass. Then if q is in motion, there will, as we have seen, be a force on the solenoid. The solenoid will now acquire a velocity. And there will be an electric field at q given by (9-119):

$$E = - \frac{\partial A}{\partial t} \tag{147}$$

Thus we see that there is the possibility of momentum exchange between q and the solenoid as long as the solenoid is permitted to move.

If we imagine a classical limit in which $\hbar$ goes to zero, the particle wave length goes to zero, and the interference pattern is washed out by the finite scale on which physical measurements must be made. At the same time in this limit, the coherent exchange of momentum between q and the solenoid is precluded by extraneous fields acting on q or mechanical vibration of the solenoid, the things that plague experiment-alists even in the limit of $\hbar$ "large."

Exercise

24. By using (6-56) for the external current in (129) obtain (143) directly.

18 **[Angular Momentum].** From (109) we write, for the angular momentum,

$$L = \int r \times g \, dV = \int r \times (D \times B) \, dV \tag{148}$$

where g, D, and B are all functions of r. As an example of interest, we find the field angular momentum of a static distribution of electric and magnetic charge. We write, for the electric displacement,

$$D = - \epsilon_0 \, \nabla \phi \tag{149}$$

where the potential ϕ is the usual electrostatic potential:

$$\phi(r) = \frac{1}{4\pi\epsilon_0} \int_{V_1} \frac{\rho(r_1)}{r_{01}} \, dV_1 \tag{150}$$

and for the magnetic field:

$$\mathbf{B} = -\nabla\overline{\phi} \tag{151}$$

where the magnetic scalar potential is given by:

$$\overline{\phi}(\mathbf{r}) = \frac{\mu_0}{4\pi}\int_{V_2}\frac{\overline{\rho}(\mathbf{r}_2)}{r_{02}}\,d\mathbf{V}_2 \tag{152}$$

Substituting from **(149)** and **(151)** we may rewrite **(148)** as

$$\mathbf{L} = \epsilon_0\int \mathbf{r}\times(\nabla\phi\times\nabla\overline{\phi})\,dV \tag{153}$$

We now define the vector:

$$\mathbf{C} = \tfrac{1}{2}\epsilon_0(\phi\nabla\overline{\phi} - \overline{\phi}\nabla\phi) \tag{154}$$

which has the property from **(B-3)**:

$$\nabla\times\mathbf{C} = \epsilon_0\,\nabla\phi\times\nabla\overline{\phi} \tag{155}$$

Substituting **(155)** into **(153)** we have:

$$\mathbf{L} = \int \mathbf{r}\times(\nabla\times\mathbf{C})\,dV \tag{156}$$

In what follows we make use of a number of tensor identities to transform **(156)** into a tractable integral. In the process we obtain two surface integrals that we will want to set to zero in the limit that the surface goes to infinity. [Since $\nabla\phi$ and $\nabla\overline{\phi}$ go to zero as $1/r^2$, we might expect $\mathbf{C}$ to go to zero as $1/r^3$. Then contributions to **(156)** from large r might be expected to diverge logarithmically requiring us to retain the surface contributions. Actually however, because $\mathbf{C}$ is antisymmetric in the interchange of ϕ and $\overline{\phi}$, it falls off at large r as $1/r^4$ rather than as $1/r^3$, and we can properly neglect the surface integrals for large r.]

Expanding the integrand of **(156)** we write in tensor form **(C-1)**:

$$\mathbf{r}\times(\nabla\times\mathbf{C}) = (\nabla\mathbf{C})\cdot\mathbf{r} - (\mathbf{r}\cdot\nabla)\mathbf{C} \tag{157}$$

We introduce the additional identities from **(C-4)**:

$$\nabla\cdot(\mathbf{r}\mathbf{C}) = 3\mathbf{C} + (\mathbf{r}\cdot\nabla)\mathbf{C} \tag{158}$$

$$\nabla\cdot(\mathbf{C}\mathbf{r}) = \mathbf{r}(\nabla\cdot\mathbf{C}) + \mathbf{C} \tag{159}$$

and from **(C-3)**:

$$\mathbf{V}(\mathbf{C} \cdot \mathbf{r}) = (\mathbf{V}\mathbf{C}) \cdot \mathbf{r} + \mathbf{C} \tag{160}$$

By eliminating $(\mathbf{V}\mathbf{C}) \cdot \mathbf{r}$, $(\mathbf{r} \cdot \mathbf{V})\mathbf{C}$, and $\mathbf{C}$ we obtain:

$$\mathbf{r} \times (\mathbf{V} \times \mathbf{C}) = \mathbf{V}(\mathbf{C} \cdot \mathbf{r}) - \mathbf{V} \cdot (\mathbf{r}\mathbf{C}) + 2\mathbf{V} \cdot (\mathbf{C}\mathbf{r}) - 2\mathbf{r}(\mathbf{V} \cdot \mathbf{C}) \tag{161}$$

Substituting into **(156)** we may convert $\mathbf{V}(\mathbf{C} \cdot \mathbf{r})$ to a surface integral by **(B-12)** and $\mathbf{V} \cdot (\mathbf{r}\mathbf{C})$ and $\mathbf{V} \cdot (\mathbf{C}\mathbf{r})$ to surface integrals by **(C-6)**. By letting the surface go to infinity these integrals vanish, and we have:

$$\mathbf{L} = -2 \int \mathbf{r}(\mathbf{V} \cdot \mathbf{C}) \, dV \tag{162}$$

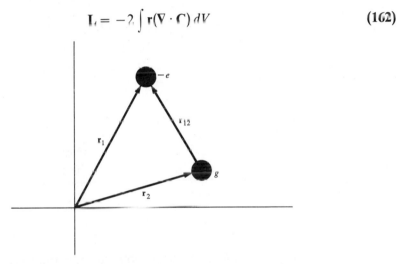

Figure 9 Field angular momentum. An electric charge $-e$ and a magnetic charge g generate an electromagnetic field with angular momentum.

Taking the divergence of **(154)** we obtain:

$$-2\mathbf{V} \cdot \mathbf{C} = -\epsilon_0(\phi\nabla^2\overline{\phi} - \overline{\phi}\nabla^2\phi) = \frac{1}{c^2} \, \phi\overline{\rho} - \overline{\phi}\rho \tag{163}$$

so that **(162)** becomes:

$$\mathbf{L} = \int \left[\frac{1}{c^2} \, \phi\overline{\rho}(\mathbf{r}) - \overline{\phi}\rho(\mathbf{r}) \right] \mathbf{r} \, dV \tag{164}$$

As a specific example we find the angular momentum of an electric charge $-e$ at $\mathbf{r}_1$ and a magnetic charge g at $\mathbf{r}_2$ as shown in Figure 9. The contributions to **(164)** arise from integrating over the electric and magnetic charges so that we may write for the field angular momentum:

$$\mathbf{L} = \frac{1}{c^2} \, g\mathbf{r}_2 \, \phi(\mathbf{r}_2) - e\mathbf{r}_1\phi(\mathbf{r}_1) \tag{165}$$

where from **(150)** we have for the electric potential:

$$\phi(\mathbf{r}_2) = -\frac{1}{4\pi\epsilon_0}\frac{e}{r_{12}} \tag{166}$$

and from **(152)** for the magnetic potential:

$$\overline{\phi}(\mathbf{r}_1) = \frac{\mu_0}{4\pi}\frac{g}{r_{12}} \tag{167}$$

Substituting from **(166)** and **(167)** into **(165)** we obtain the simple result:

$$\mathbf{L} = \frac{\mu_0}{4\pi}\, ge\hat{\mathbf{r}}_{12} \tag{168}$$

As discussed in Section D-5, Dirac obtained the result **(D-31)**:

$$\mu_0\, ge = 2\pi n\hbar \tag{169}$$

by requiring that electronic wave functions be single-valued in the presence of magnetic charge. Substituting into **(168)** we have for the field angular momentum:

$$\mathbf{L} = \frac{n}{2}\,\hbar\hat{\mathbf{r}}_{12} \tag{170}$$

For an electron that passes a magnetic monopole the total *change* in field angular momentum is $n\hbar$ and is directed along the path of the electron.[12]

Exercises

---

25. Verify from **(148)** and Figure 9 that the field angular momentum is directed *from* a positively charged monopole *toward* a negatively charged electron.
26. An electron moves past a monopole at rest. Show that the angular momentum acquired by the electron is the negative of **(168)** so that the sum of external and field angular momentum is a constant.
27. A magnetic monopole moves past an electron at rest. Show that the angular momentum acquired by the monopole is the negative of **(168)** so that the sum of external and field angular momentum is a constant. [A. S. Goldhaber, *Phys. Rev.*, **140**, B1407 (1965).]
28. Verify that **C (154)** falls off as $1/r^4$ for a magnetic charge q at the origin and an electric charge q at **a**. Write the potential of q as a moment expansion about the origin.

---

[12] For alternate derivations of **(168)**, see I. Adawi, *Am. J. Phys.*, **44**, 762 (1976).

29. A charge q moves slowly with velocity $\mathbf{v}$ in the magnetic field of steady solenoidal currents. From **(148)** show that the field angular momentum about the origin is given by:

$$\mathbf{L} \simeq q\mathbf{r} \times \mathbf{A}(\mathbf{r})$$

Use the method leading to **(120)**.

VECTOR AND SCALAR POTENTIALS

In our discussion of static charge distributions we found it useful to introduce the scalar potential ϕ where the electric field was given by **(1-22)**:

$$\mathbf{E} = -\nabla\phi \tag{171}$$

Similarly, for steady currents we were able to introduce a vector potential for the magnetic field **(6-54)**:

▶
$$\mathbf{B} = \nabla \times \mathbf{A} \tag{172}$$

We recall that **(171)** is a consequence of **(1-28)**, which states that the curl of an electrostatic field is zero while **(172)** is a consequence of **(6-59)**, which states that the divergence of the field of steady currents is zero. With time-varying fields the curl of the electric field is no longer zero according to **(39)** so that **(171)** is no longer tenable. In the absence of magnetic charge we may retain **(172)** as a consequence of ME IV with the understanding that both $\mathbf{B}$ and $\mathbf{A}$ may be functions of time as well as position.

19   Scalar Potential. Substituting **(172)** into ME II we obtain:

$$\nabla \times \left(\mathbf{E} + \frac{\partial \mathbf{A}}{\partial t}\right) = 0 \tag{173}$$

as the generalization of **(1-28)** to time-varying fields. From **(B-8)** we have

$$\nabla \times (\nabla\phi) = 0 \tag{174}$$

for any scalar, permitting us to *define* the scalar potential by the relation:

$$-\nabla\phi = \mathbf{E} + \frac{\partial \mathbf{A}}{\partial t} \tag{175}$$

Solving **(175)** for the electric field we have:

▶
$$\mathbf{E} = -\nabla\phi - \frac{\partial \mathbf{A}}{\partial t} \tag{176}$$

In terms of the scalar and vector potentials ME I becomes on substitution of (176) into (29):

$$\nabla^2 \phi + \frac{\partial}{\partial t} \nabla \cdot \mathbf{A} = -\frac{1}{\epsilon_0} \rho \tag{177}$$

Substituting (172) and (176) into ME III we obtain:

$$\nabla^2 \mathbf{A} - \frac{1}{c^2} \frac{\partial^2 \mathbf{A}}{\partial t^2} - \nabla \left(\nabla \cdot \mathbf{A} + \frac{1}{c^2} \frac{\partial \phi}{\partial t} \right) = -\mu_0 \mathbf{j} \tag{178}$$

Exercise

30. External currents are accelerated by an electrostatic potential. Show that the work done by the electrostatic field goes into driving the currents and supplying the energy in the magnetic field:

$$\frac{dW}{dt} = \int dV\, \mathbf{j}_{\text{ext}} \cdot \mathbf{E} + \int dV\, \mathbf{j}_{\text{ext}} \cdot \frac{d\mathbf{A}}{dt}$$

Use (176). Compare with Exercise 10-7.

20 **Lorentz and Coulomb Gauges.** Because the magnetic field **B** is determined by the curl of **A** and is unaffected by the divergence of **A**, we are free to choose the divergence of **A** so long as we make a corresponding change in ϕ as indicated by (175). A choice that brings (177) and (178) to symmetric form is the Lorentz condition:

▶
$$\nabla \cdot \mathbf{A} + \frac{1}{c^2} \frac{\partial \phi}{\partial t} = 0 \tag{179}$$

which has the form of a continuity equation. Substituting (179) into (177) and (178) we obtain a pair of fully decoupled wave equations:

▶
$$\nabla^2 \phi - \frac{1}{c^2} \frac{\partial^2 \phi}{\partial t^2} = -\frac{1}{\epsilon_0} \rho \tag{180}$$

▶
$$\nabla^2 \mathbf{A} - \frac{1}{c^2} \frac{\partial^2 \mathbf{A}}{\partial t^2} = -\mu_0 \mathbf{j} \tag{181}$$

Vector and scalar potentials related by (179) are said to be in the *Lorentz gauge*.

A choice of guage which is employed in the theory of superconductivity is the *Coulomb gauge* (8-97):

$$\nabla \cdot \mathbf{A}_L = 0 \tag{182}$$

In this gauge (177) becomes:

$$\nabla^2 \phi = -\frac{1}{\epsilon_0} \rho \tag{183}$$

and ϕ is formally equivalent to the Coulomb potential of a static charge distribution. We may write the London equation (8-105):

$$\mu_0 \mathbf{j} = -\frac{1}{\lambda_L^2} \mathbf{A}_L \tag{184}$$

for slowly varying as well as for static fields. Substituting (184) into (178) we obtain:

$$\nabla^2 \mathbf{A}_L + \frac{1}{\lambda_L^2} \mathbf{A}_L = -\frac{1}{c^2} \frac{\partial \mathbf{E}}{\partial t} \tag{185}$$

as the equation connecting the vector potential and the electric field within a super-conductor. Because of its use in superconductivity, (182) is also called the *London gauge*.

Exercise

---

31. From (180) and (181) use (172) and (176) to obtain the wave equations for the electric and magnetic fields in vacuum:

$$\nabla^2 \mathbf{E} - \frac{1}{c^2} \frac{\partial^2 \mathbf{E}}{\partial t^2} = 0$$

$$\nabla^2 \mathbf{B} - \frac{1}{c^2} \frac{\partial^2 \mathbf{B}}{\partial t^2} = 0$$

ELECTROMAGNETIC RADIATION

We next consider the radiation of electromagnetic energy by time-varying charges and currents. We introduce retarded solutions for the vector and scalar potentials. We then develop the multipole expansion of the potentials and obtain expressions for radiation from electric and magnetic dipoles and from electric quadrupoles. Finally we discuss as an example radiation by charges moving rapidly on a circular orbit—synchrotron radiation.

21 Retarded Potentials. We discuss briefly the solutions to (180) and (181), the equations for the scalar and vector potential of an arbitrary charge and current distribution:

$$\nabla^2 \phi - \frac{1}{c^2} \frac{\partial^2 \phi}{\partial t^2} = -\frac{1}{\epsilon_0} \rho \tag{186}$$

$$\nabla^2 \mathbf{A} - \frac{1}{c^2} \frac{\partial^2 \mathbf{A}}{\partial t^2} = -\mu_0 \mathbf{j} \tag{187}$$

If for the moment we disregard the time derivative, we recognize **(186)** and **(187)** as **(1-19)** and **(6-49)** for which we have had the solutions:

$$\phi(\mathbf{r}) = \frac{1}{4\pi\epsilon_0} \int \frac{\rho(\mathbf{r}_1)}{r_{01}} dV_1 \tag{188}$$

$$\mathbf{A}(\mathbf{r}) = \frac{\mu_0}{4\pi} \int \frac{\mathbf{j}(\mathbf{r}_1)}{r_{01}} dV_1 \tag{189}$$

The effect of the second derivative with respect to time in **(186)** and **(187)** is to propagate the scalar and vector potentials away from the charge and current sources at the velocity of light. This means that the potentials at position $\mathbf{r}$ and time t are determined by the charge and current densities at position $\mathbf{r}_1$ but at an earlier time $t - r_{01}/c$ where r_{01}/c is the time for the disturbance to propagate from $\mathbf{r}_1$ to $\mathbf{r}$. We may write then for the scalar and vector potentials:

▶
$$\phi(\mathbf{r},t) = \frac{1}{4\pi\epsilon_0} \int \frac{\rho(\mathbf{r}_1, t - r_{01}/c)}{r_{01}} dV_1 \tag{190}$$

▶
$$\mathbf{A}(\mathbf{r},t) = \frac{\mu_0}{4\pi} \int \frac{\mathbf{j}(\mathbf{r}_1, t - r_{01}/c)}{r_{01}} dV_1 \tag{191}$$

It is worth emphasizing that the sources of the scalar and vector potentials are charges and currents and not fields. The charge and current densities $\rho(\mathbf{r}_1, t - r_{01}/c)$ and $\mathbf{j}(\mathbf{r}_1, t - r_{01}/c)$ are called the *retarded* charge and current densities and **(190)** and **(191)** are called the retarded solutions of **(186)** and **(187)**. The notation may be simplified by writing the retarded time in brackets:

$$\rho(\mathbf{r}_1, t - r_{01}/c) = \rho(\mathbf{r}_1, [t]) \qquad [t] = t - r_{01}/c \tag{192}$$

Exercises

32. Show that the expressions for the electric and magnetic fields of a uniformly moving charge **(6-21)** and **(6-22)** may be written as:

$$\mathbf{E} = \frac{q}{4\pi\epsilon_0} \frac{[\mathbf{r}] + \mathbf{v}[r]/c}{\gamma^2 s^3}$$

$$\mathbf{B} = \frac{1}{c^2} \mathbf{v} \times \mathbf{E}$$

where the distance s is given by

$$s = [r] + \frac{1}{c}\mathbf{v} \cdot [\mathbf{r}]$$

Here $[\mathbf{r}]$ is the position of q at the retarded time and $[r]$ is the distance to the charge at the retarded time.

33. Verify **(190)** and **(191)** by substitution into **(186)** and **(187)**. For the Laplacian of $1/r_{01}$ see **(B-71)**.

34. Show that **(190)** and **(191)** satisfy **(179)**.

35. Show that a charge q moving at a speed less than c contributes at only one position to **(190)** and **(191)**. If the speed is greater than c, show that the charge makes two contributions to each of the integrals.

22 Multipole Radiation. We have already discussed multipole expansions of static charge and steady solenoidal current distributions. In Section 1-10 we wrote the potential of a limited charge distribution as a moment expansion **(1-70)**:

$$\phi(\mathbf{r}) = \phi_{E0}(\mathbf{r}) + \phi_{E1}(\mathbf{r}) + \phi_{E2}(\mathbf{r}) + \cdots \tag{193}$$

The electric monopole term was given by:

$$\phi_{E0}(\mathbf{r}) = \frac{1}{4\pi\epsilon_0} \frac{q}{r} \tag{194}$$

where:

$$q = \int_{V_1} \rho(\mathbf{r}_1) \, dV_1 \tag{195}$$

is the total enclosed charge and is called the electric monopole moment. The electric dipole term was given by:

$$\phi_{E1}(\mathbf{r}) = \frac{1}{4\pi\epsilon_0} \frac{\hat{\mathbf{r}} \cdot \mathbf{p}}{r^2} \tag{196}$$

where

$$\mathbf{p} = \int_{V_1} \mathbf{r}_1 \rho(\mathbf{r}_1) \, dV_1 \tag{197}$$

is called the electric dipole moment. The electric quadrupole term was given by

$$\phi_{E2}(\mathbf{r}) = \frac{1}{8\pi\epsilon_0} \frac{\hat{\mathbf{r}} \cdot \mathbf{Q} \cdot \hat{\mathbf{r}}}{r^3} \tag{198}$$

where

$$\mathbf{Q} = \int_{V_1} [3\mathbf{r}_1\mathbf{r}_1 - r_1^2 \mathbf{I}]\rho(\mathbf{r}_1) \, dV_1 \tag{199}$$

is the electric quadrupole moment and is a tensor.

In Chapter 6 we expanded the vector potential of a distribution of steady currents (6-103):

$$\blacktriangleright \qquad \mathbf{A}(\mathbf{r}) = \mathbf{A}_{M0}(\mathbf{r}) + \mathbf{A}_{M1}(\mathbf{r}) + \mathbf{A}_{M2}(\mathbf{r}) + \cdots \qquad (200)$$

The first term, which would arise from a magnetic monopole moment, was shown to be zero for steady solenoidal currents. The second term is from the magnetic dipole:

$$\mathbf{A}_{M1}(\mathbf{r}) = \frac{\mu_0}{4\pi} \frac{\boldsymbol{\mu} \times \hat{\mathbf{r}}}{r^2} \qquad (201)$$

where

$$\boldsymbol{\mu} = \tfrac{1}{2} \int_{V_1} \mathbf{r}_1 \times \mathbf{j}(\mathbf{r}_1)\, dV_1 \qquad (202)$$

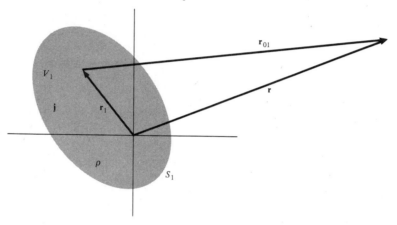

Figure 10   Retarded potential expansion. We assume that we are well removed from a bounded distribution of currents and charges.

is called the magnetic dipole moment. The next term, which is the magnetic quadrupole moment, is rarely of importance in classical radiation problems and will not be further considered.

We now wish to expand the retarded scalar and vector potentials (187) and (188) for a general distribution of charges and currents. We will assume that we are well removed from a bounded distribution as shown in Figure 10. We expand r_{01} and $1/r_{01}$:

$$r_{01} = (r^2 - 2\mathbf{r} \cdot \mathbf{r}_1 + r_1{}^2)^{1/2} \simeq r - \hat{\mathbf{r}} \cdot \mathbf{r}_1 + \cdots \qquad (203)$$

$$\frac{1}{r_{01}} = (r^2 - 2\mathbf{r} \cdot \mathbf{r}_1 + r_1{}^2)^{-1/2} \simeq \frac{1}{r} + \frac{\hat{\mathbf{r}} \cdot \mathbf{r}_1}{r^2} + \cdots \qquad (204)$$

In order to obtain the leading terms in the potentials at large distance we retain only the first term in $1/r_{01}$ but the first two terms in the expansion of r_{01}. We may then write (173) as

$$\mathbf{A}(\mathbf{r},t) = \frac{\mu_0}{4\pi} \frac{1}{r} \int_{V_1} \mathbf{j}(\mathbf{r}_1, t - r/c + \hat{\mathbf{r}} \cdot \mathbf{r}_1/c)\, dV_1 \qquad (205)$$

We next expand the current as a function of time:

$$\mathbf{j}(\mathbf{r}_1, t - r/c + \hat{\mathbf{r}} \cdot \mathbf{r}_1/c) = \mathbf{j}(\mathbf{r}_1, t - r/c) + \frac{1}{c} \hat{\mathbf{r}} \cdot \mathbf{r}_1 \frac{\partial}{\partial t} \mathbf{j}(\mathbf{r}_1, t - r/c)$$

$$+ \frac{1}{2c^2} (\hat{\mathbf{r}} \cdot \mathbf{r}_1)^2 \frac{\partial^2}{\partial t^2} \mathbf{j}(\mathbf{r}_1, t - r/c) + \cdots \quad (206)$$

The first term in **(206)** makes a contribution to the vector potential:

$$\mathbf{A}_{E1}(\mathbf{r}) = \frac{\mu_0}{4\pi} \frac{1}{r} \int_{V_1} \mathbf{j}(\mathbf{r}_1, t - r/c) \, dV_1 \quad (207)$$

We make use of the identity **(C-4)**:

$$\mathbf{j} = \mathbf{\nabla}_1 \cdot (\mathbf{j}\mathbf{r}_1) - \mathbf{r}_1 \mathbf{\nabla} \cdot \mathbf{j} \quad (208)$$

to transform **(207)**. If there are no currents on the bounding surface the divergence term integrates to zero and we have:

$$\mathbf{A}_{E1}(\mathbf{r}) = -\frac{\mu_0}{4\pi} \frac{1}{r} \int_{V_1} \mathbf{r}_1 \mathbf{\nabla} \cdot \mathbf{j}(\mathbf{r}_1, t - r/c) \, dV_1 \quad (209)$$

From the equation of continuity **(32)** we may write:

$$\mathbf{\nabla} \cdot \mathbf{j}(\mathbf{r}_1, t - r/c) = -\frac{\partial}{\partial t} \rho(\mathbf{r}_1, t - r/c) \quad (210)$$

We have then for **(209)**:

$$\mathbf{A}_{E1}(\mathbf{r}) = \frac{\mu_0}{4\pi r} \frac{\partial}{\partial t} \mathbf{p}(t - r/c) \quad (211)$$

where **p** is the electric dipole moment and is given by **(197)**. We could have designated this contribution as arising from the magnetic monopole moment of a divergent current although it is customary to designate it as arising from the time derivative of the electric dipole moment.

We develop the second term of **(206)** by writing $\mathbf{r}_1\mathbf{j}$ as the sum of an antisymmetric and a symmetric tensor:

$$\mathbf{r}_1\mathbf{j} = \tfrac{1}{2}(\mathbf{r}_1\mathbf{j} - \mathbf{j}\mathbf{r}_1) + \tfrac{1}{2}(\mathbf{r}_1\mathbf{j} + \mathbf{j}\mathbf{r}_1) \quad (212)$$

The vector potential of the antisymmetric part may be written as

$$\mathbf{A}_{M1}(\mathbf{r}) = \frac{\mu_0}{4\pi} \frac{\hat{\mathbf{r}}}{rc} \cdot \frac{\partial}{\partial t} \int_{V_1} \tfrac{1}{2}(\mathbf{r}_1 \mathbf{j} - \mathbf{j}\mathbf{r}_1) \, dV_1$$

$$= \frac{\mu_0}{4\pi} \frac{1}{rc} \frac{\partial}{\partial t} \left(\frac{1}{2} \int_{V_1} \mathbf{r}_1 \times \mathbf{j}(\mathbf{r}_1, t - r/c) \, dV_1 \right) \times \hat{\mathbf{r}} \qquad (213)$$

From **(202)** we write **(213)** as

$$\mathbf{A}_{M1}(\mathbf{r}) = \frac{\mu_0}{4\pi} \frac{1}{rc} \frac{\partial}{\partial t} \boldsymbol{\mu}(t - r/c) \times \hat{\mathbf{r}} \qquad (214)$$

which is the contribution to the vector potential from the time variation of the magnetic dipole moment.

The symmetric part of **(212)** contributes to the vector potential a term:

$$\mathbf{A}_{E2}(\mathbf{r}) = \frac{\mu_0}{8\pi} \frac{\hat{\mathbf{r}}}{rc} \frac{\partial}{\partial t} \int_{V_1} (\mathbf{r}_1 \mathbf{j} + \mathbf{j}\mathbf{r}_1) \, dV_1 \qquad (215)$$

We transform the integrand of **(215)** by making use of the tensor identity **(C-8)**:

$$\nabla_1 \cdot (\mathbf{j}\,\mathbf{r}_1 \mathbf{r}_1) = \mathbf{r}_1 \mathbf{r}_1 \nabla_1 \cdot \mathbf{j} + (\mathbf{r}_1 \mathbf{j} + \mathbf{j}\,\mathbf{r}_1) \qquad (216)$$

Substituting into **(215)**, the left side of **(216)** integrates to zero over S_1 and we are left with:

$$\mathbf{A}_{E2}(\mathbf{r}) = -\frac{\mu_0}{8\pi} \frac{\hat{\mathbf{r}}}{rc} \frac{\partial}{\partial t} \int \mathbf{r}_1 \mathbf{r}_1 \, \nabla_1 \cdot \mathbf{j}(\mathbf{r}_1, t - r/c) \, dV_1 \qquad (217)$$

Using the equation of continuity **(207)** we have, finally,

$$\mathbf{A}_{E2}(\mathbf{r}) = \frac{\mu_0}{8\pi} \frac{\hat{\mathbf{r}}}{rc} \cdot \frac{\partial^2}{\partial t^2} \int \mathbf{r}_1 \mathbf{r}_1 \rho(\mathbf{r}_1, t - r/c) \, dV_1 \qquad (218)$$

Note that the integrand of **(218)** is the quadrupole moment **(199)** apart from a multiple of the unit tensor.

The expansion of the scalar potential **(190)** is much simpler. By the same argument that led to **(205)** we write:

$$\phi(\mathbf{r},t) = \frac{1}{4\pi\epsilon_0} \frac{1}{r} \int_{V_1} \rho(\mathbf{r}_1, t - r/c + \hat{\mathbf{r}} \cdot \mathbf{r}_1/c) \, dV_1 \qquad (219)$$

We expand the charge density as a function of time to obtain:

$$\blacktriangleright \qquad \phi(\mathbf{r},t) = \phi_{E0}(\mathbf{r},t) + \phi_{E1}(\mathbf{r},t) + \phi_{E2}(\mathbf{r},t) + \cdots \qquad (220)$$

where we have

$$\phi_{E0}(\mathbf{r}) = \frac{1}{4\pi\epsilon_0} \frac{1}{r} \int_{V_1} \rho(\mathbf{r}_1, t - r/c) \, dV_1 = \frac{1}{4\pi\epsilon_0} \frac{q}{r} \tag{221}$$

$$\phi_{E1}(\mathbf{r}) = \frac{1}{4\pi\epsilon_0} \frac{\hat{\mathbf{r}}}{rc} \cdot \frac{\partial}{\partial t} \int_{V_1} \mathbf{r}_1 \rho(\mathbf{r}_1, t - r/c) \, dV_1$$

$$= \frac{1}{4\pi\epsilon_0} \frac{\hat{\mathbf{r}}}{rc} \cdot \frac{\partial}{\partial t} \mathbf{p}(t - r/c) \tag{222}$$

$$\phi_{E2}(\mathbf{r}) = \frac{1}{8\pi\epsilon_0} \frac{\hat{\mathbf{r}}}{rc^2} \cdot \left[\frac{\partial^2}{\partial t^2} \int_{V_1} \mathbf{r}_1 \mathbf{r}_1 \rho(\mathbf{r}_1, t - r/c) \, dV_1 \right] \cdot \hat{\mathbf{r}} \tag{223}$$

Note that in the Lorentz gauge the magnetic dipole moment does not contribute to the scalar potential.

We now obtain expressions for the radiation field from **(172)** and **(176)**:

$$\mathbf{B}(\mathbf{r}) = \nabla \times \mathbf{A}(\mathbf{r}) \tag{224}$$

$$\mathbf{E}(\mathbf{r}) = -\nabla \phi(\mathbf{r}) - \frac{\partial}{\partial t} \mathbf{A}(\mathbf{r}) \tag{225}$$

For the electric monopole moment we obtain:

$$\mathbf{E}_{E0}(\mathbf{r}) = \frac{1}{4\pi\epsilon_0} \hat{\mathbf{r}} \frac{q}{r^2} \qquad \mathbf{B}_{E0}(\mathbf{r}) = 0 \tag{226}$$

This is just the electrostatic field since q must be constant for a bounded distribution. The electric E1 field is computed by substituting **(211)** and **(222)** into **(225)**:

$$\mathbf{E}_{E1}(\mathbf{r}) = -\frac{1}{4\pi\epsilon_0} \nabla \left[\frac{\hat{\mathbf{r}}}{rc} \cdot \frac{\partial}{\partial t} \mathbf{p}(t - r/c) \right] - \frac{1}{4\pi\epsilon_0} \frac{1}{rc^2} \frac{\partial^2}{\partial t^2} \mathbf{p}(t - r/c) \tag{227}$$

The leading term involves the gradient of $\hat{\mathbf{r}} \cdot \mathbf{p}$ which gives by **(B-6)**:

$$\mathbf{E}_{E1}(\mathbf{r}) = -\frac{1}{4\pi\epsilon_0} \frac{1}{rc} \left[(\hat{\mathbf{r}} \cdot \nabla) \frac{\partial[\mathbf{p}]}{\partial t} + \hat{\mathbf{r}} \times \left(\nabla \times \frac{\partial[\mathbf{p}]}{\partial t} \right) + \frac{1}{c} \frac{\partial^2[\mathbf{p}]}{\partial t^2} \right] \tag{228}$$

We use the identities:

$$(\hat{\mathbf{r}} \cdot \nabla)[\mathbf{p}] = \left(\frac{x}{r} \frac{\partial}{\partial x} + \frac{y}{r} \frac{\partial}{\partial y} + \frac{z}{r} \frac{\partial}{\partial z} \right)[\mathbf{p}] = \frac{\partial}{\partial r} \mathbf{p}(t - r/c) = -\frac{1}{c} \frac{\partial[\mathbf{p}]}{\partial t} \tag{229}$$

$$\nabla \times \mathbf{p}(t - r/c) = \hat{\mathbf{r}} \times \frac{\partial}{\partial r} \mathbf{p}(t - r/c) = -\frac{1}{c} \hat{\mathbf{r}} \times \frac{\partial[\mathbf{p}]}{\partial t} \tag{230}$$

By **(229)** we see that the first and third terms in **(228)** cancel and we are left with:

$$\mathbf{E}_{E1}(\mathbf{r}) = \frac{1}{4\pi\epsilon_0} \frac{\hat{\mathbf{r}}}{rc^2} \times \left(\hat{\mathbf{r}} \times \frac{\partial^2 [\mathbf{p}]}{\partial t^2} \right) \tag{231}$$

To obtain the $E1$ magnetic field we substitute **(211)** into **(224)**:

$$\mathbf{B}_{E1}(\mathbf{r}) = \frac{\mu_0}{4\pi} \nabla \times \left[\frac{1}{r} \frac{\partial}{\partial t} \mathbf{p}(t - r/c) \right] \tag{232}$$

To obtain the leading term we take out the $1/r$:

$$\mathbf{B}_{E1}(\mathbf{r}) = \frac{\mu_0}{4\pi} \frac{1}{r} \nabla \times \left[\frac{\partial}{\partial t} \mathbf{p}(t - r/c) \right] = \frac{\mu_0}{4\pi} \frac{\hat{\mathbf{r}}}{r} \times \left[\frac{\partial^2}{\partial r \partial t} \mathbf{p}(t - r/c) \right]$$

$$= -\frac{\mu_0}{4\pi} \frac{\hat{\mathbf{r}}}{rc} \times \frac{\partial^2}{\partial t^2} \mathbf{p}(t - r/c) \tag{233}$$

Comparing **(231)** with **(233)** we have:

$$\mathbf{E}_{E1}(\mathbf{r}) = -c\hat{\mathbf{r}} \times \mathbf{B}_{E1}(\mathbf{r}) \tag{234}$$

Although we have obtained **(234)** for the special case of electric dipole radiation it applies generally to radiation fields.

To obtain the energy flow we compute the Poynting vector **(95)**:

$$\mathbf{\Pi} = \mathbf{E}(\mathbf{r}) \times \mathbf{H}(\mathbf{r}) = \frac{1}{\mu_0} \mathbf{E}(\mathbf{r}) \times \mathbf{B}(\mathbf{r}) = -\frac{c}{\mu_0} (\hat{\mathbf{r}} \times \mathbf{B}) \times \mathbf{B}$$

$$= -\frac{c}{\mu_0} [\mathbf{B}(\hat{\mathbf{r}} \cdot \mathbf{B}) - \hat{\mathbf{r}}B^2] = \frac{c}{\mu_0} \hat{\mathbf{r}}B^2 \tag{235}$$

We show in Figure 11 the relative directions of $\hat{\mathbf{r}}$, $\mathbf{E}$, $\mathbf{B}$, and $\partial^2 \mathbf{p}/\partial t^2$. We see that the magnitude of $\mathbf{B}$ is proportional to $\sin\theta$ and we may write for the magnitude of the Poynting vector:

$$\Pi = \frac{1}{16\pi^2\epsilon_0} \frac{1}{r^2c^3} \left(\frac{\partial^2 [p]}{\partial t^2} \right)^2 \sin^2\theta \tag{236}$$

The total rate of energy flow across a sphere of radius r is given by:

$$\frac{dU}{dt} = \oint_S \mathbf{\Pi} \cdot d\mathbf{S} = \frac{2}{3c^3} \frac{1}{4\pi\epsilon_0} \left(\frac{\partial^2 [p]}{\partial t^2} \right)^2 \tag{237}$$

The radiation pattern of a radiating system is a polar plot in which Π is plotted versus θ as shown in Figure 12a for $E1$ radiation.

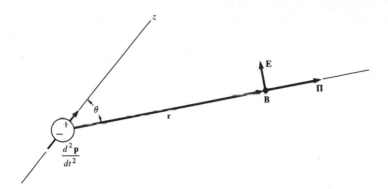

Figure 11 Electric dipole radiation. The relative directions of **E** and **B** are shown with respect to the oscillating dipole.

The electric and magnetic fields and radiation patterns may be obtained in a similar way for $M1$ and $E2$ radiation. The pattern for $M1$ radiation is the same as for $E1$ and is also shown in Figure 12a. The pattern for $E2$ radiation is shown in Figure 12b.

23 **[Synchrotron Radiation].** We discuss here a very striking example of the theory that we have developed, the radiation from charged particles moving at speeds close to that of light.[13] Such radiation is observed from highly energetic electrons moving on curved paths as in the electron synchrotron. It is observed now with much greater intensity from electrons in storage rings. And there is increasing evidence for such radiation from interstellar sources. The characteristics of this radiation, as we will see,

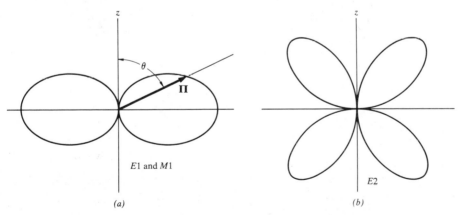

Figure 12 Radiation pattern. (a) Electric dipole and magnetic dipole radiation have the same pattern with two cusps. (b) Electric quadrupole radiation has a pattern with four cusps.

[13] The present discussion follows R. P. Feynman, R. B. Leighton, and M. Sands, *The Feynman Lectures,* Addison-Wesley, 1963; Section I-34-3. See also M. L. Perlman, E. M. Rose, and R. E. Watson, *Phys. Today,* **27**(7), 30 (1974) and letters of comment by G. C. Baldwin and D. W. Kerst, *Phys. Today,* **28**(1), 9 (1975). The first discussion of synchrotron radiation was that of G. A. Schott, *Electromagnetic Radiation,* Cambridge, 1912.

are a continuous spectrum, polarization of the electric field in the plane of the orbit, and radiation in the direction of the charged particle's motion.

Synchrotron radiation is strongly oriented in the forward direction. This means that an observer will observe radiation only from those charges directly approaching. This is the case for only a very short segment of the path of the particle so that the distance to the observer will always be large compared with the effective size of the source. For this reason, we may regard the radiation as arising through the electric dipole moment associated with the radial acceleration of individual charges. We have for the radiation fields from **(233)** and **(234)**:

$$\mathbf{B}(\mathbf{r}) = -\frac{\mu_0}{4\pi} \frac{\hat{\mathbf{r}}_{01}}{c r_{01}} \times \frac{\partial^2}{\partial t^2} \mathbf{p}(t - r_{01}/c) \tag{238}$$

$$\mathbf{E}(\mathbf{r}) = -c\hat{\mathbf{r}}_{01} \times \mathbf{B}(\mathbf{r}) \tag{239}$$

where we regard $\mathbf{p}$ as the electric dipole moment associated with a charge q at $\mathbf{r}_1$:

$$\mathbf{p}(t - r_{01}/c) = q\mathbf{r}_1(t - r_{01}/c) \tag{240}$$

We show in Figure 13a the path of a particle moving with angular velocity ω on a circular path of radius R. To the right we plot the transverse displacement of the charge versus the retarded time $[t]$. To understand why the charge radiates it is

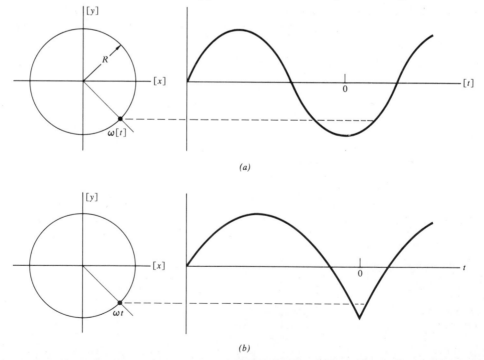

(a)

(b)

Figure 13   Synchrotron radiation. (a) A charge moves on a circular path of radius R. The transverse displacement is a sinusoidal function of retarded time. (b) When the transverse displacement is plotted as a function of present time, a cusp is developed. This cusp has an extremely high second derivative and thus produces intense radiation.

useful to think of the retarded displacement as a function of present time t since the field is given by the second derivative with respect to t. We write, from **(192)**,

$$[t] = t - \frac{r_{01}}{c} = t + \frac{[x]}{c} \tag{241}$$

where we have arbitrarily set the zero of $[t]$ and t at $[x] = 0$. Since we are only concerned with derivatives with respect to time we are free to do this.

We show in Figure 13b a plot of $[y]$ versus t. The total period of the motion is the same but the portion in which the particle is moving toward the observer is highly compressed while the portion in which the particle is moving away is expanded. We get in this way a cusp in $[y]$, which gives an extremely high second derivative and thus intense electromagnetic radiation. We may do this problem analytically by writing

$$[x] = R \sin \omega[t] \tag{242}$$

$$[y] = -R \cos \omega[t] \tag{243}$$

$$[t] = t + \frac{[x]}{c} = t + \frac{R}{c} \sin \omega[t] \tag{244}$$

Taking the differential of **(244)** we obtain

$$dt = (1 - \beta \cos \omega[t]) \, d[t] \tag{245}$$

where $\beta = \omega R/c$ is the ratio of the speed of the charge to that of light. Taking the second derivative of **(243)** with respect to t we obtain

$$\frac{d^2[y]}{dt^2} = \omega^2 R \frac{\cos \omega[t]}{(1 - \beta \cos \omega[t])^3} \tag{246}$$

We may obtain an approximate expression for **(246)** in the vicinity of the cusp by expanding:

$$\cos \omega[t] \simeq 1 - \tfrac{1}{2}(\omega[t])^2 \simeq 1 - \frac{1}{2}\left(\frac{\omega t}{1 - \beta}\right)^2 \tag{247}$$

Substituting into **(246)** we have, approximately,

$$\frac{d^2[y]}{dt^2} \simeq \frac{\omega^2 R/(1 - \beta)^3}{1 + \tfrac{3}{2}\beta(\omega t)^2/(1 - \beta)^3} \tag{248}$$

This shape is known as a *Lorentz line*. The standard normalized form of such a line is plotted in Figure 14a:

$$f(t) = \frac{1}{\pi} \frac{\tau}{\tau^2 + t^2} \tag{249}$$

Comparing **(248)** and **(249)** we obtain:

$$\omega^2\tau^2 = \frac{2}{3\beta}(1-\beta)^3 \tag{250}$$

For $\beta \sim 1$ we may rewrite **(250)** as

▶
$$\omega\tau = \frac{\sqrt{3}}{6\gamma^3} \tag{251}$$

A pulse of duration τ will excite oscillators with frequencies up to $1/\tau$ that from **(251)** is γ^3 times the orbital frequency. A more satisfactory way of discussing such problems is to imagine the pulse as a superposition of waves, all of which come into phase at $t = 0$. This is discussed in Section I-6 where it is shown that for the line shape of Figure 14a the wave amplitudes fall off exponentially with frequency as shown in Figure 14b.

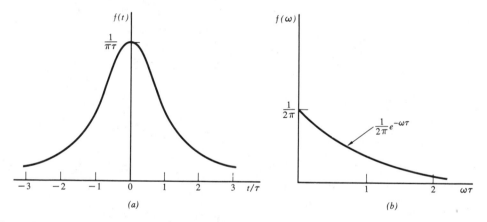

Figure 14   Radiation spectrum. (a) The transverse acceleration as a function of time gives a Lorentz line. (b) The wave amplitudes are expected to fall off exponentially with frequency.

QUASI-STATIONARY APPROXIMATION

There are many situations where the full apparatus of the Maxwell equations is not needed. Generally the condition is that the change in the distribution of currents and charges be small during the time that is required for the electromagnetic fields from any of the sources to propagate to the field position. We show that by neglecting vacuum displacement current we are able to write expressions for the scalar and vector potential without retardation. Finally, we discuss the problem of steady currents as a limiting case of the motion of individual charges.

24   Quasi-Stationary Currents. In the quasi-stationary approximation we continue to write ME I **(29)**:

$$\nabla \cdot \mathbf{E}(\mathbf{r},t) = \frac{1}{\epsilon_0}\rho(\mathbf{r},t) \tag{252}$$

and ME II **(31)**:

$$\nabla \times \mathbf{E}(\mathbf{r},t) = -\frac{\partial}{\partial t} \mathbf{B}(\mathbf{r},t) \tag{253}$$

and ME IV **(42)**:

$$\nabla \cdot \mathbf{B}(\mathbf{r},t) = 0 \tag{254}$$

but we approximate ME III **(38)** by assuming that the current $\mathbf{j}(\mathbf{r},t)$ is everywhere much greater than $\epsilon_0(\partial/\partial t)\mathbf{E}(\mathbf{r},t)$ so that we have essentially Ampère's law **(6-48)**:

$$\nabla \times \mathbf{B}(\mathbf{r},t) = \mu_0 \mathbf{j}(\mathbf{r},t) \tag{255}$$

Note that **(255)** requires that $\mathbf{j}(\mathbf{r},t)$ be approximately solenoidal. We continue to write scalar and vector potentials with

$$\mathbf{E}(\mathbf{r},t) = -\nabla \phi(\mathbf{r},t) - \frac{\partial}{\partial t} \mathbf{A}(\mathbf{r},t) \tag{256}$$

$$\mathbf{B}(\mathbf{r},t) = \nabla \times \mathbf{A}(\mathbf{r},t) \tag{257}$$

We also assume consistent with **(255)** that the vector potential is approximately solenoidal:

$$\nabla \cdot \mathbf{A}(\mathbf{r},t) = 0 \tag{258}$$

Substituting **(256)** and **(257)** into the modified Maxwell equations with **(258)** we obtain:

$$\nabla^2 \phi(\mathbf{r},t) \simeq \frac{1}{\epsilon_0} \rho(\mathbf{r},t) \tag{259}$$

$$\nabla^2 \mathbf{A}(\mathbf{r},t) \simeq \mu_0 \mathbf{j}(\mathbf{r},t) \tag{260}$$

which are **(186)** and **(187)** without the time derivative terms. As we have discussed, these equations may be integrated to give **(188)** and **(189)**

$$\phi(\mathbf{r},t) \simeq \frac{1}{4\pi\epsilon_0} \int_{V_1} \frac{\rho(\mathbf{r}_1,t)}{r_{01}} dV_1 \tag{261}$$

$$\mathbf{A}(\mathbf{r},t) = \frac{\mu_0}{4\pi} \int_{V_1} \frac{\mathbf{j}(\mathbf{r}_1,t)}{r_{01}} dV_1 \tag{262}$$

which are quasi-stationary potentials and do not exhibit retardation.

Exercises

37. Obtain the magnetic field $\mathbf{B}(\mathbf{r},t)$ by taking the curl of **(262)**. Show that the Biot-Savart law **(6-40)** is obtained within the quasi-stationary approximation.
38. Integrate ME III **(38)** in the quasi-stationary approximation. Show that the displacement current does not contribute to the magnetic field in this approximation. [See A. P. French and J. R. Tessman, *Am. J. Phys.*, **31**, 201 (1963) and W. G. V. Rosser, *Am. J. Phys.*, **44**, 1221 (1976).]

25   Current Ring. We obtained in Section 6-6 the magnetic field of steady currents by the following line of argument:

1. By Lorentz transformation we obtained the electric and magnetic fields of a uniformly moving charge.

2. We integrated over a line of such charges to obtain the magnetic field of a current filament.

3. We used the result that any distribution of currents may be decomposed into filaments to obtain Ampere's law.

Now that we have expressions for the radiation fields of a single *accelerating* charge **(238)** and **(239)**, we wish to reexamine the problem of steady currents. It is clear from **(186)** and **(187)** that if the charge and current densities are *time independent* we must have a solution in which the scalar and vector potentials are time independent and thus there is no radiation. Similarly, looking at **(190)** and **(191)**, if the charge and current are independent of present time, they must be independent of retarded time as well and thus lead to **(188)** and **(189)**.

Although obtaining this same result from the fields is hardly required, it should give us some new insight. One may ask, for example, why it is that electrons in a storage ring (Section 23) radiate while electrons moving on a circular wire do not. We discuss this question by examining the conditions under which it is possible to obtain the magnetic field of a distribution of uniformly moving charges.

We show in Appendix H that the electric and magnetic fields of a steady charge distribution are the same as for static charge **(1-6)** and steady currents **(6-40)**, even though individual charges are in accelerated motion. Then charge acceleration is not sufficient for radiation. We must be sure in addition that there are no other charges accelerating *into* the states being vacated.

It is instructive to reexamine synchrotron radiation from this point of view. We wish to look at the change in the charge distribution over a time interval τ, which is the duration of a synchrotron pulse **(251)**. Can we say that in this interval the distribution has not changed? Let us imagine that the charges move on the storage ring separated by a mean distance a, with the current given by

$$I = \frac{e}{a^3} vS \simeq \frac{e}{a^3} cS \tag{263}$$

where S is the beam cross section.

We take as a radiation parameter

$$\eta = \frac{a}{\lambda} \tag{264}$$

where $\lambda = c\tau$ is the critical wavelength. If this parameter is large, the charges are far apart and their radiation fields may be computed independently. If the parameter is small, however, the pulses will interfere reducing the amount of radiation.

It is interesting to estimate **(264)** for an electron storage ring. The rest energy of the electron is:

$$U_0 = m_0 c^2 = 0.511 \text{ MeV} \tag{265}$$

and we have:

$$\gamma = \frac{U}{U_0} = \frac{m}{m_0} \tag{266}$$

The frequency of rotation around the ring is given by

$$\omega = \frac{qB}{m} = 1.76 \times 10^{11} \frac{B}{\gamma} \tag{267}$$

For the Stanford positron-electron accelerator ring (SPEAR) the peak energy is 4.5 GeV, which from **(266)** gives $\gamma = 8800$. The orbital frequency is given by $\omega/2\pi = 1.28$ MHz. The beam is 3.22 mm wide and 1.59 mm high for a cross-sectional area $S = 5.1 \times 10^{-6}$ m². The mean current is 45 mA. But since the beam is filled over a pulse time $\Delta t = 10^{-10}$ s, the maximum current I_m is given by

$$I_m = \frac{T}{\Delta t} I = 350 \text{ A} \tag{268}$$

where $T = 0.78$ μs is the orbital period. Substituting into **(250)** we obtain:

$$a = (ecS/I_m)^{1/3} = 0.9 \ \mu\text{m} \tag{269}$$

for the mean charge separation. The critical wavelength is given by:

$$\lambda = c\tau = \frac{c/\omega}{2\sqrt{3}\gamma^3} = 1.6 \times 10^{-11} \text{ m} \tag{270}$$

Thus the separation between electrons is large compared with the wavelength radiated, by nearly five orders of magnitude, and we can neglect interference.

SUMMARY

1. The force on a charge q moving with velocity $\mathbf{v}$ is given by:

$$\mathbf{F} = q(\mathbf{E} + \mathbf{v} \times \mathbf{B}) \tag{1}$$

where $\mathbf{E}$ and $\mathbf{B}$ are the electric and magnetic fields appropriate to the particular frame. The magnitude of the charge q is invariant.

2. The transformation properties of the electric and magnetic fields may be obtained from the Lorentz force and the transformation properties of forces.

3. Under a Lorentz transformation current density transforms like a displacement and charge density transforms like a time.

4. The first two Maxwell equations give the sources of flux and circulation of the electric field. The first is a generalization of Gauss's law:

$$\text{ME I} \qquad\qquad \mathbf{V} \cdot \mathbf{E} = \frac{1}{\epsilon_0}\, \rho \tag{29}$$

5. The second Maxwell equation is Faraday's law:

$$\text{ME II} \qquad\qquad \mathbf{V} \times \mathbf{E} = -\frac{\partial \mathbf{B}}{\partial t} \tag{31}$$

6. The third Maxwell equation expresses the sources of circulation of the magnetic field and may be obtained as a consequence of the Lorentz invariance of ME I:

$$\text{ME III} \qquad\qquad \mathbf{V} \times \mathbf{B} = \mu_0\!\left(\mathbf{j} + \epsilon_0 \frac{\partial \mathbf{E}}{\partial t}\right) \tag{38}$$

7. The fourth Maxwell equation expresses the sources of flux of the magnetic field and may be obtained as a consequence of the Lorentz invariance of ME II:

$$\text{ME IV} \qquad\qquad \mathbf{V} \cdot \mathbf{B} = 0 \tag{42}$$

8. To describe the behavior of fields near magnetic charge both ME II and ME IV must be modified. The source of flux of $\mathbf{B}$ is enclosed magnetic charge. There will be a contribution to the circulation of $\mathbf{E}$ from magnetic charge current, which modifies Faraday's law.

9. In the absence of magnetic charge it is still possible to express the magnetic field as the curl of a vector potential:

$$\mathbf{B}(\mathbf{r},t) = \mathbf{V} \times \mathbf{A}(\mathbf{r},t) \tag{172}$$

The electric field must be written as the sum of two terms:

$$\mathbf{E}(\mathbf{r},t) = -\mathbf{V}\phi(\mathbf{r},t) - \frac{\partial}{\partial t}\mathbf{A}(\mathbf{r},t) \tag{176}$$

10. The divergence of the vector potential is specified by the gauge, which gives an additional subsidiary condition. In the Lorentz gauge we write:

$$\mathbf{V} \cdot \mathbf{A}(\mathbf{r},t) = -\frac{1}{c^2}\frac{\partial}{\partial t}\phi(\mathbf{r},t) \tag{179}$$

In this gauge we obtain two wave equations:

$$\nabla^2\phi - \frac{1}{c^2}\frac{\partial^2\phi}{\partial t^2} = -\frac{1}{\epsilon_0}\rho \tag{180}$$

$$\nabla^2\mathbf{A} - \frac{1}{c^2}\frac{\partial^2\mathbf{A}}{\partial t^2} = -\mu_0\mathbf{j} \tag{181}$$

11. The energy flux in a medium of permittivity ϵ and permeability μ is given by the Poynting vector:

$$\mathbf{\Pi} = \mathbf{E} \times \mathbf{H} = \frac{1}{\mu}\mathbf{E} \times \mathbf{B} \tag{95}$$

12. The solutions of the wave equations for the scalar and vector potentials are the retarded potentials:

$$\phi(\mathbf{r},t) = \frac{1}{4\pi\epsilon_0}\int\frac{\rho(\mathbf{r}_1, t - r_{01}/c)}{r_{01}}dV_1 \tag{190}$$

$$\mathbf{A}(\mathbf{r},t) = \frac{\mu_0}{4\pi}\int\frac{\mathbf{j}(\mathbf{r}_1, t - r_{01}/c)}{r_{01}}dV_1 \tag{191}$$

13. The radiation fields from a bounded distribution of currents and charges may be expressed in terms of an expansion of the vector potential:

$$\mathbf{A}(\mathbf{r},t) = \mathbf{A}_{M0}(\mathbf{r},t) + \mathbf{A}_{M1}(\mathbf{r},t) + \mathbf{A}_{E2}(\mathbf{r},t) + \cdots \tag{200}$$

and of the scalar potential:

$$\phi(\mathbf{r},t) = \phi_{E0}(\mathbf{r},t) + \phi_{E1}(\mathbf{r},t) + \phi_{E2}(\mathbf{r},t) + \cdots \tag{220}$$

The first term in the scalar potential is just the potential of the total enclosed charge and gives no radiation. The second term together with $\mathbf{A}_{M0}$ leads to the electric dipole or magnetic monopole fields, where the two contributions are coupled by the equation of continuity. The total energy radiated across a sphere of radius r by an electric dipole is given by:

$$\frac{dU}{dt} = \frac{2}{3c^3}\frac{1}{4\pi\epsilon_0}\left(\frac{\partial^2[p]}{\partial t^2}\right)^2 \tag{237}$$

14. Electric and magnetic dipoles have their most intense radiation in a plane transverse to dipole axis. The radiation pattern from an electric quadrupole has nodes along the principal axes of the quadrupole tensor with maxima in the diagonal direction.

15. Intense radiation is obtained from charged particles moving on a curved path at velocities close to c. The radiation is transmitted in the forward direction with each charge radiating a Lorentz pulse of half width

$$\tau = \sqrt{3}/6\omega\gamma^3 \tag{251}$$

The pulse may also be regarded as a superposition of sinusoidal waves whose amplitudes fall off exponentially with frequency.

16. A steady current is the limiting case of a distribution of moving charges where the total distribution of charge and current is independent of the time. Beginning with the general expression for the electric and magnetic fields of an accelerating charge it is possible to obtain Coulomb's law and the Biot-Savart law for the electric and magnetic fields even though individual charges may be in accelerated motion.

PROBLEMS

1. **Displacement current.** The plates of a parallel plane capacitor are discs of radius R separated by a distance d. The volume between the plates is filled with an insulating medium of permittivity ϵ. Across the plates is connected a resistor and a chemical battery that provides an electromotive force $\mathscr{E}$. At $t = 0$ the battery is disconnected leaving the capacitor and resistor connected. Find the displacement current in the dielectric as a function of time. Find the magnetic field between the plates.

2. **Conducting dielectric.** The insulating dielectric described in Problem 1 is replaced by a conducting dielectric of real permittivity ϵ and conductivity σ. We dispense with the resistor. Find the potential across the plates as a function of time, the current through the dielectric, and the *magnetic field* between the plates.

3. **Electromagnetic induction.** An alternating voltage is applied across a pair of plane parallel discs of radius R and separation d. A Rogowski coil (see Problem 9-13) is placed between the plates and coaxial with them. Find the electromotive force induced in the coil. [See T. R. Carver and J. Rajhel, *Am. J. Phys.*, **42**, 246 (1974).] In a realization of this situation for a lecture demonstration, the radius of the plates was 0.4 m and the separation d was 0.1 m. The plates were driven at a frequency of 20 kHz with a peak voltage of 100 V. The Rogowski coil had a circumference of 2.38 m with 14,000 turns wound on a tube of radius 1.27 cm. What voltage should have been obtained? [See also A. G. Klein, *Am. J. Phys.*, **43**, 368 (1975).]

4. **Maxwell equations.** By decomposing the charge density into external and polarization charge:

$$\rho = \rho_{\text{ext}} - \nabla \cdot \mathbf{P}$$

and the current density into external, polarization, and magnetization current:

$$\mathbf{j} = \mathbf{j}_{ext} + \frac{\partial \mathbf{P}}{\partial t} + \nabla \times \mathbf{M}$$

write ME I and ME III in terms of $\mathbf{D}$ and $\mathbf{H}$. Show that the equations written in this form require the conservation of external charge.

5. **Induced current.** A circular ring of radius r and resistance R is placed so that its plane is normal to a uniform time-varying magnetic field $B(t)$.
   (a) Find the current induced in the ring.
   (b) Show that the electrostatic potential is zero.
   A new ring is constructed such that half the ring is of resistance R_1 and other half of resistance R_2.
   (c) Find the current in the new ring.
   (d) Find the electrostatic potential on the ring.

6. **Hertz potential.** Let the vector and scalar potentials be given in terms of the Hertz electric potential [E. A. Essex, *Am. J. Phys.* **45**, 1099 (1977)]:

$$\mathbf{A} = \frac{1}{c^2} \frac{\partial}{\partial t} \mathbf{\Pi}_e \qquad \psi = -\nabla \cdot \mathbf{\Pi}_e$$

   (a) Show that the Lorentz condition (**179**) is automatically satisfied.
   (b) By writing

$$\rho = -\nabla \cdot \mathbf{P} \qquad \mathbf{j} = \frac{\partial \mathbf{P}}{\partial t}$$

   which neglects magnetization current, obtain the differential equation:

$$\nabla^2 \mathbf{\Pi}_e - \frac{1}{c^2} \frac{\partial^2}{\partial t^2} \mathbf{\Pi}_e = -\frac{1}{\epsilon_0} \mathbf{P}(\mathbf{r},t)$$

   (c) Obtain the retarded solution for $\mathbf{\Pi}_e$.

7. **Debye potential.** A vector $\mathbf{Z}$ is defined by the relation

$$\mathbf{Z} = \phi(\mathbf{r},t)\hat{\mathbf{z}}$$

where the Debye potential $\phi(\mathbf{r},t)$ is a solution of the wave equation:

$$\nabla^2 \phi(\mathbf{r},t) = \frac{1}{c^2} \frac{\partial^2}{\partial t^2} \phi(\mathbf{r},t)$$

   (a) Show that

$$\mathbf{E} = \nabla \times (\nabla \times \mathbf{Z})$$

is a possible electric field in vacuum.

(b) Within a perfectly conducting spherical shell show that the Debye potential may be of the form:

$$\phi(\mathbf{r},t) = \frac{1}{r} \sin \frac{\omega r}{c} \sin \omega t$$

8. Complex fields.[14] An interesting way of showing the symmetry of the Maxwell equations is through the introduction of a complex field:

$$\mathbf{F}(\mathbf{r},t) = \mathbf{E}(\mathbf{r},t) + ic\mathbf{B}(\mathbf{r},t)$$

where $\mathbf{E}(\mathbf{r},t)$ and $\mathbf{B}(\mathbf{r},t)$ are real physical fields. We introduce a *complex* charge density:

$$R(\mathbf{r},t) = \rho(\mathbf{r},t) + \frac{i}{c} \bar{\rho}(\mathbf{r},t)$$

where the real part is electric charge and the imaginary part is magnetic charge. Similarly we may write a *complex* current density:

$$\mathbf{J}(\mathbf{r},t) = \mathbf{j}(\mathbf{r},t) + \frac{i}{c} \bar{\mathbf{j}}(\mathbf{r},t)$$

(a) By equating real and imaginary parts write **(32)** and **(57)** as a single conservation equation:

$$\nabla \cdot \mathbf{J}(\mathbf{r},t) + \frac{\partial}{\partial t} R(\mathbf{r},t) = 0$$

(b) Show that the first and fourth Maxwell equations **(29)** and **(55)** may be written as a single complex equation:

$$\nabla \cdot \mathbf{F}(\mathbf{r},t) = \frac{1}{\epsilon_0} R(\mathbf{r},t)$$

(c) Show that the second and third Maxwell equations may be similarly written as a single equation:

$$\nabla \times \mathbf{F} = i\mu_0 c \left[\mathbf{J}(\mathbf{r},t) + \epsilon_0 \frac{\partial}{\partial t} \mathbf{F}(\mathbf{r},t) \right]$$

[14] Problems 8, 9, and 10 on complex fields were suggested by D. Frieden.

9. **Lorentz transformation of complex fields.**[14] (a) For an infinitesimal Lorentz transformation, show that **(11)**, **(12)**, **(13)**, and **(14)** are equivalent to:

$$F' = F - \frac{i}{c} d\mathbf{V} \times F$$

(b) Show that a finite Lorentz transformation may be regarded as a rotation of $F(\mathbf{r},t)$ about $\hat{\mathbf{V}}$ through an angle θ with $\tan \theta = -iV/c$ which may be written as the transformation:

$$F' = F + \hat{\mathbf{V}} \times F \sin \theta + (1 - \cos \theta)\hat{\mathbf{V}} \times (\hat{\mathbf{V}} \times F)$$

See Exercise A-24.

10. **Complex field energy.**[14] Using the complex fields introduced in Problem 8 show:
(a) that the Poynting vector may be written as:

$$\Pi = \frac{i}{2\mu_0 c} F(\mathbf{r},t) \times F^*(\mathbf{r},t)$$

(b) and that the energy density may be written as

$$u = \tfrac{1}{2}\epsilon_0 F(\mathbf{r},t) \cdot F^*(\mathbf{r},t)$$

(c) Obtain the statement of conservation of energy:

$$\nabla \cdot \Pi + \frac{\partial u}{\partial t} + \mathrm{Re}\{F \cdot J^*\} = 0$$

11. **Electric dipole radiation.** Find the electric and magnetic fields of a harmonically oscillating charge.
(a) Show that the electric field is the sum of a zero frequency term which falls as $1/r^2$ and a term at frequency ω that falls as $1/r$.
(b) Show that there is *no* dipolar field that falls as $1/r^3$.
(c) Show that the magnetic field has a term that falls as $1/r^2$ at frequency ω and a term, also at frequency ω that falls as $1/r$.

12. **Radiation from a long wire.** A very long wire carries a current $I \cos \omega t$. By integrating **(223)** and **(234)** over an *infinite* wire show that the radiation fields fall off as $r^{-1/2}$. Show by means of a Poynting vector argument that this result is to be expected.

13. **Field energy.** Compute the energy density in the electromagnetic field for the oscillating charge of Problem 11. Identify terms which are at zero frequency, which are at frequency ω, and which are at frequency 2ω.

14. **Poynting vector.** Compute the Poynting vector of an oscillating electric dipole.
(a) Which terms provide for the change in stored energy and average to zero over a cycle?

(b) Which term gives the radiation field?

(c) Identify terms which fall as $1/r^3$ at ω and 2ω.

(d) Which terms fall as $1/r^2$ at zero frequency and at 2ω?

15. **Electromagnetic mass.** A spherical shell of radius R, with a charge q uniformly distributed over its surface, moves with velocity v much less than c.

(a) Obtain the relation between field momentum and field energy:

$$\mathbf{p}_{\text{field}} = \frac{4}{3c^2} U_{\text{field}} \mathbf{v} = \frac{\mu_0}{6\pi} \frac{q^2 \mathbf{v}}{R}$$

(b) If we define a field mass by the relation

$$\mathbf{p}_{\text{field}} = m_{\text{field}} \mathbf{v}$$

obtain for the electromagnetic field mass

$$m_{\text{field}} = \frac{4}{3c^2} U_{\text{field}}$$

16. **Image radiation.** A charge q moves with nearly uniform velocity $\mathbf{v}$ parallel to the plane of a diffraction grating. Assume that the grating depth is sinusoidal and that the charge moves transverse to the rulings.

(a) Obtain an expression for the radiated magnetic field.

(b) How does the radiated frequency vary with the direction to the field position?

(c) What is the source of the radiated energy? What is the mechanism for delivering this energy? [S. J. Smith and E. M. Purcell, *Phys. Rev.*, **92**, 1069 (1953).]

17. **Magnetic pressure.** A sphere of permeability μ and radius R is introduced into a magnetic field $\mathbf{B}_{\text{ext}}$. Show that the mean pressure within the sphere is given by:

$$P = \langle \hat{\mathbf{n}} \cdot \mathbf{i} \times \mathbf{B} \rangle_S = \langle (\hat{\mathbf{n}} \cdot \mathbf{M})(\hat{\mathbf{n}} \cdot \mathbf{B}) \rangle_S - \langle \mathbf{M} \cdot \mathbf{B} \rangle$$

where $\mathbf{B}$ is the mean field through the surface, and $\mathbf{i}$ is the surface current density. Compare with **(111)**.

18. **Field angular momentum.** A charge q moves slowly with velocity $\mathbf{v}$ past a magnetic moment $\boldsymbol{\mu}$ located at the origin.

(a) From the force on q in the field of a magnetic dipole, obtain the torque on q.

(b) From the magnetic field of q at the dipole obtain the torque on $\boldsymbol{\mu}$ by **(6-134)**.

(c) Compare your answers to (a) and (b). What does this result suggest for the field angular momentum of a charge moving past a magnetic moment?

19. **Helicity.** A circularly polarized electromagnetic wave is of the form

$$E_x = E(x,y)\cos(kz - \omega t)$$

$$E_y = \pm E(x,y)\sin(kz - \omega t)$$

where $E(x,y)$ is slowly varying.

(a) Find E_z and the components of the magnetic field.

(b) Find the energy U per unit length associated with the wave.

(c) Use **(148)** to obtain the angular momentum L per unit length of the wave and obtain the relation:

$$L = \pm U/\omega$$

20. **van der Waals interaction.** The interaction between a pair of polarizable fluctuating dipoles is modified when the fluctuation period is short compared with the time required for a light signal to pass between the molecules. Quantum electrodynamics gives in this limit for the energy of interaction:

$$U_{12} = -\frac{\alpha_1 \alpha_2}{(4\pi\epsilon_0 \, r_{12}{}^3)^2} \cdot \frac{23\hbar c}{4\pi r_{12}}$$

Compare this expression with the classical van der Waals interaction (Problem 4-7) and show that it is smaller by a factor of the order of $1/\omega\tau$ where ω is an atomic frequency and τ is the time for light to propagate between the molecules. [H. B. G. Casimir and D. Polder, "The Influence of Retardation on the London-van der Waals Forces," *Phys. Rev.*, **73**, 360 (1948).]

REFERENCES

Ashkin, A., "The Pressure of Laser Light," *Scientific American*, **226**(2), 62 (February 1972).

Henry, G. E., "Radiation Pressure," *Scientific American*, **196**(6), 99 (June 1957).

Rosser, W. G. V., *Classical Electromagnetism via Relativity*, Plenum, 1968.

Rowe, E. M. and J. H. Weaver, "The Uses of Synchrotron Radiation," *Scientific American*, **236**(6), 32 (June 1977).

Tricker, R. A. R., *The contributions of Faraday and Maxwell to Electrical Science*, Pergamon, 1966.

12 ELECTROMAGNETIC MEDIA I

We begin the discussion of the propagation of electromagnetic waves through material media by reviewing the Maxwell equations. The first equation (11-29) is the differential statement of Gauss's law:

ME I
$$\nabla \cdot \mathbf{E} = \frac{1}{\epsilon_0} \rho \tag{1}$$

where ρ is the *total* charge density, $\mathbf{E}$ is the microscopic local electric field and is defined from the force on a charge (11-1), and ϵ_0 is a constant of proportionality called the permittivity of vacuum and equal in the SI system of units to $1/\mu_0 c^2$ farads per metre.

The second Maxwell equation is the differential statement of Faraday's law (11-31):

ME II
$$\nabla \times \mathbf{E} = -\frac{\partial \mathbf{B}}{\partial t} \tag{2}$$

The third Maxwell equation is a generalization of Ampère's law (6-48) and follows from the Lorentz invariance of (1) as we saw in Section 11-8 (11-38):

ME III
$$\nabla \times \mathbf{B} = \mu_0\left(\mathbf{j} + \epsilon_0 \frac{\partial \mathbf{E}}{\partial t}\right) \tag{3}$$

Here $\mathbf{j}$ is the *total* current density, $\mathbf{B}$ is the local microscopic magnetic field as defined by the force on a moving charge (11-1), and μ_0 is a constant of proportionality called the permeability of vacuum and equal in the SI system to $4\pi/10^7$ henrys per metre.

Maxwell obtained (3) from a quite separate and ingenious line of argument. Using (B-9) we have:

$$\nabla \cdot (\nabla \times \mathbf{B} - \mu_0 \mathbf{j}) = -\mu_0 \nabla \cdot \mathbf{j} \tag{4}$$

435

From conservation of charge **(11-32)** and **(1)** we have:

$$\mathbf{V} \cdot \mathbf{j} = -\frac{\partial \rho}{\partial t} = -\mathbf{V} \cdot \left(\epsilon_0 \frac{\partial \mathbf{E}}{\partial t} \right) \tag{5}$$

which on substitution into **(4)** yields

$$\mathbf{V} \cdot \left[\mathbf{V} \times \mathbf{B} - \mu_0 \left(\mathbf{j} + \mu_0 \frac{\partial \mathbf{E}}{\partial t} \right) \right] = 0 \tag{6}$$

Maxwell's contribution was to take the bracketed quantity in **(6)** to be identically equal to zero, which is then **(3)**. Put in a slightly different way, if we *assume* that the curl of the bracketed quantity is everywhere equal to zero, then by Helmholtz's theorem (Section B-15), we are led to **(3)**.

Finally, as we saw in Section 11-8 the invariance of **(2)** under a Lorentz transformation requires that there be no magnetic charge

ME IV $$\mathbf{V} \cdot \mathbf{B} = 0 \tag{7}$$

This means, of course, that should magnetic charge be established, Faraday's law would have to be modified together with **(7)** becoming the magnetic equivalent of **(1)** as discussed in Section 11-10.

In describing the propagation of fields in material media we find it convenient to work with *macroscopic* fields which, as we have seen, are space averages of the local microscopic fields. We introduce the macroscopic displacement field through the relation **(4-1)**:

$$\mathbf{D} = \epsilon_0 \mathbf{E} + \mathbf{P} \tag{8}$$

where **P** is the macroscopic polarization **(1-126)**:

$$\mathbf{P} = \frac{1}{V} \int_V (\rho - \rho_{\text{ext}}) \mathbf{r} \, dV \tag{9}$$

where ρ_{ext} is what we have called the *external* charge density and is not to be included in the polarization. The electric field **E** as used in **(8)** is also a macroscopic quantity. Taking the divergence of **(9)** leads to **(4-6)**:

$$\rho = \rho_{\text{ext}} - \mathbf{V} \cdot \mathbf{P} \tag{10}$$

In a similar way we introduce the macroscopic field **H** **(10-1)**:

$$\mathbf{H} = \frac{\mathbf{B}}{\mu_0} - \mathbf{M} \tag{11}$$

where $\mathbf{M}$ is the macroscopic magnetization density **(6-143)**:

$$\mathbf{M} = \frac{1}{2V} \int_V \mathbf{r} \times \left(\mathbf{j} - \mathbf{j}_{ext} - \frac{\partial \mathbf{P}}{\partial t} \right) dV \tag{12}$$

where $\mathbf{j}_{ext}$ is what we have called the *external* current and $\partial \mathbf{P}/\partial t$ is the polarization current. Taking the *curl* of **(12)** we obtain **(11-52)**:

$$\mathbf{j} = \mathbf{j}_{ext} + \frac{\partial \mathbf{P}}{\partial t} + \nabla \times \mathbf{M} \tag{13}$$

Now, taking the divergence of **(8)** and using **(1)** and **(10)** we obtain:

$$\nabla \cdot \mathbf{D} = \epsilon_0 \nabla \cdot \mathbf{E} + \nabla \cdot \mathbf{P} = \rho + (\rho_{ext} - \rho) = \rho_{ext} \tag{14}$$

which is to say that the divergence of the displacement is the *external* charge density.[1]
Taking the curl of **(11)** we obtain using **(3)** and **(13)** and **(8)**:

$$\nabla \times \mathbf{H} = \frac{1}{\mu_0} \nabla \times \mathbf{B} - \nabla \times \mathbf{M} = \mathbf{j} + \epsilon_0 \frac{\partial \mathbf{E}}{\partial t} - \left(\mathbf{j} - \mathbf{j}_{ext} - \frac{\partial \mathbf{P}}{\partial t} \right) = \mathbf{j}_{ext} + \frac{\partial \mathbf{D}}{\partial t} \tag{15}$$

which is close to the form in which Maxwell wrote the generalization of Ampère's law **(11-46)**.

In discussing the propagation of fields in material media, we use **(14)** and **(15)** in place of **(1)** and **(3)** so that our field equations become:

$$\nabla \cdot \mathbf{D} = \rho_{ext} \tag{16}$$

$$\nabla \times \mathbf{E} = -\frac{\partial \mathbf{B}}{\partial t} \tag{17}$$

$$\nabla \times \mathbf{H} = \mathbf{j}_{ext} + \frac{\partial \mathbf{D}}{\partial t} \tag{18}$$

$$\nabla \cdot \mathbf{B} = 0 \tag{19}$$

To solve these equations we shall require, in addition, statements of the properties of the medium, which may be expressed in terms of the relation between $\mathbf{D}$ and $\mathbf{E}$ and the relation between $\mathbf{H}$ and $\mathbf{B}$.

---

[1] Most authors call our ρ_{ext} the free charge, regarding the charge in a conductor as free and the charge in an insulator as bound. We have tried to avoid this distinction, which has no fundamental basis in modern solid state theory. Our ρ_{ext} and $\mathbf{j}_{ext}$ are *external* in the sense that they are applied by external agents, although they need not be physically external to the medium being investigated.

Exercise

1. Show that the field equations in material media imply local conservation of *external* charge **(11-48)**:

$$\mathbf{V} \cdot \mathbf{j}_{ext} + \frac{\partial}{\partial t} \rho_{ext} = 0$$

PROPAGATION IN LINEAR MEDIA

In discussing the propagation in linear media we will for the present limit ourselves to magnetically isotropic systems. Then **(11)** may be written **(10-8)** as:

$$\mathbf{H} = \frac{1}{\mu_0}(1 - \chi)\mathbf{B} = \frac{1}{\mu}\mathbf{B} \tag{20}$$

and we may concentrate on the relationship between **D** and **E**.

1 **Equations of Motion.** If we have some way of relating **P** to the macroscopic field **E**, then through **(8)** we have the required relation between **D** and **E**. In our discussion of dielectric insulators we could have written equations of motion for the several contributions to the molecular polarization **(3-63)**. Similarly, in our discussion of plasmas, we obtained an equation **(5-132)** that related the current and charge densities to the electric field. One approach, then, is to augment **(16)**, **(17)**, **(18)**, and **(19)** by equations of motion describing the behavior of the polarization in a general applied field. The limitation of such an approach is that, as we have seen, an equation of motion may be obtained only for model systems where we have made drastic simplifications. The more usual situation is that we wish a description of propagation in a real system for which we have no complete theory and wish only a semiempirical scheme for relating the propagation to other properties of the system.

2 **Dielectric Function.** A semiempirical approach, which is more generally useful, is to relate the displacement to the electric field by the dielectric function. The only requirement on the system is that of linearity. Let us *assume* an electric field of the form:

$$\mathbf{E}_1(\mathbf{r},t) = \mathbf{E}e^{-\mathbf{k}'' \cdot \mathbf{r}} \cos(\mathbf{k}' \cdot \mathbf{r} - \omega t) \tag{21}$$

where **E** is a constant electric field at some general orientation with respect to the propagation vectors **k'** and **k''**. The induced polarization may be written as:

$$\mathbf{P}_1(\mathbf{r},t) = \epsilon_0 \overline{\overline{\chi}}'(\omega,\mathbf{k}',\mathbf{k}'') \cdot \mathbf{E}e^{-\mathbf{k}'' \cdot \mathbf{r}} \cos(\mathbf{k}' \cdot \mathbf{r} - \omega t)$$
$$- \epsilon_0 \overline{\overline{\chi}}''(\omega,\mathbf{k}',\mathbf{k}'') \cdot \mathbf{E}e^{-\mathbf{k}'' \cdot \mathbf{r}} \sin(\mathbf{k}' \cdot \mathbf{r} - \omega t) \tag{22}$$

where $\overline{\chi}'$ and $\overline{\chi}''$ are tensors that give the in-phase and out-of-phase components of the polarization.

We may considerably simplify the description of the medium by introducing a complex susceptibility tensor:

$$\underline{\overline{\chi}} = \overline{\chi}' + i\overline{\chi}'' \tag{23}$$

and a complex propagation vector:

▶ $$\underline{k} = k' + ik'' \tag{24}$$

The technique that we use is similar to that employed in Section 5-7 where we introduced the complex conductivity. We define a second field $E_2(r,t)$, which is shifted in time with respect to (21):

$$E_2(r,t) = Ee^{-k''\cdot r}\sin(k'\cdot r - \omega t) \tag{25}$$

To be consistent with (22) the induced polarization must be given by

$$P_2(r,t) = \epsilon_0\,\overline{\chi}'(\omega,k',k'')\cdot Ee^{-k''\cdot r}\sin(k'\cdot r - \omega t)$$
$$+ \epsilon_0\,\overline{\chi}''(\omega,k',k'')\cdot Ee^{-k''\cdot r}\cos(k'\cdot r - \omega t) \tag{26}$$

Now, in a linear medium any linear combination of (21) and (25) must produce the same linear combination of (22) and (26). We introduce the *complex* field:

▶ $$\underline{E}(r,t) = E_1(r,t) + iE_2(r,t) = \underline{E}e^{i(\underline{k}\cdot r - \omega t)} \tag{27}$$

where $\underline{k}$ is given by (24). Although (27) is not a physical field, nonetheless linearity requires a corresponding polarization:

$$\underline{P}(r,t) = P_1(r,t) + iP_2(r,t) = \epsilon_0\,\underline{\overline{\chi}}(\omega,\underline{k})\cdot \underline{E}e^{i(\underline{k}\cdot r - \omega t)} \tag{28}$$

where $\underline{\overline{\chi}}(\omega,\underline{k})$ is given by (23). Corresponding to $E(r,t)$ and $P(r,t)$ we have a displacement:

▶ $$\underline{D}(r,t) = \epsilon_0\,\underline{E}(r,t) + \underline{P}(r,t) = \underline{\overline{\epsilon}}(\omega,\underline{k})\cdot \underline{E}(r,t) \tag{29}$$

where $\underline{\overline{\epsilon}}(\omega,\underline{k})$ is the complex permittivity tensor and is given by:

$$\underline{\overline{\epsilon}}(\omega,\underline{k}) = \epsilon_0[I + \underline{\overline{\chi}}(\omega,\underline{k})] \tag{30}$$

Although we will work with complex electric and magnetic fields we must understand that we are carrying the complex combination of two real fields shifted in phase by $\pi/2$ and that the application of a *real* electric field (21) must yield a displacement which is the real part of (29).

We now use **(20)** and **(29)** to rewrite **(17)** and **(18)** as:

$$\nabla \times \underline{\mathbf{E}} = -\mu \frac{\partial \underline{\mathbf{H}}}{\partial t} \tag{31}$$

$$\nabla \times \underline{\mathbf{H}} = \underline{\mathbf{j}}_{\text{ext}} + \bar{\bar{\epsilon}} \cdot \frac{\partial \underline{\mathbf{E}}}{\partial t} \tag{32}$$

Taking the *curl* of **(31)** and using **(B-10)** we obtain:

$$\nabla \times (\nabla \times \underline{\mathbf{E}}) = \nabla(\nabla \cdot \underline{\mathbf{E}}) - \nabla^2 \underline{\mathbf{E}} = -\mu \frac{\partial}{\partial t} \nabla \times \underline{\mathbf{H}} \tag{33}$$

Substituting from **(32)** and regrouping terms, we have, as the wave equation for the electric field,

$$\nabla^2 \underline{\mathbf{E}} - \nabla(\nabla \cdot \underline{\mathbf{E}}) - \mu \bar{\bar{\epsilon}} \cdot \frac{\partial^2 \underline{\mathbf{E}}}{\partial t^2} = \mu \frac{\partial \underline{\mathbf{j}}_{\text{ext}}}{\partial t} \tag{34}$$

where $\underline{\mathbf{j}}_{\text{ext}}$ is a complex periodic current of the form:

$$\underline{\mathbf{j}}_{\text{ext}}(\mathbf{r},t) = \underline{\mathbf{j}}_{\text{ext}}\, e^{i(\underline{\mathbf{k}} \cdot \mathbf{r} - \omega t)} \tag{35}$$

We take the magnetic field **H** to be of the form:

$$\underline{\mathbf{H}}(\mathbf{r},t) = \underline{\mathbf{H}} e^{i(\underline{\mathbf{k}} \cdot \mathbf{r} - \omega t)} \tag{36}$$

where we allow $\underline{\mathbf{H}}$ as well as $\underline{\mathbf{E}}$ in **(27)** to contain a complex phase factor.

Substituting **(27)** and **(35)** into **(34)** we obtain for the wave equation:

$$(\underline{k}^2 \mathbf{I} - \underline{\mathbf{k}}\underline{\mathbf{k}} - \omega^2 \mu \bar{\bar{\epsilon}}) \cdot \underline{\mathbf{E}} = i\omega \mu \underline{\mathbf{j}}_{\text{ext}} \tag{37}$$

Similarly, **(19)** and **(31)** become

$$\underline{\mathbf{k}} \cdot \underline{\mathbf{H}} = 0 \tag{38}$$

$$\underline{\mathbf{H}} = \frac{1}{\omega \mu} \underline{\mathbf{k}} \times \underline{\mathbf{E}} \tag{39}$$

where

$$\underline{\rho}_{\text{ext}} = \frac{1}{\omega} \underline{\mathbf{k}} \cdot \underline{\mathbf{j}}_{\text{ext}} \tag{40}$$

is established by the conservation of external charge. Equations **37**, **38**, **39** and **40** give expressions for the complex fields $\underline{\mathbf{E}}$ and $\underline{\mathbf{H}}$ in a medium characterized by a complex permittivity $\underline{\epsilon}(\omega,\mathbf{k})$ and in the presence of charges and currents ρ_{ext} and $\mathbf{j}_{ext}$. Although **(39)** *appears* to give $\underline{\mathbf{H}}$ normal to both $\underline{\mathbf{E}}$ and $\underline{\mathbf{k}}$, this will not be true of the *real* fields $\mathbf{H}$ and $\mathbf{E}$ unless $\mathbf{k}'$ and $\mathbf{k}''$ are parallel.

3 Poynting Vector. In Section 11-15 we introduced the Poynting vector **(11-95)**:

$$\mathbf{\Pi} = \mathbf{E} \times \mathbf{H} \tag{41}$$

as giving the flux of energy in nonabsorbing media. We extend the argument here to absorbing media. Taking the scalar product of $\mathbf{H}$ with **(17)** we obtain

$$\mathbf{H} \cdot \nabla \times \mathbf{E} = -\mathbf{H} \cdot \frac{\partial \mathbf{B}}{\partial t} \tag{42}$$

and taking the scalar product of $\mathbf{E}$ with **(18)** we have

$$\mathbf{E} \cdot \nabla \times \mathbf{H} = \mathbf{E} \cdot \mathbf{j}_{ext} + \mathbf{E} \cdot \frac{\partial \mathbf{D}}{\partial t} \tag{43}$$

Subtracting **(42)** from **(43)** and using **(B-4)** we write:

$$-\nabla \cdot (\mathbf{E} \times \mathbf{H}) = \mathbf{H} \cdot \frac{\partial \mathbf{B}}{\partial t} + \mathbf{E} \cdot \frac{\partial \mathbf{D}}{\partial t} + \mathbf{E} \cdot \mathbf{j}_{ext} \tag{44}$$

where the fields in **(42)**, **(43)**, and **(44)** are all real. The left side is interpreted as the inward flow of energy per unit volume. The first term on the right is the rate at which energy enters the magnetic field *plus* the rate at which energy is dissipated via the magnetic field. The second term is similarly the rate at which energy enters the electric field *plus* the rate at which energy is dissipated via the electric field. The third term is the rate at which work is done by the electric field on external currents. Thus, even with dissipation **(41)** continues to describe the energy flow where $\mathbf{E}$ and $\mathbf{H}$ are real fields.

It is also possible to write the *average* Poynting vector in terms of the complex fields **(27)** and **(36)** in the following way. We have from **(41)**:

$$\langle \mathbf{\Pi} \rangle = \langle \mathbf{E}_1(\mathbf{r},t) \times \mathbf{H}_1(\mathbf{r},t) \rangle = \langle \mathbf{E}_2(\mathbf{r},t) \times \mathbf{H}_2(\mathbf{r},t) \rangle \tag{45}$$

where the average is over a period $T = 2\pi/\omega$. Combining terms we may write:

$$\langle \mathbf{\Pi} \rangle = \tfrac{1}{2}\langle \mathbf{E}_1(\mathbf{r},t) \times \mathbf{H}_1(\mathbf{r},t) + \mathbf{E}_2(\mathbf{r},t) \times \mathbf{H}_2(\mathbf{r},t) \rangle \tag{46}$$

By substitution, this is equivalent to

$$\blacktriangleright \qquad \langle \mathbf{\Pi} \rangle = \tfrac{1}{2} \operatorname{Re} \langle \underline{\mathbf{E}}(\mathbf{r},t) \times \underline{\mathbf{H}}^*(\mathbf{r},t) \rangle \tag{47}$$

where $\underline{\mathbf{H}}^*$ is the complex conjugate of $\underline{\mathbf{H}}$ and the symbol Re means the real part.

Exercise

---

2. A propagating electric field is of the form

$$\mathbf{E}(\mathbf{r},t) = \mathbf{E}e^{i(\mathbf{k}\cdot\mathbf{r}-\omega t)}$$

Obtain explicitly the relations

$$\nabla \cdot \mathbf{E}(\mathbf{r},t) = i\mathbf{k}\cdot\mathbf{E}(\mathbf{r},t) \qquad \nabla \times \mathbf{E}(\mathbf{r},t) = i\mathbf{k}\times\mathbf{E}(\mathbf{r},t)$$

---

4 Energy Loss. We may examine more closely the expressions for energy loss via electric and magnetic fields. We discuss here only the electric field problem. A very similar discussion may be constructed for magnetic fields. As observed following **(44)**, the rate at which energy enters the electric field plus the rate at which energy is dissipated is given by

$$\frac{dw}{dt} = \mathbf{E}\cdot\frac{d\mathbf{D}}{dt} \tag{48}$$

Averaging over some integral number of cycles we are left with the dissipative part only:

$$\left\langle\frac{dw}{dt}\right\rangle = \left\langle\mathbf{E}_1(\mathbf{r},t)\cdot\frac{d}{dt}\mathbf{D}_1(\mathbf{r},t)\right\rangle = \left\langle\mathbf{E}_2(\mathbf{r},t)\cdot\frac{d}{dt}\mathbf{D}_2(\mathbf{r},t)\right\rangle \tag{49}$$

We may write **(49)** in terms of complex fields:

$$\frac{dw}{dt} = \tfrac{1}{2}\,\mathrm{Re}\left\langle\underline{\mathbf{E}}(\mathbf{r},t)\cdot\frac{d}{dt}\underline{\mathbf{D}}^*(\mathbf{r},t)\right\rangle = -\tfrac{1}{2}\omega\,\mathrm{Im}\langle\underline{\mathbf{E}}(\mathbf{r},t)\cdot\underline{\mathbf{D}}^*(\mathbf{r},t)\rangle \tag{50}$$

Writing

$$\underline{\mathbf{D}}(\mathbf{r},t) = (\bar{\boldsymbol{\epsilon}}' + i\bar{\boldsymbol{\epsilon}}'')\cdot\underline{\mathbf{E}}(\mathbf{r},t) \tag{51}$$

we obtain for the rate of dissipation:

$$\blacktriangleright \qquad\qquad \frac{dw}{dt} = \tfrac{1}{2}\omega\underline{\mathbf{E}}\cdot\bar{\boldsymbol{\epsilon}}''\cdot\underline{\mathbf{E}}^* \tag{52}$$

The imaginary part of the complex permittivity gives the out-of-phase component of the displacement. Another way of representing the phase difference between **D** and **E** and its relation to energy loss may be shown graphically from **(48)**. The energy dissipated over one cycle may be written from **(48)** as

$$\Delta w = \int_0^T \frac{dw}{dt}\,dt = \int_0^T \mathbf{E}\cdot\frac{d\mathbf{D}}{dt}\,dt = \oint \mathbf{E}\cdot d\mathbf{D} \tag{53}$$

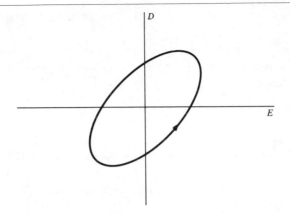

Figure 1 Contour in the *E-D* plane. The curve is an ellipse and the energy dissipated over a cycle is just the area of the ellipse.

Now if **E** and **D** are sinusoidally varying, the integral given in **(53)** may be written over a contour in the *E-D* plane as shown schematically in Figure 1. The curve is an ellipse and the energy dissipated over a cycle is just the area of the ellipse.

ISOTROPIC MEDIA

In an isotropic medium **P** will be parallel to **E** so that we may write in place of **(29)**:

$$\underline{\mathbf{D}}(\mathbf{r},t) = \underline{\epsilon}(\omega,\underline{\mathbf{k}})\underline{\mathbf{E}}(\mathbf{r},t) \tag{54}$$

that is, the permittivity tensor is a multiple of the identity tensor:

$$\underline{\underline{\boldsymbol{\epsilon}}}(\omega,\underline{\mathbf{k}}) = \underline{\epsilon}(\omega,\underline{\mathbf{k}})\mathbf{I} \tag{55}$$

Then **(16)** becomes

$$\boldsymbol{\nabla} \cdot \mathbf{E} = i\underline{\mathbf{k}} \cdot \underline{\mathbf{E}} = -\frac{1}{\underline{\epsilon}}\,\rho_{\text{ext}} = \frac{1}{\omega\underline{\epsilon}}\,\underline{\mathbf{k}} \cdot \mathbf{j}_{\text{ext}} \tag{56}$$

and **(34)** takes the form

$$\nabla^2\mathbf{E} - \mu\underline{\epsilon}\,\frac{\partial^2\mathbf{E}}{\partial t^2} = -(\underline{k}^2 - \omega^2\mu\underline{\epsilon})\mathbf{E} = -i(\omega\mu\mathbf{j}_{\text{ext}} - \underline{\mathbf{k}}\rho_{\text{ext}}/\underline{\epsilon}) = \frac{i}{\omega\epsilon}(\underline{\mathbf{k}}\underline{\mathbf{k}} - \omega^2\mu\underline{\epsilon}\mathbf{I}) \cdot \mathbf{j}_{\text{ext}} \tag{57}$$

5 Free Propagation. Here we are primarily concerned with situations in which an electromagnetic wave propagates through a passive medium, that is, there are no mechanisms within the medium for feeding energy into the electromagnetic field. This means that we must take $\mathbf{j}_{\text{ext}}$ and ρ_{ext} equal to zero. In addition, this requirement imposes some restrictions on the form of the complex permittivity as discussed in Appendix J.

As long as the permittivity $\underline{\epsilon}(\omega,\mathbf{k})$ is not equal to zero, the right side of (57) vanishes and the condition that there be a nonzero propagating electric field is that the coefficient of **E** must be equal to zero:

$$\underline{k}^2 = \underline{\mathbf{k}} \cdot \underline{\mathbf{k}} = \omega^2 \mu \underline{\epsilon} \tag{58}$$

where for simplicity we take $\mathbf{k}'$ and $\mathbf{k}''$ to be parallel. From (56) **E** must then be perpendicular to **k**. From (39) **H** is similarly perpendicular to **k** as shown in Figure 2. The ratio of the electric field to the magnetic field is called the intrinsic impedance and is given from (39) and (58) by

$$\blacktriangleright \qquad \underline{Z} = Z' + iZ'' = \frac{E_x}{\underline{H}_y} = -\frac{E_y}{\underline{H}_x} = \frac{\omega\mu}{\underline{k}} = \left(\frac{\mu}{\underline{\epsilon}}\right)^{1/2} \tag{59}$$

In general, there will be dielectric relaxation and the permittivity will be complex, implying through (59) that **H** and **E** will be out of phase.

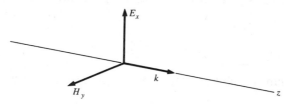

Figure 2   Vector relations in an isotropic medium. The electric and magnetic fields and the wavevector are mutually perpendicular.

It is possible, as we will see in Section 13-4, to excite a wave such that $\mathbf{k}'$ and $\mathbf{k}''$ are not parallel. This is true even for an isotropic medium. Under these circumstances either or both **E** and **H** will have a longitudinal component. In general, **E**, **H**, and $\mathbf{k}'$ need not be mutually orthogonal.

We now use (58) to obtain an expression for the attenuation in the medium. We assume that $\mathbf{k}'$ and $\mathbf{k}''$, which are the real and imaginary components of **k** (24), are parallel. Then substituting into (58) and writing for the complex permittivity:

$$\underline{\epsilon} = \epsilon' + i\epsilon'' \tag{60}$$

we obtain:

$$k'^2 - k''^2 = \omega^2 \mu \epsilon' \tag{61}$$

$$2k'k'' = \omega^2 \mu \epsilon'' \tag{62}$$

Solving (61) and (62) we obtain:

$$k'^2 = \tfrac{1}{2}\omega^2 \mu [\epsilon' + (\epsilon'^2 + \epsilon''^2)^{1/2}] \tag{63}$$

$$k''^2 = \tfrac{1}{2}\omega^2 \mu [-\epsilon' + (\epsilon'^2 + \epsilon''^2)^{1/2}] \tag{64}$$

Exercises

3. A plane elliptically polarized wave is generated by the superposition of two perpendicular electric fields shifted in phase by some angle ϕ:

$$\underline{E}_x(z,t) = E_x\,e^{i(kz - \omega t)} \qquad \underline{E}_y(z,t) = E_y\,e^{i(kz - \omega t + \phi)}$$

where E_x and E_y may be taken to be real. Obtain the components of the magnetic field **H**. Show that the condition for the *real* electric and magnetic fields to be orthogonal is either $\phi = 0$ (linear polarization) or Z as given by **(59)** real.

4. An electromagnetic field penetrates into a lossless medium. Show that either k' or k'' must be zero. Describe the spatial variation in field for the two cases. How does the field vary when ϵ'' is much greater than ϵ' as for a plasma at frequencies for which $\omega\tau \ll 1$?

5. Show that **(59)** is equivalent to the vector relation:

$$\mathbf{H} = \frac{1}{Z}\,\hat{\mathbf{k}} \times \mathbf{E}$$

---

6 **Phase Velocity.** We write for the real part of a wave propagating along $\hat{z}$:

$$E_1(z,t) = Ee^{-k''z}\cos(k'z - \omega t) \tag{65}$$

Alternatively, we may write **(65)** as:

$$\mathbf{E}_1(z,t) = Ee^{-k''z}\cos k'(z - v_p t) \tag{66}$$

where v_p is called the phase velocity and is the velocity with which a section of the wave with fixed phase propagates. Comparing **(65)** and **(66)** we have for the phase velocity:

▶
$$v_p = \frac{\omega}{k'} \tag{67}$$

7 **Energy Density and Flow.** Because the energy density and flow are quadratic in the electric and magnetic fields, we reduce the complex fields **(27)** and **(36)** to one of the solutions **(21)** or **(25)** and the corresponding solution for the magnetic field. [It is possible to use **(47)** and similar expressions for the energy density as well.]

We begin by writing **(27)** in the form:

$$\mathbf{E} = \hat{x}E_x\,e^{i(kz - \omega t)} + \hat{y}E_y\,e^{i(kz - \omega t + \phi)} \tag{68}$$

where we choose both E_x and E_y to be real and introduce ϕ explicitly in order to represent the difference in phase between the components of the field along $\hat{x}$ and $\hat{y}$.

From (39) we obtain for **H**:

$$\mathbf{H} = -\hat{\mathbf{x}} \frac{k}{\omega\mu} E_y e^{i(kz - \omega t + \phi)} + \hat{\mathbf{y}} \frac{k}{\omega\mu} E_x e^{i(kz - \omega t)} \tag{69}$$

Taking just the real parts of (68) and (69) we have:

$$\mathbf{E}_1 = \hat{\mathbf{x}} E_x e^{-k''z} \cos(k'z - \omega t) + \hat{\mathbf{y}} E_y e^{-k''z} \cos(k'z - \omega t + \phi) \tag{70}$$

$$\mathbf{H}_1 = -\frac{1}{\omega\mu} \hat{\mathbf{x}} E_y e^{-k''z}[k' \cos(k'z - \omega t + \phi) - k'' \sin(k'z - \omega t + \phi)]$$

$$+ \frac{1}{\omega\mu} \hat{\mathbf{y}} E_x e^{-k''z}[k' \cos(k'z - \omega t) - k'' \sin(k'z - \omega t)] \tag{71}$$

We write the energy density in the electric field from (11-68) as

$$u_{\text{elec}} = \tfrac{1}{2}\mathbf{E} \cdot \mathbf{D} \tag{72}$$

and the energy stored in the magnetic field from (11-70) as:

$$u_{\text{mag}} = \tfrac{1}{2}\mathbf{B} \cdot \mathbf{H} \tag{73}$$

We write for the displacement the real part of (54):

$$\mathbf{D}_1 = \hat{\mathbf{x}} E_x e^{-k''z}[\epsilon' \cos(k'z - \omega t) - \epsilon'' \sin(k'z - \omega t)]$$

$$+ \hat{\mathbf{y}} E_y e^{-k''z}[\epsilon' \cos(k'z - \omega t + \phi) - \epsilon'' \sin(k'z - \omega t + \phi)] \tag{74}$$

Substituting (70) and (74) into (72) and averaging over one cycle we obtain:

$$\langle u_{\text{elec}} \rangle = \frac{\epsilon'}{4} (E_x{}^2 + E_y{}^2)e^{-2k''z} \tag{75}$$

Similarly using (71) and (20) we obtain for the average energy density in the magnetic field:

$$\langle u_{\text{mag}} \rangle = \tfrac{1}{4}(\epsilon'^2 + \epsilon''^2)^{1/2}(E_x{}^2 + E_y{}^2)e^{-2k''z} \tag{76}$$

Adding (75) and (76) and using (63) we write for the total average energy density:

$$\langle u \rangle = \frac{1}{2} \frac{k'^2}{\omega^2\mu} (E_x{}^2 + E_y{}^2)e^{-2k''z} \tag{77}$$

It is instructive to verify the time average of the quantities given in (44) assuming free propagation with $\mathbf{j}_{\text{ext}} = 0$. From (20) the first term on the right of (44) must average to zero over a cycle, that is, there is no energy dissipation in the magnetic field. From (70) and (74) we obtain for the time average of the second term:

$$\left\langle \mathbf{E} \cdot \frac{\partial \mathbf{D}}{\partial t} \right\rangle = \tfrac{1}{2}\omega\epsilon'' e^{-2k''z}(E_x^2 + E_y^2) \tag{78}$$

From (70) and (71) we obtain for the Poynting vector:

$$\mathbf{\Pi} = \langle \mathbf{E} \times \mathbf{H} \rangle = \frac{1}{2}\frac{k'}{\omega\mu} e^{-2k''z}(E_x^2 + E_y^2)\hat{\mathbf{z}} \tag{79}$$

Taking the divergence of (79) we obtain:

$$-\mathbf{\nabla} \cdot \langle \mathbf{E} \times \mathbf{H} \rangle = \frac{k'k''}{\omega\mu} e^{-2k''z}(E_x^2 + E_y^2) \tag{80}$$

From (62) we see that (78) and (80) are equivalent.

8 Group Velocity.

To signal from one position to another via electromagnetic waves we must vary either the amplitude (or what is equivalent, the polarization) or the phase (or frequency) of the wave. Ideally we would like to transmit a very short pulse. The problem with the analysis of such a situation is that the shape of the pulse will change as the pulse propagates through the medium and it may not be clear how the velocity is to be defined.

This problem is avoided if we signal by modulating the electric field with a limited number of sinusoidal signals. The simplest case is to take for the electric field at $z = 0$:

$$\mathbf{E}_1 = \hat{\mathbf{x}} E_x \cos \omega_m t \cos \omega t \tag{81}$$

where ω_m is some modulation frequency as shown in Figure 3a. Now, we may rewrite (81) as the sum of two sinusoidal signals:

$$\mathbf{E}_1 = \tfrac{1}{2}\hat{\mathbf{x}} E_x [\cos(\omega + \omega_m)t + \cos(\omega - \omega_m)t] \tag{82}$$

as shown in Figures 3b and c. In a linear medium the signals propagate independently, and we expect

$$\mathbf{E}_1(z,t) = \tfrac{1}{2}\hat{\mathbf{x}} E_x [\cos(k_1 z - \omega_1 t) + \cos(k_2 z - \omega_2 t)] \tag{83}$$

where the frequencies are given by

$$\omega_1 = \omega + \omega_m \tag{84}$$

$$\omega_2 = \omega - \omega_m \tag{85}$$

and k_1 and k_2, which for simplicity we assume to be real[2], are given by (58).

[2] For a discussion of propagation in lossy media, see D. Rader, *Am. J. Phys.* **41**, 420 (1973); **44**, 1005 (1976) and R. A. Fisher, *Am. J. Phys.* **44**, 1002 (1976). Also see S. C. Bloch, *Am. J. Phys.* **45**, 538 (1977).

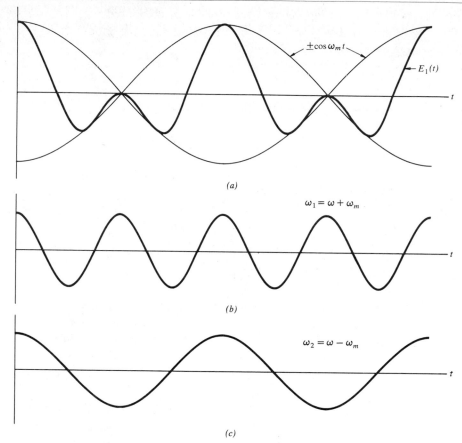

Figure 3. Electric field modulated by a single sinusoid. (*a*) The electric field $E_1(t)$ as given by **(81)**. The modulating function $\cos \omega_m t$ is the envelope. (*b*) The upper sideband is at the sum of the carrier and modulation frequencies. (*c*) The lower sideband is at the difference of the carrier and modulation frequencies.

Corresponding to **(83)** we have a magnetic field:

$$\mathbf{H}_1(z,t) = \tfrac{1}{2}\hat{\mathbf{y}}E_x\left[\frac{k_1}{\omega_1 \mu}\cos(k_1 z - \omega_1 t) + \frac{k_2}{\omega_2 \mu}\cos(k_2 z - \omega_2 t)\right] \qquad \textbf{(86)}$$

Computing the Poynting vector, we obtain

$$\boldsymbol{\Pi} = \mathbf{E}_1 \times \mathbf{H}_1 = \frac{1}{4\mu}\hat{\mathbf{z}}E_x{}^2\left[\frac{k_1}{\omega_1}\cos^2(k_1 z - \omega_1 t) + \frac{k_2}{\omega_2}\cos^2(k_2 z - \omega_2 t)\right.$$

$$\left. + \left(\frac{k_1}{\omega_1} + \frac{k_2}{\omega_2}\right)\cos(k_1 z - \omega_1 t)\cos(k_2 z - \omega_2 t)\right] \qquad \textbf{(87)}$$

We rewrite the third term as

$$\cos(k_1 z - \omega_1 t)\cos(k_2 z - \omega_2 t) = \tfrac{1}{2}\cos[(k_1 + k_2)z - (\omega_1 + \omega_2)t]$$

$$+ \tfrac{1}{2}\cos[(k_1 - k_2)z - (\omega_1 - \omega_2)t] \qquad \textbf{(88)}$$

We now average the Poynting vector over times long compared with $1/\omega_1$ and $1/\omega_2$ but short compared with $1/(\omega_1 - \omega_2)$. This assumes that the modulation frequency ω_m is low compared with ω. We obtain for the average:

$$\langle \Pi \rangle = \frac{E_x^2}{8\mu} \left(\frac{k_1}{\omega_1} + \frac{k_2}{\omega_2} \right) [1 + \cos(\Delta kz - \Delta \omega t)] \tag{89}$$

where we have:

$$\Delta k = k_1 - k_2 \tag{90}$$

$$\Delta \omega = \omega_1 - \omega_2 = 2\omega_m \tag{91}$$

The expression given by (89) describes a sinusoidally modulated flow of energy as shown in Figure 4. The maxima move with a velocity that, in the limit that ω_m is much smaller than ω approaches what is called the group velocity v_g:[3]

$$v_g = \lim \frac{\Delta \omega}{\Delta k} = \frac{d\omega}{dk} \tag{92}$$

As we show in Section 9, the group velocity v_g is always less than c.

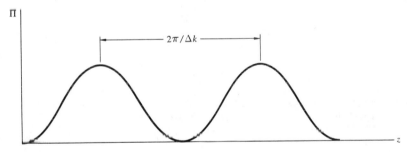

Figure 4   Sinusoidally modulated flow of energy. The maxima move with the group velocity.

9   Dispersion in Ionic Solids. In this section we discuss effects associated with the frequency dependence of the phase velocity in ionic solids. As a simple model for the development of dispersion, we imagine an ionic polarization density $\mathbf{P}(\mathbf{r},t)$, which satisfies the equation (Section 4-10):

$$\frac{d^2}{dt^2} \mathbf{P}(\mathbf{r},t) + \omega_T^2 \mathbf{P}(\mathbf{r},t) = \epsilon_{core} \omega_p^2 \mathbf{E}(\mathbf{r},t) \tag{93}$$

where ω_T is the resonance frequency and

$$\omega_p^2 = \frac{nq^2}{\epsilon_{core} M} \tag{94}$$

[3] For a discussion of group velocity based on a single-frequency wave, see D. F. Nelson, "Group Velocity in Crystal Optics," *Am. J. Phys.* **45**, 1187 (1977).

is the ion plasma frequency. Here ϵ_{core} is the contribution to the permittivity from the polarization of the ion cores, and q and M are representative ionic charges and masses. Substituting into **(93)** the electric field:

$$\mathbf{E}(\mathbf{r},t) = \mathbf{E}e^{i(\mathbf{k}\cdot\mathbf{r}-\omega t)} \tag{95}$$

and polarization

$$\mathbf{P}(\mathbf{r},t) = \mathbf{P}e^{i(\mathbf{k}\cdot\mathbf{r}-\omega t)} \tag{96}$$

we obtain:

$$\mathbf{P} = \epsilon_{core}\,\frac{\omega_p^{\,2}}{\omega_T^{\,2}-\omega^2}\,\mathbf{E} \tag{97}$$

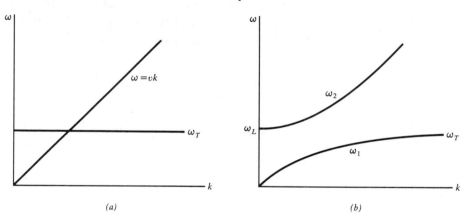

(a)   *(b)*

Figure 5   Dispersion relation for an ionic solid. (*a*) When the polarization can be neglected, a wave propagates at v without dispersion. (*b*) When polarization can not be neglected, the wave propagates on two branches, both of which exhibit dispersion.

The permittivity of the medium is given by:

$$\epsilon = \epsilon_{core} + \frac{P}{E} = \epsilon_{core}\left(1 + \frac{\omega_p^{\,2}}{\omega_T^{\,2}-\omega^2}\right) \tag{98}$$

The connection between the frequency and the wave vector k for a propagating wave is called the *dispersion relation*. The dispersion relation for **(97)** is simply given by the resonance denominator:

$$\omega = \omega_T \tag{99}$$

and is plotted as the horizontal line in Figure 5*a*.

Combining **(17)** and **(18)** we have in the absence of external charges and currents:

$$\nabla \times (\nabla \times \mathbf{E}) = -\mu\,\frac{\partial^2 D}{\partial t^2} \tag{100}$$

Writing for the displacement from (98):

$$\mathbf{D} = \epsilon\mathbf{E} = \epsilon_{core}\,\mathbf{E} + \mathbf{P} \tag{101}$$

we have in the absence of external charge:

$$\mathbf{V} \cdot \mathbf{E} = \frac{1}{\epsilon}\mathbf{V} \cdot \mathbf{D} = 0 \tag{102}$$

Using (101) and (102) we may rewrite (100) as:

$$\nabla^2\mathbf{E}(\mathbf{r},t) - \mu\epsilon_{core}\frac{\partial^2}{\partial t^2}\mathbf{E}(\mathbf{r},t) = \mu\frac{\partial^2}{\partial t^2}\mathbf{P}(\mathbf{r},t) \tag{103}$$

Substituting from (95) and (96) the form of the electric field we have:

$$(k^2 - \omega^2/v^2)\mathbf{E} = \omega^2\mu\mathbf{P} \tag{104}$$

with

$$v = (\mu\epsilon_{core})^{-1/2} \tag{105}$$

Neglecting the polarization density $\mathbf{P}$ in (104) we obtain free propagation as given by the straight line with slope v in Figure 5a.

We now couple (97) and (104) to obtain the dispersion relation:

$$v^2k^2 - \omega^2 = \frac{\omega^2\omega_p{}^2}{\omega_T{}^2 - \omega^2} \tag{106}$$

We clear the denominator on the right side of (106), obtaining a quadratic equation in ω^2. We assume for simplicity that $\omega_p{}^2$, which plays the role of a coupling constant is small. To first order in $\omega_p{}^2$ we obtain two branches:

$$\omega \simeq \omega_T\left(1 + \frac{1}{2}\frac{\omega_p{}^2}{\omega_T{}^2 - v^2k^2}\right) \tag{107}$$

$$\omega \simeq vk\left(1 - \frac{1}{2}\frac{\omega_p{}^2}{\omega_T{}^2 - v^2k^2}\right) \tag{108}$$

which are sketched in Figure 5b.

For vk *small* compared with ω_T we obtain, for the lower branch from (108),

$$\omega_1 = vk\left(1 - \frac{1}{2}\frac{\omega_p{}^2}{\omega_T{}^2}\right) \tag{109}$$

and for the upper branch from **(107)** a quadratic dispersion relation

$$\omega_2 = \omega_L + \frac{1}{2}\frac{\omega_p^2}{\omega_T^3}v^2k^2 \tag{110}$$

with

$$\omega_L \simeq \omega_T\left(1 + \frac{1}{2}\frac{\omega_p^2}{\omega_T^2}\right) \tag{111}$$

For *vk large* compared with ω_T we obtain for the lower branch from **(107)**:

$$\omega_1 = \omega_T\left(1 - \frac{1}{2}\frac{\omega_p^2}{v^2k^2}\right) \tag{112}$$

and for the upper branch from **(108)**:

$$\omega_2 = vk\left(1 + \frac{1}{2}\frac{\omega_p^2}{v^2k^2}\right) \tag{113}$$

Low frequency waves propagate on the lower branch with velocity

$$v_p = v_g = v\left(1 - \frac{1}{2}\frac{\omega_p^2}{\omega_T^2}\right) \tag{114}$$

Very high frequency waves propagate on the upper branch with velocity

$$v_p = v_g \simeq v \tag{115}$$

At intermediate frequencies near ω_T there is substantial dispersion and the phase and group velocities will differ considerably. We observe from Figure 5*b* that the group velocity is *always* less than v and approaches v only in the limit of very high frequencies.

Finally we discuss very qualitatively propagation through a dispersive medium of an electric field of the form shown in Figure 6 where an electric field that oscillates at frequency ω is turned on sharply at time zero.[4] The very sharp leading edge of the signal propagates with velocity v as expected for very high frequencies on the upper branch of Figure 5*b*. This signal, which is called the Sommerfeld, or first, precurser, is also very similar to the "whistlers" that we will discuss in connection with plasmas. The highest frequencies arrive first and lower frequencies progressively later. A second signal, the Brillouin, or second, precurser, travels with velocity $v(1 - \chi/2)$ and represents the very low frequencies that are present because the electric field starts with definite sign and the average field in the wave falls off as $1/t$. The shape of the second

---

[4] J. D. Jackson, *Classical Electrodynamics*, Second Edition, Wiley, 1975; striking photographs of the Sommerfeld and Brillouin precursers are included in a very extensive discussion of the subject in Section 7.11. For a microscopic model of dispersion, see K. S. Kunz and E. Gemoets, *Am. J. Phys.* **44**, 264 (1976).

precurser is rather like the diffraction pattern from a sharp edge. Finally the main body of the wave, which propagates at the group velocity appropriate to the driving frequency ω, arrives.

10 **Plasmas.** Although all directions of propagation within a plasma are equivalent, the permittivity must still be regarded as a tensor quantity with a principle axis given by the direction of the propagation vector **k**. According to the phenomenological theory of Section 5-17, the polarization satisfies an equation of the form **(5-153)**

$$\frac{d^2\mathbf{P}}{dt^2} + \frac{1}{\tau}\frac{d\mathbf{P}}{dt} - s^2\nabla(\nabla \cdot \mathbf{P}) = \epsilon_{\text{ion}}\,\omega_p{}^2\mathbf{E} \tag{116}$$

where

$$\omega_p{}^2 = nq^2/\epsilon_{\text{ion}}\,m \tag{117}$$

is the plasma frequency of the mobile carriers. The quantity s in **(116)** is an acoustic wave velocity.

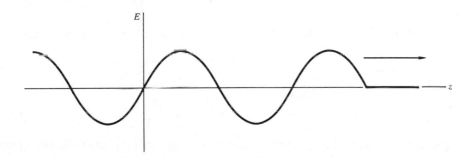

Figure 6 Propagation of an electric field at frequency ω. The field is turned on sharply at $t = 0$.

For an electric field of the form of **(27)** and a polarization of the form of **(28)** we obtain:

$$[-\omega(\omega + i/\tau)\mathbf{I} + s^2\mathbf{k}\mathbf{k}] \cdot \mathbf{P} = \epsilon_{\text{ion}}\,\omega_p{}^2\mathbf{E} \tag{118}$$

We relate **P** to **E** by the susceptibility tensor:

$$\mathbf{P} = \epsilon_0 \bar{\chi} \cdot \mathbf{E} \tag{119}$$

where $\bar{\chi}$ has the form for **k** along the $\hat{z}$ direction:

$$\bar{\chi} = \begin{pmatrix} \chi_T & 0 & 0 \\ 0 & \chi_T & 0 \\ 0 & 0 & \chi_L \end{pmatrix} \tag{120}$$

From **(118)** we obtain:

$$\chi_T = -\frac{\epsilon_{\text{ion}}}{\epsilon_0}\frac{\omega_p{}^2}{\omega(\omega + i/\tau)} \tag{121}$$

$$\chi_L = -\frac{\epsilon_{\text{ion}}}{\epsilon_0}\frac{\omega_p{}^2}{\omega(\omega + i/\tau) - s^2 k^2} \tag{122}$$

To obtain the dispersion of a freely propagating wave we write for the permittivity tensor:

$$\bar{\bar{\epsilon}} = \epsilon_{\text{ion}}\, \mathbf{I} + \epsilon_0\, \bar{\bar{\chi}} \tag{123}$$

and substitute into **(37)** written in the absence of driving currents:

$$[(k^2 - \epsilon_{\text{ion}}\,\mu\omega^2)\mathbf{I} - \omega^2 \epsilon_0\, \mu\bar{\bar{\chi}} - \mathbf{kk}] \cdot \mathbf{E} = 0 \tag{124}$$

We now discuss separately the two solutions to **(124)** for transverse and longitudinal waves.

1. For transverse waves **E** and **P** are both orthogonal to **k**, and we have from **(124)**:

$$k^2 = \epsilon_{\text{ion}}\,\mu\omega^2\left(1 - \frac{\omega_p{}^2}{\omega(\omega + i/\tau)}\right) \tag{125}$$

At frequencies sufficiently high that relaxation may be ignored we may write from **(125)**:

$$\omega^2 = \omega_p{}^2 + (ck'/n)^2 \tag{126}$$

where

$$n = (\epsilon_{\text{ion}}/\epsilon_0)^{1/2} \tag{127}$$

is the high frequency refractive index and is determined by the polarizability of the ion background. The dispersion relation for this wave is plotted in Figure 7.

2. The second possibility is **E** parallel to **k**, which is the condition for a *longitudinal mode* with zero permittivity and thus zero displacement as discussed in Section 5-17. This is what is generally meant by a plasma wave. From **(124)** we have for the dispersion relation of the longitudinal mode:

$$\omega(\omega + i/\tau) = \omega_p{}^2 + s^2 k^2 \tag{128}$$

For frequencies sufficiently high that relaxation may be neglected, **(128)** may be plotted as shown in Figure 7.

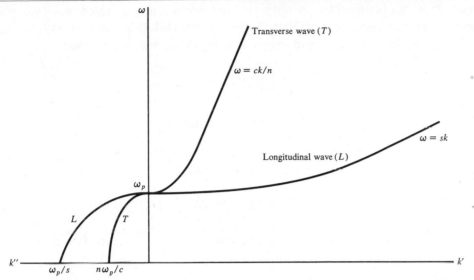

Figure 7   Dispersion relations for electromagnetic wave propagation in a plasma. The real part k' of the wavevector is plotted to the right and the imaginary part k'' to the left.

TRANSMISSION LINES

We discuss in this topic the guided propagation of transverse electromagnetic fields between conductors in a medium of permittivity ϵ and permeability μ. We obtain expressions for the inductance and capacitance per unit length, for the phase velocity, and for the characteristic impedance of the line. Although we can find waves which are transverse, they will not in general be uniform in the transverse plane.

11 Propagation of Transverse Waves. We begin by discussing the propagation of waves guided by a pair of conductors whose cross section is independent of z as shown in Figure 8. We further assume that the intervening medium is isotropic and homogeneous

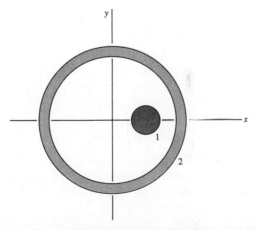

Figure 8   Cross section of a transmission line. The configuration of the line is independent of distance along the line.

and may be characterized by permittivity ϵ and permeability μ, which may be explicit functions of frequency, but not of wavevector. From **(16)**, **(17)**, **(18)**, and **(19)** we have the equations for the electric and magnetic fields in the intervening space:

$$\nabla \cdot \mathbf{D}(\mathbf{r},t) = 0 \tag{129}$$

$$\nabla \times \mathbf{E}(\mathbf{r},t) = -\frac{\partial}{\partial t} \mathbf{B}(\mathbf{r},t) \tag{130}$$

$$\nabla \times \mathbf{H}(\mathbf{r},t) = \frac{\partial}{\partial t} \mathbf{D}(\mathbf{r},t) \tag{131}$$

$$\nabla \cdot \mathbf{B}(\mathbf{r},t) = 0 \tag{132}$$

where we have taken the external charge and current to be zero. For a transverse guided wave we assume solutions of the form:

$$\mathbf{E}(\mathbf{r},t) = \mathbf{E}(x,y)e^{i(kz-\omega t)} \tag{133}$$

$$\mathbf{H}(\mathbf{r},t) = \mathbf{H}(x,y)e^{i(kz-\omega t)} \tag{134}$$

with

$$\mathbf{D}(\mathbf{r},t) = \epsilon\mathbf{E}(\mathbf{r},t) \tag{135}$$

$$\mathbf{B}(\mathbf{r},t) = \mu\mathbf{H}(\mathbf{r},t) \tag{136}$$

where the fields are all in the transverse plane. From **(129)** we have for the electric field:

$$\frac{\partial E_x}{\partial x} + \frac{\partial E_y}{\partial y} = 0 \tag{137}$$

and from the $\hat{\mathbf{z}}$ component of **(130)**:

$$\frac{\partial E_y}{\partial x} = \frac{\partial E_x}{\partial y} \tag{138}$$

Combining **(137)** and **(138)** we obtain

$$\left(\frac{\partial^2}{\partial x^2} + \frac{\partial^2}{\partial y^2}\right)\mathbf{E}(x,y) = 0 \tag{139}$$

which is the equation of a cylindrical electrostatic field. Similarly we may obtain from **(132)** and the z component of **(131)**

$$\left(\frac{\partial^2}{\partial x^2} + \frac{\partial^2}{\partial y^2}\right)\mathbf{H}(x,y) = 0 \tag{140}$$

which is the equation of a cylindrical magnetostatic field. Taking the curl of (130) and using (129) and (131) we have

$$\mathbf{V}^2 \mathbf{E}(\mathbf{r},t) = \left(\frac{\partial^2}{\partial x^2} + \frac{\partial^2}{\partial y^2} + \frac{\partial^2}{\partial z^2} \right) \mathbf{E}(\mathbf{r},t) = -\mu\epsilon\omega^2 \mathbf{E}(\mathbf{r},t) \tag{141}$$

Substituting from (133) for the form of $\mathbf{E}(\mathbf{r},t)$ and using (139) we obtain the dispersion relation for guided transverse waves

$$\omega = k/(\mu\epsilon)^{1/2} \tag{142}$$

which is exactly the same as for a *plane* wave. Although it is not obvious from our treatment we require perfect conductors in order to be able to assume transverse waves. We will return to this problem in Section 13-8 when we discuss surface electromagnetic waves.

12 Wave Impedance. We had in Section 5 the relation for a transverse plane wave (59)

$$\mathbf{H} = \frac{1}{Z} \hat{z} \times \mathbf{E} \tag{143}$$

where

$$Z = (\mu/\epsilon)^{1/2} \tag{144}$$

is the intrinsic impedance of the medium. We write from (131)

$$-i\omega\epsilon E_x = -\frac{\partial H_y}{\partial z} = -ikH_y \tag{145}$$

$$-i\omega\epsilon E_y = \frac{\partial H_x}{\partial z} = ikH_x \tag{146}$$

We see that (145) and (146) are consistent with (143) and with the dispersion relation (142) give a wave impedance equal to the intrinsic impedance of the medium (144).

13 Capacitance and Inductance. The equations 137 and 138 permit us to write in the transverse plane

$$\mathbf{E}(x,y) = -\mathbf{V}_T \phi(x,y) \tag{147}$$

where $\mathbf{V}_T$ is the gradient with respect to the transverse coordinates only. We write for the capacitance per unit length between the two conductors (Section 4-13):

$$C' = \frac{\lambda}{V} \tag{148}$$

where $V = \phi_1 - \phi_2$ is the difference in potential and λ is the charge per unit length on the conductors. The charge within a cylinder of height dz is given by Gauss's law:

$$\lambda \, dz = \epsilon \oint_S \mathbf{E} \cdot d\mathbf{S} = \epsilon \, dz \oint_C \mathbf{E} \cdot (d\mathbf{r} \times \hat{\mathbf{z}}) \tag{149}$$

where C is a contour in the transverse plane as shown in Figure 9a. From (149) and (143) we may write

$$\lambda = \epsilon Z \oint_C \mathbf{H} \cdot d\mathbf{r} = \frac{1}{v_p} \oint_C \mathbf{H} \cdot d\mathbf{r} \tag{150}$$

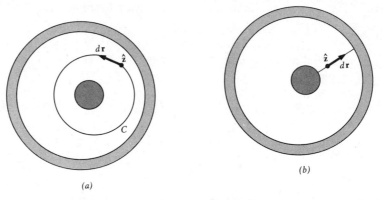

(a)

(b)

Figure 9 Capacitance and inductance of a transmission line. (a) The contour is that used in (149). (b) The contour is that used in (152).

Now the integral of $\mathbf{H}$ is the current I flowing through the contour so that we have the simple expression for the current

$$\blacktriangleright \qquad\qquad I = \lambda v_p \tag{151}$$

where the phase velocity v_p is given by (67).

The inductance per unit length may be written as (Section 10-11)

$$L = \frac{1}{I} \frac{d\Phi}{dz} = \frac{\mu}{I} \int_C \mathbf{H} \cdot (\hat{\mathbf{z}} \times d\mathbf{r}) \tag{152}$$

where the integral is between the two conductors as shown in Figure 9b. Substituting from (143) we obtain:

$$L = \frac{\mu}{IZ} \int_C \mathbf{E} \cdot d\mathbf{r} = \frac{\mu}{Z} \frac{V}{I} = \frac{1}{v_p} \frac{V}{I} \tag{153}$$

Taking the product of **(148)** and **(153)** we obtain

$$\blacktriangleright \qquad L'C' = \mu\epsilon = 1/v_p{}^2 \qquad (154)$$

where we have used **(144)**, **(151)**, and **(67)**.

Taking the ratio of **(153)** and **(148)** we obtain

$$\frac{L'}{C'} = (V/I)^2 = Z_0{}^2 \qquad (155)$$

where

$$Z_0 = V/I \qquad (156)$$

is called the characteristic impedance of the line. Comparing **(154)** and **(155)** we write for the inductance per unit length

$$L' = Z_0/v_p \qquad (157)$$

and for the capacitance per unit length

$$C' = 1/Z_0 v_p \qquad (158)$$

Exercises

6. A twin lead transmission line has a characteristic impedance Z_0 of 300 Ω and a velocity factor $v_p/c = 0.8$. Find the capacitance in picofarads per metre and the inductance in microhenrys per metre.
7. RG 174/U coaxial cable has a characteristic impedance Z_0 of 50 Ω and a capacitance of 100 pF/m. Find the inductance per metre and the velocity factor v_p/c. What is the relative permittivity ϵ/ϵ_0 of the insulating dielectric?

ANISOTROPIC MEDIA

Electromagnetic media are characterized as isotropic, uniaxial, and biaxial. For isotropic media, as we have discussed, all directions of propagation and transverse polarization are equivalent. For a medium to be uniaxial, there must be a single axis about which there is threefold or fourfold symmetry. For propagation along the symmetry axis, the medium *appears* isotropic. In the absence of *any* axis of threefold or higher symmetry the medium is biaxial: there are *two* directions of propagation for which the medium *appears* to be isotropic. The relation between crystal symmetry and optical properties is best discussed in terms of the properties of the permittivity tensor.[5]

---

[5] For a readable account of the relation between symmetry and optical properties, see E. A. Wood, *Crystals and Light*, Van Nostrand, 1964.

14 **Permittivity Tensor.** We begin by establishing the well-known result that for sufficiently low frequencies, the permittivity tensor of a dielectric is symmetric. We begin with **(44)**, which may be written in the absence of external currents as:

$$du = \mathbf{H} \cdot d\mathbf{B} + \mathbf{E} \cdot d\mathbf{D} \tag{159}$$

Now, if there are no losses in the medium, which will be the case at least for dielectrics if the fields are varied sufficiently slowly, then du should be a perfect differential and it should be possible to express the energy of the system as a quadratic function of the components of the electric and magnetic field. But for u to be a perfect differential, the combination

$$E_i \epsilon_{ij} \, dE_j + E_j \epsilon_{ji} \, dE_i \tag{160}$$

must be a perfect differential and this requires

$$\epsilon_{ij} = \epsilon_{ji} \tag{161}$$

that is, the permittivity tensor must be symmetric. This derivation strictly applies to the zero frequency limit of the permeability tensor. This limit excludes dielectric loss and it also excludes optical rotation, which we will discuss separately.

As shown in Section C-8 any symmetric tensor may be brought to diagonal form by means of a coordinate transformation. This means that for every crystal there exists a set of orthogonal axes $\hat{\mathbf{x}}$, $\hat{\mathbf{y}}$, and $\hat{\mathbf{z}}$ such that the permittivity tensor may be written as

$$\bar{\bar{\epsilon}} = \epsilon_{xx} \hat{\mathbf{x}}\hat{\mathbf{x}} + \epsilon_{yy} \hat{\mathbf{y}}\hat{\mathbf{y}} + \epsilon_{zz} \hat{\mathbf{z}}\hat{\mathbf{z}} \tag{162}$$

15 **[Index Ellipsoid].** A simple physical interpretation of the solution of **(37)** for free propagation in an anisotropic medium was first given by Fresnel.[6] Following his argument we introduce the tensor inverse to **(162)** and write in place of **(29)**:

$$\mathbf{E} = \bar{\bar{\epsilon}}^{-1} \cdot \mathbf{D} \tag{163}$$

Then we may write **(37)** as an equation for the normal modes of $\mathbf{D}$:

$$(k^2 \bar{\bar{\epsilon}}^{-1} - \mathbf{k}\mathbf{k} \cdot \bar{\bar{\epsilon}}^{-1} - \omega^2 \mu \mathbf{I}) \cdot \mathbf{D} = 0 \tag{164}$$

---

[6] For a discussion closer to that of Fresnel, see M. Born and E. Wolf, *Principles of Optics*, Pergamon, Fifth Edition, 1975; Chapter 14. K. S. Kunz, *Am. J. Phys.* **45**, 267 (1977) presents a discussion in terms of an eigenvalue problem.

If we take the scalar product with **k** from the left, the first terms cancel and we are left with

$$\mathbf{k} \cdot \mathbf{I} \cdot \mathbf{D} = \mathbf{k} \cdot \mathbf{D} = 0 \tag{165}$$

which is to say that **D** is perpendicular to **k**, which is simply a consequence of **(16)**. Next we take the scalar product of **(164)** from the left with **D**, obtaining the equation

$$\mathbf{D} \cdot \bar{\epsilon}^{-1} \cdot \mathbf{D} = \frac{\omega^2 \mu}{k^2} D^2 \tag{166}$$

which is the equation of an ellipsoid as may be seen by writing **(166)** in terms of the components of **D** along axes for which the permittivity tensor is diagonal:

$$\frac{D_x^2}{\epsilon_{xx}} + \frac{D_y^2}{\epsilon_{yy}} + \frac{D_z^2}{\epsilon_{zz}} = \frac{\omega^2 \mu}{k^2} D^2 \tag{167}$$

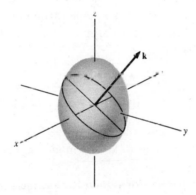

Figure 10   The index ellipsoid. The surface shown is defined by **(169)**.

We next introduce a vector **r** parallel to **D** and defined by

$$\mathbf{r} = \frac{k}{\omega \mu^{1/2}} \hat{\mathbf{D}} \tag{168}$$

In terms of the components of **r** we write **(130)** as

$$\frac{x^2}{\epsilon_{xx}} + \frac{y^2}{\epsilon_{yy}} + \frac{z^2}{\epsilon_{zz}} = 1 \tag{169}$$

The surface defined by **(169)** is called the index ellipsoid and is shown in Figure 10.

We use the index ellipsoid in finding the solution of **(164)** by first observing that **(165)** defines a plane normal to **k** and passing through the origin. Then **r** must lie on the intersection of the ellipsoid and this plane, which is an ellipse and is shown in

Figure 10. Finally, we must find the direction of **D**. To do this we take the cross product of **(164)** from the left with **k** to obtain:

$$\mathbf{k} \times \bar{\epsilon}^{-1} \cdot \mathbf{D} = \frac{\omega^2 \mu}{k^2} \mathbf{k} \times \mathbf{D} \tag{170}$$

We now wish to argue that **(170)** implies that **D** be along one of the axes of the ellipse. To make the argument let us imagine that we rotate into a new coordinate system with $\hat{\mathbf{z}}$ along **k**. Now as long as **D** is contained in the transverse plane we may write **(170)** as:

$$\bar{\epsilon}^{-1} \cdot \mathbf{D} = \frac{\omega^2 \mu}{k^2} \mathbf{D} \tag{171}$$

where we only use the components of the reciprocal permittivity tensor that are in the plane. The condition on **D** given by **(171)** is that **D** be parallel to the principal axes within the plane, that is, along one of the axes of the ellipse.

To summarize, we have shown that for some particular direction of **k** there are two orthogonal solutions of **(164)** along the principal axes of the ellipse formed by the intersection of the index ellipsoid and a plane through the origin normal to **k**. If the lengths of the axes of the ellipse are $\epsilon_1^{1/2}$ and $\epsilon_2^{1/2}$, the phase velocities of the two solutions are given from **(168)** by:

$$v_1 = \frac{\omega}{k} = \frac{1}{(\mu\epsilon_1)^{1/2}} \qquad v_2 = \frac{1}{(\mu\epsilon_2)^{1/2}} \tag{172}$$

The electric fields for the two waves are given by **(163)**. The two fields need not be orthogonal nor need they be normal to **k**. The magnetic fields as given by **(39)** will be normal to **k**.

It is clear from Figure 10 that for $\epsilon_{xx} < \epsilon_{yy} < \epsilon_{zz}$ there will be two orientations of **k** in the xz plane for which the intersection of the ellipsoid and the plane of propagation will be circular with $\epsilon_1 = \epsilon_2$. These are the two axes for a general biaxial crystal.

If there is an axis of threefold or higher symmetry, then we must have $\epsilon_{xx} = \epsilon_{yy}$ or $\epsilon_{yy} = \epsilon_{zz}$. Then there will be only one optic axis, either the $\hat{\mathbf{z}}$ or the $\hat{\mathbf{x}}$ axis, depending on whether the ellipsoid is prolate or oblate. One may think of the uniaxial crystal as the limit of a biaxial crystal for which the two optic axes have coalesced.

16 **[Electrooptic Effect].** Applying a strong electric field to an isotropic medium lowers the optical symmetry to uniaxial. Applying an electric field off the optic axis of a uniaxial crystal makes the crystal biaxial. For crystals with a center of symmetry the anisotropy in the permittivity tensor is quadratic in E since reversing the field cannot change the permittivity tensor. If the crystal is not centrosymmetric, the electrically induced anisotropy may be linear in the electric field.

17 **[Optical Activity].** The preceding theory predicts that in an isotropic medium like a liquid or a glass a linearly polarized wave should propagate without change in polarization. It is, in fact, observed that through solutions of molecules that have *handedness*[7] there may be a considerable rotation of the plane of polarization. The most common example of this phenomenon, which is called optical activity, is given by sugar solutions. Crystals such as quartz, which lack a center of symmetry, also show rotation. Our main interest here will be with liquids.

Our model of an organic molecule exhibiting handedness is shown in Figure 11. We imagine that the molecule has polarizability α along the axis of each of its parts and negligible transverse polarizability.[8]

It is at first sight surprising that a liquid containing randomly oriented molecules will still exhibit anisotropic behavior. This occurs because *handedness* is not averaged out. The molecule shown in Figure 11 is right handed no matter what its orientation and no matter how we label the two constituents.

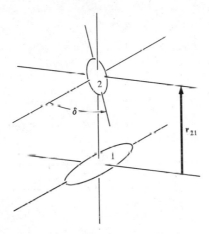

Figure 11   Model of an organic molecule exhibiting handedness. The molecule has a polarizability α along the axis of each of its parts.

In order to properly average over all possible orientations of the molecule we characterize the molecule by the Euler angles shown in Figure 12.

If we let the vectors $\hat{\mathbf{a}}_1$ and $\hat{\mathbf{a}}_2$ lie along the polarization axes, then we may write from **(A-3)**, **(A-4)**, and **(A-5)**:

$$\hat{\mathbf{a}} = \hat{\mathbf{x}}' = \hat{\mathbf{x}}(\cos \psi \cos \phi - \sin \psi \sin \phi \cos \theta)$$
$$+ \hat{\mathbf{y}}(\cos \psi \sin \phi + \sin \psi \cos \phi \cos \theta) - \hat{\mathbf{z}} \sin \psi \sin \theta \quad \textbf{(173)}$$

For a general orientation of the electric field **E**, we may write the induced moment as:

$$\mathbf{p} = \alpha \hat{\mathbf{a}}(\hat{\mathbf{a}} \cdot \mathbf{E}) = \alpha(\hat{\mathbf{a}}\hat{\mathbf{a}}) \cdot \mathbf{E} \quad \textbf{(174)}$$

and we may regard $\alpha \hat{\mathbf{a}}\hat{\mathbf{a}}$ as a polarizability tensor.

---

[7] For a general account of rotational symmetry see M. Gardner, *The Ambidextrous Universe*, Mentor, 1964.

[8] For a rather different model, see G. E. Desobry and P. K. Kabir, *Am. J. Phys.* **41**, 1350 (1973) and **42**, 790 (1974). For a general discussion, see R. M. Peterson, *Am. J. Phys.* **43**, 969 (1975).

We first regard the two parts as noninteracting and write:

$$\mathbf{p} = \mathbf{p}_1 + \mathbf{p}_2 = \alpha[e^{-i\eta}\hat{\mathbf{a}}_1\hat{\mathbf{a}}_1 + e^{+i\eta}\hat{\mathbf{a}}_2\hat{\mathbf{a}}_2] \cdot \mathbf{E} \qquad (175)$$

where we have taken the phase of $\mathbf{E}$ to be that midway between the parts. We now must average the polarizability tensor:

$$\bar{\bar{\alpha}} = \alpha(e^{-i\eta}\hat{\mathbf{a}}_1\hat{\mathbf{a}}_1 + e^{+i\eta}\hat{\mathbf{a}}_2\hat{\mathbf{a}}_2) \qquad (176)$$

over the Euler angles. Performing the average over ϕ and ψ only we obtain

$$\langle\bar{\bar{\alpha}}\rangle_{\phi\psi} = \frac{\alpha}{2}(1 + \cos^2\theta)\cos\eta\,\mathbf{I} + \frac{\alpha}{2}(1 - 3\cos^2\theta)\cos\eta\,\hat{\mathbf{z}}\hat{\mathbf{z}} \qquad (177)$$

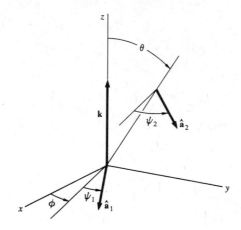

Figure 12 Euler angles for two-component molecule. An average is made over all possible orientations of the molecule.

where the phase difference is given by

$$2\eta = \mathbf{k} \cdot \mathbf{r}_{21} = kr_{21}\cos\theta \qquad (178)$$

and $\mathbf{k}$ is along the $\hat{\mathbf{z}}$ direction. The polarizability given by (177) will be isotropic in the transverse plane, even after averaging over θ, and thus can produce no rotation even though there are some interference effects.

We introduce now an interaction between the parts by means of the classical dipolar field. We write for the fields at $\mathbf{p}_1$ and $\mathbf{p}_2$:

$$\mathbf{E}_1 = \mathbf{E}e^{-i\eta} - \beta\mathbf{p}_2 \qquad (179)$$

$$\mathbf{E}_2 = \mathbf{E}e^{+i\eta} - \beta\mathbf{p}_1 \qquad (180)$$

Substituting (179) and (180) into (174) we obtain:

$$\mathbf{p}_1 = \alpha\hat{\mathbf{a}}_1\hat{\mathbf{a}}_1 \cdot (\mathbf{E}e^{-i\eta} - \beta\mathbf{p}_2) \qquad (181)$$

$$\mathbf{p}_2 = \alpha\hat{\mathbf{a}}_2\hat{\mathbf{a}}_2 \cdot (\mathbf{E}e^{+i\eta} - \beta\mathbf{p}_1) \qquad (182)$$

Eliminating $\mathbf{p}_2$ between **(181)** and **(182)** we may write

$$\mathbf{p}_1 = \alpha \hat{\mathbf{a}}_1 \hat{\mathbf{a}}_1 \cdot (e^{-i\eta}\mathbf{I} - \alpha\beta e^{+i\eta}\hat{\mathbf{a}}_2\hat{\mathbf{a}}_2) \cdot \mathbf{E} + \alpha^2\beta^2\hat{\mathbf{a}}_1\hat{\mathbf{a}}_1 \cdot \hat{\mathbf{a}}_2\hat{\mathbf{a}}_2 \cdot \mathbf{p}_1 \tag{183}$$

with a similar expression for $\mathbf{p}_2$. We may perform the two scalar products indicated in **(183)**:

$$\hat{\mathbf{a}}_1\hat{\mathbf{a}}_1 \cdot \hat{\mathbf{a}}_2\hat{\mathbf{a}}_2 \cdot \mathbf{p}_1 = \cos \delta \, \hat{\mathbf{a}}_1\hat{\mathbf{a}}_2 \cdot \mathbf{p}_1 = \cos^2 \delta \, \hat{\mathbf{a}}_1 p_1 = \mathbf{p}_1 \cos^2 \delta \tag{184}$$

We may then write for $\mathbf{p}_1$:

$$\mathbf{p}_1 = \frac{\alpha}{1 - (\alpha\beta\cos\delta)^2} (e^{-i\eta}\,\hat{\mathbf{a}}_1\hat{\mathbf{a}}_1 - \alpha\beta e^{+i\eta}\cos\delta\,\hat{\mathbf{a}}_1\hat{\mathbf{a}}_2) \cdot \mathbf{E} \tag{185}$$

and a similar expression for $\mathbf{p}_2$. We have for the polarizability tensor the expression:

$$\bar{\bar{\alpha}} = \frac{\alpha}{1 - (\alpha\beta\cos\delta)^2} (e^{-i\eta}\,\hat{\mathbf{a}}_1\hat{\mathbf{a}}_1 + e^{+i\eta}\,\hat{\mathbf{a}}_2\hat{\mathbf{a}}_2 - \alpha\beta\cos\delta\,e^{i\eta}\,\hat{\mathbf{a}}_1\hat{\mathbf{a}}_2 - \alpha\beta\cos\delta\,e^{i\eta}\,\hat{\mathbf{a}}_2\hat{\mathbf{a}}_1) \tag{186}$$

Averaging only over ϕ and ψ we obtain:

$$\langle\hat{\mathbf{a}}_1\hat{\mathbf{a}}_2\rangle_{\phi\psi} = \tfrac{1}{4}\cos\delta\,(1 + \cos^2\theta)\mathbf{I} + \tfrac{1}{4}\cos\delta\,(1 - 3\cos^2\theta)\hat{\mathbf{z}}\hat{\mathbf{z}} + \tfrac{1}{2}\sin\delta\cos\theta(\hat{\mathbf{x}}\hat{\mathbf{y}} - \hat{\mathbf{y}}\hat{\mathbf{x}}) \tag{187}$$

with a similar expression for $\langle\hat{\mathbf{a}}_2\hat{\mathbf{a}}_1\rangle_{\phi\psi}$ except that the sign of δ is reversed. Substituting into **(186)** we obtain finally for the average polarizability tensor:

$$\langle\bar{\bar{\alpha}}\rangle = \frac{\alpha}{2}\frac{1 - \alpha\beta\cos^2\delta}{1 - (\alpha\beta\cos\delta)^2} \langle(1 + \cos^2\theta)\cos\eta\,\mathbf{I} + (1 - 3\cos^2\theta)\cos\eta\,\hat{\mathbf{z}}\hat{\mathbf{z}}\rangle_\theta$$

$$- i\alpha\frac{\alpha\beta}{1 - (\alpha\beta\cos\delta)^2}\sin 2\delta\,\langle\sin\eta\cos\theta\rangle_\theta(\hat{\mathbf{x}}\hat{\mathbf{y}} - \hat{\mathbf{y}}\hat{\mathbf{x}}) \tag{188}$$

If we assume that the wavelength is long compared with the separation between parts of the molecule we may make the approximations:

$$\cos\eta = \cos\left(\tfrac{1}{2}kr_{12}\cos\theta\right) \simeq 1 \qquad \sin\eta \simeq \tfrac{1}{2}kr_{12}\cos\theta \tag{189}$$

With this simplification we can do the averages over θ and obtain:

$$\langle\bar{\bar{\alpha}}\rangle = \tfrac{2}{3}\alpha\frac{1 - \alpha\beta\cos^2\delta}{1 - (\alpha\beta\cos\delta)^2}\mathbf{I} - i\frac{k}{6}\frac{\alpha^2\beta r_{12}\sin 2\delta}{1 - (\alpha\beta\cos\delta)^2}(\hat{\mathbf{x}}\hat{\mathbf{y}} - \hat{\mathbf{y}}\hat{\mathbf{x}}) \tag{190}$$

Disregarding the local field correction (Section 3-7), we may write for the permittivity tensor:

$$\bar{\bar{\epsilon}} = \epsilon_0(\mathbf{I} + n\langle\bar{\bar{\alpha}}\rangle) \tag{191}$$

where n is the number of molecules per unit volume. From **(190)** we may write the permittivity as:

$$\bar{\bar{\epsilon}} = \epsilon\mathbf{I} - i\gamma k(\hat{\mathbf{x}}\hat{\mathbf{y}} - \hat{\mathbf{y}}\hat{\mathbf{x}}) = \epsilon\mathbf{I} + i\gamma\mathbf{k} \times \mathbf{I} \tag{192}$$

where $\mathbf{k}$ is along $\hat{\mathbf{z}}$ and ϵ and γ may be obtained by comparison with **(190)**.

Substituting into **(29)** we obtain:

$$\mathbf{D}(\mathbf{r},t) = \epsilon\mathbf{E}(\mathbf{r},t) + i\gamma\mathbf{k} \times \mathbf{E}(\mathbf{r},t) \tag{193}$$

Using the expression for the curl of a plane wave:

$$\nabla \times \mathbf{E}(\mathbf{r},t) = i\mathbf{k} \times \mathbf{E}(\mathbf{r},t) \tag{194}$$

we may if we wish rewrite **(193)** as:

$$\mathbf{D}(\mathbf{r},t) = \epsilon\mathbf{E}(\mathbf{r},t) + \gamma\nabla \times \mathbf{E}(\mathbf{r},t) \tag{195}$$

It remains to show that **(193)** or **(195)** actually leads to a rotation of the plane of polarization. This was first shown by Fresnel who demonstrated that when **(192)** is substituted into **(37)** the solutions are right and left circularly polarized waves with slightly different phase velocities. A linearly polarized incident wave may be regarded as a superposition of the two circularly polarized waves with equal amplitude. Because of the different phase velocities, the relative phase between the circular waves will be a function of z, leading to rotation of the plane of polarization. This is the standard derivation to be given in Section 19. An alternative derivation, which avoids normal modes, may be given using **(195)**.

From **(17)**, **(18)**, and **(195)** we write:

$$\nabla^2\mathbf{E}(\mathbf{r},t) - \mu\,\epsilon\,\frac{d^2}{dt^2}\,\mathbf{E}(\mathbf{r},t) = \mu\gamma\,\frac{d^2}{dt^2}\,\nabla \times \mathbf{E}(\mathbf{r},t) \tag{196}$$

where we have used **(B-10)** to expand the double curl and have set the divergence of $\mathbf{E}$ to zero using **(195)**. We now write $\mathbf{E}(\mathbf{r},t)$ in the form:

$$\mathbf{E}(\mathbf{r},t) = \mathbf{E}(z)\cos(kz - \omega t) \tag{197}$$

Substituting **(197)** into **(196)** we obtain two equations by equating separately the coefficients of $\cos(kz - \omega t)$ and $\sin(kz - \omega t)$:

$$\frac{d}{dz}\mathbf{E} = -\tfrac{1}{2}\omega^2\mu\gamma\hat{\mathbf{z}} \times \mathbf{E} \qquad (198)$$

$$\frac{d^2}{dz^2}\mathbf{E} + (\omega^2\mu\epsilon - k^2)\mathbf{E} = -\omega^2\mu\gamma\hat{\mathbf{z}} \times \frac{d}{dz}\mathbf{E} \qquad (199)$$

The solution of **(198)** is a vector that rotates uniformly with z:

$$\theta = -\tfrac{1}{2}\omega^2\mu\gamma z \qquad (200)$$

Substituting **(200)** into **(199)** we obtain for the wave vector:

$$k^2 = \omega^2\mu\epsilon + \tfrac{1}{4}(\omega^2\mu\gamma)^2 \qquad (201)$$

Note that the plane of polarization is a function of z alone and not of the time, a result that is particularly clear from the above derivation.

Exercise

---

8. Born has demonstrated that a molecule composed of four identical polarizable spheres, interacting through Coulomb fields, can produce optical rotation. Would this be true if the spheres were at the corners of a tetrahedron? Can any arrangement for which the atoms are all in the same plane produce rotation? Might one expect rotation from a molecule built of three atoms? [M. Born, *Proc. Roy. Soc. London*, **150**, 84 (1935); B. Y. Oke, *Proc. Roy. Soc. London*, **153**, 339 (1936).]

MAGNETOOPTIC EFFECTS IN PLASMAS

The application of a magnetic field to an isotropic medium produces a rotation of the plane of polarization for propagation along the field (the Faraday effect) and double refraction (the Voigt effect) for propagation transverse to the field. The sense of rotation is determined by the direction of the magnetic field and not by the direction of propagation. We discuss two examples of magnetooptic effects: in this section plasmas and in the next section transparent ferromagnetic insulators.

To treat the effect of a magnetic field on a plasma we replace the force $q\mathbf{E}$ by the full Lorentz force:

$$\mathbf{F} = q(\mathbf{E} + \mathbf{v} \times \mathbf{B}) \qquad (202)$$

which we must rewrite in terms of the polarization current: ·

$$\mathbf{j}_p = \frac{d\mathbf{P}}{dt} = nq\mathbf{v} \qquad (203)$$

For a charge q moving in a magnetic field we have:

$$m \frac{d\mathbf{v}}{dt} = m\boldsymbol{\omega}_c \times \mathbf{v} = q\mathbf{v} \times \mathbf{B} \tag{204}$$

where $\boldsymbol{\omega}_c$ is the cyclotron frequency and is given from (205) by

$$\boldsymbol{\omega}_c = -\frac{q}{m}\mathbf{B} \tag{205}$$

We may now write (202) as

$$\mathbf{F} = q\left(\mathbf{E} + \frac{m}{nq^2}\boldsymbol{\omega}_c \times \frac{d\mathbf{P}}{dt}\right) = q\left(\mathbf{E} + \frac{1}{\omega_p{}^2 \epsilon_{\text{ion}}}\boldsymbol{\omega}_c \times \frac{d\mathbf{P}}{dt}\right) \tag{206}$$

where (117)

$$\omega_p{}^2 = nq^2/\epsilon_{\text{ion}}\, m \tag{207}$$

is the expression for the square of the plasma frequency. Next we rewrite (116) as:

$$\frac{d^2\mathbf{P}}{dt^2} - \boldsymbol{\omega}_c \times \frac{d\mathbf{P}}{dt} + \frac{1}{\tau}\frac{d\mathbf{P}}{dt} - s^2\boldsymbol{\nabla}(\boldsymbol{\nabla}\cdot\mathbf{P}) = \epsilon_{\text{ion}}\,\omega_p{}^2\mathbf{E} \tag{208}$$

Allowing $\mathbf{E}(\mathbf{r},t)$ to vary as (27) and $\mathbf{P}(\mathbf{r},t)$ to vary as (28) we obtain:

$$[-\omega(\omega + i/\tau)\mathbf{I} + i\omega\boldsymbol{\omega}_c \times \mathbf{I} + s^2\mathbf{k}\mathbf{k}]\cdot\mathbf{P} = \epsilon_{\text{ion}}\,\omega_p{}^2\mathbf{E} \tag{209}$$

18 Circular Polarization. Writing from (123)

$$\bar{\bar{\epsilon}} = \epsilon_{\text{ion}}\,\mathbf{I} + \epsilon_0\,\bar{\bar{\chi}} \tag{210}$$

where we have assumed ϵ_{ion} isotropic, we may use (209) to obtain an expression for the dielectric function. As a simple example, we assume the propagation vector $\mathbf{k}$ parallel to the magnetic field, which is along the z direction. From (209) and (210) we obtain, for the longitudinal dielectric function,

$$\epsilon_L(k,\omega) = \epsilon_{\text{ion}}\left(1 - \frac{\omega_p{}^2}{\omega(\omega + i/\tau) - s^2 k^2}\right) \tag{211}$$

which is just the same as that obtained from (122) in the absence of a magnetic field. This is because the field exerts no force on parallel polarization current.

For a transverse wave we have, writing the components of (209),

$$-\omega(\omega + i/\tau)P_x - i\omega\omega_c P_y = \epsilon_{\text{ion}}\,\omega_p{}^2 E_x \tag{212}$$

$$i\omega\omega_c P_x - \omega(\omega + i/\tau)P_y = \epsilon_{\text{ion}}\,\omega_p{}^2 E_y \tag{213}$$

To solve **(212)** and **(213)** we write the electric field in terms of circularly polarized components. If we imagine circularly polarized fields E_r and E_l as shown in Figure 13, we may write for the real fields

$$E_{1x}(z,t) = E_r\,e^{-k''z}\cos(k'z - \omega t) - E_l\,e^{-k''z}\sin(k'z - \omega t + \phi) \qquad (214)$$

$$E_{1y}(z,t) = -E_r\,e^{-k''r}\sin(k'z - \omega t) + E_l\,e^{-k''z}\cos(k'z - \omega t + \phi) \qquad (215)$$

Complex fields are formed in the usual way by shifting the time by a quarter period to obtain a second set of solutions:

$$E_{2x}(z,t) = E_r\,e^{-k''z}\sin(k'z - \omega t) + E_l\,e^{-k''z}\cos(k'z - \omega t + \phi) \qquad (216)$$

$$E_{2y}(z,t) = E_r\,e^{-k''z}\cos(k'z - \omega t) + E_l\,e^{-k''z}\sin(k'z - \omega t + \phi) \qquad (217)$$

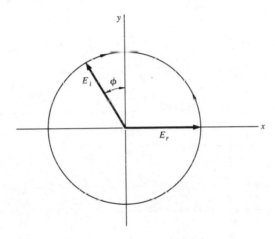

Figure 13 Circularly polarized fields. The two components rotate uniformly in opposite directions.

Taking the complex combination of the two sets of solutions we write:

$$\underline{E}_x(z,t) = E_{1x}(z,t) + iE_{2x}(z,t) = (E_r + iE_l)e^{i(kz - \omega t)} \qquad (218)$$

$$\underline{E}_y(z,t) = E_{1y}(z,t) + iE_{2y}(z,t) = (E_l + iE_r)e^{i(kz - \omega t)} \qquad (219)$$

where $\underline{k}$ is the usual complex wave vector. The phase shift ϕ is absorbed into a complex E_l. Similarly, the polarization may be written in terms of circularly polarized components:

$$\underline{P}_x(z,t) = (P_r + iP_l)e^{i(kz - \omega t)} \qquad (220)$$

$$\underline{P}_y(z,t) = (P_l + iP_r)e^{i(kz - \omega t)} \qquad (221)$$

Substituting the circularly polarized field and polarization into (212) and (213) we obtain

$$P_r = \epsilon_0 \chi_r E_r = -\frac{\epsilon_{ion}\,\omega_p{}^2}{\omega(\omega - \omega_c + i/\tau)}E_r \tag{222}$$

$$P_l = \epsilon_0 \chi_l E_l = -\frac{\epsilon_{ion}\,\omega_p{}^2}{\omega(\omega + \omega_c + i/\tau)}E_l \tag{223}$$

From (222) and (223) we write for the permittivity of right (r) and left (l) circularly polarized waves:

$$\epsilon_r = \epsilon_{ion} + \epsilon_0 \chi_r = \epsilon_{ion}\left[1 - \frac{\omega_p{}^2}{\omega(\omega - \omega_c + i/\tau)}\right] \tag{224}$$

$$\epsilon_l = \epsilon_{ion} + \epsilon_0 \chi_l = \epsilon_{ion}\left[1 - \frac{\omega_p{}^2}{\omega(\omega + \omega_c + i/\tau)}\right] \tag{225}$$

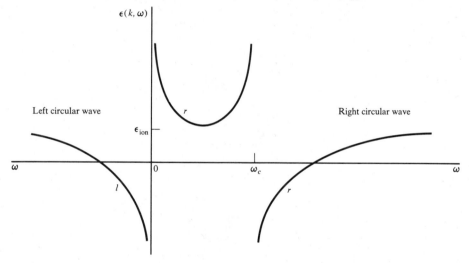

Figure 14   Dielectric function of an uncompensated electron plasma. The right side corresponds to right circular waves and the left side to left circular waves.

In Figure 14 we plot the dielectric function of an uncompensated electron plasma as a function of ω. The right side corresponds to r waves and the left to l waves. We neglect relaxation and take ω_c positive as would be the case for *negative* carriers.

As we have seen (5-159) in the absence of a magnetic field, the permittivity becomes positive only at frequencies above the plasma frequency ω_p. The presence of a magnetic field introduces an additional resonance at ω_c. Below this resonance the permittivity is positive again, allowing propagation through the plasma by helicons as is discussed in Section 20.

Exercises

9. From (212) and (213) obtain the elements of the susceptibility tensor χ_{xx}, χ_{xy}, χ_{yx}, and χ_{yy}.

10. From the form of **(224)** and **(225)** show that Figure 14 is symmetric with respect to $\omega = \frac{1}{2}\omega_c$.

11. Equations **224** and **225** have resonances at $\omega = 0$ and $\omega = \omega_c$. Describe the charge motion in the vicinity of these resonances.

---

19   Faraday Effect. We begin by considering propagation with **k** and **B** both parallel to $\hat{z}$ as shown in Figure 15. As Faraday showed, the plane of polarization of the electric field may be observed to rotate as the wave propagates along the magnetic field.

From **(37)** we write for the wavevector of right and left polarized waves:

$$k_r{}^2 = \omega^2 \mu \epsilon_r \qquad k_l{}^2 = \omega^2 \mu \epsilon_l \tag{226}$$

For ω_c small compared with ω we may write **(226)** approximately as

$$k_r = k - \Delta k \qquad k_l = k + \Delta k \tag{227}$$

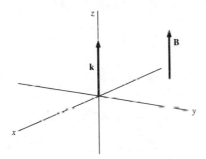

Figure 15 Wave propagation along a magnetic field. The plane of polarization of the wave is observed to rotate as the wave propagates along the field.

with

$$k^2 = \omega^2 \mu \epsilon_{\text{ion}}(1 - \omega_p{}^2/\omega^2) \tag{228}$$

and

$$\Delta k \simeq \frac{\omega_p{}^2 \omega_c}{2\omega(\omega^2 - \omega_p{}^2)} k \tag{229}$$

where we have expanded **(224)** and **(225)**.

In place of **(218)** and **(219)** we write freely propagating circular waves:

$$\underline{E}_x(z,t) = E_r\, e^{i(k_r z - \omega t)} + iE_l\, e^{i(k_l z - \omega t)} \tag{230}$$

$$\underline{E}_y(z,t) = iE_r\, e^{i(k_r z - \omega t)} + E_l\, e^{i(k_l z - \omega t)} \tag{231}$$

We set the relative phase between right and left circular waves by assuming $E_y = 0$ at $z = 0$, which gives

$$E_l = iE_r = -iE \tag{232}$$

Substituting (227) and (232) into (230) and (231) we obtain finally for the components of the electric field:

$$E_x(z,t) = 2E \cos \Delta kz\, e^{i(kz-\omega t)} \tag{233}$$

$$E_y(z,t) = 2E \sin \Delta kz\, e^{i(kz-\omega t)} \tag{234}$$

We see from (233) and (234) that the plane of polarization of the electric field rotates with z through an angle:

$$\theta = \Delta kz = VBz \tag{235}$$

where V is called the *Verdet constant*. From (229) and (205) we obtain for the Verdet constant of a plasma:

$$V = \frac{q/2mv_p}{(\omega/\omega_p)^2 - 1} \tag{236}$$

where v_p is the phase velocity of the wave.

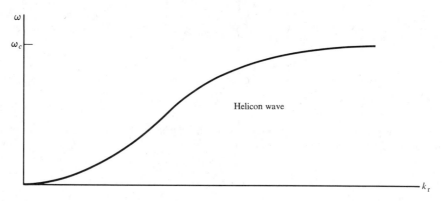

Figure 16   Helicon wave. The plasma supports a right circular wave at frequencies below the cyclotron frequency.

20   [Helicons]. For frequencies below ω_c and small compared with ω_p we may write (226) as

$$k_r^2 \simeq \omega^2 \mu\epsilon_{\text{ion}} \frac{\omega_p^2}{\omega(\omega_c - \omega)} \tag{237}$$

$$k_l^2 \simeq -\omega^2 \mu\epsilon_{\text{ion}} \frac{\omega_p^2}{\omega(\omega_c + \omega)} \tag{238}$$

Since (237) is positive the plasma will support a propagating right circular mode up to the cyclotron frequency as shown in Figure 16. The wave vector of the left circular mode is pure imaginary, and there is no propagation. For ω small compared with ω_c we write from (237)

$$\omega \simeq \frac{\omega_c}{\mu\epsilon_{\text{ion}}\,\omega_p^2} k_r^2 \tag{239}$$

which is the dispersion relation for a free particle. Note that the group velocity $d\omega/dk$ goes to zero at low frequency and increases with frequency.

The excitation of helicon modes in the magnetosphere leads to the observation of "whistlers," sound bursts that start at high frequency and move toward low frequency.[9]

The helicon is a circularly polarized wave with the sense of polarization the same as that of the cyclotron motion of the carriers. The positive contribution to the permittivity arises from carriers driven below their cyclotron frequency. Thus, the frequency for cyclotron resonance ω_c is a limiting frequency for this mode.

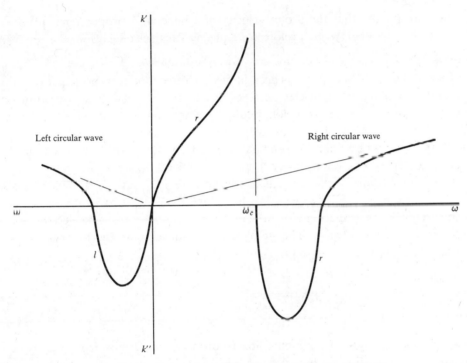

Figure 17 Dispersion relation for propagation along a magnetic field. The wave propagates without attenuation where the dielectric function is positive. Where the dielectric function is negative, the wave is exponentially attenuated.

An alternative way of looking at helicon propagation is in terms of the transverse dielectric function (224). From (57) we have, as the condition for free wave propagation,

$$\epsilon(\underline{k},\omega) = \epsilon_0 \, \frac{c^2 \underline{k}^2}{\omega^2} = \epsilon_0 \, \frac{c^2}{\omega^2} \, (k' + ik'')^2 \tag{240}$$

From Figure 14 for $\epsilon(\underline{k},\omega)$ positive we expect propagation, and for $\varepsilon(\underline{k},\omega)$ negative we expect pure attenuation as shown in Figure 17.

[9] R. Helliwell, *Whistlers and Related Ionspheric Phenomena*, Stanford, 1965. For a discussion of very similar acoustic dispersion, see F. S. Crawford, "Culvert Whistlers," *Am. J. Phys.*, **39**, 610 (1971).

Exercises

12. Show that **(239)** is equivalent to:

$$\omega = \epsilon_0\, RBc^2 k_r{}^2$$

where $R = 1/nq$ is the Hall constant.

13. Show from **(237)** that the phase velocity of a helicon is maximum at half the cyclotron frequency and that the maximum is equal to $(c/2n)(\omega_c/\omega_p)$. Neglect relaxation.

14. Show from **(239)** that the group velocity of a helicon is proportional to $\omega^{1/2}$ at frequencies well below the cyclotron frequency. Neglect relaxation.

21 **[Alfvén Waves].** Our discussion has been limited to three modes of propagation, which are admixtures of the three modes obtained in the absence of a field: two transverse electromagnetic modes and one longitudinal electric wave. Had we considered in Section 5-16 the motion of the positive charge carriers as well as the electrons, we would have obtained a second longitudinal electric wave, the positive ion wave, in which the electrons and positive ions move together (Problems 5-24 and 5-25).

In the presence of a magnetic field the positive ion wave is admixed into the other waves producing what at very low frequencies are called hydromagnetic waves. For low frequency propagation along the magnetic field the transverse electromagnetic waves are called *Alfvén waves*. The simplest way of developing the theory is to write from **(222)** for the electron susceptibility:

$$\epsilon_0\, \chi_{\text{eln}}(k,\omega) = -\frac{\omega_{\text{ep}}{}^2}{\omega(\omega - \omega_{\text{ec}} + i/\tau_{\text{eln}})}\,\epsilon_{\text{core}} \tag{241}$$

where ϵ_{core} includes only the contribution to the ion permittivity from the ion core. The electron plasma frequency is defined as:

$$\omega_{\text{ep}}{}^2 = \frac{n_{\text{eln}}\, e^2}{\epsilon_{\text{core}}\, m_{\text{eln}}} \tag{242}$$

with the electron cyclotron frequency given by

$$\omega_{\text{ec}} = \frac{eB}{m_{\text{eln}}} \tag{243}$$

We have an expression similar to **(241)** for the susceptibility associated with displacement of the ions:

$$\epsilon_0\, \chi_{\text{ion}}(k,\omega) = -\frac{\omega_{\text{ip}}{}^2}{\omega(\omega - \omega_{\text{ic}} + i/\tau_{\text{ion}})}\,\epsilon_{\text{core}} \tag{244}$$

The ion plasma frequency is defined by:

$$\omega_{ip}{}^2 = \frac{n_{ion}\, q^2}{\epsilon_{core}\, m_{ion}} \tag{245}$$

with the ion cyclotron frequency given by:

$$\omega_{ic} = -\frac{qB}{m_{ion}} \tag{246}$$

Adding contributions from ion cores and electron and ion displacement we obtain for the dielectric function:

$$\epsilon(\underline{k},\omega) = \epsilon_{core}\left(1 - \frac{\omega_{ep}{}^2}{(\omega - \omega_{ec} + i/\tau_{eln})} - \frac{\omega_{ip}{}^2}{\omega(\omega - \omega_{io} + i/\tau_{ion})}\right) \tag{247}$$

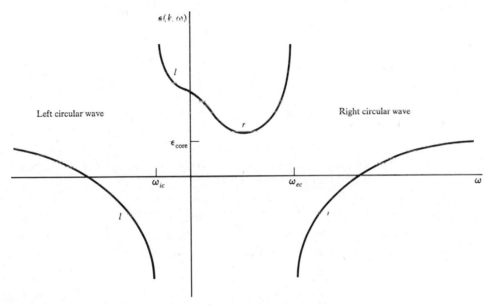

Figure 18 Dielectric function of a compensated plasma. For the electron and ion concentrations equal, the singularity at zero frequency has disappeared.

We plot in Figure 18 the dielectric function of a compensated plasma for which the electron and ion concentrations and charges are equal leading to:

$$\frac{\omega_{ep}{}^2}{\omega_{ec}} + \frac{\omega_{ip}{}^2}{\omega_{ic}} = 0 \tag{248}$$

Comparing Figure 18 with Figure 17 we see that the presence of ion cyclotron motion allows for the propagation of *l* waves up to the ion cyclotron frequency. The helicon wave, which was associated with the singularity in $\varepsilon(k,\omega)$ at zero frequency, is absent in a compensated electron-ion plasma.

Exercise

15. Plot the real and imaginary parts of the wavevector as a function of ω for right and left circular waves in the region of Alfvén wave propagation. You may wish to represent all four plots on the same graph in the way shown in Figure 18.

22 [**Voigt Effect**]. The Voigt effect is the appearance of double refraction for propagation transverse to a magnetic field. We take **k** along the z direction as before but with **B** in the x direction as shown in Figure 19. We again may use (**209**). We infer a susceptibility tensor of the form:

$$\overline{\overline{\chi}} = \begin{pmatrix} \chi_x & 0 & 0 \\ 0 & \chi_y & i\beta \\ 0 & -i\alpha & \chi_z \end{pmatrix} \tag{249}$$

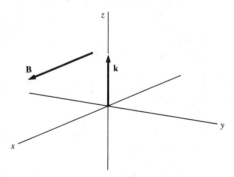

Figure 19 Propagation transverse to a magnetic field. The dispersion relation will depend on whether the wave is polarized parallel or perpendicular to the field.

The justification for this form is that polarization along the field is not carried into the other two directions while polarization along $\hat{y}$ and $\hat{z}$ are expected to be mixed. Substituting into (**209**) we obtain for the diagonal components:

$$\chi_x = -\frac{\epsilon_{\text{ion}}}{\epsilon_0} \frac{\omega_p^2}{\omega(\omega + i/\tau)} \tag{250}$$

$$\chi_y = -\frac{\epsilon_{\text{ion}}}{\epsilon_0} \frac{\omega_p^2}{\omega^2} \frac{\omega(\omega + i/\tau)}{(\omega + i/\tau)^2 - \omega_c^2} \tag{251}$$

$$\chi_z = (1 + k^2 s^2/\omega_p^2)\chi_y \tag{252}$$

For the off-diagonal elements we obtain

$$\alpha = \frac{\omega_c}{\omega + i/\tau} \chi_y \tag{253}$$

$$\beta = \frac{\omega_c}{\omega + i/\tau} \chi_z = \frac{\omega_c}{\omega + i/\tau} (1 + k^2 s^2/\omega_p^2)\chi_y \tag{254}$$

A wave polarized with **E** along $\hat{x}$ has the same velocity as in the absence of a magnetic field and is called the *ordinary wave*. The square of the wavevector is given at frequencies for which relaxation can be neglected by:

$$k^2 = \omega^2 \mu \epsilon_{\mathrm{ion}} \left(1 - \frac{\omega_p^2}{\omega^2} \right) \tag{255}$$

The two other modes are obtained on substitution of the $\hat{y}$ and $\hat{z}$ parts of **(249)** into **(37)**. We obtain

$$k^2 E_y - \omega^2 \mu [(\epsilon_{\mathrm{ion}} + \epsilon_0 \chi_y) E_y + \epsilon_0 \, l\beta E_z] = 0 \tag{256}$$

$$-i\alpha\epsilon_0 \, E_y + (\epsilon_{\mathrm{ion}} + \epsilon_0 \chi_z) E_z = 0 \tag{257}$$

From the vanishing of the determinant of the coefficients of E_x and E_y we obtain:

$$(\epsilon_{\mathrm{ion}} + \epsilon_0 \chi_z) \left[(\epsilon_{\mathrm{ion}} + \epsilon_0 \chi_y) - \frac{k^2}{\omega^2 \mu} \right] - \epsilon_0^2 \alpha\beta \tag{258}$$

By neglecting carrier relaxation and pressure we obtain

$$\frac{k^2}{\omega^2 \mu \epsilon_{\mathrm{ion}}} = 1 - \frac{\omega_p^2}{\omega^2} \frac{\omega_p^2 - \omega^2}{\omega_p^2 + \omega_c^2 - \omega^2} \tag{259}$$

This equation has two branches, which we discuss below.

For zero applied field ($\omega_c = 0$) the transverse and longitudinal waves were shown in Figure 7. The threshold for the transverse wave is the plasma frequency. At frequencies well above the plasma frequency the mobile carriers are no longer able to follow the field and the wave propagates with a permittivity given by the ion cores. The threshold for the longitudinal wave is also the plasma frequency. At high frequencies the phase velocity approaches s, which is a longitudinal acoustic wave velocity as discussed in Section 5-16.

For propagation *along* a magnetic field the longitudinal wave is unaffected. The transverse waves are circularly polarized as will be discussed in detail for magnetic insulators. For propagation *across* the field the wave which is polarized along the field is called the ordinary wave and as we have seen has the same dispersion as the zero-field transverse wave. The mode that is polarized perpendicular to the field is mixed with the longitudinal wave by the field as shown in Figure 20. The splitting at $k = 0$ is just equal to the cyclotron frequency. From **(259)**, which neglects pressure, we expect the lower wave to asymptotically approach $(\omega_p^2 + \omega_c^2)^{1/2}$. Including pressure leads to $\omega = sk$ at high frequencies as we have seen.

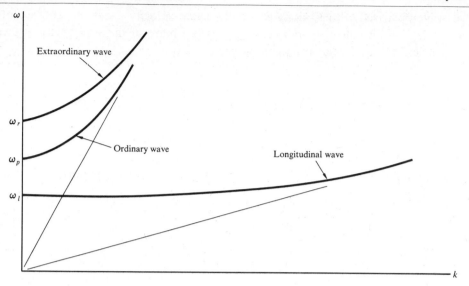

Figure 20   Dispersion relation for transverse propagation. Three modes of propagation may be obtained.

Exercise

16. Show that at $k = 0$ the extraordinary wave is purely electrostatic.

MAGNETOOPTIC EFFECTS IN MAGNETIC INSULATORS

We discuss here optical effects in transparent magnetic insulators. Our model will be the same as that introduced in Section 3-2 in the discussion of atomic polarizability except that we add an effective magnetic field $\mathbf{B}_{eff}$ as shown in Figure 21.

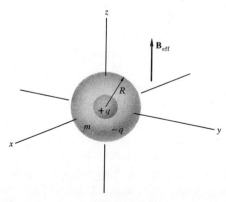

Figure 21   Model of a magnetic insulator. A polarizable atom experiences an effective magnetic field.

23 [Effective Field]. The equation of motion that we will use is:

$$m\frac{d^2\mathbf{p}}{dt^2} + \frac{m}{\tau}\frac{d\mathbf{p}}{dt} - q\frac{d\mathbf{p}}{dt} \times \mathbf{B}_{\text{eff}} + \frac{q^2}{\alpha_0}\mathbf{p} = q^2\mathbf{E} \tag{260}$$

where the polarizability is given from **(3-12)** as:

$$\alpha_0 = 4\pi\epsilon_0\,R^3 \tag{261}$$

The effective field is equal to the local magnetic field plus a term proportional to the mean atomic magnetization:

$$\mathbf{B}_{\text{eff}} = \mathbf{B} + \beta\mu_0\,\mathbf{M} \tag{262}$$

The origin of the term in the magnetization is the spin-orbit coupling, which we discussed in Section 9-2, where we saw **(9-26)** that a moving magnetic moment $\boldsymbol{\mu}'$ experiences a force:

$$\mathbf{F} = \frac{1}{c^2}(\mathbf{v} \times \boldsymbol{\mu}' \cdot \nabla)\mathbf{E} \tag{263}$$

With the electric field given by

$$\mathbf{E} = \frac{q}{4\pi\epsilon_0}\frac{\hat{\mathbf{r}}}{r^2} \tag{264}$$

the field gradient tensor is given by:

$$\nabla\mathbf{E} = \frac{q}{4\pi\epsilon_0}\left(\frac{\mathbf{I}}{r^3} - 3\frac{\hat{\mathbf{r}}\hat{\mathbf{r}}}{r^3}\right) \tag{265}$$

Substituting **(265)** into **(263)** and averaging over angles we obtain:

$$\mathbf{F} = -\frac{\mu_0}{2\pi r^3}\,q\mathbf{v} \times \boldsymbol{\mu}' \tag{266}$$

which leads to an effective field when averaged over spin orientation of:

$$\beta\mathbf{M} = \beta n\langle\boldsymbol{\mu}'\rangle = \frac{1}{2\pi r^3}\langle\boldsymbol{\mu}'\rangle \tag{267}$$

The factor β is then approximately the ratio of the volume of the atomic cell to the actual volume occupied by the atomic magnetization. This factor is between 10 and several hundred, depending primarily on the structure of the magnetic medium.

We wish to discuss the response of the medium to a circularly polarized electric field of the form:

$$\mathbf{E}_1(z,t) = E[\hat{\mathbf{x}} \cos(kz - \omega t) \mp \hat{\mathbf{y}} \sin(kz - \omega t)] \tag{268}$$

The minus sign is for right circular polarization and the plus sign for left circular polarization. By delaying the zero of time by a quarter cycle we obtain a second field:

$$\mathbf{E}_2(z,t) = E[\hat{\mathbf{x}} \sin(kz - \omega t) \pm \hat{\mathbf{y}} \cos(kz - \omega t)] \tag{269}$$

As before, we construct a complex field that is a linear combination of (268) and (269):

$$\mathbf{E}(z,t) = \mathbf{E}_1(z,t) + i\mathbf{E}_2(z,t) = E(\hat{\mathbf{x}} \pm i\hat{\mathbf{y}})e^{i(kz - \omega t)} \tag{270}$$

We understand that the real part of the polarization computed from (270) is the polarization that would result from (268). This is the same argument that we used earlier for linearly polarized fields.

By analogy with (268) we may expect, as a response to that field, a polarization of the form

$$\mathbf{p}_1(z,t) = p[\hat{\mathbf{x}} \cos(kz - \omega t - \phi) \mp \hat{\mathbf{y}} \sin(kz - \omega t - \phi)] \tag{271}$$

and to (180) a polarization of the form:

$$\mathbf{p}_2(z,t) = p[\hat{\mathbf{x}} \sin(kz - \omega t - \phi) \pm \hat{\mathbf{y}} \cos(kz - \omega t - \phi)] \tag{272}$$

and where we understand that p and ϕ will depend on frequency as well as on the sense of polarization. We develop a complex polarization corresponding to (270) by taking a complex linear combination of (271) and (272):

$$\mathbf{p}(z,t) = \mathbf{p}_1(z,t) + i\mathbf{p}_2(z,t) = p(\hat{\mathbf{x}} \pm i\hat{\mathbf{y}})e^{i(kz - \omega t)} \tag{273}$$

To simplify the notation, the phase angle ϕ has been absorbed into p, which is to be regarded as a complex number.

From the form of (273) we may write, for the cross product in (260),

$$-q\frac{d\mathbf{p}}{dt} \times \mathbf{B}_{\text{eff}} = m\boldsymbol{\omega}_c \times \frac{d\mathbf{p}}{dt} = -i\omega m\boldsymbol{\omega}_c \times \mathbf{p} = -i\omega\omega_c\, mp(\hat{\mathbf{y}} \mp i\hat{\mathbf{x}})e^{i(kz - \omega t)} \tag{274}$$

where we have taken for a negative charge $\boldsymbol{\omega}_c = (q/m)\mathbf{B}_{\text{eff}}$. But (274) is equal to:

$$\mp \omega\omega_c\, m\, p(z,t) \tag{275}$$

Finally, substituting into (251), we have

$$[-\omega(\omega \pm \omega_c + i/\tau) + \omega_0^2]\mathbf{p}(z,t) = \alpha_0\,\omega_0^2\mathbf{E}(z,t) \tag{276}$$

where the square of the electronic zero field resonance frequency is given by

$$\omega_0{}^2 = \frac{q^2}{m\alpha_0} \tag{277}$$

The complex atomic polarizability is then given by:

$$\alpha = \frac{p}{E} = \frac{\alpha_0\,\omega_0{}^2}{\omega_0{}^2 - \omega(\omega \pm \omega_c + i/\tau)} \tag{278}$$

The complex permittivity is written as

$$\epsilon = \epsilon' + i\epsilon'' = \epsilon_0 + n\alpha \tag{279}$$

and the real and imaginary parts from **(234)** and **(235)** as:

$$\epsilon' = \epsilon_0 + n\alpha_0\,\omega_0{}^2\,\frac{\omega_0{}^2 - \omega(\omega \pm \omega_c)}{[\omega_0{}^2 - \omega(\omega \pm \omega_c)]^2 + \omega^2/\tau^2} \tag{280}$$

$$\epsilon'' = n\alpha_0\,\omega_0{}^2\,\frac{\alpha/\tau}{[\omega_0{}^2 - \omega(\omega + \omega_c)]^2 + \omega^2/\tau^2} \tag{281}$$

The expressions given by **(280)** and **(281)** are for circularly polarized electric fields with the plus sign corresponding to right circular polarization and the minus sign to left circular polarization. We now apply these expressions to propagation through a magnetic medium.

24 **[Optical Rotatory Dispersion]**. An important technique for studying the electronic structure of magnetically active media is to measure the frequency dependence of Faraday rotation. This technique is called optical rotatory dispersion. As we saw in Section 19, for propagation along a magnetic field in an otherwise isotropic medium, the right and left circularly polarized waves propagate with wave vectors:

$$k_r = k - \Delta k \qquad k_l = k + \Delta k \tag{282}$$

and the plane of polarization rotates as

$$\theta = \Delta k z \tag{283}$$

From **(282)** we obtain

$$\Delta k = \tfrac{1}{2}(k_l - k_r) \qquad k = \tfrac{1}{2}(k_l + k_r) \tag{284}$$

and we may write **(283)** as:

$$\theta = \tfrac{1}{2}(k_l - k_r)z \tag{285}$$

Neglecting absorption, (285) may be written as:

$$\theta = \tfrac{1}{2}\omega\mu^{1/2}z(\epsilon_l'^{1/2} - \epsilon_r'^{1/2}) = \tfrac{1}{2}\omega\,\frac{\mu^{1/2}z}{\epsilon_l'^{1/2} + \epsilon_r'^{1/2}}\,(\epsilon_l' - \epsilon_r') \qquad (286)$$

Although we should examine the full expression, the main features of the rotatory dispersion are given by looking at the term $\epsilon_l' - \epsilon_r'$, which is shown in Figure 22.

25 [**Magnetic Circular Dichroism**]. A second technique for studying the electronic structure of magnetic media involves looking at the difference in absorption of the right and

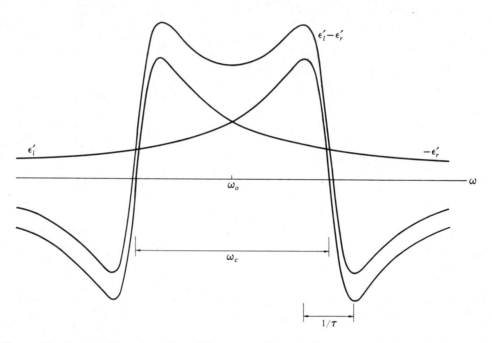

Figure 22 Rotary dispersion. The difference in dielectric function for left and right circular waves is shown as a function of frequency.

left circularly polarized components of the wave. If we include absorption, (230) and (231) should be written as

$$\mathbf{E}(z,t) = E_r(\hat{\mathbf{x}} + i\hat{\mathbf{y}})e^{-k_r''z}e^{i(k_r'z - \omega t)} + E_l(\hat{\mathbf{y}} + i\hat{\mathbf{x}})e^{-k_l''z}e^{i(k_l'z - \omega t)} \qquad (287)$$

The real part of (287) will be an elliptically polarized wave with the major axis given by (286) and the ellipticity determined by the ratio of the axes

$$\tanh(k_l'' - k_r'')z = \frac{e^{-k_l''z} - e^{-k_r''z}}{e^{-k_l''z} + e^{-k_r''z}} \qquad (288)$$

As a measure of the ellipticity we plot in Figure 23 the quantity $\epsilon_l'' - \epsilon_r''$.

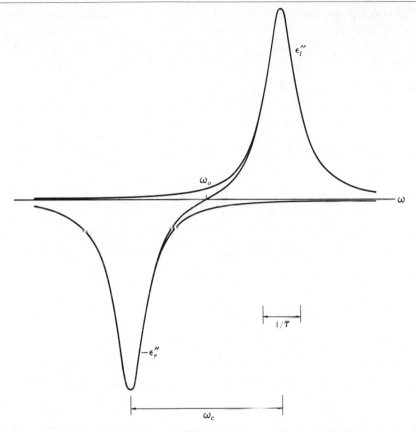

Figure 23 Magnetic circular dichroism. The ellipticity of the wave depends on the difference in the imaginary parts of the dielectric function for left and right circular waves.

SUMMARY

1. The electric field and displacement are related in a linear medium by

$$\underline{\mathbf{D}}(\mathbf{r},t) = \underline{\bar{\bar{\epsilon}}}(\omega,\mathbf{k}) \cdot \underline{\mathbf{E}}(\mathbf{r},t) \tag{29}$$

where $\underline{\mathbf{D}}(\mathbf{r},t)$ and $\underline{\mathbf{E}}(\mathbf{r},t)$ are complex sinusoidally varying fields:

$$\underline{\mathbf{E}}(\mathbf{r},t) = \underline{\mathbf{E}}e^{i(\mathbf{k}\cdot\mathbf{r}-\omega t)} \tag{27}$$

and $\underline{\mathbf{k}}$ is a complex vector

$$\underline{\mathbf{k}} = \mathbf{k}' + i\mathbf{k}'' \tag{24}$$

2. The Poynting vector is written for complex fields as

$$\boldsymbol{\Pi} = \tfrac{1}{2}\mathrm{Re}\langle \underline{\mathbf{E}}(\mathbf{r},t) \times \underline{\mathbf{H}}^*(\mathbf{r},t)\rangle \tag{47}$$

where $\underline{\mathbf{H}}^*(\mathbf{r},t)$ is the complex conjugate of the magnetic field $\underline{\mathbf{H}}(\mathbf{r},t)$. The average is to be taken over one cycle and Re indicates the real part of the bracketed expression.

3. The rate of energy dissipation may be written in terms of the imaginary part of the permittivity:

$$\frac{dw}{dt} = \tfrac{1}{2}\omega\underline{\mathbf{E}}\cdot\bar{\bar{\epsilon}}''\cdot\underline{\mathbf{E}}^* \tag{52}$$

4. The electric and magnetic fields in an isotropic medium are related by the intrinsic impedance:

$$Z = \frac{E_x}{\underline{H}_y} = -\frac{E_y}{\underline{H}_x} = \frac{\omega\mu}{\underline{k}} = (\mu/\underline{\epsilon})^{1/2} \tag{59}$$

5. The velocity with which a position of definite phase advances is given by the phase velocity:

$$v_p = \frac{\omega}{k'} \tag{67}$$

where k' is the real part of the complex wavevector.

6. The group velocity is the velocity with which information is transmitted and is most easily pictured as the velocity of propagation of a wave envelope:

$$v_g = \frac{d\omega}{dk} \tag{92}$$

7. By examining a simple model for a dispersive medium, an oscillator at a single frequency coupled to the electromagnetic field, it may be seen that the group velocity is less than c over the entire spectrum.

8. An electromagnetic wave propagates through a plasma either as a longitudinal electric wave with

$$\omega^2 = \omega_p{}^2 + s^2k^2 \tag{128}$$

or as a transverse electromagnetic wave with

$$\omega^2 = \omega_p{}^2 + k^2/\mu\epsilon_{\text{ion}} \tag{126}$$

9. In an anisotropic medium there are two directions, the optic axes, for which the two transverse modes have the same properties. For a medium with an axis of threefold or higher symmetry, the two optic axes merge into a single axis.

10. In a medium where the individual components have "handedness" the plane of polarization of an electromagnetic wave will rotate even though the components are randomly oriented. This rotation is called optical activity. A simple model is given by two polarizable dipoles coupled through their dipolar fields.

11. For propagation of an electromagnetic wave parallel to a magnetic field and through a plasma, the longitudinal dielectric function is the same as without the field. The transverse dielectric function shows a resonance at the carrier cyclotron frequency, and is positive for frequencies below the cyclotron frequency.

12. For transverse electromagnetic waves propagated through a plasma and along a magnetic field there is a rotation of the plane of polarization, called the Faraday effect.

13. Associated with the fact that the transverse permittivity is positive at low frequency in a magnetoplasma, there is a new mode of propagation, the helicon, which extends up to the carrier cyclotron frequency.

14. Where both positive and negative carriers are mobile and of equal charge and concentration, propagation is possible from negative circular frequencies equal to the ion cyclotron frequency to positive circular frequencies equal to the electron cyclotron frequency. The singularity at zero frequency that characterizes the helicon is not present in a compensated plasma where the low frequency waves are called Alfvén waves.

15. Electromagnetic waves may also be propagated across a magnetic field. The transverse wave polarized along the magnetic field is unaffected by the field and is the same as in zero field. The other transverse wave mixes with the longitudinal wave.

16. The propagation of a wave through a magnetic insulator may be strongly affected by a magnetic field. These field effects are generally associated with the sensitivity of the permittivity of the medium to the field. A strong enhancement of the coupling to the field may exist in media which are also magnetic and arises from spin-orbit coupling. A single resonance frequency may be split by the magnetic field and observed either through the associated Faraday rotation (optical rotatory dispersion) or through the preferential absorption of one of the components (magnetic circular dichroism).

PROBLEMS

1. **Lorentz force.** A charge q travels with velocity $\mathbf{v}$ in the same direction as the propagation vector $\mathbf{k}$ of an electromagnetic wave in vacuum. Obtain an expression for the Lorentz force acting on the particle. Explain why the force goes to zero as v approaches c.

2. **Optical path.** Assume solutions of the Maxwell equations of **(16)**, **(17)**, **(18)**, and **(19)** of the form:

$$\underline{\mathbf{E}}(\mathbf{r},t) = \mathbf{E}(\mathbf{r})e^{i\omega(\tau - t)}$$

$$\underline{\mathbf{H}}(\mathbf{r},t) = \mathbf{H}(\mathbf{r})e^{i\omega(\tau - t)}$$

where ρ_{ext} and $\mathbf{j}_{\text{ext}}$ are zero and where τ is a real function of position.
(a) Show that $\mathbf{E}(\mathbf{r})$, $\mathbf{H}(\mathbf{r})$, and $\nabla\tau$ are mutually orthogonal.
(b) Obtain the relation with μ and ϵ real

$$\nabla\tau = (\mu\epsilon)^{1/2}\hat{\Pi}$$

3. **Lyddane-Sachs-Teller relation.** Show that in a dispersive medium as discussed in Section 9 electromagnetic waves will not propagate for frequencies between ω_T and ω_L where ω_L is given by

$$\omega_L{}^2 = \omega_T{}^2 + \omega_p{}^2$$

4. **Propagation through a plasma.** A transverse electromagnetic wave propagates in a plasma characterized by a permittivity:

$$\epsilon = \epsilon_{\text{ion}}\left(1 - \frac{\omega_p{}^2}{\omega(\omega + i/\tau)}\right)$$

Assuming $\omega_p > 1/\tau$, obtain the form of the wavevector and intrinsic impedance for the three limiting cases:
(a) Low frequencies: $\omega < 1/\tau < \omega_p$
(b) Intermediate frequencies: $1/\tau < \omega < \omega_p$
(c) High frequencies: $1/\tau < \omega_p < \omega$

5. **Propagation through a lossy medium.** An electromagnetic wave propagates in an isotropic but lossy dielectric medium. The electric field is of the form

$$\mathbf{E}(z,t) = E\hat{\mathbf{x}}e^{i(kz - \omega t)}$$

where both $\underline{\mathbf{E}}(z,t)$ and $\underline{k}$ are complex quantities.
(a) Obtain the form of $\underline{\mathbf{H}}(z,t)$.
(b) Obtain the expression for the Poynting vector using

$$\mathbf{\Pi} = \tfrac{1}{2}\,\text{Re}[\underline{\mathbf{E}}(z,t) \times \underline{\mathbf{H}}^*(z,t)]$$

(c) Obtain the expression for the complex wave inpedance $\underline{Z} = \underline{E}/\underline{H}$. Write the real and imaginary parts, Z' and Z'' separately.

6. **Propagation through sea water.** The electric conductivity of sea water is nominally 4.3 siemens per metre (reciprocal ohm metre). Assume a relative permittivity ϵ'/ϵ_0 of 81 as for pure water.
(a) Find the phase velocity v_p and attenuation distance $\delta = 1/k''$ at frequencies $\omega/2\pi$ of 20 kHz and 10 MHz. These are typical very low frequency (VLF) and radio frequencies (RF).
(b) At the frequency of light the relative permittivity has dropped to $n^2 = 1.777$. Assume that the conductivity arises entirely from Na^+ ions, which are present to 1 percent concentration. Find $\omega\tau$ in this frequency range.
(c) Find the attenuation distance at the frequency of light.

7. **Eddy current damping.** An electromagnetic wave is propagated through a highly conducting medium. For what frequencies may the propagation be treated as in Section 9-12 where the term $\partial\mathbf{D}/\partial t$ in **(18)** does not appear.

8. **Nonlocal permittivity.** A linear medium may be characterized at zero frequency by a permittivity $\bar{\epsilon}(\mathbf{k})$:

$$\mathbf{D}(\mathbf{k}) = \bar{\epsilon}(\mathbf{k}) \cdot \mathbf{E}(\mathbf{k})$$

(a) Show that the displacement at position $\mathbf{r}$ is given by:

$$\mathbf{D}(\mathbf{r}) = \int_{V_1} \bar{\epsilon}(\mathbf{r} - \mathbf{r}_1) \cdot \mathbf{E}(\mathbf{r}_1) \, dV_1$$

where we have

$$\bar{\epsilon}(\mathbf{r} - \mathbf{r}_1) = (2\pi)^{-3} \int \bar{\epsilon}(\mathbf{k}) e^{i\mathbf{k} \cdot (\mathbf{r} - \mathbf{r}_1)} \, dV_1$$

(b) Show that for $\bar{\epsilon}(\mathbf{k}) = \bar{\epsilon}$ we obtain

$$\mathbf{D}(\mathbf{r}) = \bar{\epsilon} \cdot \mathbf{E}(\mathbf{r})$$

9. **Carrier relaxation.** By including carrier relaxation obtain the low frequency propagation modes of an electromagnetic wave directed along a magnetic field. Find the attenuation of the helicon and Alfvén waves.

10. **Propagation of light through a plasma.** An electromagnetic wave is propagated through a plasma. Obtain the expression for the group velocity. Find the time for a pulse to be propagated through a plasma of extent L and integrated density $N = \int n \, dz$. Show that for light pulses differing in wave length by $\Delta\lambda$, the difference in arrival times is given by

$$\Delta T = \frac{e^2 N}{4\pi^2 \sqrt{\epsilon_0} \, mc^3} \lambda \Delta\lambda$$

where we have assumed that the light frequency $\omega = 2\pi c/\lambda$ is much higher than the plasma frequency.

11. **Extreme ultraviolet filters.** Metallic foils of aluminum, indium, and tin are used as filters that pass ultraviolet radiation. Foil thickness is between 150 and 170 nm. The low frequency or long wavelength cutoff is determined by the plasma frequency. The high frequency cutoff is determined by core absorption. Locate the pass band for the three metals.

12. **Perfect conductor.** A perfect conductor is an idealization in the limit that the carrier concentration n goes to infinity while the relaxation rate $1/\tau$ remains finite. A perfectly conducting body is placed in an oscillating electromagnetic field.
    (a) Show that the electric field is zero everywhere within the conductor.
    (b) Show that the magnetic field is zero everywhere within the conductor.
    (c) Show that the current density is zero everywhere *within* the conductor.

13. **Intrinsic admittance.** A medium may be characterized by a complex admittance:

$$\underline{Y} = \frac{1}{\underline{Z}}$$

which is the reciprocal of the intrinsic impedance. Write the real and imaginary parts of $\underline{Y}$ in terms of ϵ' and ϵ''.

14. **Microwave Faraday rotation.** Linearly polarized microwave radiation propagates through a magnetic rod parallel to an applied magnetic field **B**. Use **(7-63)** to obtain the permeability tensor and the specific rotation. [K. L. Yan and W. P. Lonc, *Am. J. Phys.*, **43**, 718 (1975).]

REFERENCES

Baldwin, G. C., *An Introduction to Nonlinear Optics*, Plenum, 1969.

Born, M. and E. Wolf, *Principles of Optics*, Fifth Edition, Pergamon, 1975.

Bowers, R., "Plasmas in Solids," *Scientific American*, **209** (5), 46 (November, 1963).

Chen, F. F., *Introduction to Plasma Physics*, Plenum, 1974.

Clemmow, P. C. and J. P. Dougherty, *Electrodynamics of Particles and Plasmas*, Addison-Wesley, 1969.

Ferro, U. C. A. and C. Plumpton, *Magneto-fluid Mechanics*, Oxford, 1967.

Haerendel, G. and R. Lust, "Artificial Plasma Clouds in Space," *Scientific American*, **219** (5), 80 (November, 1968).

Ichimaru, S., *Basic Principles of Plasma Physics*, Benjamin, 1973.

Jenkins, F. A., and H. E. White, *Fundamentals of Optics*, Fourth Edition, McGraw-Hill, 1976.

Kittel, C., *Introduction to Solid State Physics*, Fifth Edition, Wiley, 1976.

Lilley, A. E., "The Absorption of Radio Waves in Space," *Scientific American*, **197** (1), 48 (July, 1957).

Pierce, J. R., *Almost All about Waves*, MIT, 1974.

Spitzer, L., *Physics of Fully Ionized Gases*, Second Edition, Interscience, 1962.

Wooten, F. O., *Optical Properties of Solids*, Academic, 1972.

Zernike, F. and J. E. Midwinter, *Applied Nonlinear Optics*, Wiley, 1973.

13 ELECTROMAGNETIC MEDIA II

We begin this chapter by discussing the reflection and refraction of electromagnetic radiation at the interface between media of differing permittivity or permeability. To get insight into the problem we first look at the radiation from the interfacial polarization that is excited by the wave. In this way we are able to obtain the directions of the reflected and refracted wave and to understand the phenomenon of total internal reflection. The equivalent result is obtained from the continuity of the fields by requiring that the boundary conditions be satisfied in an invariant way over the interface. In later sections we apply this theory to guided waves and to recent developments in forming what are called integrated optical devices.

REFLECTION AND REFRACTION

We show in Figure 1a a plane wave with propagation vector $\mathbf{k}_1$ incident on the interface between media of permittivity ϵ_1 and ϵ_2.

If the incident wave is written as

$$\mathbf{E} = \mathbf{E}_1 \cos(\mathbf{k}_1 \cdot \mathbf{r} - \omega t) \tag{1}$$

there will be a charge (or current) at the interface of the form:

$$\rho = \rho_1 \cos(\mathbf{k}_1 \cdot \mathbf{r} - \omega t) = \rho_1 \cos(k_1 x \sin \theta_1 - \omega t) \tag{2}$$

The charges radiate back into the first medium along a direction such that the radiation from the interfacial charge interferes constructively. If the medium is homogeneous, then the only direction for which there is constructive interference is for the angle of reflection θ_3 equal to the angle of incidence θ_1 as shown in Figure 1b.

1   Snell's Law and Refractive Index. Because the phase velocity in the second medium is different from that in the first, the direction for constructive interference will be different. If we write for the transmitted wave in the interfacial plane as shown in Figure 1c:

$$E = E_2 \cos(\mathbf{k}_2 \cdot \mathbf{r} - \omega t) = E_2 \cos(k_2 x \sin \theta_2 - \omega t) \tag{3}$$

then **(2)** and **(3)** must have the same periodicity from which we must conclude

$$k_1 \sin \theta_1 = k_2 \sin \theta_2 \tag{4}$$

which is Snell's law with

$$k = \omega \sqrt{\mu \epsilon} = \omega/v_p \tag{5}$$

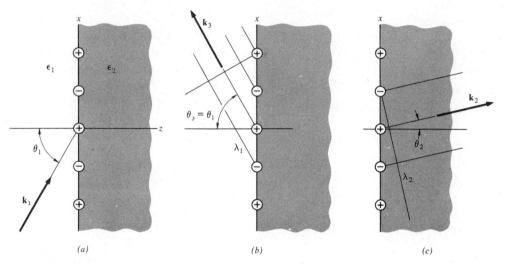

(a)   (b)   (c)

Figure 1 Reflection at an interface. (*a*) A plane wave is incident on the interface between two media. (*b*) The wave is reflected with the angle of reflection equal to the angle of incidence. (*c*) The wave is partially transmitted at an angle given by Snell's law **(4)**.

Combining **(4)** and **(5)**, Snell's law may also be expressed in terms of the phase velocities in the two media:

$$\frac{v_{p1}}{\sin \theta_1} = \frac{v_{p2}}{\sin \theta_2} \tag{6}$$

The index of refraction of a medium is the ratio of c to the phase velocity in the medium:

$$n = \frac{c}{v_p} = \frac{ck}{\omega} \tag{7}$$

In terms of the refractive indices of the two media, **(4)** becomes:

$$n_1 \sin \theta_1 = n_2 \sin \theta_2 \tag{8}$$

In Table 1 we give the refractive indices of a number of selected materials. Refractive indices are for sodium light ($\lambda = 589.3$ nm) and at room temperature. For materials that are cubic or amorphous, only one index is given. For materials with a single optic axis, indices are given along and transverse to the optic axis. For biaxial crystals, the indices are given for the principal directions of the permittivity tensor.

Table 1   Refractive Indices of Selected Materials.

Crystalline	n	Glass	n
C (diamond)	2.4173	Zinc crown	1.517
NaI	1.7745	Light flint	1.575
S	1.957, 2.0377, 2.2454	Heavy flint	1.650
ZnCl	1.687, 1.713		

Plastic	n
Methyl methacrylate—acrylic (lucite, plexiglas)	1.491
Methyl methacrylate styrene copolymer (nas)	1.563
Polycarbonate (lexan, merlon)	1.586
Polystyrene—styrene (dylene, styron)	1.590

Exercise

1. Light of wavelength λ is refracted by a thin prism. Show that near normal incidence the angle of refraction is approximately

$$\theta \simeq (n - 1)\alpha$$

where n is the refractive index and α is the prism angle.

2   Total Internal Reflection. An interesting situation exists when medium 1 has higher permittivity than medium 2 and the angle of incidence θ_1 is not too small. Rewriting **(4)** as

$$\sin \theta_2 = \frac{k_1}{k_2} \sin \theta_1 \tag{9}$$

we see there is a critical angle θ_c determined by the relation:

▶
$$\frac{k_1}{k_2} \sin \theta_c = 1 \tag{10}$$

such that for θ_1 greater than θ_c the right side of **(9)** is greater than 1, and it is not possible to satisfy **(9)** with θ_2 real. At the critical angle of incidence the refracted wave has its wavevector in the interfacial plane. If the angle of incidence θ_1 is increased beyond θ_c, there is *no* direction for which we can have constructive interference, and there can be no refracted wave.

Under conditions of total internal reflection the electric field in medium 1 is given by

$$E_1(r,t) = 2E_1 \cos(k_1 z \cos\theta_1 + \phi)\cos(k_1 x \sin\theta_1 - \omega t) \tag{11}$$

which has the form of a wave traveling along the x direction and sinusoidally modulated along the z direction as a result of interference between the incident and reflected wave. The electric field in medium 2 is of the form

$$E_2(r,t) = E_2\, e^{-\beta z}\cos(k_1 x \sin\theta_1 - \omega t) \tag{12}$$

Substituting **(12)** into the wave equation **(12-57)** in the absence of external currents we obtain for the attenuation constant

$$\beta = \omega(\mu_1\epsilon_1 \sin^2\theta_1 - \mu_2\epsilon_2)^{1/2} \tag{13}$$

By calculating the Poynting vector for **(11)** and **(12)** it may be seen that the flow of energy in both media is entirely along $\hat{x}$. Thus, once the wave is established, there is no continuing flow of energy from medium 1 into medium 2.

3 Normal Incidence. In order to obtain the relative intensities of the reflected and refracted waves it is necessary to match the fields at the boundary. We first examine the problem for normal incidence ($\sin\theta_1 = 0$).

We take the incident electric field to be of the form:

$$\underline{\mathbf{E}}_1(\mathbf{r},t) = E_1\hat{x}e^{i(k_1 z - \omega t)} \tag{14}$$

Then the incident magnetic field will be from **(12-39)**:

$$\underline{\mathbf{H}}_1(\mathbf{r},t) = \frac{E_1}{Z_1}\hat{y}e^{i(k_1 z - \omega t)} \tag{15}$$

where Z is the intrinsic impedance of the medium and is given **(12-59)** by

$$\blacktriangleright \qquad Z = \frac{\omega\mu}{k} = \mu v_p = \left(\frac{\mu}{\epsilon}\right)^{1/2} \tag{16}$$

For a medium with both the permeability and permittivity real positive, the intrinsic impedance Z will be real. If either μ or ϵ is real negative and the other real positive, then Z will be pure imaginary. In general, both μ and ϵ will be complex as a result of energy loss and Z will also be a complex quantity. The transmitted electric field, as shown in Figure 2 will be given by

$$\mathbf{E}_2(\mathbf{r},t) = E_2\,\hat{x}e^{i(k_2z-\omega t)} \tag{17}$$

and the magnetic field by

$$\mathbf{H}_2(\mathbf{r},t) = \frac{E_2}{Z_2}\,\hat{y}e^{i(k_2z-\omega t)} \tag{18}$$

where Z_2 is the intrinsic impedance of the second medium. The wave reflected back into the first medium will have an electric field:

$$\mathbf{E}_3(\mathbf{r},t) = E_3\,\hat{x}e^{-i(k_1z+\omega t)} \tag{19}$$

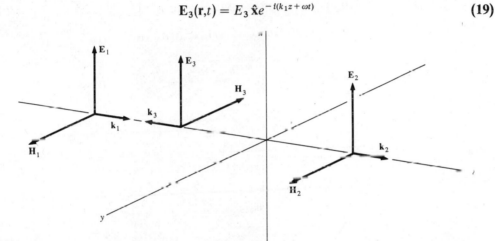

Figure 2 Reflection and transmission at normal incidence. The electric and magnetic fields are related by the intrinsic impedance of the medium.

and a magnetic field:

$$\mathbf{H}_3(\mathbf{r},t) = -\frac{E_3}{Z_1}\,\hat{y}e^{-i(k_1z+\omega t)} \tag{20}$$

At the boundary the transverse electric field must be continuous **(4-35)** and the transverse magnetic field **H** must be continuous as well **(10-24)**. These conditions lead to the equations:

$$E_1 + E_3 = E_2 \quad \text{or} \quad Z_1H_1 + Z_1H_3 = Z_2H_2 \tag{21}$$

$$\frac{E_1}{Z_1} - \frac{E_3}{Z_1} = \frac{E_2}{Z_2} \quad \text{or} \quad H_1 - H_3 = H_2 \tag{22}$$

for which the solutions are:

$$E_2 = \frac{2Z_2}{Z_1 + Z_2}\,E_1 \qquad E_3 = \frac{Z_2 - Z_1}{Z_1 + Z_2}\,E_1 \tag{23}$$

We define a complex amplitude reflection coefficient:

$$\blacktriangleright \qquad r = \rho e^{i\phi} = \frac{Z_2 - Z_1}{Z_2 + Z_1} = \frac{(\mu_2/\mu_1)^{1/2} - (\epsilon_2/\epsilon_1)^{1/2}}{(\mu_2/\mu_1)^{1/2} + (\epsilon_2/\epsilon_1)^{1/2}} \qquad (24)$$

If the real part of the incident field is written as:

$$\mathrm{Re}\ \mathbf{E}_1(\mathbf{r},t) = E_1 \hat{\mathbf{x}}\ \cos(k_1 z - \omega t) \qquad (25)$$

then the real part of the reflected field will be written as:

$$\mathrm{Re}\ \mathbf{E}_3(\mathbf{r},t) = \rho E_1 \hat{\mathbf{x}}\ \cos(k_1 z + \omega t - \phi) \qquad (26)$$

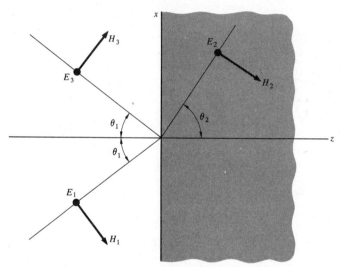

Figure 3   Oblique incidence of the transverse electric (TE) wave. The electric field for the incident, the reflected, and the transmitted wave is normal to the scattering plane.

and the power reflection coefficient will be given by:

$$R(\omega) = \rho^2 = \left| \frac{Z_2 - Z_1}{Z_2 + Z_1} \right|^2 = \left| \frac{(\mu_2/\mu_1)^{1/2} - (\epsilon_2/\epsilon_1)^{1/2}}{(\mu_2/\mu_1)^{1/2} + (\epsilon_2/\epsilon_1)^{1/2}} \right|^2 \qquad (27)$$

The relations **(23)** are the same as for reflection and transmission at a junction between transmission lines of characteristic impedance Z_1 and Z_2 as shown in Section E-7.

4 Oblique Incidence of the Transverse Electric (TE) Wave. We show in Figure 3 the situation for nonvanishing angle of incidence and **E** normal to the scattering plane, which is the plane determined by the incident wave vector $\mathbf{k}_1$ and the normal $\hat{\mathbf{n}}$ to the interface. Equating the tangential components of the electric field we obtain:

$$E_1 e^{i(k_1 x \sin \theta_1)} + E_3\, e^{i(k_1 x \sin \theta_1)} = E_2\, e^{i(k_2 x \sin \theta_2)} \qquad (28)$$

For k_1 and k_2 real, the interpretation of (28) is simple enough. For the boundary condition to be satisfied for all values of x, the two phase factors must be identical, which leads to Snell's law (4). But what if either or both k_1 and k_2 are complex? If k_1 is complex the intensity of E_1 will drop along the interface, leading to a decrease in the amplitude of the refracted wave along a direction other than $\hat{\mathbf{k}}_2$. This may be handled in a quite formal way by regarding $\sin \underline{\theta}_2$ as a complex quantity determined by (9). Similarly, if k_2 is complex, the wave amplitude will not be uniform over the wavefront and again this may be incorporated into a complex $\sin \underline{\theta}_2$.

For $\underline{\theta}_2$ complex, the propagating wave will not be strictly transverse. We may decompose $\underline{\theta}_2$ into real and imaginary parts:

$$\underline{\theta}_2 = \theta_2' + i\theta_2'' \tag{29}$$

We have then for the components of the complex propagation vector $\underline{\mathbf{k}}_2$:

$$\underline{k}_{2x} = k_2 \sin \underline{\theta}_2 = (k_2 \cosh \theta_2'') \sin \theta_2' + i(k_2 \sinh \theta_2'') \cos \theta_2' \tag{30}$$

$$\underline{k}_{2z} = k_2 \cos \underline{\theta}_2 = (k_2 \cosh \theta_2'') \cos \theta_2' - i(k_2 \sinh \theta_2'') \sin \theta_2' \tag{31}$$

where we have assumed k_2 real. It is apparent from (30) and (31) that under this assumption the real part of $\underline{\mathbf{k}}_2$ is oriented at an angle θ_2' with respect to $\hat{\mathbf{z}}$ while the imaginary part of $\underline{\mathbf{k}}_2$ is normal to the real part. With this understanding we may write (28) as:

$$E_1 + E_3 = E_2 \tag{32}$$

Equating the tangential components of **H** leads to:

$$\frac{E_1}{Z_1} \cos \theta_1 - \frac{E_3}{Z_1} \cos \theta_1 = \frac{E_2}{Z_2} \cos \underline{\theta}_2 \tag{33}$$

where $\underline{\theta}_2$ is determined from θ_1 by (9). The equations are formally equivalent to (21) and (22) except that Z_1 is replaced by $Z_1/\cos \theta_1$ and Z_2 by $Z_2/\cos \underline{\theta}_2$. We may then write from (24) for the reflection coefficient of the TE wave:

$$\blacktriangleright \qquad r = \frac{Z_2/\cos \underline{\theta}_2 - Z_1/\cos \theta_1}{Z_2/\cos \underline{\theta}_2 + Z_1/\cos \theta_1} \tag{34}$$

5 Oblique Incidence of the Transverse Magnetic (TM) Wave. The transverse magnetic wave exhibits a feature not shown by the transverse electric wave: there is an angle (the Brewster angle) for which the reflection coefficient is zero and the wave is fully transmitted. For an unpolarized wave incident at this angle, the reflected wave is completely polarized.[1] We show the geometry in Figure 4. Writing the boundary conditions in terms of H we have:

$$H_1 - H_3 = H_2 \tag{35}$$

$$H_1 Z_1 \cos\theta_1 + H_3 Z_1 \cos\theta_1 = H_2 Z_2 \cos\theta_2 \tag{36}$$

Comparing with **(21)** and **(22)** we see that Z_1 is replaced by $Z_1 \cos\theta_1$ and Z_2 by $Z_2 \cos\theta_2$. In particular, the reflection coefficient will be given by:

$$r = \frac{Z_2 \cos\theta_2 - Z_1 \cos\theta_1}{Z_1 \cos\theta_1 + Z_2 \cos\theta_2} \tag{37}$$

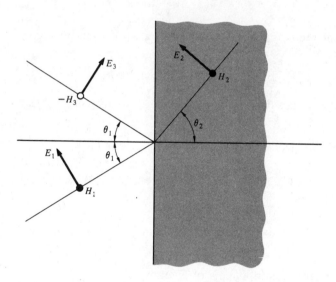

Figure 4   Oblique incidence of the transverse magnetic (TM) wave. The magnetic field for the incident, the reflected, and the transmitted wave is normal to the scattering plane.

Using **(4)** and **(16)** we may rewrite **(37)** as:

$$r = \frac{\mu_2 \sin 2\theta_2 - \mu_1 \sin 2\theta_1}{\mu_1 \sin 2\theta_1 + \mu_2 \sin 2\theta_2} \tag{38}$$

Note that for $\mu_1 = \mu_2$ and

$$\theta_1 + \theta_2 = \frac{\pi}{2} \tag{39}$$

---

[1] For a description of polarization in terms of the Stokes parameters, see E. Collett, *Am. J. Phys.* **36**, 713 (1968); **39**, 517 (1971). An error in these papers is corrected by R. Rimon and V. Srinivasan, *Am. J. Phys.* **45**, 1223 (1977).

we have $r = 0$. The angle of incidence θ_1 for which r vanishes is called the *Brewster angle*. The tangent of this angle is given by:

$$\blacktriangleright \qquad \tan \theta_1 = \frac{\sin \theta_1}{\cos \theta_1} = \frac{\sin \theta_1}{\sin \theta_2} = \frac{k_2}{k_1} = \frac{n_2}{n_1} \qquad (40)$$

Exercise

2. Show that if the Brewster condition (39) is satisfied for the incident wave in medium 1 at angle of incidence θ_1, it is also satisfied for the incident wave in medium 2 at incident angle θ_2.

6 Metallic Reflection. In this section we discuss as an example the reflection of light from a metallic surface. We make the simplifying assumption of treating the metal as an electron plasma whose permittivity is given from (12-121) by

$$\epsilon = \epsilon_{\text{ion}} + \epsilon_0 \chi_T = \epsilon_{\text{ion}}[1 - \omega_p^2/\omega(\omega + i/\tau)] \qquad (41)$$

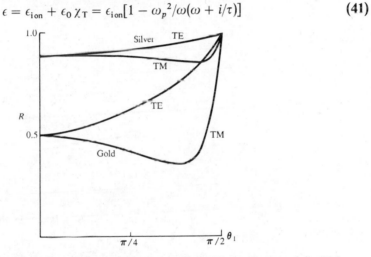

Figure 5 Reflectance of silver and gold for plane polarized light. The reflection of the TM wave is less than that of the TE wave for all angles of incidence.

For metals like aluminum and silver we may regard ϵ_{ion} as real and independent of frequency. For metals like copper and gold, which are absorbing in the blue, we may within the context of (41) associate the additional absorption with the ion cores. (Although such a point of view is convenient, it is not, in fact, possible to make a fundamental distinction between core and conduction electrons.) From (16) we write for the intrinsic impedance of the metal:

$$Z = (\mu/\epsilon)^{1/2} = (\mu/\epsilon_{\text{ion}})^{1/2}[1 - \omega_p^2/\omega(\omega + i/\tau)]^{-1/2} \qquad (42)$$

The reflectivity for TE waves as a function of the angle of incidence may be obtained from (34) and for TM waves from (37). We show in Figure 5 the observed reflectance R of silver and gold for plane polarized white light.

SURFACE ELECTROMAGNETIC WAVES

In discussing total internal reflection (Section 2) we saw that for sufficiently large angles of incidence θ_1, an incident wave is totally reflected off the interface with a second medium in which the phase velocity is lower. Under these conditions the wave decays exponentially into the second medium. In this part we examine the conditions for a wave to be concentrated at the interface between two media, which is to say that the wave must decay exponentially into *both* media.

We take the interface between the media to be the xz plane as shown in Figure 6. The first medium, which is characterized by μ_1 and ϵ_1, is to the left of the interface with y negative. The second medium, characterized by μ_2 and ϵ_2 is at positive y. We assume an electric field in medium 1 of the form:

$$\mathbf{E}(\mathbf{r},t) = \mathbf{E}_1 e^{\alpha y} e^{i(kx - \omega t)} \tag{43}$$

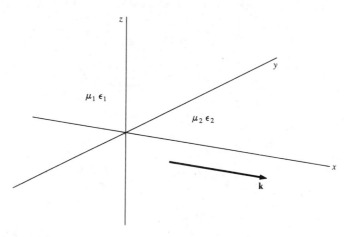

Figure 6 Surface electromagnetic wave. The xz plane is taken as the interface between the two media.

For a homogeneous, isotropic, and linear medium we have, from **(12-16)** in the absence of external charge,

$$\nabla \cdot \mathbf{D} = \epsilon_1 \, \nabla \cdot \mathbf{E} = \epsilon_1 (ikE_{1x} + \alpha E_{1y}) e^{\alpha y} e^{i(kx - \omega t)} = 0 \tag{44}$$

which leads to a field in the xy plane elliptically polarized in the *positive* direction.

$$E_{1x} = i \frac{\alpha}{k} E_{1y} \tag{45}$$

Substituting **(43)** into the wave equation **(12-57)** we have from the condition for free propagation in the absence of external currents:

$$\alpha^2 = k^2 - \omega^2 \mu_1 \epsilon_1 \tag{46}$$

The magnetic field in medium 1 is obtained from **(12-17)**:

$$\mathbf{H}(\mathbf{r},t) = \frac{1}{i\omega\mu_1} \nabla \times \mathbf{E}(\mathbf{r},t) = \left(-\frac{E_{1z}}{\omega\mu_1} (i\alpha\hat{\mathbf{x}} + k\hat{\mathbf{y}}) + \frac{\omega\epsilon_1}{k} \hat{\mathbf{z}}E_{1y} \right) e^{\alpha y} e^{i(kx-\omega t)} \qquad (47)$$

where we have used **(45)** and **(46)** to simplify the coefficient of E_{1y}.

In the second medium we must assume a solution of the form:

$$E(\mathbf{r},t) = E_2 e^{-\beta y} e^{i(kx-\omega t)} \qquad (48)$$

Corresponding to **(45)** we have:

$$E_{2x} = -i\frac{\beta}{k} E_{2y} \qquad (49)$$

and the fields in the xy plane are elliptically polarized in the *negative* direction. Corresponding to **(46)** we have in the second medium:

$$\beta^2 = k^2 - \omega^2\mu_2\epsilon_2 \qquad (50)$$

and a magnetic field:

$$\mathbf{H}(\mathbf{r},t) = \left(-\frac{E_{2z}}{\omega\mu_2} (-i\beta\hat{\mathbf{x}} + k\hat{\mathbf{y}}) + \frac{\omega\epsilon_2}{k} \hat{\mathbf{z}}E_{2y} \right) e^{-\beta y} e^{i(kx-\omega t)} \qquad (51)$$

7 Boundary Conditions. As we have discussed, **(12-16)** when applied at an interface requires in the absence of *external* surface charge:

$$D_{1y} = D_{2y} \quad \text{or} \quad \epsilon_1 E_{1y} = \epsilon_2 E_{2y} \qquad (52)$$

Faraday's law **(12-17)** requires when applied at an interface that the transverse components of the electric field are continuous:

$$E_{1x} = E_{2x} \qquad E_{1z} = E_{2z} \qquad (53)$$

That there are no sources of **B (12-19)** requires, at the interface,

$$B_{1y} = B_{2y} \quad \text{or} \quad \mu_1 H_{1y} = \mu_2 H_{2y} \qquad (54)$$

Ampère's law **(12-18)** requires at an interface in the absence of external surface currents that the transverse components of **H** are continuous:

$$H_{1x} = H_{2x} \quad \text{and} \quad H_{1z} = H_{2z} \qquad (55)$$

Applying **(53)** and using **(45)** and **(49)** gives:

$$\alpha E_{1y} = -\beta E_{2y} \tag{56}$$

Comparing **(52)** and **(56)** we must have either:

$$\alpha \epsilon_2 + \beta \epsilon_1 = 0 \tag{57}$$

or

$$E_{1y} = E_{2y} = 0 \quad \text{and} \quad E_{1x} = E_{2x} = 0 \tag{58}$$

where the second equality follows from **(45)** and **(49)**. We note that since we require both α and β to be positive, **(57)** requires that the permittivity of *one* of the two media be negative.

From **(55)** we have:

$$\frac{\alpha}{\mu_1} E_{1z} = -\frac{\beta}{\mu_2} E_{2z} \tag{59}$$

Comparing **(59)** with **(53)** we must have either:

$$\alpha \mu_2 + \beta \mu_1 = 0 \tag{60}$$

or

$$E_{1z} = E_{2z} = 0 \tag{61}$$

We note that **(60)** requires that μ_1 and μ_2 be opposite in sign.

It is necessary only that either **(57)** or **(60)** be satisfied in order to have a surface wave. It is not necessary that both conditions be satisfied. If neither condition is satisfied, then from **(58)** and **(61)** all field components must be zero, and there is no possibility of a surface wave.

8   Metals and Insulators. Although it is possible to satisfy **(60)** just below ferromagnetic resonance (see Problem 10-16) such circumstances are rather special. On the other hand, as we have seen (Section 12-10) the transverse permittivity of a plasma:

$$\epsilon = \epsilon_{\text{ion}}[1 - \omega_p^2/\omega(\omega + i/\tau)] \tag{62}$$

is negative for frequencies up to the plasma frequency. Then, one way to satisfy **(57)** is for one of the media to have positive permittivity, while the other is a plasma driven below its plasma frequency.

A second situation for which we expect negative permittivity is just above a strong resonance in a dielectric as discussed in Section 12-9. From **(12-98)** we have for the permittivity of an ionic solid:

$$\epsilon = \epsilon_{\text{core}}\left(1 + \frac{\omega_p^2}{\omega_T^2 - \omega^2}\right) \tag{63}$$

where ϵ_{core} is the contribution to the permittivity of core electrons, ω_p is the ion plasma frequency, and ω_T is the resonance frequency.

9 **Magnetosurface Waves.** We show in Figure 7 the electric fields associated with a surface wave for permittivities of opposite sign. The fields are circularly polarized in opposite directions. For propagation in the positive x direction the wave in medium 1 rotates counterclockwise at fixed z, while the wave in medium 2 rotates clockwise. Reversing the direction of propagation reverses the sense of rotation. If one of the media is a plasma, and we apply a magnetic field B along z, then we might expect for one direction of propagation to couple to the cyclotron motion of the carriers, but not for the other direction of propagation. The dispersion relation will then be different for the two directions of propagation.[2] This is called *non-reciprocal dispersion*.

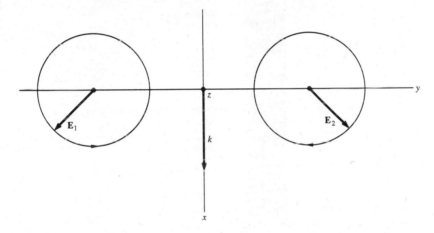

Figure 7 Electric fields associated with surface waves. The two media have permittivities of opposite sign.

MULTIPLE REFLECTION

We begin this part by considering the general problem of reflection from a pair of parallel surfaces. As an example we discuss the Fabry-Perot interferometer.

We show in Figure 8 a dielectric slab of permittivity ϵ and thickness d separating media of permittivity ϵ_1 and ϵ_2. We assume an electric field at normal incidence. For $z < 0$ the electric field may then be written as:

$$E(z,t) = E_1 e^{i(k_1 z - \omega t)} + E_3 e^{-i(k_1 z + \omega t)} \tag{64}$$

The transmitted field has the form:

$$E(z,t) = E_2 e^{i(k_2 z - \omega t)} \tag{65}$$

[2] J. J. Brion, R. F. Wallis, A. Harts, and E. Burstein, *Phys. Rev. Lett.*, **28**, 1455 (1972).

In the intermediate region we write the electric field as:

$$E(z,t) = E_+ \, e^{i(kz - \omega t)} + E_- \, e^{-i(kz + \omega t)} \tag{66}$$

To obtain the transmitted and reflected fields we assume continuity of transverse electric and magnetic fields at the two boundaries. At $z = 0$ we write:

$$E_1 + E_3 = E_+ + E_- \tag{67}$$

$$E_1/Z_1 - E_3/Z_1 = E_+/Z - E_-/Z \tag{68}$$

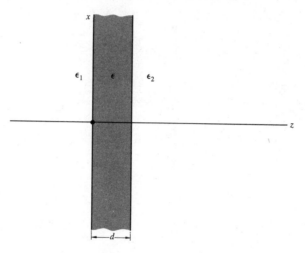

Figure 8 Dielectric slab. A slab of permittivity ϵ separates media of permittivity ϵ_1 and ϵ_2.

At $z = d$ the continuity of fields leads to

$$E_2 \, e^{ik_2 d} = E_+ \, e^{ik d} + E_- \, e^{-ik d} \tag{69}$$

$$(E_2/Z_2)e^{ik d} = (E_+/Z)e^{ik d} - (E_-/Z)e^{-ik d} \tag{70}$$

Eliminating E_+ and E_- we obtain for the equations connecting E_2 and E_3 with E_1:

$$E_1 + E_3 = [(Z_2 - Z)/2Z_2]E_2 \, e^{i(k_2 + k) d} + [(Z_2 + Z)/2Z_2]E_2 \, e^{i(k_2 - k) d} \tag{71}$$

$$(E_1/Z_1) - (E_3/Z_1) = [(Z_2 + Z)/2Z](E_2/Z_2)e^{i(k_2 - k) d}$$

$$+ [(Z - Z_2)/2Z](E_2/Z_2)e^{i(k_2 + k) d} \tag{72}$$

Note that if d goes to zero, then (71) and (72) approach (21) and (22). Similarly if Z_2 and Z are equal we obtain the same result.

10 Fabry-Perot Interferometer. In the Fabry-Perot interferometer the width of the intermediate region is nearly an integral number of half wavelengths. Solving **(71)** and **(72)** for E_2 we obtain

$$E_2 = \frac{2Z_2 Z}{(Z_1 + Z_2)Z \cos kd + i(Z^2 + Z_1 Z_2)\sin kd} E_1 e^{-ik_2 d} \tag{73}$$

For kd an odd number of quarter wavelengths, and $Z^2 = Z_1 Z_2$ which is the condition for Z to form a match between Z_1 and Z_2 **(E-42)**, we obtain from **(73)**

$$E_2 = (Z_2/Z_1)^{1/2} E_1 e^{-i(k_2 d \pm \pi/4)} \tag{74}$$

which gives all the incident energy flowing into the second medium:

$$\Pi_2 = \tfrac{1}{2} E_2 E_2^* / Z_2 = \tfrac{1}{2} E_1 E_1^* / Z_1 = \Pi_1 \tag{75}$$

For kd an integral number of half wavelengths, which is the resonance condition, and $Z_1 = Z_2$ we similarly have all the incident power transmitted into the second medium. At other frequencies the transmitted power may be substantially reduced, and the interferometer acts as a narrow-band filter. In general we write from **(74)** for the transmittance:

$$T = \Pi_2/\Pi_1 = \frac{4Z^2 Z_1 Z_2}{Z^2(Z_1 + Z_2)^2 + (Z^2 - Z_1^2)(Z^2 - Z_2^2)\sin^2 kd} \tag{76}$$

For Z large compared with Z_1 and Z_2 the ratio of the transmittance on-resonance to off-resonance is equal to $[Z/(Z_1 + Z_2)]^2$.

PROPAGATION BETWEEN PLANE PARALLEL CONDUCTORS

We begin the discussion of the propagation of bounded electromagnetic waves by considering the propagation of radiation in the region between two parallel conducting plates as shown in Figure 9. We take the separation between the plates to be a and assume further that the plates are perfect conductors.

11 Boundary Conditions. In the interior of the conducting plates we write:

$$\mathbf{j}(\mathbf{r}) = \sigma \mathbf{E}(\mathbf{r}) \tag{77}$$

For a perfect conductor σ is infinite. With $\mathbf{j}(\mathbf{r})$ finite we must have everywhere within the conductor $\mathbf{E} = 0$. Now, from Faraday's law **(12-17)**:

$$\nabla \times \mathbf{E} = -\frac{\partial \mathbf{B}}{\partial t} \tag{78}$$

we obtain the continuity of the *transverse* component of **E** at the conducting surface. But since the electric field is zero within the conductor, we have the boundary condition on **E**:

$$\hat{\mathbf{n}} \times \mathbf{E} = 0 \tag{79}$$

From the fact that the electric field is zero within a perfect conductor we have from **(78)** that the magnetic field must also be zero. From the fourth Maxwell equation **(12-19)**:

$$\nabla \cdot \mathbf{B} = 0 \tag{80}$$

we obtain the continuity of the normal component of **B** leading to the boundary condition on **B**:

$$\hat{\mathbf{n}} \cdot \mathbf{B} = 0 \tag{81}$$

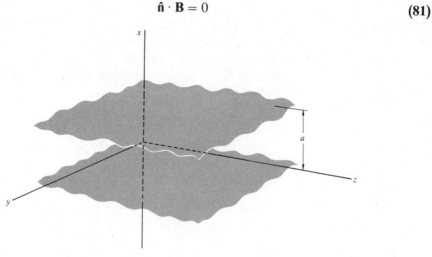

Figure 9   Parallel conducting plates. Electromagnetic radiation propagates in the region bounded by the plates.

We may now use **(79)** and **(81)** together with the Maxwell equations outside the conductor to obtain derived boundary conditions. In the absence of *external* currents we write outside the conductor from **(12-18)**

$$\nabla \times \mathbf{H} = \epsilon \frac{\partial \mathbf{E}}{\partial t} \tag{82}$$

Crossing **(82)** with the surface normal and using **(79)** we have, at the surface, the derived boundary condition

$$\hat{\mathbf{n}} \times (\nabla \times \mathbf{H}) = 0 \tag{83}$$

Taking the scalar product of **(78)** with the surface normal and using **(80)** we obtain the second derived boundary condition:

$$\hat{n} \cdot \nabla \times \mathbf{E} = 0 \tag{84}$$

Exercises

3. Obtain from **(83)** at the surface of a perfect conductor:

$$\nabla(\hat{n} \cdot \mathbf{H}) - (\hat{n} \cdot \nabla)\mathbf{H} = 0$$

4. Obtain from **(84)** at the surface of a perfect conductor:

$$(\hat{n} \times \nabla) \cdot \mathbf{E} = 0$$

12 Transverse Electromagnetic (TEM) Waves. There are three modes of propagation, transverse electromagnetic (TEM), transverse electric (TE), and transverse magnetic (TM), which we consider separately. As long as **k** is parallel to the conducting plates and **E** is normal to the plates (with **H** as a consequence parallel to the plates), we may propagate a purely transverse electromagnetic wave of the form:

$$\mathbf{E}(z,t) = E\hat{x}e^{i(kz - \omega t)} \tag{85}$$

$$\mathbf{H}(z,t) = \frac{E}{Z} \hat{y}e^{i(kz - \omega t)} \tag{86}$$

where we have taken **k** along the z direction and Z is the wave impedance of the medium between the plates. The energy flux is given by:

$$\Pi = \tfrac{1}{2}EH = \tfrac{1}{2}E^2/Z \tag{87}$$

The fields given by **(85)** and **(86)** are exactly the same as for waves in free space except that they are restricted by current and charge in the conducting plates to the region between the plates as discussed in Section 12-11. As long as the plates are regarded as ideal conductors they have no effect on the fields except to restrict them. It is clear that we cannot propagate a transverse wave polarized with the electric field along $\hat{y}$ since the field must vanish at both plates and (for a TEM wave) must be uniform across the wavefront.

13 Transverse Electric (TE) Waves. It is possible to propagate a wave with the electric field polarized along $\hat{y}$ by lifting the restriction that both the electric and magnetic fields be transverse. A way of visualizing this mode of propagation is shown in Figure 10. We propagate a wave with **E** polarized along $\hat{y}$ and with **k** at an angle θ to the axis. The wave propagates until it is reflected specularly from the top surface with a phase

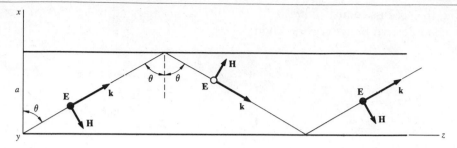

Figure 10   Transverse electric (TE) wave. The wave propagates by multiple reflection from the top and bottom surfaces.

shift of π. The wave continues to propagate until it is reflected from the bottom surface, and so on.

The representation shown in Figure 10 is somewhat misleading in that it appears that energy flows diagonally. Actually, at each position between the plates there is an interference between the incident and reflected wavefronts so that the energy flow for perfectly reflecting walls is always along $\hat{z}$.

Not all values of θ make it possible to simultaneously satisfy **(79)** at both surfaces. To find the condition on θ we assume as shown in Figure 11 a wave reflected specularly (and with a phase change of π) from the upper surface. The electric field at the lower surface will be a superposition of fields produced by the incident and reflected waves. For the electric field to be zero at the lower surface, we must have:

$$d_1 + d_2 = n\lambda \tag{88}$$

From Figure 11 we have the relations:

$$d_1 = \frac{a}{\cos\theta} - z_1 \sin\theta \tag{89}$$

$$d_2 = \frac{a}{\cos\theta} - z_2 \sin\theta \tag{90}$$

$$z_1 + z_2 = 2a \tan\theta \tag{91}$$

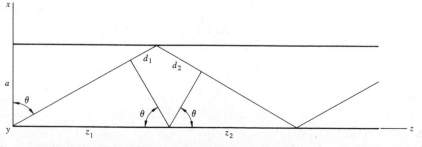

Figure 11   Destructive interference. The condition that the electric field vanish at the lower surface as well as at the upper surface is given by **(88)**.

Adding **(89)** and **(90)** and using **(91)** and requiring **(88)** we obtain:

$$\frac{2a}{\cos\theta} - \frac{2a\sin^2\theta}{\cos\theta} = n\lambda \tag{92}$$

which leads to

$$\blacktriangleright \qquad \qquad \cos\theta = \frac{n\lambda}{2a} \tag{93}$$

Notice that if λ is larger than $2a$ there is no nonvanishing solution to **(93)**. The distance $2a$ is called the cutoff wavelength for propagation. Since energy flows at angles $\pm(\pi/2 - \theta)$ to the z axis, we have, for the Poynting vector,

$$\Pi = EH\sin\theta \tag{94}$$

Note from Figure 10 that although the electric field is transverse, the magnetic field is not, having a component along z.

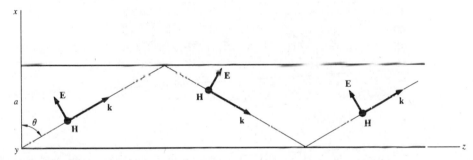

Figure 12 Transverse magnetic (TM) wave. The wave propagates by multiple reflection from the top and bottom surfaces

14 Transverse Magnetic (TM) Waves. We may also propagate a wave with **H** polarized along $\hat{y}$ and with **k** at an angle θ to $\hat{x}$ as shown in Figure 12. There is no phase shift on reflection, but because the tangential component of **E** still changes sign, the condition for cancellation at the opposite face is still given by **(92)**, which leads to **(93)**. The difference here is that $n = 0$, which corresponds to propagation along z is allowed and gives the TEM wave described by **(85)** and **(86)**. Although there can be no longitudinal component of **E** at the conducting surface, there will be such a component between the plates except for the $n = 0$ (TEM) mode.

PROPAGATION IN A DIELECTRIC SLAB

We now consider the general problem of propagation in a dielectric slab of permittivity ϵ_1 and thickness a and bounded by a medium of permittivity ϵ_2. By letting ϵ_2 go to ϵ_0 we should have the case of propagation in a dielectric slab in vacuum. Going to the other limit of ϵ_2 approaching minus infinity we should have the case already considered of propagation between conducting plates. We show in Figure 13 a section through the xz plane. We consider TE waves and TM waves separately.

15   Transverse Electric (TE) Waves. We assume that the electric field is of the form:

$$\underline{\mathbf{E}}(x,z,t) = E(x)\hat{\mathbf{y}}e^{i(kz - \omega t)} \tag{95}$$

where instead of writing $\mathbf{E}$ as a superposition of plane waves we take $\mathbf{k}$ to be along z and allow the magnitude of $\mathbf{E}$ to be a function of the transverse coordinate. We obtain the magnetic field from **(12-31)**:

$$\nabla \times \underline{\mathbf{E}} = -\mu \frac{\partial \underline{\mathbf{H}}}{\partial t} \tag{96}$$

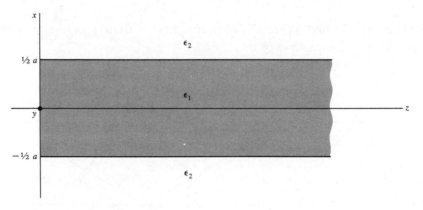

Figure 13 Propagation in a dielectric slab. Propagation in a slab in vacuum and between conductors are limiting cases.

which gives on substitution of **(95)** into **(96)**:

$$\underline{\mathbf{H}}(x,z,t) = \frac{1}{\omega\mu}\left[k\hat{\mathbf{x}}E(x) + i\hat{\mathbf{z}}\frac{\partial}{\partial x}E(x)\right]e^{i(kz - \omega t)} \tag{97}$$

We next substitute **(97)** into **(12-32)** written in the absence of *external* currents and for an isotropic medium:

$$\nabla \times \underline{\mathbf{H}} = \epsilon \frac{\partial \underline{\mathbf{E}}}{\partial t} \tag{98}$$

Substituting **(97)** into **(98)** we obtain for the electric field the equation:

$$\frac{\partial^2}{\partial x^2}E(x) + (\omega^2\mu\epsilon - k^2)E(x) = 0 \tag{99}$$

Our problem is to obtain general solutions for $E(x)$ in both media and from the boundary conditions to obtain discrete solutions.

The electric field $E(x)$ may be either an even or an odd function of x. We consider first the even case and write for the electric field within the slab:

$$E_1(x) = E_1 \cos \alpha x \tag{100}$$

where on substitution into **(99)** we have:

$$\alpha^2 = \omega^2 \mu_1 \epsilon_1 - k^2 \tag{101}$$

In the medium and for x positive we look for a solution of the form:

$$E_2(x) = E_2 e^{-\beta x} \tag{102}$$

Since we wish to restrict the wave to the slab we are interested only in solutions for exponentially decreasing $E_2(x)$. Substituting **(102)** into **(99)** we have:

$$\beta^2 = k^2 - \omega^2 \mu_2 \epsilon_2 \tag{103}$$

Now, if both media are lossless dielectrics with $\mu_1 = \mu_2$ the only way in which we may satisfy both **(101)** and **(103)** with both ϵ_1 and ϵ_2 real is for the permittivity of the slab to be greater than that of the surrounding medium.

If the bounding medium is a conducting plasma, we have from **(12-121)** for transverse propagation:

$$\epsilon_2 = \epsilon_{\text{ion}} \left[1 - \frac{\omega_p{}^2}{\omega(\omega + i/\tau)} \right] \tag{104}$$

where ϵ_{ion} represents the permittivity of the background, ω_p is the plasma frequency of the carriers, and τ is their relaxation time. If we make the assumption that $\omega\tau$ is large and ω is less than the plasma frequency, then ϵ_2 is negative and α, β, and k may all be real. The effect of including carrier relaxation is that the penetration into medium 2 will be a damped sinusoid rather than a pure exponential, k will be complex representing attenuation from carrier losses, and α will be complex representing the flow of energy into the plasma.

The boundary conditions are that $\mathbf{E}$, which is tangential, be continuous across the boundary and that the tangential component of $\mathbf{H}$ to be continuous, which as may be seen from **(97)** is equivalent for $\mu_1 = \mu_2$ to requiring that the normal derivative of $\mathbf{E}$ be continuous. The boundary conditions give:

$$E_1 \cos \tfrac{1}{2}\alpha a = E_2 e^{-(1/2)(\beta a)} \tag{105}$$

$$-\alpha E_1 \sin \tfrac{1}{2}\alpha a = -\beta E_2 e^{-(1/2)(\beta a)} \tag{106}$$

Taking the ratio and eliminating β we obtain:

$$\tan \tfrac{1}{2}\alpha a = \frac{\beta}{\alpha} = \left(\frac{\omega^2(\mu_1 \epsilon_1 - \mu_2 \epsilon_2)}{\alpha^2} - 1 \right)^{1/2} \tag{107}$$

The solutions of **(107)** give the allowed modes of propagation.

For $E(x)$ an odd function of x we write:

$$E_1(x) = E_1 \sin \alpha x \tag{108}$$

In the surrounding medium we assume a solution of the form of **(102)** for positive x and the negative of this field for negative x. The boundary conditions give in place of **(105)** and **(106)**:

$$E_1 \sin \tfrac{1}{2}\alpha a = E_2 \, e^{-(1/2)(\beta a)} \tag{109}$$

$$\alpha E_1 \cos \tfrac{1}{2}\alpha a = -\beta E_2 \, e^{-(1/2)(\beta a)} \tag{110}$$

Taking the ratio of **(109)** to **(110)** and eliminating β we obtain:

$$-\cot \tfrac{1}{2}\alpha a = \frac{\beta}{\alpha} = \left(\frac{\omega^2(\mu_1 \epsilon_1 - \mu_2 \epsilon_2)}{\alpha^2} - 1 \right)^{1/2} \tag{111}$$

The simplest way to represent the solutions of **(107)** and **(111)** is to graph both the trigonometric functions and the right side as functions of α and find the intersections as shown in Figure 14. For a given mode the dispersion relation **(101)** is given in Figure 15.

In drawing Figure 15, we have assumed that the medium itself is nondispersive in the frequency range of interest. Then the phase velocity in the waveguide:

$$v_p = \omega/k \tag{112}$$

is greater than the phase velocity in an infinite medium, while the group velocity:

$$v_g = \frac{d\omega}{dk} = \frac{k}{(\omega \mu_1 \epsilon_1)} = \frac{k v^2}{\omega} \tag{113}$$

is always less than the phase velocity in an infinite medium $v = (\mu_1 \epsilon_1)^{-1/2}$. Taking the product of **(112)** and **(113)** we have the relation:

$$v_p v_g = v^2 \tag{114}$$

The right side of **(107)** and **(111)** intersect the axis at the critical value

$$\alpha_c = \omega(\mu_1 \epsilon_1 - \mu_2 \epsilon_2)^{1/2} \tag{115}$$

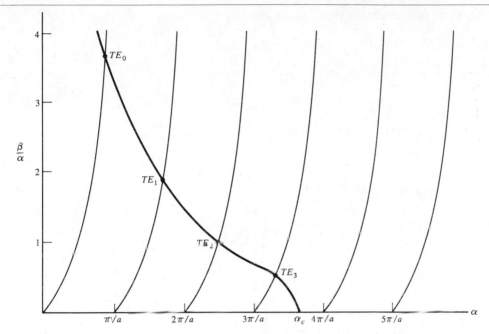

Figure 14   Modes of propagation. The allowed modes are obtained by graphical solution of **(107)** and **(111)**.

We observe that no matter how close the permittivities of the two media may be, there is one mode that is concentrated in the slab. As the difference in permittivities increases, the intersection with the axis moves to larger values of α and more modes become possible. The index with which the modes are labeled is the number of nodes of $E(x)$ as shown in Figure 16 for the TE_0, TE_1, and TE_2 modes.

One may see that **(115)** is just the condition for total internal reflection at the interface between the two dielectrics, as suggested by the requirement that β be real. We have from **(10)** the condition for total internal reflection:

$$k_1 \sin \theta > k_2 \tag{116}$$

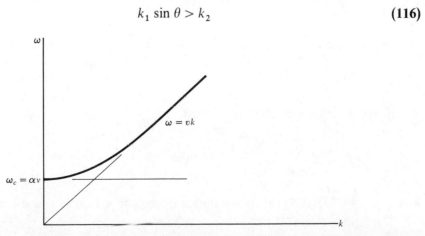

Figure 15   Dispersion relation. The assumption is made that the medium is nondispersive. The dispersion arises entirely from the boundaries.

If we write for the propagation vector along z:

$$k = k_1 \sin \theta \tag{117}$$

then **(116)** may be written as

$$k^2 > k_2^2 = \omega^2 \mu_2 \epsilon_2 \tag{118}$$

Substituting **(118)** into **(101)** we have:

$$\alpha^2 < \omega^2 (\mu_1 \epsilon_1 - \mu_2 \epsilon_2) \tag{119}$$

which is equivalent to **(115)**.

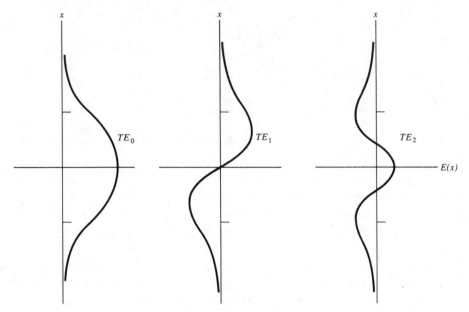

Figure 16   Transverse electric field. The index with which the modes are labeled is the number of nodes of $E(x)$.

16   Transverse Magnetic (TM) Waves.

We assume a magnetic field of the form:

$$\mathbf{H}(x,z,t) = H(x)\hat{\mathbf{y}}e^{i(kz-\omega t)} \tag{120}$$

From **(98)** we obtain for the electric field:

$$\mathbf{E}(x,z,t) = \frac{1}{\omega\epsilon}\left[k\hat{\mathbf{x}}H(x) + i\hat{\mathbf{z}}\frac{\partial}{\partial x}H(x)\right]e^{i(kz-\omega t)} \tag{121}$$

Substituting **(121)** into **(96)** we obtain for the magnetic field the equation:

$$\frac{\partial^2}{\partial x^2}H(x) + (\omega^2\mu\epsilon - k^2)H(x) = 0 \tag{122}$$

This equation is the same as **(99)** and we expect to obtain modes similar to those for a TE wave. The boundary conditions are slightly different however. Assuming solutions to **(122)** of the form:

$$H_1(x) = H_1 \sin \alpha x \tag{123}$$

$$H_2(x) = H_2 e^{-\beta |x|} \tag{124}$$

we obtain from the continuity of H_y and E_z:

$$H_1 \sin \tfrac{1}{2}\alpha a = H_2 e^{-(1/2)(\beta a)} \tag{125}$$

$$\frac{\alpha H}{\epsilon_1} H_1 \cos \tfrac{1}{2}\alpha a = -\frac{\beta}{\epsilon_2} H_2 e^{-(1/2)(\beta a)} \tag{126}$$

Taking the ratio of **(126)** to **(125)**:

$$\cot \tfrac{1}{2}\alpha a = -\frac{\epsilon_1}{\epsilon_2}\frac{\beta}{\alpha} \tag{127}$$

which differs from **(111)** by the factor ϵ_1/ϵ_2. If ϵ_1 and ϵ_2 are similar, the TE and TM modes will have very nearly the same propagation vectors. A similar result is obtained for even waves with $H(x) = H_1 \cos \alpha x$.

PROPAGATION THROUGH CONDUCTING WAVEGUIDES

In the examples just considered of the propagation of electromagnetic waves between conducting plates and within a dielectric slab, the wave was bounded along x but not along y. We now consider propagation through a waveguide where the radiation is fully confined in the transverse plane. We first consider the case where the bounding walls are planar and the cross section is rectangular. Then we consider the case of propagation through a round conducting pipe.

We write Maxwell's equations for linear isotropic media in the absence of external charge or current as:

ME I	$\nabla \cdot \mathbf{E} = 0$	(128)

$$\text{ME II} \qquad \nabla \times \mathbf{E} = -\mu \frac{\partial \mathbf{H}}{\partial t} \tag{129}$$

$$\text{ME III} \qquad \nabla \times \mathbf{H} = \epsilon \frac{\partial \mathbf{E}}{\partial t} \tag{130}$$

$$\text{ME IV} \qquad \nabla \cdot \mathbf{H} = 0 \tag{131}$$

Taking the curl of **(129)** and using **(128)** we obtain:

$$\nabla^2 \mathbf{E} = \mu\epsilon \frac{\partial^2 \mathbf{E}}{\partial t^2} \tag{132}$$

Similarly, taking the curl of **(130)** and using **(131)** we have:

$$\nabla^2 \mathbf{H} = \mu\epsilon \frac{\partial^2 \mathbf{H}}{\partial t^2} \tag{133}$$

For propagation through a cylindrical wave guide we expect solutions to **(132)** and **(133)** of the form:

$$\mathbf{E}(x,y,z,t) = \mathbf{E}(x,y)e^{i(kz-\omega t)} \tag{134}$$

$$\mathbf{H}(x,y,z,t) = \mathbf{H}(x,y)e^{i(kz-\omega t)} \tag{135}$$

Substituting **(134)** into **(132)** and **(135)** into **(133)** we have:

$$\nabla_T^2 \mathbf{E} = -(\omega^2\mu\epsilon - k^2)\mathbf{E} \tag{136}$$

$$\nabla_T^2 \mathbf{H} = -(\omega^2\mu\epsilon - k^2)\mathbf{H} \tag{137}$$

where ∇_T^2 is the Laplacian with respect to the *transverse* coordinates only:

$$\nabla_T^2 = \frac{\partial^2}{\partial x^2} + \frac{\partial^2}{\partial y^2} \tag{138}$$

It is not at all necessary to solve **(136)** and **(137)** for all three components of **E** and of **H**. It is sufficient to find the z components of **E** and of **H** and then to use **(129)** and **(130)** to obtain:

$$i\omega\mu H_x = \frac{\partial E_z}{\partial y} - ikE_y \tag{139}$$

$$i\omega\mu H_y = -\frac{\partial E_z}{\partial x} + ikE_x \tag{140}$$

$$i\omega\epsilon E_x = -\frac{\partial H_z}{\partial y} + ikH_y \tag{141}$$

$$i\omega\epsilon E_y = \frac{\partial H_z}{\partial x} - ikH_x \tag{142}$$

These equations may be solved for the transverse fields:

$$(\omega^2\mu\epsilon - k^2)H_x = -i\omega\epsilon\frac{\partial E_z}{\partial y} + ik\frac{\partial H_z}{\partial x} \tag{143}$$

$$(\omega^2\mu\epsilon - k^2)H_y = i\omega\epsilon\frac{\partial E_z}{\partial x} + ik\frac{\partial H_z}{\partial y} \tag{144}$$

$$(\omega^2\mu\epsilon - k^2)E_x = i\omega\mu\frac{\partial H_z}{\partial y} + ik\frac{\partial E_z}{\partial x} \tag{145}$$

$$(\omega^2\mu\epsilon - k^2)E_y = -i\omega\mu\frac{\partial H_z}{\partial x} + ik\frac{\partial E_z}{\partial y} \tag{146}$$

By specifying E_z and H_z, the transverse field components are determined. The boundary condition (79) requires at the conducting boundary:

$$E_z = 0 \tag{147}$$

We have from the z component of (83):

$$(\hat{\mathbf{n}} \cdot \nabla)H_z = 0 \tag{148}$$

We note that if (147) and (148) are satisfied at the boundary for E_z and H_z, then the boundary conditions (79) and (81) are automatically satisfied through (143), (144), (145), and (146) for the transverse components as well.

For any perfectly conducting boundary the solutions to (136) and (137) take two forms. First, we may take $H_z(x,y) = 0$ everywhere [which automatically satisfies (148)] and take for $E_z(x,y)$ a form that satisfies (147). This gives transverse magnetic (TM) waves. Conversely, we may take $E_z(x,y) = 0$ and take for $H_z(x,y)$ a form that satisfies (148), generating the transverse electric (TE) waves.

This separation follows from the separate form of the boundary conditions (79) and (81). If instead the conductivity of the surrounding conductor is finite, the TE and TM modes will in general be mixed.

17 Transverse Magnetic (TM) Waves in a Rectangular Pipe. We show in Figure 17 the geometry for which we wish to find the propagating electric and magnetic fields. We take $H_z = 0$ and write for the longitudinal electric field:

$$E_z(x,y) = E\sin\frac{m\pi x}{a}\sin\frac{n\pi y}{b} \tag{149}$$

where (149) is a solution of (136) giving:

$$\omega^2 = \omega_{mn}{}^2 + \frac{k^2}{\mu\epsilon} \tag{150}$$

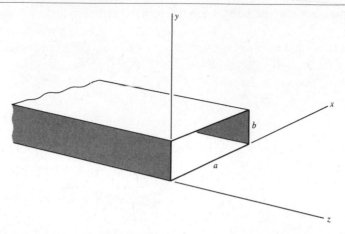

Figure 17 Transverse magnetic (TM) waves in a rectangular pipe. The z direction is the direction of propagation with the reflecting surfaces in the xz and yz planes.

and where ω_{mn} is called the cutoff frequency given by:

$$\blacktriangleright \qquad \mu\epsilon\omega_{mn}{}^2 = \left(\frac{m\pi}{a}\right)^2 + \left(\frac{n\pi}{b}\right)^2 \qquad (151)$$

We observe that **(150)** has the same form as **(12-126)**, which is the dispersion relation in a plasma. We plot in Figure 18 the form of **(150)**. The form of **(149)** automatically satisfies the boundary condition that the tangential component of E_z be zero for m and n integral. From **(143)**, **(144)**, **(145)**, and **(146)** we obtain for the transverse fields:

$$H_x = \frac{-i\omega\epsilon(n\pi/b)E}{(m\pi/a)^2 + (n\pi/b)^2}\, E \sin\frac{m\pi x}{a}\cos\frac{n\pi y}{b} = -\frac{\omega\epsilon}{k}E_y \qquad (152)$$

$$H_y = \frac{i\omega\epsilon(n\pi/a)E}{(m\pi/a)^2 + (n\pi/b)^2}\, E \cos\frac{m\pi x}{a}\sin\frac{n\pi y}{b} = \frac{\omega\epsilon}{k}E_x \qquad (153)$$

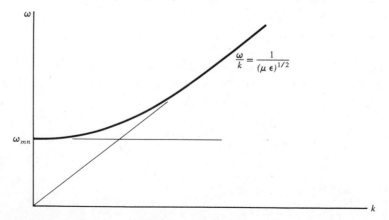

Figure 18 Dispersion relation. The dispersion relation has the same form as for propagation in a plasma.

These waveguide modes are designated as TM_{mn}. The lowest frequency mode is for m and n both equal to one and is shown in Figure 19. For lower frequencies we see from (150) that k will be pure imaginary and the wave will be exponentially attenuated.

Note that (152) and (153) automatically satisfy the boundary condition that the tangential component of **E** be zero.

18 **Transverse Electric (TE) Waves in a Rectangular Pipe.** We now consider propagating waves with $E_z = 0$. The requirement that the tangential component of **E** be zero at the boundaries requires from (79) with (78) that the normal gradient of **H** be zero at the conducting surfaces leading to a longitudinal magnetic field of the form:

$$H_z(x,y) = H \cos \frac{m\pi x}{a} \cos \frac{n\pi y}{b} \tag{154}$$

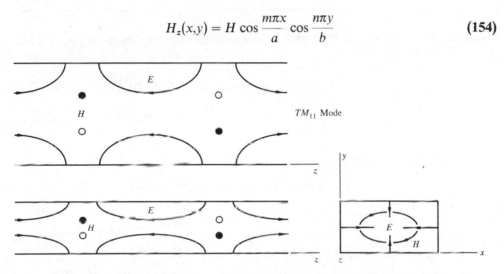

Figure 19 Lowest frequency mode. The indices m and n are both equal to 1.

Substituting (154) into (137) again gives (150) and (151). From (143), (144), (145), and (146) we obtain for the transverse fields:

$$E_x = \frac{-i\omega\mu(n\pi/b)}{(m\pi/a)^2 + (n\pi/b)^2} H \cos \frac{mx}{a} \sin \frac{n\pi y}{b} = \frac{\omega\mu}{k} H_y \tag{155}$$

$$E_y = \frac{i\omega\mu(m\pi/z)}{(m\pi/a)^2 + (n\pi/b)^2} H \sin \frac{m\pi x}{a} \cos \frac{n\pi y}{b} = -\frac{\omega\mu}{k} H_x \tag{156}$$

Note that as for the TM modes, the boundary conditions on **E** are automatically satisfied. For a greater than b the lowest frequency mode is for $m = 1$ and $n = 0$, and is shown in Figure 20.

19 **[Propagation in a Round Pipe].** To treat the propagation of electromagnetic waves in a round pipe it is necessary to obtain the form of the Laplacian in cylindrical coordinates as shown in Figure 21a.

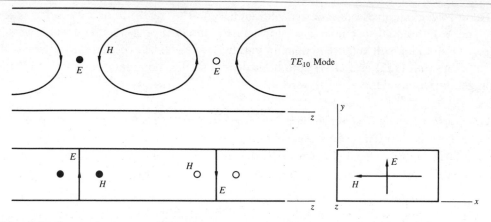

Figure 20   Transverse electric (TE) wave in a rectangular pipe. The lowest frequency mode is for $m = 1$ and $n = 0$.

The gradient operator in cylindrical coordinates is from **(B-83)**:

$$\mathbf{\nabla} = \hat{\boldsymbol{\rho}}\,\frac{\partial}{\partial \rho} + \hat{\boldsymbol{\phi}}\,\frac{1}{\rho}\,\frac{\partial}{\partial \phi} + \hat{\mathbf{z}}\,\frac{\partial}{\partial z} \tag{157}$$

To form the Laplacian of a vector we write:

$$\nabla^2 \mathbf{E} = (\mathbf{\nabla} \cdot \mathbf{\nabla})\mathbf{E} = \mathbf{\nabla} \cdot \mathbf{\nabla}\mathbf{E} \tag{158}$$

Using the relations **(B-86)** we write:

$$\mathbf{\nabla E} = \hat{\boldsymbol{\rho}}\hat{\boldsymbol{\rho}}\,\frac{\partial E_\rho}{\partial \rho} + \hat{\boldsymbol{\rho}}\hat{\boldsymbol{\phi}}\,\frac{\partial E_\phi}{\partial \rho} + \hat{\boldsymbol{\rho}}\hat{\mathbf{z}}\,\frac{\partial E_z}{\partial \rho} + \hat{\boldsymbol{\phi}}\hat{\boldsymbol{\rho}}\,\frac{1}{\rho}\left(\frac{\partial E_\rho}{\partial \phi} - E_\phi\right) + \hat{\boldsymbol{\phi}}\hat{\boldsymbol{\phi}}\,\frac{1}{\rho}\left(E_\rho + \frac{\partial E_\phi}{\partial \phi}\right)$$

$$+ \hat{\boldsymbol{\phi}}\hat{\mathbf{z}}\,\frac{1}{\rho}\,\frac{\partial E_z}{\partial \phi} + \hat{\mathbf{z}}\hat{\boldsymbol{\rho}}\,\frac{\partial E_\rho}{\partial z} + \hat{\mathbf{z}}\hat{\boldsymbol{\phi}}\,\frac{\partial E_\phi}{\partial z} + \hat{\mathbf{z}}\hat{\mathbf{z}}\,\frac{\partial E_z}{\partial z} \tag{159}$$

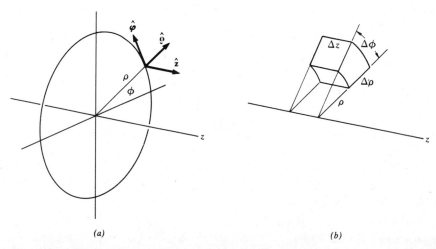

(a)   (b)

Figure 21   Propagation of electromagnetic waves in a round pipe. (*a*) The unit vectors for cylindrical coordinates are shown. (*b*) A volume element in cylindrical coordinates is shown.

Taking the divergence of **(159)** we obtain finally:

$$\nabla^2 \mathbf{E} = \hat{\mathbf{\rho}} \left[\frac{1}{\rho} \frac{\partial}{\partial \rho} \left(\rho \frac{\partial E_\rho}{\partial \rho} \right) + \frac{1}{\rho^2} \left(\frac{\partial^2 E_\rho}{\partial \phi^2} - E_\rho + 2 \frac{\partial E_\phi}{\partial \phi} \right) + \frac{\partial^2 E_\rho}{\partial z^2} \right]$$

$$+ \hat{\mathbf{\phi}} \left[\frac{1}{\rho} \frac{\partial}{\partial \rho} \left(\rho \frac{\partial E_\phi}{\partial \rho} \right) + \frac{1}{\rho^2} \left(\frac{\partial^2 E_\phi}{\partial \phi^2} - E_\phi + 2 \frac{\partial E_\rho}{\partial \phi} \right) + \frac{\partial^2 E_\phi}{\partial z^2} \right]$$

$$+ \hat{\mathbf{z}} \left[\frac{1}{\rho} \frac{\partial}{\partial \rho} \left(\rho \frac{\partial E_z}{\partial \rho} \right) + \frac{1}{\rho^2} \frac{\partial^2 E_z}{\partial \phi^2} + \frac{\partial^2 E_z}{\partial z^2} \right] \qquad (160)$$

We first consider the propagation of a transverse magnetic (TM) wave in a round conducting pipe. Applying **(136)** to the longitudinal component of the electric field we have, from **(160)**,

$$\frac{1}{\rho} \frac{\partial}{\partial \rho} \rho \frac{\partial E_z}{\partial \rho} + \frac{1}{\rho^2} \frac{\partial^2 E_z}{\partial \phi^2} = -(\omega^2 \mu \epsilon - k^2) E_z \qquad (161)$$

We write E_z as a product function:

$$E_z(\rho, \phi) = R(\rho) \Phi(\phi) \qquad (162)$$

Substituting **(162)** into **(161)** leads to the two equations:

$$\frac{d^2}{d\phi^2} \Phi + n^2 \Phi = 0 \qquad (163)$$

$$\frac{d^2 R}{d\rho^2} + \frac{1}{\rho} \frac{dR}{d\rho} + \left(\alpha^2 - \frac{n^2}{\rho^2} \right) R = 0 \qquad (164)$$

with

$$\alpha = (\omega^2 \mu \epsilon - k^2)^{1/2} \qquad (165)$$

The solution of **(163)** is either of the forms:

$$\Phi = \cos n\phi \qquad \Phi = \sin n\phi \qquad (166)$$

where n must be integral for the solutions to be single valued. The radial equation **(164)** may be put in standard form by introducing the variable:

$$x = \alpha \rho \qquad (167)$$

in terms of which we write **(164)** as:

$$\frac{d^2 R}{dx^2} + \frac{1}{x} \frac{dR}{dx} + \left(1 - \frac{n^2}{x^2} \right) R = 0 \qquad (168)$$

This is *Bessel's equation* for which the solutions are called Bessel functions of order n. Solutions of the *first kind* converge for all finite values of x:

$$J_n(x) = \left(\frac{x}{2}\right)^n \sum_{p=0}^{\infty} \frac{(-1)^p}{p!(p+n)!} \left(\frac{x}{2}\right)^{2p} \tag{169}$$

From **(169)** we may write the recurrence relations for $J_n(x)$:

$$J_{n+1}(x) = \frac{n}{x} J_n(x) - \frac{d}{dx} J_n(x) \tag{170}$$

$$J_{n-1}(x) = \frac{n}{x} J_n(x) + \frac{d}{dx} J_n(x) \tag{171}$$

$$J_{-n}(x) = (-1)^n J_n(x) \tag{172}$$

Also we have the relation for small x:

$$J_n \sim \left(\frac{x}{2}\right)^n \tag{173}$$

A solution of the *second kind*, called the *Neumann function*, is obtained in the *limit* that n approaches an integer:

$$N_n(x) = \frac{J_n(x)\cos n\pi - J_{-n}(x)}{\sin n\pi} \tag{174}$$

The Neumann function diverges at small x and is not an acceptable solution within the interior of a waveguide.

In order to study the asymptotic behavior at large x it is useful to rewrite **(168)** as an equation in $x^{1/2}R$:

$$\frac{d^2}{dx^2}(x^{1/2}R) + \left(1 - \frac{n^2 - \frac{1}{4}}{x^2}\right) x^{1/2}R = 0 \tag{175}$$

In the limit of x large compared with n, the solutions of **(175)** are trigonometric functions that are the asymptotic forms of **(169)** and **(174)**:

$$J_n(x) \simeq \left(\frac{2}{\pi x}\right)^{1/2} \cos(x - n\pi/2 - \pi/4) \tag{176}$$

$$N_n(x) \simeq \left(\frac{2}{\pi x}\right)^{1/2} \sin(x - n\pi/2 - \pi/4) \tag{177}$$

If we replace x by ix in **(168)** we obtain the *modified Bessel's equation*:

$$\frac{d^2R}{dx^2} + \frac{1}{x}\frac{dR}{dx} - \left(1 + \frac{n^2}{x^2}\right)R = 0 \tag{178}$$

Solutions of this equation that converge for large x are of the form:

$$K_n(x) = \frac{\pi}{2} i^{n+1}[J_n(ix) + iN_n(ix)] \tag{179}$$

The asymptotic form of **(179)** may be obtained from **(176)** and **(177)**:

$$K_n(x) = \left(\frac{\pi}{2x}\right)^{1/2} e^{-x} \tag{180}$$

The recurrence relations for $K_n(x)$ are from **(179)** and **(170)** and **(171)**:

$$K_{n+1}(x) = \frac{n}{x} K_n(x) - \frac{d}{dx} K_n(x) \tag{181}$$

$$K_{n-1}(x) = -\frac{n}{x} K_n(x) - \frac{d}{dx} K_n(x) \tag{182}$$

$$K_{-n}(x) = K_n(x) \tag{183}$$

The asymptotic form of $K_0(x)$ is for x small:

$$K_0(x) \simeq -\left(\ln\frac{x}{2} + 0.5772\right) \tag{184}$$

For $n > 1$ we have:

$$K_n(x) \simeq \tfrac{1}{2}(n-1)!\left(\frac{2}{x}\right)^n \tag{185}$$

The requirement that E_z be zero on the cylindrical surface leads through **(162)** to the condition that R vanish at $x = \alpha a$. We list in Table 2 the first few zeros of $J_n(x)$.

Table 2   Zeros of Bessel Functions $J_n(x)$

	$m = 1$	$m = 2$	$m = 3$
$n = 0$	$x = 2.405$	$x = 5.520$	$x = 8.654$
$n = 1$	$x = 0$	$x = 3.832$	$x = 7.016$
$n = 2$	$x = 0$	$x = 5.136$	$x = 8.417$

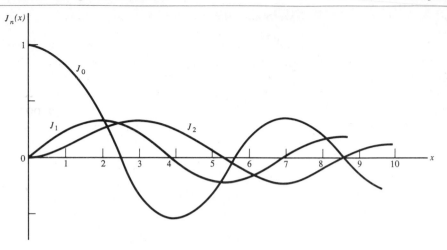

Figure 22   Solutions of Bessel's equation. The first few Bessel functions are shown.

From **(165)** the frequency of propagation is given by:

$$\omega^2 \mu \epsilon = k^2 + \frac{x^2}{a^2} \tag{186}$$

where x is given in Table 2. The modes are designated as circular TM_{nm} and the lowest frequency mode corresponds to the first zero of $J_0(x)$.

For transverse electric (TE) waves we write:

$$H_z(r,\phi) = R(r)\Phi(\phi) \tag{187}$$

The boundary conditions on H_z are that the normal derivative vanish at the boundary, which leads to

$$\frac{dR}{dx} = 0 \tag{188}$$

at $x = a$. We list in Table 3 the first few zeros of $J_n'(x)$. The frequency of the wave is given by **(186)** where the values of x are given in Table 3. The lowest frequency wave is then a TE_{11} and is followed by the TM_{01}. The first few Bessel functions are plotted in Figure 22.

Table 3   Zeros of $J_n'(x)$

	$m = 1$	$m = 2$	$m = 3$
$n = 0$	$x = 0$	$x = 3.832$	$x = 7.016$
$n = 1$	$x = 1.841$	$x = 5.331$	$x = 8.536$
$n = 2$	$x = 3.054$	$x = 6.706$	$x = 9.970$

PROPAGATION THROUGH DIELECTRIC WAVEGUIDES

The recent development of fused silica optical fibers of high purity and perfection has made possible communication by the transmission of modulated light through such fibers. The fundamental limit on propagation distance is light scattering by the irregularities characteristic of any amorphous material. Such scattering decreases with increasing wavelength and for fused silica is less than 3 dB (half the power) per kilometre in the range of one micrometre wavelength. This kind of scattering, which also accounts for the blue color of the sky, was first explained by Rayleigh and will be discussed in the next chapter.

A number of solid-state light sources, operating in the one micrometre wavelength range—both coherent and incoherent—are available and will be discussed in Section 15-10.

Our interest here is in the design of optical waveguides so as to minimize dispersion, either from a broad spectral source or from the propagation of a narrow spectral source in more than one mode. The design that we will discuss is shown in Figure 23. Single-mode fibers have an inner core of pure fused silica a few micrometres

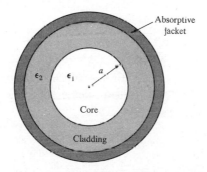

Figure 23 Optical waveguide. An inner core is surrounded by a cladding of lower permittivity. The cladding is covered with an absorptive jacket.

in diameter. Multimode fibers may be one hundred micrometres in diameter. Surrounding the inner core is a cladding of borosilicate glass of permittivity only slightly smaller than that of the core. As we have seen in our discussion of propagation in a dielectric slab, the permittivity of the surrounding medium must be lower if the radiation is to be contained. As we have also seen from **(115)** and Figure 14, the number of modes carried is of the order of:

$$n \simeq \alpha_c/(\pi/a) = (2a/\lambda)(\Delta\epsilon/\epsilon)^{1/2} \tag{189}$$

where λ is the wavelength in the medium. As we may see from **(189)** for single mode operation the fractional difference in permittivity will be inversely proportional to the square of the number of wavelengths across the slab. A similar condition operates for circular fibers.

Finally, the cladding is surrounded by a jacket that absorbs most of the light reaching it through the cladding. This jacket is necessary for the isolation of the fibers within a single bundle. Our analysis here will simply assume that the cladding extends to infinity, which is nearly equivalent to a perfectly absorbing jacket.

For a TM mode we write the longitudinal field within the core in the form of (162):

$$E_z(\rho,\phi) = E_1 J_n(\alpha\rho) \cos n\phi \tag{190}$$

where α is given by:

$$\alpha = (\omega^2 \mu_1 \epsilon_1 - k^2)^{1/2} \tag{191}$$

If the radiation is to be concentrated within the core, the solution outside must be exponentially decreasing and of the form:

$$E_z(\rho,\phi) = E_2 K_n(\beta\rho) \cos n\phi \tag{192}$$

where β is given by:

$$\beta = (k^2 - \omega^2 \mu_2 \epsilon_2)^{1/2} \tag{193}$$

The boundary conditions on E_z are continuity of the field and of its radial derivative, which is equivalent to continuity of the transverse components of $\mathbf{H}$ as may be seen from (143) and (144):

$$E_1 J_n(\alpha a) = E_2 K_n(\beta a) \tag{194}$$

$$\frac{\epsilon_1}{\alpha^2} E_1 \frac{d}{dr} J_n(\alpha a) = -\frac{\epsilon_2}{\beta^2} E_2 \frac{d}{dr} K_n(\beta a) \tag{195}$$

By using the recurrence relations for $J_n(x)$ and $K_n(x)$ we may write (195) as:

$$\frac{\epsilon_1}{\alpha^2} E_1[J_{n-1}(\alpha a) - J_{n+1}(\alpha a)] = \frac{\epsilon_2}{\beta^2} E_2[K_{n-1}(\beta a) + K_{n+1}(\beta a)] \tag{196}$$

Dividing (196) by (194) we have

$$\frac{\epsilon_1}{\alpha} \frac{J_{n-1}(\alpha a) - J_{n+1}(\alpha a)}{J_n(\alpha a)} = \frac{\epsilon_2}{\beta} \frac{K_{n-1}(\beta a) + K_{n+1}(\beta a)}{K_n(\beta a)} \tag{197}$$

20 Single-Mode Fibers. For a single-mode fiber the diameter $2a$ is made sufficiently small that (197) has only a single solution corresponding to $n = 0$. Writing (197) for $n = 0$ we have

$$\frac{\epsilon_1}{\alpha} \frac{J_1(\alpha a)}{J_0(\alpha a)} = -\frac{\epsilon_2}{\beta} \frac{K_1(\beta a)}{K_0(\beta a)} \tag{198}$$

which must be solved together with the subsidiary relation obtained from **(191)** and **(197)**:

$$\alpha^2 + \beta^2 = \omega^2(\mu_1\epsilon_1 - \mu_2\epsilon_2) \tag{199}$$

For a given mode we have from **(191)** a dispersion relation of the form shown in Figure 15. Although we may operate in a region of appreciable dispersion, the fact that sources are of narrow band width means that there will be little loss of phase coherence.

21 **Multimode Fibers.** For a multimode fiber the radius of curvature is large compared with a wavelength and the boundary conditions are quite similar to those for the dielectric slab. For the radius large and ϵ_1 and ϵ_2 not too close we may use the asymptotic forms of $J_n(x)$ and $K_n(x)$ and write **(197)** as:

$$\frac{\epsilon_1}{\alpha} \tan\left(\alpha a - \frac{n\pi}{2} - \frac{\pi}{4}\right) = \frac{\epsilon_2}{\beta} \tag{200}$$

which, except for the phase shift, is of the same form as obtained earlier for the dielectric slab.

If we are to minimize dispersion, however, we must not permit α to become large. In the limit that ϵ_1 and ϵ_2 are very close there is only a single allowed mode for each value of n as may be seen by writing **(197)** in the limiting form for αa and βa small. We assume for simplicity that n is greater than 2 so as to avoid the logarithmic term in the asymptotic limit of $K_1(x)$. We obtain from **(197)**:

$$\frac{1}{\alpha^2} - \frac{1}{\beta^2} \simeq \frac{a^2}{2} \frac{1}{n^2 - 1} \tag{201}$$

Taken together with the relation between α and β obtained from **(191)** and **(193)**:

$$\alpha^2 + \beta^2 = \omega^2(\mu_1\epsilon_1 - \mu_2\epsilon_2) \tag{202}$$

we may obtain α and β. We show in Figure 24 the graphical solution of **(201)** and **(202)**. From **(191)** we have

$$\omega^2\mu_1\epsilon_1 = k^2 + \alpha^2 \tag{203}$$

For ϵ_1 and ϵ_2 close, α^2 will be small and there will be very little dispersion of any of the propagating modes.

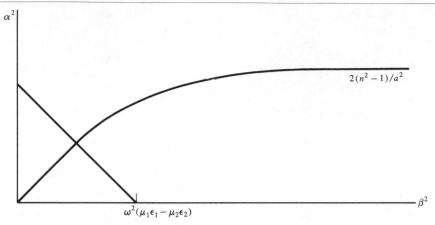

Figure 24   Graphical solution. The parameters α and β are determined from the simultaneous solution of **(201)** and **(202)**.

INTEGRATED OPTICS

With the development of coherent light sources and improvements in the fabrication of uniform thin films it has become possible to fabricate thin film lenses and prisms as well as polarizers, gratings, light modulators, and thin film lasers. We limit ourselves here to a brief discussion of thin film lenses and prisms.

We saw in the discussion of propagation in a dielectric slab that the dispersion relation is given by:

$$\omega^2 = \frac{1}{\mu_1 \epsilon_1} (\alpha^2 + k^2) \tag{204}$$

where αa has a fixed value for a particular propagating mode and is independent of wavelength and film thickness. We write αa in terms of the parameter x:

$$\alpha a = x \tag{205}$$

From **(204)** and **(205)** we may write the phase velocity as

$$v = \frac{\omega}{k} = \frac{1}{(\mu_1 \epsilon_1)^{1/2}} \left(1 - \frac{x^2}{\omega^2 \mu_1 \epsilon_1 a^2}\right)^{-1/2} \tag{206}$$

For film thicknesses a few wavelengths of light we may approximate **(206)**:

$$v \simeq v_0 \left(1 + \frac{x^2}{8\pi^2} \frac{\lambda^2}{a^2}\right) \tag{207}$$

where λ is the wavelength in the medium. We observe that the larger a the smaller the phase velocity, which is equivalent to increasing the refractive index.

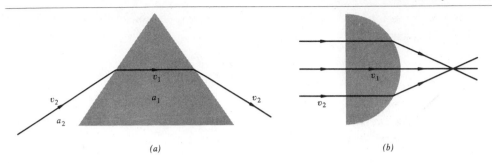

(a)   (b)

Figure 25 Integrated optics. (a) A prism may be formed by evaporating through a mask a slightly thicker dielectric slab. (b) A lens may be formed similarly.

We show in Figure 25a a prism formed by first evaporating a layer of a dielectric like zinc sulfide a few microns thick on a glass substrate. A second layer is evaporated through a triangular mask. Typical transverse dimensions are about one millimetre. The boundaries of the prism are tapered to minimize reflection and mode mixing. The refractive index of zinc sulfide is approximately 2.4 while that of the glass substrate is around 1.5. Thus the zinc sulfide film acts as a dielectric waveguide. The relative index of the prism, or of the lens shown in Figure 25b may be written as:

$$n = \frac{v_2}{v_1} = 1 + \frac{1}{2}\left(\frac{\lambda x}{2\pi}\right)^2 \left(\frac{1}{a_2{}^2} - \frac{1}{a_1{}^2}\right) \tag{208}$$

SUMMARY

1. For a wave incident on the interface between two media, the angle of incidence θ_1 is related to the angle of refraction θ_2 by Snell's law:

$$n_1 \sin \theta_1 = n_2 \sin \theta_2 \tag{8}$$

where n_1 and n_2 are the refractive indices of the two media.

2. When the permittivity (and thus the refractive index) of medium 1 is higher than that of medium 2, there will be a critical angle θ_c for total internal reflection given by:

$$\sin \theta_c = \frac{n_2}{n_1} \tag{10}$$

3. A medium may be characterized by an intrinsic impedance

$$Z = \frac{\omega\mu}{k} = (\mu/\varepsilon)^{1/2} \tag{16}$$

4. In terms of the intrinsic impedances, the complex reflection coefficient for normal incidence is given by

$$r = \frac{Z_2 - Z_1}{Z_2 + Z_1} \tag{24}$$

5. At oblique incidence and with the electric field transverse to the scattering plane we obtain for the reflection coefficient of the TE wave:

$$r = \frac{Z_2/\cos \underline{\theta}_2 - Z_1/\cos \theta_1}{Z_2/\cos \underline{\theta}_2 + Z_1/\cos \theta_1} \tag{34}$$

where the complex angle $\underline{\theta}_2$ is obtained from Snell's law.

6. At oblique incidence and with the magnetic field transverse to the scattering plane, the reflection coefficient of the TM wave is given by:

$$r = \frac{Z_2 \cos \underline{\theta}_2 - Z_1 \cos \theta_1}{Z_2 \cos \underline{\theta}_2 + Z_1 \cos \theta_1} \tag{37}$$

7. At an angle of incidence given by

$$\tan \theta_1 = \frac{Z_1}{Z_2} \tag{40}$$

the reflection coefficient vanishes for the TM wave. Then for initially unpolarized radiation, the reflected radiation will be polarized with the electric field normal to the plane of incidence. This angle is called the Brewster angle.

8. An electromagnetic wave may be propagated in the space between plane parallel conductors separated by a distance a. There exist discrete modes characterized by an index n where the angle of incidence with the planes is given by:

$$\cos \theta = n\lambda/2a \tag{93}$$

This relation holds for both TE and TM waves. For the TE wave there is no $n = 0$ wave. For the TM wave, $n = 0$ gives a TEM wave.

9. TE and TM waves may also be propagated in a dielectric slab. The condition for propagation is equivalent to the condition for total internal reflection at the boundaries of the dielectric.

10. Both TE and TM (but not TEM) waves may be propagated through a conducting pipe or waveguide. The transverse fields are entirely determined by the longitudinal field components E_z and H_z.

11. The waves are characterized by a pair of indices m and n with a dispersion relation

$$\omega^2 = \omega_{mn}^{\ 2} + k^2/\mu\epsilon \tag{150}$$

12. For a rectangular pipe of sides a and b the frequencies of the modes are given by

$$\mu\epsilon\omega_{mn}^{2} = (m\pi/a)^2 + (n\pi/b)^2 \tag{151}$$

13. Within a round pipe there similarly exist both TM and TE modes characterized by indices n and m. The index n is associated with the azimuthal behavior of the wave and leads to a radial variation of E_z or H_z with the Bessel function $J_n(\rho)$. For the TM wave the allowed modes correspond to the zeros of J_n. For the TE wave the allowed modes correspond to the zeros of $dJ_n/d\rho$.

14. It is possible to propagate TE and TM waves through dielectric fibers. For a multimode fiber the condition for a mode is that the wave be totally internally reflected at the boundary of the fiber.

15. By varying the thickness of a dielectric slab, it is possible to vary the effective refractive index within the slab. In this way lenses and prisms as well as other optical devices may be synthesized.

PROBLEMS

1. **Surface polarization current.** An electromagnetic field is incident normally on the interface with an absorbing medium. Integrate **(12-18)** in the absence of external currents to obtain at the interface:

$$\hat{n} \times \mathbf{H} = \mathbf{i}_p$$

where

$$\mathbf{i}_p = \int_0^\infty \frac{\partial}{\partial t} \mathbf{P}(z)\, dz$$

is the effective surface polarization current density.

2. **Fresnel biprism.** A monochromatic plane wave is divided by a Fresnel biprism as shown in the following diagram.

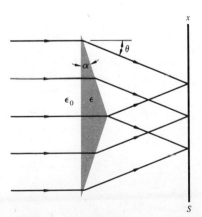

(a) For the wave incident normal to the base of the prism obtain an expression for the deflection angle θ, where α is the prism angle.

(b) The waves refracted through the two halves of the prism are detected at a screen. Obtain an expression for the detected intensity as a function of x.

(c) How large is the fringe separation d for soda glass ($n = 1.5$) with $\alpha = 1°$ and green light ($\lambda = 560$ nm)?

3. **Lloyd's mirror.** A Lloyd's mirror produces interference between the light reflected from a long mirror and the light coming directly from the source as shown in the following diagram. Assume that the source is sufficiently far distant that the incident and reflected waves may be regarded as plane.

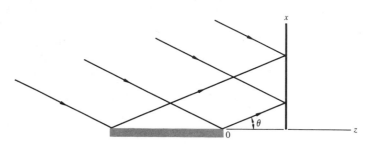

(a) Using complex notation write the expression for the incident electric field. Taking the mirror to be an unsilvered glass plate write an expression for the reflected electric field.

(b) Locate the positions of the dark fringes at a screen placed at z.

(c) When the screen is placed at the end of the mirror at $z = 0$, does $x = 0$ correspond to a dark fringe or a light fringe?

4. **Atmospheric refraction.** A luminous sphere of radius R (e.g., the sun) is observed at a distance $r \gg R$. The sphere is surrounded by an atmosphere of refractive index n and radius R'.

(a) Show that for $R' > nR$ the apparent radius of the sphere is nR.

(b) Show that for $R' < nR$ the apparent radius is R'.

5. **Ray equation.** A wave travels in a medium of variable refractive index n. At position $\mathbf{r}$ choose the $\hat{z}$ direction parallel to the gradient of n and let θ be the angle between the wavevector $\mathbf{k}$ and $\hat{z}$.

(a) Obtain the relation:

$$\frac{d\theta}{ds} = -\frac{1}{n}\frac{dn}{dz}\sin\theta$$

(b) Obtain the ray equation:

$$\frac{d}{ds}\mathbf{k} = \nabla k$$

where ds is along the direction of $\mathbf{k}$.

6. **Inhomogeneous medium.** An electromagnetic wave propagates in a medium of permeability μ_0 and permittivity $\epsilon(\mathbf{r})$.

(a) Show that the wave equations for **E** and **H** are of the form:

$$\nabla^2 \mathbf{E} - \nabla\left(\epsilon \mathbf{E} \cdot \nabla \frac{1}{\epsilon}\right) = \mu_0 \epsilon \frac{\partial^2}{\partial t^2} \mathbf{E}$$

$$\nabla^2 \mathbf{H} - \epsilon\left(\nabla \frac{1}{\epsilon}\right) \times (\nabla \times \mathbf{H}) = \mu_0 \epsilon \frac{\partial^2}{\partial t^2} \mathbf{H}$$

(b) For a wave propagating along the $\hat{z}$ direction in a medium where the permittivity is a slowly varying function of z, show that a possible solution is of the form:

$$\mathbf{E}(z,t) = \mathbf{E}_1(\epsilon/\epsilon_0)^{1/2} e^{i(\int k\,dz - \omega t)}$$

(c) What is the condition on the variation of ϵ for the wave to propagate without reflection as in (b)?

7. **Fresnel drag.** Fresnel's drag law states that the velocity of light in a body of refractive index n and moving with velocity V is given by:

$$v - \frac{c}{n} + (1 - 1/n^2)V$$

Derive this equation using the Einstein velocity addition law (Section F-2) under the assumption that V is much less than c. [I. Lerche, "The Fizeau Effect," *Am. J. Phys.* **45**, 1154 (1977.]

8. **Fermat's principle.** Fermat's principle states that the time required for a light ray to transverse the actual path is an extremum as compared to the times for all infinitesimally close paths. Use this principle to obtain Snell's law.

9. **Total internal reflection.** An electromagnetic wave is totally reflected from a medium of lower permittivity. Take the electric field in the second medium to be of the form:

$$\mathbf{E}_2(\mathbf{r},t) = E\hat{y}e^{-\kappa z}e^{i(kx \sin \theta - \omega t)}$$

(a) Find the magnetic field in the second medium.
(b) Show that the energy flow in the second medium is parallel to the interface and obtain an expression for the Poynting vector.

10. **Resonant absorption.** A wave is incident normally on a lossy dielectric of thickness a backed by a perfect conductor. Show that the condition for total absorption of the wave is given by:

$$\frac{\omega \mu_2}{i k_2 \cot k_2 a} = Z_1$$

where Z_1 is the intrinsic impedance in the incident medium, and k_2 is the wavevector in the lossy dielectric.

11. **Nonreflective coating.** An electromagnetic wave in a medium of intrinsic impedance Z_1 is incident normally on a medium of intrinsic impedance Z_2. Show that if the two media are separated by a film one-quarter wavelength thick and of intrinsic impedance $(Z_1 Z_2)^{1/2}$, then there will be no reflection.

12. **Salisbury screen.** An absorber of microwave radiation is constructed by placing a thin absorbing layer of resistivity ρ and thickness d at one-quarter wavelength from a conducting surface.
    (a) Find the optimum resistance ρ/d in ohms of the absorbing layer. Does this value depend on the permittivity ϵ or permeability μ of the intermediate layer?
    (b) How does the reflection coefficient vary with frequency where the separation between the absorber and the reflector is fixed?
    (c) Obtain an expression for the reflection coefficient of a Salisbury screen as a function of angle of incidence.

13. **Nonresonant absorber.** A broad-band absorber is constructed of a lossy dielectric with ϵ increasing gradually from ϵ_0 to some maximum value. Use the result of Problem 6(b) to obtain an expression for the electric and magnetic fields within the absorbing medium. Find an expression for the wave impedance within the medium. Nonresonant absorbers are also constructed from uniform material formed into tapered pyramids as on the walls of an anechoic chamber. Explain how such an arrangement accomplishes low reflectivity.

14. **Reflecting metal films.** A partially reflecting and partially transmitting mirror is made by evaporating a metallic film of controlled thickness.
    (a) Obtain expressions for the reflection coefficient R and the transmission coefficient T as a function of film thickness.
    (b) At what thickness are R and T equal for aluminum? At this thickness what fraction of the incident light is absorbed?
    (c) Obtain an expression for the absorption relative to transmission. How does this quantity vary with increasing film thickness?
    (d) Obtain an expression for R in the limit of infinite film thickness. Obtain a numerical value for aluminum.

15. **Highly reflecting surfaces.** In order to make glass $(\epsilon = 2.25\epsilon_0)$ surfaces highly reflecting they are coated with a quarter wavelength film of silicon monoxide $(\epsilon = 4.0\epsilon_0)$ or titanium dioxide $(\epsilon = 6.9\epsilon_0)$.
    (a) Obtain numerical values for the intensity reflection coefficient R from glass, bulk silicon monoxide, and bulk titanium dioxide.
    (b) Obtain an expression for the reflection coefficient of a quarter-wavelength coated glass surface and obtain numerical values for glass coated with silicon monoxide and with titanium dioxide.

16. **Anodization of aluminum.** A thick porous layer of amorphous alumina (Al_2O_3) will form on an aluminum anode in an electrolyte of sulfuric acid. Such layers are widely used for decorative purposes.
    (a) Taking the refractive index of alumina as 1.7, what is the thickness of a film that is one-quarter wavelength at 560 nm?
    (b) Discuss the reflection coefficient of the film described in (a) as a function of optical wavelength.

17. **Lambert's law.** A body that absorbs all incident radiation is described as a "black body."

(a) Show that for such a body to be in equilibrium with isotropic radiation, the energy radiated into a solid angle $d\Omega$ at an angle θ with the surface normal must be proportional to $\cos\theta$. This is Lambert's law.

(b) Show that for a radiant surface obeying Lambert's law the apparent brightness of the surface is independent of the angle from which it is viewed.

18. **Diffuse reflection.** A surface is described as a diffuse reflector if the intensity of reflected radiation is independent of the direction of incident radiation and varies as the cosine of the angle of reflection. Consider a surface that is composed of small segments of uniform size and randomly oriented.

(a) Show that such a surface will give diffuse reflection for normally incident radiation.

(b) Explain why, for grazing incidence, the reflection is specular.

19. **Brewster condition.** A TM wave is incident on the interface between media of differing permeabilities and permittivities.

(a) Show that the Brewster condition is given by:

$$\tan^2\theta_1 = \frac{(\epsilon_2/\epsilon_1)(\epsilon_2/\epsilon_1 - \mu_2/\mu_1)}{(\mu_2/\mu_1)(\epsilon_2/\epsilon_1 - \mu_1/\mu_2)}$$

(b) Show that the existence of the Brewster condition for the TM wave requires that ϵ_2/ϵ_1 be outside the region between μ_2/μ_1 and μ_1/μ_2.

(c) Find the Brewster condition for the TE wave and show that if (a) cannot be met for the TM wave, then it will be met for the TE wave and conversely.

20. **Reflection boundary conditions.** A wave is reflected at normal incidence along z and off the interface between a medium of intrinsic impedance Z_1 and a medium of intrinsic impedance Z_2. Show that the boundary conditions on the *transverse* components of **E** and **D** take the form:

$$\frac{\partial \mathbf{D}}{\partial t} = -\frac{Z_2}{Z_1{}^2}\frac{\partial \mathbf{E}}{\partial z} \qquad \frac{\partial \mathbf{B}}{\partial t} = -\frac{Z_1{}^2}{Z_2}\frac{\partial \mathbf{H}}{\partial z}$$

Use **(14)**, **(15)**, **(19)**, and **(20)**.

21. **Ellipsometry.** The thickness and refractive index of thin films on dielectric or metallic surfaces may be measured by a technique called ellipsometry. For an incident angle θ let the electric field be polarized at an angle ϕ to the scattering plane. Obtain an expression for the electric field of the reflected wave. Show that the reflected wave is elliptically polarized. What is the orientation of the major axis with respect to the scattering plane? What is the ratio of the major axis to the minor axis of the ellipse?

22. **Metallic reflection.** Light is reflected at normal incidence from a metal of low intrinsic impedance.

(a) Show that in the visible the intrinsic impedance may be approximated by:

$$Z = [1/(2\omega_p\tau) - i\omega/\omega_p](\mu/\epsilon_{\text{ion}})^{1/2}$$

(b) Show that for the intrinsic impedance Z much less than Z_0 the impedance of vacuum, the intensity reflection coefficient is approximately:

$$R \simeq 1 - \frac{4}{Z_0} \text{Re}\{Z\}$$

23. **Waveguide charges and currents.** An electromagnetic wave propagates down a rectangular waveguide of perfectly conducting walls.
    (a) Find the current and charge density on the walls in terms of the electric and magnetic fields at the walls.
    (b) Show that the *longitudinal* current density is equal to the surface charge density times the phase velocity.
    (c) Obtain explicit expressions for the charge and current density for the TE_{01} mode.

24. **Waveguide attenuation.** A rectangular waveguide contains a slightly lossy dielectric. Show that for frequencies above cutoff the fields are slightly attenuated. Find the attenuation constant k''. Show that below cutoff there is a change in phase along the guide and obtain the expression for k'.

25. **Energy transfer.** Electromagnetic waves are reflected from the interface with a medium of intrinsic impedance Z. Show that under the assumption that the wave-vector in the second medium is nearly normal to the interface, the rate of energy transfer into the second medium may be written as:

$$\Pi = \tfrac{1}{2} |\hat{\mathbf{n}} \times \mathbf{H}|^2 \, \text{Re}\{Z\}$$

where $\mathbf{H}$ is the magnetic field at the interface. Explain how this theorem might be used to compute the attenuation in a waveguide as a result of the finite conductivity of the walls.

26. **Resonant cavity.** A rectangular box with sides a, b, and l has perfectly conducting walls and is filled with a medium of permittivity ϵ.
    (a) Assuming $l > a > b$ find the lowest frequency standing wave. When excited in this way, this structure is called a resonant cavity.
    (b) Assume $\epsilon = \epsilon' + i\epsilon''$ to be complex. Find the decay time of the excited oscillations.

27. **Microwave horn.** A transverse magnetic (TM) wave propagates between conducting plates of variable separation $a(z)$. Assume a longitudinal electric field of the form:

$$E_z(x,z,t) = E \cos \frac{\pi x}{a} \, e^{i(\int k \, dz - \omega t)}$$

with $k = k(z) = (\omega^2 \mu \epsilon - \pi^2/a^2)^{1/2}$.
(a) Show that the condition that $E_z(x,z,t)$ be a solution of the wave equation is

$$\frac{d}{dz} \frac{1}{k} \ll 1$$

(b) interpret this condition.

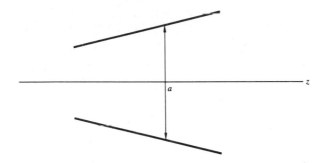

28. Waveguide transients. A waveguide is excited by a transverse electric field with the values at $z = 0$:

$$t < 0 \qquad E_y = 0$$

$$t > 0 \qquad E_y = E \sin \omega t$$

Assuming that propagation is in the fundamental mode, obtain an expression for $E_y(z,t)$. Obtain the form of this expression in the limit that ω is much higher and much lower than the cutoff frequency for the fundamental mode. See Section I-1.

REFERENCES

Arnaud, J. A., *Beam and Fiber Optics*, Academic, 1976.

Barnoski, M., editor, *Introduction to Integrated Optics*, Plenum, 1976.

Clarricoats, P. J. B., "Optical Fibre Waveguides—a Review" in *Progress in Optics*, XIV, E. Wolfe, editor, North-Holland, 1976.

Cook, J. S., "Communication by Optical Fiber," *Scientific American*, **229**(5), 28 (November 1973).

Henry, G. E., "Radiation Pressure," *Scientific American*, **196**(6), 99 (June, 1957).

Hertzberger, M., *Modern Geometrical Optics*, Interscience, 1958.

Jenkins, F. A., and H. E. White, *Fundamentals of Physical Optics*, Fourth Edition, McGraw-Hill, 1976.

Kapany, N. S., *Fiber Optics*, Academic, 1967.

Kapany, N., and J. J. Burke, *Optical Waveguides*, Academic, 1972.

Kline, M., and I. W. Kay, *Electromagnetic Theory and Geometrical Optics*, Interscience, 1965.

Levin, L., *Theory of Waveguides*, Halsted, 1975.

Marcuse, D., *Light Transmission Optics*, Van Nostrand Reinhold, 1972.

Marcuse, D., *Theory of Dielectric Optical Waveguides*, Academic, 1974.

Pratt, W. K., *Laser Communication Systems*, Wiley, 1969.

Tien, P. K., "Integrated Optics," *Scientific American*, **230**(4), 28 (April 1974).

von Hippel, A., *Dielectrics and Waves*, Wiley, 1954.

Zimmer, H. G., *Geometrical Optics*, Springer, 1970.

14 ELECTROMAGNETIC MEDIA III

In this chapter we are concerned with the scattering of electromagnetic radiation. We begin with the scattering by a single free charge $-e$ and describe what is called Thomson scattering. We next discuss scattering by a dielectric sphere and then by periodic slits and apertures in a thin dielectric screen. We next discuss scattering by media in which the dielectric susceptibility is a periodic function of position. Next, we discuss Rayleigh scattering, which arises from aperiodic variations in susceptibility. Finally, we examine the Cherenkov radiation of a rapidly moving charge.

THOMSON SCATTERING

We first discuss the scattering of radiation by a particle of charge $-e$ and mass m under conditions that the field is sufficiently weak that the particle velocity is much less than the velocity of light. We write, for the equation of motion of the particle,

$$m \frac{d\mathbf{v}}{dt} = -e\mathbf{E}(\mathbf{r},t) \tag{1}$$

where $\mathbf{E}(\mathbf{r},t)$ is the total electric field acting on the particle, and we have neglected the force of the magnetic field.

1   Radiation Fields. The further approximation that we make is to assume that the scattering is sufficiently weak that the electric field acting on the particle is predominantly the incident field:

$$\underline{\mathbf{E}}(\mathbf{r},t) = \mathbf{E}_0 \, e^{i(\mathbf{k} \cdot \mathbf{r} - \omega t)} \tag{2}$$

and that the scattered field may be neglected as far as the motion of the particle is concerned.[1] For the charge $-e$ located at the origin and assuming that its displacement

---

[1] For a treatment of scattering that includes the radiation reaction force, see K. Hagenbuch, *Am. J. Phys.*, **45**, 693 (1977).

is small compared with a wavelength, we obtain, for the induced moment,

$$\underline{p}(t) = e\underline{r}_1(t) = -\frac{e^2}{m\omega^2} \mathbf{E}_0 \, e^{-i\omega t} \tag{3}$$

We have obtained **(11-211)** and **(11-222)** for the vector and scalar potentials of an oscillating electric dipole:

$$\mathbf{A}(\mathbf{r},t) = \frac{\mu_0}{4\pi r} \frac{\partial}{\partial t} \underline{p}(t - r/c) \tag{4}$$

$$\phi(\mathbf{r},t) = \frac{1}{4\pi\epsilon_0} \frac{\hat{\mathbf{r}}}{rc} \cdot \frac{\partial}{\partial t} \underline{p}(t - r/c) \tag{5}$$

which leads to the radiation field:

$$\underline{\mathbf{B}}(\mathbf{r},t) = -\frac{\mu_0}{4\pi} \frac{\hat{\mathbf{r}}}{rc} \times \frac{\partial^2}{\partial t^2} \underline{p}(t - r/c) \tag{6}$$

Using **(3)** we may write the scattered magnetic field as:

$$\underline{\mathbf{B}}(\mathbf{r},t) = -\frac{\mu_0}{4\pi rc} \frac{e^2}{m} \hat{\mathbf{r}} \times \mathbf{E}_0 \, e^{i(kr - \omega t)} \tag{7}$$

From **(11-234)** the electric field is given by:

$$\underline{\mathbf{E}}(\mathbf{r},t) = -c\hat{\mathbf{r}} \times \underline{\mathbf{B}}(\mathbf{r},t) \tag{8}$$

The Poynting vector for the energy flow is given by:

$$\mathbf{\Pi}(\mathbf{r},t) = \mathbf{E}(\mathbf{r},t) \times \mathbf{H}(\mathbf{r},t) \tag{9}$$

where the fields given in **(9)** are the *real* physical fields. Substituting from **(6)** and **(8)** and averaging over a cycle we obtain for the scattered radiation:

$$\mathbf{\Pi}(\mathbf{r}) = \frac{1}{32\pi^2\epsilon_0} \frac{\omega^4}{r^2 c^3} \hat{\mathbf{r}} p^2 \sin^2 \psi \tag{10}$$

where ψ is the angle between $\mathbf{p}$ and $\hat{\mathbf{r}}$ as shown in Figure 1 and where p is the peak value of the moment. From **(6)** and **(8)** it is apparent that the scattered radiation is partially polarized. Writing the Poynting vector as the sum of a linearly polarized and an unpolarized component:

$$\mathbf{\Pi}(\mathbf{r}) = \mathbf{\Pi}_p(\mathbf{r}) + \mathbf{\Pi}_u(\mathbf{r}) \tag{11}$$

the degree of polarization V is defined by the relation:

$$V = \frac{\Pi_p}{\Pi_u + \Pi_p} \tag{12}$$

Exercise

1. An electromagnetic wave is scattered off a free charge. Show that radiation scattered through $90°$ is linearly polarized even though the incident wave is unpolarized.

2 **Thomson Cross Section.** From Figure 1 we have the relation:

$$\sin^2 \psi = 1 - \cos^2 \psi = 1 - \cos^2 \phi \sin^2 \theta \tag{13}$$

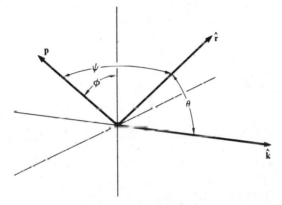

Figure 1 Thomson scattering. The wave is incident along $\mathbf{k}$ and the field position is along $\mathbf{r}$.

Averaging (13) over ϕ, which is the polarization angle of the incident field, we obtain:

$$\langle \sin^2 \psi \rangle = 1 - \tfrac{1}{2}\sin^2 \theta = \tfrac{1}{2}(1 + \cos^2 \theta) \tag{14}$$

Integrating (10) over a sphere of radius r and using (14) we obtain, for the total rate of energy radiation,

$$\frac{dU}{dt} = \oint_S \Pi_1(\mathbf{r}) \cdot d\mathbf{S} = \frac{1}{12\pi\epsilon_0} \frac{\omega^4}{c^3} p^2 = \frac{1}{12\pi\epsilon_0} \frac{e^4}{m^2 c^3} E_0{}^2 \tag{15}$$

where we have substituted from (3) for p. We note that the rate at which energy is radiated is independent of frequency.

It is useful to characterize the scattering by the area that would intercept an energy flow from the incident beam equal to (15). We write for the Poynting vector of the incident radiation:

$$\Pi_0 = \tfrac{1}{2}E_0 H_0 \mathbf{k} = \frac{E_0{}^2}{2Z_0} \mathbf{k} = \tfrac{1}{2}\epsilon_0 c E_0{}^2 \mathbf{k} \tag{16}$$

We define the Thomson cross section by the relation:

$$\frac{dU}{dt} = \Pi_0 \, \sigma_T \tag{17}$$

Substituting from **(15)** and **(16)** we obtain:

$$\sigma_T = \frac{1}{6\pi\epsilon_0{}^2} \frac{e^4}{m^2 c^4} \tag{18}$$

If we define a characteristic length by the relation:

$$\frac{e^2}{4\pi\epsilon_0 r_0} = mc^2 \tag{19}$$

then we may write **(18)** as:

$$\blacktriangleright \qquad\qquad \sigma_T = \frac{8\pi}{3} r_0{}^2 \tag{20}$$

where the distance $r_0 = \mu_0 \, e^2/4\pi m = 2.8178 \times 10^{-15}$ m is called the classical radius of the electron.

SCATTERING BY A DIELECTRIC SPHERE

We next consider the situation shown in Figure 2 where an electromagnetic wave described by **(2)** is incident on a dielectric sphere of susceptibility $\chi(\omega)$ and radius R.

3 **Fraunhofer Diffraction.** We assume that $\chi(\omega)$ is sufficiently small that we may neglect the scattered field and write for the polarization density:

$$\mathbf{P}(\mathbf{r}_1, t) = \epsilon_0 \, \chi \mathbf{E}_0 \, e^{i(\mathbf{k} \cdot \mathbf{r}_1 - \omega t)} \tag{21}$$

We expect to be able to write the radiation fields in the form of **(6)** and **(8)** only in the limit that the radius of the sphere is small compared with the wavelength of the radiation. Otherwise we must allow for the possibility of destructive interference between

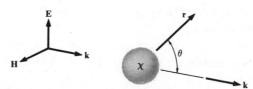

Figure 2 Scattering by a dielectric sphere. A dielectric sphere of radius R is characterized by a susceptibility $\chi(\omega)$.

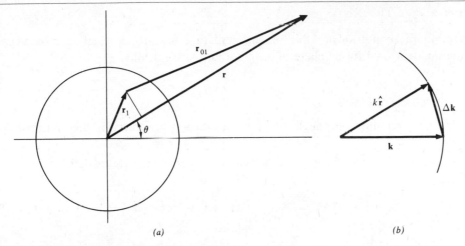

(a)   *(b)*

Figure 3 Fraunhofer scattering. (*a*) The distance to the field position is assumed large compared with the size of the scatterer. (*b*) The change in wavevector is the final wavevector less the initial wavevector.

radiating regions of the sphere. Given the approximation that the scattered radiation is weak, the simplest way to proceed is to write (**6**) as an integral over the sphere:

$$\mathbf{B}(\mathbf{r},t) = \frac{\omega^2 \mu_0}{4\pi c} \int_{V_1} \frac{\hat{\mathbf{r}}_{01} \times \mathbf{P}(\mathbf{r}_1, t - r_{01}/c)}{r_{01}} \, dV_1 \tag{22}$$

As long as the distance to the field position is large compared with the size of the sphere we may make the approximation (**11-203**) as shown in Figure 3*a*:

$$r_{01} \simeq r - \hat{\mathbf{r}} \cdot \mathbf{r}_1 \tag{23}$$

In writing (**22**) from (**11-233**) we have already assumed that the distance from the original source is large compared with a wavelength. Making in addition the assumption that the distance to the field position is large compared with the size of the source, we have what is called the *Fraunhofer condition*. With these approximations (**22**) becomes:

$$\mathbf{B}(\mathbf{r},t) = \frac{\mu_0}{4\pi} \frac{\omega^2 \hat{\mathbf{r}}}{rc} \times \int_{V_1} \mathbf{P}(\mathbf{r}_1, t - r/c + \hat{\mathbf{r}} \cdot \mathbf{r}_1/c) \, dV_1 \tag{24}$$

Finally, substituting from (**21**) for $\mathbf{P}(\mathbf{r}_1, t)$, we obtain:

$$\mathbf{B}(\mathbf{r},t) = \frac{1}{4\pi} \frac{\omega^2}{rc^3} \hat{\mathbf{r}} \times \mathbf{E}_0 \int_{V_1} \chi(\mathbf{r}) e^{-i(\Delta \mathbf{k} \cdot \mathbf{r}_1 + \omega t - kr)} \, dV_1 \tag{25}$$

where

$$\Delta \mathbf{k} = k\hat{\mathbf{r}} - \mathbf{k} \tag{26}$$

as shown in Figure 3*b*, is the difference between the outgoing wavevector along **r** and the incident wavevector along **k**.

Exercise

2. Obtain the condition for the forward radiating field to be weak compared with the incident field for a sphere of radius R and susceptibility χ:

$$\chi < 3/(kR)^2$$

This is the condition for neglecting the polarization produced by the scattered field.

4 Form Factor.

The form factor for a body with susceptibility $\chi(\mathbf{r})$ is defined as

$$f(\Delta \mathbf{k}) = -\frac{\epsilon_0}{e^2/m\omega^2} \int_{V_1} \chi(\mathbf{r}_1) e^{-i\,\Delta \mathbf{k} \cdot \mathbf{r}_1}\, dV_1 \tag{27}$$

In terms of the form factor we have, for the scattered magnetic field (25),

$$\mathbf{B}(\mathbf{r},t) = -\frac{\mu_0}{4\pi rc}\frac{e^2}{m} f(\Delta \mathbf{k})\hat{\mathbf{r}} \times \mathbf{E}_0\, e^{i(kr - \omega t)} \tag{28}$$

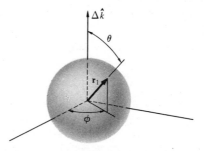

Figure 4 Scattering sphere. The form factor is obtained by integrating over the scattering sphere in spherical coordinates with the z axis along $\Delta \mathbf{k}$.

The form factor for $\Delta \mathbf{k} = 0$ is the *effective* number of free charges:

$$\blacktriangleright \qquad f(0) = -\frac{\epsilon_0}{e^2/m\omega^2} \int_{V_1} \chi(\mathbf{r}_1)\, dV_1 = N_{\text{eff}} \tag{29}$$

Note that for χ positive, N_{eff} is actually negative because the scattered field is π out of phase with the field scattered from a free charge. For the special case that χ is uniform throughout the sample, we may write:

$$f(\Delta \mathbf{k}) = \frac{N_{\text{eff}}}{V_1} \int_{V_1} e^{-i\,\Delta \mathbf{k} \cdot \mathbf{r}_1}\, dV_1 \tag{30}$$

Comparing (28) with (7) we see that the form factor $f(\Delta \mathbf{k})$ is the ratio of the field amplitude for an extended scatterer with N_{eff} electrons to the field scattered by a single electron.

We may transform **(30)** by taking the z axis along $\Delta \mathbf{k}$ as shown in Figure 4. Performing the integration over θ and ϕ we have:

$$f(\Delta k) = \frac{4\pi N_{eff}}{V_1} \int_0^R \frac{\sin r\,\Delta k}{r\,\Delta k} r^2\, dr \tag{31}$$

The integral **(31)** may be performed by integration by parts and yields:

▶

$$f(\Delta k) = 3N_{eff}\left(\frac{\sin R\,\Delta k}{(R\,\Delta k)^3} - \frac{\cos R\,\Delta k}{(R\,\Delta k)^2}\right) \tag{32}$$

The function $f(\Delta k)$ is plotted in Figure 5.

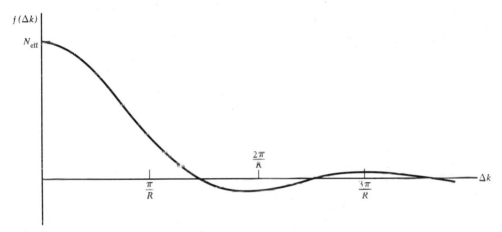

Figure 5   Form factor. The form factor, which is a measure of the scattering, with wavevector change Δk is shown as a function of Δk.

An alternate way of regarding **(27)** is suggested in Section I-1 where it is shown that given a function $\chi(\mathbf{r})$, the Fourier transform of this function is given by **(I-20)**:

$$\bar{\chi}(\mathbf{k}) = \frac{1}{(2\pi)^3} \int \chi(\mathbf{r}) e^{-i\mathbf{k}\cdot\mathbf{r}}\, dV \tag{33}$$

Thus, the form factor **(27)** is proportional to the Fourier transform of the susceptibility function:

$$f(\Delta \mathbf{k}) = -\frac{(2\pi)^3 \epsilon_0}{e^2/m\omega^2}\, \bar{\chi}(\Delta \mathbf{k}) \tag{34}$$

Exercises

3. An electromagnetic wave is scattered off a thin dielectric shell of thickness t, much less than the wavelength of the radiation, and radius R. Find the form factor.

4. Show that the condition of Exercise 2 may be written in terms of the effective number of scatterers N_{eff} and the classical radius of the electron r_0 as

$$N_{eff} < R/r_0$$

5. Show that for $kR \ll 1$ the radiated field has the same form factor as a point scatterer while for $kR \gg 1$ the scattering is largely contained within a cone of half angle $\theta \simeq 1/kR$.

6. From **(32)** obtain an expression for the zeros of $f(\Delta k)$. Obtain an approximate graphical solution and compare with Figure 5.

SCATTERING BY APERTURES IN A DIELECTRIC SCREEN

We wish in this section to discuss the scattering by various kinds of apertures in a two dimensional dielectric screen. We begin with the scattering by a circular aperture following which we discuss single slit and then multiple slit diffraction.

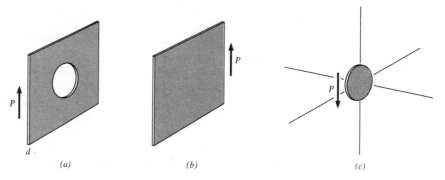

(a)   (b)   (c)

Figure 6 Circular aperture. (*a*) We wish the scattering from a circular aperture in a uniformly polarized screen. (*b*) The aperture may be replaced by a uniformly polarized screen and (*c*) a disc of reversed polarization.

5 **Circular Aperture.** We wish to discuss the diffraction of radiation by a circular aperture of radius R in a dielectric screen as shown in Figure 6a. Since we assume that the scattered wave is weak, we may replace the scatterer by a uniformly polarized screen as shown in Figure 6b and a disc of reversed polarization as shown in Figure 6c. Except for the forward and reverse directions, the scattered wave from (*a*) must be the same as from (*c*), which is Babinet's principle as applied to a dielectric screen.[2] To find the scattered wave we integrate **(30)** over the thickness of the screen replacing **P** by $-\mathbf{P}$ to obtain the form factor:

$$f(\Delta\mathbf{k}) = -\frac{N_{eff}}{S_1} \int_{S_1} e^{-i\,\Delta\mathbf{k}\cdot\mathbf{r}_1}\,dS_1 \tag{35}$$

[2] For a discussion of Babinet's principle as applied to thin *conducting* screens, see C. L. Andrews and D. P. Margolis, *Am. J. Phys.*, **43**, 672 (1975).

where we have performed the integration normal to the disc of thickness d assumed thin compared with a wavelength. We show in Figure 7 an expanded view of the disc to indicate the angles over which we integrate.

Since $\mathbf{r}_1$ is in the plane of the disc, only the transverse component of $\Delta\mathbf{k}$ appears in the exponential and we have:

$$f(\Delta\mathbf{k}) = -\frac{N_{\text{eff}}}{S_1} \int_0^R r_1 \, dr_1 \int_0^{2\pi} e^{-ikr_1 \sin\theta \cos\phi} \, d\phi \tag{36}$$

The integral over ϕ may be performed by using the integral representation of the Bessel function of zero order:

$$J_0(x) = \frac{1}{2\pi} \int_0^{2\pi} e^{ix \cos\phi} \, d\phi \tag{37}$$

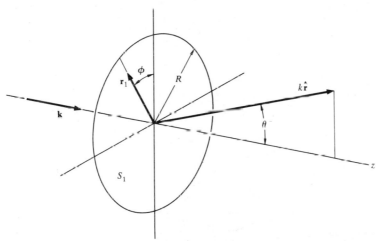

Figure 7 Scattering disc. The scattered wave is obtained by integrating over the disc.

Using (37) we have, for (36),

$$f(\Delta\mathbf{k}) = -\frac{2\pi N_{\text{eff}}}{S_1} \int_0^R J_0(kr_1 \sin\theta) r_1 \, dr_1 \tag{38}$$

We may integrate (38) by using the recurrence relation for $J_n(x)$ **(13-171)** which may be written in the form:

$$xJ_n(x) = x^{-n} \frac{d}{dx} \left[x^{n+1} J_{n+1}(x) \right] \tag{39}$$

Writing (39) for $n = 0$, (38) becomes:

$$f(\Delta\mathbf{k}) = -N_{\text{eff}} \frac{J_1(kR \sin\theta)}{\tfrac{1}{2}kR \sin\theta} \tag{40}$$

This quantity is plotted in Figure 8. Most of the radiation falls within a half angle:

$$\blacktriangleright \qquad \theta < \sin^{-1}\frac{3.832}{kR} = \sin^{-1}\frac{1.220\lambda}{2R} \qquad\qquad (41)$$

which defines what is called the *Airy disc*. This theory was first applied to diffraction by circular lenses. The *Rayleigh* criterion for the resolution of two sources that subtend an angle θ is that the center of the Airy disc of one source should be outside the Airy disc of the other source:

$$\sin\theta > 1.220\,\frac{\lambda}{D} \qquad\qquad (42)$$

where $D = 2R$ is the diameter of the aperture.

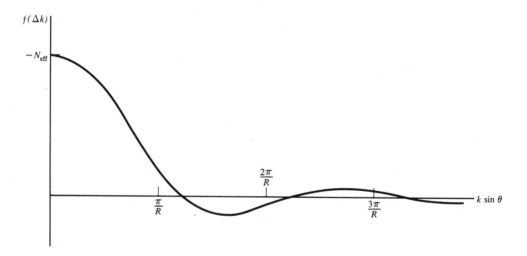

Figure 8   Form factor. The form factor is shown as a function of $k \sin\theta$.

Exercise

7. Show that the condition that the dielectric screen affects the fields very little is

$$k_1\, dZ_1 < Z_0$$

where $Z_1 = (\mu_1/\epsilon_1)^{1/2}$ is the intrinsic impedance of the dielectric medium, and Z_0 is the impedance of vacuum.

6 Opaque Screen. Although the approximations of the above theory restrict our results to transparent screens, it is possible with additional restrictions to extend the theory to the case where the screen is opaque. We use **(13-23)** to write for the field within the dielectric:

$$E_1(r,t) = \frac{2Z_1}{Z_0 + Z_1} E_0\, e^{i(k_1 z - \omega t)} \tag{43}$$

We write for the polarization within the dielectric:

$$P(r,t) = \epsilon_0\, \chi E_1(r,t) = (\epsilon_1 - \epsilon_0) E_1(r,t) \tag{44}$$

Now, the total polarization per unit area within the dielectric may be written as:

$$\int_0^d P\, dz = \epsilon_0\, \chi E_1 \int_0^d e^{i(k_1 z - \omega t)}\, dz = -\frac{\epsilon_1 - \epsilon_0}{ik_1} E_1 e^{-i\omega t} = -\frac{1}{i\omega Z_1}\frac{\epsilon_1 - \epsilon_0}{\epsilon_1} E_1 e^{-i\omega t} \tag{45}$$

where we have assumed that the screen is sufficiently thick that the electromagnetic field does not penetrate to $z = d$. Substituting from (43) for $E_1(r,t)$ and using the definition of the intrinsic impedance:

$$Z_0 = (\mu_0/\epsilon_0)^{1/2} \qquad Z_1 = (\mu_1/\epsilon_1)^{1/2} \tag{46}$$

we obtain for (45):

$$\int_0^d P\, dz = -\frac{2}{i\omega\epsilon_1}\frac{\epsilon_1}{Z_0 + Z_1}\frac{\epsilon_0}{} E_0\, e^{-i\omega t} \tag{47}$$

Since the effective scatterer is the *negative* of (47), we have, from (24) for the scattered magnetic field,

$$B(r,t) = -\frac{iZ_0}{\lambda r c \epsilon_1}\frac{\epsilon_1 - \epsilon_0}{Z_0 + Z_1} e^{i(kr - \omega t)}(\hat{r} \times E_0) \int_{S_1} e^{-i\,\Delta k \cdot r_1}\, dS_1 \tag{48}$$

Comparing (48) with (7) we have for the effective number of scatterers:

$$\blacktriangleright \qquad N_{\text{eff}} = -\frac{4\pi i}{\mu_0}\frac{S_1}{\lambda e^2/m}\frac{Z_0(\epsilon_1 - \epsilon_0)}{\epsilon_1(Z_1 + Z_0)} \tag{49}$$

The fact that N_{eff} is imaginary simply means that the scattered field is $\pi/2$ out of phase with the field scattered by a single free charge. If the screen is highly reflecting as well as being opaque, we may neglect ϵ_0 as compared with ϵ_1, and Z_1 compared with Z_0 and we have finally for the scattered electric and magnetic fields:

$$B(r,t) = -\frac{1}{c}\frac{i}{\lambda r}\hat{r} \times E_0\, e^{i(kr - \omega t)} \int_{S_1} e^{-i\,\Delta k \cdot r_1}\, dS_1 \tag{50}$$

$$E(r,t) = \frac{i}{\lambda r}\hat{r} \times (\hat{r} \times E_0)e^{i(kr - \omega t)} \int_{S_1} e^{-i\,\Delta k \cdot r_1}\, dS_1 \tag{51}$$

Our final expression for the scattered electric field, written as an integral over the incident electric field within the aperture of a highly reflecting screen is known as the *Fresnel-Kirchoff diffraction formula*.

In the foregoing discussion we have assumed the polarization in the rest of the screen to be unaffected by the presence of the aperture. Only in this way can we treat the radiated fields as arising from a polarization density $-\mathbf{P}(z)$ within the aperture. This assumption requires that the fields diffracted by the aperture affect very little the fields and resulting polarization in the rest of the screen. This requires that there be very little radiation scattered at 90° and, consequently, that the aperture be *large* compared with a wavelength.[3] It should be emphasized that we have made no assumption about the fields in the aperture itself, but only about the fields in the screen.

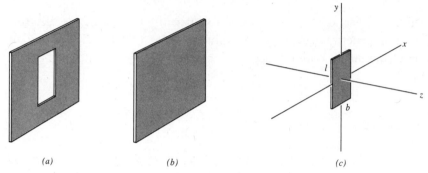

Figure 9 Rectangular aperture. (*a*) We wish the scattering from a rectangular aperture in a uniformly polarized screen. (*b*) The aperture may be replaced by a uniformly polarized screen and (*c*) a rectangle of reversed polarization.

7 Scattering by a Rectangular Aperture. We next discuss the scattering of radiation by a slit of length l and of width b in a screen as shown in Figure 9. We use the same argument as for the circular aperture and compute the field radiated by the oppositely polarized strip, which would fill the slit. The form factor may be written from (**34**) as

$$f(\Delta\mathbf{k}) = -N_{\text{eff}} \int_{-1/2b}^{1/2b} e^{-ix\,\Delta k_x}\,\frac{dx}{b} \times \int_{-l/2}^{l/2} e^{-iy\,\Delta k_x}\,\frac{dy}{l} \tag{52}$$

We may perform the integrals given in (**48**) to obtain

$$f(\Delta\mathbf{k}) = -N_{\text{eff}}\,\frac{\sin \tfrac{1}{2}b\,\Delta k_x}{\tfrac{1}{2}b\,\Delta k_x}\,\frac{\sin \tfrac{1}{2}l\,\Delta k_y}{\tfrac{1}{2}l\,\Delta k_y} \tag{53}$$

In the limit that l becomes large, $f(\Delta\mathbf{k})$ goes to zero very rapidly with increasing Δk_y.

[3] For an experimental test of the theory, see P. M. Rinard, *Am. J. Phys.*, **44**, 70 (1976).

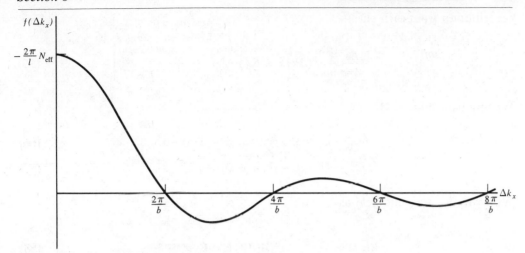

Figure 10 Form factor. The form factor of a rectangular aperture is shown as a function of Δk_x.

We may define a form factor for Δk_x alone as the integral over Δk_y:

$$\blacktriangleright \qquad f(\Delta k_x) = \int f(\Delta \mathbf{k})\, dk_y = -\frac{2\pi N_{\mathrm{eff}}}{l}\frac{\sin \frac{1}{2}b\, \Delta k_x}{\frac{1}{2}b\, \Delta k_x} \qquad (54)$$

This function is plotted in Figure 10.

8 Double-Slit Diffraction. We next consider the problem of two slits of width b separated by a displacement a as shown in Figure 11. The scattered field **(25)** will have a

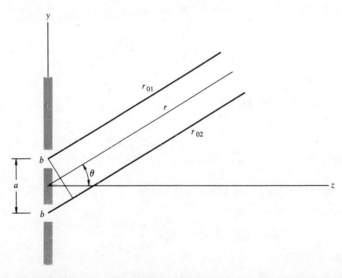

Figure 11 Double-slit diffraction. The single-slit form factor is an envelope for the double-slit interference function.

contribution from each slit:

$$\mathbf{B}(\mathbf{r},t) = -\frac{\mu_0}{4\pi c}\frac{e^2}{m} f(\Delta\mathbf{k})\,\hat{\mathbf{r}}\times\mathbf{E}_0\left[\frac{e^{i(kr_{01}-\omega t)}}{r_{01}} + \frac{e^{i(kr_{02}-\omega t)}}{r_{02}}\right] \tag{55}$$

Writing from Figure 11:

$$kr_{01} = kr - \tfrac{1}{2}ka\sin\theta = kr - \tfrac{1}{2}\Delta\mathbf{k}\cdot\mathbf{a} \tag{56}$$

$$kr_{02} = kr + \tfrac{1}{2}ka\sin\theta = kr + \tfrac{1}{2}\Delta\mathbf{k}\cdot\mathbf{a} \tag{57}$$

We write the magnetic field as

$$\mathbf{B}(\mathbf{r},t) = -\frac{\mu_0}{4\pi rc}\frac{e^2}{m} A(\Delta\mathbf{k})\,\hat{\mathbf{r}}\times\mathbf{E}_0\,e^{i(kr-\omega t)} \tag{58}$$

where $A(\Delta\mathbf{k})$ is called the scattering amplitude. Using (56) and (57) we may write:

▶
$$A(\Delta\mathbf{k}) = 2f(\Delta\mathbf{k})\cos\tfrac{1}{2}\mathbf{a}\cdot\Delta\mathbf{k} \tag{59}$$

where $f(\Delta\mathbf{k})$ is the form factor for a single slit. Since the separation between slit centers must be wider than the slit width b, the form factor shown in Figure 10 provides an envelope for the interference between slits. For long slits we integrate (59) over Δk_y to obtain:

$$A(\Delta k_x) = 2f(\Delta k_x)\cos\tfrac{1}{2}a\,\Delta k_x \tag{60}$$

9 **Multiple-Slit Diffraction.** For an array of N slits we extend (60) to obtain for the scattering amplitude:

$$A(\Delta k_x) = -f(\Delta k_x)\sum_{n=-(N-1)/2}^{(N-1)/2} e^{ina\,\Delta k_x} \tag{61}$$

To evaluate the summation we use

$$\sum_{n=-(N-1)/2}^{(N-1)/2} e^{in\phi} = \frac{\sin\tfrac{1}{2}N\phi}{\sin\tfrac{1}{2}\phi} \tag{62}$$

With (62) we obtain for the scattering amplitude:

▶
$$A(\Delta k_x) = f(\Delta k_x)\frac{\sin\tfrac{1}{2}Na\,\Delta k_x}{\sin\tfrac{1}{2}a\,\Delta k_x} \tag{63}$$

which is the familiar result for a diffraction grating with N slits. The coefficient of $f(\Delta k_x)$ in (63) is plotted in Figure 12a for $N = 5$. In Figure 12b we plot the form

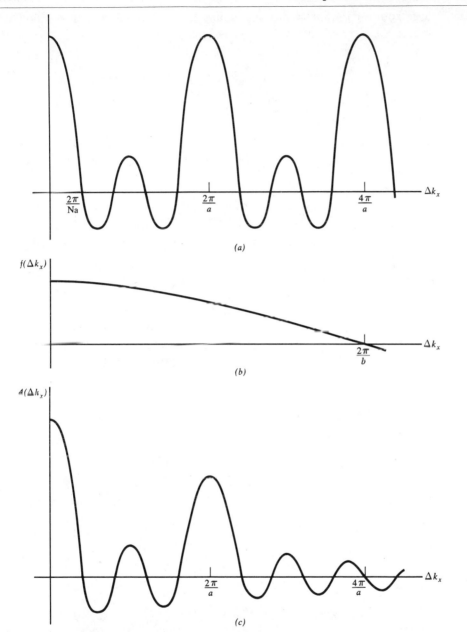

Figure 12 Multiple-slit diffraction. (*a*) The interference between slits is shown for five slits. (*b*) The form factor of a single slit of width $b = a/2$ is shown. (*c*) The scattering amplitude is the product of the interference function and the form factor.

factor $f(\Delta k_x)$ for $b = a/2$. Finally in Figure 12*c* we plot the scattering amplitude **(63)**, which is the product of the quantities shown in Figures 12*a* and 12*b*. The peaks in the scattering amplitude are for values of Δk_x for which the denominator goes to zero:

$$\Delta k_x = \frac{2n\pi}{a} \qquad\qquad\qquad \textbf{(64)}$$

For N large and in the vicinity of a maximum, the denominator changes slowly and we may make the approximation for the coefficient of the form factor in (63):

$$\frac{\sin \frac{1}{2}Na\,\Delta k_x}{\sin \frac{1}{2}a\,\Delta k_x} \simeq \frac{\sin \frac{1}{2}Na\,\Delta k_x}{\frac{1}{2}a\,\Delta k_x - n\pi} \tag{65}$$

The right side of (65) is the same as the form factor for a single slit and integrates over Δk_x to $\pm 2\pi/a$. We may then describe the scattering amplitude as a set of discrete values:

$$A_{\Delta k_x} = (-1)^{n(N-1)}\frac{2\pi}{a}\,f(\Delta k_x) \tag{66}$$

where Δk_x is given by (64).

10 **Aperture Function.** Instead of having an aperture cut off sharply at its edges, it is possible to taper the aperture from completely open to opaque. By using photographic techniques an aperture with any taper may be generated. To describe the taper we introduce the aperture function $F(\mathbf{r}_1)$ and we write (50) and (51) in the form

$$\mathbf{B}(\mathbf{r},t) = -\frac{1}{c}\frac{i}{\lambda r}\,\hat{\mathbf{r}} \times \mathbf{E}_0\,e^{i(kr-\omega t)}\int_{S_1} F(\mathbf{r}_1)e^{-i\,\Delta\mathbf{k}\cdot\mathbf{r}_1}\,dS_1 \tag{67}$$

$$\mathbf{E}(\mathbf{r},t) = \frac{1}{\lambda r}\,\hat{\mathbf{r}} \times (\hat{\mathbf{r}} \times \mathbf{E}_0)e^{i(kr-\omega t)}\int_{S_1} F(\mathbf{r}_1)e^{-i\,\Delta\mathbf{k}\cdot\mathbf{r}_1}\,dS_1 \tag{68}$$

The form factor (30) takes the form

$$f(\Delta\mathbf{k}) = f(0)\frac{\int_{S_1} F(\mathbf{r}_1)e^{-i\,\Delta\mathbf{k}\cdot\mathbf{r}_1}\,dS_1}{\int_{S_1} F(\mathbf{r}_1)\,dS_1} \tag{69}$$

For a long slit parallel to the y axis, we integrate (69) over y to obtain:

$$f(\Delta k_x) = \frac{2\pi f(0)}{l}\frac{\int F(x_1)e^{-ix_1\,\Delta k_x}\,dx_1}{\int F(x_1)\,dx_1} \tag{70}$$

From Section I-1 it is apparent that the form factor is proportional to the Fourier transform of the aperture function.

We may also use the aperture function to describe scattering by a transparent screen of variable thickness. Where the screen thickness approaches a wavelength, retardation effects may be treated by the use of a complex aperture function.

HOLOGRAPHY

As Dennis Gabor showed, it is possible to store on a photographic plate both amplitude and phase information with respect to the light reaching the plate from a source. As he further showed, by diffracting light through the developed plate one obtains both a virtual image at the position of the original source and real image at the mirror position.

We first consider the simple example of an incident *plane* wave, which we call the object wave as shown in Figure 13*a*:

$$E_{obj}(r,t) = E e^{i(k_1 \cdot r - \omega t)} \tag{71}$$

In addition we illuminate the screen with a reference wave

$$E_{ref}(r,t) = E_0 \, e^{i(kz - \omega t)} \tag{72}$$

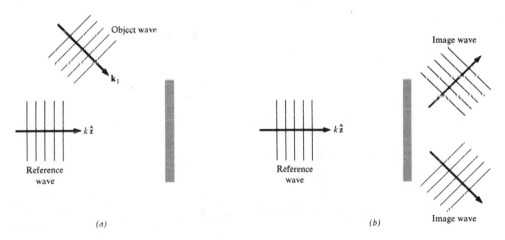

(a)   (b)

Figure 13 Holography. (*a*) A photographic plate is exposed to an object wave and a reference wave. The exposed plate is then developed. (*b*) When the developed plate is illuminated by the reference wave, two image waves are produced.

The darkening of the developed photographic plate and thus the aperture function will vary with the light intensity in the plane of the screen

$$I(r_1) = \tfrac{1}{2}|E_{obj} + E_{ref}|^2 = \tfrac{1}{2}E_0^2 + \tfrac{1}{2}E^2 + EE_0 \cos k_1 \cdot r_1 \tag{73}$$

where r_1 is a vector in the plane of the screen.

From (73) we write for the aperture function of the photographic plate

$$F(r_1) = F_0 + F_1 \cos k_1 \cdot r_1 \tag{74}$$

Following the development of the plate we illuminate with the reference wave as shown in Figure 13*b* obtaining a diffracted electric field of the form of **(68)**. Substituting **(74)** into **(69)** we obtain for the form factor:

$$f(\Delta k) = \frac{f(0)}{S_1} \left\{ \int_{S_1} e^{-i\,\Delta k \cdot r_1}\, dS_1 + \frac{1}{2}\frac{F_1}{F_0} \left[\int_{S_1} e^{i(k_1 - \Delta k) \cdot r_1}\, dS_1 + \int_{S_1} e^{-i(k_1 + \Delta k) \cdot r_1}\, dS_1 \right] \right\} \tag{75}$$

Using the results of Section I-1 we may write the integrals in **(75)** in terms of Dirac delta functions as long as plate dimensions are large compared with a wavelength:

$$f(\Delta k) = \frac{(2\pi)^2}{S_1} f(0)\left\{ \delta(\Delta k)_T + \frac{1}{2}\frac{F_1}{F_0}\left[\delta(\Delta k - k_1)_T + \delta(\Delta k + k_1)_T \right] \right\} \tag{76}$$

Note that since the integrals are restricted to the plane of the screen the delta functions apply only to the transverse components of Δk. We identify in **(76)** three contributions to the form factor:

1. The first term is simply the reference wave with transverse wave vector

$$k_T = \Delta k_T = 0 \tag{77}$$

2. The second term is the original object wave with wave vector

$$k_T = \Delta k_T = k_{1T} \tag{78}$$

3. And the third term is a second image wave reflected through z

$$k_T = -\Delta k_T = -k_{1T} \tag{79}$$

As we have seen the interference condition establishes only the transverse component of Δk. But since the magnitude of k must be the same in the scattered wave as in the incident wave, we must have corresponding to **(77)**, **(78)**, and **(79)**:

$$k = k\hat{z} \qquad k = k_1 \qquad k = k_1 - 2k_{1T} \tag{80}$$

11   Image Recording. In this section we find the aperture function when monochromatic light from a point source is mixed with a reference wave onto a photographic plate as shown in Figure 14*a*. Taking the source to be a dipole, the electric field is, from **(7)** and **(8)**,

$$E_{obj}(r,t) = -\frac{\omega^2 \mu_0}{4\pi r_{01}}\, \hat{f}_{01} \times (\hat{f}_{01} \times p) e^{i(kr_{01} - \omega t)} \tag{81}$$

where r_{01} is the distance from the position r of the dipole to a general position r_1

on the screen. Mixing the object wave with a reference wave we obtain an aperture function:

$$F(\mathbf{r}_1) = F_0 + F_1 \cos k r_{01} \qquad (82)$$

The aperture function as shown in Figure 14b is a zone plate centered at the minimum distance from **p** to the plate.

12 **Image Reconstruction.** To reconstruct the image of Figure 14a, the reference wave is projected through the zone plate shown in Figure 14b. Substituting **(82)** into **(69)** we obtain for the form factor of the scattered wave:

$$f(\Delta k) = \frac{f(0)}{S_1} \left\{ (2\pi)^2 \, \delta(\Delta k)_T + \frac{1}{2} \frac{F_1}{F_0} \left[\int_{S_1} e^{i(k r_{01} - \Delta \mathbf{k} \cdot \mathbf{r}_1)} \, dS_1 + \int_{S_1} e^{-i(k r_{01} + \Delta \mathbf{k} \cdot \mathbf{r}_1)} \, dS_1 \right] \right\} \qquad (83)$$

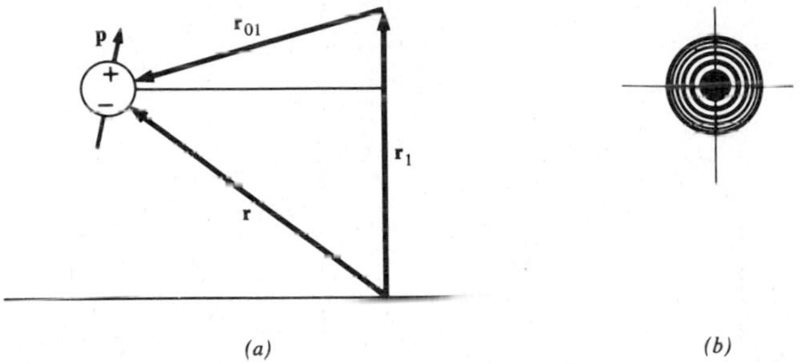

(a)   (b)

Figure 14 Image recording. (a) Monochromatic light from a dipole **p** is mixed with a reference wave on a photographic plate. (b) The aperture function produced is a zone plate.

and where

$$\Delta \mathbf{k} = \mathbf{k} - k \hat{z}$$

is the momentum difference between the outgoing wave and the incoming wave. As for a plane wave hologram we may identify three contributions to the form factor:

1. The first contribution is for $\Delta \mathbf{k}_T = 0$ and is simply the reference wave partially transmitted through the plate.

2. The second term has its maximum contribution for directions of **k** such that the phase of the exponential is independent of $\mathbf{r}_1$, and all contributions to the integral interfere constructively. We show that this will be the case for **k** parallel to $\mathbf{r}_{01}$ as shown in Figure 15a so that all wavevectors diverge from the object position, which constitutes a virtual image. We require

$$k r_{01} - \mathbf{k} \cdot \mathbf{r}_1 = k r \qquad (84)$$

constant independent of $\mathbf{r}_1$. For **k** antiparallel to $\mathbf{r}_{01}$ we may write **(84)** as

$$r = r_{01} + \hat{\mathbf{r}}_{01} \cdot \mathbf{r}_1 \qquad (85)$$

Now from Figure 15a we have the identity

$$\mathbf{r} = \mathbf{r}_{01} + \mathbf{r}_1 \tag{86}$$

Squaring (86) we have

$$r^2 = r_{01}{}^2 + 2\mathbf{r}_{01} \cdot \mathbf{r}_1 + r_1{}^2 \tag{87}$$

For r_1 small compared with r_{01} we obtain approximately, from (87),

$$r \simeq r_{01} + \hat{\mathbf{f}}_{01} \cdot \mathbf{r}_1 \tag{88}$$

which is the requirement of (85). We observe, though, that (85) is satisfied only approximately in the limit that the size of the plate is small compared with the distance to the object.

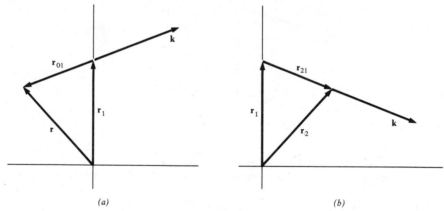

(a)   (b)

Figure 15   Image reconstruction. (a) For $\mathbf{k}$ parallel to $\mathbf{r}_{01}$ we obtain a virtual source at the position of the original source. (b) For $\mathbf{k}$ parallel to $\mathbf{r}_{21}$, where $\mathbf{r}_2$ is the mirror position of $\mathbf{r}$, we obtain a real image.

3. The third term similarly will be a maximum for directions of $\mathbf{k}$ such that the phase of the exponential is independent of $\mathbf{r}_1$. We show that this will be the case for $\mathbf{k}$ parallel to $\mathbf{r}_{21}$ as shown in Figure 15b where $\mathbf{r}_2$ is the mirror position of $\mathbf{r}$. Since all wave vectors must actually pass through $\mathbf{r}_2$, a real image is formed at this position. We require

$$kr_{01} + \mathbf{k} \cdot \mathbf{r}_1 = kr \tag{89}$$

constant independent of $\mathbf{r}_1$. For $\mathbf{k}$ parallel to $\mathbf{r}_{21}$ we may write (89) as

$$r = r_{01} + \hat{\mathbf{f}}_{21} \cdot \mathbf{r}_1 \tag{90}$$

We have from Figure 15b

$$\mathbf{r}_2 = \mathbf{r}_{21} + \mathbf{r}_1 \tag{91}$$

with

$$r_2 = r \qquad r_{21} = r_{01} \tag{92}$$

Obtaining from **(91)** the magnitude of r_2 and using **(92)** we have approximately

$$r \simeq r_{01} + \hat{\mathbf{r}}_{21} \cdot \mathbf{r}_1 \tag{93}$$

which is the requirement of **(90)** and establishes to the approximation of **(93)** that $\mathbf{r}_2$ is a real image.

Finally, the fact that **(85)** and **(90)** can be satisfied only approximately requires some comment. Although this requirement might appear to cause problems for large holograms, the fact is that for any given position of a detector (such as the eye of an observer) the amount of the hologram intercepted is relatively small. What the large hologram makes possible is the observation of the source from a wide range of directions.

CRYSTAL DIFFRACTION

We are concerned here with the scattering of electromagnetic radiation by three-dimensional periodic structures. We begin with the derivation of Bragg's law, which gives the condition for constructive interference between scattered waves. Following this, we discuss the Laue derivation of the scattering amplitude, which is closer to the kind of analysis that we have been pursuing.

13 Bragg Reflection. We wish to find the condition for the scattering of radiation by a regular periodic three dimensional crystal such as that shown in part in Figure 16. For a general orientation of the incident wave with respect to the crystal axes, we will

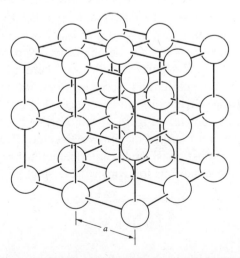

Figure 16 Bragg reflection. Radiation is scattered by a regular periodic three-dimensional crystal.

obtain a wave specularly reflected from the surface of the crystal and a transmitted wave that is refracted from the direction of the incident wave. These waves have been discussed for a model in which the medium is regarded as homogeneous, which is perfectly appropriate as long as the wavelength of the radiation is long compared with the distance between atoms.

A few years after Roentgen's discovery of X rays in 1895, the diffraction of these rays by a narrow slit was observed. From the magnitude of the diffraction the wave length was estimated to be of the order of 10^{-10} m. Laue recognized that with wavelengths this short it should be possible to observe diffraction from a crystal in much the way that diffraction is observed from a grating. Diffraction of the sort predicted by Laue was first observed in 1912, establishing that X rays were electromagnetic waves of definite wavelength, and that crystalline materials have a regular periodic structure.

Laue recognized that for waves to be diffracted along a particular direction, the waves scattered by individual lattice points must interfere constructively. A simple physical interpretation of this requirement, first given by W. L. Bragg in 1912, treats the crystalline solid as a set of lattice planes. For example, the simple cubic crystal shown

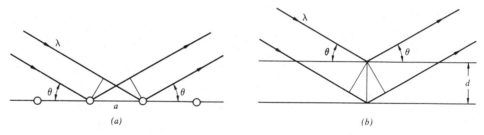

Figure 17 Bragg plane. (*a*) Scattering is first considered from a plane of atoms. (*b*) Next, interference between scattering from adjacent planes is considered.

in Figure 16 may be decomposed into planes of atoms. There are many ways in which this may be done. The simplest is to take planes parallel to the crystal faces and separated by the lattice constant *a*. A second possibility is to take planes normal to a face diagonal and separated by $a/\sqrt{2}$. Or we might decompose the lattice into planes normal to a body diagonal and separated by $a/\sqrt{3}$. There are an infinite number of possible planes, which become closer and closer together. Rather than apply the Laue condition all at once, Bragg divided it into two stages. First he considered diffraction from a plane of atoms as shown in Figure 17*a* and argued that the wave specularly reflected from the plane must represent constructive interference in zero order of scattered waves. Next we must consider the interference between planes. As may be seen from Figure 17*b* the condition for constructive interference between planes is given by *Bragg's law*:

$$2d \sin \theta = n\lambda \tag{94}$$

where *n* must be an integer. Although **(94)** is stated in terms of constructive interference between planes, it actually is the condition for constructive interference between atoms since it assumes specular reflection from lattice planes. It should be noted that in order to observe Bragg reflection at all we must have, from **(94)**

$$\lambda < 2d \qquad (95)$$

because $\sin \theta$ must be less than one. Even for the wavelength sufficiently short, diffraction takes place only for certain well-defined angles θ and not for a general incident direction. This is in contrast with diffraction from a planar grating or array, which is allowed for all incident directions.

14 **Scattering of Radiation.** We now discuss the general theory of the scattering of electromagnetic radiation from a three-dimensional periodic structure. We characterize the crystal by a dielectric susceptibility $\chi(\mathbf{r})$, which has the periodicity of the lattice. As shown in Section I-7, we may write $\chi(\mathbf{r})$ as a Fourier series:

$$\chi(\mathbf{r}) = \sum_{\mathbf{G}} \chi_{\mathbf{G}} e^{i\mathbf{G}\cdot\mathbf{r}} \qquad (96)$$

where the vectors $\mathbf{G}$ form a lattice that is reciprocal to the crystal lattice. Conversely, if we know $\chi(\mathbf{r})$ we may obtain $\chi_{\mathbf{G}}$ from the inverse transform:

$$\chi_{\mathbf{G}} = \frac{1}{V_c} \int_{V_c} \chi(\mathbf{r}) e^{-i\mathbf{G}\cdot\mathbf{r}} \, dV \qquad (97)$$

where the integral is over the unit cell.

From **(25)** we have for the scattered magnetic field:

$$\mathbf{B}(\mathbf{r},t) = \frac{1}{4\pi} \frac{\omega^2}{rc^3} \hat{\mathbf{r}} \times \mathbf{E}_0 \, e^{i(kr-\omega t)} \int_{V_1} \chi(\mathbf{r}_1) e^{-i\Delta\mathbf{k}\cdot\mathbf{r}_1} \, dV_1 \qquad (98)$$

where V_1 is the volume of the entire sample. Substituting from **(96)** for $\chi(\mathbf{r})$ we obtain:

$$\mathbf{B}(\mathbf{r},t) = \frac{1}{4\pi} \frac{\omega^2}{rc^3} \hat{\mathbf{r}} \times \mathbf{E}_0 \, e^{i(kr-\omega t)} \sum_{\mathbf{G}} \chi_{\mathbf{G}} \int_{V_1} e^{i(\mathbf{G}-\Delta\mathbf{k})\cdot\mathbf{r}_1} \, dV_1 \qquad (99)$$

Now, the integral in **(99)** is zero unless there happens to be a reciprocal lattice vector $\mathbf{G}$, which equals $\Delta\mathbf{k}$, where $\Delta\mathbf{k}$ is given by **(26)**. Under these conditions the integral becomes the volume of the sample and we have:

$$\mathbf{B}(\mathbf{r},t) = \frac{1}{4\pi} \frac{\omega^2 V_1 \chi_{\mathbf{G}}}{rc^3} \hat{\mathbf{r}} \times \mathbf{E}_0 \, e^{i(kr-\omega t)} \qquad (100)$$

where $\mathbf{G}$ *must* satisfy the condition:

$$\mathbf{G} = k\hat{\mathbf{r}} - \mathbf{k} \qquad (101)$$

15   Form Factor. We may define the form factor as the integral over a unit cell:

$$f(\Delta \mathbf{k}) = \frac{\epsilon_0}{e^2/m\omega^2} \int_{V_c} \chi(\mathbf{r}_1) e^{-i\,\Delta\mathbf{k}\cdot\mathbf{r}_1}\, dV_1 \tag{102}$$

Comparing (97) and (102) we have the relationship between the form factor and the Fourier component of the susceptibility:

$$f(\mathbf{G}) = -\frac{\epsilon_0 V_c}{e^2/m\omega^2}\, \chi_{\mathbf{G}} \tag{103}$$

Finally, we may write the scattered magnetic field in terms of the form factor as

$$\mathbf{B}(\mathbf{r},t) = -\frac{\mu_0}{4\pi r c}\frac{V}{V_c}\frac{e^2}{m}\, f(\mathbf{G})\,\hat{\mathbf{r}} \times \mathbf{E}_0\, e^{i(kr-\omega t)} \tag{104}$$

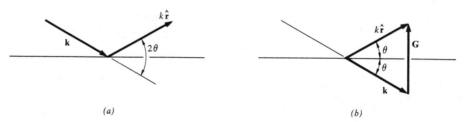

Figure 18   Scattering angle. (*a*) The scattering angle is given by 2θ. (*b*) The reciprocal lattice vectors **G** are normal to Bragg planes.

Using (8) for the electric field:

$$\mathbf{E}(\mathbf{r},t) = -c\hat{\mathbf{r}} \times \mathbf{B}(\mathbf{r},t) \tag{105}$$

we obtain for the Poynting vector where we have averaged over polarization:

$$\Pi(\theta) = \frac{3c}{8}\frac{\sigma_T}{4\pi r^2}\,\epsilon_0 E_0{}^2 \left(\frac{V}{V_c}\right)^2 |f(\mathbf{G})|^2 (1 + \cos^2 2\theta) \tag{106}$$

where σ_T is the Thomson cross section as given by (18) and 2θ is the scattering angle as shown in Figure 18*a*.

Comparing Figure 18*b* with Figure 17*b* it is apparent that the reciprocal lattice vectors **G** are normal to Bragg planes. From Figure 18*b* we have, for the magnitude of **G**,

$$G = 2k \sin \theta \tag{107}$$

which when compared with **(94)** gives:

$$G = \frac{2\pi n}{d} \tag{108}$$

which relates the magnitude of the reciprocal lattice vectors to the distance between Bragg planes.

Exercises

8. Show that at very high frequency the form factor is given by

$$f(G) = V_c \rho(G)$$

where $\rho(G)$ is the Fourier component with wave vector G of the electron charge density.

RAYLEIGH SCATTERING

16 Gases. The origin of the blue color of the sky was first explained by Lord Rayleigh as arising from the fact that gas molecules are random in position rather than distributed strictly uniformly. If we regard a gas molecule as a scatterer of polarizability α and diameter small compared with a wavelength, we have, from **(25)**,

$$\mathbf{B}(\mathbf{r},t) = \frac{\mu_0}{4\pi} \frac{\alpha\omega^2}{rc} \,\hat{\mathbf{r}} \times \mathbf{E}_0 \, e^{i(kr - \omega t)} \tag{109}$$

For a distribution of molecules we have, for the Poynting vector at θ for unpolarized radiation **(14)**,

$$\Pi(\theta) = nV \frac{k^4}{r^2} \left(\frac{\alpha}{4\pi\epsilon_0}\right)^2 \tfrac{1}{2}(1 + \cos^2 \theta)\Pi_0(z) \tag{110}$$

where V is the volume out of which radiation is scattered with n molecules per unit volume and we have assumed that interference between scatterers cancels out on the average. By integrating **(110)** we obtain for the total energy scattered per unit volume:

$$-\frac{d\Pi_0(z)}{dz} = \frac{1}{V} \int \Pi(\theta) r^2 \, d\Omega = \frac{8\pi}{3} nk^4 \left(\frac{\alpha}{4\pi\epsilon_0}\right)^2 \Pi_0(z) \tag{111}$$

We may define the scattering coefficient out of the incident beam by writing for the Poynting vector of the incident radiation:

$$\Pi_0(z) = \Pi_0(0)e^{-\kappa z} \tag{112}$$

We have from **(111)** and **(112)** for the scattering coefficient:

$$\blacktriangleright \qquad \kappa = \frac{8\pi}{3} n k^4 \left(\frac{\alpha}{4\pi\epsilon_0}\right)^2 \tag{113}$$

The fourth power of the wavevector in **(113)** arises from the assumption that α is independent of frequency, which is the case as long as the radiation frequency is low compared with that for resonant absorption.

Writing the molecular polarizability in terms of the permittivity:

$$\alpha = \frac{\epsilon - \epsilon_0}{n} \tag{114}$$

we may write the scattering coefficient as:

$$\kappa = \frac{k^4}{6\pi n} \left(\frac{\epsilon}{\epsilon_0} - 1\right)^2 \tag{115}$$

Thus, it is possible to use Rayleigh scattering to determine n and then Avogadro's number. A related use is the determination of the concentration of organic molecules in solution and from this their molecular weight.

As the diameter of the scattering particle approaches the wavelength of light, the backward scattering begins to drop off as shown in Figure 5. In the limit of very large particles, the form factor falls off as $1/(R\,\Delta k)^2 \sim 1/R^2 k^2$ as may be seen from **(29)** and **(32)**. Since the intensity of the scattered radiation is proportional to k^4 times the square of the form factor, the scattering coefficient will be independent of wavelength in this limit. This is why, for example, clouds are white whereas the sky is blue.

17 [**Liquids**]. Rayleigh scattering from pure liquids is dominated by the presence of density fluctuations associated with thermal excitations in the medium. Such fluctuations are also present in glass and provide the principal long wavelength mechanism for light scattering in pure fused quartz fibers, as indicated in our discussion of fiber optics.

From **(25)**, **(8)**, and **(9)** we have, for the light scattered out of a volume V_1 as shown in Figure 19,

$$\Pi(\theta) = \frac{\pi}{2} V_1 \frac{k^4}{r^2} \tfrac{1}{2}(1 + \cos^2 \theta) \frac{1}{(2\pi)^3 V_1} \left| \int_{V_1} \chi(\mathbf{r}_1) e^{-i\mathbf{r}_1 \cdot \Delta\mathbf{k}} \, dV_1 \right|^2 \Pi_0 \tag{116}$$

with $\sin \theta = \Delta k / 2k$.

We now must develop the modulus-squared integral in **(116)**, which we denote by $\overline{X}(\Delta\mathbf{k})$ and call the power scattering coefficient:

$$\overline{X}(\Delta\mathbf{k}) = \frac{1}{(2\pi)^3 V_1} \int_{V_1} dV_1 \int_{V_2} \chi(\mathbf{r}_1)\chi^*(\mathbf{r}_2) e^{i\Delta\mathbf{k} \cdot (\mathbf{r}_2 - \mathbf{r}_1)} \, dV_2 \tag{117}$$

We introduce the change of variables:

$$r_2 = r_1 + \rho \tag{118}$$

as a result of which we have for (117):

$$\overline{X}(\Delta k) = \frac{1}{(2\pi)^3 V_1} \int_{V_1} dV_1 \int_{V_\rho} \chi(r_1)\chi^*(r_1 + \rho)e^{i \Delta k \cdot \rho} dV_\rho \tag{119}$$

The function:

$$X(\rho) = \frac{1}{V_1} \int_{V_1} \chi(r_1)\chi^*(r_1 + \rho) \, dV_1 \tag{120}$$

is called the cross-correlation function between $\chi(r_1)$ and $\chi(r_1 + \rho)$. In terms of (120) we now have for (119):

$$\overline{X}(\Delta k) = \frac{1}{(2\pi)^3} \int_{V_\rho} X(\rho)e^{i \Delta k \cdot \rho} dV_\rho \tag{121}$$

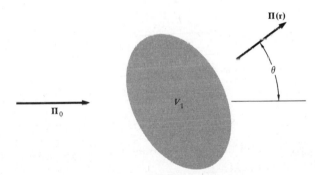

Figure 19   Scattering from liquids. Light is scattered by density fluctuations within a volume V_1.

We observe from (121) that the power scattering coefficient $\overline{X}(\Delta k)$ is the Fourier transform (see Section I-1) of the cross-correlation function.

From (121) we may develop the inverse relation:

$$X(\rho) = \int_{V_k} \overline{X}(\Delta k)e^{-i \Delta k \cdot \rho} dV_k \tag{122}$$

The equations (121) and (122) are called the *Wiener-Khintchine relations*.

We now want to apply this theory to liquids. But before doing so we discuss as an example Rayleigh scattering from a gas of independent molecules. We take the molecules to be spherical and of volume $V_s = 4\pi R^3/3$. The cross-correlation function (120) for $\rho = 0$ will simply be:

$$X(0) = \langle \chi^2 \rangle = \frac{n\alpha^2}{\epsilon_0{}^2 V_s} \tag{123}$$

For ρ much larger than R and in the absence of any correlation between molecules, we expect:

$$X(\rho) = \langle\chi\rangle^2 = \frac{n^2\alpha^2}{\epsilon_0{}^2} \tag{124}$$

The difference between **(123)** and **(124)** is the mean square fluctuation or dispersion:

$$(\Delta\chi)^2 = \langle\chi^2\rangle - \langle\chi\rangle^2 = n\left(\frac{1}{V_s} - n\right)\frac{\alpha^2}{\epsilon_0{}^2} \tag{125}$$

The form of $X(\rho)$ away from $\rho = 0$, as shown in Figure 20, must simply be proportional to the overlap of a pair of spheres of radius R and separation between centers ρ:

$$X(\rho) = \langle\chi\rangle^2 + (\Delta\chi)^2\left(1 - \frac{3}{4}\frac{\rho}{R} + \frac{1}{16}\frac{\rho^3}{R^3}\right) \tag{126}$$

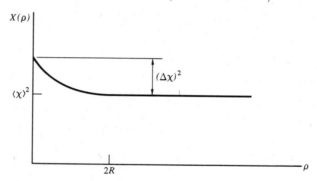

Figure 20   Cross-correlation function. At large distances the function must approach the square of the mean susceptibility.

Substituting into **(121)** and assuming $R\,\Delta k$ small, we obtain:

$$\overline{X}(\Delta\mathbf{k}) = \frac{(\Delta\chi)^2}{(2\pi)^3} \cdot V_s \tag{127}$$

Substituting **(125)** into **(127)** we obtain:

$$\overline{X}(\Delta k) = \frac{n(1 - nV_s)}{(2\pi)^3}\frac{2}{\epsilon_0{}^2} \tag{128}$$

Substituting **(128)** into **(116)** leads to **(110)**, the Rayleigh result in the limit that nV_s is small.

Now we are prepared to consider more complex fluctuations in density. Our problem is to determine the mean square fluctuation level:

$$(\Delta\chi)^2 = \langle\chi^2\rangle - \langle\chi\rangle^2 \tag{129}$$

and the way in which $X(\rho)$ behaves with distance. We assume the form:

$$X(\rho) = \langle\chi\rangle^2 + (\Delta\chi)^2 e^{-\rho/\rho_c} \tag{130}$$

where we may not necessarily assume that ρ_c is short compared with a wavelength. Substituting (130) into (121) we have, after integrating over the polar angle,

$$\overline{X}(\Delta k) = -\frac{(\Delta\chi)^2}{4\pi^2} \int_{-1}^{1} d\cos\theta \int_{0}^{\infty} e^{-\rho(1/\rho_c + i\,\Delta k\,\cos\theta)}\rho^2\,d\rho \tag{131}$$

Integrating over ρ we obtain

$$\overline{X}(\Delta k) = \frac{(\Delta\chi)^2}{2\pi^2} \int_{-1}^{1} \frac{\rho_c^3\,d\cos\theta}{(1 + i\rho_c\,\Delta k\,\cos\theta)^3} \tag{132}$$

Finally, completing the integral over θ, we have

$$\overline{X}(\Delta k) = \frac{(\Delta\chi)^2}{\pi^2} \frac{\rho_c^3}{(1 + \rho_c^2\,\Delta k^2)^2} \tag{133}$$

Substituting into (116) we obtain for the scattered radiation:

$$\Pi(\theta) = \frac{\Pi_0}{4\pi} \frac{k^4}{r^2} V_1 (\Delta\chi)^2 \rho_c^3 \frac{1 + \cos^2\theta}{(1 + \rho_c^2\,\Delta k^2)^2} \tag{134}$$

At long wavelengths, we have $\rho_c\,\Delta k < 1$ and we may approximate (134) by

$$\Pi(\theta) \simeq \frac{\Pi_0}{4\pi} \frac{k^4}{r^2} V_1 (\Delta\chi)^2 \rho_c^3 (1 + \cos^2\theta) \tag{135}$$

which has the same form as molecular scattering. At short wavelengths, however, the scattering is reduced. Writing for Δk^2:

$$\Delta k^2 = 2k^2(1 - \cos\theta) \tag{136}$$

we have from (134) in the limit of short wavelengths:

$$\Pi(\theta) = \frac{\Pi_0}{16\pi} \frac{V_1}{\rho_c r^2} (\Delta\chi)^2 \frac{1 + \cos^2\theta}{(1 - \cos\theta)^2} \tag{137}$$

We note that in this limit the scattering is independent of wavelength and is sharply peaked in the forward direction just as for a large dielectric sphere.

Assuming that the variations in susceptibility are associated with density fluctuations we have:[4]

$$(\Delta\chi)^2 = \frac{k_B T}{B V_c}\chi^2 \tag{138}$$

where $k_B = 1.38 \times 10^{-23}$ J/K is the Boltzmann constant, T is the absolute temperature, B is the bulk modulus, and $V_c = 4\pi\rho_c^3/3$ is the volume over which the fluctuation is determined. Substituting **(135)** into **(111)** we have, for the scattering coefficient at long wavelengths,

$$\kappa = \frac{8\pi}{3}\frac{r^2}{V}\frac{\Pi(0)}{\Pi_0} = \frac{1}{\pi}k^4\chi^2\frac{k_B T}{B} \tag{139}$$

For clear fused quartz $B = 3.6 \times 10^{10}$ J/m³. At the softening temperature $T = 1940$ K, we have $k_B T/B = 7.5 \times 10^{-31}$ m³. Taking $\chi = 1.13$ as determined from the refractive index at optical frequencies, we have:

$$\kappa = 3.0 \times 10^{-31} k^4 \tag{140}$$

where k is 2π times the wavenumber in reciprocal metres. At a wavelength of one micrometre we expect, in agreement with present observations,[5]

$$\frac{1}{\kappa} \simeq 2100 \text{ m} \tag{141}$$

Exercise

9. From **(140)** compute the attenuation in decibels per kilometre at a wavelength of 1.51 micrometres. Compare with the value of 0.46 db/km reported[5] on very high purity optical fibers.

18 **[Glasses].** In addition to scattering from density fluctuations we may expect in fused quartz and other glassy materials that there will be scattering from the random bonding network characteristic of such media.[6] The density correlation function for such scattering will be of short range, leading to additional scattering above the k^4 Rayleigh scattering at short wavelengths.

For a glass, the random bonding network leads to a correlation function of the form shown in Figure 21. Since all the bonds must be saturated, the integral of $X(\rho)$ **(106)** over space must give

$$\int X(\rho)\, dV_\rho = \int_V \langle\chi\rangle^2\, dV_\rho \tag{142}$$

---

[4] F. Reif, *Statistical and Thermal Physics*, McGraw-Hill, 1965, pp. 300–301.

[5] M. Horiguchi and H. Osani, *Electron Lett.* **12**, 310 (1976). For discussions of the ultimate lower limits of attenuation in optical fibers, see D. B. Keck, R. D. Maurer, and P. G. Schultz, *Appl. Phys. Lett.* **22**, 307 (1973) and D. A. Pinnow, T. C. Rich, F. W. Ostermayer, Jr. and M. Didomenico, Jr., *Appl. Phys. Lett.* **22**, 527 (1973).

[6] E. U. Condon, "Physics of the Glassy State I," *Am. J. Phys.*, **22**, 43 (1954).

When substituted into **(121)** we have for the leading term:

$$\overline{X}(\Delta k) \simeq (\Delta \chi)^2 \rho_c{}^3 (\rho_c{}^2 \, \Delta k^2 + \cdots) \tag{143}$$

Using **(136)** we have:

$$\overline{X}(\Delta k) \sim (\Delta \chi)^2 k^2 \rho_c{}^5 (1 - \cos \theta) \tag{144}$$

Substituting into **(92)** we obtain a scattering intensity that varies as k^6.

It is interesting to examine at what wavelength we expect the scattering from density fluctuations and from random bonding to be comparable. Assuming $(\Delta \chi)^2 \sim \chi^2$ and equating **(133)** and **(143)** we obtain $2\pi/k \sim 1$ micrometre for ρ_c a few atomic distances.

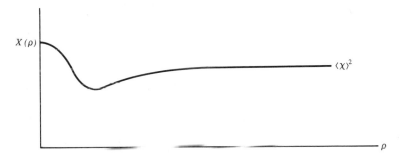

Figure 21 Cross-correlation function of a glass. The random bonding network of a glass leads to an oscillatory function.

CHERENKOV RADIATION

As we have seen in Section 6-1, the electric and magnetic fields of a uniformly moving charge fall off as $1/r^2$, and there is no electromagnetic radiation. In a material medium, as we saw in Section 12-9, the velocity of propagation may be substantially less than c. Now if a charged particle moves with speed v greater than the speed of propagation in the medium s, we may expect a transfer of energy from the particle to the electromagnetic field as first observed in 1934 by P. A. Cherenkov. This situation is something like the excitation of a bow wave by a ship. As discussed in introductory physics texts, the wave front makes an angle θ_C with the velocity of the ship given by the relation:

$$\sin \theta_C = \frac{s}{v} \tag{145}$$

and wave excitation is detected only at positions within a cone of half angle θ_C.

19 Tachyons. We begin the discussion of Cherenkov radiation by considering a very simple but highly hypothetical situation, the fields of a charged particle moving in vacuum at velocity v greater than c. Although it is evidently not possible to excite a particle to velocities in excess of c, the special theory of relativity is not incompatible with the *existence* of such particles.[7]

Our reason for considering such a situation is that we wish first to discuss radiation in a medium that is entirely nondispersive with the permittivity $\epsilon(\omega,k)$ simply equal to ϵ_0. To simplify the discussion we assume an observer on the x axis and a charge q moving uniformly along the z axis with velocity $\mathbf{v}$ as shown in Figure 22. As we saw in Section 11-21 the scalar and vector potentials of an arbitrarily moving charge may be written from (11-190) and (11-191) as:

$$\phi(\mathbf{r},t) = \frac{1}{4\pi\epsilon_0} \int_{V_1} \frac{\rho(\mathbf{r}_1,[t])}{r_{01}} dV_1 \tag{146}$$

$$\mathbf{A}(\mathbf{r},t) = \frac{\mu_0}{4\pi} \int_{V_1} \frac{\mathbf{j}(\mathbf{r}_1,[t])}{r_{01}} dV_1 \tag{147}$$

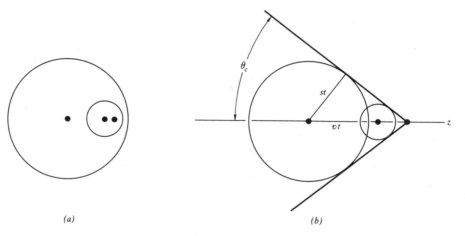

(a)   (b)

Figure 22   Cherenkov radiation. (a) A charge moving with velocity v less than the speed s of wave propagation in the medium does not radiate. (b) When the velocity v of the charge exceeds the speed of wave propagation s, a wavefront is developed.

where $[t] = t - r_{01}/c$ is the retarded time. We show in Figure 23 the retarded position $[z]$ of the charge and the retarded distance $[r]$ to the field position for z positive and for a charge moving with $\beta = v/c$ less than 1.

We find the retarded position $[z]$ from the condition that the time required for radiation to travel from $[z]$ to the field position is equal to the time for the charge to travel from $[z]$ to the present position:

$$\frac{[r]}{c} = -\frac{[z]}{v} \tag{148}$$

[7] G. Feinberg, *Phys. Rev.*, **159**, 1089 (1967). See also L. Parker, *Phys. Rev.*, **188**, 2287 (1969). This subject has been considered from a similar point of view by G. A. Schott, *Electromagnetic Radiation*, Cambridge, 1912. For a short bibliography, see L. M. Feldman, *Am. J. Phys.*, **42**, 179 (1974).

Writing:

$$[r]^2 = (z - [z])^2 + x^2 \tag{149}$$

and substituting into **(148)** we obtain

$$[z] = \frac{\beta}{1 - \beta^2} \left[-\beta z \pm (r^2 - \beta^2 x^2)^{1/2} \right] \tag{150}$$

The negative solution is the usual retarded solution shown in Figure 23. The positive solution is called the advanced solution and places the charge *ahead* of its present position as a consequence of the inequality

$$r^2 - \beta^2 x^2 > \beta^2 z^2 \tag{151}$$

for $\beta < 1$. Although the advanced position leads to a potential that is formally a solution of the Maxwell equations, we have to reject it on the grounds of causality—a particle cannot contribute to the potential from a position that it has not yet reached.

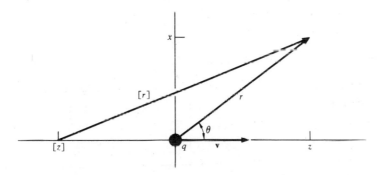

Figure 23   Retarded position. The retarded position $[z]$ of the charge is shown for $\beta = v/c$ less than. 1.

For β greater than unity we rewrite **(150)** as

$$[z] = \frac{\beta}{\beta^2 - 1} \left[\beta z \pm (r^2 - \beta^2 x^2)^{1/2} \right] \tag{152}$$

In Figure 24 we plot $[z]$ horizontally and β vertically over the range from $\beta = 0$ to the limiting value $\beta = r/x$. The plot is shown for z *negative*. We observe that for β small the retarded and advanced positions are symmetrical with respect to the present position. As the velocity v approaches c the advanced solution (for z negative) moves to positive infinity while the retarded position remains finite. For β greater than unity the "advanced" solution comes in from minus infinity so that in this range both solutions are acceptable. The condition $\beta < r/x$ or $x < r/\beta$ places the field position within the Cherenkov cone.

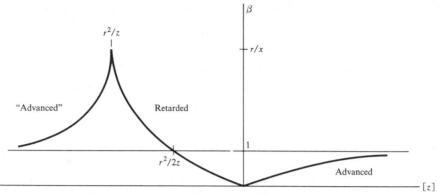

Figure 24   Retarded and advanced solutions. For $\beta < 1$ only the retarded solution is physical. For $\beta > 1$ both the retarded and "advanced" solutions are physical.

We obtain in Section H-1 **(H-17)** for the retarded potential of a charge with velocity v *less than* c and moving away from the observer

$$\phi_-(\mathbf{r},t) = \frac{1}{4\pi\epsilon_0} \frac{q}{r_- + \beta z_-} \tag{153}$$

where $r_- = [r_{01}]$ is the distance from the retarded position of the charge to the field position and $z_- = [\hat{\mathbf{v}} \cdot \mathbf{r}_{01}]$ is the projection of $[\mathbf{r}_{01}]$ onto the trajectory of the charge. For $\beta = v/c > 1$ we must integrate over the contour shown in Figure 25a in order to obtain the contribution to the potential from the charge at the retarded position. Comparing Figure 25a with Figure H-2b, we observe that the contour has slope less than that of the lines bounding the charge. This interchanges the limits on the path integral and reverses the sign of the potential:

$$\phi_-(\mathbf{r},t) = -\frac{1}{4\pi\epsilon_0} \frac{q}{\beta z_- + r_-} \tag{154}$$

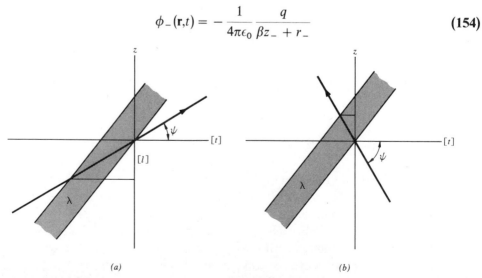

(a)   (b)

Figure 25   Retarded potential. (a) For $\beta > 1$ we integrate over the contour shown. (b) For $\beta > 1$ we must also integrate over the "advanced" contour.

Now, in addition, we must include the contribution to the potential from the "advanced" position of the charge. From **(H-10)** we also write for the advanced contour

$$t = [t] + r_{01}/c = \text{const} \tag{155}$$

since the "advanced" solution is actually retarded for $\beta > 1$. Taking the differential of **(155)** we obtain:

$$c\, dt = c\, d[t] + dr_{01} = c\, dt + dz \cos \theta = 0 \tag{156}$$

where the geometry is that of Figure 23. From **(156)** the contour $t = \text{const}$ has slope:

$$\tan \psi = \frac{1}{c} \frac{dz}{dt} = -\frac{1}{\cos \theta} \tag{157}$$

as shown in Figure. 25*b*. The contribution to the advanced potential is then:

$$\phi_+(\mathbf{r},t) = \frac{1}{4\pi\epsilon_0} \frac{q}{\beta z_+ + r_+} \tag{158}$$

Adding **(154)** and **(158)** and writing a similar expression for the vector potential we obtain within the Cherenkov cone:

$$\phi(\mathbf{r},t) = -\frac{1}{4\pi\epsilon_0} \frac{q}{\beta z_- + r_-} + \frac{1}{4\pi\epsilon_0} \frac{q}{\beta z_+ + r_+} \tag{159}$$

$$\mathbf{A}(\mathbf{r},t) = -\frac{\mu_0}{4\pi} \frac{q\mathbf{v}}{\beta z_- + r_-} + \frac{\mu_0}{4\pi} \frac{q\mathbf{v}}{\beta z_+ + r_+} \tag{160}$$

Outside the Cherenkov cone both the scalar and vector potential are zero.
From **(148)** and **(152)** we write:

$$r_\pm + \beta z_\pm = (z - z_\pm)/\beta + \beta z_\pm = [z + (\beta^2 - 1)z_\pm]/\beta = \mp(r^2 - \beta^2\rho^2)^{1/2} \tag{161}$$

Then, within the Cherenkov cone the contributions from the advanced and retarded positions are equal and we have

$$\phi(\mathbf{r},t) = -\frac{1}{4\pi\epsilon_0} \frac{2q}{r(1 - \beta^2 \sin^2 \theta)^{1/2}} \tag{162}$$

$$\mathbf{A}(\mathbf{r},t) = -\frac{\mu_0}{4\pi} \frac{2q\mathbf{v}}{r(1 - \beta^2 \sin^2 \theta)^{1/2}} \tag{163}$$

The electric and magnetic fields are obtained from **(11-176)** and **(11-172)**:

$$\mathbf{E}(\mathbf{r},t) = -\nabla\phi(\mathbf{r},t) - \frac{\partial}{\partial t}\mathbf{A}(\mathbf{r},t) \tag{164}$$

$$\mathbf{B}(\mathbf{r},t) = \nabla \times \mathbf{A}(\mathbf{r},t) \tag{165}$$

Substituting from (162) and (163) we obtain:

$$E(r,t) = \frac{1}{4\pi\epsilon_0} \frac{2(\beta^2 - 1)q}{r^2(1 - \beta^2 \sin^2 \theta)^{3/2}} \, \hat{r} \tag{166}$$

$$B(r,t) = \frac{\mu_0}{4\pi} \frac{2(\beta^2 - 1)q}{r^2(1 - \beta^2 \sin^2 \theta)^{3/2}} \, v \times \hat{r} = \frac{1}{c^2} v \times E \tag{167}$$

These expressions are just twice the corresponding expressions (6-21) and (6-22) for the fields of a charge moving at less than c. The Poynting vector (9) is then:

$$\Pi(r,t) = E(r,t) \times H(r,t) = \frac{1}{4\pi^2\epsilon_0} \frac{(\beta^2 - 1)^2}{r^4(1 - \beta^2 \sin^2 \theta)^3} \, \hat{r} \times (v \times \hat{r}) \tag{168}$$

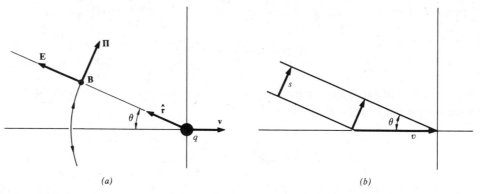

(a)   (b)

Figure 26   Cherenkov cone. (*a*) The directions of the electric and magnetic fields are as shown in the surface of the cone. (*b*) Energy flow is normal to the surface of the cone.

The energy density is given by

$$u(r,t) = \tfrac{1}{2}\epsilon_0 E^2 + \tfrac{1}{2}\mu_0 H^2 = \frac{(\beta^2 - 1)^2 q^2}{8\pi^2\epsilon_0} \frac{1 + \beta^2 \sin^2 \theta}{r^3(1 - \beta^2 \sin^2 \theta)^3} \tag{169}$$

The *directions* of the electric and magnetic fields and the Poynting vector are shown in Figure 26a. The effective velocity for energy flow is given by:

$$s = \frac{\Pi}{u} = \frac{2\beta \sin \theta}{1 + \beta^2 \sin^2 \theta} c \tag{170}$$

At the edge of the Cherenkov cone this velocity is equal to c and is less than c within the cone. From Figure 26b, we have, for the energy flow,

$$s = v \sin \theta = c \tag{171}$$

Note from (168) that the Poynting vector falls off with distance as $1/r^4$. More important, there is no radial component of energy flux for finite distance. This is quite different from radiation from an accelerated charge **(11-235)** that is radial and falls off as $1/r^2$ so that the energy flow across a sphere of radius r is independent of r. In the case of Cherenkov radiation the energy is transferred from the uniformly moving charge into the Cherenkov cone. And the energy flow at finite r serves only to maintain a constant cone angle.

Exercises

10. Show by integration of **(166)** that the flux of **E** is q/ϵ_0. Use **(6-15)**.
11. Show that for velocities greater than c the retarded potential alone leads to a normal and a forward-directed Cherenkov cone. Show that the "advanced" solution cancels the forward-directed cone and doubles the flux through the normal cone.

20 [**Dispersive Media**]. In this section we discuss the more realistic situation where a charge moves with velocity v less than c through a dispersive medium. The dispersion of the medium will have the effect of broadening the Cherenkov cone leading to finite electric and magnetic fields.[8]

We write the Maxwell equations in a dispersive medium from **(12-16)**, **(12-17)**, **(12-18)**, and **(12-19)** as

$$\mathbf{V} \cdot \mathbf{D} = \rho_{\text{ext}} \tag{172}$$

$$\mathbf{V} \times \mathbf{E} = -\frac{\partial \mathbf{B}}{\partial t} \tag{173}$$

$$\mathbf{V} \times \mathbf{H} = \mathbf{j}_{\text{ext}} + \frac{\partial \mathbf{D}}{\partial t} \tag{174}$$

$$\mathbf{V} \cdot \mathbf{B} = 0 \tag{175}$$

We write the external charges and currents as Fourier integrals:

$$\rho_{\text{ext}}(\mathbf{r},t) = \int \rho_{\text{ext}}(\mathbf{k},\omega) e^{i(\mathbf{k} \cdot \mathbf{r} - \omega t)} \, dV_k \, d\omega \tag{176}$$

$$\mathbf{j}_{\text{ext}}(\mathbf{r},t) = \int \mathbf{j}_{\text{ext}}(\mathbf{k},\omega) e^{i(\mathbf{k} \cdot \mathbf{r} - \omega t)} \, dV_k \, d\omega \tag{177}$$

with, as a consequence of local charge conservation,

$$\mathbf{k} \cdot \mathbf{j}_{\text{ext}}(\mathbf{k},\omega) = \omega \rho_{\text{ext}}(\mathbf{k},\omega) \tag{178}$$

[8] G. P. Sastry, *Am. J. Phys.*, **44**, 707 (1976). See also the earlier discussion for nondispersive media by G. M. Volkoff, *Am. J. Phys.*, **31**, 601 (1963).

The electric field and displacement may be written as:

$$\mathbf{D}(\mathbf{r},t) = \int \mathbf{D}(\mathbf{k},\omega)e^{i(\mathbf{k}\cdot\mathbf{r}-\omega t)} \, dV_k \, d\omega \tag{179}$$

$$\mathbf{E}(\mathbf{r},t) = \int \mathbf{E}(\mathbf{k},\omega)e^{i(\mathbf{k}\cdot\mathbf{r}-\omega t)} \, dV_k \, d\omega \tag{180}$$

where we write, for a linear isotropic medium,

$$\mathbf{D}(\mathbf{k},\omega) = \epsilon(k,\omega)\mathbf{E}(\mathbf{k},\omega) \tag{181}$$

Finally, by assuming for simplicity that the magnetic permeability is a constant, we may write

$$\mathbf{B}(\mathbf{r},t) = \mu\mathbf{H}(\mathbf{r},t) = \mu \int \mathbf{H}(\mathbf{k},\omega)e^{i(\mathbf{k}\cdot\mathbf{r}-\omega t)} \, dV_k \, d\omega \tag{182}$$

The Maxwell equations may be written in terms of Fourier components as

$$i\epsilon(k,\omega)\mathbf{k}\cdot\mathbf{E}(\mathbf{k},\omega) = \rho_{\text{ext}}(\mathbf{k},\omega) \tag{183}$$

$$i\mathbf{k}\times\mathbf{E}(\mathbf{k},\omega) = i\omega\mu\mathbf{H}(\mathbf{k},\omega) \tag{184}$$

$$i\mathbf{k}\times\mathbf{H}(\mathbf{k},\omega) = \mathbf{j}_{\text{ext}}(\mathbf{k},\omega) - i\omega\epsilon(k,\omega)\mathbf{E}(\mathbf{k},\omega) \tag{185}$$

$$i\mathbf{k}\cdot\mathbf{H}(\mathbf{k},\omega) = 0 \tag{186}$$

To obtain the electric field we cross $\mathbf{k}$ into (184) to obtain:

$$\mathbf{k}\times[\mathbf{k}\times\mathbf{E}(\mathbf{k},\omega)] = \omega\mu\mathbf{k}\times\mathbf{H}(\mathbf{k},\omega) \tag{187}$$

Expanding the triple product and substituting from (183) and (185) we write:

$$\mathbf{k}\rho_{\text{ext}}(\mathbf{k},\omega)/i\epsilon(k,\omega) - k^2\mathbf{E}(\mathbf{k},\omega) = \omega\mu\mathbf{j}_{\text{ext}}(\mathbf{k},\omega) - \omega^2\mu\epsilon(k,\omega)\mathbf{E}(\mathbf{k},\omega) \tag{188}$$

Solving for $\mathbf{E}(\mathbf{k},\omega)$ and using (178) we have finally for the electric field:

$$\mathbf{E}(\mathbf{k},\omega) = \frac{\omega\mu\mathbf{j}_{\text{ext}}(\mathbf{k},\omega) + \mathbf{k}\mathbf{k}\cdot\mathbf{j}_{\text{ext}}(\mathbf{k},\omega)/i\omega\epsilon(k,\omega)}{k^2 - \omega^2\mu\epsilon(k,\omega)} \tag{189}$$

We may invert (177) to write for the Fourier transform of a current density:

$$\mathbf{j}_{\text{ext}}(\mathbf{k},\omega) = \frac{1}{(2\pi)^4} \int \mathbf{j}_{\text{ext}}(\mathbf{r},t)e^{-i(\mathbf{k}\cdot\mathbf{r}-\omega t)} \, dV \, dt \tag{190}$$

For a uniformly moving charge we write $j_{ext}(r,t)$ in terms of the Dirac delta function:

$$j_{ext}(r,t) = qv\,\delta(r - vt) \tag{191}$$

We have then from (190) after performing the space integration:

$$j_{ext}(k,\omega) = \frac{1}{(2\pi)^4}\,qv\int e^{-i(k\cdot v - \omega)t}\,dt \tag{192}$$

But (192) is the integral expansion of the delta function so that we have finally:

$$j_{ext}(k,\omega) = \frac{1}{(2\pi)^3}\,qv\,\delta(k\cdot v - \omega) \tag{193}$$

We next substitute (193) into (189) and (189) into (180) to obtain for the electric field:

$$E(r,t) = \frac{1}{(2\pi)^3}\,q\int \frac{\omega\mu v + k(k\cdot v)/i\omega\epsilon(k,\omega)}{k^2 - \omega^2\mu\epsilon(k,\omega)}\,e^{i(k\cdot r - \omega t)}\,\delta(k\cdot v - \omega)\,dV_k\,d\omega \tag{194}$$

Carrying out the integration over frequency and using the delta function we have:

$$E(r,t) = \frac{1}{(2\pi)^3}\,q\int \frac{\mu v(k\cdot v) + k/i\epsilon(k,\omega)}{k^2 - (k\cdot v)^2\mu\epsilon(k,\omega)}\,e^{ik\cdot(r - vt)}\,dV_k \tag{195}$$

with $\omega = k\cdot v$. Writing with $\hat{v}$ as the polar axis:

$$k\cdot v = kv\sin\theta \quad\text{and}\quad dV_k = k^2\,dk\,\cos\theta\,d\theta\,d\phi \tag{196}$$

where θ is the angle between the wave front and v, we obtain, for (193),

$$E(r,t) = \frac{1}{8\pi^2}\,q\int \frac{v\mu v\sin\theta + \hat{k}/i\epsilon(k,\omega)}{1 - \mu\epsilon(k,\omega)v^2\sin^2\theta}\,e^{i(k\cdot r - kvt\cos\theta)}\frac{1}{k}\,dV_k \tag{197}$$

The principal contributions to (197) arise from values of k and $\sin\theta$ for which the denominator vanishes. The fields are concentrated in a thin conical shell whose half angle[9] is given by

$$\tan\theta = \frac{(v^2/v_p^2 - 1)^{1/2}}{v^2/v_p v_g - 1} \tag{198}$$

where v_g is the group velocity (12-92):

$$v_g = \frac{d\omega}{dk} \tag{199}$$

[9] I. Tamm, *J. Phys. U.S.S.R.*, **1**, 439 (1939).

The Poynting vector still makes an angle θ_C with the perpendicular as shown in Figure 27a with

$$\sin \theta_C = \frac{v_p}{v} \tag{200}$$

If the medium is dispersive with the group velocity v_g less than the phase velocity v_p, then the angle θ is less than θ_C and there is a component of energy flow in the surface of the cone.

The velocity at which energy flows may be obtained from Figure 27b from which we write the identity:

$$s \cos(\theta_C - \theta) = v \sin \theta \tag{201}$$

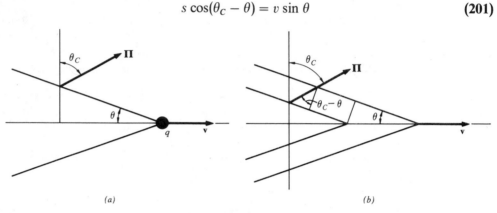

(a)   (b)

Figure 27   Poynting vector. (a) The Poynting vector makes an angle θ_C with the perpendicular to the direction of charge velocity. (b) For the group velocity less than the phase velocity, there is a component of energy flow in the surface of the cone.

Substituting from **(198)** and **(200)** we have for the velocity of energy flow[10]

$$s = \frac{v \sin \theta}{\cos(\theta_C - \theta)} = v_g \tag{202}$$

We emphasize that the energy flow is not normal to the cone, but has a component in the surface of the cone as well.

SUMMARY

1. A free charge driven by a uniform electric field produces electric dipole radiation. The rate at which energy is radiated is independent of frequency and is characterized for an electron by the Thomson cross section:

$$\sigma_T = \frac{8\pi}{3} r_0{}^2 \tag{20}$$

where $r_0 = \mu_0 e^2/4\pi m$ is the classical radius of the electron.

[10] H. Motz and L. I. Schiff, *Am. J. Phys.*, **21**, 258 (1953).

2. The radiation fields of a dielectric sphere of radius R are the same as those for a free electron but multiplied by the form factor

$$f(\Delta k) = 3N_{\text{eff}}\left(\frac{\sin R\,\Delta k}{(R\,\Delta k)^3} - \frac{\cos R\,\Delta k}{(R\,\Delta k)^2}\right) \tag{32}$$

where

$$N_{\text{eff}} = -\frac{\epsilon_0}{e^2/m\omega^2}\int_{V_1}\chi(\mathbf{r}_1)\,dV_1 \tag{29}$$

is the equivalent number of free electrons that would produce the same forward scattering. For $\chi(\mathbf{r}_1)$ positive, N_{eff} is negative, signifying that the scattered radiation is π out of phase with that of free charges.

3. The form factor of an aperture in a dielectric screen may be taken to be the negative of that for the object filling the aperture as long as the screen is thin. The form factor of a circular aperture is:

$$f(\Delta k) = -N_{\text{eff}}\frac{J_1(kR\sin\theta)}{\tfrac{1}{2}kR\sin\theta} \tag{40}$$

where $J_1(x)$ is the Bessel function of order one. The first zero of $J_1(x)$ occurs for $x = kR\sin\theta = 3.832$. A circular image extending out to the first zero of $J_1(x)$ is called the Airy disc. Points within the disc satisfy the condition:

$$\sin\theta < 1.220\frac{\lambda}{D} \tag{41}$$

4. For an aperture in an absorbing screen of permittivity ϵ_1 the effective number of radiating charges is given by

$$N_{\text{eff}} = -\frac{4\pi i}{\mu_0}\frac{S_1}{\lambda e^2/m}\frac{(\epsilon_1 - \epsilon_0)Z_0}{(Z_0 + Z_1)\epsilon_1} \tag{49}$$

where S_1 is the area of the aperture. The fact that N_{eff} is complex indicates that the radiation fields may be of intermediate phase. This result is a generalization of the Fresnel-Kirchoff diffraction theory.

5. The form factor of a rectangular aperture of width b and length l is:

$$f(\Delta\mathbf{k}) = -N_{\text{eff}}\frac{\sin\tfrac{1}{2}b\,\Delta k_x}{\tfrac{1}{2}b\,\Delta k_x}\frac{\sin\tfrac{1}{2}l\,\Delta k_y}{\tfrac{1}{2}l\,\Delta k_y} \tag{53}$$

6. A slit is the limit of a long rectangular aperture for which the form factor is given by:

$$f(\Delta k_x) = -\frac{2\pi}{l}N_{\text{eff}}\frac{\sin\tfrac{1}{2}b\,\Delta k_x}{\tfrac{1}{2}b\,\Delta k_x} \tag{54}$$

7. For a pair of slits the scattered radiation is that of a single elementary charge multiplied by the scattering amplitude:

$$A(\Delta k) = 2f(\Delta k) \cos \tfrac{1}{2}\mathbf{a} \cdot \Delta \mathbf{k} \tag{59}$$

where $f(\Delta k)$ is the form factor of single slit and $\mathbf{a}$ is the displacement between slits.

8. For N slits the scattering amplitude is given by

$$A(\Delta k_x) = f(\Delta k_x) \frac{\sin \tfrac{1}{2}N \, \Delta k_x}{\sin \tfrac{1}{2}a \, \Delta k_x} \tag{63}$$

9. The condition for constructive interference of waves diffracted from a periodic lattice is the *Bragg condition*:

$$2d \sin \theta = n\lambda \tag{94}$$

where d is the distance between planes, and θ is the angle between the wave vector and the lattice planes.

10. The scattering from randomly distributed polarizable scatterers is called Rayleigh scattering and explains the blue color of the sky. As long as the molecular polarizability is independent of frequency and the scatterers are small, the scattering intensity is proportional to the fourth power of the frequency. The attenuation constant of the incident radiation is given by:

$$\kappa = \frac{8\pi}{3} \, nk^4 (\alpha/4\pi\epsilon_0)^2 \tag{113}$$

For extended scatterers the form factor falls off as $(\omega/R \, \Delta k)^2$ which is independent of frequency. Thus, large water droplets as in clouds are expected to scatter uniformly at all wavelengths, explaining why clouds are white.

PROBLEMS

1. **Scattering by a bound charge.** A molecule may be characterized by a set of electronic resonance frequencies ω_i and oscillator strengths f_i (see Section 3-11).
   (a) Obtain the form of the scattered radiation in the vicinity of a resonance.
   (b) Where the lifetime of the excited state is finite the resonance frequency may be treated as complex $\underline{\omega}_i = \omega_i - i/\tau_i$. How is the scattering cross section affected?

2. **Gaussian aperture function.** A single slit has an aperture function of the form

$$F(x) = e^{-x^2/2b^2}$$

Obtain the form factor for the slit.

3. **Ronchi ruling.** A transmission grating is composed of N slits of width b separated by opaque regions of width $a-b$ as shown in the following diagram.

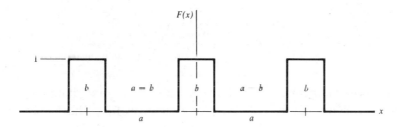

(a) Find the form factor $f(\Delta k)$ for $N = 1$.

(b) Find the scattering amplitude for some general value of N. Take $b = a/m$ with m integral. What orders are missing in diffraction?

4. **Infinite Ronchi ruling.** Letting N go to infinity in Problem 3, Δk takes on discrete values.

(a) For $b = a/m$ with m integral, find the allowed values of Δk and $f(\Delta k)$. Show that the allowed values of Δk correspond to the orders of diffraction.

(b) Letting m go to infinity makes what change in $f(\Delta k)$ for the allowed values of Δk?

5. **Triangular diffraction grating.** A transmission grating is constructed with a triangular aperture function as shown in the following diagram.

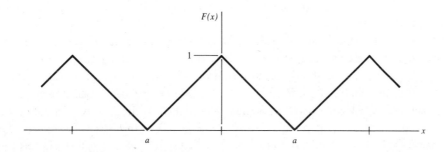

(a) Obtain the form factor $f(\Delta k)$ for a single aperture.

(b) Obtain the scattering amplitude $A(\Delta k)$ for a diffraction grating with an infinite number of slits.

6. **Sinusoidal transmission grating.** A transmission grating is constructed photographically with a sinusoidal variation in transmission as shown in the following

diagram. Find the form factor $f(\Delta k)$ and the scattering amplitude $A(\Delta k)$. What orders appear in diffraction? Explain.

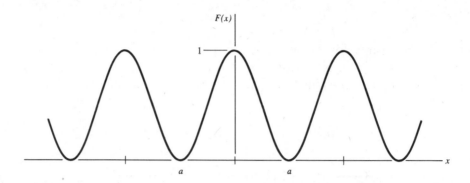

7. **Apodization.** By modifying the aperture function of a single slit the energy going into subsidiary maxima may be reduced. (Apodization is from the Greek and means without feet.)

   (a) Compute the fraction of energy in the central maximum for a slit that cuts off sharply at $x = \pm\frac{1}{2}d$.

   (b) What fraction of the incident energy is directed into the central maximum for a slit with an aperture function $F(x) = \cos(\pi x/d)$. Compare with (a).

8. **Beam pattern.** Show that the beam pattern of an antenna array is the Fraunhofer diffraction pattern of the aperture of the antenna.

9. **Very long baseline array.** A radio telescope has been proposed with eight antenna elements spanning North America along the 40° parallel of latitude. With the array operating at a wavelength of two centimeters, find the east-west angular resolution.

10. **Fresnel diffraction.** Where the distance from the source to the scatterer or from the scatterer to the receiver is comparable with the size of the scatterer, it is not possible to assume that all regions of the scatterer contribute to the received field simply in proportion to their susceptibility. Under such conditions obtain an expression for the interference between a pair of widely separated but very narrow slits excited by a coherent line source.

11. **Waveguide Cherenkov radiation.** Radiation may be generated by an electron beam moving down a waveguide at a velocity greater than the phase velocity in the waveguide. Show that the detected radiation will consist of a discrete rather than a continuous spectrum. [J. R. Pierce, *J. Appl. Phys.*, **26**, 627 (1955), Section 13.]

12. **Surface-generated Cherenkov radiation.** Cherenkov radiation is generated by an electron beam moving near the surface of a dielectric. For titanium dioxide $(\epsilon/\epsilon_0 \simeq 100)$ what is the minimum beam energy? [M. Danos and H. Lashinsky, *Trans. IRE*, **MTT-2**, 21 (1954).]

13. **Cherenkov emission of plasma waves.** A charged particle moving through a plasma with velocity **v** excites a plasma wave of angular frequency ω and wave vector **k**. Show that the condition for Cherenkov radiation is:

$$\omega = \mathbf{k} \cdot \mathbf{v}$$

14. Patterson function. Show that the modulus-squared form factor at wave vector **G** for a periodic charge distribution is given by the Fourier transform of the Patterson function

$$|\rho(\mathbf{G})|^2 = (2\pi)^{-6} \int \mathbf{P}(\mathbf{u}) e^{i\mathbf{G}\cdot\mathbf{u}} \, dV_u$$

where the Patterson function is given by

$$\mathbf{P}(\mathbf{u}) = \int \rho(\mathbf{r})\rho(\mathbf{r}+\mathbf{u}) \, dV = (2\pi)^{-3} \int |\rho(\mathbf{G})|^2 e^{-i\mathbf{G}\cdot\mathbf{u}} \, dV_G$$

Note that the Patterson function has the form of a charge correlation function.

REFERENCES

Bragg, L., "X-ray Crystallography," *Scientific American,* **219**(1), 58 (July 1968). Reprinted in *Lasers and Light*, W. H. Freeman, 1969.

Brillouin, L., *Wave Propagation in Periodic Structures*, Second Edition, McGraw-Hill, 1946; Dover reprint edition, 1953.

Cathey, W. T., *Optical Information Processing and Holography*, Wiley, 1974.

Collier, R. J., C. B. Burckhardt, and L. H. Lin, *Optical Holography*, Academic, 1977.

Cowley, J. M., *Diffraction Physics*, North-Holland, 1975.

Fabelinskii, I. L., *Molecular Scattering of Light*, Plenum, 1968.

Francon, M., *Holography*, Academic, 1974.

Germer, L. H., "The Structure of Crystal Surfaces," *Scientific American*, **212**(3), 32 (March 1965).

Jelley, J. V., *Cerenkov Radiation and Its Applications*, Pergamon, 1958.

Kittel, C., *Introduction to Solid State Physics*, Fifth Edition, Wiley, 1976.

Kock, W. E., *Lasers and Holography*, Doubleday, 1969.

Leith, E. N., "White-Light Holograms," *Scientific American*, **235**(4), 80 (October 1976).

Leith, E. N., and J. Upatnieks, "Laser Photography," *Scientific American*, **212**(6), 24 (June 1965).

Marcuse, D., *Light Transmission Optics*, Van Nostrand Reinhold, 1972.

Schmahl, G., and D. Rudolph, "Optical Gratings Produced Holographically," *Progress in Optics*, Volume 14, E. Wolfe, Editor, North-Holland, 1976.

Smith, H. W., *Principles of Holography*, Second Edition, Wiley, 1975.

Stroke, G., *Introduction to Coherent Optics and Holography*, Academic, 1969.

Zachariasen, W. H., *Theory of X-ray Diffraction in Crystals*, Wiley, 1945; Dover reprint edition, 1967.

15 ACTIVE MEDIA

In this final chapter we discuss the amplification of propagating electromagnetic waves. The term *active* refers to a system that shows power gain and is to be contrasted with a *passive* system. The power delivered to the outgoing signal comes from some source other than the incoming signal. For a lumped device like a transistor, the source of power may be a battery. For systems that we consider here, the sources of energy, which are stored in the medium, are quite varied and will be discussed for each individual case.

Where the amplification is more than sufficient to make up for system losses and the radiation is bounded, we can expect sustained oscillations at some natural resonance of the system. Some of our discussion can be phenomenological and macroscopic while much of it must be more fundamental and microscopic.

The acronyms MASER, for microwave amplification by stimulated emission of radiation, and LASER, for light amplification by stimulated emission of radiation, refer to *quantum* processes. After discussing the connection between classical and quantum models of radiation, we consider microwave amplification in magnetic media because it can be treated classically. Then we consider the ammonia MASER as a model for dielectric media. Finally, we discuss qualitatively the amplification by conducting media followed by a survey of light amplification.

Our discussion up to now of media has been almost entirely phenomenological and classical. We have used the structure of quantum mechanics but only indirectly. For example, quantization was required as a mechanism for stabilizing matter and entered also into our discussion of paramagnetism and ferromagnetism and especially superconductivity.

In discussing the response of dielectric media, we were able to describe excitation processes by classical harmonic oscillators. A specific atomic absorption is strictly associated with the excitation between *two* atomic levels. The quantum analog of the harmonic oscillator, on the other hand, has an infinite number of uniformly spaced levels. If the excitation out of the ground state is weak, there is not much excitation above the first excited state, and the two systems respond in the same way. But for a situation where the population in the excited state of a two-level system is higher than in the ground state, the behavior of the system can *not* be described by the classical harmonic oscillator. As we will see, the equations of a spinning top *are* analogous to   **583** the two-level system, and we will use them at least for the description of magnetic systems. For other systems where angular momentum is not involved, we prefer to work directly with the equations for the wave amplitudes in the two states.

CLASSICAL OSCILLATION

1 Radiative Damping. We have seen **(11-237)** that the instantaneous rate of *radiation* from an oscillating electric dipole may be written as:

$$\frac{dW}{dt} = \frac{1}{6\pi\epsilon_0 c^3}\left(\frac{d^2p}{dt^2}\right)^2 \tag{1}$$

Writing the energy of the dipole as

$$U = \tfrac{1}{2}mv^2 + \frac{1}{2}\frac{p^2}{\alpha} = \frac{m}{2e^2}\left[\left(\frac{dp}{dt}\right)^2 + \omega_0{}^2 p^2\right] \tag{2}$$

we have, on substitution into **(1)**,

$$\frac{dU}{dt} = \frac{m}{e^2}\left(\frac{d^2p}{dt^2} + \omega_0{}^2 p\right)\frac{dp}{dt} = -\frac{dW}{dt} \tag{3}$$

We define a relaxation rate by the ratio:

$$\frac{1}{\tau} = \frac{\mu_0 e^2}{12\pi mc}\left(\frac{d^2p/dt^2}{dp/dt}\right)^2 \tag{4}$$

Substituting **(4)** into **(3)** and dropping the common factor of dp/dt we obtain:

$$\frac{d^2p}{dt^2} + \frac{2}{\tau}\frac{dp}{dt} + \omega_0{}^2 p = 0 \tag{5}$$

which we recognize as the equation of a damped harmonic oscillator. In general the damping constant $1/\tau$ is a function of the motion as given by **(4)**. If the motion is periodic or nearly periodic, we may average **(4)** over a cycle of the motion. For oscillation at frequency ω we write on the average

$$\frac{1}{\tau} = \frac{r_0}{3c}\frac{\langle(d^2p/dt^2)^2\rangle}{\langle(dp/dt)^2\rangle} = \frac{\omega^2 r_0}{3c} \tag{6}$$

where $r_0 = \mu_0 e^2/4\pi m$ is the classical radius of the electron. In order to average in this way we require:

$$\omega\tau = \frac{3c}{\omega r_0} = \frac{3}{2\pi}\frac{\lambda}{r_0} \gg 1 \tag{7}$$

From (7) the wavelength emitted by the oscillator must be greater than $r_0 = 2.8178 \times 10^{-15}$ m.

Our picture of a classical oscillator, then, is that the oscillator emits an exponentially decaying wavetrain extending over a mean time τ. From (6) we may write the relaxation time as:

$$\tau = \frac{3}{4\pi^2} \frac{\lambda^2}{r_0 c} \tag{8}$$

For a wavelength of 0.5 micrometre, (8) gives:

$$\tau \simeq 2 \times 10^{-8} \text{ s} \tag{9}$$

2 Absorption. We now write (5) in the presence of a driving force

$$F(t) = -eE(t) \tag{10}$$

as

$$\frac{d^2 p}{dt^2} + \frac{2}{\tau} \frac{dp}{dt} + \omega_0^2 p = \frac{e^2}{m} E(t) \tag{11}$$

The energy of the oscillator is given by (2). The rate at which energy is radiated may be written from (1) and (4) as:

$$\frac{dW}{dt} = \frac{2m}{e^2 \tau} \left(\frac{dp}{dt}\right)^2 \tag{12}$$

Adding (3) and (12) and comparing with (11) we have

$$\frac{dU}{dt} + \frac{dW}{dt} = E \frac{dp}{dt} \tag{13}$$

The right side of (13) is the rate at which the field does work on the polarization. From the left side of (13), part of this energy is absorbed by the oscillator and goes to increase its kinetic and potential energy, and part is radiated. This is no more than a statement of conservation of energy. In the absence of a driving field we have, from (13) or (3),

$$\frac{dW}{dt} = -\frac{dU}{dt} \tag{14}$$

that is, energy is radiated at the expense of the internal energy of the oscillator.

MICROWAVE AMPLIFICATION

The first microwave amplifying medium was developed by Townes and independently by Basov and Prokhorov and utilized a natural vibrational mode of the ammonia molecule at approximately 24 GHz. More recently, microwave amplifiers have been built utilizing magnetic resonance at a frequency determined by the strength of an external magnetic field. We first discuss magnetic media followed by dielectric and then conducting media.

3 Magnetic Media. We discussed in Section 7-7 the excitation of magnetic resonance and obtained the equation **(7-47)** for the response of the angular momentum **J** of an atom to an applied magnetic field **B**:

$$\frac{d\mathbf{J}}{dt} = \gamma(\mathbf{J} \times \mathbf{B}) \tag{15}$$

where γ is the magnetomechanical ratio of the atom. For a system of n atoms per unit volume we write, for the magnetization density,

$$\mathbf{M} = n\gamma\mathbf{J} \tag{16}$$

In terms of the magnetization density, **(15)** becomes

$$\frac{d\mathbf{M}}{dt} = \gamma(\mathbf{M} \times \mathbf{B}) \tag{17}$$

If we understand that **B** is the macroscopic field then we must augment **(17)** by a term that brings the magnetization to equilibrium. Such terms are strictly phenomenological and there are several choices. If we wish to regard **M** as a vector that, for a distribution of moments relaxes exponentially along both the longitudinal and transverse directions, we may write

$$\frac{d\mathbf{M}}{dt} = \gamma(\mathbf{M} \times \mathbf{B}) - \frac{1}{\tau}(\mathbf{M} - M\hat{\mathbf{B}}) \tag{18}$$

where τ is called the relaxation time. We show in Figure 1 the direction of the two terms in **(18)**. We take γ to be positive and **B** along the $\hat{\mathbf{z}}$ direction. The first term provides a precession of **M** in the negative direction. The second term relaxes **M** into the direction of the field. Writing **(18)** in component form we have

$$\frac{dM_x}{dt} + \frac{M_x}{\tau} = \gamma B_z M_y - \gamma M_z B_y \tag{19}$$

$$\frac{dM_y}{dt} + \frac{M_y}{\tau} = -\gamma B_z M_x + \gamma M_z B_x \tag{20}$$

$$\frac{dM_z}{dt} + \frac{M_z}{\tau} = \frac{M}{\tau} + \gamma B_y M_x - \gamma B_x M_y \tag{21}$$

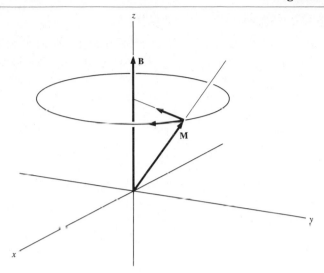

Figure 1 Magnetic resonance. The magnetization processes on a cone about the magnetic field.

We take the sample to be a thin sheet lying in the x-z plane as shown in Figure 2. Applying an oscillating field only along x, we have:

$$B_y = 0 \qquad\qquad (22)$$

The reason for this geometry is that later we will consider propagation of an electro-magnetic wave along the $\hat{y}$ direction. Then (22) is a consequence of the Maxwell equation,

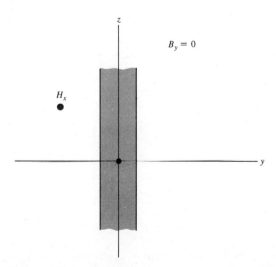

Figure 2 Magnetic sample. The sample is taken to be a thin sheet lying in the x-z plane. An oscillating field is applied along x.

$\nabla \cdot \mathbf{B} = 0$. With (22) we have, for (19) and (20)[1],

$$\frac{dM_x}{dt} + \frac{M_x}{\tau} = \gamma B_z M_y \tag{23}$$

$$\frac{dM_y}{dt} + \frac{M_y}{\tau} = -\gamma B_z M_x + \gamma M_z B_x \tag{24}$$

Writing

$$B_x = \mu_0(H_x + M_x) \tag{25}$$

and substituting into (24) we obtain

$$\frac{dM_y}{dt} + \frac{M_y}{\tau} = -\gamma \mu_0 H_z M_x + \gamma \mu_0 M_z H_x \tag{26}$$

Eliminating M_y between (23) and (26) we obtain for the equation of motion of M_x:

$$\frac{d^2 M_x}{dt^2} + \frac{2}{\tau}\frac{dM_x}{dt} + (\omega_0{}^2 + 1/\tau^2)M_x = \omega_0{}^2 \frac{M_z}{H_z} H_x \tag{27}$$

where $\omega_0{}^2$ is given by:

$$\omega_0{}^2 = \gamma^2 \mu_0 B_z H_z \tag{28}$$

Taking the driving field H_x of the form:

$$H_x(t) = H_x e^{-i\omega t} \tag{29}$$

we obtain for M_x:

$$M_x = \frac{\omega_0{}^2}{\omega_0{}^2 + 1/\tau^2 - \omega(\omega + 2i/\tau)} \frac{M_z}{H_z} H_x \tag{30}$$

The complex permeability may be written as:

$$\underline{\mu} = \mu' + i\mu'' = \frac{B_x}{H_x} = \mu_0 \left(1 + \frac{M_x}{H_x}\right) \tag{31}$$

Substituting from (27) we obtain, for the permeability,

$$\frac{\underline{\mu}}{\mu_0} = 1 + \frac{\omega_0{}^2}{\omega_0{}^2 + 1/\tau^2 - \omega(\omega + 2i/\tau)} \frac{M_z}{H_z} \tag{32}$$

[1] We may note the similarity of (23) and (24) to (D-72) and (D-74) where M_x is analogous to p and M_y to q. The frequency ω_0 corresponds to γB_z and M_z/M corresponds to $(a_1 a_1{}^* - a_2 a_2{}^*)$. This is to justify the use of (23) and (24) to describe quantum oscillators. See R. P. Feynman, F. L. Vernon, Jr., and R. W. Hellwarth, *Am. J. Phys.*, **28**, 49 (1957).

We now consider propagating fields and write

$$\underline{H}_x(y,t) = H_x\, e^{i(\underline{k}y - \omega t)} = H_x\, e^{-k''z} e^{i(k'y - \omega t)} \tag{33}$$

As we have seen (12-58) the wave equation gives:

$$\underline{k} = k' + ik'' = \omega[\underline{\mu}(\omega)\epsilon]^{1/2} \tag{34}$$

Squaring both sides of (34) we obtain for ϵ real:

$$k'^2 - k''^2 = \omega\mu'\epsilon \tag{35}$$

$$2k'k'' = \omega\mu''\epsilon \tag{36}$$

For propagation in the positive y direction k' must be positive. Then from (36) we observe that k'' will have the same sign as μ''. From (32) we write for the real and imaginary parts of the permeability:

$$\frac{\mu'}{\mu_0} = 1 + \frac{\omega_0^2(\omega_0^2 + 1/\tau^2 - \omega^2)}{(\omega_0^2 + 1/\tau^2 - \omega^2)^2 + 4\omega^2/\tau^2}\frac{M_z}{H_z} \tag{37}$$

$$\frac{\mu''}{\mu_0} = \frac{\omega_0^2(2\omega/\tau)}{(\omega_0^2 + 1/\tau^2 - \omega^2)^2 + 4\omega^2/\tau^2}\frac{M_z}{H_z} \tag{38}$$

and we observe from (38) that as long as M_z/H_z is positive then μ'' will be positive, and the wave will be attenuated.

If, on the other hand, we are able to establish and maintain a reversed longitudinal magnetization, then (38) will be negative rather than positive. This means that the wave, instead of decaying exponentially, will grow exponentially. The energy, of course must come from the longitudinal energy:

$$u = -M_z \mu_0 H_z \tag{39}$$

which is positive for a reversed magnetization.

An alternative way of regarding (34) is to imagine that k is real as it would be for a resonant mode between a pair of reflecting plates and the frequency is regarded as complex. If we write:

$$\underline{\omega} = \omega' + i\omega'' \tag{40}$$

then we have, instead of (33),

$$H_x(z,t) = H_x\, e^{i(kz - \underline{\omega} t)} = H_x\, e^{\omega'' t} e^{i(kz - \omega' t)} \tag{41}$$

and in place of **(35)** and **(36)**:

$$\frac{\omega'^2 - \omega''^2}{(\omega'^2 + \omega''^2)^2} = \frac{\mu'\epsilon}{k} \tag{42}$$

$$\frac{2\omega'\omega''}{(\omega'^2 + \omega''^2)^2} = -\frac{\mu''\epsilon}{k} \tag{43}$$

The standard form of **(41)** requires ω' positive. Then from **(43)** ω'' must be opposite in sign from μ''. For μ'' positive ω'' will be negative, and the fields will decay in time as may be observed from **(41)**. On the other hand, for μ'' negative as the result of reversed magnetization, ω'' will be positive and **(41)** leads to exponential growth of the fields. This growth will not be unlimited but will approach a level at which the energy supplied to maintain the reversed magnetization is just balanced by the losses.

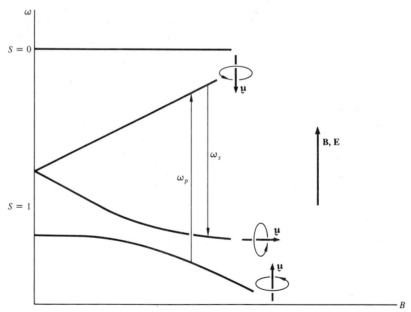

Figure 3   Three-level system. In thermal equilibrium most atoms will be in the lowest energy state with their magnetic moments along the applied field.

4 **Three-Level System.** To understand how it is possible to maintain a reverse magnetization we imagine that each of the atoms contributing to the magnetization contains a pair of equivalent electrons. The lowest energy state will be a triplet for which the electron spins are parallel. Much higher in energy is a singlet for which the spins are antiparallel. If the crystal has an axial electric field the spin triplet will be split into a twofold degenerate level corresponding to spin either parallel or antiparallel to the electric field and a level with the spin normal to the electric field. If we now impose a magnetic field along the electric field the energies of the states will be modified as shown in Figure 3. In thermal equilibrium most atoms will be in the lowest energy state, which gives a magnetic moment along the applied field. By exciting

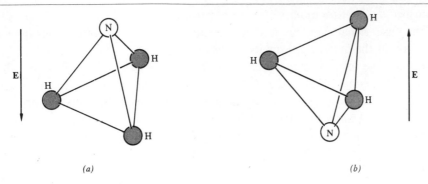

(a)   (b)

Figure 4 Ammonia molecule. (*a*) The nitrogen is stable above the plane of the three hydrogens. (*b*) The inverted position of the nitrogen is also stable.

at ω_p, which reverses the magnetization, it is possible to obtain amplification at a frequency corresponding to a moment turning from antiparallel to normal to the field. The energy that is required for amplification clearly comes from the source at the pump frequency ω_p and is delivered at the signal frequency ω_s. The difference in energy is dissipated as heat.

5 Dielectric Media—Ammonia. We describe next the use of molecular vibrational modes such as those of the ammonia molecule NH_3 to produce microwave amplification. The ideas that we present here have their basis in the quantum nature of molecular vibrational levels and have no classical analog as does the magnetic problem.

We are used to thinking of the ammonia molecule as shown in Figure 4a or 4b where the stable orientation is as indicated. If we measure the polarization as a function of field, however, we observe the behavior shown in Figure 5 with

$$p_0 = 0.306 \times 10^{-10} \text{ electron metre} \tag{44}$$

the ionic dipole moment of the ammonia molecule. The field necessary for saturation of the moment is of the order of 10^6 volt per metre. Although one might think that the

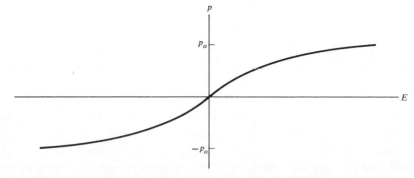

Figure 5 Field-induced polarization. The field necessary for the saturation of the ionic dipole moment of the ammonia molecule is of the order of 10^6 volt per metre.

mechanism by which the polarization is reversed is for the molecule to flop over, this is not the case. What happens is that the nitrogen passes through the plane of the hydrogens and the molecule turns itself "inside out."

How are we to understand a dipole moment less than the saturation value p_0? One way is to regard Figure 4a and 4b as "resonating structures." In the absence of an electric field the molecule spends equal amounts of time in the two structures and the moment is zero. Imposing a field stabilizes one of the structures over the other, the molecule spends more time in that state, and a moment develops. For fields sufficiently small that we are in the linear regime we may write a classical equation of motion for the polarization:

$$\frac{d^2p}{dt^2} + \frac{2}{\tau}\frac{dp}{dt} + \omega_0{}^2 p = \alpha\omega_0{}^2 E \tag{45}$$

This equation has the same form as (27) with α taking the place of M_z/B_z. To obtain amplification we must reverse the sign of α, which means that a polarization must be developed in opposition to the field. To understand how this is accomplished we show

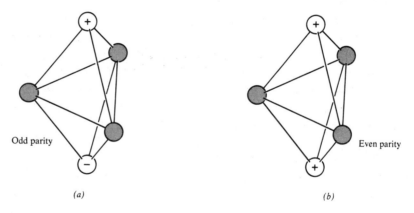

(a)   *(b)*

Figure 6 States of the ammonia molecule. (*a*) The state of odd parity has slightly lower energy. (*b*) The state of even parity is more energetic because of the possibility of finding the nitrogen in the plane of hydrogens.

in Figures 6a and 6b a more fundamental way of regarding the states of the ammonia molecule. By the plus and minus signs we indicate the sign of the wavefunction characterizing the nitrogen atom. The configuration shown in Figure 6a is described as having odd parity and has slightly lower energy than the configuration with even parity shown in Figure 6b. In neither state of the system is there a polarization. If we have a molecule in state (*a*) and apply an electric field we develop a polarization that lowers the energy at first quadratically and then linearly with E as shown in Figure 7. Since the energy decreases linearly with E for large E, we must have an induced polarization that is along the field. On the other hand, applying a field to the even state raises the energy, which must mean that a moment is developed opposite to the field as needed for amplification.

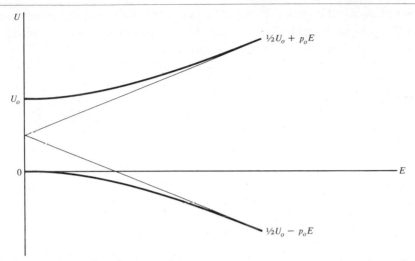

Figure 7 Field energy of ammonia. Molecules in the lower energy state have their energy reduced still further by an electric field.

The problem is one of separating molecules with even parity from those with odd parity. The energy difference is quite small and we expect that in thermal equilibrium both states would be present in almost equal concentration. The method developed for separating the two states takes advantage of the difference in behavior shown in Figure 7. We show in Figure 8 an end-on view of the arrangement used. A beam of ammonia molecules is surrounded by rods of alternating potential. Molecules in the lower energy state have their energy lowered still further by the field and are drawn out of the beam. Molecules in the higher energy state have their energy raised by the field and so are driven into a central core. In this way, the final beam consists of molecules predominantly in the higher energy state and amplification results.

6   **[Coupled Circuit Representation].** It is possible to represent the equation for a two-state system such as that of the ammonia molecule by the equivalent circuit shown in Figure 9a. Here a_1 and a_2 are the amplitudes of the configurations shown in Figure 9b.

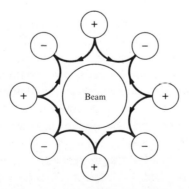

Figure 8 Separation of ammonia molecules. The beam of ammonia molecules is surrounded by rods of alternating potential.

The coupling between the two circuits is by a capacitor $2/\omega_0$, and the perturbation of the system by an applied field is represented by the capacitors $\pm 1/\omega_e$. Note that the resistors have the value $-i$. The circuit equations are:

$$-i\frac{da_1}{dt} + \tfrac{1}{2}\omega_0(a_1 + a_2) + \omega_e a_1 = 0 \tag{46}$$

$$-i\frac{da_2}{dt} + \tfrac{1}{2}\omega_0(a_1 + a_2) - \omega_e a_2 = 0 \tag{47}$$

Assuming a time dependence for a_1 and a_2 of the form $e^{-i\omega t}$ we obtain for the normal mode frequencies:

$$\omega = \tfrac{1}{2}\omega_0 \pm (\tfrac{1}{4}\omega_0{}^2 + \omega_e{}^2)^{1/2} \tag{48}$$

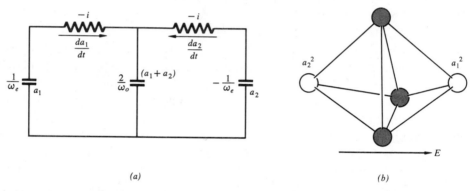

(a)   (b)

Figure 9   Coupled circuit representation. (a) The equation for a two-state system may be represented by an equivalent circuit. (b) The charges a_1 and a_2 correspond to the amplitudes of the two nitrogen positions.

Comparing **(48)** with Figure 7 we may make the identification of ω_0 with the zero field energy separation and ω_e with the perturbation energy $p_0 E$. Varying the electric field corresponds in Figure 9a to modulating the capacitors $\pm 1/\omega_e$.

The frequencies of the normal modes are shown in Figure 10. The amplitudes of the modes are shown in Figure 11a and 11b.

7   Conducting Media.

A number of electrical devices have characteristics of the form shown in Figure 12. Examples of the characteristic shown in Figure 12a are:

1. Glow discharge, such as in a neon indicator lamp.
2. The mineral thyrite (SiC), which is used in lightening arrestors.
3. The thyratron, a triode or tetrode gas discharge tube.
4. Thyristors, which are solid state junction devices with characteristics quite similar to those of thyratrons.

Examples of Figure 12b are:

1. The anode and screen grid characteristic in vacuum tetrodes.

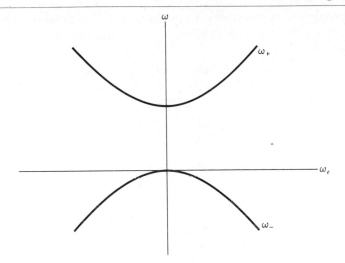

Figure 10   Normal modes. The frequencies of the normal modes are given as a function of the frequency ω_c.

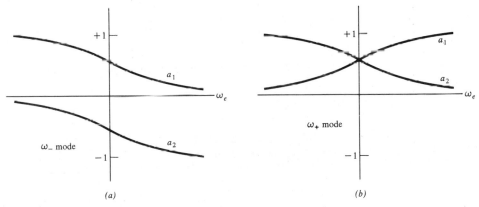

(a)   (b)

Figure 11   Mode amplitudes. (a) The amplitudes of the odd mode are given as a function of the frequency ω_e. (b) The amplitudes of the even mode are given as a function of the frequency ω_e.

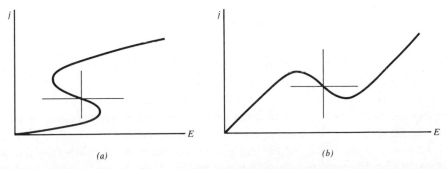

(a)   (b)

Figure 12   Negative differential conductivity. (a) Glow discharge devices show current that is a double-valued function of field. (b) Tunnel diodes require fields that are a double-valued function of current.

2. The screen grid-suppressor grid transconductance characteristic in vacuum pentodes.

3. Tunnel diodes.

4. Gunn effect semiconductors such as gallium arsenide.

To achieve amplification in a medium with the characteristics of Figure 12a or 12b, we apply a static voltage that biases the medium in a region where the slope of the current density versus field characteristic is negative. Writing the current as a rate of change of polarization:

$$j = \frac{dP}{dt} \tag{49}$$

and assuming a time dependence $e^{-i\omega t}$ we have:

$$\epsilon'' = \epsilon_0 \chi'' = \frac{\sigma}{\omega} \tag{50}$$

A negative differential conductivity thus means that the imaginary part of the permittivity is negative and we may expect gain.

The problem in achieving gain at microwave frequencies is in finding a medium with sufficiently short relaxation time that it can respond in times of 10^{-10} seconds to an alternating electric field. The semiconductor gallium arsenide shows the characteristic of Figure 12b and does appear to respond sufficiently quickly. The principal use of gallium arsenide is as a microwave oscillator rather than as an amplifier. The oscillations obtained result from the Gunn effect, which is a charge instability associated with the negative differential conductivity. We have seen **(5-133)** that the equation for charge density in a plasma is of the form:

$$\frac{\partial^2 \rho}{\partial t^2} + \frac{1}{\tau} \frac{\partial \rho}{\partial t} + \frac{\sigma}{\epsilon \tau} \rho = s^2 \nabla^2 \rho \tag{51}$$

If σ is positive the solutions of **(51)** are sinusoidal waves. If however, σ is negative the solutions are exponentially growing and the medium breaks up into domains of charge. These domains drift in an applied field, generating relaxation oscillations with a period of the order of the time for charge to be swept across the sample.

LIGHT AMPLIFICATION

The principles underlying amplification at optical frequencies are the same as we have discussed at microwave frequencies. By some process an excess population must be established in a higher energy level. An incoming signal at the difference frequency stimulates a transition to the state of lower energy with the radiated energy added coherently to the signal field. An equivalent way of viewing the situation is in terms of a reversed electric or magnetic susceptibility for a transition for which the excess population is in the higher state. In the following we discuss briefly the characteristics of various types of LASERS.

8   Ion Lasers. The first light amplifier was built by Maiman in 1960 and utilized synthetic ruby, which is alumina (Al_2O_3) together with a fraction of a percent of Cr^{3+} ion. Ruby is red because of a broad absorption band in the yellow and green. It was known that with excitation of ruby anywhere in this band a sharp fluorescence at a wavelength of 0.6943 micrometre could be observed. The energy level structure of ruby is shown in Figure 13. Electrons excited into the absorption band of ruby relax to a long lived and thus very sharp level from which they then radiate to the ground state.

What Maiman was able to do by very intense excitation was to establish a sufficiently high population in the long lived level that the rate of growth of a coherent wave was greater than the rate of decay by fluorescence. Since 1960, laser action has been obtained with a large number of transition metal ions in ionic crystals, in glasses, and in liquids where the ion is protected.

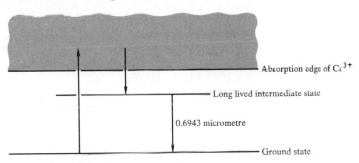

Figure 13   Energy level structure of ruby. Electrons excited into the absorption band relax to a long-lived level from which they radiate to the ground state.

9   Gas Discharge Lasers. The first gas discharge laser, which was proposed by Javan in 1959 and successfully operated in 1961, used a mixture of helium and neon gas. Helium atoms are directly excited to a metastable state by electrons in the discharge. Collisions between excited helium and unexcited neon atoms results in a transfer of excitation from helium to neon. As the excited neon atoms lose energy by radiation there are intermediate states of sufficient lifetime that laser action is possible. The radiation observed in the visible at 0.6328 micrometre is from a level directly excited by energy transfer from helium to a long-lived intermediate level. A large number of atomic gas discharge lasers have now been developed from a wavelength of 0.2 micrometre in the ultraviolet to 133 micrometre in the far infrared. In addition, molecular lasers utilizing vibrational-rotational transitions of the electronic ground state of carbon dioxide and other molecules have been obtained. Such lasers have produced continuous powers of thousands of watts in the infrared.

10   Semiconductor Injection Lasers. It was discovered in 1962 that a gallium arsenide *pn* junction is an efficient emitter of radiation at a wavelength of 0.85 micrometre and with virtually all of the electrical energy going into light. We show in Figure 14 an energy band diagram at a *pn* junction. In the absence of an applied voltage very few electrons have sufficient energy to diffuse into the *p*-type region where they can recombine with holes. Similarly, holes in the *p*-type region are prevented from diffusing into the *n*-type region. Now, if a forward bias is applied across the junction there will be a current, primarily of electrons crossing from the *n*-type into the *p*-type region where they recombine radiatively with holes, which are the majority carriers.

Shortly after the demonstration of such high efficiency (something like 20 percent of the light is reabsorbed), it was demonstrated that by polishing two faces of a gallium arsenide crystal, laser action could be achieved with some 10 W of emitted coherent radiation. Current densities required for laser action were of the order of 25,000 amperes per square centimeter. For a junction 5×10^{-4} cm^2 in area, which is a typical value, the peak current was 12.5 A. Because it is not possible to efficiently conduct away watts

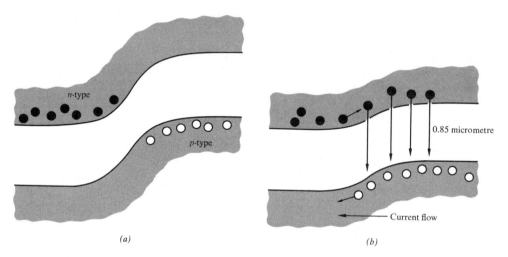

(a)   (b)

Figure 14   Energy bands at a *p-n* junction. (*a*) Few carriers have sufficient energy to cross the junction and recombine. (*b*) When a forward bias is applied across the junction, electrons will cross from the *n*-type region into the *p*-type region to recombine with holes.

of power from such a small volume, such diodes had to be operated in pulses of something like one microsecond duration with a millisecond between pulses.

Recent development has been directed toward diodes that "lase" at sufficiently low current densities that they may be safely operated continuously at room temperature. It has been recognized that if the light could be both generated and contained in a narrow region less than a micrometre in width that the current density could be substantially reduced. The injected electrons must be prevented from diffusing any appreciable distance into the indefinitely extended *p*-type region and some way must be found to contain the generated light within the junction. With junctions of the kind shown in Figure 15, threshold currents have been reduced to 1000 amperes per square centimetre and continuous operation at room temperature has become possible.

The active lasing region of the junction is gallium arsenide about half a micrometre across. On either side of the gallium arsenide is aluminum gallium arsenide, which has a larger band gap and thus a lower refractive index. This means that light generated in the gallium arsenide can be contained by total internal reflection. The regions are doped as shown with n-type aluminum gallium arsenide on the left, p-type gallium arsenide in the center, and p-type aluminum gallium arsenide on the right. With the difference in doping of the two aluminum gallium arsenide regions, electrons injected from the left and holes injected from the right are both contained within the gallium arsenide.

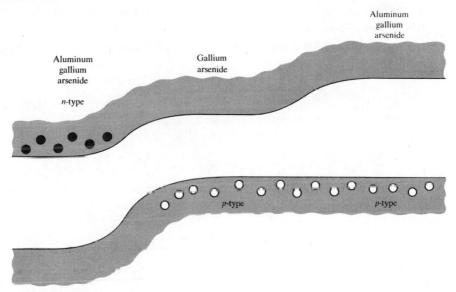

Figure 15   Semiconductor injection laser. Injected electrons are prevented from diffusing out of the junction. Also, light is contained within the junction because of the lower permittivity of aluminium gallium arsenide.

REFERENCES

"Annual Instrumentation Issue: Lasers," *Phys. Today*, **24**(3), 23 (March (1971).

Bowers, R., "A Solid-State Source of Microwaves," *Scientific American*, **215**(2), 22 (August 1966).

Brown, R. Hanbury, *The Intensity Interferometer, Its Application to Astronomy*, Halsted, 1974.

Gordon, J. P., "The Maser," *Scientific American*, **199**(6), 42 (December 1958).

Heitler, W., *The Quantum Theory of Radiation*, Third Edition, Oxford, 1954.

Lindsay, P. A., *Introduction to Quantum Electronics*, Halsted, 1975.

Mandel, L. and E. Wolf, editors, *Coherence and Quantum Optics*, Plenum, 1973.

Morehead, F. F., Jr., "Light-Emitting Semiconductors," *Scientific American*, **216**(5), 108 (May 1967).

Nussenzveig, H. M., *Introduction to Quantum Optics*, Gordon and Breach, 1973.

Panish, M. B. and I. Hayashi, "A New Class of Diode Lasers," *Scientific American*, **225**(1), 32 (July 1971).

Pankove, J. I., *Optical Processes in Semiconductors*, Prentice-Hall, 1971; Dover reprint edition, 1976.

Sargent, M., III, M. O. Scully, and W. E. Lamb, Jr., *Laser Physics*, Addison-Wesley, 1975.

Schawlow, A. L., *Lasers and Light*, Readings from the *Scientific American*, Freeman 1969.

Siegman, A. E., *Microwave Solid State Masers*, McGraw-Hill, 1964.

Siegman, A. E., *An Introduction to Lasers and Masers*, McGraw-Hill, 1971.

Silfvast, W. T., "Metal-Vapor Lasers," *Scientific American*, **228**(2), 88 (February 1973).

Troup, G. J. F., *Masers and Lasers*, Methuen, 1963.

Yariv, A., *Introduction to Optical Electronics*, Second Edition, Holt, Rinehart and Winston, 1976.

APPENDICES

A VECTOR ALGEBRA

The laws of physics are generally invariant under the translation, rotation, and inversion of coordinate axes. Thus, it should be possible to write these laws without reference to a specific coordinate system; it is for this purpose that we introduce the language of vector analysis.[1]

TRANSFORMATION OF COORDINATES

A *vector* is a directed quantity such as, for example, a displacement; a *scalar* is an undirected quantity such as the magnitude of a displacement. A more precise definition of vector and scalar quantities may be made in terms of their invariance under the rotation of coordinate axes. In Figure 1a we represent a vector $\mathbf{A}$ together with orthogonal unit vectors $\hat{\mathbf{x}}$, $\hat{\mathbf{y}}$, and $\hat{\mathbf{z}}$. (In type, vectors are represented by boldface; a unit vector is written in boldface with a caret. In handwritten form an arrow or caret is used, respectively.) In Figure 1b we show a second set of orthogonal vectors $\hat{\mathbf{x}}'$, $\hat{\mathbf{y}}'$, and $\hat{\mathbf{z}}'$. The orientation of the primed axes with respect to the unprimed axes is characterized by the Euler angles ϕ, θ, and ψ. To rotate $\hat{\mathbf{x}}\hat{\mathbf{y}}\hat{\mathbf{z}}$ into $\hat{\mathbf{x}}'\hat{\mathbf{y}}'\hat{\mathbf{z}}'$ we first rotate through ϕ about $\hat{\mathbf{z}}$ to produce $\hat{\mathbf{x}}''$ and $\hat{\mathbf{y}}''$. Next we rotate about $\hat{\mathbf{x}}''$ through θ to produce $\hat{\mathbf{z}}'$ and $\hat{\mathbf{y}}'''$. Finally, we rotate about $\hat{\mathbf{z}}'$ through ψ to produce $\hat{\mathbf{x}}'$ and $\hat{\mathbf{y}}'$. (To simplify Figure 1 we omit the intermediate vectors $\hat{\mathbf{x}}''$, $\hat{\mathbf{y}}''$ **and** $\hat{\mathbf{y}}'''$.)

The transformation that takes the primed axes back into the orientation of the unprimed axes is first a rotation through $-\psi$ about $\hat{\mathbf{z}}'$ followed by a rotation through $-\theta$ about $\hat{\mathbf{x}}''$ and finally a rotation through $-\phi$ about $\hat{\mathbf{z}}$. Thus, the transformation which is inverse to (ϕ,θ,ψ) is $(-\psi,-\theta,-\phi)$.

We may write $\mathbf{A}$ in terms of its components along $\hat{\mathbf{x}}$, $\hat{\mathbf{y}}$, and $\hat{\mathbf{z}}$:

$$\mathbf{A} = A_x\hat{\mathbf{x}} + A_y\hat{\mathbf{y}} + A_z\hat{\mathbf{z}} \tag{1}$$

---

[1] For a discussion of symmetry in physical laws, see R. P. Feynman, R. B. Leighton, and Matthew Sands, *The Feynman Lectures on Physics*, Addison-Wesley, 1963, I-52.

or in terms of its components along $\hat{x}'$, $\hat{y}'$, and $\hat{z}'$:

$$\mathbf{A} = A_{x'}\hat{\mathbf{x}}' + A_{y'}\hat{\mathbf{y}}' + A_{z'}\hat{\mathbf{z}}' \tag{2}$$

The relations connecting the primed and unprimed components are

$$A_x = A_{x'}(\cos\psi\cos\phi - \sin\psi\sin\phi\cos\theta)$$
$$- A_{y'}(\cos\psi\cos\theta\sin\phi + \sin\psi\cos\phi) - A_{z'}\sin\phi\sin\theta \tag{3}$$

$$A_y = A_{x'}(\cos\psi\sin\phi + \sin\psi\cos\phi\cos\theta)$$
$$+ A_{y'}(\cos\psi\cos\phi\cos\theta - \sin\psi\sin\phi) + A_{z'}\cos\phi\sin\theta \tag{4}$$

$$A_z = - A_{x'}\sin\psi\sin\theta - A_{y'}\cos\psi\sin\theta + A_{z'}\cos\theta \tag{5}$$

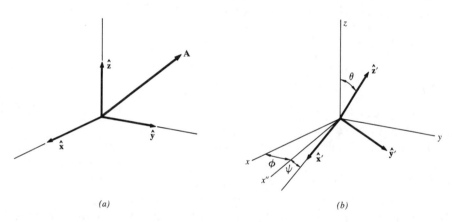

(a)   (b)

Figure 1   Orthogonal coordinates. (*a*) A vector **A** is shown together with a set of orthogonal unit vectors. (*b*) A second set of unit vectors is shown, connected to the first set by rotation.

There are two ways of regarding the transformation properties of a vector:

1. We take the point of view that **A** is a fixed vector and the coordinate axes are arbitrary. Then the components of **A** for two sets of coordinate axes related by ϕ, θ, and ψ are given by (3), (4), and (5).

2. An alternative point of view is that **A** is characterized by its components A_x, A_y, A_z or $A_{x'}$, $A_{y'}$, $A_{z'}$. The operation of rotating the unprimed coordinates into the orientation of the primed coordinates is called a coordinate transformation. The relations between the primed and unprimed components of **A** describe the transformation properties of **A**. For components related by (3), (4), and (5) we say that **A** is invariant under coordinate rotation.

In what follows we adopt the first point of view, that is, that vectors and vector operations have meaning regardless of the particular coordinates that may be chosen. Thus, any expression that is coordinate free must be invariant.

1 **Direction Cosines.** A more compact way of writing the components of a vector is provided by the use of direction cosines. We write

$$A_x = A \cos \theta_{Ax} = Al_{Ax} \tag{6}$$

$$A_y = A \cos \theta_{Ay} = Al_{Ay} \tag{7}$$

$$A_z = A \cos \theta_{Az} = Al_{Az} \tag{8}$$

where θ_{Ax}, θ_{Ay}, and θ_{Az} are the angles between $\mathbf{A}$ and $\hat{\mathbf{x}}$, $\hat{\mathbf{y}}$, and $\hat{\mathbf{z}}$. The quantities l_{Ax}, l_{Ay}, and l_{Az} are called the direction cosines of $\mathbf{A}$ with respect to $\hat{\mathbf{x}}$, $\hat{\mathbf{y}}$, and $\hat{\mathbf{z}}$. Note that direction cosine of $\mathbf{A}$ with respect to $\hat{\mathbf{x}}$ is equal to the direction cosine of $\hat{\mathbf{x}}$ with respect to $\mathbf{A}$ and similarly for the other direction cosines:

$$l_{Ax} = l_{xA} \qquad l_{Ay} = l_{yA} \qquad l_{Az} = l_{zA} \tag{9}$$

Similarly, we may introduce the nine direction cosines of $\hat{\mathbf{x}}$, $\hat{\mathbf{y}}$, and $\hat{\mathbf{z}}$ with respect to $\hat{\mathbf{x}}'$, $\hat{\mathbf{y}}'$, and $\hat{\mathbf{z}}'$. In this way we may write (3), (4), and (5) as:

$$A_x = l_{xx'} A_{x'} + l_{xy'} A_{y'} + l_{xz'} A_{z'} \tag{10}$$

$$A_y = l_{yx'} A_{x'} + l_{yy'} A_{y'} + l_{yz'} A_{z'} \tag{11}$$

$$A_z = l_{zx'} A_{x'} + l_{zy'} A_{y'} + l_{zz'} A_{z'} \tag{12}$$

Exercises

1. Express $A_{x'}$, $A_{y'}$, and $A_{z'}$ in terms of A_x, A_y, and A_z using the form of (3), (4), and (5) but with the transformation inverse to (ϕ,θ,ψ). Alternatively, use the form of (10), (11), and (12) together with the property (9) of the direction cosines.
2. Show that the direction cosines have the property

$$l_{Ax}^2 + l_{Ay}^2 + l_{Az}^2 = 1$$

3. Two vectors $\mathbf{A}$ and $\mathbf{B}$ are separated by an angle γ. Obtain the relation:

$$\cos \gamma = l_{Ax} l_{Bx} + l_{Ay} l_{By} + l_{Az} l_{Bz}$$

4. Let the vectors $\hat{\mathbf{x}}$, $\hat{\mathbf{y}}$, and $\hat{\mathbf{z}}$ define a unit cube. Find the four body diagonals. Show that the angle between pairs of body diagonals is equal to the tetrahedral angle 109.471° or its complement.
5. Find the six face diagonals of the unit cube described in Exercise 4. Find the angles between all pairs of face diagonals and show that the face diagonals form the edges of a tetrahedron.

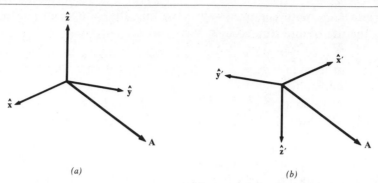

Figure 2   Coordinate inversion. (*a*) A vector **A** is shown together with a set of orthogonal coordinates. (*b*) A second set of coordinates is shown, related to the first set by inversion.

2   Polar and Axial Vectors.

A second kind of coordinate transformation is shown in Figure 2. The operation that takes the unprimed into the primed coordinates is called inversion. If **A** is invariant under inversion, its components will be related by

$$A_x = -A_{x'} \qquad A_y = -A_{y'} \qquad A_z = -A_{z'} \tag{13}$$

A vector that transforms according to **(13)** is called a polar vector, the simplest example of which is a displacement.

It may be surprising that there exists a second class of vectors that do not transform in this way. The simplest example of this class is a rotation.

We show in Figure 3*a* a particle moving counterclockwise on a circle of radius *R* in the *xy* plane. It is convenient to characterize the motion of the particle by an angular frequency $\omega = v/R$.

We regard **ω** as a vector along the axis about which the particle rotates. In order to relate a vector to a rotation we must use the "handedness" of the coordinate

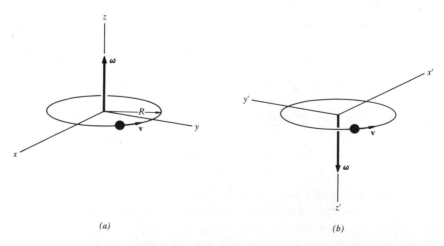

Figure 3   Axial vector. (*a*) The vector **ω** represents positive (or counterclockwise) rotation in the *x-y* plane. (*b*) the vector **ω** now represents positive (or *clockwise*) rotation in the *x'-y'* plane.

representation. Thus, in Figure 3*a* if we orient the fingers of our right hand in the counterclockwise direction (from $\hat{\mathbf{x}}$ to $\hat{\mathbf{y}}$), our thumb points along $\hat{\mathbf{z}}$. For $\hat{\mathbf{x}}$, $\hat{\mathbf{y}}$, and $\hat{\mathbf{z}}$ in cyclic order, such a coordinate system is called righthanded. In order to generate $\hat{\mathbf{z}}'$ from $\hat{\mathbf{x}}'$ and $\hat{\mathbf{y}}'$ as shown in Figure 3*b*, we must use our left hand and such a coordinate system is called lefthanded.

Now, the direction that we take for $\boldsymbol{\omega}$ is dictated by the handedness of the coordinate representation. In Figure 3*a* we would take $\boldsymbol{\omega}$ in the positive $\hat{\mathbf{z}}$ direction:

$$\boldsymbol{\omega} = \omega\hat{\mathbf{z}} \tag{14}$$

In Figure 3*b*, however, where the coordinate representation is lefthanded, we must take $\boldsymbol{\omega}$ in the positive $\hat{\mathbf{z}}'$ direction:

$$\boldsymbol{\omega} = \omega\hat{\mathbf{z}}' \tag{15}$$

For $\boldsymbol{\omega}$ in some general direction we have the transformation:

$$\omega_x = \omega_{x'} \qquad \omega_y = \omega_{y'} \qquad \omega_z = \omega_{z'} \tag{16}$$

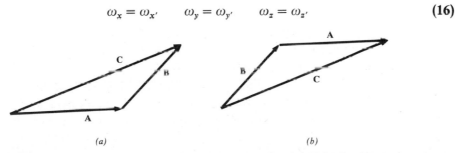

(a)   (b)

Figure 4 Law of triangles. (*a*) A pair of vectors **A** and **B** are added by forming a triangle as shown. (*b*) The resultant vector **C** is the same when **A** and **B** are interchanged, establishing that vector addition is commutative.

A vector that transforms according to **(16)** under coordinate inversion is called an axial vector. The laws of physics may be expressed in terms of both polar and axial vectors although there are some restrictions imposed by the requirement that the laws themselves be invariant under coordinate transformation.[2]

3 Addition of Vectors. Two vectors **A** and **B** are added by the geometrical construction shown in Figure 4*a*. The result of summing **A** and **B** is written as:

▶ $$\mathbf{C} = \mathbf{A} + \mathbf{B} \tag{17}$$

Note that the vector formed in Figure 4*b* by adding **A** onto **B** is also **C**. We write this construction as:

▶ $$\mathbf{C} = \mathbf{B} + \mathbf{A} \tag{18}$$

---

[2] A special situation is evidently presented by beta decay. See footnote 1.

Equations **17** and **18** taken together establish the commutativity of vector addition. (Note that vectors **A** and **B** in Figures 4*a* and *b* are taken to be equivalent even though they have been relatively displaced. That is, a vector is characterized by magnitude and direction *only*—not by position.)

Multiplying a vector by a positive scalar leaves its direction unchanged while multiplying its magnitude by the value of the scalar. Multiplication by a negative scalar reverses the direction of the vector. The subtraction of two vectors may then be regarded as the sum of one vector and the negative of the second vector.

Finally, vector addition is associative so that vectors may be manipulated under addition and subtraction in the same way as numbers:

$$\mathbf{A} + (\mathbf{B} + \mathbf{C}) = \mathbf{A} + \mathbf{B} + \mathbf{C} = (\mathbf{A} + \mathbf{B}) + \mathbf{C} \tag{19}$$

Vectors **B** and **C** may be decomposed just as **A** in **(1)** and **(2)** giving:

$$C_x = A_x + B_x \qquad C_y = A_y + B_y \qquad C_z = A_z + B_z \tag{20}$$

or

$$C_{x'} = A_{x'} + B_{x'} \qquad C_{y'} = A_{y'} + B_{y'} \qquad C_{z'} = A_{z'} + B_{z'} \tag{21}$$

The same result **(21)** holds under coordinate inversion as long as **A** and **B** are either both polar vectors or both axial vectors. We cannot, in general, have a physical quantity in which a polar vector is added to an axial vector since the magnitude of that quantity would not be invariant under coordinate inversion.[2]

MULTIPLICATION OF VECTORS

One may think of various ways of taking the product of two vectors. We are interested here in only those products that have the distributive property. In this part we discuss the scalar product of two polar vectors, which is a scalar, and the vector product of two polar vectors, which is an axial vector. In the next part of this appendix we discuss tensor products of vectors.

4 **Scalar Product.** The scalar product of two vectors is the number that is obtained by taking the product of the magnitudes of the two vectors with the cosine of the intermediate angle. If the vectors are **A** and **B** and the intermediate angle is θ as shown in Figure 5*a*, then the scalar (or dot) product is written as:

$$\mathbf{A} \cdot \mathbf{B} = AB \cos \theta = \mathbf{B} \cdot \mathbf{A} \tag{22}$$

It is clear from **(22)** that scalar multiplication is commutative since $\cos \theta = \cos(-\theta)$. We now show that the scalar product is distributive:

$$\mathbf{D} \cdot (\mathbf{A} + \mathbf{B}) = \mathbf{D} \cdot \mathbf{A} + \mathbf{D} \cdot \mathbf{B} \tag{23}$$

We rewrite the left side of **(23)**:

$$\mathbf{D} \cdot (\mathbf{A} + \mathbf{B}) = \mathbf{D} \cdot \mathbf{C} = D \cos \phi (A \cos \beta + B \cos \alpha) \tag{24}$$

where the angles $\alpha,\, \beta,\, \gamma,\, \theta,$ and ϕ are shown in Figure 5b.
Now we write **(24)** as:

$$\mathbf{D} \cdot \mathbf{C} = DA \cos(\phi + \beta) + DB \cos(\phi - \alpha) + DA \sin \phi \sin \beta - DB \sin \phi \sin \alpha \tag{25}$$

where we have used the identity:

$$\cos(\phi + \psi) = \cos \phi \cos \psi - \sin \phi \sin \psi \tag{26}$$

Now by the law of sines we have the identity:

$$\frac{\sin \alpha}{A} = \frac{\sin \beta}{B} \tag{27}$$

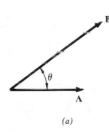

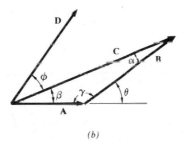

(a) (b)

Figure 5 Scalar product. (*a*) The scalar product of two vectors **A** and **B**, separated by the angle θ, is given by **(22)**. (*b*) By introducing an arbitrary vector **D**, it may be demonstrated that the scalar product is distributive.

by which the final two terms in **(25)** cancel. But $\phi + \beta$ is the angle between **D** and **A**, and $\phi - \alpha$ is the angle between **D** and **B**, leading to **(23)**.

We may use the distributive property to write the scalar product of **A** and **B** in terms of their rectangular components:

$$\mathbf{A} \cdot \mathbf{B} = A_x B_x + A_y B_y + A_z B_z \tag{28}$$

where we have used $\hat{\mathbf{x}} \cdot \hat{\mathbf{x}} = \hat{\mathbf{y}} \cdot \hat{\mathbf{y}} = \hat{\mathbf{z}} \cdot \hat{\mathbf{z}} = 1$, and $\hat{\mathbf{x}} \cdot \hat{\mathbf{y}} = \hat{\mathbf{y}} \cdot \hat{\mathbf{z}} = \hat{\mathbf{z}} \cdot \hat{\mathbf{x}} = 0$. What can we say about the product:

$$A_{x'} B_{x'} + A_{y'} B_{y'} + A_{z'} B_{z'} \tag{29}$$

where the primed coordinates are related to the unprimed coordinates by a rotation? Since the scalar product is defined without reference to any particular coordinate system it is clear that **(28)** must hold for any coordinate system and thus **(28)** and **(29)** must be equal to the same scalar.

Exercises

---

6. Obtain the law of cosines:

$$C^2 = A^2 + B^2 - 2AB \cos \gamma$$

(where the angle γ is shown in Figure 5*b*) by taking the scalar product of **(18)** with itself.

7. Show that the equation of a plane may be written as:

$$\mathbf{r} \cdot \hat{\mathbf{n}} = R$$

where **r** is a general position vector, $\hat{\mathbf{n}}$ is the unit normal to the plane, and R is the distance from the origin to the plane.

8. The vector $\mathbf{S} = \hat{\mathbf{n}}S$ represents a planar surface where S is the magnitude of the area and $\hat{\mathbf{n}}$ is normal to the surface. If the surface moves with velocity **v**, show that $\mathbf{v} \cdot \mathbf{S}$ is equal to the rate at which volume is swept out.

9. Show that the scalar product of **A** and **B** is equivalent to the indicated product of **A** and **B** written as row and column vectors:

$$\mathbf{A} \cdot \mathbf{B} = (A_x\, A_y\, A_z)\begin{pmatrix} B_x \\ B_y \\ B_z \end{pmatrix}$$

---

5 Vector Product. The vector product of two vectors is written as:

$$\mathbf{B} \times \mathbf{C} \tag{30}$$

and has the magnitude:

$$BC \sin \phi \tag{31}$$

where ϕ is the smaller angle through which **B** may be turned to bring it into the direction of **C**.

The vector $\mathbf{B} \times \mathbf{C}$ is along the axis about which **B** must be turned through less than π to bring it into alignment with **C**. The sense of this vector depends on the handedness of the space as discussed earlier for the rotation vector. Thus, for a right-handed system and $\phi < \pi$ as shown in Figure 6*a*, **(30)** will be in the $\hat{\mathbf{z}}$ direction. For a left-handed system, as shown in Figure 6*b*, **(30)** will be in the $\hat{\mathbf{z}}'$ direction. It is clear that **(30)** is an axial vector.

Whatever the handedness of the coordinate system, the rotation that takes **B** into **C** is just the negative of that which takes **C** into **B**. Thus the vector product is anticommutative:

▶ $$\mathbf{B} \times \mathbf{C} = -\mathbf{C} \times \mathbf{B} \tag{32}$$

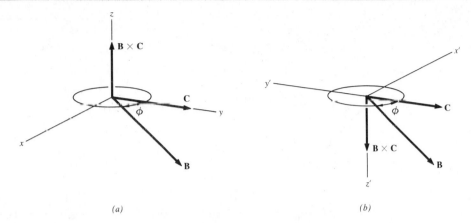

(a)   (b)

Figure 6 Vector product. (*a*) For a right-handed coordinate system and $\phi < \pi$, the cross product of **B** and **C** is along the positive z direction. (*b*) For a left-handed coordinate system and $\phi < \pi$, the cross product of **B** and **C** is along the positive z' direction.

Exercises

10. For three vectors **A**, **B**, and **C**, which satisfy **(17)**, obtain the result

$$\mathbf{A} \times \mathbf{B} = \mathbf{A} \times \mathbf{C} - \mathbf{C} \times \mathbf{B}$$

11. By using the vector identity of Exercise 10 obtain the law of sines:

$$\frac{\sin \alpha}{A} = \frac{\sin \beta}{B} = \frac{\sin \gamma}{C}$$

where α, β, and γ are the opposite angles as shown in Figure 5*b*.

12. Show that the vector product of **A** and **B** gives both the area enclosed by the parallelogram determined by **A** and **B** and the direction of the surface normal.

6   Triple Scalar Product.

The scalar product of **(30)** with a vector is called the triple scalar product:

$$\mathbf{A} \cdot (\mathbf{B} \times \mathbf{C}) \tag{33}$$

and has the value:

$$\mathbf{A} \cdot (\mathbf{B} \times \mathbf{C}) = \pm ABC \sin \phi \cos \theta \tag{34}$$

where the angles are shown in Figure 7*a*. The sign ambiguity arises because of the handedness of the space. For a right-handed coordinate system we have the positive sign. For a left-handed system, we have a negative sign. This discussion assumes that **A**, **B**, and **C** are all polar vectors. Because the triple scalar product changes sign under coordinate inversion it cannot be an ordinary scalar and is called a pseudoscalar.

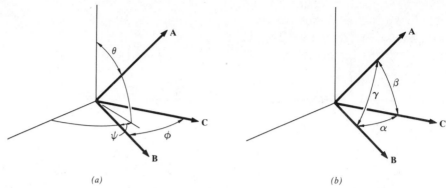

Figure 7   Triple scalar product. (*a*) With the angles ϕ, θ, and ψ as indicated, the triple scalar product is given by **(34)**. (*b*) Alternatively, the triple scalar product may be expressed in terms of the intermediate angles α, β, and γ.

We demonstrate now that in the triple scalar product **(33)** the dot and cross product may be interchanged without affecting the result:

$$\blacktriangleright \qquad\qquad \mathbf{A} \cdot (\mathbf{B} \times \mathbf{C}) = (\mathbf{A} \times \mathbf{B}) \cdot \mathbf{C} \qquad\qquad (35)$$

We show in Figure 7*b* a second way of labeling the angles. Comparing the two figures we obtain

$$\cos \alpha = \cos \phi \qquad\qquad (36)$$

$$\cos \beta = \sin \theta \cos(\phi - \psi) \qquad\qquad (37)$$

$$\cos \gamma = \sin \theta \cos \psi \qquad\qquad (38)$$

Eliminating ψ between **(37)** and **(38)** we obtain:

$$\sin \phi \cos \theta = \pm [1 + 2 \cos \alpha \cos \beta \cos \gamma - (\cos^2 \alpha + \cos^2 \beta + \cos^2 \gamma)]^{1/2} \qquad (39)$$

where the plus sign is for α, β, and γ in cyclic order as shown in Figure 7*b* for a right-handed coordinate system. The minus sign is for anticyclic order, which is obtained from cyclic order by an odd number of interchanges of pairs of angles. We observe that **(39)** is invariant under cyclic permutation of α, β, and γ and thus of **A**, **B**, and **C**. By permuting the three vectors and commuting the scalar product we obtain **(35)**.

Exercises
___

13. Show that the triple scalar product **(35)** may be written as the determinant of the matrix[1]

$$\mathbf{M} = \begin{pmatrix} A_x & A_y & A_z \\ B_x & B_y & B_z \\ C_x & C_y & C_z \end{pmatrix}$$

[1] In type matrices are customarily set in boldface block capital letters (*sans serif*).

14. A parallelepiped is defined by the vectors **A**, **B**, and **C**. Show that the enclosed volume is plus or minus (depending on whether **A**, **B**, and **C** are in cyclic or anticyclic order) the triple scalar product.

15. Verify that the vectors **A**, **B**, and **C** in **(33)** may be cyclically permuted:

$$\mathbf{A} \cdot (\mathbf{B} \times \mathbf{C}) = \mathbf{B} \cdot (\mathbf{C} \times \mathbf{A}) = \mathbf{C} \cdot (\mathbf{A} \times \mathbf{B})$$

7 Reciprocal Vectors. Suppose that **a**, **b**, and **c** are three non-coplanar vectors. We may introduce a set of reciprocal vectors **A**, **B**, and **C** that satisfy the following conditions:

$$
\begin{array}{lll}
\mathbf{A} \cdot \mathbf{a} = 1 & \mathbf{B} \cdot \mathbf{a} = 0 & \mathbf{C} \cdot \mathbf{a} = 0 \\
\mathbf{A} \cdot \mathbf{b} = 0 & \mathbf{B} \cdot \mathbf{b} = 1 & \mathbf{C} \cdot \mathbf{b} = 0 \\
\mathbf{A} \cdot \mathbf{c} = 0 & \mathbf{B} \cdot \mathbf{c} = 0 & \mathbf{C} \cdot \mathbf{c} = 1
\end{array}
\tag{40}
$$

These equations may be solved to give:

$$\mathbf{A} = \frac{\mathbf{b} \times \mathbf{c}}{\mathbf{a} \cdot \mathbf{b} \times \mathbf{c}} \qquad \mathbf{B} = \frac{\mathbf{c} \times \mathbf{a}}{\mathbf{a} \cdot \mathbf{b} \times \mathbf{c}} \qquad \mathbf{C} = \frac{\mathbf{c} \times \mathbf{b}}{\mathbf{a} \cdot \mathbf{b} \times \mathbf{c}} \tag{41}$$

Note that **(41)** gives **A** orthogonal to both **b** and **c** although not in general parallel to **a**. Corresponding relations hold for **B** and C.

As an example of the use of reciprocal vectors, let us imagine that we are given a fixed vector:

$$\mathbf{F} = F_1\mathbf{a} + F_2\mathbf{b} + F_2\mathbf{c} \tag{42}$$

[As long as **a**, **b**, and **c** do not all lie in the same plane, the expansion **(42)** is possible.] We wish now to find F_1, F_2, and F_3. Taking the scalar product of **(42)** with **A**, **B**, and **C** we obtain:

$$F_1 = \mathbf{A} \cdot \mathbf{F} \qquad F_2 = \mathbf{B} \cdot \mathbf{F} \qquad F_3 = \mathbf{C} \cdot \mathbf{F} \tag{43}$$

Exercises

16. Given that **a**, **b**, and **c** are non-coplanar, show that **A**, **B**, and **C** are also non-coplanar.

17. Given a set of vectors **A**, **B**, and **C** described by **(41)**, express the vectors reciprocal to **A**, **B**, and **C** in terms of **a**, **b**, and **c**.

8 Distributivity. We show next that the vector product is distributive. To do this we form the product:

$$(\mathbf{A} + \mathbf{B}) \times \mathbf{C} \cdot \mathbf{D} = (\mathbf{A} + \mathbf{B}) \cdot (\mathbf{C} \times \mathbf{D}) \tag{44}$$

where **D** is some general vector and we have used **(35)** to obtain the above equality. Since the scalar product is distributive **(33)** we may expand the right side of **(44)** to obtain:

$$\mathbf{A} \cdot (\mathbf{C} \times \mathbf{D}) + \mathbf{B} \cdot (\mathbf{C} \times \mathbf{D}) = (\mathbf{A} \times \mathbf{C}) \cdot \mathbf{D} + (\mathbf{B} \times \mathbf{C}) \cdot \mathbf{D} \tag{45}$$

where again we have used **(35)**. But if **D** is a general vector we must have from **(44)** and **(45)**:

$$\blacktriangleright \qquad (\mathbf{A} + \mathbf{B}) \times \mathbf{C} = \mathbf{A} \times \mathbf{C} + \mathbf{B} \times \mathbf{C} \tag{46}$$

which is the desired statement of the distributivity of the vector product. We have used the result that if $\mathbf{F} \cdot \mathbf{D}$ and $\mathbf{G} \cdot \mathbf{D}$ are equal for *all* **D**, then **F** and **G** must be equal.

Making use of distributivity one may write the vector product in terms of the rectangular components of **A** and **B**:

$$\mathbf{A} \times \mathbf{B} = \hat{\mathbf{x}}(A_y B_z - A_z B_y) + \hat{\mathbf{y}}(A_z B_x - A_x B_z) + \hat{\mathbf{z}}(A_x B_y - A_y B_x) \tag{47}$$

Exercise

18. Show that **(47)** may be written as the determinant of the matrix:

$$\mathbf{M} = \begin{pmatrix} \hat{\mathbf{x}} & \hat{\mathbf{y}} & \hat{\mathbf{z}} \\ A_x & A_y & A_z \\ B_x & B_y & B_z \end{pmatrix}$$

9 Triple Vector Product. The triple vector product is written as:

$$\mathbf{A} \times (\mathbf{B} \times \mathbf{C}) \tag{48}$$

where the parentheses are necessary in order to indicate the order in which the cross products are to be taken. As we have seen, the vector $\mathbf{B} \times \mathbf{C}$ is normal to the plane of **B** and **C** and has magnitude $BC \sin \phi$. Since the triple vector product must be perpendicular to $\mathbf{B} \times \mathbf{C}$, it must lie *in* the plane of **B** and **C** and be perpendicular to **A** as shown in Figure 8.

Since **(48)** is in the plane of **B** and **C** it can be written as a linear combination of **B** and **C**:

$$\mathbf{A} \times (\mathbf{B} \times \mathbf{C}) = b\mathbf{B} + c\mathbf{C} \tag{49}$$

Taking the scalar product of **(49)** with **A** we obtain:

$$b(\mathbf{A} \cdot \mathbf{B}) + c(\mathbf{A} \cdot \mathbf{C}) = 0 \tag{50}$$

so that we may write **(49)** as:

$$\mathbf{A} \times (\mathbf{B} \times \mathbf{C}) = a[\mathbf{B}(\mathbf{A} \cdot \mathbf{C}) - \mathbf{C}(\mathbf{A} \cdot \mathbf{B})] \tag{51}$$

We now wish to evaluate a. The magnitude of the left side of **(51)** is from Figure 8 given by:

$$|\mathbf{A} \times (\mathbf{B} \times \mathbf{C})| = ABC \sin \theta \sin \phi \tag{52}$$

The magnitude of the bracketed quantity in **(51)** is given by:

$$|\mathbf{B}(\mathbf{A} \cdot \mathbf{C}) - \mathbf{C}(\mathbf{A} \cdot \mathbf{B})| = ABC(\cos^2 \beta + \cos^2 \gamma - 2 \cos \alpha \cos \beta \cos \gamma)^{1/2} \tag{53}$$

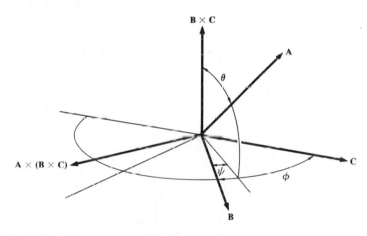

Figure 8 Triple vector product. The triple vector product of the three vectors **A**, **B**, and **C** lies in the plane of **B** and **C** as shown.

where α, β, and γ are shown in Figure 7b. From **(36)**, **(37)**, and **(38)** we obtain the identity:

$$(\cos^2 \beta + \cos^2 \gamma - 2 \cos \alpha \cos \beta \cos \gamma)^{1/2}$$
$$= \sin \theta[\cos^2 \psi - 2 \cos \phi \cos \psi \cos(\phi - \psi) + \cos^2(\phi - \psi)]^{1/2} = \sin \theta \sin \phi \tag{54}$$

which gives $a = 1$ and permits us to write the triple vector product as:

$$\blacktriangleright \qquad \mathbf{A} \times (\mathbf{B} \times \mathbf{C}) = \mathbf{B}(\mathbf{A} \cdot \mathbf{C}) - \mathbf{C}(\mathbf{A} \cdot \mathbf{B}) \tag{55}$$

Because of the presence of two vector products the direction of **(55)** is independent of the handedness of the coordinate system and thus is an ordinary polar vector as long as **A**, **B**, and **C** are polar vectors.

Exercises

19. Prove Lagrange's identity:

$$(\mathbf{A} \times \mathbf{B}) \cdot (\mathbf{C} \times \mathbf{D}) = (\mathbf{A} \cdot \mathbf{C})(\mathbf{B} \cdot \mathbf{D}) - (\mathbf{A} \cdot \mathbf{D})(\mathbf{B} \cdot \mathbf{C})$$

20. Prove the vector identity:

$$(\mathbf{A} \times \mathbf{B}) \times (\mathbf{C} \times \mathbf{D}) = [\mathbf{A} \cdot (\mathbf{B} \times \mathbf{D})]\mathbf{C} - [\mathbf{A} \cdot (\mathbf{B} \times \mathbf{C})]\mathbf{D}$$

21. Prove the vector identity:

$$(\mathbf{A} \times \mathbf{B}) \times (\mathbf{C} \times \mathbf{D}) = (\mathbf{A} \times \mathbf{C})(\mathbf{B} \cdot \mathbf{D}) - (\mathbf{A} \times \mathbf{D})(\mathbf{B} \cdot \mathbf{C})$$

22. Prove the following identity:

$$\mathbf{A} \times (\mathbf{B} \times \mathbf{C}) + \mathbf{B} \times (\mathbf{C} \times \mathbf{A}) + \mathbf{C} \times (\mathbf{A} \times \mathbf{B}) \equiv 0$$

23. Prove the identity:

$$\mathbf{A} \cdot [\mathbf{B} \times (\mathbf{C} \times \mathbf{D})] = [(\mathbf{A} \times \mathbf{B}) \times \mathbf{C}] \cdot \mathbf{D}$$

24. A vector $\mathbf{A}$ lying in the xy plane is rotated about the $\hat{z}$ direction through an angle θ. Show that the transformed vector may be written in the form:

$$\mathbf{A}' = \mathbf{A} + (\hat{z} \times \mathbf{A})\sin\theta + (1 - \cos\theta)\hat{z} \times (\hat{z} \times \mathbf{A})$$

TENSOR PRODUCT

The tensor product of two vectors $\mathbf{A}$ and $\mathbf{B}$ is written by mathematicians as:

$$\mathbf{A} \otimes \mathbf{B} \tag{56}$$

The meaning of the tensor product of $\mathbf{A}$ with $\mathbf{B}$, which we write simply as the tensor $\mathbf{AB}$, is established by taking the scalar and vector product of a third vector $\mathbf{C}$ with the tensor product. The scalar product of a tensor product and a vector may be taken either from the right:

$$(\mathbf{A} \otimes \mathbf{B}) \cdot \mathbf{C} \equiv (\mathbf{AB}) \cdot \mathbf{C} = \mathbf{A}(\mathbf{B} \cdot \mathbf{C}) \tag{57}$$

or from the left:

▶ $$\mathbf{C} \cdot (\mathbf{A} \otimes \mathbf{B}) \equiv \mathbf{C} \cdot (\mathbf{AB}) = (\mathbf{C} \cdot \mathbf{A})\mathbf{B} \tag{58}$$

and is a vector.

Similarly, the vector product may be taken either from the right:

$$\blacktriangleright \qquad (\mathbf{A} \otimes \mathbf{B}) \times \mathbf{C} \equiv (\mathbf{AB}) \times \mathbf{C} = \mathbf{A}(\mathbf{B} \times \mathbf{C}) \tag{59}$$

or from the left:

$$\mathbf{C} \times (\mathbf{A} \otimes \mathbf{B}) \equiv \mathbf{C} \times (\mathbf{AB}) = (\mathbf{C} \times \mathbf{A})\mathbf{B} \tag{60}$$

and is a tensor product.

Note that as a consequence of (57) and (59) we have the identities:

$$(\mathbf{AB}) \cdot \mathbf{C} = (\mathbf{AC}) \cdot \mathbf{B} \qquad (\mathbf{AB}) \times \mathbf{C} = -(\mathbf{AC}) \times \mathbf{B} \tag{61}$$

Exercise
_____

25. Prove the identity:

$$[\mathbf{A}(\mathbf{B} \times \mathbf{C})] \cdot \mathbf{D} = (\mathbf{AB}) \cdot (\mathbf{C} \times \mathbf{D})$$

---

10 Distributivity. We now show that the tensor product is distributive:

$$\blacktriangleright \qquad \mathbf{A}(\mathbf{B} + \mathbf{C}) = \mathbf{AB} + \mathbf{AC} \tag{62}$$

To do this we take the scalar product of (62) with some general vector **D**:

$$[\mathbf{A}(\mathbf{B} + \mathbf{C})] \cdot \mathbf{D} = \mathbf{A}[(\mathbf{B} + \mathbf{C}) \cdot \mathbf{D}] = \mathbf{A}(\mathbf{B} \cdot \mathbf{D}) + \mathbf{A}(\mathbf{C} \cdot \mathbf{D}) \tag{63}$$

where the first equality follows from (57) and the second from the distributivity of the scalar product. Using (57) again we may rewrite the right side of (63) to obtain:

$$\mathbf{A}(\mathbf{B} + \mathbf{C}) \cdot \mathbf{D} = (\mathbf{AB}) \cdot \mathbf{D} + (\mathbf{AC}) \cdot \mathbf{D} \tag{64}$$

But if **D** is a general vector we must have (62) as a consequence of (64).

Exercise
_____

26. Show that the vector triple product $(\mathbf{A} \times \mathbf{B}) \times \mathbf{C}$ may be written in the form:

$$(\mathbf{A} \times \mathbf{B}) \times \mathbf{C} = \mathbf{C} \cdot (\mathbf{AB} - \mathbf{BA})$$

which is the difference between (58) and (57).

---

11 Rectangular Coordinates. We have defined the properties of the tensor product without reference to any particular coordinate system and thus may conclude that the scalar and vector products are invariant under coordinate transformation. For some purposes it may be convenient to write **A** and **B** in a specific representation:

$$\mathbf{A} = A_x \hat{\mathbf{x}} + A_y \hat{\mathbf{y}} + A_z \hat{\mathbf{z}} \tag{65}$$

$$\mathbf{B} = B_x \hat{\mathbf{x}} + B_y \hat{\mathbf{y}} + B_z \hat{\mathbf{z}} \tag{66}$$

Using the distributive property **(62)** we write the tensor product in a particular representation as:

$$\begin{aligned}
\mathbf{AB} = \; & A_x B_x \,\hat{\mathbf{x}}\hat{\mathbf{x}} + A_x B_y \,\hat{\mathbf{x}}\hat{\mathbf{y}} + A_x B_z \,\hat{\mathbf{x}}\hat{\mathbf{z}} \\
& + A_y B_x \,\hat{\mathbf{y}}\hat{\mathbf{x}} + A_y B_y \,\hat{\mathbf{y}}\hat{\mathbf{y}} + A_y B_z \,\hat{\mathbf{y}}\hat{\mathbf{z}} \\
& + A_z B_x \,\hat{\mathbf{z}}\hat{\mathbf{x}} + A_z B_y \,\hat{\mathbf{z}}\hat{\mathbf{y}} + A_z B_z \,\hat{\mathbf{z}}\hat{\mathbf{z}}
\end{aligned} \tag{67}$$

The coefficients of the tensor products of the unit vectors thus form a 3×3 matrix. The addition of tensor products is defined and is equivalent to the addition of matrices. A sum of tensor products is called a tensor and cannot in general be written as a *single* tensor product. This is because the tensor product of two vectors has only five independent components while a 3×3 matrix has nine. We return to a more extended discussion of tensors in Appendix C.

Exercise

27. Prove the above statement that the tensor product of two vectors has five independent components (and not six). As a corollary show that **A** and **B** are not uniquely determined by the tensor product **AB**.

SUMMARY

1. Vector addition is commutative and associative:

$$\mathbf{A} + \mathbf{B} = \mathbf{B} + \mathbf{A} \tag{17}\;(18)$$

$$\mathbf{A} + (\mathbf{B} + \mathbf{C}) = (\mathbf{A} + \mathbf{B}) + \mathbf{C} \tag{19}$$

2. The scalar product is commutative and distributive:

$$\mathbf{A} \cdot \mathbf{B} = \mathbf{B} \cdot \mathbf{A} \tag{22}$$

$$\mathbf{A} \cdot (\mathbf{B} + \mathbf{C}) = \mathbf{A} \cdot \mathbf{B} + \mathbf{A} \cdot \mathbf{C} \tag{23}$$

3. The vector product is anticommutative and distributive:

$$\mathbf{A} \times \mathbf{B} = -\mathbf{B} \times \mathbf{A} \tag{32}$$

$$\mathbf{A} \times (\mathbf{B} + \mathbf{C}) = \mathbf{A} \times \mathbf{B} + \mathbf{A} \times \mathbf{C} \tag{46}$$

4. The triple scalar product has the property:

$$\mathbf{A} \cdot (\mathbf{B} \times \mathbf{C}) = (\mathbf{A} \times \mathbf{B}) \cdot \mathbf{C} \tag{35}$$

5. The triple vector product may be written as:

$$\mathbf{A} \times (\mathbf{B} \times \mathbf{C}) = \mathbf{B}(\mathbf{A} \cdot \mathbf{C}) - \mathbf{C}(\mathbf{A} \cdot \mathbf{B}) \tag{55}$$

6. The tensor product is distributive but not commutative:

$$\mathbf{A}(\mathbf{B} + \mathbf{C}) = \mathbf{A}\mathbf{B} + \mathbf{A}\mathbf{C} \tag{62}$$

$$\mathbf{A}\mathbf{B} \neq \mathbf{B}\mathbf{A}$$

7. The triple tensor-scalar product gives the vector:

$$\mathbf{A} \cdot (\mathbf{B}\mathbf{C}) = (\mathbf{A} \cdot \mathbf{B})\mathbf{C} \tag{59}$$

8. The triple tensor-vector product gives the tensor product:

$$\mathbf{A} \times (\mathbf{B}\mathbf{C}) - (\mathbf{A} \times \mathbf{B})\mathbf{C} \tag{60}$$

REFERENCES

Arfken, G. B., *Mathematical Methods for Physicists*, Second Edition, Academic, 1970.

Chambers, Ll. G., *A Course in Vector Analysis*, Chapman and Hall, 1969.

Hoffmann, B., *About Vectors*, Prentice-Hall, 1966; Dover, 1975.

Kemmer, N., *Vector Analysis*, Cambridge, 1977.

Marion, J. B., *Principles of Vector Analysis*, Academic 1965.

B VECTOR CALCULUS

In this appendix we extend the techniques of differentiation and integration to vectors. We obtain a number of vector identities, which may be expressed in terms of the gradient operator. We regularly refer to these identities by number in the body of the text and list them here for convenient reference.

VECTOR IDENTITIES

$$\mathbf{\nabla}(\phi\psi) = \phi\mathbf{\nabla}\psi + \psi\mathbf{\nabla}\phi \tag{1}$$

$$\mathbf{\nabla}\cdot(\phi\mathbf{A}) = (\mathbf{\nabla}\phi)\cdot\mathbf{A} + \phi(\mathbf{\nabla}\cdot\mathbf{A}) \tag{2}$$

$$\mathbf{\nabla}\times(\phi\mathbf{A}) = (\mathbf{\nabla}\phi)\times\mathbf{A} + \phi(\mathbf{\nabla}\times\mathbf{A}) \tag{3}$$

$$\mathbf{\nabla}\cdot(\mathbf{A}\times\mathbf{B}) = \mathbf{B}\cdot(\mathbf{\nabla}\times\mathbf{A}) - \mathbf{A}\cdot(\mathbf{\nabla}\times\mathbf{B}) \tag{4}$$

$$\mathbf{\nabla}\times(\mathbf{A}\times\mathbf{B}) = \mathbf{A}(\mathbf{\nabla}\cdot\mathbf{B}) - (\mathbf{A}\cdot\mathbf{\nabla})\mathbf{B} - \mathbf{B}(\mathbf{\nabla}\cdot\mathbf{A}) + (\mathbf{B}\cdot\mathbf{\nabla})\mathbf{A} \tag{5}$$

$$\mathbf{\nabla}(\mathbf{A}\cdot\mathbf{B}) = (\mathbf{A}\cdot\mathbf{\nabla})\mathbf{B} + \mathbf{A}\times(\mathbf{\nabla}\times\mathbf{B}) + (\mathbf{B}\cdot\mathbf{\nabla})\mathbf{A} + \mathbf{B}\times(\mathbf{\nabla}\times\mathbf{A}) \tag{6}$$

$$\mathbf{\nabla}(\mathbf{A}\cdot\mathbf{B}) = \mathbf{A}(\mathbf{\nabla}\cdot\mathbf{B}) + (\mathbf{A}\times\mathbf{\nabla})\times\mathbf{B} + \mathbf{B}(\mathbf{\nabla}\cdot\mathbf{A}) + (\mathbf{B}\times\mathbf{\nabla})\times\mathbf{A} \tag{7}$$

$$\mathbf{\nabla}\times(\mathbf{\nabla}\phi) = 0 \tag{8}$$

$$\mathbf{\nabla}\cdot(\mathbf{\nabla}\times\mathbf{A}) = 0 \tag{9}$$

$$\mathbf{\nabla}\times(\mathbf{\nabla}\times\mathbf{A}) = \mathbf{\nabla}(\mathbf{\nabla}\cdot\mathbf{A}) - \nabla^2\mathbf{A} \tag{10}$$

1 Gauss's theorems

$$\int_V (\mathbf{\nabla}\cdot\mathbf{A})\,dV = \oint_S d\mathbf{S}\cdot\mathbf{A} \tag{11}$$

$$\int_V \mathbf{\nabla}\phi\,dV = \oint_S \phi\,d\mathbf{S} \tag{12}$$

$$\int_V (\mathbf{\nabla}\times\mathbf{A})\,dV = \oint_S d\mathbf{S}\times\mathbf{A} \tag{13}$$

619

2   Stokes's theorems

$$\int_S d\mathbf{S} \cdot (\nabla \times \mathbf{A}) = \oint_C d\mathbf{r} \cdot \mathbf{A} \tag{14}$$

$$\int_S d\mathbf{S} \times \nabla\phi = \oint_C \phi \, d\mathbf{r} \tag{15}$$

$$\int_S (d\mathbf{S} \times \nabla) \times \mathbf{A} = \oint_C d\mathbf{r} \times \mathbf{A} \tag{16}$$

3   Green's theorems

$$\int_V (\phi\nabla^2\psi - \psi\nabla^2\phi) \, dV = \oint_S d\mathbf{S} \cdot (\phi\nabla\psi - \psi\nabla\phi) \tag{17}$$

$$\int_V (\nabla\phi \times \nabla\psi) \, dV = \tfrac{1}{2}\oint_S d\mathbf{S} \times (\phi\nabla\psi - \psi\nabla\phi) \tag{18}$$

$$\int_S d\mathbf{S} \cdot (\nabla\phi \times \nabla\psi) = \tfrac{1}{2}\oint_C d\mathbf{r} \cdot (\phi\nabla\psi - \psi\nabla\phi) \tag{19}$$

DIFFERENTIATION OF VECTORS

We begin by considering vectors $\mathbf{A}(t)$ and $\mathbf{B}(t)$, which are explicit functions of time. The time derivative as for scalars is defined in the limit as Δt approaches zero:

$$\frac{d\mathbf{A}}{dt} = \lim \frac{\mathbf{A}(t + \Delta t) - \mathbf{A}(t)}{\Delta t} \tag{20}$$

The following relations follow from this definition:

$$\frac{d}{dt}\,\phi\mathbf{A} = \frac{d\phi}{dt}\,\mathbf{A} + \phi\,\frac{d\mathbf{A}}{dt} \tag{21}$$

$$\frac{d}{dt}\,\mathbf{A} \cdot \mathbf{B} = \frac{d\mathbf{A}}{dt} \cdot \mathbf{B} + \mathbf{A} \cdot \frac{d\mathbf{B}}{dt} \tag{22}$$

$$\frac{d}{dt}\,\mathbf{A} \times \mathbf{B} = \frac{d\mathbf{A}}{dt} \times \mathbf{B} + \mathbf{A} \times \frac{d\mathbf{B}}{dt} \tag{23}$$

Since the above relations may be defined without respect to a particular coordinate system, it is clear that they must be invariant under coordinate transformation.

We next consider scalar and vector functions of position, where we may imagine that the position $\mathbf{r}$ is an implicit function of the time $\mathbf{r} = \mathbf{r}(t)$. We first consider $\phi[\mathbf{r}(t)]$

and we wish to develop an expression for $d\phi/dt$, which we define in the limit as Δt goes to zero:

$$\frac{d\phi}{dt} = \lim \frac{\phi[\mathbf{r}(t + \Delta t)] - \phi[\mathbf{r}(t)]}{\Delta t} \tag{24}$$

4 Gradient of a Scalar.

We may write **(24)** in the form of the chain rule through the introduction of a vector called the gradient of ϕ, or **grad** ϕ. This vector is defined with reference to Figure 1*a* where we show surfaces of constant ϕ together with a contour which is everywhere orthogonal to these surfaces. The gradient of ϕ is *defined* in the limit that $\Delta \mathbf{r}$ goes to zero:

$$\mathbf{grad}\ \phi = \lim \frac{\phi(\mathbf{r} + \Delta \mathbf{r}) - \phi(\mathbf{r})}{\hat{\mathbf{n}} \cdot \Delta \mathbf{r}} \hat{\mathbf{n}} \tag{25}$$

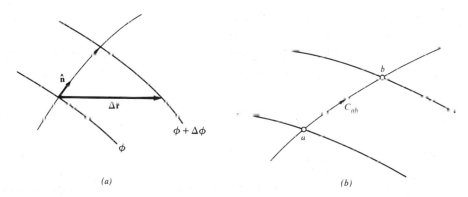

(a)   (b)

Figure 1 Gradient of a scalar. (*a*) Surfaces of constant ϕ are shown together with a contour that is everywhere orthogonal to these surfaces. The gradient is directed along this contour. (*b*) The difference in ϕ at points *a* and *b* may be obtained by integrating the gradient over an arbitrary contour connecting the two points as given by **(29)**.

where $\hat{\mathbf{n}}$ is a unit vector orthogonal to the surface of constant ϕ. If we write $\mathbf{r}(t + \Delta t)$ as $\mathbf{r} + \Delta \mathbf{r}$, we may rewrite **(24)** as:

$$\frac{d\phi}{dt} = \lim \frac{\phi(\mathbf{r} + \Delta \mathbf{r}) - \phi(\mathbf{r})}{\hat{\mathbf{n}} \cdot \Delta \mathbf{r}} \hat{\mathbf{n}} \cdot \frac{\Delta \mathbf{r}}{\Delta t} = \mathbf{v} \cdot \mathbf{grad}\ \phi \tag{26}$$

where $\mathbf{v} = \lim \Delta \mathbf{r}/\Delta t$. From the definitions that we have given, it is clear that **grad** ϕ is a vector that is invariant under coordinate transformation.

Although **(24)** is expressed as a derivative with respect to time, in fact any parameter will do, making it possible to write **(26)** as a relation between differentials:

$$d\phi = d\mathbf{r} \cdot \mathbf{grad}\ \phi \tag{27}$$

Integrating over any contour C_{ab} as shown in Figure 1b we have

$$\phi_a = \phi_b + \int_{C_{ab}} d\phi \tag{28}$$

Substituting from (27) for $d\phi$ we may write:

$$\phi_a = \phi_b + \int_{C_{ab}} d\mathbf{r} \cdot \mathbf{grad}\ \phi \tag{29}$$

Exercise

1. Show that the direction of **grad** ϕ is that of the maximum rate of change of ϕ and that the magnitude of **grad** ϕ is the value of the maximum rate of change.

5 Gradient Operator in Rectangular Coordinates. We may extend the interpretation of the gradient by writing (27) in a specific coordinate representation. Although we may consider any set of coordinates, the discussion is considerably simplified by taking rectangular coordinates x, y, and z. For any function $\phi(x,y,z)$ we may write:

$$d\phi = \frac{\partial \phi}{\partial x}\, dx + \frac{\partial \phi}{\partial y}\, dy + \frac{\partial \phi}{\partial z}\, dz \tag{30}$$

Writing $d\mathbf{r}$ as

$$d\mathbf{r} = \hat{\mathbf{x}}\, dx + \hat{\mathbf{y}}\, dy + \hat{\mathbf{z}}\, dz \tag{31}$$

We see that (27) and (30) are equivalent if

$$\blacktriangleright \qquad\qquad \mathbf{grad}\ \phi = \hat{\mathbf{x}}\, \frac{\partial \phi}{\partial x} + \hat{\mathbf{y}}\, \frac{\partial \phi}{\partial y} + \hat{\mathbf{z}}\, \frac{\partial \phi}{\partial z} \tag{32}$$

We may regard (32) as defining a vector differential operator, which is called *del*, and is defined as

$$\nabla = \hat{\mathbf{x}}\, \frac{\partial}{\partial x} + \hat{\mathbf{y}}\, \frac{\partial}{\partial y} + \hat{\mathbf{z}}\, \frac{\partial}{\partial z} \tag{33}$$

We may now use the definition of the gradient—or del—operator to write in compact form the expression for the differential of a vector. We imagine that the vector **A** is an explicit function of x, y, and z. Then we have:

$$d\mathbf{A} = dx\, \frac{\partial \mathbf{A}}{\partial x} + dy\, \frac{\partial \mathbf{A}}{\partial y} + dz\, \frac{\partial \mathbf{A}}{\partial z} \tag{34}$$

From **(31)** and **(32)** we may write **(34)** as:

▶
$$dA = (dr \cdot \nabla)A \tag{35}$$

The gradient operator may be defined in an analogous way for curvilinear coordinates. We must recognize, however, that the gradient operates on the unit coordinate vectors as well as on their coefficients. For this reason the decomposition of **(35)** has a more complex form in curvilinear than in rectangular coordinates.

Exercises

2. Let the potential ϕ be expressed in terms of the relative position with respect to $\mathbf{r}_1$:

$$\phi = \phi(\mathbf{r}_{01})$$

with $\mathbf{r}_{01} = \mathbf{r} - \mathbf{r}_1$. Introducing the gradient operator with respect to $\mathbf{r}_1$:

$$\nabla_1 = \hat{\mathbf{x}}\frac{\partial}{\partial x_1} + \hat{\mathbf{y}}\frac{\partial}{\partial y_1} + \hat{\mathbf{z}}\frac{\partial}{\partial z_1}$$

obtain the identity:

$$\nabla\phi = -\nabla_1\phi$$

3. For Λ a fixed vector, obtain the identity:

$$\nabla(\mathbf{r} \cdot \mathbf{A}) = (\mathbf{A} \cdot \nabla)\mathbf{r} = \mathbf{A}$$

4. Obtain the result:

$$\nabla\frac{1}{r} = -\frac{\hat{\mathbf{r}}}{r^2}$$

5. Obtain the relation:

$$\hat{\mathbf{r}} = \nabla r$$

INTEGRAL RELATIONS

6 **Flux and Divergence.** We represent an increment of surface of area dS and normal $\hat{\mathbf{n}}$ by a vector $d\mathbf{S} = \hat{\mathbf{n}}\, dS$. The differential flux of some vector $\mathbf{A}$ through the surface dS is defined and written as

$$d\Phi = \mathbf{A} \cdot d\mathbf{S} \tag{36}$$

The total flux through a surface is an invariant and is written as:

$$\Phi = \int_S \mathbf{A} \cdot d\mathbf{S} \tag{37}$$

We now consider the flux through a surface that entirely surrounds a point $\mathbf{r}$ as shown in Figure 2. We further imagine that the total volume V is divided into cells of volume ΔV_i. Then, since the fluxes across common internal boundaries cancel, we may write for the total flux through the *closed* surface S;

$$\Phi = \sum_i \Delta\Phi_i = \sum_i \frac{\Delta\Phi_i}{\Delta V_i} \Delta V_i \tag{38}$$

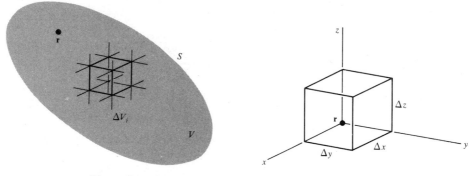

Figure 2   **Figure 3**

Figure 2   Flux. A surface S encloses a point $\mathbf{r}$. The included volume V may be divided into cells ΔV_i.

Figure 3   Divergence. The divergence is the flux per unit volume for an incremental volume.

We define the divergence of $\mathbf{A}$ at any point within V as

▶
$$\text{div } \mathbf{A} = \lim \frac{\Delta\Phi}{\Delta V} \tag{39}$$

Finally, we may write for the total flux by equating (37) and (38):

▶
$$\Phi = \oint_S \mathbf{A} \cdot d\mathbf{S} = \int_V \text{div } \mathbf{A}\, dV \tag{40}$$

The equation (40) is called Gauss' theorem and requires only that the limit expressed by (39) exist.

We now show how div $\mathbf{A}$ may be written explicitly in terms of the gradient operator. Let $\mathbf{A}$ be a function of x, y, and z. We expand $\mathbf{A}$ about $\mathbf{r}$ as

$$\mathbf{A} = \mathbf{A}(\mathbf{r}) + (\Delta\mathbf{r} \cdot \nabla)\mathbf{A} \tag{41}$$

from **(35)**. Integrating over the surface shown in Figure 3 we obtain by **(37)** for the flux from ΔV:

$$\Delta \Phi = \left(\frac{\partial A_x}{\partial x}\,\Delta x\right)\Delta y\,\Delta z + \left(\frac{\partial A_y}{\partial y}\,\Delta y\right)\Delta x\,\Delta z + \left(\frac{\partial A_z}{\partial z}\,\Delta z\right)\Delta x\,\Delta y \tag{42}$$

or by **(39)**:

$$\text{div } \mathbf{A} = \lim \frac{\Delta \Phi}{\Delta V} = \frac{\partial A_x}{\partial x} + \frac{\partial A_y}{\partial y} + \frac{\partial A_z}{\partial z} \tag{43}$$

Note from the definition **(33)** of the gradient operator that we may write the divergence of **A** as:

▶
$$\text{div } \mathbf{A} = \mathbf{V} \cdot \mathbf{A} \tag{44}$$

Similarly, the divergence may be written in curvilinear coordinates as the scalar product with the gradient operator, remembering that differential operations are performed *before* taking the scalar products. Using **(44)**, we may write Gauss's theorem as **(11)**.

Exercise

6. By applying Gauss's theorem to the vector **r**, show that the volume bounded by an open surface S, and the conical surface formed by tracing **r** around the contour C, which bounds S, is given by

$$V = \frac{1}{3}\int_S \mathbf{r} \cdot d\mathbf{S}$$

7 **Gradient Theorem.** An alternative definition of the gradient of a scalar may be made by introducing a vector function $\mathbf{\Lambda}$, which is the integral of the potential over a *closed* surface:

$$\mathbf{\Lambda} = \oint_S \phi\, d\mathbf{S} \tag{45}$$

where S surrounds a volume V as shown in Figure 2. We may now regard the volume V as divided into cells of volume ΔV_i, each of which is surrounded by a surface ΔS_i. Since the contributions to **(45)** from internal surfaces cancel, we may write **(45)** as a sum over cells:

$$\mathbf{\Lambda} = \sum_i \Delta \mathbf{\Lambda}_i \tag{46}$$

where $\Delta\Lambda_i$ is defined as in **(45)**. We define the gradient of ϕ at any point within V as the limit as ΔV goes to zero:

$$\textbf{grad } \phi = \lim \frac{\Delta\Lambda}{\Delta V} \tag{47}$$

We must now demonstrate the equivalence of **(32)** and **(47)**. For the cell shown in Figure 3 we have:

$$\Delta\Lambda = \left[\left(\phi + \frac{\partial\phi}{\partial x}\Delta x\right) - \phi\right]\hat{\mathbf{x}}\,\Delta y\,\Delta z + \left[\left(\phi + \frac{\partial\phi}{\partial y}\Delta y - \phi\right)\right]\hat{\mathbf{y}}\,\Delta z\,\Delta x$$
$$+ \left[\left(\phi + \frac{\partial\phi}{\partial z}\Delta z - \phi\right)\right]\hat{\mathbf{z}}\,\Delta x\,\Delta y \tag{48}$$

The increment of volume is given by $\Delta V = \Delta x\,\Delta y\,\Delta z$. Substituting into **(47)** and taking the limit we obtain **(32)**, which establishes the equivalence.

From **(46)** and **(47)** we have:

$$\Lambda = \int \textbf{grad }\phi\, dV \tag{49}$$

Equating **(45)** and **(49)** yields **(12)**.

Exercises

7. Let ϕ represent the negative of a hydrostatic pressure p. Interpret **(12)** as giving two equivalent descriptions of the force on a volume V.
8. If S is a closed surface that bounds V, prove the relation:

$$\oint_S d\mathbf{S} = 0$$

8 Circulation and Curl. The circulation of a vector $\mathbf{A}$ around a closed contour C is defined and written as

$$\blacktriangleright \qquad\qquad \Gamma = \oint_C \mathbf{A}\cdot d\mathbf{r} \tag{50}$$

and is an invariant. Dividing the surface S, which is bounded by the contour C into increments ΔS_i, we may write (Figure 4):

$$\Gamma = \sum_i \Gamma_i \tag{51}$$

since the contributions from internal boundaries cancel.

We introduce the vector **curl A** (which is sometimes given the symbol Ω and called the vorticity of **A**) through the relation

$$\hat{\mathbf{n}} \cdot \mathbf{curl\ A} = \text{limit}\ \frac{\Delta\Gamma}{\Delta S} \tag{52}$$

where $\hat{\mathbf{n}}$ is the normal to the limiting ΔS. From **(51)** and **(52)** we obtain:

▶ $$\Gamma = \int_S \hat{\mathbf{n}} \cdot \mathbf{curl\ A}\ dS = \int_S \mathbf{curl\ A} \cdot d\mathbf{S} \tag{53}$$

Equating **(50)** and **(53)** we obtain Stokes's theorem:

$$\Gamma = \oint_C \mathbf{A} \cdot d\mathbf{r} = \int_S \mathbf{curl\ A} \cdot d\mathbf{S} \tag{54}$$

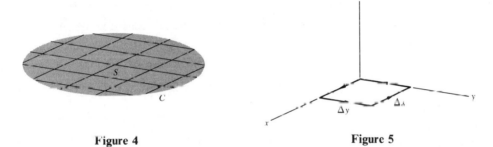

Figure 4   **Figure 5**

Figure 4 Circulation. The surface S bounded by the contour C may be divided into increments.
Figure 5 Curl. The normal component of the curl is the circulation per unit area of an incremental surface.

We now show how **curl A** may be written explicitly in terms of the del operator. Let **A** be a function of x, y, and z. We expand **A** about **r** as in **(41)**. Integrating over the contour shown in Figure 5 we obtain:

$$\Delta\Gamma_z = \left(\frac{\partial A_y}{\partial x}\,\Delta x\right)\Delta y - \left(\frac{\partial A_x}{\partial y}\,\Delta y\right)\Delta x \tag{55}$$

From **(52)** we have:

$$\hat{\mathbf{z}} \cdot \mathbf{curl\ A} = \lim \frac{\Delta\Gamma_z}{\Delta S} = \frac{\partial A_y}{\partial x} - \frac{\partial A_x}{\partial y} \tag{56}$$

In a similar way we may take contours in the yz and zx planes to obtain:

$$\mathbf{curl\ A} = \hat{\mathbf{x}}\left(\frac{\partial A_z}{\partial y} - \frac{\partial A_y}{\partial z}\right) + \hat{\mathbf{y}}\left(\frac{\partial A_x}{\partial z} - \frac{\partial A_z}{\partial x}\right) + \hat{\mathbf{z}}\left(\frac{\partial A_y}{\partial x} - \frac{\partial A_x}{\partial y}\right) \tag{57}$$

Finally, from **(33)**, we may write **curl A** as:

$$\mathbf{curl\ A} = \nabla \times \mathbf{A} \tag{58}$$

The sign of Γ depends on convention concerning the positive direction for contour integration. This is taken as shown in Figure 5 for a right-handed space. Under coordinate inversion, the space would become "left handed," reversing the sign of Γ. For this reason Γ is called a pseudoscalar and **curl A** is an axial vector. The appearance of the cross product in **(58)** is thus to be expected. Using **(58)** we write Stokes's theorem **(54)** as **(14)**.

In curvilinear coordinates we may use **(52)** to express the curl. Alternatively, we may continue to use **(58)** as long as we remember that

1. The gradient operates on the unit coordinate vectors as well as on their coefficients.

2. The differential operations are to be performed *before* the vector cross products are taken.

An alternative way of defining the *curl* of a vector follows from the introduction of the vector function:

$$\mathbf{\Omega} = \oint_S d\mathbf{S} \times \mathbf{A} \tag{59}$$

where S is a closed surface.

Again, we may divide V into incremental volumes ΔV_i bounded by surfaces ΔS_i as in Figure 2. Contributions to $\mathbf{\Omega}$ from interior surfaces will again cancel so that we may write:

$$\mathbf{\Omega} = \sum_i \Delta\mathbf{\Omega}_i \tag{60}$$

Now, we define **curl A** at a point by the relation:

$$\blacktriangleright \qquad \mathbf{curl\ A} = \lim \frac{\Delta\mathbf{\Omega}}{\Delta V} \tag{61}$$

From **(61)** we may write:

$$\blacktriangleright \qquad \mathbf{\Omega} = \int_V \mathbf{curl\ A}\ dV \tag{62}$$

Equating **(59)** and **(62)** we have:

$$\int_V \mathbf{curl\ A}\ dV = \oint_S d\mathbf{S} \times \mathbf{A} \tag{63}$$

Employing **(58)** we obtain **(13)**.

Exercises

---

9. Prove the identity:

$$(\mathbf{A} \cdot \nabla)\mathbf{A} = \tfrac{1}{2}\nabla A^2 - \mathbf{A} \times (\nabla \times \mathbf{A})$$

10. Obtain the integral theorem:

$$\oint_C \mathbf{r} \times d\mathbf{r} = 2 \int_S d\mathbf{S}$$

   Use Stokes's theorem with $\mathbf{A} = \mathbf{C} \times \mathbf{r}$ where $\mathbf{C}$ is a constant vector.

11. Use Stokes's theorem to prove $\nabla \times (\nabla\phi) = 0$.

12. Use Gauss's and Stokes's theorems to prove $\nabla \cdot (\nabla \times \mathbf{A}) = 0$.

13. Obtain the relation (15):

$$\oint_C \phi \, d\mathbf{r} = \int_S d\mathbf{S} \times \nabla\phi$$

   Use Stokes's theorem with $\mathbf{A} = \phi\mathbf{C}$ where $\mathbf{C}$ is a constant vector.

14. Show that **curl A** may be written formally in rectangular coordinates as the determinant of the matrix:

$$\mathbf{M} = \begin{pmatrix} \hat{\mathbf{x}} & \hat{\mathbf{y}} & \hat{\mathbf{z}} \\ \dfrac{\partial}{\partial x} & \dfrac{\partial}{\partial y} & \dfrac{\partial}{\partial z} \\ A_x & A_y & A_z \end{pmatrix}$$

15. Prove the relation

$$\nabla \times \hat{\mathbf{r}} \equiv 0$$

16. Verify (5). Expand the triple product assuming **B** constant and add to this the triple product expansion with **A** assumed constant.

17. From (54) obtain the relation:

$$\oint_C \mathbf{A} \cdot d\mathbf{r} = \int_S (d\mathbf{S} \times \nabla) \cdot \mathbf{A}$$

18. From (54) obtain the identity (16):

$$\oint_C d\mathbf{r} \times \mathbf{A} = \int_S (d\mathbf{S} \times \nabla) \times \mathbf{A}$$

   by replacing **A** by $\mathbf{A} \times \mathbf{C}$ where C is a constant vector.

19. Obtain **(15)** from **(54)** by replacing **A** by **A**ϕ where **A** is a constant vector.
20. Demonstrate the equivalence of **(52)** and **(61)**.
21. For S a closed surface that bounds V, prove the following:

$$\oint_S \mathbf{r} \times d\mathbf{S} = 0$$

22. Obtain **(63)** from **(40)** by replacing **A** by **A** $\times$ **C** where **C** is a constant vector.

---

9   Combinations of Gradient, Curl, and Divergence. Operating on a scalar ϕ we may take the divergence of the gradient of ϕ or the **curl** of the gradient of ϕ:
    1. The first is called the Laplacian of ϕ and is written as:

$$\blacktriangleright \qquad \nabla^2\phi = \text{div } \mathbf{grad }\phi = \mathbf{V}\cdot(\nabla\phi) = (\nabla\cdot\nabla)\phi \tag{64}$$

    2. The second is identically zero as may be seen from the following:

$$\blacktriangleright \qquad \mathbf{curl\ grad }\phi = \nabla\times(\nabla\phi) = (\nabla\times\nabla)\phi \equiv 0 \tag{65}$$

because the operator $(\nabla\times\nabla)$ is identically zero.
    Operating on a vector, we may take the gradient of the divergence of **A**, the divergence of the **curl** of **A**, or the **curl** of the **curl** of **A**:
    1. The first is written as:

$$\mathbf{grad} \text{ div } \mathbf{A} = \nabla(\nabla\cdot\mathbf{A}) \tag{66}$$

    By careful manipulation of the del operator we may show that the second is zero and write the third in terms of previously defined derivatives:
    2. We write the divergence of the **curl** of **A** as:

$$\blacktriangleright \qquad \text{div }\mathbf{curl\ A} = \nabla\cdot(\nabla\times\mathbf{A}) = (\nabla\times\nabla)\cdot\mathbf{A} \equiv 0 \tag{67}$$

where we have interchanged the cross and dot product. Since the cross product must be taken before the dot product, the operator is identically zero as in **(65)**.
    3. We may expand **curl (curl A)** as in **(A-51)** to obtain:

$$\blacktriangleright \qquad \nabla\times(\nabla\times\mathbf{A}) = \nabla(\nabla\cdot\mathbf{A}) - (\nabla\cdot\nabla)\mathbf{A} = \nabla(\nabla\cdot\mathbf{A}) - \nabla^2\mathbf{A} \tag{68}$$

    Particular care must be exercised when **(68)** is applied in curvilinear coordinates. This is because the Laplacian ∇^2 operates not only on the components of **A** along coordinate directions but on the unit coordinate vectors as well. Although it is occasionally suggested that **(68)** be regarded as the definition of $\nabla^2\mathbf{A}$ in curvilinear coordinates, **(10)** stands by itself with the Laplacian understood as the divergence of the gradient.

Exercises

23. Prove the identity

$$\nabla^2(\phi\psi) = \phi\nabla^2\psi + \psi\nabla^2\phi + 2\nabla\phi \cdot \nabla\psi$$

24. Given the vector

$$A(r) = Ae^{ik \cdot r}$$

where **A** and **k** are constant vectors, show the following:

$$\nabla \cdot A(r) = ik \cdot Ae^{ik \cdot r}$$
$$\nabla \times A(r) = ik \times Ae^{ik \cdot r}$$
$$\nabla \times [\nabla \times A(r)] = -k \times (k \times A)e^{ik \cdot r}$$
$$\nabla[\nabla \cdot A(r)] = -k(k \cdot A)e^{ik \cdot r}$$
$$\nabla^2 A(r) = \quad k^2 Ae^{ik \cdot r}$$

25. Show that if a vector $A(r)$ is proportional to its curl:

$$\nabla \times A(r) = kA(r)$$

then $A(r)$ must be a solution of the differential equation

$$\nabla^2 A(r) = -k^2 A(r)$$

[See A. Barnes, *Am. J. Phys.* **45**, 371 (1977) for application to force-free fields.]

26. Use Gauss's theorem and Stokes's theorem to establish **(67)**. Divide the closed surface S into two open surfaces sharing a common contour C.

27. Use Stokes's theorem to establish **(65)**.

28. Obtain the identity:

$$\nabla \times [(A \cdot \nabla)A] = (A \cdot \nabla)(\nabla \times A) + (\nabla \cdot A)(\nabla \times A) - [(\nabla \times A) \cdot \nabla]A$$

Use **(3)** and then **(2)**.

29. Obtain the identity:

$$\nabla \times (C \times \nabla\phi) = -(C \times \nabla) \times \nabla\phi$$

where **C** is a constant vector.

10 Laplacian of $1/r$. We imagine a velocity field $v(r)$, which has a source at the origin from which the flow is spherically symmetric. Then the flux of an incompressible fluid is given from **(37)** by:

$$\Phi_0 = 4\pi r^2 v(r) \tag{69}$$

and the velocity must then be given by:

$$\mathbf{v}(\mathbf{r}) = \frac{\Phi_0}{4\pi r^2}\,\hat{\mathbf{r}} \tag{70}$$

The divergence of $\mathbf{v}(\mathbf{r})$, which is the flux per unit volume, is zero except at the origin where, as may be seen from the following argument, it has an integrable singularity. From **(40)** we write

$$\Phi = \int_S \mathbf{v} \cdot d\mathbf{S} = \int_V \boldsymbol{\nabla} \cdot \mathbf{v}\, dV \tag{71}$$

where the integral is equal to Φ_0 if V includes the origin and zero otherwise. The Dirac delta function has this same property. That is, the integral

$$\int \delta(\mathbf{r})\, dV \tag{72}$$

is equal to unity if V includes the origin and zero otherwise. We may then write:

$$\boldsymbol{\nabla} \cdot \mathbf{v} = \Phi_0\, \delta(\mathbf{r}) \tag{73}$$

We next write the velocity field **(70)** as a constant times the gradient of $1/r$:

$$\mathbf{v}(\mathbf{r}) = -\frac{\Phi_0}{4\pi}\boldsymbol{\nabla}\frac{1}{r} \tag{74}$$

Finally, by substituting **(74)** into **(73)** we obtain:

$$\nabla^2\frac{1}{r} = -4\pi\delta(\mathbf{r}) \tag{75}$$

CURVILINEAR COORDINATES

Rectangular coordinates have the special property that the orientation of the unit vectors $\hat{\mathbf{x}}$, $\hat{\mathbf{y}}$, and $\hat{\mathbf{z}}$ is independent of the coordinates x, y, and z as shown in Figure 6a. As we have seen, we may decompose a vector according to **(A-1)**:

$$\mathbf{A} = A_x\hat{\mathbf{x}} + A_y\hat{\mathbf{y}} + A_z\hat{\mathbf{z}} \tag{76}$$

where the coefficients of $\hat{\mathbf{x}}$, $\hat{\mathbf{y}}$, and $\hat{\mathbf{z}}$ are given by

$$A_x = \mathbf{A} \cdot \hat{\mathbf{x}} \qquad A_y = \mathbf{A} \cdot \hat{\mathbf{y}} \qquad A_z = \mathbf{A} \cdot \hat{\mathbf{z}} \tag{77}$$

The divergence of **A**, for example, may be written as the scalar product with the gradient operator:

$$\text{div }\mathbf{A} = \nabla \cdot \mathbf{A} = \left(\hat{\mathbf{x}}\frac{\partial}{\partial x} + \hat{\mathbf{y}}\frac{\partial}{\partial y} + \hat{\mathbf{z}}\frac{\partial}{\partial z}\right)\cdot\mathbf{A} = \frac{\partial A_x}{\partial x} + \frac{\partial A_y}{\partial y} + \frac{\partial A_z}{\partial z} \tag{78}$$

and it does not matter in which order we perform the differentiation and the scalar product.

By contrast, if we decompose a vector in curvilinear coordinates, as shown in Figure 6b for cylindrical coordinates:

$$\mathbf{A} = A_\rho\hat{\boldsymbol{\rho}} + A_\phi\hat{\boldsymbol{\phi}} + A_z\hat{\mathbf{z}} \tag{79}$$

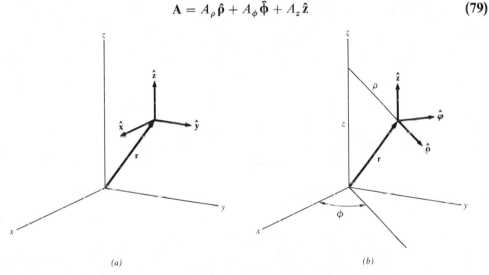

(a)　　　　　　　　　　(b)

Figure 6 Coordinate systems. (a) For rectangular coordinates, the orientation of the coordinate unit vectors is independent of the coordinate values. (b) For cylindrical coordinates, for example, the orientation of the coordinate unit vectors depends on the coordinate values.

the orientation of $\hat{\boldsymbol{\rho}}$ and $\hat{\boldsymbol{\phi}}$ are functions of ϕ. Although it is possible to develop general expressions for divergence and curl in curvilinear coordinates, we instead work explicitly with the two most commonly used curvilinear systems, cylindrical and spherical coordinates.

11 Cylindrical Coordinates. For a function $\psi(\rho,\phi,z)$ we write

$$d\psi = \frac{\partial\psi}{\partial\rho}\,d\rho + \frac{\partial\psi}{\partial\phi}\,d\phi + \frac{\partial\psi}{\partial z}\,dz \tag{80}$$

Writing an incremental displacement as

$$d\mathbf{r} = \hat{\boldsymbol{\rho}}\,d\rho + \hat{\boldsymbol{\phi}}\rho\,d\phi + \hat{\mathbf{z}}\,dz \tag{81}$$

we may write

$$d\psi = d\mathbf{r}\cdot\mathbf{grad}\,\psi \tag{82}$$

with

$$\mathbf{grad}\ \psi = \hat{\mathbf{\rho}}\,\frac{\partial\psi}{\partial\rho} + \hat{\mathbf{\phi}}\,\frac{1}{\rho}\frac{\partial\psi}{\partial\phi} + \hat{\mathbf{z}}\,\frac{\partial\psi}{\partial z} \tag{83}$$

where the gradient operator then takes the form:

$$\mathbf{\nabla} = \hat{\mathbf{\rho}}\,\frac{\partial}{\partial\rho} + \hat{\mathbf{\phi}}\,\frac{1}{\rho}\frac{\partial}{\partial\phi} + \hat{\mathbf{z}}\,\frac{\partial}{\partial z} \tag{84}$$

The increment of a vector may be written as:

$$d\mathbf{A} = d\rho\,\frac{\partial\mathbf{A}}{\partial\rho} + d\phi\,\frac{\partial\mathbf{A}}{\partial\phi} + dz\,\frac{\partial\mathbf{A}}{\partial z} = (d\mathbf{r}\cdot\mathbf{\nabla})\mathbf{A} \tag{85}$$

If we wish to write the decomposition of (85) we must make use of the relations[1]

$$\frac{\partial\hat{\mathbf{\rho}}}{\partial\rho} = 0 \qquad \frac{\partial\hat{\mathbf{\phi}}}{\partial\rho} = 0 \qquad \frac{\partial\hat{\mathbf{z}}}{\partial\rho} = 0$$

$$\frac{\partial\hat{\mathbf{\rho}}}{\partial\phi} = \hat{\mathbf{\phi}} \qquad \frac{\partial\hat{\mathbf{\phi}}}{\partial\phi} = -\hat{\mathbf{\rho}} \qquad \frac{\partial\hat{\mathbf{z}}}{\partial\phi} = 0 \tag{86}$$

$$\frac{\partial\hat{\mathbf{\rho}}}{\partial z} = 0 \qquad \frac{\partial\hat{\mathbf{\phi}}}{\partial z} = 0 \qquad \frac{\partial\hat{\mathbf{z}}}{\partial z} = 0$$

to obtain

$$
\begin{aligned}
d\mathbf{A} = \hat{\mathbf{\rho}}&\left(\frac{\partial A_\rho}{\partial\rho}\,d\rho + \frac{\partial A_\rho}{\partial\phi}\,d\phi + \frac{\partial A_\rho}{\partial z}\,dz - A_\phi\,d\phi\right) \\
+ \hat{\mathbf{\phi}}&\left(\frac{\partial A_\phi}{d\rho}\,d\rho + \frac{\partial A_\phi}{\partial\phi}\,d\phi + \frac{\partial A_\phi}{\partial z}\,dz + A_\rho\,d\phi\right) \\
+ \hat{\mathbf{z}}&\left(\frac{\partial A_z}{\partial\rho}\,d\rho + \frac{\partial A_z}{\partial\phi}\,d\phi + \frac{\partial A_z}{\partial z}\,dz\right)
\end{aligned}
\tag{87}
$$

To obtain the divergence of **A** we integrate the flux of **A** over the surface shown in Figure 7 to obtain:

$$\Delta\Phi = [(\rho + \Delta\rho)A_\rho(\rho + \Delta\rho) - \rho A_\rho(\rho)]\,\Delta\phi\,\Delta z + [A_\phi(\phi + \Delta\phi) - A_\phi(\phi)]\,\Delta\rho\,\Delta z$$

$$+ [A_z(z + \Delta z) - A_z(z)]\rho\,\Delta\phi\,\Delta\rho \quad \textbf{(88)}$$

---

[1] For a similar approach see P. D. Gupta, *Am. J. Phys.* **44**, 888 (1976); K. Srinivasan, *Am. J. Phys.* **45**, 767 (1977), V. Namias, *Am. J. Phys.* **45**, 773 (1977), D. P. Prato, *Am. J. Phys.* **45**, 1003 (1977).

We obtain for the divergence of **A**:

$$\mathbf{div}\ \mathbf{A} = \lim \frac{\Delta\Phi}{\Delta V} = \frac{1}{\rho}\frac{\partial}{\partial\rho}(\rho A_\rho) + \frac{1}{\rho}\frac{\partial}{\partial\phi}A_\phi + \frac{\partial}{\partial z}A_z \tag{89}$$

We next obtain the expression for the **curl** of **A** in cylindrical coordinates. With reference to Figure 7, we may write for the incremental circulation in planes normal to $\hat{\rho}$, $\hat{\phi}$, and $\hat{z}$:

$$\Delta\Gamma_\rho = \Delta z[A_z(\phi + \Delta\phi) - A_z(\phi)] - \rho\,\Delta\phi[A_\phi(z + \Delta z) - A_\phi(z)] \tag{90}$$

$$\Delta\Gamma_\phi = \Delta\rho[A_\rho(z + \Delta z) - A_\rho(z)] - \Delta z[A_z(\rho + \Delta\rho) - A_z(\rho)] \tag{91}$$

$$\Delta\Gamma_z = \Delta\phi[(\rho + \Delta\rho)A_\phi(\rho + \Delta\rho) - \rho A_\phi(\rho)] - \Delta\rho[A_\rho(\phi + \Delta\phi) - A_\rho(\phi)] \tag{92}$$

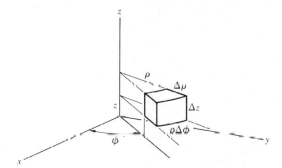

Figure 7 Cylindrical coordinates. The coordinate values and an incremental volume are as shown in cylindrical coordinates.

Substituting into **(52)** and taking the limit as $\Delta\rho$, $\Delta\phi$, and Δz go to zero, we obtain

$$\mathbf{curl}\ \mathbf{A} = \hat{\rho}\left(\frac{1}{\rho}\frac{\partial A_\phi}{\partial\phi} - \frac{\partial A_\phi}{\partial z}\right) + \hat{\phi}\left(\frac{\partial A_\rho}{\partial z} - \frac{\partial A_z}{\partial\rho}\right) + \frac{\hat{z}}{\rho}\left(\frac{\partial}{\partial\rho}\rho A_\phi - \frac{\partial A_\rho}{\partial\phi}\right) \tag{93}$$

Exercises

30. Demonstrate in cylindrical coordinates the relation

$$\mathrm{div}\ \mathbf{A} = \nabla \cdot \mathbf{A}$$

where the gradient operator is given by **(84)** and the differentiation is performed before the scalar product is taken.

31. Obtain the form of the acceleration $\mathbf{a} = d^2\mathbf{r}/dt^2$ in cylindrical coordinates.
32. Demonstrate in cylindrical coordinates the relation:

$$\mathrm{curl}\ \mathbf{A} = \nabla \times \mathbf{A}$$

where the gradient operator is given by **(84)** and it is understood that the differentiation is performed *before* the cross product is taken.

33. Show explicitly in cylindrical coordinates the identity:

$$\mathbf{V} \times \mathbf{V} \equiv 0$$

34. Obtain the expression for the Laplacian in cylindrical coordinates:

$$\nabla^2 \phi = \mathbf{V} \cdot (\mathbf{V}\phi) = \text{div } \mathbf{grad} \ \phi$$

12 Spherical Coordinates. We show in Figure 8 the unit vectors $\hat{\mathbf{r}}$, $\hat{\boldsymbol{\theta}}$, and $\hat{\boldsymbol{\phi}}$ in spherical coordinates. The incremental displacement has the form:

$$d\mathbf{r} = \hat{\mathbf{r}} \, dr + \hat{\boldsymbol{\theta}} r \, d\theta + \hat{\boldsymbol{\phi}} r \sin\theta \, d\phi \tag{94}$$

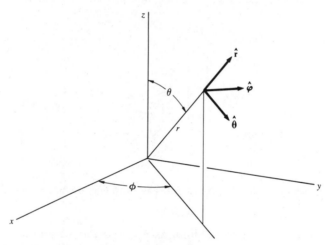

Figure 8 Spherical coordinates. The coordinate values and the unit vectors are as shown in spherical coordinates.

and the gradient operator may be written as

$$\mathbf{V} = \hat{\mathbf{r}} \frac{\partial}{\partial r} + \hat{\boldsymbol{\theta}} \frac{1}{r} \frac{\partial}{\partial \theta} + \hat{\boldsymbol{\phi}} \frac{1}{r \sin\theta} \frac{\partial}{\partial \phi} \tag{95}$$

Differentiation of the unit vectors with respect to the coordinates takes the form

$$\frac{\partial \hat{\mathbf{r}}}{\partial r} = 0 \qquad\qquad \frac{\partial \hat{\boldsymbol{\theta}}}{\partial r} = 0 \qquad\qquad \frac{\partial \hat{\boldsymbol{\phi}}}{\partial r} = 0$$

$$\frac{\partial \hat{\mathbf{r}}}{\partial \theta} = \hat{\boldsymbol{\theta}} \qquad\qquad \frac{\partial \hat{\boldsymbol{\theta}}}{\partial \theta} = -\hat{\mathbf{r}} \qquad\qquad \frac{\partial \hat{\boldsymbol{\phi}}}{\partial \theta} = 0 \tag{96}$$

$$\frac{\partial \hat{\mathbf{r}}}{\partial \phi} = \hat{\boldsymbol{\phi}} \sin\theta \qquad \frac{\partial \hat{\boldsymbol{\theta}}}{\partial \phi} = \hat{\boldsymbol{\phi}} \cos\theta \qquad \frac{\partial \hat{\boldsymbol{\phi}}}{\partial \phi} = -\hat{\mathbf{r}} \sin\theta - \hat{\boldsymbol{\theta}} \cos\theta$$

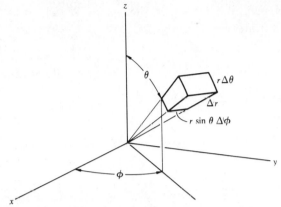

Figure 9 Spherical coordinates. The coordinate values and an incremental volume are as shown in spherical coordinates.

By integrating the flux of **A** over the spherical segment shown in Figure 9 and taking the limiting flux per unit volume we obtain:

$$\textbf{div A} = \nabla \cdot \textbf{A} = \frac{1}{r^2}\frac{\partial}{\partial r}\left(r^2 A_r\right) + \frac{1}{r\sin\theta}\frac{\partial}{\partial\theta}\left(\sin\theta A_\theta\right) + \frac{1}{r\sin\theta}\frac{\partial}{\partial\phi}A_\phi \qquad (97)$$

Integrating over contours normal to $\hat{\textbf{r}}$, $\boldsymbol{\theta}$, and $\hat{\boldsymbol{\phi}}$ and taking the limiting circulation per unit area we obtain for the curl in spherical coordinates:

$$\textbf{curl A} = \nabla \times \textbf{A} = \hat{\textbf{r}}\,\frac{1}{r\sin\theta}\left[\frac{\partial}{\partial\theta}\left(\sin\theta A_\phi\right) - \frac{\partial}{\partial\phi}A_\theta\right]$$
$$+ \boldsymbol{\theta}\,\frac{1}{r}\left[\frac{1}{\sin\theta}\frac{\partial}{\partial\phi}A_r - \frac{\partial}{\partial r}\left(rA_\phi\right)\right]$$
$$+ \hat{\boldsymbol{\phi}}\,\frac{1}{r}\left[\frac{\partial}{\partial r}\left(rA_\theta\right) - \frac{\partial}{\partial\theta}A_r\right] \qquad (98)$$

Exercises

35. Obtain the expression for the Laplacian in spherical coordinates:

$$\nabla^2\phi = \nabla \cdot (\nabla\phi) = \text{div } \textbf{grad } \phi$$

36. Obtain the form of the acceleration $\textbf{a} = d\textbf{r}/dt$ in spherical coordinates.

37. Show explicitly in spherical coordinates the identity:

$$\nabla \times \nabla \equiv 0$$

GREEN'S THEOREMS

We derive here an identity attributed to George Green, a nineteenth-century British mathematician. We also derive two other closely related identities, all of which we loosely call Green's theorems.

13 [**Gauss-Based Theorems**]. We introduce the vector **A**, which is an antisymmetric combination derived from two scalars ϕ and ψ:

$$\mathbf{A} = \phi\nabla\psi - \psi\nabla\phi \tag{99}$$

Substituting into Gauss's theorem (**11**) we obtain

$$\int_V (\phi\nabla^2\psi - \psi\nabla^2\phi)\, dV = \oint_S d\mathbf{S} \cdot (\phi\nabla\psi - \psi\nabla\phi) \tag{100}$$

where we have used the relation:

$$\nabla \cdot (\phi\nabla\psi - \psi\nabla\phi) = \phi\nabla^2\psi - \psi\nabla^2\phi \tag{101}$$

The equation (**100**) is usually called Green's second identity, the first identity being obtained with $\mathbf{A} = \phi\nabla\psi$.

By substituting (**99**) into (**13**) we obtain the identity:

$$\int_V (\nabla\phi \times \nabla\psi)\, dV = \tfrac{1}{2}\oint_S d\mathbf{S} \times (\phi\nabla\psi - \psi\nabla\phi) \tag{102}$$

where we have used the result:

$$\nabla \times (\phi\nabla\psi - \psi\nabla\phi) = 2\nabla\phi \times \nabla\psi \tag{103}$$

Exercise

38. Derive the vector Green's theorem:

$$\int_V \{\mathbf{A} \cdot [\nabla \times (\nabla \times \mathbf{B})] - \mathbf{B} \cdot [\nabla \times (\nabla \times \mathbf{A})]\}\, dV = \oint_S d\mathbf{S} \cdot [\mathbf{B} \times (\nabla \times \mathbf{A}) - \mathbf{A} \times (\nabla \times \mathbf{B})]$$

14 [**Stokes-Based Theorem**]. By substituting (**99**) into (**14**) we obtain the identity:

$$\int_S d\mathbf{S} \cdot (\nabla\phi \times \nabla\psi) = \tfrac{1}{2}\oint_C d\mathbf{r} \cdot (\phi\nabla\psi - \psi\nabla\phi) \tag{104}$$

and where we used (**103**).

HELMHOLTZ'S THEOREM

Helmholtz showed in 1858 in developing the theory of fluid mechanics that any vector function **W** whose divergence and **curl** have potentials may be expressed uniquely as the sum of an irrotational and solenoidal part as defined below.

15 Vector Decomposition. We write $\mathbf{W}$ as the sum of two vectors:

$$\mathbf{W} = \Lambda + \Omega \tag{105}$$

where Ω is taken to be solenoidal, which is equivalent to:

$$\operatorname{div} \Omega = 0 \tag{106}$$

and Λ is taken to be irrotational, which is equivalent to:

$$\operatorname{curl} \Lambda = 0 \tag{107}$$

Writing:

$$\Omega = \operatorname{curl} \mathbf{A} \tag{108}$$

where we assume for simplicity:

$$\operatorname{div} \mathbf{A} = 0 \tag{109}$$

and writing

$$\Lambda = -\operatorname{\mathbf{grad}} \phi \tag{110}$$

then **(106)** follows from **(108)** by **(67)**; and **(107)** follows from **(110)** through **(65)**. Note that **(109)** does not affect **(108)**.

From **(105)**, **(106)**, **(110)**, and **(64)** we obtain

$$\nabla \cdot \mathbf{W} = \nabla \cdot \Lambda = -\nabla^2 \phi \tag{111}$$

From **(105)**, **(107)**, **(108)**, and **(109)** with **(68)** we obtain

$$\nabla \times \mathbf{W} = \nabla \times \Omega = \nabla \times (\nabla \times \mathbf{A}) = -\nabla^2 \mathbf{A} \tag{112}$$

Now **(111)** and **(112)** may be regarded as differential equations for ϕ and $\mathbf{A}$. As we discuss in the text, these equations are called Poisson's equation and have a solution if the inhomogeneous part is specified *everywhere*. From this it follows that $\mathbf{A}$ and ϕ exist and that Ω and Λ, which are given by **(108)** and **(110)**, exist as well.

16 Uniqueness Theorem. As a corollary to Helmholtz's theorem we have the result that if the divergence and **curl** of a vector are given *everywhere*, the vector is determined to within a constant vector. This result follows from the *uniqueness* of the solution for $\mathbf{A}$ and ϕ from Poisson's equation and thus the uniqueness of Λ and Ω.

BOUNDARY CONDITIONS

If the sources of divergence and **curl** of a vector are given only within some limited region, V, then we must know something about the vector or its potentials on S in order to be able to determine the vector uniquely within V. We first look at the problem for irrotational vectors, then for solenoidal vectors, and then for a general vector field.

17 **[Irrotational Vectors].** We consider a vector Λ, which has the property *everywhere*:

$$\mathbf{V} \cdot \mathbf{\Lambda} = \rho \qquad \mathbf{V} \times \mathbf{\Lambda} = 0 \tag{113}$$

Let us imagine that we are given ρ only within some limited volume V as shown in Figure 10a and that we have obtained a vector field Λ which satisfies **(113)** within V. What additional conditions must be imposed on Λ to establish its uniqueness?

As a result of **(113)** we may write:

$$\mathbf{\Lambda} = -\mathbf{V}\phi \qquad \mathbf{V}^2\phi = -\rho \tag{114}$$

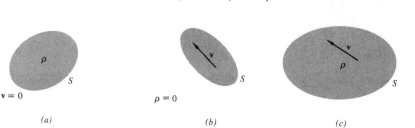

| (a) | (b) | (c) |

Figure 10   Boundary conditions. (*a*) An irrotational vector has sources of flux ρ, but its sources of circulation $\mathbf{v}$ are zero. (*b*) A solenoidal vector has sources of circulation $\mathbf{v}$, but its sources of flux ρ are zero. (*c*) A general vector has both sources of flux ρ and sources of circulation $\mathbf{v}$.

Now let us imagine that we have two solutions to **(113)** and **(114)** that we designate as ϕ_1, Λ_1 and ϕ_2, Λ_2. We form the vector difference:

$$\mathbf{D} = (\phi_1 - \phi_2)(\mathbf{\Lambda}_1 - \mathbf{\Lambda}_2) \tag{115}$$

Substituting into Gauss's theorem **(11)** we obtain:

$$\int_V \mathbf{V} \cdot \mathbf{D}\, dV = -\int_V (\mathbf{\Lambda}_1 - \mathbf{\Lambda}_2) \cdot (\mathbf{\Lambda}_1 - \mathbf{\Lambda}_2)\, dV = \oint_S (\phi_1 - \phi_2)(\mathbf{\Lambda}_1 - \mathbf{\Lambda}_2) \cdot d\mathbf{S} \tag{116}$$

where we have used **(2)** and **(114)**.

If we are able to specify ϕ on S, which is called the Dirichlet condition, the difference between ϕ_1 and ϕ_2 *must* be zero on S and the right side **(116)** vanishes. But since the left side is everywhere positive we must have Λ_1 and Λ_2 identical.

Alternatively, we may specify $\hat{\mathbf{n}} \cdot \mathbf{\Lambda}$ on S, which is called the Neumann condition. Similarly, the right side of **(116)** vanishes and Λ_1 and Λ_2 must be identical. We may, of course, specify ϕ on part of S and $\hat{\mathbf{n}} \cdot \mathbf{\Lambda}$ on the remaining part and this will determine Λ uniquely as well.

18 [Solenoidal Vectors]. We next consider a vector $\boldsymbol{\Omega}$ that has the property everywhere:

$$\mathbf{V} \cdot \boldsymbol{\Omega} = 0 \qquad \mathbf{V} \times \boldsymbol{\Omega} = \mathbf{v} \tag{117}$$

Let us imagine that we are given $\mathbf{v}$ only within some limited volume V as shown in Figure 10*b* and that we have obtained a vector field $\boldsymbol{\Omega}$, which satisfies **(117)** within V. What additional conditions must be imposed on $\boldsymbol{\Omega}$ to establish its uniqueness?

As a result of **(108)** we may write:

$$\boldsymbol{\Omega} = \mathbf{V} \times \mathbf{A} \qquad \nabla^2 \mathbf{A} = -\mathbf{v} \tag{118}$$

where we have used **(10)** and have the simplifying assumption $\mathbf{V} \cdot \mathbf{A} = 0$. We imagine that we have two solutions that we designate as $\boldsymbol{\Omega}_1$, $\mathbf{A}_1$ and $\boldsymbol{\Omega}_2$, $\mathbf{A}_2$. We form the vector

$$\mathbf{D} = (\mathbf{A}_1 - \mathbf{A}_2) \times (\boldsymbol{\Omega}_1 - \boldsymbol{\Omega}_2) \tag{119}$$

Substituting into Gauss's theorem we may write:

$$\int_V \mathbf{V} \cdot \mathbf{D} \, dV = \int_V (\boldsymbol{\Omega}_1 - \boldsymbol{\Omega}_2) \times (\boldsymbol{\Omega}_1 - \boldsymbol{\Omega}_2) \, dV = \oint_S (\mathbf{A}_1 - \mathbf{A}_2) \times (\boldsymbol{\Omega}_1 - \boldsymbol{\Omega}_2) \cdot d\mathbf{S} \tag{120}$$

The analog of the Dirichlet condition for a solenoidal vector is the specification of the tangential component of $\mathbf{A}$ on S. The analog of the Neumann condition is the specification of the tangential component of $\boldsymbol{\Omega}$. It is also possible to show that the specification of $(\hat{\mathbf{n}} \cdot \mathbf{V})\mathbf{A}$ on S establishes $\boldsymbol{\Omega}$ uniquely in V.

19 [General Vector Fields]. Finally, we consider the problem of a vector $\mathbf{W}$, which has the property:

$$\mathbf{V} \cdot \mathbf{W} = \rho \qquad \mathbf{V} \times \mathbf{W} = \mathbf{v} \tag{121}$$

As we have seen, if ρ and $\mathbf{v}$ are given everywhere, then $\mathbf{W}$ may be written uniquely as:

$$\mathbf{W} = \boldsymbol{\Lambda} + \boldsymbol{\Omega} \tag{122}$$

where **(113)** and **(117)** are satisfied. Let us imagine that ρ and $\mathbf{v}$ are given only within a limited volume. What properties of $\mathbf{W}$ must be specified on S in order to obtain $\mathbf{W}$ uniquely? We imagine again two sets of solutions $\mathbf{W}_1$, ϕ_1, $\mathbf{A}_1$ and $\mathbf{W}_2$, ϕ_2, $\mathbf{A}_2$ which satisfy **(121)**. We form the difference vector:

$$\mathbf{D} = (\mathbf{W}_1 - \mathbf{W}_2)(\phi_1 - \phi_2) + (\mathbf{W}_1 - \mathbf{W}_2) \times (\mathbf{A}_1 - \mathbf{A}_2) \tag{123}$$

Substituting into Gauss's theorem we obtain:

$$\int_V \mathbf{V} \cdot \mathbf{D} \, dV = -\int_V (\mathbf{W}_1 - \mathbf{W}_2) \cdot (\mathbf{W}_1 - \mathbf{W}_2) \, dV = \oint_S \mathbf{D} \cdot d\mathbf{S} \tag{124}$$

Now if **W** is specified on S, then **D** is zero on S and **W** is determined uniquely within V. Alternatively, if we know both ϕ and the tangential component of **A** on S, **W** is determined. But as **(124)** demonstrates, it is not necessary to know the nature of the sources of **W** outside V as long as we have both the normal and tangential components of **W** on S.

20 **[Connectivity]**. A problem that may arise is the following: let us imagine that **W** has sources of circulation **v** outside V but only sources of flux within V so that we have *within* V

$$\nabla \cdot \mathbf{W} = \rho \qquad \nabla \times \mathbf{W} = 0 \tag{125}$$

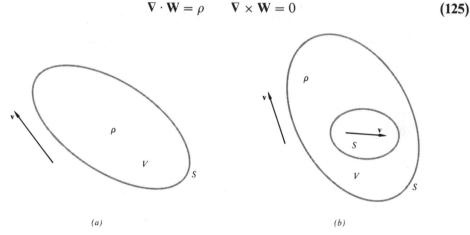

(a)   (b)

Figure 11 Connectivity. (*a*) A vector has sources of circulation **v** outside V but only sources of flux ρ within V and V is simply connected. (*b*) Again, the vector has only sources of flux ρ within V, but the volume is multiply connected.

as shown in Figure 11*a*. May we as a consequence of **(125)** write within V:

$$\mathbf{W} = \mathbf{\Lambda} \tag{126}$$

where $\mathbf{\Lambda}$ satisfies **(107)**.

The answer is that we may always write **(126)** if the volume is *simply connected*. If the volume is *multiply connected*, that is, if there are nonequivalent contours within V as shown in Figure 11*b*, and if **v** is nonzero in the holes, then we cannot assume **(126)** that **W** is irrotational within V as clearly, it is not.

Since we will know **W** on S but not the external currents, the question as to whether or not **W** is irrotational within V must be established in terms of the boundary conditions. In order to write **(126)** we must have

$$\oint_C \mathbf{W} \cdot d\mathbf{r} = 0 \tag{127}$$

for any closed contour *on* S. If **W** as given on S meets this condition, then we may assume **(126)**. If however, **W** does not satisfy **(127)** we must write **W** within V as **(122)**.

SUMMARY

1. Vectors and vector products may be differentiated with respect to a parameter such as the time in the same way as scalars and scalar products.

2. The differential of a scalar that is a function of position may be written as:

$$d\phi = d\mathbf{r} \cdot \mathbf{grad}\ \phi \tag{27}$$

3. The gradient of a scalar may be written in rectangular coordinates in terms of the del operator:

$$\mathbf{grad}\ \psi = \nabla\psi = \left(\hat{\mathbf{x}}\frac{\partial}{\partial x} + \hat{\mathbf{y}}\frac{\partial}{\partial y} + \hat{\mathbf{z}}\frac{\partial}{\partial z} \right) \phi \tag{32}$$

4. The differential of a vector $\mathbf{A(r)}$ may be written as:

$$d\mathbf{A} = (d\mathbf{r} \cdot \nabla)\mathbf{A} \tag{35}$$

5. The flux of a vector $\mathbf{A}$ through a *closed* surface may be written as

$$\Phi = \oint_S \mathbf{A} \cdot d\mathbf{S} = \int_V \text{div}\ \mathbf{A}\ dV \tag{40}$$

where we have

$$\text{div}\ \mathbf{A} = \lim \frac{\Delta\Phi}{\Delta V} = \nabla \cdot \mathbf{A} \tag{39)(44}$$

6. The circulation of a vector around a *closed* contour may be written as:

$$\Gamma = \oint_C \mathbf{A} \cdot d\mathbf{r} = \int_S \mathbf{curl}\ \mathbf{A} \cdot d\mathbf{S} \tag{50)(53}$$

7. The volume integral of **curl** A may be written as:

$$\Omega = \int_V \mathbf{curl}\ A\ dV = \oint_S d\mathbf{S} \times \mathbf{A} \tag{62)(63}$$

from which we have:

$$\mathbf{curl}\ A = \text{limit}\ \frac{\Delta\Omega}{\Delta V} \tag{61}$$

8. The operator ∇ is a differential operator leading to the following identities:

$$\nabla^2 = \nabla \cdot \nabla \tag{64}$$

$$\nabla \times \nabla = 0 \tag{65}$$

$$\nabla \cdot \nabla \times \quad = 0 \tag{67}$$

$$\nabla \times (\nabla \times \quad) = \nabla\nabla \cdot \; - \nabla^2 \tag{68}$$

9. Any vector $\mathbf{W}$, provided that its divergence and curl are derivable from potentials, may be written as the sum of an irrotational part (zero **curl**) and a solenoidal part (zero divergence).

10. If the divergence and **curl** of a vector are everywhere specified, the vector is uniquely determined.

REFERENCES

Arfken, G. B., *Mathematical Methods for Physicists*, Second Edition, Academic, 1970.

Chambers, Ll. G., *A course in Vector Analysis*, Chapman and Hall, 1969.

Green, B. A., Jr., *Vector Calculus*, Appleton-Century-Crofts, 1967.

Kemmer, N., *Vector Analysis*, Cambridge, 1977.

Schey, H. M., *Div, Grad, Curl, and All That, an Informal Text on Vector Calculus*, Norton, 1973.

Wills, A. P., *Vector Analysis with an Introduction to Tensor Analysis*, Dover, 1958.

C TENSORS AND MATRICES

In the development of electromagnetic theory it is useful to have a compact way of describing a linear relationship between two vectors that are not necessarily parallel. One may write such a relationship through the use of tensors, which may be represented (see Appendix A) by the sum of tensor products of vectors.

An alternative to the tensor product representation requires that one work in a specific coordinate system and relate the components of the two vectors by means of a matrix. The advantage of the tensor product approach is that it is compatible with vector analysis and the relations that we obtain must be invariant under coordinate transformation.

Most readers will be familiar with matrix algebra but will have had little experience with tensors. Our approach here will be to review briefly a particular property of matrices and then to describe the corresponding property of tensors.

TENSOR IDENTITIES

We list for convenient reference, a number of the tensor identities[1] that are developed in this appendix and to which we refer in the text.

$$\mathbf{A} \times (\nabla \times \mathbf{B}) = (\nabla \mathbf{B}) \cdot \mathbf{A} - (\mathbf{A} \cdot \nabla)\mathbf{B} \tag{1}$$

$$(\mathbf{A} \times \nabla) \times \mathbf{B} = (\nabla \mathbf{B}) \cdot \mathbf{A} - \mathbf{A}(\nabla \cdot \mathbf{B}) \tag{2}$$

$$\nabla(\mathbf{A} \cdot \mathbf{B}) = (\nabla \mathbf{A}) \cdot \mathbf{B} + (\nabla \mathbf{B}) \cdot \mathbf{A} \tag{3}$$

$$\nabla \cdot (\mathbf{AB}) = (\nabla \cdot \mathbf{A})\mathbf{B} + (\mathbf{A} \cdot \nabla)\mathbf{B} \tag{4}$$

$$\nabla \times (\mathbf{AB}) = (\nabla \times \mathbf{A})\mathbf{B} - (\mathbf{A} \times \nabla)\mathbf{B} \tag{5}$$

$$\nabla(\mathbf{A} \times \mathbf{B}) = (\nabla \mathbf{A}) \times \mathbf{B} - (\nabla \mathbf{B}) \times \mathbf{A} \tag{6}$$

$$\nabla \times (\mathbf{A} \times \mathbf{B}) = \nabla \cdot (\mathbf{BA} - \mathbf{AB}) \tag{7}$$

$$\nabla \cdot (\mathbf{ABC}) = \mathbf{BC}(\nabla \cdot \mathbf{A}) + [(\mathbf{A} \cdot \nabla)\mathbf{B}]\mathbf{C} + \mathbf{B}(\mathbf{A} \cdot \nabla)\mathbf{C} \tag{8}$$

[1] In type matrices and tensors are customarily set in boldface block capital letters (*sans serif*). The notation used in this text is close to that devised jointly by E. A. Milne and S. Chapman in 1926. See S. Chapman and T. G. Cowling, *The Mathematical Theory of Non-Uniform Gases*, Third Edition, Cambridge, 1970.

1   Gauss's Theorems

$$\int_V \nabla \cdot \mathbf{T}\, dV = \oint_S d\mathbf{S} \cdot \mathbf{T} \tag{9}$$

$$\int_V \nabla \mathbf{A}\, dV = \oint_S d\mathbf{S}\mathbf{A} \tag{10}$$

$$\int_V \nabla \times \mathbf{T}\, dV = \oint_S d\mathbf{S} \times \mathbf{T} \tag{11}$$

2   Stokes's Theorems

$$\int_S d\mathbf{S} \cdot \nabla \times \mathbf{T} = \oint_C d\mathbf{r} \cdot \mathbf{T} \tag{12}$$

$$\int_S d\mathbf{S} \times \nabla \mathbf{A} = \oint_C d\mathbf{r}\mathbf{A} \tag{13}$$

$$\int_S (d\mathbf{S} \times \nabla) \times \mathbf{T} = \oint_C d\mathbf{r} \times \mathbf{T} \tag{14}$$

ROTATION OF A VECTOR

As an example of the use of matrices and then of tensors we first discuss the rotation of a vector. We wish to find an operation that takes a vector $\mathbf{A}$ into a vector $\mathbf{A}'$, as shown in Figure 1, where the transformation involves a rotation of the *vector* through an angle $\Delta\phi$ about the z axis.

From Figure 1, the components of $\mathbf{A}'$ and $\mathbf{A}$ are related by the expressions:

$$A_x' = A \sin\theta \cos(\phi + \Delta\phi) = A_x \cos\Delta\phi - A_y \sin\Delta\phi \tag{15}$$
$$A_y' = A \sin\theta \sin(\phi + \Delta\phi) = A_x \sin\Delta\phi + A_y \cos\Delta\phi \tag{16}$$
$$A_z' = A \cos\theta = A_z \tag{17}$$

3   Rotation Matrix. We represent $\mathbf{A}$ and $\mathbf{A}'$ by column vectors

$$\mathbf{A} = \begin{pmatrix} A_1 \\ A_2 \\ A_3 \end{pmatrix} \qquad \mathbf{A}' = \begin{pmatrix} A_1' \\ A_2' \\ A_3' \end{pmatrix} \tag{18}$$

Then, any linear transformation, including rotation, may be represented by the multiplication of $\mathbf{A}$ by a matrix $\mathbf{M}$:

$$\mathbf{A}' = \mathbf{M}\mathbf{A} \tag{19}$$

In particular, for rectangular coordinates, A_1, A_2, and A_3 are just A_x, A_y, and A_z and the elements of **M** are given from **(15)**, **(16)**, and **(17)** by

$$\mathbf{M} = \begin{pmatrix} \cos \Delta\phi & -\sin \Delta\phi & 0 \\ \sin \Delta\phi & \cos \Delta\phi & 0 \\ 0 & 0 & 1 \end{pmatrix} \tag{20}$$

The inverse matrix $\mathbf{M}^{-1}$ rotates $\mathbf{A'}$ back into $\mathbf{A}$:

$$\mathbf{A} = \mathbf{M}^{-1}\mathbf{A'} \tag{21}$$

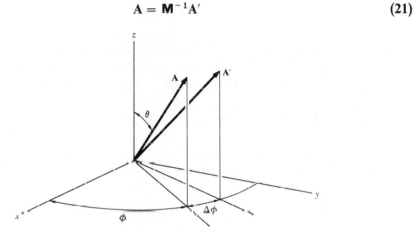

Figure 1 Rotation of a vector. A rotation through $\Delta\phi$ about the z axis takes a vector **A** into a vector **A'**.

and must be given by

$$\mathbf{M}^{-1} = \begin{pmatrix} \cos \Delta\phi & \sin \Delta\phi & 0 \\ -\sin \Delta\phi & \cos \Delta\phi & 0 \\ 0 & 0 & 1 \end{pmatrix} \tag{22}$$

If we define $\tilde{\mathbf{M}}$ (read as the transpose of **M**) as the matrix obtained from **M** by interchanging its rows and columns, we have, from **(20)** and **(22)**,

$$\tilde{\mathbf{M}} = \mathbf{M}^{-1} \tag{23}$$

Matrices that have the property **(23)** are described as orthogonal. Taking the transpose of a column vector as in **(18)** converts the vector to row form as $\tilde{\mathbf{A}} = (A_1 A_2 A_3)$.

Exercises

1. Obtain the relation:

$$\tilde{\mathbf{A}}\,\mathbf{M}\mathbf{B} = \tilde{\mathbf{B}}\,\tilde{\mathbf{M}}\mathbf{A}$$

where **M** is any matrix.

2. Show that the product of **A** with the transpose of **B** generates a matrix whose elements are the products of the components of **A** and **B**:

$$\mathbf{A\tilde{B}} = \begin{pmatrix} A_x \\ A_y \\ A_z \end{pmatrix}^{(B_x \ B_y \ B_z)} = \begin{pmatrix} A_x B_x & A_x B_y & A_x B_z \\ A_y B_x & A_y B_y & A_y B_z \\ A_z B_x & A_z B_y & A_z B_z \end{pmatrix}$$

4 Rotation Tensor. The tensor equivalent of **(19)** is the expression[2]

▶
$$\mathbf{A'} = \mathbf{T} \cdot \mathbf{A} \tag{24}$$

where **T** is a sum of tensor products as described in Appendix A:

$$\mathbf{T} = \mathbf{BC} + \mathbf{DE} + \mathbf{FG} + \cdots \tag{25}$$

A tensor of the form of **(25)** is described as being of second rank. A tensor of third rank would contain terms of the form **BCD** where each term involves two tensor products. The extension to a tensor of rank n follows directly. We may write the particular tensor for the rotation of Figure 1 as

$$\mathbf{T} = \mathbf{\hat{x}\hat{x}} \cos \Delta\phi - \mathbf{\hat{x}\hat{y}} \sin \Delta\phi + \mathbf{\hat{y}\hat{x}} \sin \Delta\phi + \mathbf{\hat{y}\hat{y}} \cos \Delta\phi + \mathbf{\hat{z}\hat{z}} \tag{26}$$

Comparing **(26)** with **(20)** we see that there is a one-to-one correspondence between the elements of the matrix **M** and the rectangular tensor products.

The transpose of **(25)** is obtained simply by interchanging the vectors in the individual tensor products:

$$\mathbf{\tilde{T}} = \mathbf{CB} + \mathbf{ED} + \mathbf{GF} + \cdots \tag{27}$$

Exercises

3. Show that a general second rank tensor may always be written in the form:

$$\mathbf{T} = \mathbf{\hat{x}A} + \mathbf{\hat{y}B} + \mathbf{\hat{z}C}$$

or in the form:

$$\mathbf{T} = \mathbf{D\hat{x}} + \mathbf{E\hat{y}} + \mathbf{F\hat{z}}$$

4. Show that the tensor given by **(26)** is orthogonal:

$$\mathbf{\tilde{T}} = \mathbf{T}^{-1}$$

---

[2] For a discussion of the connection between geometric language and component language, see C. W. Misner, K. S. Thorne, and J. A. Wheeler, *Gravitation*, Freeman, 1973, Chapter 3.

5. Prove the identity:

$$\mathbf{A} \cdot (\mathbf{T} \cdot \mathbf{B}) = (\mathbf{A} \cdot \mathbf{T}) \cdot \mathbf{B}$$

which may be written consequently without parentheses:

$$\mathbf{A} \cdot \mathbf{T} \cdot \mathbf{B}$$

6. Prove the identity

$$\mathbf{A} \cdot \mathbf{T} \cdot \mathbf{B} = \mathbf{B} \cdot \tilde{\mathbf{T}} \cdot \mathbf{A}$$

5 Unit Tensor. Since the rotation shown in Figure 1 is characterized by the *direction* of the z axis and the rotation angle $\Delta\phi$, it should be possible to write **(26)** without reference to a specific coordinate representation. We may do this by using the unit tensor **I**, which has the properties

▶
$$\mathbf{A} = \mathbf{I} \cdot \mathbf{A} \qquad \mathbf{A} = \mathbf{A} \cdot \mathbf{I} \tag{28}$$

In rectangular coordinates we must have

▶
$$\mathbf{I} = \hat{x}\hat{x} + \hat{y}\hat{y} + \hat{z}\hat{z} \tag{29}$$

From **(26)** and **(29)** and the properties of the cross product we may write:

$$\mathbf{T} = \mathbf{I} \cos \Delta\phi + \hat{z}\hat{z}(1 - \cos \Delta\phi) + \hat{z} \times \mathbf{I} \sin \Delta\phi \tag{30}$$

which as required involves only $\hat{z}$ and $\Delta\phi$ since **I** may be defined by **(28)** without reference to a particular coordinate representation.

Exercises

7. Obtain the identity:

$$\nabla \mathbf{r} = \mathbf{I}$$

8. Obtain the identity:

$$\nabla \hat{\mathbf{r}} = \frac{1}{r} (\mathbf{I} - \hat{\mathbf{r}}\hat{\mathbf{r}})$$

9. Let **A**, **B**, and **C** be a set of vectors reciprocal to **a**, **b**, and **c**. (See Section A-7.) Show that the sum of the tensor products is equal to the unit tensor:

$$\mathbf{A}\mathbf{a} + \mathbf{B}\mathbf{b} + \mathbf{C}\mathbf{c} = \mathbf{I}$$

6 **Scalar and Vector of a Tensor.** Writing a tensor as the sum of tensor products:

$$\mathbf{T} = \mathbf{AB} + \mathbf{CD} + \cdots \tag{31}$$

the scalar of **T** is obtained by taking the sum of the scalar products of the pairs of vectors:

$$T_S = \mathbf{A} \cdot \mathbf{B} + \mathbf{C} \cdot \mathbf{D} + \cdots \tag{32}$$

The vector of **T** is defined as:

$$\mathbf{T}_V = \mathbf{A} \times \mathbf{B} + \mathbf{C} \times \mathbf{D} + \cdots \tag{33}$$

Since the vectors **A**, **B**, **C**, **D**, ... are all invariant under coordinate rotation, it is clear that both T_S and $\mathbf{T}_V$ are invariant as well.

Exercises

10. By writing **(32)** in rectangular coordinates, show that the scalar of **T** is equal to the sum of the diagonal elements of the corresponding matrix.

11. Obtain the identity:

$$\mathbf{A} \cdot (\mathbf{T} \times \mathbf{B}) = (\mathbf{A} \cdot \mathbf{T}) \times \mathbf{B} = \mathbf{A} \cdot \mathbf{T} \times \mathbf{B}$$

MULTIPLICATION OF TENSORS OR MATRICES

Let us imagine that we transform a column vector **A** into a column vector **A**′ by matrix multiplication with $\mathbf{M}_1$:

$$\mathbf{A}' = \mathbf{M}_1 \mathbf{A} \tag{34}$$

The matrix $\mathbf{M}_1$ may generate a single rotation but this is not necessary. Following this operation we transform **A**′ into **A**″ with a matrix $\mathbf{M}_2$:

$$\mathbf{A}'' = \mathbf{M}_2 \, \mathbf{A}' = \mathbf{M}_2(\mathbf{M}_1 \mathbf{A}) = \mathbf{M}\mathbf{A} \tag{35}$$

Now, as may be readily demonstrated, the single transformation that takes **A** into **A**″ is multiplication by a matrix **M**, which is the matrix product of $\mathbf{M}_2$ and $\mathbf{M}_1$:

$$\mathbf{M} = \mathbf{M}_2 \, \mathbf{M}_1 \tag{36}$$

It is well known that finite rotations about nonparallel axes do not commute and we should expect that applying the transformations in opposite order will not generate the same transformation, that is, matrix multiplication is *not* commutative.

Exercise

12. Show that if $\mathbf{M}_1$ and $\mathbf{M}_2$ have the orthogonality property **(23)** then their product also has this property.

We may obtain quite analogous results for tensors. The transformation from $\mathbf{A}$ to $\mathbf{A}'$ is given by:

$$\mathbf{A}' = \mathbf{T}_1 \cdot \mathbf{A} \tag{37}$$

and from $\mathbf{A}'$ to $\mathbf{A}''$ by:

$$\mathbf{A}'' = \mathbf{T}_2 \cdot \mathbf{A}' = \mathbf{T}_2 \cdot (\mathbf{T}_1 \cdot \mathbf{A}) = \mathbf{T} \cdot \mathbf{A} \tag{38}$$

where the equivalent single transformation is simply the scalar product of the individual transformations:

$$\blacktriangleright \qquad \mathbf{T} = \mathbf{T}_2 \cdot \tilde{\mathbf{T}}_1 \tag{39}$$

SYMMETRIC TENSORS AND MATRICES

The requirement that a matrix be symmetric is given by the expression:

$$M_{ij} = M_{ji} \tag{40}$$

As we have seen, the matrix that is the transpose of $\mathbf{M}$ is obtained by interchanging the rows and columns of $\mathbf{M}$ and is written as $\tilde{\mathbf{M}}$. Then **(40)** implies for a *symmetric* matrix:

$$\mathbf{M} = \tilde{\mathbf{M}} \tag{41}$$

A symmetric 3×3 matrix has six independent components and may be written out as:

$$\mathbf{M} = \begin{pmatrix} M_{11} & M_{12} & M_{13} \\ M_{12} & M_{22} & M_{23} \\ M_{13} & M_{23} & M_{33} \end{pmatrix} \tag{42}$$

In a similar way we may develop a symmetric tensor. Since we have six independent components we might expect to generate a general symmetric tensor from two vectors. A more satisfactory way of accomplishing this is to begin by writing a general tensor in what is called normal form. Texts in tensor analysis show that for any tensor it is possible to find *two sets* of rectangular coordinates $\hat{\mathbf{x}}$, $\hat{\mathbf{y}}$, $\hat{\mathbf{z}}$ and $\hat{\mathbf{x}}'$, $\hat{\mathbf{y}}'$, $\hat{\mathbf{z}}'$ such that the tensor may be written as:

$$\blacktriangleright \qquad \mathbf{T} = A_x \hat{\mathbf{x}}\hat{\mathbf{x}}' + A_y \hat{\mathbf{y}}\hat{\mathbf{y}}' + A_z \hat{\mathbf{z}}\hat{\mathbf{z}}' \tag{43}$$

The orientation of each of the coordinate systems requires three angles, as shown in Appendix A. With coefficients A_x, A_y, A_z we have in all nine components. One may show fairly easily that if the tensor **T** is symmetric, then the two coordinate systems must be parallel and we have:

▶
$$\mathbf{T} = A_x \,\hat{\mathbf{x}}\hat{\mathbf{x}} + A_y \,\hat{\mathbf{y}}\hat{\mathbf{y}} + A_z \,\hat{\mathbf{z}}\hat{\mathbf{z}} \tag{44}$$

which requires six components, three for the orientation of what are called the principal axes $\hat{\mathbf{x}}$, $\hat{\mathbf{y}}$, and $\hat{\mathbf{z}}$ and three for the principal values A_x, A_y, A_z. The tensor **(44)** is described as being in diagonal form.

Exercises

13. Show that $\mathbf{T}_V$ **(33)** is zero for a symmetric tensor.
14. Show that the matrix corresponding to the tensor

$$\mathbf{T} = \mathbf{AB} + \mathbf{BA}$$

is symmetric.
15. Show that a tensor in normal form **(43)** and symmetric takes the form of **(44)**.

7 Moment of Inertia Tensor. We imagine a rigid body as shown in Figure 2 rotating with angular velocity **ω** about an axis through the origin. An increment of volume ΔV and density ρ at **r** has linear momentum:

$$\Delta\mathbf{p} = \rho\mathbf{v}\,\Delta V = \rho\,\Delta V\boldsymbol{\omega} \times \mathbf{r} \tag{45}$$

The contribution of ΔV to the angular momentum is:

$$\Delta\mathbf{L} = \mathbf{r} \times \Delta\mathbf{p} = \mathbf{r} \times (\boldsymbol{\omega} \times \mathbf{r})\rho\,\Delta V \tag{46}$$

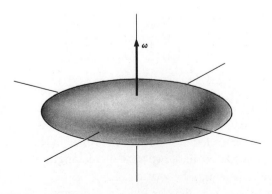

Figure 2   Rigid body. A body rotates with angular velocity **ω** about an axis through the origin.

The total angular momentum, which is obtained by integrating over V, has the form:

$$\mathbf{L} = \int_V \mathbf{r} \times (\boldsymbol{\omega} \times \mathbf{r})\rho \, dV \tag{47}$$

We transform the vector triple product:

$$\mathbf{r} \times (\boldsymbol{\omega} \times \mathbf{r}) = \boldsymbol{\omega} r^2 - (\boldsymbol{\omega} \cdot \mathbf{r})\mathbf{r} = [\boldsymbol{\omega} - (\boldsymbol{\omega} \cdot \hat{\mathbf{r}})\hat{\mathbf{r}}]r^2 \tag{48}$$

By use of the identity tensor $\mathbf{I}$ and the tensor product $\hat{\mathbf{r}}\hat{\mathbf{r}}$ we may write **(48)** as

$$\mathbf{r} \times (\boldsymbol{\omega} \times \mathbf{r}) = \boldsymbol{\omega} \cdot (\mathbf{I} - \hat{\mathbf{r}}\hat{\mathbf{r}})r^2 \tag{49}$$

Then **(47)** becomes:

$$\mathbf{L} = \boldsymbol{\omega} \cdot \mathbf{K} \tag{50}$$

where $\mathbf{K}$ is the moment of inertia tensor and is given by

$$\mathbf{K} = \int_V (\mathbf{I} - \hat{\mathbf{r}}\hat{\mathbf{r}})\rho r^2 \, dV \tag{51}$$

It is clear from the form of **(51)** that $\mathbf{K}$ is a symmetric tensor. As has been indicated by **(44)** there exists for every symmetric tensor a set of rectangular axes for which the tensor may be written in diagonal form **(44)**:

$$\mathbf{K} = K_x \hat{\mathbf{x}}\hat{\mathbf{x}} + K_y \hat{\mathbf{y}}\hat{\mathbf{y}} + K_z \hat{\mathbf{z}}\hat{\mathbf{z}} \tag{52}$$

The axes x, y, and z are called the principal axes of the moment of inertia tensor and K_x, K_y, and K_z are its principal values.

Exercises

16. Show that **(51)** is equivalent to the expression for the moment of inertia matrix:

$$K_{ij} = \int_V (\delta_{ij} \sum x_k^2 - x_i x_j)\rho \, dV$$

where δ_{ij} is the Kronecker delta and the x_i are the components of $\mathbf{r}$ along rectangular axes.

17. Show that the trace of the moment of inertia matrix is given by

$$K_{xx} + K_{yy} + K_{zz} = 2 \int_V \rho r^2 \, dV$$

and thus is invariant under coordinate transformation. Obtain the same result for the scalar of (51) as defined by (32).

---

8 Tensor Ellipsoid. A procedure that helps to visualize the properties of symmetric tensors is given by their representation as ellipsoids. For a symmetric tensor **T**, the equation:

$$\mathbf{r} \cdot \mathbf{T} \cdot \mathbf{r} = 1 \tag{53}$$

is that of an ellipsoidal surface as may be seen by writing out the terms:

$$x^2 T_{xx} + 2xy T_{xy} + y^2 T_{yy} + 2yz T_{yz} + z^2 T_{zz} + 2xz T_{xz} = 1 \tag{54}$$

A rotation of coordinates about the z axis, followed by rotations about $\hat{\mathbf{x}}$ and $\hat{\mathbf{y}}$ brings the major axes of the ellipsoid along the rectangular axes and thus the tensor **T** into diagonal form.

The intersection of (54) with the xy plane is obtained by setting $z = 0$:

$$x^2 T_{xx} + 2xy T_{xy} + y^2 T_{yy} = 1 \tag{55}$$

This is the equation of an ellipse and is obtained from (54) by projecting **r** onto the xy plane.

9 Elastic Compliance Tensor. As a second example of the use of tensors we imagine a particle bound to a point by *harmonic forces* and under action of an external force **F**. For a force **F** in some general direction there will be a displacement **r**, where the relationship between **r** and **F** depends on the nature of the internal restoring forces. We establish a coordinate system and write for the displacements along x, y, and z:

$$x = S_{xx} F_x + S_{xy} F_y + S_{xz} F_z \tag{56}$$
$$y = S_{yx} F_x + S_{yy} F_y + S_{yz} F_z \tag{57}$$
$$z = S_{zx} F_x + S_{zy} F_y + S_{zz} F_z \tag{58}$$

This is the most general set of equations we can write for a linear system. We may write the above equations in matrix notation by writing the position and the force as column vectors:

$$\mathbf{r} = \mathbf{SF} \tag{59}$$

where **S** is called the elastic compliance matrix. Alternatively, we may write **(59)** as a tensor equation:

$$\mathbf{r} = \mathbf{S} \cdot \mathbf{F} \tag{60}$$

where now **S** is the compliance tensor. Although the notation may be different, both formulations of the problem contain the same physics.

From the form of the energy we argue that **S** must be a symmetric tensor. The work done in displacing the point of application is given for any force by:

$$dW = \mathbf{F} \cdot d\mathbf{r} \tag{61}$$

From **(60)** we may write the displacement as:

$$d\mathbf{r} = \mathbf{S} \cdot d\mathbf{F} \tag{62}$$

Substituting into **(61)** we obtain:

$$dW = \mathbf{F} \cdot \mathbf{S} \cdot d\mathbf{F} \tag{63}$$

We write out the terms in rectangular coordinates.

$$dW = S_{xx} F_x \, dF_x + S_{xy} F_x \, dF_y + S_{xz} F_x \, dF_z + S_{yx} F_y \, dF_x + S_{yy} F_y \, dF_y$$
$$+ S_{yz} F_y \, dF_z + S_{zx} F_z \, dF_x + S_{zy} F_z \, dF_y + S_{zz} F_z \, dF_z \tag{64}$$

Now, if the work done to displace the point to **r** is independent of the path taken, then dW must be a perfect differential and we must have:

$$S_{xy} = S_{yx} \qquad S_{xz} = S_{zx} \qquad S_{yz} = S_{zy} \tag{65}$$

which is the requirement for **S** to be a symmetric tensor.

We may now use the tensor ellipsoid to show that for any harmonically bound system there exist three mutually perpendicular directions such that a force along any of the three directions produces a displacement only along that direction. In such a representation the compliance matrix is diagonal:

$$\mathbf{S} = \begin{pmatrix} S_x & 0 & 0 \\ 0 & S_y & 0 \\ 0 & 0 & S_z \end{pmatrix} \tag{66}$$

or, equivalently, the compliance tensor may be written as the sum of tensor products:

$$\mathbf{S} = S_x \hat{\mathbf{x}}\hat{\mathbf{x}} + S_y \hat{\mathbf{y}}\hat{\mathbf{y}} + S_z \hat{\mathbf{z}}\hat{\mathbf{z}} \tag{67}$$

The proof of the above consists in writing out the equation for the tensor ellipsoid:

$$\mathbf{r} \cdot \mathbf{S} \cdot \mathbf{r} = 1 \tag{68}$$

Now, we can always rotate to principal axes, for which **(68)** has the form

$$x^2 S_x + y^2 S_y + z^2 S_z = 1 \tag{69}$$

It follows, then, that the same rotation must bring the compliance matrix to the form **(66)** or the compliance tensor to **(67)**.

ANTISYMMETRIC TENSORS AND MATRICES

The requirement that a matrix be antisymmetric is given by the expression:

$$M_{ij} = -M_{ji} \tag{70}$$

In particular, setting $i = j$ we see that the diagonal elements must be zero:

$$M_{ii} = -M_{ii} = 0 \tag{71}$$

As we have seen, the matrix which is the transpose of **M** is obtained by interchanging the rows and columns of **M** and is written as $\tilde{\mathbf{M}}$. Then **(70)** implies for an antisymmetric matrix:

$$\mathbf{M} = -\tilde{\mathbf{M}}$$

An antisymmetric 3×3 matrix has only three independent components and may be written out as:

$$\mathbf{M} = \begin{pmatrix} 0 & M_{12} & M_{13} \\ -M_{12} & 0 & M_{23} \\ -M_{13} & -M_{23} & 0 \end{pmatrix} \tag{72}$$

In the same way we may develop an antisymmetric tensor. Since we expect only three independent components, an antisymmetric tensor can be generated from a vector. The simplest way of doing this is take the cross product of the identity tensor with some vector **C**. We write:

▶ $$\mathbf{T} = \mathbf{I} \times \mathbf{C} = (\hat{x}\hat{x} + \hat{y}\hat{y} + \hat{z}\hat{z}) \times (\hat{x}C_x + \hat{y}C_y + \hat{z}C_z) \tag{73}$$

Expanding out the cross products we obtain:

$$\mathbf{T} = -\hat{x}\hat{y}C_z + \hat{x}\hat{z}C_y - \hat{y}\hat{z}C_x + \hat{y}\hat{x}C_z + \hat{z}\hat{y}C_x - \hat{z}\hat{x}C_y \tag{74}$$

Comparing **(74)** with **(72)** we see that the coefficients of the tensor products form an antisymmetric matrix.

Exercises

18. Show that the matrix corresponding to the tensor:

$$\mathbf{T} = \mathbf{AB} - \mathbf{BA}$$

is antisymmetric.

19. Show that the vector product of **A** and **B** may be associated with an antisymmetric tensor by the relation

$$(\mathbf{A} \times \mathbf{B}) \times \mathbf{I} = \mathbf{BA} - \mathbf{AB}$$

20. Prove the identity:

$$\mathbf{I} \times \mathbf{C} = \mathbf{C} \times \mathbf{I}$$

21. Prove the identity:

$$\mathbf{A} \cdot (\mathbf{I} \times \mathbf{B}) = \mathbf{A} \cdot (\mathbf{B} \times \mathbf{I}) = \mathbf{A} \times \mathbf{B}$$

22. Prove the identity:

$$\mathbf{A} \times (\mathbf{B} \times \mathbf{I}) - (\mathbf{I} \times \mathbf{A}) \times \mathbf{B} = \mathbf{BA} - \mathbf{I}(\mathbf{A} \cdot \mathbf{B})$$

23. For **T** antisymmetric prove the identity:

$$\mathbf{T} \cdot \mathbf{D} = -\mathbf{D} \cdot \mathbf{T}$$

where **D** is any vector.

24. For **T** antisymmetric and **A** any vector, prove the identity:

$$\mathbf{A} \cdot \mathbf{T} = \tfrac{1}{2}\mathbf{T}_V \times \mathbf{A}$$

where $\mathbf{T}_V$ is the vector of **T**.

25. Show that the vector product of **A** and **B** may be written as the scalar product of a tensor with **A**:

$$\mathbf{A} \times \mathbf{B} = \mathbf{T} \cdot \mathbf{A}$$

Let **B** be written as the vector product of **C** and **D**. Show that **T** is the antisymmetric tensor:

$$\mathbf{T} = \mathbf{CD} - \mathbf{DC}$$

26. Show that if **T** is an antisymmetric tensor expressed in terms of **(73)**, the vector of **T (33)** is given by:

$$\mathbf{T}_V = -2\mathbf{C}$$

27. For **T** antisymmetric prove the identity:

$$\mathbf{D} \cdot \mathbf{T} \cdot \mathbf{D} = 0$$

where **D** is any vector.

28. For **T** a symmetric tensor obtain the relation

$$\mathbf{T} \times \mathbf{A} = -\widetilde{\mathbf{A} \times \mathbf{T}}$$

DIFFERENTIAL RELATIONS

The differential relations developed in Appendix B for vectors may also be applied to tensors. As we saw, the gradient operator ∇ has the properties of a vector operator. Taking the scalar product of ∇ with a vector gives the divergence of the vector while taking the vector product gives the **curl**. With these properties we may utilize ∇ in forming and operating on tensors.

10   Gradient of a Vector. The gradient of a vector may be regarded as the tensor product of ∇ with the vector and is thus a tensor:

$$\mathbf{T} = \nabla \mathbf{A} \tag{75}$$

The scalar of **T** (32) is the divergence of **A**:

$$T_S = \nabla \cdot \mathbf{A} \tag{76}$$

and the vector of **T** (33) is the **curl** of **A**:

$$T_V = \nabla \times \mathbf{A} \tag{77}$$

We may take the scalar product of (75) from the left with a second vector according to (A-54):

$$\mathbf{B} \cdot \mathbf{T} = \mathbf{B} \cdot \nabla \mathbf{A} = (\mathbf{B} \cdot \nabla)\mathbf{A} \tag{78}$$

We may also take the scalar product on the right, but we must be sure that ∇ does not operate on the second vector:

$$\mathbf{T} \cdot \mathbf{B} = (\nabla \mathbf{A}) \cdot \mathbf{B} \tag{79}$$

Exercises

29. Show that the gradient of the scalar product of two vectors (B-6) or (B-7) may be expressed simply through the use of the tensor product where it is understood that ∇ operates only on the vector within parentheses:

$$\nabla(\mathbf{A} \cdot \mathbf{B}) = (\nabla \mathbf{A}) \cdot \mathbf{B} + (\nabla \mathbf{B}) \cdot \mathbf{A}$$

30. Verify **(B-6)** by expanding the two vector triple products and using the result of Exercise 30.

31. Verify **(B-7)** by expanding the two vector triple products and using the result of Exercise 30.

32. Obtain the identity:

$$\mathbf{A} \times (\nabla \times \mathbf{B}) = \mathbf{A} \cdot (\widetilde{\nabla \mathbf{B}} - \nabla \mathbf{B})$$

33. Obtain the identity:

$$\nabla\nabla \frac{1}{r} = \frac{3\hat{\mathbf{r}}\hat{\mathbf{r}} - \mathbf{I}}{r^3}$$

---

11 Divergence and Curl of a Tensor. To obtain the divergence of a tensor **AB** we take the scalar product of ∇ with the tensor just as for vectors:

$$\nabla \cdot \mathbf{T} = \nabla \cdot (\mathbf{AB}) = (\nabla \cdot \mathbf{A})\mathbf{B} + (\mathbf{A} \cdot \nabla)\mathbf{B} \tag{80}$$

Note that both the vector and operator properties are preserved in **(80)**.

Similarly we may expand the **curl** of the tensor **AB**:

$$\nabla \times \mathbf{T} = \nabla \times (\mathbf{AB}) = (\nabla \times \mathbf{A})\mathbf{B} - (\mathbf{A} \times \nabla)\mathbf{B} \tag{81}$$

Exercises

34. Obtain the identity for a third-rank tensor:

$$\nabla \cdot (\mathbf{ABC}) = \mathbf{BC}(\nabla \cdot \mathbf{A}) + (\mathbf{A} \cdot \nabla\mathbf{B})\mathbf{C} + \mathbf{BA} \cdot \nabla\mathbf{C}$$

35. Obtain the identity:

$$\nabla \cdot (\mathbf{BA} - \mathbf{AB}) = \nabla \times (\mathbf{A} \times \mathbf{B})$$

and compare with **(B-5)**.

36. Obtain the identity:

$$\nabla \cdot (\mathbf{I}\phi) = \nabla\phi$$

37. For **T** independent of **r** obtain the identity:

$$\nabla^2(\mathbf{r} \cdot \mathbf{T} \cdot \mathbf{r}) = 2T_S$$

38. For **T** independent of **r** obtain the identity:

$$\mathbf{r} \times [\nabla \times (\mathbf{T} \cdot \mathbf{r})] = \mathbf{r} \cdot \mathbf{T} - \mathbf{T} \cdot \mathbf{r}$$

39. For **T** independent of **r** obtain the identity:

$$\nabla \cdot (\mathbf{r} \cdot \mathbf{T}) = T_S$$

40. For **T** independent of **r** obtain the identity:

$$\nabla \times (\mathbf{r} \cdot \mathbf{T}) = T_V$$

41. For **T** independent of **r** obtain the identity:

$$\nabla(\mathbf{r} \cdot \mathbf{T} \cdot \mathbf{r}) = \mathbf{T} \cdot \mathbf{r} + \mathbf{r} \cdot \mathbf{T}$$

INTEGRAL RELATIONS

The integral relations developed in Appendix B for vectors may also be applied to tensors. Thus, as the equivalent of **(B-11)** we have Gauss's theorem for a tensor function of position:

$$\int_V (\nabla \cdot \mathbf{T}) \, dV = \oint_S d\mathbf{S} \cdot \mathbf{T} \tag{82}$$

If we write **T** as a tensor product, the divergence is given by terms of the form of **(80)**. The tensor equivalent of the curl theorem **(B-13)** is:

$$\int_V \nabla \times \mathbf{T} \, dV = \oint_S d\mathbf{S} \times \mathbf{T} \tag{83}$$

If we write **T** as a tensor product, the curl may be written as terms of the form of **(81)**. Similarly, Stokes theorem **(B-14)** may be written as:

$$\int_S d\mathbf{S} \cdot (\nabla \times \mathbf{T}) = \oint_C d\mathbf{r} \cdot \mathbf{T} \tag{84}$$

Note in **(84)** as well as in **(82)** and **(83)** that vector operations must consistently be taken on the same member of the tensor product.

As an application of **(84)** we let **T** equal the antisymmetric tensor:

$$\mathbf{T} = \mathbf{I} \times \mathbf{A} = \mathbf{A} \times \mathbf{I} \tag{85}$$

Substituting into **(84)** we obtain for the right side:

$$\oint_C d\mathbf{r} \cdot (\mathbf{A} \times \mathbf{I}) = \oint_C d\mathbf{r} \times \mathbf{A} \cdot \mathbf{I} = \oint_C d\mathbf{r} \times \mathbf{A} \tag{86}$$

For the left side we have:

$$\int_S d\mathbf{S} \cdot (\nabla \times \mathbf{T}) = \int_S (d\mathbf{S} \times \nabla) \cdot (\mathbf{A} \times \mathbf{I}) = \int_S (d\mathbf{S} \times \nabla) \times \mathbf{A} \cdot \mathbf{I} \qquad (87)$$

Equating **(86)** and **(87)** we obtain **(B-16)** as a consequence of Stokes's theorem:

$$\int_S (d\mathbf{S} \times \nabla) \times \mathbf{A} = \oint_C d\mathbf{r} \times \mathbf{A} \qquad (88)$$

Exercise

42. Using **(82)** and **(80)** obtain the identity:

$$\oint_S (d\mathbf{S} \cdot \mathbf{A})\mathbf{B} = \int_V [\mathbf{B}(\nabla \cdot \mathbf{A}) + (\mathbf{A} \cdot \nabla)\mathbf{B}]\, dV$$

SUMMARY

1. Any linear transformation that connects two vectors may be written as a second rank tensor:

$$\mathbf{A}' = \mathbf{T} \cdot \mathbf{A} \qquad (24)$$

2. There exists a unit tensor with the property

$$\mathbf{A} = \mathbf{I} \cdot \mathbf{A} = \mathbf{A} \cdot \mathbf{I} \qquad (28)$$

In rectangular coordinates the unit tensor may be written as

$$\mathbf{I} = \hat{\mathbf{x}}\hat{\mathbf{x}} + \hat{\mathbf{y}}\hat{\mathbf{y}} + \hat{\mathbf{z}}\hat{\mathbf{z}} \qquad (29)$$

3. Multiplication of tensors is defined and is simply the scalar product:

$$\mathbf{T} = \mathbf{T}_2 \cdot \mathbf{T}_1 \qquad (39)$$

where **T** is the single transformation that is equivalent to applying $\mathbf{T}_1$ and then $\mathbf{T}_2$.
4. Any tensor may be brought to normal form:

$$\mathbf{T} = A_x \hat{\mathbf{x}}\hat{\mathbf{x}}' + A_y \hat{\mathbf{y}}\hat{\mathbf{y}}' + A_z \hat{\mathbf{z}}\hat{\mathbf{z}}' \qquad (43)$$

where x, y, and z and x', y', and z' are two sets of rectangular coordinates.

5. For a symmetric tensor the two sets of coordinates are parallel and the tensor may be written as:

$$\mathbf{T} = A_x \hat{x}\hat{x} + A_y \hat{y}\hat{y} + A_z \hat{z}\hat{z} \tag{44}$$

where x, y, and z are the principal axes of the tensor and A_x, A_y and A_z are the corresponding principal values.

6. An antisymmetric tensor may be characterized by a single vector and written as:

$$\mathbf{T} = \mathbf{I} \times \mathbf{C} \tag{73}$$

REFERENCES

Arfken, G. B., *Mathematical Methods for Physicists*, Second Edition, Academic, 1970.

Chambers, Ll. G., *A Course in Vector Analysis*, Chapman and Hall, 1969.

Chorlton, F., *Vector and Tensor Methods*, Halsted, 1975.

Coburn, N., *Vector and Tensor Analysis*, Macmillan, 1955.

Joshi, A. W., *Matrices and Tensors in Physics*, Wiley, 1975.

Marion, J. B., *Principles of Vector Analysis*, Academic, 1965.

Symon, K. R., *Mechanics*, Third Edition, Addison-Wesley, 1971. Chapter 10 of this text contains an introduction to tensor algebra.

Wills, A. P., *Vector Analysis with an Introduction to Tensor Analysis*, Dover, 1958.

D QUANTIZATION

The approach of this text is phenomenological, that is, we are not primarily concerned with fundamental theories of matter. We may, for example, discuss certain charge or current distributions that are *known* to exist without proving from "first principles" that such distributions are stable.

In our discussion of dielectric media, we are satisfied to establish in terms of simple models that the presence of kinetic energy together with the constancy of angular momentum stabilizes the energy. But why should angular momentum be fixed? We suggest here the answer to such questions in terms of the wave nature of matter.

Second, how is it that matter is stable against electromagnetic radiation? If there is kinetic energy, charge must be in motion and there should be changing magnetic as well as electric fields. An answer may be given to this question based on the Schrödinger approach to wave mechanics. A result of this theory is that for a state of definite energy the current and charge distributions are steady and there are no time-varying fields.

There are many good books that discuss these questions at an introductory level. Some are listed as references at the end of this appendix. We do not aim here to provide a comprehensive discussion of the subject but only wish to point out the quantization conditions that make possible an essentially classical discussion of the interaction between fields and matter.

STABILITY AND SIZE OF ATOMS

In terms of purely classical theory there is no reason why we should have atoms of unique structure and size. Let us imagine, for example, an atom with a nucleus of charge $+Ze$ about which move Z electrons of charge $-e$. The equations of motion are of the form:

$$m\frac{d^2}{dt^2}\mathbf{r}_i = -e\mathbf{E}(\mathbf{r}_i,t) \tag{1}$$

Now, if we make the change of scale:

$$\mathbf{r}_i = u^2\mathbf{r}_i' \qquad t = u^3 t' \tag{2}$$

we obtain an equation of the same form as **(1)** but in $\mathbf{r}'$ and t' since $E(\mathbf{r})$ varies as $1/u^4$. The period of the motion, as for any inverse-square force, is proportional to the three-halves power of size of the orbit. Just as for planetary orbits there is no reason to expect that atomic orbits need be unique.

1   Quantization of Angular Momentum. A solution to this problem was first given for the hydrogen atom by Niels Bohr. Bohr showed that by assuming an electron moving on a circular orbit with angular momentum $m\hbar$ one obtains radii $R = n^2 a_0$ with

$$\blacktriangleright \qquad a_0 = \frac{4\pi\epsilon_0}{e^2}\frac{\hbar^2}{m} = 5.2917706 \times 10^{-11} \text{ m} \qquad (3)$$

Bohr's theory gave the right answer for the size of the hydrogen atom and was in *quantitative* agreement with the observed spectra of hydrogen. Elliptical orbits were treated later by Sommerfeld, but his theory could not be readily extended to multi-electron atoms. Worse, an orbiting electron should generate electric and magnetic fields varying at the frequency of rotation and such fields should carry away electromagnetic energy. Somehow, Bohr supposed, the atom was prevented from radiating in the $n = 1$ state. Radiation was allowed from other states at frequencies

$$\hbar\omega = U_n - U_{n'} \qquad (4)$$

A more satisfactory theory of atomic structure was formulated by Erwin Schrödinger in 1926. In the Schrödinger theory, instead of having solutions of **(1)**, a stationary state of an atom is characterized by a wave function

$$\psi_n(\mathbf{r}_1, \mathbf{r}_2, \ldots \mathbf{r}_Z) \qquad (5)$$

where ψ_n is one of a number of discrete solutions of a wave-type equation, called the Schrödinger equation. In this theory the Z electrons are regarded as indistinguishable so that **(5)** must look exactly the same if we rearrange the electron labels. Quantization of angular momentum enters the theory in a natural way through the requirement that acceptable solutions of the wave equation be single valued functions of the coordinates of the electrons. For example, a traveling free particle might be represented by a wave function:

$$\psi(\mathbf{r}) = e^{i\mathbf{k} \cdot \mathbf{r}} \qquad (6)$$

where

$$\mathbf{k} = (m/\hbar)\mathbf{v} \qquad (7)$$

is the propagation vector. For a particle moving in an axially symmetric potential we have:

$$\psi(\mathbf{r}) = \psi(r,\theta)e^{il_z\phi} \qquad (8)$$

where l_z is the z component of the vector:

$$\mathbf{l} = \mathbf{r} \times \mathbf{k} = (m/\hbar)\mathbf{r} \times \mathbf{v} \tag{9}$$

Now, increasing ϕ by 2π represents the same electron position. If ψ represents a state with definite angular momentum, l_z must be integral, which through (9) leads to quantization.

The electron is observed to have spin angular momentum:

$$\blacktriangleright \qquad\qquad\qquad S = \tfrac{1}{2}\hbar \tag{10}$$

and an intrinsic magnetic moment:

$$\blacktriangleright \qquad\qquad\qquad \mu = \tfrac{1}{2}g\mu_B \tag{11}$$

where μ_B is called the Bohr magnetron and is the quantum of orbital magnetic moment in the Bohr atom:

$$\mu_B = \tfrac{1}{2}eva_0 = \frac{e}{2m}\hbar = 9.274078 \times 10^{-24}\ \text{J/T} \tag{12}$$

The quantity g is called the spectroscopic splitting factor and is approximately equal to 2. Radiative corrections lead to a slightly larger value:

$$g = 2\left(1 + \frac{v}{2c} + \cdots\right) \simeq 2.00231931 \tag{13}$$

A physical interpretation of the wave function was given by Born, who proposed that the square of the absolute value of (5) represents a probability density. With this interpretation one may compute from (5) charge densities and current densities, both of which are independent of the time. Thus, electric and magnetic fields are independent of time and the absence of radiation is explained in a natural way as long as the system is *entirely* in a single well-defined state.

2 **Field Energy.** The Schrödinger theory has given us a fully adequate description of atomic and molecular structure. Yet even this description is phenomenological in character and is not a complete theory. We must be particularly careful in problems where we are concerned with field energies. Here there are difficulties even for the hydrogen atom, for example. Let us imagine that we wish to find the ionization energy of the hydrogen atom. Our prescription is to find the wave function of the lowest energy state of hydrogen and of a state in which the electron is ionized with vanishingly small kinetic energy. Having found the wave functions, the energies are the eigenvalues of the Schrödinger equation and the ionization energy is their difference.

An alternate way of computing energies might be in terms of electric and magnetic fields. In the ground state of hydrogen there is no mean current and thus no magnetic field. Our problem, then, should be a purely electrostatic one. Given the wave function, we presumably know the charge density from which we may compute fields and energies. In the ionized state, the electron is fully dispersed and has according to the Schrödinger theory no field energy. Then the only field energy is that stored in the field generated by the proton. In bound states this field is shielded by the electron and so the energy should be lower. The values that we calculate, though, are not in agreement with those obtained correctly from the Schrödinger equation. Attempting to compute field energies in the way that we have described makes a serious error. A computation of field energy as we have described it involves coulomb interaction between different parts of the electron charge distribution, which amounts to an interaction of the electron with itself. The energy computed correctly from the Schrödinger equation contains no such terms and so we must be careful in our use of charge density on an atomic scale. Current density presents a similar problem.

Any fundamental theory of atoms and molecules must recognize the particle character of fields as well as the wave character of particles. The Schrödinger theory is an approximation to a more general quantum field theory in the limit that the charges move slowly compared with c. We may estimate the electron speed in the hydrogen ground state by writing for a body moving on a circular orbit:

$$U_P = -2U_K \tag{14}$$

which leads on substitution from (3) to:

$$\tfrac{1}{2}mv^2 = \frac{1}{2}\frac{1}{4\pi\epsilon_0}\frac{e^2}{a_0} = \left(\frac{e^2}{4\pi\epsilon_0}\right)^2\frac{m}{2\hbar^2} \tag{15}$$

We obtain for the ratio of the electron velocity to the speed of light:

▶
$$\frac{v}{c} = \frac{1}{4\pi\epsilon_0}\frac{e^2}{\hbar c} = \alpha = \frac{1}{137.03604} \tag{16}$$

where α is called the fine-structure constant. The velocities of *valence* electrons at least are relatively small so that the Schrödinger equation should give a good description of atomic and molecular processes.

INTERACTION WITH MAGNETIC FIELDS

The Schrödinger formulation of quantum mechanics is based on the classical Hamiltonian with the momenta acting as operators in the conjugate coordinates. For an electron moving in a uniform magnetic field **B**, the momentum conjugate to the position coordinate **r** is given as shown in Section G-1 by:

$$\mathbf{p} = m\mathbf{v} + q\mathbf{A} \tag{17}$$

where for a uniform field we have:

$$A = \tfrac{1}{2}B \times r \tag{18}$$

3 Diamagnetism. The orbital angular momentum is given by:

$$L = r \times p = mr \times v + qr \times A = mr \times v + \tfrac{1}{2}qr \times (B \times r) = n\hbar\hat{z} \tag{19}$$

We write for the enclosed flux:

$$\Phi = \pi r^2 B \tag{20}$$

and for the orbital magnetic moment of an electron with $q = -e$:

$$\mu = -\tfrac{1}{2}e(r \times v) \cdot \hat{z} \tag{21}$$

Substituting into **(19)** we obtain for the moment:

$$\mu = -n\,\frac{e\hbar}{2m} - \frac{e^2}{4\pi m}\,\Phi \tag{22}$$

For a given quantum state the first term is a constant, so that **(22)** provides a relation between the induced diamagnetic moment and the contained flux.

4 Flux Quantization. In a completely diamagnetic plasma, the magnetic moment, as shown in Section 8-4, is given by **(8-71)**:

$$\mu = -\frac{q^2}{2\pi m}\,\Phi \tag{23}$$

where q is the charge per carrier. Substituting into **(22)** we obtain for a charge q:

$$\Phi = -\frac{nh}{q} \tag{24}$$

The fluxoid h/q is the quantum of magnetic flux and depends on the magnitude of charge. For a superconductor, electrons are paired and we must take $q = -2e$ so that the fluxoid for a superconductor is:

$$\blacktriangleright \qquad \Phi_0 = \frac{h}{2e} = 2.0678506 \times 10^{-15}\ \text{W} \tag{25}$$

5 Magnetic Monopoles. In 1931, Dirac[1] showed that solutions of the Schrödinger equation could be constructed under conditions of flux divergence as long as the total divergent flux was quantized. We show in Figure 1 a source of total flux $\mu_0 \bar{q}$. From the fact that the integral of the vector potential over a closed contour is the enclosed flux **(6-57)** we may write:

$$2\pi r \sin \theta A_\phi = \frac{\Omega}{4\pi} \mu_0 \bar{q} \tag{26}$$

where Ω is the solid angle subtended by the contour. But Ω is undetermined to within a multiple of 4π. For example, it is not clear from Figure 1 whether the subtended angle is Ω or $-(4\pi - \Omega) = \Omega - 4\pi$. We write from **(26)** for the arbitrariness in A_ϕ:

$$r \sin \theta \, \Delta A_\phi = \frac{\mu_0 \bar{q}}{2\pi} \tag{27}$$

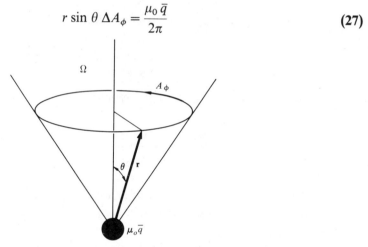

Figure 1   Magnetic charge. The total flux from a magnetic charge $\bar{q}$ is $\mu_0 \bar{q}$.

From **(27)** we have an indeterminacy of the z component of angular momentum of **(19)**:

$$L_z = -er \sin \theta \, \Delta A_\phi \tag{28}$$

If the solutions of the Schrödinger equation are to be single valued, the indeterminacy in L_z must be a multiple of $\hbar$, which gives from **(28)**:

$$r \sin \theta \, \Delta A_\phi = \frac{n\hbar}{e} \tag{29}$$

Comparing **(27)** and **(29)** we obtain:

$$\mu_0 \bar{q} = \mu_0 g = \frac{nh}{e} \tag{30}$$

[1] P. A. M. Dirac, *Proc. Roy. Soc.* (London), **A133**, 60 (1931) and *Phys. Rev.*, **74**, 817 (1948). For additional references, see P. Hrasko, *Am. J. Phys.* **45**, 838 (1977).

The quantum of magnetic charge is then given by:

$$g_0 = \frac{g}{n} = \frac{h}{\mu_0 e} = \frac{1}{2}\left(\frac{4\pi\epsilon_0}{e^2}\hbar c\right)ce = \frac{ce}{2\alpha} = 3.29109 \times 10^{-9}\ \text{A m} \tag{31}$$

Comparing **(31)** and **(25)** we have:

$$\mu_0 g = 2n\Phi_0 \tag{32}$$

Thus, in an experiment utilizing superconducting elements as detectors, each monopole should generate $2n$ fluxoids each time it passes through a multiply connected superconducting element.[2]

QUANTUM OSCILLATOR

A quantum oscillator at frequency ω_0 is characterized by two or more states separated in energy by U_0 where the frequency and energy are related by **(4)**:

$$U_0 = \hbar\omega_0 \tag{33}$$

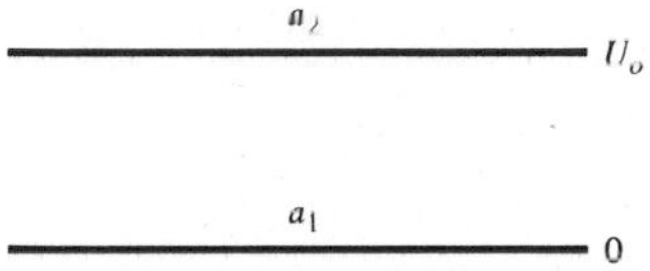

Figure 2 Two-level system. We assume two states separated by energy U_0 and with amplitudes a_1 and a_2.

We assume for simplicity that we have just two states as shown in Figure 2. We take the lower state to be at the zero of energy and the upper state to be at U_0. We represent the states by wave functions ψ_1 and ψ_2. The wave function of the system is written as:

$$\psi = a_1\psi_1 + a_2\psi_2 \tag{34}$$

where a_1 and a_2 are the amplitudes of the states. The probability that the system is in the ground state is given by $a_1 a_1{}^*$ and the probability that the system is in the excited state by $a_2 a_2{}^*$. The energy of the system is then expected to be:

$$U = U_0 a_2 a_2{}^* \tag{35}$$

with:

$$a_1 a_1{}^* + a_2 a_2{}^* = 1 \tag{36}$$

[2] L. W. Alvarez et al., *Rev. Sci. Instr.*, **42**, 326 (1971). For an alternative development of **(31)** see L. J. Garwin and R. L. Garwin, *Am. J. Phys.*, **45**, 164 (1977).

6   Forced Oscillation. A fundamental difference between the quantum oscillator and the classical oscillator is that the quantum oscillator may have positive energy without an oscillating polarization. This would be the case, as we will see, for the oscillator entirely in the upper state. How then, are we to describe stimulated emission or absorption? The external field must somehow develop a polarization that then leads to a rate of change of energy. This process is described by a pair of equations which give the rate of change of the *amplitudes* of the states:

$$i\hbar \frac{d}{dt} a_1 = -p_0 E(t) a_2 \tag{37}$$

$$i\hbar \frac{d}{dt} a_2 - U_0 a_2 = -p_0 E(t) a_1 \tag{38}$$

where p_0 is a polarization that gives the strength of the coupling to the electric field. These equations follow from the Schrödinger equation and will not be derived here. The reader is referred to an introductory text on quantum physics.[3]

As an example of the solution of (37) and (38) we assume a classical driving field of the form:

$$E(t) = 2E_1 \cos \omega_0 t = E_1 \left(e^{i\omega_0 t} + e^{-i\omega_0 t} \right) \tag{39}$$

Substituting into (37) and (38) we obtain:

$$\frac{d}{dt} a_1 = \frac{i}{2} \omega_n a_2{}^0 \left(1 + e^{-2i\omega_0 t} \right) \tag{40}$$

$$\frac{d}{dt} a_2{}^0 = \frac{i}{2} \omega_n a_1 \left(1 + e^{2i\omega_0 t} \right) \tag{41}$$

where ω_n is the nutation frequency and is given by:

$$\hbar \omega_n = -2p_0 E_1 \tag{42}$$

and where we have taken:

$$a_2 = a_2{}^0 e^{-i\omega_0 t} \tag{43}$$

If the nutation frequency ω_n is much smaller than the resonance frequency ω_0 we may neglect the oscillating terms in (40) and (41) to obtain, as solutions for a_1 and a_2,

$$a_1 = \cos \tfrac{1}{2}\omega_n t \tag{44}$$

$$a_2 = i \sin \tfrac{1}{2}\omega_n t \, e^{-i\omega_0 t} \tag{45}$$

[3] See, for example, R. P. Feynman, R. B. Leighton, and M. Sands, *The Feynman Lectures*, II-8-5, Addison-Wesley, 1965.

For the probability that the oscillator is in the ground state we obtain

$$a_1 a_1{}^* = \cos^2 \tfrac{1}{2}\omega_n t = \tfrac{1}{2}(1 + \cos \omega_n t) \tag{46}$$

and for the excited state:

$$a_2 a_2{}^* = \sin^2 \tfrac{1}{2}\omega_n t = \tfrac{1}{2}(1 - \cos \omega_n t) \tag{47}$$

where we have assumed that at $t = 0$ the oscillator is entirely in the ground state. These functions are plotted in Figure 3.

Our result is that for the oscillator driven at resonance the energy varies sinusoidally at the nutation frequency ω_n:

$$U = U_0 a_2 a_2{}^* = \tfrac{1}{2}U_0(1 - \cos \omega_n t) \tag{48}$$

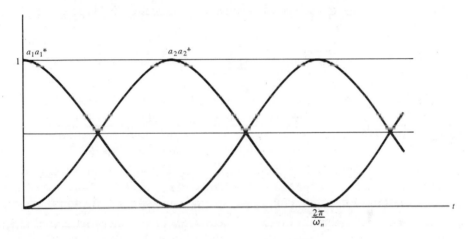

Figure 3 Forced oscillation. The probabilities of the ground and excited states are shown as a function of time.

Next we wish to obtain from **(38)** and **(39)** a general expression for the rate at which the energy of the oscillator changes. From **(37)** and **(38)** we may write:

$$\frac{d}{dt}\,|a_1|^2 = \frac{1}{i\hbar}\,(a_1 a_2{}^* - a_1{}^* a_2)p_0\, E(t) \tag{49}$$

$$\frac{d}{dt}\,|a_2|^2 = \frac{1}{i\hbar}\,(a_1{}^* a_2 - a_1 a_2{}^*)p_0\, E(t) \tag{50}$$

Using **(35)** for the energy and **(15-13)** for the rate at which work is done, we obtain for the rate of change of the polarization:

$$\frac{dp}{dt} = -i\omega_0\, p_0(a_1{}^* a_2 - a_1 a_2{}^*) \tag{51}$$

Substituting into **(37)** and **(38)** we can rewrite **(51)** as:

$$\frac{dp}{dt} = p_0 \frac{d}{dt}(a_1{}^*a_2 + a_1 a_2{}^*) \tag{52}$$

which leads by integration to the expression for the expected polarization:

$$p = p_0(a_1{}^*a_2 + a_1 a_2{}^*) \tag{53}$$

Taking the time derivative of **(51)** and using **(53)** we obtain:

$$\frac{d^2 p}{dt^2} + \omega_0{}^2 p = f \frac{e^2}{m}(a_1 a_1{}^* - a_2 a_2{}^*)E(t) \tag{54}$$

where f is the oscillator strength of the transition (see Section 3-11) and is given by:

$$f = \frac{2m\omega_0}{\hbar e^2} p_0{}^2 \tag{55}$$

7   Induced Emission. In comparing **(54)** with **(15-11)** we note that in addition to the presence of the oscillator strength we also have the factor $(a_1 a_1{}^* - a_2 a_2{}^*)$. For the oscillator in the ground state this factor is plus one and a quantum oscillator initially builds up just as does a classical oscillator. The energy of the oscillator must increase from which we infer that the average value of $E\,dp/dt$ is positive. If, on the other hand, the oscillator is in the excited state, then the factor $(a_1 a_1{}^* - a_2 a_2{}^*)$ is equal to minus one and the polarization developed by the field is the negative of that for the previous case. This means that the average value $E\,dp/dt$ must be negative and by **(15-13)** the oscillator must deliver energy to the field. This process is called induced emission.[4]

8   Spontaneous Emission. For exponential decay the equations for the populations in the ground and excited states take the form:

$$\frac{d}{dt}(a_1 a_1{}^*) = \frac{2}{\tau}(1 - a_1 a_1{}^*) \tag{56}$$

$$\frac{d}{dt}(a_2 a_2{}^*) = -\frac{2}{\tau}(a_2 a_2{}^*) \tag{57}$$

Adding **(56)** and **(57)** gives a result consistent with **(36)**. Although we would like to be able to write equations for the rate of change of a_1 and a_2, there is no simple way of doing this that gives exponential decay at the same time. The description of radiation by a single oscillator requires a quantum description not only of the oscillator, but of the electromagnetic field and is the subject of what is called quantum electrodynamics.

[4] For a discussion of the induced radiation, see A. V. Durrant, *Am. J. Phys.*, **44**, 630 (1976), **45**, 752 (1977).

If we take the initial energy of the oscillator to be given by:

$$U(0) = U_0 a_2 a_2{}^* \tag{58}$$

we have from **(56)** or **(57)** for the energy of the oscillator:

▶
$$U(t) = U(0)e^{-2t/\tau} \tag{59}$$

9 Superradiance. We next wish to describe the coherent emission of radiation by a system of N oscillators in the absence of a driving field.[5] The equations that we write are phenomenological and are constructed to have the corresponding classical properties. Since the amplitude in the ground state can build up only if there is some energy in the system, we write

$$\frac{d}{dt} a_1 = (a_2 a_2{}^*) \frac{a_1}{\tau} \tag{60}$$

By a similar argument we have, for the amplitude of the excited state,

$$\frac{d}{dt} a_2 + i\omega_0 a_2 = -(a_1 a_1{}^*) \frac{a_2}{\tau} \tag{61}$$

Multiplying **(60)** by $a_1{}^*$ and then adding the conjugate equation we have

$$\frac{d}{dt} (a_1 a_1{}^*) = \frac{2}{\tau} (a_1 a_1{}^*)(a_2 a_2{}^*) \tag{62}$$

Similarly, we obtain, from **(61)**,

$$\frac{d}{dt} (a_2 a_2{}^*) = -\frac{2}{\tau} (a_1 a_1{}^*)(a_2 a_2{}^*) \tag{63}$$

We interpret **(62)** and **(63)** as giving a relaxation rate proportional to the product $a_2 a_2{}^*$, which is the probability that the system is in the excited state, times $a_1 a_1{}^*$, the probability that the system is in the ground state. We note that adding **(62)** and **(63)** gives zero as required by **(36)**.

The reason for the form of **(63)** may be seen from **(15-12)**, which gives a rate of energy radiation proportional to the square of the polarization. Neglecting the radiative terms we have, from **(60)** and **(61)**,

$$a_1 = (a_1 a_1{}^*)^{1/2} e^{-i\phi} \qquad a_2 = (a_2 a_2{}^*)^{1/2} e^{-i\omega_0 t} \tag{64}$$

[5] This subject was first discussed by R. H. Dicke, *Phys. Rev.*, **93**, 90 (1954). See also R. P. Feynman, F. L. Vernon, and R. W. Hellwarth, *J. Appl. Phys.*, **28**, 212 (1957).

Substituting into **(53)** the polarization is given by:

$$p(t) = p_0(a_1 a_1 {}^* a_2 a_2 {}^*)^{1/2} \cos(\omega_0 t + \phi) \tag{65}$$

and the square of the polarization is proportional to the right side of **(63)**.

As an example we solve **(63)** for a system of N oscillators radiating coherently. At $t = 0$ we imagine that the system has energy:

$$U(0) = NU_0 \, a_2 a_2 {}^* \tag{66}$$

Then, integrating **(63)** we obtain for the energy of the system:

▶
$$U(t) = \frac{NU_0 \, U(0)}{U(0) + [NU_0 - U(0)]e^{2t/\tau}} \tag{67}$$

We observe that if all the oscillators are in the excited state at $t = 0$, there is no coherent radiation. The relaxation time τ that appears in **(60)** and **(61)** is not the relaxation time of a single oscillator, but is the relaxation time for N oscillators and is $1/N^2$ times the classically computed single oscillator relaxation time.

We may simplify **(67)** by taking $t = 0$ to be the time at which the energy is just $\frac{1}{2}NU_0$. We have then, for $a_2 a_2 {}^*$,

$$a_2 a_2 {}^* = \frac{1}{1 + e^{2t/\tau}} \tag{68}$$

The oscillating polarization is obtained from **(65)** and takes the form:

▶
$$p = p_0 \operatorname{sech} \frac{t}{\tau} \cos(\omega_0 t + \phi) \tag{69}$$

The energy and polarization are sketched in Figures 4a and 4b.

It is clear that **(59)** and **(67)** describe opposite regimes. The first describes the behavior of a single quantum oscillator and neglects interference between oscillators. The second describes N oscillators radiating coherently. For superradiance we require:

1. The number of oscillators N must be large.

2. The oscillators must be interacting in such a way that their coherence is maintained.

10 Equation of Motion. Finally, we combine **(37)** and **(38)** with **(60)** and **(61)** to obtain equations for the amplitudes in the presence of a driving force and superradiance:

$$\frac{d}{dt} a_1 - (a_2 a_2 {}^*)\frac{a_1}{\tau} = \frac{i}{\hbar} p_0 E(t)a_2 \tag{70}$$

$$\frac{d}{dt} a_2 + (a_1 a_1 {}^*)\frac{a_2}{\tau} + i\omega_0 a_2 = \frac{i}{\hbar} p_0 E(t)a_1 \tag{71}$$

From **(70)** and **(71)** we may write for the rate of change of polarization:

$$\frac{dp}{dt} + (a_1 a_1{}^* - a_2 a_2{}^*)\frac{p}{\tau} = \omega_0 q \tag{72}$$

where $\omega_0 q$ is given by **(51)** with:

$$q = ip_0(a_1{}^* a_2 - a_1 a_2{}^*) \tag{73}$$

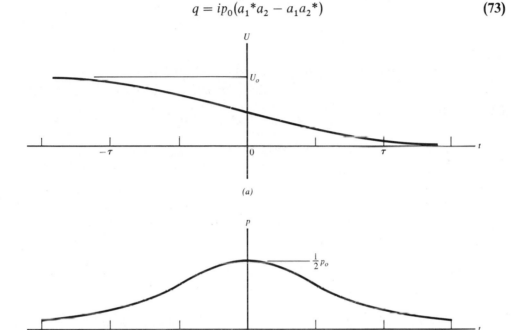

(a)

(b)

Figure 4 Superradiance. (a) The energy of the superradiant state is shown as a function of time. (b) The polarization of the superradiant state is shown as a function of time.

Similarly, we obtain for the rate of change of q:

$$\frac{dq}{dt} + (a_1 a_1{}^* - a_2 a_2{}^*)\frac{q}{\tau} = -\omega_0 p + \frac{fe^2}{\omega_0 m}(a_1 a_1{}^* - a_2 a_2{}^*)E(t) \tag{74}$$

Combining **(73)** and **(74)** and assuming that $\omega_0 \tau$ is large we obtain:

$$\frac{d^2 p}{dt^2} + \frac{2}{\tau}(a_1 a_1{}^* - a_2 a_2{}^*)\frac{dp}{dt} + \omega_0{}^2 p = f\frac{e^2}{m}(a_1 a_1{}^* - a_2 a_2{}^*)E(t) \tag{75}$$

We may write **(75)** in the form of **(15-11)** by introducing an effective mass

$$\frac{1}{m^*} = (a_1 a_1{}^* - a_2 a_2{}^*)\frac{1}{m} \tag{76}$$

and an effective relaxation rate

$$\frac{1}{\tau^*} = (a_1 a_1{}^* - a_2 a_2{}^*)\frac{1}{\tau} = \frac{m}{m^*}\frac{1}{\tau} \tag{77}$$

in terms of which the equation of motion for the polarization becomes

$$\frac{d^2 p}{dt^2} + \frac{2}{\tau^*}\frac{dp}{dt} + \omega_0{}^2 p = f\frac{e^2}{m^*}E(t) \tag{78}$$

For the system in the ground state $1/m^*$ and $1/\tau^*$ are positive while for the system in the excited state they are negative. It is this property of a quantum oscillator which makes power gain possible.

We observe from (54) and (74):

$$p = iq = 2a_1{}^* a_2 p_0 \tag{79}$$

Taking the modulus square of (79) we obtain:

$$|p + iq|^2 = 4a_1 a_1{}^* a_2 a_2{}^* p_0{}^2 = [1 - (a_1 a_1{}^* - a_2 a_2{}^*)^2]p_0{}^2 \tag{80}$$

where we have added and subtracted the square of (36). As shown in Section 15-3, (72) and (74) when taken together with (80) are analogous to the equation of a gyroscope.

SUMMARY

1. Angular momentum is quantized in units of

$$\hbar = 1.0545887 \times 10^{-34} \text{ J/s}$$

2. Quantization of angular momentum leads to hydrogenic radii $R = n^2 a_0$ with:

$$a_0 = \frac{4\pi\epsilon_0}{e^2}\frac{\hbar^2}{m} = 5.2917706 \times 10^{-11} \text{ m} \tag{3}$$

3. The electron is observed to have a spin angular momentum:

$$S = \tfrac{1}{2}\hbar \tag{10}$$

and a magnetic moment:

$$\mu = \tfrac{1}{2}g\mu_B \tag{11}$$

where μ_B is the Bohr magneton and is given by:

$$\mu_B = \frac{e\hbar}{2m_e} = 9.274078 \times 10^{-24} \text{ J/T} \tag{12}$$

and g is the spectroscopic splitting factor, which is equal to 2.00231931.

4. The velocity in the first Bohr orbit as compared with the speed of light is equal to the fine structure constant:

$$\frac{v}{c} = \frac{1}{4\pi\epsilon_0} \frac{e^2}{\hbar c} = \alpha = \frac{1}{137.03604} \tag{16}$$

5. Magnetic flux is quantized and the quantum of flux or *fluxoid* is given by

$$\Phi_0 = \frac{h}{2e} = 2.0678506 \times 10^{-15} \text{ W} \tag{25}$$

6. Magnetic charge is similarly quantized with the quantum of magnetic charge:

$$g_0 = \frac{ce}{2\alpha} = 3.29109 \times 10^{-9} \text{ A m} \tag{31}$$

7. Spontaneous emission of energy by a quantum oscillator may on the average be described as exponential decay:

$$U(t) = U(0)e^{-2t/\tau} \tag{59}$$

8. For a large number N coherently interacting oscillators the energy decays instead as:

$$U(t) = \frac{NU_0 \, U(0)}{U(0) + [NU_0 - U(0)]e^{2t/\tau}} \tag{67}$$

where U_0 is the energy separation.

9. The initial polarization decays as the hyperbolic secant:

$$p = p_0 \operatorname{sech} \frac{t}{\tau} \cos(\omega_0 t + \phi) \tag{69}$$

REFERENCES

Feynman, R. P., R. B. Leighton, and M. Sands, *The Feynman Lectures on Physics*, Volume III, Addison-Wesley, 1965.

Mandel, L., "The Case for and against Semiclassical Radiation Theory," *Progress in Optics*, Volume 13, E. Wolf, editor, North Holland, 1975.

Parks, R. D., "Quantum Effects in Superconductors," *Scientific American*, **213**(4), 57 (October 1965).

Wichmann, E. H., *Quantum Physics, Berkeley Physics Course*, Volume IV, McGraw-Hill, 1971.

E NETWORK REPRESENTATIONS

It is extremely helpful in physics to have *graphical representations* for mathematical quantities. Thus we represent a force or a velocity by a vector, which we draw as a directed line segment.

Similarly, we find it useful to have a geometrical representation for differential equations, particularly for coupled equations. The representation that we introduce uses electrical circuit elements to represent physical quantities (which in themselves may have little or nothing to do with electricity). For example, mass is represented by inductance, restoring force by capacitance, and forces that dissipate energy by resistance. An external force is represented by a voltage generator. All this works, of course, because the differential equations that describe the motion of particles are analogous to the differential equations that describe networks. This applies to transient as well as periodic behavior. One can even, and again this is a consequence of the equations, make an analogy between the energy stored in the magnetic field of an inductor and the kinetic energy of motion of a mass. Similarly the potential energy stored in a spring is the analogue of the energy stored in the electric field of a capacitor.

The danger in using equivalent networks in a book on electromagnetism is that the physics of the system being studied may be confused with the physics of the circuit elements. So one must be careful to limit the analogy to the form of the equations and stop there.

PARTICLES

The equation of motion of a free particle of mass m acted on by a force F is

$$F = m \frac{dv}{dt} \tag{1}$$

This equation is represented in Figure 1 by a voltage generator driving an inductance. The analogue of particle velocity is the current through the inductance. **679**

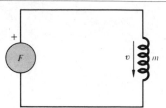

Figure 1   Free particle. A free particle, whose equation of motion is given by **(1)**, is represented by a voltage generator driving an inductance.

The rate at which work is done on the particle:

$$\frac{dW}{dt} = Fv \tag{2}$$

is the analogue of the voltage current product. The total work done on the particle is equal to its kinetic energy:

$$U_K = \tfrac{1}{2}mv^2 \tag{3}$$

The analogue of this energy is the field energy stored in the inductor, which is $\tfrac{1}{2}LI^2$.

1   **Restoring Force.** Work done against a restoring force is represented by a generator charging a capacitor as shown in Figure 2. Note that F is the force exerted *on* the spring and is the negative of the restoring force exerted *by* the spring:

$$F = kx \tag{4}$$

The analogue of particle position is charge on the capacitor. The work done on the particle is equal to the stored potential energy:

$$U_P = \tfrac{1}{2}kx^2 \tag{5}$$

The analogue of this energy is the energy stored in the capacitor $\tfrac{1}{2}Q^2/C$. The capacitance and $1/k$ the spring compliance (or reciprocal stiffness) are thus analogues.

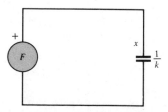

Figure 2   Restoring force. A harmonic restoring force of the form given by **(4)** is represented by a capacitor.

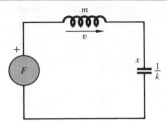

Figure 3 Harmonic oscillator. A harmonic oscillator is represented by a voltage generator driving an inductor in series with a capacitor.

2 Harmonic Oscillator. The harmonic oscillator is a particle of mass m and restoring force linearly proportional and opposite to the displacement. The equivalent circuit representation is given in Figure 3 and the equation of motion in **(6)**.

$$F = m\frac{dv}{dt} + kx \qquad (6)$$

The energy stored is the sum of **(3)** and **(5)**:

$$U = U_K + U_P \qquad (7)$$

3 Damped Particle. The relaxation of particle motion by a damping force is often represented by a force proportional to and opposite to the velocity. The equation of motion of a damped particle of mass m is given by **(8)** and shown in Figure 4. The time τ is the time for decay of the velocity to $1/e$ of its initial value in the absence of a driving force.

$$F = m\frac{dv}{dt} + \frac{m}{\tau}v \qquad (8)$$

The rate at which energy is dissipated under these circumstances is equal to the rate of loss of kinetic energy and is the analogue of I^2R:

$$\frac{dU_K}{dt} = mv\frac{dv}{dt} = -v^2\frac{m}{\tau} \qquad (9)$$

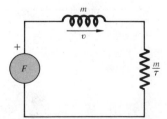

Figure 4 Damped particle. A damped particle is represented by a voltage generator driving an inductor in series with a resistor.

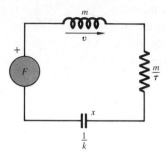

Figure 5   Damped harmonic oscillator. A damped harmonic oscillator is represented by a voltage generator driving an inductor in series with a resistor and a capacitor.

4   Damped Harmonic Oscillator. Finally, in the damped harmonic oscillator we combine the three circuit elements which we have introduced. The equation of the damped oscillator is given by **(10)** and the equivalent circuit shown in Figure 5.

$$F = m\frac{dv}{dt} + \frac{m}{\tau}v + kx \tag{10}$$

The rate at which work is done on the system may be obtained by multiplying **(10)** by v:

$$\frac{dW}{dt} = Fv = \frac{d}{dt}(U_K + U_P) + v^2\frac{m}{\tau} \tag{11}$$

where U_K and U_P are given by **(3)** and **(5)**.

5   Coupled Oscillators. We now examine the circuit representation of a pair of coupled oscillators. The equations are given by **(12)** and **(13)** and the equivalent circuit is shown in Figure 6.

$$F_1 = m_1\frac{dv_1}{dt} + k_1 x_1 + k_{12}(x_1 - x_2) \tag{12}$$

$$F_2 = m_2\frac{dv_2}{dt} + k_2 x_2 + k_{12}(x_2 - x_1) \tag{13}$$

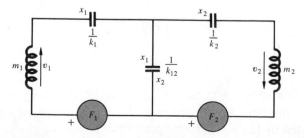

Figure 6   Coupled oscillators. A pair of oscillators coupled by their relative displacement is represented by the circuit shown.

6 Normal Modes. The particular advantage of the circuit representation is in the analysis of normal modes. To find the normal modes of the circuit of Figure 6, we assume a sinusoidal time dependence of x_1 and x_2 with $F_1 = F_2 = 0$. We obtain on substitution into **(12)** and **(13)**:

$$(-m_1\omega^2 + k_1 + k_{12})x_1 - k_{12}x_2 = 0 \tag{14}$$

$$-k_{12}x_1 + (-m_2\omega^2 + k_2 + k_{12})x_2 = 0 \tag{15}$$

Solving the determinantal equation for ω^2 we obtain two solutions:

$$\omega^2 = \frac{1}{2}\left[\frac{k_1}{m_1} + \frac{k_2}{m_2} + \left(\frac{1}{m_1} + \frac{1}{m_2}\right)k_{12}\right] \pm \frac{1}{2}\left[\left(\frac{k_1}{m_1}\right)^2 + \left(\frac{k_2}{m_2}\right)^2 + \left(\frac{1}{m_1} + \frac{1}{m_2}\right)^2 k_{12}{}^2\right.$$

$$\left. + 2\left(\frac{1}{m_1} - \frac{1}{m_2}\right)\left(\frac{k_1}{m_1} - \frac{k_2}{m_2}\right)k_{12} - 2\left(\frac{k_1}{m_1}\right)\left(\frac{k_2}{m_2}\right)\right] \tag{16}$$

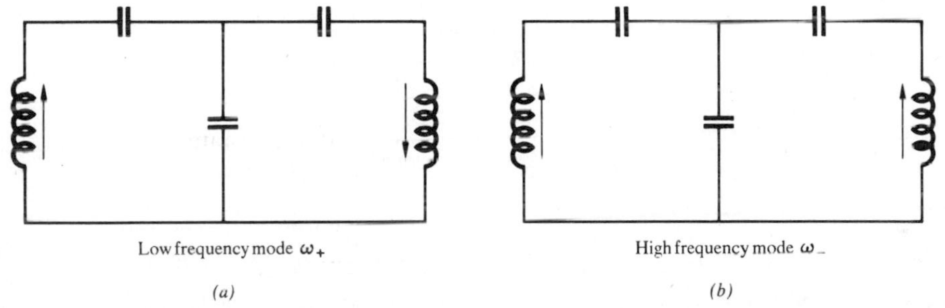

Low frequency mode ω_+ High frequency mode ω_-

(a)   (b)

Figure 7 Normal modes. (a) For the low-frequency mode the circulation has the same sense in both circuits. (b) For the high-frequency mode the circulation is opposite in sense in the two circuits.

The ratio of the velocities in the normal modes is obtained by substitution into **(14)** or **(15)**:

$$\frac{v_2}{v_1} = \frac{x_2}{x_1} = 1 + \frac{k_1 - \omega^2 m_1}{k_{12}} \tag{17}$$

The low frequency mode, which corresponds to the minus sign in **(16)** and which we call ω_+, gives a *positive* ratio for v_2/v_1. The high frequency mode ω_- gives a *negative* ratio for v_2/v_1. These modes are shown in Figures 7a and 7b.

WAVES

The transmission line serves as a network representation for wave propagation. As shown in Figure 8 the field H is represented by the current down the line and the field E is represented by the voltage across the line. The permeability μ is represented by the inductance per unit length and the permittivity ϵ by the capacitance per unit length.

7   Transmission Lines Equations. The transmission line equations are then:

$$\frac{dE}{dz} = -\mu \frac{dH}{dt} \tag{18}$$

$$\frac{dH}{dz} = -\epsilon \frac{dE}{dt} \tag{19}$$

Combining **(18)** and **(19)** we have the wave equation:

$$\frac{d^2E}{dz^2} = \epsilon\mu \frac{d^2E}{dt^2} \tag{20}$$

The solution to **(20)** is a traveling wave:

$$E(z,t) = Ee^{i(kz-\omega t)} \tag{21}$$

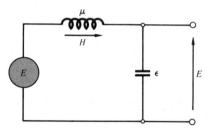

Figure 8   Transmission line. Wave propagation in space may be represented by the propagation of current and voltage on a transmission line.

where the phase velocity is given by:

$$v = \frac{\omega}{k} = (\mu\epsilon)^{-1/2} \tag{22}$$

The characteristic impedance of the line is defined and given by:

$$Z = \frac{E}{H} = \frac{k}{\omega\epsilon} = \frac{\omega\mu}{k} = \left(\frac{\mu}{\epsilon}\right)^{1/2} \tag{23}$$

We may represent the reflection and transmission of radiation by the junction of transmission lines as shown in Figure 9. The first line is characterized by $Z_1 = (\mu_1/\epsilon_1)^{1/2}$ and the second line by $Z_2 = (\mu_2/\epsilon_2)^{1/2}$.

The continuity of voltage and current give the equivalent of the boundary conditions for the extended case:

$$E_1 + E_3 = E_2 \tag{24}$$

$$H_1 - H_3 = H_2 \tag{25}$$

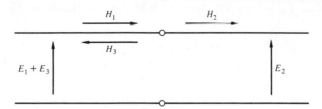

Figure 9 Reflection and transmission. The interface between two media is represented by a junction of two transmission lines.

These equations together with **(21)** may be solved to give:

$$E_2 = \frac{2Z_2}{Z_1 + Z_2} E_1 \tag{26}$$

$$E_3 = \frac{Z_2 - Z_1}{Z_2 + Z_1} E_1 \tag{27}$$

For reflection off normal incidence the effective impedances become $Z \cos \theta$ or $Z/\cos \theta$ for the TM and TE waves, respectively, as discussed in Chapter 13.

8 Matching Section. We imagine as shown in Figure 10 a medium of wave impedance Z_1 and a medium of wave impedance Z_2 joined by an intermediate region of wave impedance Z and length l. We wish to find the appropriate value of Z and l such that there is *no* reflected wave.

In medium 1 we write:

$$E_1(z,t) = E_1{}^+ e^{i(k_1 z - \omega t)} + E_1{}^- e^{i(k_1 z + \omega t)} \tag{28}$$

$$H_1(z,t) = \frac{E_1{}^+}{Z_1} e^{i(k_1 z - \omega t)} - \frac{E_1{}^-}{Z_1} e^{-i(kz + \omega t)} \tag{29}$$

In the intermediate region we write:

$$E(z,t) = E^+ e^{i(kz - \omega t)} + E^- e^{-i(kz + \omega t)} \tag{30}$$

$$H(z,t) = \frac{E^+}{Z} e^{i(kz - \omega t)} - \frac{E^-}{Z} e^{-i(kz + \omega t)} \tag{31}$$

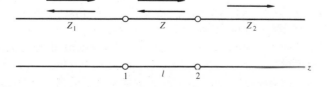

Figure 10 Intermediate region. A matching section may be used to represent an intermediate region between two semibounded regions.

In medium 2 we write:

$$E_2(z,t) = E_2\, e^{i(k_2 z - \omega t)} \tag{32}$$

$$H_2(z,t) = \frac{E_2}{Z_2}\, e^{i(k_2 z - \omega t)} \tag{33}$$

The continuity of E and H, which are analogous to the continuity of potential and current, give at $z = 0$:

$$E_1{}^+ + E_1{}^- = E^+ + E^- \tag{34}$$

$$E_1{}^+/Z_1 - E_1{}^-/Z_1 = E^+/Z - E^-/Z \tag{35}$$

At $z = l$ the continuity equations give:

$$E^+ e^{ikl} + E^- e^{-ikl} = E_2{}^+ e^{ik_2 l} \tag{36}$$

$$(E^+/Z)e^{ikl} - (E^-/Z)e^{-ikl} = (E_2{}^+/Z_2)e^{ik_2 l} \tag{37}$$

Solving for the reflected and transmitted electric fields in terms of the incident electric field we obtain:

$$E_1{}^- = \frac{(1/Z)(1/Z_1 - 1/Z_2)\cos kl - i(1/Z_1 Z_2 - 1/Z^2)\sin kl}{(1/Z)(1/Z_1 + 1/Z_2)\cos kl - i(1/Z_1 Z_2 + 1/Z^2)\sin kl}\, E_1{}^+ \tag{38}$$

$$E_2{}^+ e^{ik_2 l} = \frac{2/ZZ_1}{(1/Z)(1/Z_1 + 1/Z_2)\cos kl - i(1/Z_1 Z_2 + 1/Z^2)\sin kl}\, E_1{}^+ \tag{39}$$

The condition that there be no reflected wave is then:

$$(1/Z)(1/Z_2 - 1/Z_1)\cos kl - i(1/Z_1 Z_2 - 1/Z^2)\sin kl = 0 \tag{40}$$

To obtain a solution to **(40)** there are two conditions, that the real and imaginary parts of **(40)** vanish separately. And there are three parameters, the real and imaginary parts of the impedance Z of the matching section and the length l of the section. With the simplifying assumption that Z_1, Z_2, and Z are all real the conditions on Z and l may be written at sight:

1. Since Z_1 and Z_2 are different we must have $\cos kl = 0$ for the real part of **(40)** to vanish. This requires

$$kl = (2n + 1)\frac{\pi}{2} \tag{41}$$

that is, the matching section must be an odd number of quarter waves in length.

2. For the imaginary part of (**40**) to vanish we must have

$$Z = (Z_1 Z_2)^{1/2} \tag{42}$$

that is, the impedance of the matching section must be the geometric mean of the two sections that are to be matched.

Exercises

1. Under what conditions for $Z_1 = Z_2$ is there no reflected wave?
2. Under matching conditions obtain the expression for the transmitted wave:

$$E_2{}^+ = (Z_2/Z_1)^{1/2} E_1{}^+ e^{n\pi i/4}$$

9 Transfer Matrix. A compact way of writing the connection between the incident, reflected, and transmitted wave is through the transfer matrix. With reference to Figure 10 we write the electric and magnetic fields at the first *terminal* as:

$$E_1 = E^+ + E^- \tag{43}$$

$$H_1 = E^+/Z - E^-/Z \tag{44}$$

We have similarly, at the second terminal,

$$E_2 = E^+ e^{ikl} + E^- e^{-ikl} \tag{45}$$

$$H_2 = (E^+/Z)e^{ikl} - (E^-/Z)e^{-ikl} \tag{46}$$

Eliminating E^+ and E^- between the two pairs of equations we may write the electric and magnetic fields at the first terminal in terms of the fields at the second terminal:

$$E_1 = E_2 \cos kl - iH_2 Z \sin kl \tag{47}$$

$$H_1 = -i(E_2/Z)\sin kl + H_2 \cos kl \tag{48}$$

By constructing field column vectors we write (**47**) and (**48**) as the matrix equation:

$$\begin{pmatrix} E_1 \\ H_1 \end{pmatrix} = \begin{pmatrix} \cos kl & -iZ \sin kl \\ -(i/Z)\sin kl & \cos kl \end{pmatrix} \begin{pmatrix} E_2 \\ H_2 \end{pmatrix} \tag{49}$$

Normally, what we will want, rather than the electric and magnetic fields, are the amplitudes of the incident and reflected fields. Writing for the incident wave:

$$E_1 = E_1{}^+ + E_1{}^- \tag{50}$$

$$H_1 = E_1{}^+/Z_1 - E_1{}^-/Z_1 \tag{51}$$

We have the matrix equation:

$$\begin{pmatrix} E_1{}^+ \\ E_1{}^- \end{pmatrix} = \tfrac{1}{2} \begin{pmatrix} 1 & Z_1 \\ 1 & -Z_1 \end{pmatrix} \begin{pmatrix} E_1 \\ H_1 \end{pmatrix} \tag{52}$$

For the transmitted wave we have:

$$E_2 = E_2{}^+ e^{ikl} \tag{53}$$

$$H_2 = (E_2{}^+/Z_2)e^{ikl} \tag{54}$$

which leads to the matrix equation:

$$\begin{pmatrix} E_2 \\ H_2 \end{pmatrix} = \begin{pmatrix} 1 \\ 1/Z_2 \end{pmatrix} E_2{}^+ e^{ik_2 l} \tag{55}$$

Finally, substituting **(52)** and **(55)** into **(49)** we obtain for the incident and reflected fields:

$$\begin{pmatrix} E_1{}^+ \\ E_1{}^- \end{pmatrix} = \tfrac{1}{2} \begin{pmatrix} 1 & Z_1 \\ 1 & -Z_1 \end{pmatrix} \begin{pmatrix} \cos kl & -iZ \sin kl \\ -(i/Z)\sin kl & \cos kl \end{pmatrix} \begin{pmatrix} 1 \\ 1/Z_2 \end{pmatrix} E_2{}^+ e^{ikl} \tag{56}$$

Exercise

---

3. Perform the matrix multiplication of **(56)** to obtain $E_1{}^+$ and $E_1{}^-$ in terms of $E_2{}^+$. Compare with **(38)** and **(39)**.

---

10 **Matrix Theory of Multiple Films.** The matrix theory developed in the last section may now be used to obtain an expression for the reflection coefficient of multiple films as shown schematically in Figure 11a. Light is incident in a medium of wave impedance

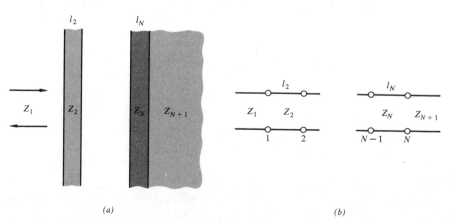

(a)   (b)

Figure 11   Multiple films. (*a*) A wave incident in a medium of wave impedance Z_1 reflects off a set of multiple films deposited on a backing of wave impedance Z_{N+1}. (*b*) The multiple films are represented by transmission line segments.

Z_1. Films of wave impedance $Z_2 \ldots Z_N$ and lengths $l_2 \ldots l_N$ are deposited on a substrate of wave impedance Z_{N+1}. The corresponding transmission line is shown in Figure 11b. In place of **(49)** we write for adjacent terminals:

$$\begin{pmatrix} E_{j-1} \\ H_{j-1} \end{pmatrix} = \begin{pmatrix} \cos k_j l_j & -iZ_j \sin k_j l_j \\ -(i/Z_j)\sin k_j l_j & \cos k_j l_j \end{pmatrix} \begin{pmatrix} E_j \\ H_j \end{pmatrix} \tag{57}$$

We have then by iteration:

$$\begin{pmatrix} E_1 \\ H_1 \end{pmatrix} = \prod_{j-2}^{N} \mathbf{T}_j \begin{pmatrix} E_N \\ H_N \end{pmatrix} \tag{58}$$

where $\mathbf{T}_j$ is the matrix represented in **(57)**. By using **(52)** and **(55)** we write for the incident and reflected fields:

$$\begin{pmatrix} E_1{}^+ \\ E_1{}^- \end{pmatrix} = \tfrac{1}{2} \begin{pmatrix} 1 & Z_1 \\ 1 & -Z_1 \end{pmatrix} \prod_{j-2}^{N} \mathbf{T}_j \begin{pmatrix} 1 \\ 1/Z_{N+1} \end{pmatrix} E_{N+1}^+ e^{ik_{N+1}l} \tag{59}$$

where l is the total film thickness. The reflection coefficient

$$r = E_1{}^- / E_1{}^+ \tag{60}$$

may be obtained by multiplying out **(59)** and taking the ratio of the fields.[1]

REFERENCES

Baggulcy, D. M. S., *Electromagnetism and Linear Circuits*, Van Nostrand Reinhold, 1973

Feynman, R. P., R. B. Leighton, and M. Sands, *The Feynman Lectures on Physics*, Addison-Wesley, 1963, I-25

[1] This analysis was first used for multiple films by P. Rouard, *Rev. d'Optique* **17**, 1, 61, 89 (1938). See A. Nussbaum and R. A. Phillips, *Contemporary Optics for Scientists and Engineers*, Prentice-Hall, 1976, Section 8-6.

F RELATIVISTIC DYNAMICS

Our purpose here is to find the relativistic transformation properties of forces. We need the form of the transformation for use in Chapter 6, where we discuss the interaction between moving charges, and in Chapter 9, where we discuss the interaction between charges and currents in relative motion.

We assume that the reader has been introduced to the special theory of relativity including a discussion of the invariance of the speed of light with respect to the velocity of the source and the observer and to the Lorentz transformation with particular discussion of length contraction and time dilation.

We begin with the *light sphere* derivation of the Lorentz transformation, followed by a discussion of momentum, energy, and force and their transformation properties. At the end of this appendix we list a number of introductory treatments of the special theory of relativity.

TRANSFORMATION OF COORDINATES

The first full theory of relativistic dynamics was given by Albert Einstein in 1905.[1] Einstein's discussion of relativistic kinematics begins with an examination of simultaneity and is followed by the transformation of coordinates and time that is required if the speed of light is to be an invariant. It is this treatment that is given in most introductory texts and that leads to the Lorentz transformation. We give here a more compact but also possibly less physical derivation of the Lorentz transformation on the assumption that the reader will already have seen the Einstein treatment.[2]

We imagine two frames of reference moving with relative velocity $\mathbf{V}$. We take the direction of $\mathbf{V}$ to be the x coordinate in one frame and the x' coordinate in the second frame. The frames are completely equivalent. Thus, as measured from the unprimed frame, the primed frame moves to the right with velocity $\mathbf{V}$. As measured from the primed frame, the unprimed frame has velocity $-\mathbf{V}$.

---

[1] *The Principle of Relativity, a Collection of Original Memoirs on the Special and General Theory of Relativity,* Methuen and Company, Ltd., 1923. Reprinted by Dover Publications, 1952.

[2] For closely related treatments see A. R. Lee and T. M. Kalotas, *Am. J. Phys.*, **43**, 434 (1975); **44**, 1000 (1976); **45**, 870 (1977) and J.-M. Levy-Leblond, *Am. J. Phys.*, **44**, 271 (1976). See also G. R. Fowles, *Am. J. Phys.*, **45**, 675 (1977). For an early and particularly full discussion, see L. A. Pars, *Phil. Mag.* **42**, 249 (1921).

We look for a *linear* transformation between the primed and unprimed frames shown in Figure 1. The reason for this is that a free particle must move in a straight line in either frame. If this were not the case Newton's first law would be violated in at least one of the frames. Furthermore, whatever transformation takes (x,y,z,t) into (x',y',z',t') must by changing the sign of V perform the inverse transformation. It follows that the transverse coordinates are unaffected by the relative motion and we have:

▶
$$y = y' \tag{1}$$

▶
$$z = z' \tag{2}$$

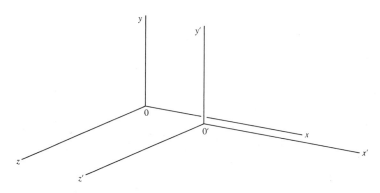

Figure 1 Frames of reference. We look for a linear transformation between the primed and unprimed frames.

The transformation of the x coordinates and the time may in general be mixed and take the form:

$$x' = a_{11}x + a_{14}t \tag{3}$$

$$t' = a_{41}x + a_{44}t \tag{4}$$

Now the requirement that the origin of the primed frame $x' = 0$ moves with velocity V as measured in the unprimed frame gives:

$$a_{14} = -Va_{11} \tag{5}$$

In place of (3) and (4) we may write a similar expression for x and t in terms of x' and t'. A way of doing this that uses the symmetry between the primed and unprimed frames is to observe that reversing the sense of time interchanges the primed and unprimed frames. This observation permits us to write:

$$x = a_{11}x' - a_{14}t' \tag{6}$$

$$-t = a_{41}x' - a_{44}t' \tag{7}$$

Substituting **(3)** and **(4)** into **(6)** and **(7)** we obtain the identities:

$$x = (a_{11}^2 - a_{14}a_{41})x + a_{14}(a_{11} - a_{44})t \tag{8}$$

$$t = -a_{41}(a_{11} - a_{44})x + (a_{44}^2 - a_{14}a_{41})t \tag{9}$$

From **(8)** and **(9)** we obtain:

$$a_{11} = a_{44} = \gamma \tag{10}$$

From **(5)** we obtain:

$$a_{14} = -\gamma V \tag{11}$$

Finally, from **(8)** and **(9)**:

$$a_{41} = (a_{44}^2 - 1)/a_{14} = (1 - \gamma^2)/\gamma V \tag{12}$$

We may then rewrite **(3)**, **(4)**, **(6)**, and **(7)** as

$$x' = \gamma(x - Vt) \qquad\qquad x = \gamma(x' + Vt') \tag{13}$$

$$t' = \gamma\left(\frac{1 - \gamma^2}{\gamma^2}\frac{x}{V} + t\right) \qquad t = \gamma\left(-\frac{1 - \gamma^2}{\gamma^2}\frac{x'}{V} + t'\right) \tag{14}$$

It remains to determine γ and for this we introduce the constancy of the speed of light as measured in both the primed and unprimed frames.

Exercise

1. Solve **(3)** and **(4)** for x and t. By comparison with **(3)** and **(4)** obtain **(12)** by using symmetry arguments.

1 **Lorentz Transformation.** We imagine that at the instant that the origins of the primed and unprimed frames coincide, a light pulse is emitted from the common origin. By requiring that the wavefront of the propagating pulse be a spherical shell centered about the origin in *both* frames, we are able to determine γ.

Let us *assume* a spherical wavefront moving out with velocity c in the unprimed frame:

$$r = ct \qquad \text{or} \qquad x^2 + y^2 + z^2 = c^2 t^2 \tag{15}$$

But if the speed of light is an invariant, the wavefront must also be spherical and centered at the origin of the primed frame:

$$r' = ct' \qquad \text{or} \qquad x'^2 + y'^2 + z'^2 = c^2 t'^2 \tag{16}$$

Thus **(15)** and **(16)** must be satisfied simultaneously together with the transformation given by **(1)**, **(2)**, **(13)**, and **(14)**. Writing **(15)** in terms of the coordinates and time in the primed frame, we obtain:

$$\frac{c^2}{V^2}\left[2 - \gamma^2\left(1 - \frac{V^2}{c^2}\right) - \frac{1}{\gamma^2}\right]x'^2 + y'^2 + z'^2$$

$$= \gamma^2\left(1 - \frac{V^2}{c^2}\right)c^2 t'^2 + \frac{2c^2}{\gamma V}\left[1 - \gamma^2\left(1 - \frac{V^2}{c^2}\right)\right]x't' \quad \text{(17)}$$

If the wavefront is to be centered about the origin in *both* frames, the coefficient of $x't'$ in **(17)** must be identically zero, which gives

$$\gamma^2 = \frac{1}{1 - V^2/c^2} \quad \text{(18)}$$

Substituting **(18)** into **(17)** gives **(16)** automatically. We may now write the transformation in a somewhat more compact form:

▶

$$x' = \gamma(x - Vt) \qquad x = \gamma(x' + Vt') \quad \text{(19)}$$

▶

$$t' = \gamma(t - Vx/c^2) \qquad t = \gamma(t' + Vx'/c^2) \quad \text{(20)}$$

The transformation given by **(1)**, **(2)**, **(19)**, and **(20)** also follows from an analysis by H. A. Lorentz based on classical electrodynamics[1] and is generally known as the Lorentz transformation.

Exercises

2. Verify from **(19)** and **(20)** the relation:

$$x^2 - c^2 t^2 = x'^2 - c^2 t'^2$$

   This quantity is called a Lorentz invariant.
3. Given the first equations in **(19)** and **(20)**, obtain the second equations by direct substitution.
4. A body of length L' moves parallel to its length with velocity V. Show that the length of the body in the laboratory frame is given by:

$$L = L'/\gamma < L'$$

   This is the Lorentz contraction.
5. Two events occurring at a given place have separation τ'. Let the frame in which the events occur at a fixed position move with velocity V with respect to the

laboratory frame. Show that the apparent separation of the events in the laboratory frame is:

$$\tau = \gamma\tau' > \tau'$$

This is called time dilation.

6. A spherical wave of angular frequency ω and wavevector k is radiated from the origin:

$$A(r,t) = A(r)\sin(kr - \omega t)$$

The wave is detected by an observer at $0'$ moving with velocity $\mathbf{V} = V\hat{\mathbf{x}}$. Show that the frequency of the wave as measured by the observer is:

$$\omega' = \gamma(\omega - kV \cos \theta)$$

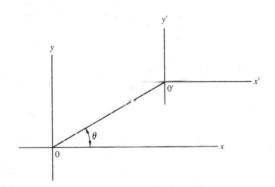

7. Use the result of the preceding exercise to obtain the longitudinal Doppler shift

$$\omega' = \gamma\omega(1 - \beta)$$

and the transverse Doppler shift

$$\omega' = \gamma\omega$$

of a light wave propagating with velocity c.

8. Show that the vector form of (1), (2), and (19) is

$$\mathbf{r}' = \mathbf{r} + (\gamma - 1)\hat{\mathbf{x}}\hat{\mathbf{x}} \cdot \mathbf{r} - \gamma V t\hat{\mathbf{x}}$$

9. Let $\mathbf{r}$ be the coordinate of a point that moves with velocity $V\hat{\mathbf{x}}$ in the unprimed frame:

$$\mathbf{r} = \mathbf{a} + V t\hat{\mathbf{x}}$$

Obtain the transformation:

$$\mathbf{r}' = \mathbf{a} + (\gamma - 1)\hat{\mathbf{x}}\hat{\mathbf{x}} \cdot \mathbf{a}$$

10. Displacements **a** and **b** are fixed in a body that moves with velocity $V\hat{\mathbf{x}}$ in the unprimed frame. The displacements define an increment of area:

$$d\mathbf{S} = \mathbf{a} \times \mathbf{b}$$

Show that the increment of area in the primed frame, which moves with the body is given by:

$$d\mathbf{S}' = \gamma d\mathbf{S} - (\gamma - 1)\hat{\mathbf{x}}\hat{\mathbf{x}} \cdot d\mathbf{S}$$

and that the magnitude of the area in the unprimed frame is given by:

$$dS = \frac{dS'}{\gamma(1 - \beta^2 \cos^2 \theta)^{1/2}}$$

where θ is the angle between $\hat{\mathbf{x}}$ and $d\mathbf{S}$ and β is V/c.

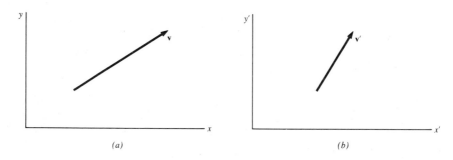

(a)   (b)

Figure 2 Transformation of velocity. (a) A point moves with velocity **v** in the unprimed frame. (b) The velocity is **v'** in the primed frame.

2 **Transformation of Velocity.** We imagine that a point moves with velocity **v** as measured in the (x,y,z,t) frame shown in Figure 2a. What velocity is measured in the (x',y',z',t') frame that has a velocity V parallel to the x coordinate as shown in Figure 2b?
   Taking the differential of **(1)**, **(19)**, and **(20)** we have:

$$dy' = dy \tag{21}$$

$$dx' = \gamma(dx - V\,dt) \tag{22}$$

$$dt' = \gamma\left(-\frac{V}{c^2}dx + dt\right) \tag{23}$$

Taking the ratio of **(21)** and **(22)** to **(23)** and dividing through by dt we obtain:

$$\blacktriangleright \qquad v_x' = \frac{dx'}{dt'} = \frac{dx - V\,dt}{dt - V\,dx/c^2} = \frac{v_x - V}{1 - v_x V/c^2} \qquad (24)$$

$$\blacktriangleright \qquad v_y' = \frac{dy'}{dt'} = \frac{dy}{\gamma(dt - V\,dx/c^2)} = \frac{v_y}{\gamma(1 - v_x V/c^2)} \qquad (25)$$

Note in particular that the ratio of v_y' to v_y depends not only on V but also on v_x as a consequence of the mixing of x and t under the Lorentz transformation. The transformation of v_z is the same as for v_y.

Exercises

11. A body moves in the primed frame with velocity c at an angle ϕ' to the x' axis. Find the magnitude and direction of its velocity in the unprimed frame. How is this result related to the postulated constancy of c in all frames?
12. A body moves with speed V about an observer. Show that the *apparent* direction of the body is ahead of its actual direction by θ with:

$$\tan\theta = \gamma V/c$$

The effect results from what is called the *aberration of light*.
13. A moving object approaches a stationary object with velocity **v**. At what velocity **V** would an observer have to move so that in the frame of the observer the two bodies have equal and opposite velocities?

3 **Gradient Transformation.** We now want to consider the general quantity $\phi(\mathbf{r},t)$, which becomes $\phi(\mathbf{r}', t')$ as a result of applying the Lorentz transformation to $\mathbf{r}$ and t. We may equate the differentials:

$$d\phi = d\mathbf{r} \cdot \nabla\phi + \frac{\partial\phi}{\partial t}\,dt = d\mathbf{r}' \cdot \nabla'\phi + \frac{\partial\phi}{\partial t'}\,dt' \qquad (26)$$

Substituting for $d\mathbf{r}'$ and dt' from **(21)**, **(22)**, and **(23)** and equating the coefficients of dx, dy, dz, and dt we obtain:

$$\frac{\partial\phi}{\partial x} = \gamma\left(\frac{\partial}{\partial x'} - \frac{V}{c^2}\frac{\partial}{\partial t'}\right)\phi \qquad (27)$$

$$\frac{\partial\phi}{\partial y} = \frac{\partial\phi}{\partial y'} \qquad \frac{\partial\phi}{\partial z} = \frac{\partial\phi}{\partial z'} \qquad (28)$$

$$\frac{\partial\phi}{\partial t} = \gamma\left(\frac{\partial}{\partial t'} - V\frac{\partial}{\partial x'}\right)\phi \qquad (29)$$

To obtain the inverse transformations we interchange the primed and unprimed coordinates and change the sign of V:

$$\frac{\partial \phi}{\partial x'} = \gamma \left(\frac{\partial}{\partial x} + \frac{V}{c^2} \frac{\partial}{\partial t} \right) \phi \tag{30}$$

$$\frac{\partial \phi}{\partial y'} = \frac{\partial \phi}{\partial y} \qquad \frac{\partial \phi}{\partial z'} = \frac{\partial \phi}{\partial z} \tag{31}$$

$$\frac{\partial \phi}{\partial t'} = \gamma \left(\frac{\partial}{\partial t} + V \frac{\partial}{\partial x} \right) \phi \tag{32}$$

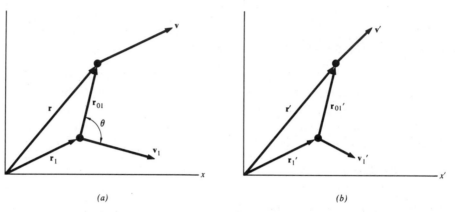

(a)   (b)

Figure 3   Transformation of relative position. (*a*) A point at **r** moves with velocity **v** and a point at $\mathbf{r}_1$ moves with velocity $\mathbf{v}_1$. (*b*) The position of the point at **r** in the unprimed frame is **r**′ in the primed frame. At the same time in the primed frame the other point is at $\mathbf{r}_1'$. In general $\mathbf{r}_1'$ is *not* related to $\mathbf{r}_1$ by a Lorentz transformation.

4 **[Transformation of Relative Position.]** We consider the situation shown in Figure 3*a* where the distance between a point at **r** and moving with constant velocity **v** and a point at $\mathbf{r}_1$ and moving with constant velocity $\mathbf{v}_1$ is $\mathbf{r}_{01} = \mathbf{r} - \mathbf{r}_1$, where the measurements of **r** and $\mathbf{r}_1$ are *simultaneous in the unprimed frame*. We wish to obtain an expression for $\mathbf{r}_{01}'$ where **r**′ and $\mathbf{r}_1'$ are simultaneous in the primed frame as shown in Figure 3*b*. We transform to a frame moving with velocity $\mathbf{V} = V\hat{\mathbf{x}}$. We obtain from **(1)**, **(2)**, and **(19)** for the components of **r**′

$$x' = \gamma(x - Vt) \qquad y' = y \qquad z' = z \tag{33}$$

and from **(20)**

$$t' = \gamma(t - Vx/c^2) \tag{34}$$

We next need the components of $\mathbf{r}_1'$ at time t'. From **(20)** the time in the unprimed frame that corresponds to t' at $\mathbf{r}_1'$ is

$$t_1 = \gamma(t' + Vx_1'/c^2) \tag{35}$$

(We emphasize that what we seek is the transformation between $\mathbf{r}_{01}$, where $\mathbf{r}_0$ and $\mathbf{r}_1$ are measured *simultaneously* in the unprimed frame, and $\mathbf{r}_{01}'$ where $\mathbf{r}_0'$ and $\mathbf{r}_1'$ are measured simultaneously in the *primed* frame.) At time t_1 the vector $\mathbf{r}_1$ has the value

$$\mathbf{r}_1(t_1) = \mathbf{r}_1(t) + (t_1 - t)\mathbf{v}_1 \tag{36}$$

Using **(1)**, **(2)**, and **(19)** we obtain for the components of $\mathbf{r}_1'$:

$$y_1' = y_1 + v_{1y}(t_1 - t) = y_1 + \gamma v_{1y} V x_{01}'/c^2 \tag{37}$$

$$z_1' = z_1 + v_{1z}(t_1 - t) = z_1 + \gamma v_{1z} V x_{01}'/c^2 \tag{38}$$

$$x_1' = \gamma[x_1 + v_{1x}(t_1 - t) - Vt_1] = \gamma x_1 + \gamma^2 V(v_{1x} - V)x_{01}'/c^2 - \gamma Vt \tag{39}$$

Subtracting **(37)**, **(38)** and **(39)** from **(33)** and transposing we obtain:

$$\gamma x_{01} = x_{01}'[1 + \gamma^2 V(v_{1x} - V)/c^2] \tag{40}$$

$$y_{01} = y_{01}' + \gamma v_{1y} V x_{01}'/c^2 \tag{41}$$

$$z_{01} = z_{01}' + \gamma v_{1z} V x_{01}'/c^2 \tag{42}$$

For the special case where we transform to a frame in which the position $\mathbf{r}_1'$ is at *rest* we have for the transformation velocity

$$v_{1x} = V \qquad v_{1y} = 0 \qquad v_{1z} = 0 \tag{43}$$

and for the transformation from **(40)**, **(41)**, and **(42)**:

$$x_{01}' = \gamma x_{01} \qquad y_{01}' = y_{01} \qquad z_{01}' = z_{01} \tag{44}$$

The magnitude of $\mathbf{r}_{01}'$ may then be written as

$$\blacktriangleright \qquad r_{01}' = (x_{01}'^2 + y_{01}'^2 + z_{01}'^2)^{1/2} = (\gamma^2 x_{01}^2 + y_{01}^2 + z_{01}^2)^{1/2}$$

$$= (\gamma^2 \cos^2 \theta + \sin^2 \theta)^{1/2} r_{01} = \gamma(1 - \beta^2 \sin^2 \theta)^{1/2} r_{01} \tag{45}$$

MOMENTUM AND ENERGY

In obtaining the Lorentz transformation and the transformation of velocity and distance we have used only kinematics. We have required that a freely moving particle move in a straight line in any frame. As a limiting case we have used also the second postulate of relativity, that the velocity of light is c in any inertial frame, independent of the velocity of the source in that frame. We now go on to consider the transformation of dynamical quantities imposed by the conservation of momentum and energy in all inertial frames.

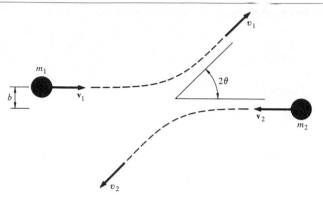

Figure 4   Collision between masses. The impact parameter is b and the scattering angle is 2θ.

5   Conservation of Momentum.

We next examine a collision between masses m_1 and m_2 first in the center of mass frame and then in a frame that moves with respect to the center of mass. We show in Figure 4 a collision in which m_1 has initial velocity $\mathbf{v}_1$ and m_2 has initial velocity $\mathbf{v}_2$. The impact parameter is b and the scattering angle is 2θ. By conservation of linear momentum, time reversal, and parity the speeds after the collision and the impact parameter are the same as before. We introduce a coordinate system as shown in Figure 5a with the x axis making an angle θ with the initial velocity directions. We show in Figure 5b the same collision as viewed from a frame moving with velocity V.

In the primed frame the mass $m_1{}'$ moves along the y' axis with velocity:

$$v_{1y}' = \frac{v_{1y}}{\gamma(1 - v_{1x} V/c^2)} = \pm \frac{v_1 \sin \theta}{\gamma(1 - v_{1x} V/c^2)} \tag{46}$$

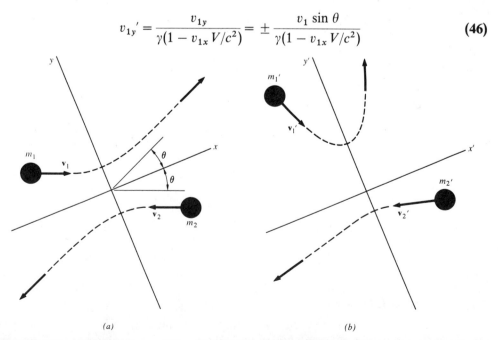

(a)   (b)

Figure 5   Conservation of momentum. (*a*) Axes are constructed such that the x axis bisects the scattering angle. (*b*) The collision is viewed from a frame moving with velocity V along x.

and the mass m_2' has a component of velocity along y':

$$v_{2y}' = \frac{v_{2y}}{\gamma(1 - v_{2x} V/c^2)} = \pm \frac{v_2 \sin \theta}{\gamma(1 - v_{2x} V/c^2)} \qquad (47)$$

Note that although we have

$$m_1 v_{1y} = -m_2 v_{2y} \qquad (48)$$

in the center of mass frame, this relation does not hold in the primed frame because v_{1y} and v_{2y} transform differently. It is for this reason that we must allow for masses m_1' and m_2' which, as we will see, differ from m_1 and m_2.

Now we have postulated that the laws of physics are identical in all frames that move with a constant relative velocity V. Then momentum must be conserved in the primed frame. We assume as the expression for the momentum transfer between the two masses.

$$m_1' v_{1y}' = -m_2' v_{2y}' \qquad (49)$$

Substituting from (46) and (47) and canceling the common factor $\sin \theta$ we obtain

$$\frac{m_1' v_1}{1 - v_{1x} V/c^2} = \frac{m_2' v_2}{1 - v_{2x} V/c^2} \qquad (50)$$

with $m_1 v_1 = m_2 v_2$ we may divide both sides of (50) by this factor to obtain, finally,

$$\frac{m_1'/m_1}{1 - v_{1x} V/c^2} = \frac{m_2'/m_2}{1 - v_{2x} V/c^2} = f(V) \qquad (51)$$

as a consequence of conservation of momentum. The second equality follows from regarding each of the other terms as independent of the particular collision.

6 Transformation of Mass. In order to determine $f(V)$ in (51) we consider a special case of the collision shown in Figure 4 with

$$V = v_{1x} = v_1 \cos \theta \qquad (52)$$

We take the particles to be identical: $m_1 = m_2 = m$. The initial velocities along x will then be given by:

$$v_{1x} = -v_{2x} = V \qquad (53)$$

and we may write (51) as

$$\frac{m_1'}{1 - V^2/c^2} = \frac{m_2'}{1 + V^2/c^2} \qquad (54)$$

Finally we let the impact. parameter b go to infinity. Then m_1' will be at rest in the primed frame of Figure 5b and we set

$$m_1' = M \qquad (55)$$

where we call M the rest mass. The second body now moves with constant velocity from (24):

$$v' = \frac{2V}{1 + V^2/c^2} \qquad (56)$$

Designating the mass of this body by m' we obtain from (54):

$$m' = \frac{1 + V^2/c^2}{1 - V^2/c^2} M \qquad (57)$$

We have, from (56),

$$1 - \frac{v'^2}{c^2} = \frac{(1 + V^2/c^2)^2 - 4V^2/c^2}{(1 + V^2/c^2)^2} = \frac{(1 - V^2/c^2)^2}{(1 + V^2/c^2)^2} \qquad (58)$$

so that we may rewrite (55) as:

$$m' = \frac{M}{(1 - v'^2/c^2)^{1/2}} \qquad (59)$$

which gives finally the mass m' of a body in a frame in which it moves with velocity $\mathbf{v}'$ in terms of the rest mass M and the magnitude of $\mathbf{v}'$.

In the unprimed frame we write the mass as:

▶
$$m = \frac{M}{(1 - v^2/c^2)^{1/2}} \qquad (60)$$

In order to obtain the transformation between m' and m we must write $\mathbf{v}'$ in terms of $\mathbf{v}$. From (24) and (25) we have

$$1 - \frac{v'^2}{c^2} = 1 - \frac{v_x^2 - 2v_x V + V^2 + v_y^2/\gamma^2}{(1 - v_x V/c^2)^2 c^2} = \frac{1 - v^2/c^2}{\gamma^2(1 - v_x V/c^2)^2} \qquad (61)$$

Substituting from (61) into (59) and (60) we obtain finally:

$$m' = \gamma(1 - v_x V/c^2)m \qquad (62)$$

Note that substituting **(62)** into **(51)** we obtain:

$$f(V) = \frac{m'/m}{1 - v_{1x}V/c^2} = \gamma \qquad (63)$$

and is as expected a function only of the transformation.

7 Potential Energy. By examining the collision of Figure 4 well before and well after the collision we have avoided any consideration of potential energy and its relation to mass. We now take up this question explicitly by considering the system shown in Figure 6, where we have two identical masses of rest mass M connected by a spring.

In the rest frame of the system we have for the velocities:

$$v_1 = v(t) \qquad v_2 = v_1 = -v(t) \qquad (64)$$

and for the masses:

$$m_1 = m_2 = m = \frac{M}{(1 - v^2/c^2)^{1/2}} \qquad (65)$$

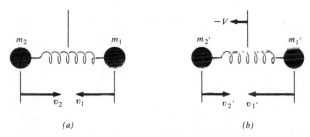

(a)   (b)

Figure 6 Potential energy. (a) Two particles of rest mass M are connected by a spring. In the rest frame the particles move with equal and opposite velocity. (b) In a frame moving with velocity $-V$ the particle velocities and masses are no longer equal.

We may now view the motion of the mass-spring system from a primed system with respect to which the center of mass moves with velocity $-V$ as shown in Figure 6b. In the primed frame we may write for the velocities from **(24)**:

$$v_1' = \frac{v - V}{1 - Vv/c^2} \qquad v_2' = \frac{-v - V}{1 + Vv/c^2} \qquad (66)$$

and from **(62)** we write the masses as:

$$m_1' = \gamma(1 - Vv/c^2)m \qquad m_2' = \gamma(1 + Vv/c^2)m \qquad (67)$$

Let us now compute the sum of the two momenta in the primed frame:

$$p_1' + p_2' = m_1'v_1' + m_2'v_2' = -2\gamma mV \qquad (68)$$

But m is a function of time and thus an observer in the primed frame would find that the momentum carried by the masses varied perodically with the time. If conservation of momentum is to apply in the primed frame we are forced to conclude that the spring has a mass which varies periodically with the time in such a way that the total momentum is constant. If m_0 is the mass of the spring in the laboratory frame we have, in the primed frame,

$$p_0' = -\gamma m_0 V \tag{69}$$

Then the constancy of the momentum, which is the sum of **(68)** and **(69)**, requires that the sum of the masses as determined in the unprimed frame be constant:

$$m_0 + 2m = m_{\text{tot}} = \text{const} \tag{70}$$

8 **Conservation of Energy.** The momentum of a mass in a frame in which it moves with velocity $\mathbf{v}$ may be written in terms of **(60)** as:

$$\mathbf{p} = m\mathbf{v} = \frac{M}{(1 - v^2/c^2)^{1/2}} \mathbf{v} \tag{71}$$

Now if a force $\mathbf{F}$ acts on m, we have, by definition,

$$\mathbf{F} = \frac{d\mathbf{p}}{dt} = M \frac{d}{dt} \frac{\mathbf{v}}{(1 - v^2/c^2)^{1/2}} = m\left(\mathbf{a} + \frac{\mathbf{v}(\mathbf{a} \cdot \mathbf{v})/c^2}{1 - v^2/c^2}\right) \tag{72}$$

where we have written $\mathbf{a} = d\mathbf{v}/dt$ and used the definition of m as given by **(60)**. Now, the rate at which work is done on the body is given by

$$\frac{dW}{dt} = \mathbf{F} \cdot \mathbf{v} = \frac{m\mathbf{a} \cdot \mathbf{v}}{1 - v^2/c^2} = \frac{d}{dt} mc^2 \tag{73}$$

where the final equality may be shown by differentiating **(60)**. Now if we interpret **(73)** as a statement of conservation of energy, we have for the kinetic energy (including the rest energy):

▶
$$U_K = mc^2 \tag{74}$$

Returning to the example of the mass-spring oscillator, we write, for the total energy,

$$U = 2U_K + U_P = 2mc^2 + U_P \tag{75}$$

where U_K is the kinetic energy of each of the masses and U_P is the potential energy stored in the spring. Comparing **(75)** with **(70)** we obtain

$$U_P = m_0 c^2 \tag{76}$$

where the potential energy stored in the spring increases its mass in just the way that kinetic energy increases the masses attached to the ends of the spring. And finally we have the relation for the total energy:

$$U = m_{tot} c^2 \tag{77}$$

which states the equivalence of mass and energy independent of whether the energy is potential or kinetic.

Exercises

14. From (60) and (71) obtain the energy-momentum relation:

$$U_K{}^2 = M^2 c^4 + p^2 c^2$$

15. Show that the velocity of a relativistic particle is given by:

$$\mathbf{v} = \frac{\mathbf{p} c^2}{U_K}$$

9 **Transformation of Momentum and Energy.** From the definition of momentum as given by (71), the transformation of mass as given by (62), the transformation of velocities as given by (24) and (25), and the definition of total energy by (77) we obtain:

$$p_x' = m'v_x' = \gamma m(v_x - V) = \gamma(p_x - UV/c^2) \tag{78}$$

$$p_y' = m'v_y' = mv_y = p_y \tag{79}$$

$$p_z' = m'v_z' = mv_z = p_z \tag{80}$$

$$U' = m'c^2 = \gamma(mc^2 - mv_x V) = \gamma(U - p_x V) \tag{81}$$

Exercises

16. Let the total momentum and energy for a two-particle system be $\mathbf{p} = \mathbf{p}_1 + \mathbf{p}_2$ and $U = U_1 + U_2$, respectively. Show explicitly that the Lorentz transformation on $\mathbf{p}$ and U are consistent with the invariance of the quantity $U^2 - p^2 c^2$.

17. Show that a free electron moving in vacuum at velocity $\mathbf{v}$ cannot emit a single light quantum without violation of the conservation laws. Show on the other hand that a hydrogen atom in an excited state can emit a light quantum. What is the difference between these two processes?

18. By treating a light flash as a particle of momentum $\mathbf{p}$ and energy U, use (78), (79), (80), and (81) to obtain the aberration formula of Exercise 12.

19. A body of rest mass M and momentum $\mathbf{p}$ approaches a similar body at rest. Find the velocity $\mathbf{V}$ of the frame in which the two bodies have equal and opposite momenta. What is the total energy of the two bodies in this frame?

20. In the primed frame an object has energy U_0 and zero momentum. Using **(78)**, **(79)**, **(80)**, and **(81)** find the momentum and energy in the laboratory frame. Compare with **(71)** and **(74)**.

21. Using **(78)**, **(79)**, **(80)**, and **(81)** obtain the energy-momentum relation:

$$U^2 = M^2 c^4 + p^2 c^2$$

10 Transformation of Force. From **(72)**, which gives the definition of force, and **(78)**–**(81)** we have

$$dp_x' = F_x'\, dt' = \gamma(dp_x - V\, dU/c^2) = \gamma(F_x - V\mathbf{F}\cdot\mathbf{v}/c^2)\, dt \tag{82}$$

$$dp_{yz}' = F_{yz}'\, dt' = dp_{yz} = F_{yz}\, dt \tag{83}$$

$$dU' = \mathbf{F}'\cdot\mathbf{v}'\, dt' = \gamma(dU - V\, dp_x) = \gamma(\mathbf{F}\cdot\mathbf{v} - VF_x)\, dt \tag{84}$$

Taking the differential of **(20)** and dividing into **(82)**, **(83)**, and **(84)**, we obtain

$$F_x' = \frac{F_x - (V/c^2)\mathbf{F}\cdot\mathbf{v}}{(1 - v_x V/c^2)} \tag{85}$$

$$F_{yz}' = \frac{F_{yz}}{\gamma(1 - v_x V/c^2)} \tag{86}$$

We find **(85)** and **(86)** to be most useful written in another form. To do this we use the following procedure. First, by expanding the scalar product we write **(85)** as:

$$F_x' = F_x - \frac{V}{c^2}\frac{F_y v_y + F_z v_z}{(1 - v_x V/c^2)} \tag{87}$$

From **(86)** we have

$$v_y F_y' + v_z F_z' = \frac{v_y F_y + v_z F_z}{\gamma(1 - v_x V/c^2)} \tag{88}$$

Now, substituting **(88)** into **(87)** we obtain:

$$F_x = F_x' + (\gamma V/c^2)(v_y F_y' + v_z F_z') \tag{89}$$

From **(86)** we may write:

$$F_y = \gamma F_y'(1 - v_x V/c^2) \tag{90}$$

$$F_z = \gamma F_z'(1 - v_x V/c^2) \tag{91}$$

By using the identity:

$$\mathbf{v}\times(\mathbf{V}\times\mathbf{F}') = \mathbf{V}(\mathbf{v}\cdot\mathbf{F}') - \mathbf{F}'(\mathbf{v}\cdot\mathbf{V}) = \hat{\mathbf{x}}V(v_y F_y' + v_z F_z') - \hat{\mathbf{y}}v_x V F_y' - \hat{\mathbf{z}}v_x V F_z' \tag{92}$$

we may write:

$$\blacktriangleright \qquad F_x = F_x' + \frac{\gamma}{c^2}[\mathbf{v} \times (\mathbf{V} \times \mathbf{F}')]_x \qquad (93)$$

$$\blacktriangleright \qquad F_{yz} = \gamma F_{yz}' + \frac{\gamma}{c^2}[\mathbf{v} \times (\mathbf{V} \times \mathbf{F}')]_{yz} \qquad (94)$$

This is the equation that we will use in analyzing the force between moving charges and between charges and currents. Note particularly that the second terms are proportional to the velocity in the *unprimed* frame and are perpendicular to $\mathbf{v}(t)$ as well.

Exercise

22. Where the velocity is zero in the primed frame obtain from **(93)** and **(94)** the transformation:

$$F_x' = F_x$$

$$F_{yz}' = \gamma F_{yz}$$

SUMMARY

1. The Lorentz transformation between an unprimed frame and a primed frame which moves with velocity $\mathbf{V}$ parallel to x is given by:

$$x' = \gamma(x - Vt) \qquad\qquad x = \gamma(x' + Vt) \qquad\qquad (19)$$

$$y' = y \qquad\qquad\qquad y = y' \qquad\qquad\qquad\quad (1)$$

$$z' = z \qquad\qquad\qquad z = z' \qquad\qquad\qquad\quad (2)$$

$$t' = \gamma(t - Vx/c^2) \qquad t = \gamma(t' + Vx'/c^2) \qquad (20)$$

2. By taking differentials of the Lorentz transformation the velocity transformation is obtained:

$$v_x' = \frac{v_x - v}{1 - v_x V/c^2} \qquad v_x = \frac{v_x' + V}{1 + v_x'V/c^2} \qquad (24)$$

$$v_{yz}' = \frac{v_{yz}}{\gamma(1 - v_x V/c^2)} \qquad v_{yz} = \frac{v_{yz}'}{\gamma(1 + v_x'V/c^2)} \qquad (25)$$

3. Positions $\mathbf{r}$ and $\mathbf{r}_1$ separated by $\mathbf{r}_{01}$ are in motion with velocities $\mathbf{v}$ and $\mathbf{v}_1$, respectively. In a frame moving with velocity $\mathbf{V} = \mathbf{v}_1$ such that $\mathbf{r}_1'$ is at rest, the relative distance between $\mathbf{r}'$ and $\mathbf{r}_1'$ is given by

$$r_{01}' = \gamma(1 + \beta^2 \sin^2 \theta)^{1/2} r_{01} \qquad (45)$$

where θ is the angle between $\mathbf{r}_{01}$ and $\mathbf{v}_1$.

4. The mass of a body moving with velocity **v** is given by:

$$m = M/(1 - v^2 c^2)^{1/2} \tag{70}$$

where M is the mass of the body at rest and $m\mathbf{v}$ is the momentum.

5. The kinetic energy of a body of rest mass M and moving with velocity **v** is given (including the rest energy) by:

$$U_K = mc^2 = Mc^2/(1 - v^2/c^2)^{1/2} \tag{74}$$

6. Where $\mathbf{F'}$ is the force on a body in a primed frame moving with velocity **V**, the components of the force in the unprimed frame are given by:

$$F_x = F_x' + \frac{\gamma}{c^2} [\mathbf{v} \times (\mathbf{V} \times \mathbf{F'})]_x \tag{93}$$

$$F_{yz} = \gamma F_{yz}' + \frac{\gamma}{c^2} [\mathbf{v} \times (\mathbf{V} \times \mathbf{F'})]_{yz} \tag{94}$$

REFERENCES

Bergmann, P. G., *Introduction to the Theory of Relativity*, Prentice-Hall, 1942.

Bohm, D., *The Special Theory of Relativity*, Benjamin, 1965.

Born, M., *Einstein's Theory of Relativity*, Dover, 1962.

French, A. P., *Special Relativity*, Norton, 1968.

Gill, T. P., *The Doppler Effect*, Academic, 1965.

Good, R. H., *Basic Concepts of Relativity*, Van Nostrand Reinhold, 1968.

Kacser, C., *Introduction to the Special Theory of Relativity*, Prentice-Hall, 1967.

Kim, S. K., *Physics: The Fabric of Reality*, Macmillan, 1975.

Mermin, N. D., *Space and Time in Special Relativity*, McGraw-Hill, 1968.

Møller, C., *The Theory of Relativity*, Second Edition, Oxford, 1972. Although this book is at a more advanced level than the others listed here, the first three chapters provide an introduction to the subject.

Resnick, R., *Relativity and Early Quantum Theory*, Wiley, 1972.

Rindler, W., *Essential Relativity*, Second Edition, Springer, 1977.

Rosser, W. G. V., *An Introduction to the Theory of Relativity*, Butterworths, 1971.

Sachs, M., *Ideas of the Theory of Relativity*, Halsted, 1974.

Sard, R. D., *Relativistic Mechanics*, Benjamin, 1970.

Shadowitz, A., *Special Relativity*, Saunders, 1968.

Smith, J. H., *Introduction to Special Relativity*, Benjamin, 1965.

Symon, *Mechanics*, Third Edition, Addison-Wesley, 1971. Chapters 13 and 14 of this text provide an introduction both to the basic postulates of the special theory of relativity and to relativistic dynamics.

Taylor, E. F. and J. A. Wheeler, *Spacetime Physics*, Freeman, 1966.

G GENERALIZED EQUATIONS OF MOTION

In this appendix we develop the Lagrangian and Hamiltonian equations for a non-relativistic charged particle and find the constants of the motion in the absence of an electric field.

1 **Lagrangian Mechanics.** The Lagrangian formulation of mechanics is based on the function of position and velocity $\mathscr{L}(\mathbf{r},\mathbf{v})$. A *generalized* momentum is defined by the gradient of the Lagrangian with respect to velocity:

$$\mathbf{p} = \nabla_{\mathbf{v}} \mathscr{L}(\mathbf{r},\mathbf{v}) \tag{1}$$

The equation of motion of the particle takes the form:

$$\frac{d\mathbf{p}}{dt} = \nabla_{\mathbf{r}} \mathscr{L}(\mathbf{r},\mathbf{v}) \tag{2}$$

where $\mathbf{p}$ is given by (1), and (2) must be equivalent to Newton's equation of motion.

As an example we take a particle of mass m and charge q moving in an electric field $\mathbf{E}$. The Newtonian equation of motion is:

$$m\frac{d\mathbf{v}}{dt} = q\mathbf{E} \tag{3}$$

By taking the Lagrangian to be:

$$\mathscr{L}(\mathbf{r},\mathbf{v}) = \tfrac{1}{2}mv^2 - q\phi \tag{4}$$

we obtain for the momentum from (1):

$$\mathbf{p} = m\mathbf{v} \tag{5}$$

709

Substituting into **(2)** we have:

$$\frac{d\mathbf{p}}{dt} = m\frac{d\mathbf{v}}{dt} = \nabla_{\mathbf{r}}\mathscr{L}(\mathbf{r},\mathbf{v}) = -q\nabla_{\mathbf{r}}\phi = q\mathbf{E} \tag{6}$$

which is the same as **(3)**.

In the presence of a magnetic field the problem is rather more complicated. In place of **(3)** we have:

$$m\frac{d\mathbf{v}}{dt} = q(\mathbf{E} + \mathbf{v} \times \mathbf{B}) \tag{7}$$

We write the Lagrangian for this problem as:

$$\mathscr{L}(\mathbf{r},\mathbf{v}) = \tfrac{1}{2}mv^2 + q\mathbf{v} \cdot \mathbf{A} - q\phi \tag{8}$$

where **A** is the vector potential. We have for the generalized momentum from **(1)**:

$$\mathbf{p} = m\mathbf{v} + q\mathbf{A} \tag{9}$$

The equation of motion is from **(2)**, **(9)**, and **(8)**:

$$\frac{d\mathbf{p}}{dt} = m\frac{d\mathbf{v}}{dt} + q\frac{d\mathbf{A}}{dt} = q\nabla_{\mathbf{r}}(\mathbf{v} \cdot \mathbf{A}) - q\nabla_{\mathbf{r}}\phi \tag{10}$$

We interpret the rate of change of the vector potential with respect to time to be taken at the position of q so that we may write:

$$\frac{d\mathbf{A}}{dt} = (\mathbf{v} \cdot \nabla_{\mathbf{r}})\mathbf{A} + \frac{\partial \mathbf{A}}{\partial t} \tag{11}$$

From **(B-6)** we have:

$$\nabla_{\mathbf{r}}(\mathbf{v} \cdot \mathbf{A}) = (\mathbf{v} \cdot \nabla_{\mathbf{r}})\mathbf{A} + \mathbf{v} \times (\nabla_{\mathbf{r}} \times \mathbf{A}) \tag{12}$$

Substituting **(11)** and **(12)** into **(10)** we obtain:

$$m\frac{d\mathbf{v}}{dt} = -q\left(\nabla_{\mathbf{r}}\phi + \frac{\partial \mathbf{A}}{\partial t}\right) + q\mathbf{v} \times (\nabla_{\mathbf{r}} \times \mathbf{A}) \tag{13}$$

Writing for the electric field **(11-114)**:

$$\mathbf{E} = -\nabla_{\mathbf{r}}\phi - \frac{\partial \mathbf{A}}{\partial t} \tag{14}$$

and for the magnetic field:

$$B = V_r \times A \tag{15}$$

we may see that **(13)** is equivalent to **(7)** as required.

2 Hamiltonian Mechanics. The Hamiltonian formulation of mechanics is based on a function of position and conjugate momentum $\mathscr{H}(r,p)$ such that the particle velocity must be given by

$$v = \nabla_p \mathscr{H}(r,p) \tag{16}$$

The equation of motion takes the form:

$$\frac{dp}{dt} = -\nabla_r \mathscr{H}(r,p) \tag{17}$$

The Hamiltonian and Langrangian are connected by the relation:

$$\mathscr{H}(r,p) = p \cdot v - \mathscr{L}(r,v) \tag{18}$$

where it is understood that **p** and **v** are related by **(1)**. For a charge q in an electric field the Hamiltonian is:

$$\mathscr{H}(r,p) = \frac{p^2}{2m} + q\phi \tag{19}$$

the velocity is from **(16)**:

$$v = \frac{p}{m} \tag{20}$$

which is in agreement with **(5)**. The equation of motion is from **(17)**

$$\frac{dp}{dt} = m\frac{dv}{dt} = -q\nabla_r \phi = qE \tag{21}$$

which is the same as **(3)**. We note that **(4)** and **(19)** are consistent with **(18)** when **(20)** is used.

We next consider the motion of q in electric and magnetic fields. We write for the Hamiltonian from **(8)** and **(18)**:

$$\mathscr{H}(r,p) = p \cdot v - \tfrac{1}{2}mv^2 - qv \cdot A + q\phi \tag{22}$$

Substituting **(9)** into **(22)** we obtain, in terms of the momentum,

$$\mathscr{H}(r,p) = \frac{1}{2m}(p - qA) \cdot (p - qA) + q\phi \tag{23}$$

Substituting **(23)** into **(16)** we obtain, for the velocity,

$$\mathbf{v} = \frac{1}{m}\,(\mathbf{p} - q\mathbf{A}) \tag{24}$$

which is consistent with **(9)**. Substituting **(23)** into **(17)** we obtain:

$$\frac{d\mathbf{p}}{dt} = \frac{q}{m}\,(\mathbf{p} - q\mathbf{A}) \cdot \mathbf{V}_{\mathbf{r}}\mathbf{A} + \frac{q}{m}\,(\mathbf{p} - q\mathbf{A}) \times (\mathbf{V}_{\mathbf{r}} \times \mathbf{A}) - q\mathbf{V}_{\mathbf{r}}\phi \tag{25}$$

Making the substitution **(24)** we see that **(25)** is equivalent to **(10)** and thus to **(7)**.

3 Constants of the Motion. Regarding the Hamiltonian as a function of position, momentum, and time, we write:

$$\frac{d}{dt}\,\mathscr{H}(\mathbf{r},\mathbf{p},t) = \mathbf{v} \cdot \mathbf{V}_{\mathbf{r}}\mathscr{H} + \frac{d\mathbf{p}}{dt} \cdot \mathbf{V}_{\mathbf{p}}\mathscr{H} + \frac{\partial}{\partial t}\,\mathscr{H} \tag{26}$$

Using **(16)** and **(17)** the first two terms cancel and we obtain the result that the Hamiltonian varies with time only if it is an *explicit* function of the time through **A** or ϕ.

We look next at the *generalized angular momentum*:

$$\mathbf{L} = \mathbf{r} \times \mathbf{p} = \mathbf{r} \times (m\mathbf{v} + q\mathbf{A}) \tag{27}$$

Differentiating with respect to the time we write:

$$\frac{d\mathbf{L}}{dt} = \mathbf{v} \times \mathbf{p} + \mathbf{r} \times \frac{d\mathbf{p}}{dt} \tag{28}$$

Substituting from **(24)** for **p** we write:

$$\frac{d\mathbf{L}}{dt} = q\mathbf{v} \times \mathbf{A} + q\mathbf{r} \times \frac{d\mathbf{A}}{dt} + m\mathbf{r} \times \frac{d\mathbf{v}}{dt} \tag{29}$$

Substituting from **(7)** for $d\mathbf{v}/dt$ we have:

$$\frac{d\mathbf{L}}{dt} = q\mathbf{v} \times \mathbf{A} + q\mathbf{r} \times \frac{d\mathbf{A}}{dt} + q\mathbf{r} \times \mathbf{E} + q\mathbf{r} \times (\mathbf{v} \times \mathbf{B}) \tag{30}$$

We now make the following simplifying assumptions:
   1. We take the magnetic field to be uniform and write:

$$\mathbf{A} = -\tfrac{1}{2}\mathbf{r} \times \mathbf{B} \tag{31}$$

2. We assume that ϕ is a central potential from the origin, which gives

$$\mathbf{r} \times \nabla\phi = 0 \tag{32}$$

3. We assume that the magnetic field is small:

$$q\mathbf{A} \ll m\mathbf{v} \tag{33}$$

Using (31) we may rewrite (30) as:

$$\frac{d\mathbf{L}}{dt} = -q\mathbf{r} \times \nabla\phi - \tfrac{1}{2}q\mathbf{B} \times (\mathbf{r} \times \mathbf{v}) \tag{34}$$

By (32) we drop the first term to obtain:

$$\frac{d\mathbf{L}}{dt} = -\tfrac{1}{2}q\mathbf{B} \times (\mathbf{r} \times \mathbf{v}) \tag{35}$$

Finally with (33) we may write:

$$\frac{d\mathbf{L}}{dt} = -\frac{q}{2m}\mathbf{B} \times (\mathbf{r} \times \mathbf{p}) \cong \boldsymbol{\omega}_L \times \mathbf{L} \tag{36}$$

with

$$\boldsymbol{\omega}_L = -\frac{q}{2m}\mathbf{B} \tag{37}$$

The relation given by (36) is called *Larmor's theorem* and (37) is called the *Larmor frequency*. We have shown under the three listed conditions that the angular momentum **L** precesses about the magnetic field **B** at the Larmor frequency (37). It is to be noted that (36) does not depend explicitly on the rate at which the magnetic field **B** is *changed*. In this respect the equation is similar to that for the response of a gyromagnetic system as discussed in Section 7-7.

Larmor's theorem is particularly useful in discussing the behavior of a system of particles interacting under central forces. It is not difficult to show where **L** is now the generalized angular momentum of the interacting particles that (36) continues to hold as long as the moving particles all have the same charge to mass ratio.

From (36) we observe that the component of **L** along the field is a constant of the motion as long as the field remains fixed in direction. It may also be shown that if the direction of the field is changed sufficiently slowly compared with the Larmor frequency **L** will follow the direction of the field at constant inclination.

Exercise

1. A system of N charges in zero field has zero total angular momentum. Use Larmor's theorem to show that in the presence of a magnetic field **B** the generalized angular momentum continues to equal zero. Show that the magnetic moment of the system is given by:

$$\boldsymbol{\mu} = \tfrac{1}{2}Nq \langle \mathbf{r} \times \mathbf{v} \rangle = \frac{Nq^2}{4m} \langle \mathbf{r} \times (\mathbf{r} \times \mathbf{B}) \rangle$$

REFERENCES

Corben, H. C. and P. Stehle, *Classical Mechanics*, Wiley, 1950.

Marion, J. B., *Classical Dynamics of Particles and Systems*, Second Edition, Academic, 1970.

Symon, K. R., *Mechanics*, Third Edition, Addison-Wesley, 1971.

H FIELDS OF MOVING CHARGES

We obtain expressions for the electric and magnetic fields of a charge q in arbitrary motion. Following this, we obtain expressions for the fields of a distribution of charges and show in the limit that the charge and current densities are independent of time that we obtain the familiar Coulomb and Biot-Savart fields.

FIELDS OF A MOVING CHARGE

Liénard-Wiechert Potentials. We show in Section 11-20, **(11-180)** and **(11 181)**, that the scalar and vector potentials are solutions of the wave equations

$$\nabla^2 \phi - \frac{1}{c^2} \frac{\partial^2 \phi}{\partial t^2} = -\frac{1}{\epsilon_0} \rho \tag{1}$$

$$\nabla^2 \mathbf{A} - \frac{1}{c^2} \frac{\partial^2 \mathbf{A}}{\partial t^2} \qquad \mu_0 \mathbf{j} \tag{2}$$

where ϕ and $\mathbf{A}$ are related by the Lorentz condition **(11-179)**:

$$\mathbf{V} \cdot \mathbf{A} + \frac{1}{c^2} \frac{\partial \phi}{\partial t} = 0 \tag{3}$$

As discussed in Section 11-21, the solutions of **(1)** and **(2)** for a general charge and current distribution take the form **(11-188)** and **(11-189)**:

$$\phi(\mathbf{r},t) = \frac{1}{4\pi\epsilon_0} \int \frac{\rho(\mathbf{r}_1,[t])}{r_{01}} dV_1 \tag{4}$$

$$\mathbf{A}(\mathbf{r},t) = \frac{\mu_0}{4\pi} \int \frac{\mathbf{j}(\mathbf{r}_1,[t])}{r_{01}} dV_1 \tag{5}$$

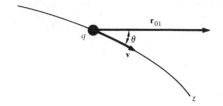

Figure 1   Liénard-Wiechert potentials. A charge q is regarded as moving on a contour z.

where

$$[t] = t - \frac{r_{01}}{c} \tag{6}$$

is called the retarded time.

   We now wish to obtain explicit solutions to **(4)** and **(5)** for a charge q that can be thought of as moving on the contour z shown in Figure 1. We discuss the scalar potential in detail; the arguments for the vector potential are quite similar. For a charge moving on a contour we write

$$\phi(\mathbf{r},t) = \frac{1}{4\pi\epsilon_0} \int \frac{\lambda(\mathbf{r}_1,[t])}{r_{01}} \, dz_1 \tag{7}$$

Now, if the lineal charge density λ were being integrated at fixed *retarded* time, we could perform the integral immediately to obtain the charge q at the retarded separation. Because the integral is at fixed *present* time, the integrated charge will differ from q.

   The argument is exceedingly tricky and we wish to pursue it with some care. Let us imagine a lineal charge density $\lambda(z,t)$, which is zero except over a finite range l, where it has the value $\lambda = q/l$. We represent in Figure 2a the motion of the charge.

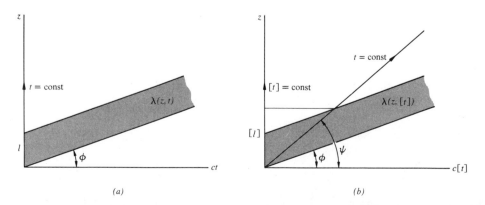

(a)

(b)

Figure 2   Charge density. (*a*) We imagine for simplicity a lineal charge density $\lambda(z,t)$, which is zero except over a finite range l, where it has the value q/l. The slope of the charge region is v/c. (*b*) The charge density at the retarded time is plotted as a function of retarded time. The slope of the line of constant present time is given by **(12)**.

The lines in Figure 2a that bound the charge have slope

$$\tan \phi = \frac{1}{c}\frac{dz}{dt} = \frac{v}{c} \tag{8}$$

Now, the total charge is obtained by integrating the charge density at fixed time:

$$q = \int \lambda(z,t)\, dz \tag{9}$$

The contour for this integral is simply the z axis shown in Figure 2a.

Let us now imagine instead a function $\lambda(z,[t])$ where the functional dependence on retarded time is exactly the same as the dependence of $\lambda(z,t)$ on present time as shown in Figure 2b. Now if we were to integrate $\lambda(z,[t])$ at constant retarded time, the contour in Figure 2b would be vertical and we would again obtain q for the total charge. The integral (7), which we must perform, is at constant *present* time, however, rather than constant retarded time. We have then the contour

$$t = [t] + r_{01}/c = \text{const} \tag{10}$$

Taking the differential of (10) we write:

$$c\,dt = c\,d[t] + dr = c\,[dt] - dz\cos\theta = 0 \tag{11}$$

where the geometry is that of Figure 1. The contour that satisfies (11) is labeled in Figure 2b as $t = $ const and has from (11) slope

$$\tan \psi = \frac{1}{c}\frac{dz}{d[t]} = \frac{1}{\cos\theta} \tag{12}$$

Integrating on this contour we obtain an amount of charge that we write as:

$$[q] = \lambda[l] \tag{13}$$

where $[l]$ is the projection from the contour $t = $ const onto the z axis as shown in Figure 2b.

From the geometry of Figure 2b we write:

$$\frac{[l]}{\sin\psi}\sin(\psi - \phi) = l\cos\phi \tag{14}$$

solving for $[l]$ we obtain:

$$[l] = \frac{l}{1 - \tan\phi\cot\psi} = \frac{l}{1 - (v/c)\cos\theta} \tag{15}$$

where we have used **(8)** and **(12)**. We then obtain for the apparent charge:

$$[q] = \lambda[l] = \frac{q}{1 - [\mathbf{v} \cdot \hat{\mathbf{f}}_{01}]/c} \tag{16}$$

And as long as the charge is reasonably localized we may write for **(7)**:

$$\phi(\mathbf{r},t) = \frac{1}{4\pi\epsilon_0} \frac{q}{[r_{01}] - [\mathbf{v} \cdot \mathbf{r}_{01}]/c} \tag{17}$$

where **v** may be quite arbitrary. In a similar way we obtain for the vector potential:

$$\mathbf{A}(\mathbf{r},t) = \frac{\mu_0}{4\pi} \frac{q[\mathbf{v}]}{[r_{01}] - [\mathbf{v} \cdot \mathbf{r}_{01}]/c} \tag{18}$$

The two expressions **(17)** and **(18)** are known as the *Liénard-Wiechert potentials* of a moving charge.

2 **Feynman's Formula.** Once we have the Liénard-Wiechert potentials for a charge q in arbitrary motion the electric and magnetic fields may be obtained by substitution into the expressions developed in Section 11-19 **(11-176)** and **(11-172)**:

$$\mathbf{E}(\mathbf{r},t) = -\nabla\phi(\mathbf{r},t) - \frac{\partial}{\partial t} \mathbf{A}(\mathbf{r},t) \tag{19}$$

$$\mathbf{B}(\mathbf{r},t) = \nabla \times \mathbf{A}(\mathbf{r},t) \tag{20}$$

R. P. Feynman has given expressions for the fields of a moving charge, written in terms of derivatives of retarded quantities with respect to *present* time[1]:

$$\mathbf{E}(\mathbf{r},t) = \frac{q}{4\pi\epsilon_0} \left(\frac{[\hat{\mathbf{f}}_{01}]}{[r_{01}]^2} + [r_{01}]\frac{1}{c}\frac{d}{dt}\frac{[\hat{\mathbf{f}}_{01}]}{[r_{01}]^2} + \frac{1}{c^2}\frac{d^2}{dt^2}[\hat{\mathbf{f}}_{01}] \right) \tag{21}$$

$$\mathbf{B}(\mathbf{r},t) = \frac{1}{c}[\hat{\mathbf{f}}_{01}] \times \mathbf{E}(\mathbf{r},t) \tag{22}$$

To obtain **(21)** and **(22)** we first reexpress the Liénard-Wiechert potentials in terms of derivatives with respect to the present time. We regard the distance from the *retarded* position of q to the field position to be a function of present time:

$$[r_{01}] = f(t) = f([t] + [r_{01}]/c) \tag{23}$$

---

[1] R. P. Feynman, R. B. Leighton, and M. Sands, *The Feynman Lectures on Physics*, Addison-Wesley, 1964; Sections I-28-1 and II-21-1.

Taking the differential of **(23)** we write:

$$d[r_{01}] = \frac{\partial f}{\partial [t]} d[t] + \frac{\partial f}{\partial [r_{01}]} d[r_{01}] = (d[t] + d[r_{01}]/c)\frac{d[r_{01}]}{dt} \tag{24}$$

Dividing through by $d[t]$ we obtain:

$$\frac{1}{[r_{01}] - [\mathbf{v} \cdot \mathbf{r}_{01}]/c} = \frac{1}{[r_{01}]}\left(1 - \frac{1}{c}\frac{d[r_{01}]}{dt}\right) \tag{25}$$

We next obtain the velocity of q at the retarded time in terms of derivatives with respect to the present time. We write for the *vector* from the retarded position of q to the field position:

$$[\mathbf{r}_{01}] = \mathbf{f}(t) \tag{26}$$

Taking the differential of **(26)** we write:

$$d[\mathbf{r}_{01}] = \frac{\partial \mathbf{f}}{\partial [t]} d[t] + \frac{\partial \mathbf{f}}{\partial [r_{01}]} d[r_{01}] = \left(d[t] + \frac{1}{c} d[r_{01}]\right)\frac{d[\mathbf{r}_{01}]}{dt} \tag{27}$$

Dividing through by $d[t]$ we have for the velocity at the retarded time:

$$[\mathbf{v}] = -\frac{d[\mathbf{r}_{01}]}{d[t]} = -\left(1 - \frac{1}{c}[\mathbf{v} \cdot \hat{\mathbf{r}}_{01}]\right)\frac{d[\mathbf{r}_{01}]}{dt} \tag{28}$$

Substituting **(25)** and **(28)** into **(17)** and **(18)** we are able to write the Liénard-Wiechert potentials as:

$$\phi(\mathbf{r},t) = \frac{1}{4\pi\epsilon_0}\frac{q}{[r_{01}]}\left(1 - \frac{1}{c}\frac{d[r_{01}]}{dt}\right) \tag{29}$$

$$\mathbf{A}(\mathbf{r},t) = -\frac{\mu_0}{4\pi}\frac{q}{[r_{01}]}\frac{d[\mathbf{r}_{01}]}{dt} \tag{30}$$

In obtaining the electric field we will need to take the gradient of **(29)**. The gradient is complicated by the fact that when we move by $d\mathbf{r}$ at fixed present time, there is a corresponding shift in the retarded source position $[\mathbf{r}_1]$. For $dt = 0$ we have

$$d[t] = -\frac{1}{c}d[r_{01}] \tag{31}$$

Then the change in $[\mathbf{r}_{01}]$ is given by:

$$d[\mathbf{r}_{01}] = d\mathbf{r} - [\mathbf{v}]\,d[t] = d\mathbf{r} + \frac{1}{c}[\mathbf{v}]\,d[r_{01}] \tag{32}$$

We write the differential of $[r_{01}]$ as:

$$d[r_{01}] = d\mathbf{r} \cdot \mathbf{V}[r_{01}] = d[\mathbf{r}_{01}] \cdot \mathbf{V}r_{01} - \frac{1}{c} d[r_{01}]\mathbf{v} \cdot \mathbf{V}[r_{01}] \qquad (33)$$

where we have substituted for $d\mathbf{r}$ from (32). We obtain from (33) for the gradient of $[r_{01}]$:

$$\mathbf{V}[r_{01}] = \frac{\hat{\mathbf{r}}_{01}}{1 - [\mathbf{v} \cdot \hat{\mathbf{r}}_{01}]/c} = \left(1 - \frac{1}{c}\frac{d[r_{01}]}{dt}\right)\hat{\mathbf{r}}_{01} \qquad (34)$$

as may be verified by substitution into (33). Substituting (29) and (30) into (19) with the gradient operation given by (34) we obtain (21).

To obtain the magnetic field (22) we take the curl of (30):

$$\mathbf{B}(\mathbf{r},t) = -\frac{\mu_0}{4\pi} q\mathbf{V} \times \frac{1}{[r_{01}]}\frac{d[\mathbf{r}_{01}]}{dt} = -\frac{\mu_0}{4\pi} q\left(\mathbf{V}\frac{1}{[r_{01}]} \times \frac{d[\mathbf{r}_{01}]}{dt} + \frac{1}{[r_{01}]}\frac{d}{dt}\mathbf{V} \times [\mathbf{r}_{01}]\right) \qquad (35)$$

From (32), (34), and (28) we obtain for the curl of $[\mathbf{r}_{01}]$:

$$\mathbf{V} \times [\mathbf{r}_{01}] = -\frac{1}{c}[\mathbf{v}] \times \mathbf{V}[r_{01}] = -\frac{1}{c}\left(1 - \frac{1}{c}\frac{d[r_{01}]}{dt}\right)[\mathbf{v}] \times [\hat{\mathbf{r}}_{01}]$$

$$= -\frac{1}{c}[\hat{\mathbf{r}}_{01}] \times \frac{d[\mathbf{r}_{01}]}{dt} \qquad (36)$$

We then have for the magnetic field:

$$\mathbf{B}(\mathbf{r},t) = \frac{\mu_0}{4\pi}\frac{q}{[r_{01}]^2}\left(1 - \frac{1}{c}\frac{d[r_{01}]}{dt}\right)\left([\hat{\mathbf{r}}_{01}] \times \frac{d[\mathbf{r}_{01}]}{dt}\right) + \frac{\mu_0}{4\pi}\frac{q}{[r_{01}]}\frac{1}{c}\frac{d}{dt}\left([\hat{\mathbf{r}}_{01}] \times \frac{d[\mathbf{r}_{01}]}{dt}\right) \qquad (37)$$

Simplifying, we obtain:

$$\mathbf{B}(r,t) = \frac{\mu_0}{4\pi}\frac{q}{[r_{01}]}[\hat{\mathbf{r}}_{01}] \times \left(\frac{d[\hat{\mathbf{r}}_{01}]}{dt} + \frac{[r_{01}]}{c}\frac{d^2[\hat{\mathbf{r}}_{01}]}{dt^2}\right) \qquad (38)$$

which is equivalent to (22).

ELECTRIC FIELD OF A CHARGE DISTRIBUTION

We now take Feynman's formula (21) for the field of a charge in arbitrary motion and sum over a distribution of such charges. We will be interested in particular in showing that when we have a steady distribution of charge and current that the equations obtained are those with which we are familiar for constant fields **(1-6)** and **(6-40)**:

$$E(r) = \frac{1}{4\pi\epsilon_0} \int_{V_1} \rho(r_1) \frac{\hat{f}_{01}}{r_{01}^2} \, dV_1 \tag{39}$$

$$B(r) = \frac{\mu_0}{4\pi} \int_{V_1} j(r_1) \times \frac{\hat{f}_{01}}{r_{01}^2} \, dV_1 \tag{40}$$

What this means is that although individual charges may be in accelerated motion, as long as the entire distribution is steady, there is no radiation.

We begin by transforming the second term in **(21)** by writing:

$$\frac{d}{dt}[\hat{r}_{01}]\frac{[\hat{f}_{01}]}{[r_{01}]^2} = \frac{[\hat{f}_{01}]}{[r_{01}]^2}\frac{d}{dt}[r_{01}] + [r_{01}]\frac{d}{dt}\frac{[r_{01}]}{[r_{01}]^2} \tag{41}$$

Substituting into **(21)** we obtain:

$$E(r) = \frac{q}{4\pi\epsilon_0}\frac{[\hat{f}_{01}]}{[r_{01}]^2}\left(1 - \frac{1}{c}\frac{d}{dt}[r_{01}]\right) + \frac{q}{4\pi\epsilon_0}\frac{1}{c}\frac{d}{dt}\left(\frac{[\hat{f}_{01}]}{[r_{01}]} + \frac{1}{c}\frac{d}{dt}[\hat{f}_{01}]\right) \tag{42}$$

We finally sum **(42)** over a distribution of charges q_i:

$$E(r) = \frac{1}{4\pi\epsilon_0}\sum_{i=1}^{N}q_i\frac{[\hat{f}_{0i}]}{[r_{0i}]^2}\left(1 - \frac{1}{c}\frac{d}{dt}[r_{0i}]\right) + \frac{1}{4\pi\epsilon_0}\frac{1}{c}\frac{d}{dt}\sum_{i=1}^{N}q_i\left(\frac{[\hat{f}_{0i}]}{[r_{0i}]} + \frac{1}{c}\frac{d}{dt}[\hat{f}_{0i}]\right) \tag{43}$$

3 Static Charge Distribution. If the distribution of charges q_i is such that the overall charge distribution is independent of time, even though individual charges are moving, then the second summation is clearly zero. The fact that the sum is over retarded rather than present positions really makes no difference. If either is independent of (present) time, the other will be.

To transform the first summation of **(43)** we examine the trajectory of charges q_i as shown in Figure 3a. The condition on the retarded vector r_{0i} is that the time taken for the charge to go from $[r_i]$ to r_i is just the same as the time taken by the radiation to go from $[r_i]$ to r:

$$\int_{[r_i]}^{r_i}\frac{dr}{v} = \int_{[r_i]}^{r}\frac{dr}{c} = \frac{[r_{0i}]}{c} \tag{44}$$

Taking the differential of **(44)** we get:

$$\frac{dr_i}{v_i} - \frac{d[r_i]}{[v_i]} = \frac{1}{c}d[r_{0i}] \tag{45}$$

Writing $dt = dr_i/v_i$ we may rewrite (45) as:

$$1 - \frac{1}{c}\frac{d}{dt}[r_{0i}] = \frac{1}{[v_i]}\frac{d}{dt}[r_i] \tag{46}$$

We show in Figure 3b the same trajectory in present time. At time t there will be at the retarded position of q_i some other charge q_j if the distribution is independent of time. The present velocity of q_j must equal the retarded velocity of q_i:

$$\mathbf{v}_j = [\mathbf{v}_i] \tag{47}$$

so that we may rewrite (46) as

$$1 - \frac{1}{c}\frac{d}{dt}[r_{0i}] = \frac{1}{v_j}\frac{d}{dt}[r_i] \tag{48}$$

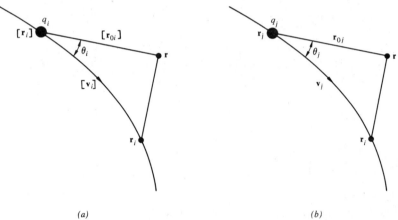

(a)   (b)

Figure 3   Charge trajectories. (a) A charge q_i with retarded velocity $[\mathbf{v}_i]$ is shown at the retarded position $[\mathbf{r}_i]$. (b) At the present time the retarded position of q_i is occupied by some other charge q_j.

This factor transforms the retarded charge density into the present charge density as may be seen by the following argument. Since we are summing over charges q_i at their retarded positions we may replace the charges q_i by their retarded density times a line element:

$$q_i \rightarrow \frac{dq}{d[r_j]}d[r_i] \tag{49}$$

Substituting (48) and (49) into (43) we have, for the electric field

$$\mathbf{E}(\mathbf{r}) = \frac{1}{4\pi\epsilon_0}\sum_{i=1}^{N}\frac{dq}{dr_j}\frac{[\hat{\mathbf{r}}_{0i}]}{[r_{0i}]^2}d[r_i] \tag{50}$$

Now, $dq/dr_j = (dq/dt)/v_j$ is the present lineal charge density at the retarded position $[r_i]$. Finally, since the distribution is independent of time we may replace $[r_{0i}]$ by the present vector r_{0j}, where the sum in (50) is to be taken over all contours. Going to the limit of a continuous distribution of charge, we clearly obtain (39).

MAGNETIC FIELD OF A CURRENT DISTRIBUTION

We take (22) for the relation between magnetic and electric field and use (21) to obtain:

$$\mathbf{B(r)} = \frac{\mu_0}{4\pi} q \left([\mathbf{r}_{01}] \times \frac{d}{dt} \frac{[\hat{\mathbf{f}}_{01}]}{[r_{01}]^2} + \frac{1}{c} [\hat{\mathbf{f}}_{01}] \times \frac{d^2}{dt^2} [\hat{\mathbf{f}}_{01}] \right) \tag{51}$$

We transform (51) by use of the identities:

$$\frac{d}{dt} \left([\mathbf{r}_{01}] \times \frac{[\hat{\mathbf{f}}_{01}]}{[r_{01}]^2} \right) = \frac{d[\mathbf{r}_{01}]}{dt} \times \frac{[\hat{\mathbf{f}}_{01}]}{[r_{01}]^2} + [\mathbf{r}_{01}] \times \frac{d}{dt} \frac{[\hat{\mathbf{f}}_{01}]}{[r_{01}]^2} = 0 \tag{52}$$

$$\frac{d}{dt} \left([\hat{\mathbf{f}}_{01}] \times \frac{d[\hat{\mathbf{f}}_{01}]}{dt} \right) = \frac{d[\hat{\mathbf{f}}_{01}]}{dt} \times \frac{d[\hat{\mathbf{f}}_{01}]}{dt} + [\hat{\mathbf{f}}_{01}] \times \frac{d^2}{dt^2} [\hat{\mathbf{f}}_{01}] \tag{53}$$

to obtain:

$$\mathbf{B(r)} = \frac{\mu_0}{4\pi} q \frac{d[\mathbf{r}_1]}{dt} \times \frac{[\hat{\mathbf{f}}_{01}]}{[r_{01}]^2} + \frac{\mu_0}{4\pi} \frac{q}{c} \frac{d}{dt} \left([\hat{\mathbf{f}}_{01}] \times \frac{d[\hat{\mathbf{f}}_{01}]}{dt} \right) \tag{54}$$

We now sum (54) over a distribution of charges q_i to obtain

$$\mathbf{B(r)} = \frac{\mu_0}{4\pi} \sum_{i=1}^{N} q_i \frac{d[\mathbf{r}_i]}{dt} \times \frac{[\hat{\mathbf{f}}_{0i}]}{[r_{0i}]^2} + \frac{\mu_0}{4\pi} \frac{d}{dt} \sum_{i=1}^{N} \frac{q_i}{c} [\hat{\mathbf{f}}_{0i}] \times \frac{d}{dt} [\hat{\mathbf{f}}_{0i}] \tag{55}$$

4 Steady Current Distribution. For a steady current it is possible to reinterpret (55) in a very simple way. First of all, the second term, which expresses the rate of change with present time of the retarded distribution, will be zero since the distribution of charges and velocities, viewed as a whole, is constant. We would like to write the first sum in terms of the present current at the retarded positions $[r_i]$. The transformation is very simple but the argument must be made carefully. We refer to Figures 4a and 4b where we show the relation between present and retarded path intervals in Figure 4a and two charge increments, both in present time, in Figure 4b.

To make the argument it is easier to think of a local charge density along the path taken by the charge q_i. We replace q_i by

$$q_i \rightarrow \frac{dq_i}{d[r_i]} d[r_i] \tag{56}$$

With reference to Figure 4a, the amount of charge dq_i in dr_i must be equal to the amount of charge within the corresponding retarded increment of path $d[r_i]$:

$$\frac{dq_i}{d[r_i]} d[r_i] = \frac{dq_i}{dr_i} dr_i \tag{57}$$

In Figure 4b we view the same path but in present time. We imagine that in an interval dt an amount of charge dq_i passes $\mathbf{r}_i$. In that same interval an equal increment of charge dq_j must pass $\mathbf{r}_j$. This is required if there is to be no accumulation of charge on the line. We must have then:

$$\frac{dq_i}{dr_i} dr_i = \frac{dq_j}{dr_j} dr_j \tag{58}$$

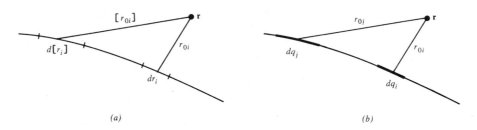

(a)   (b)

Figure 4   Current trajectories. (a) A segment of the contour of length $d[r_i]$ is located at $[\mathbf{r}_i]$. In present time this segment is of length dr_i and is located at $\mathbf{r}_i$. (b) The segments $d[r_i]$ and dr_i contain in present time charge increments dq_j and dq_i.

Now the current flowing past $\mathbf{r}_j$ in *present time* may be written as:

$$I_j = \frac{dq_j}{dt} = \frac{dq_j}{dr_j} \frac{dr_j}{dt} = \frac{dq_i}{d[r_i]} \frac{d[r_i]}{dt} \tag{59}$$

where the final equality is a consequence of (57) and (58). We have then for the magnetic field **B**:

$$\mathbf{B}(\mathbf{r}) = \frac{\mu_0}{4\pi} \sum_{i=1}^{N} I_i \, d[\mathbf{r}_i] \times \frac{[\hat{\mathbf{r}}_{0i}]}{[r_{0i}]^2} \tag{60}$$

In the limit that the increments of path go to zero we obtain:

$$\mathbf{B}(\mathbf{r}) = \frac{\mu_0}{4\pi} \sum_i I_i \int_{C_i} [d\mathbf{r}_i] \times \frac{[\hat{\mathbf{r}}_{0i}]}{[r_{0i}]^2} \tag{61}$$

where the sum is over the contours followed by the current elements. Finally we may generalize **(61)** to a general distribution of steady currents:

$$\mathbf{B}(\mathbf{r}) = \frac{\mu_0}{4\pi} \int_{V_1} \mathbf{j}(\mathbf{r}_1) \times \frac{\hat{\mathbf{r}}_{01}}{r_{01}{}^2} \, dV_1 \tag{62}$$

which is just the same as **(40)**, the Biot-Savart expression for the magnetic field from a steady current distribution. Since the currents are independent of time, it makes no difference whether we integrate over present or retarded positions, and so we have been able to drop that distinction in writing **(62)**.

I FOURIER SERIES AND COMPLEX VARIABLES

In this appendix we develop the mathematics necessary to analyze phenomena that are general functions of position or time or both in terms of sinusoidally varying functions. We may write a general function as a sum or integral of sinusoids from which we obtain integral expressions for the density of sinusoids as a function of frequency or wavenumber. To perform the integration explicitly we write the spectral density as a function of a complex frequency and/or wavevector. By extending the spectral density into the complex plane we relate the integral over real frequency or wavenumber to an integral over a contour in the complex plane.

FOURIER SERIES

In physics we commonly encounter phenomena that are periodic functions of the time. The harmonic oscillator, for example, has sinusoidal solutions of the form:

$$x(t) = x_1 \cos(\omega t + \theta) = x_1 \cos \theta \cos \omega t - x_1 \sin \theta \sin \omega t \tag{1}$$

where the angular frequency is $\omega = 2\pi/T$ and T is the period of the oscillation.

An oscillator such as the pendulum is harmonic for low amplitudes only. For higher amplitudes the displacement is still periodic but must be written as a more general series:

$$
\begin{aligned}
x(t) = x_0 &+ x_1{}' \cos \omega t + x_1{}'' \sin \omega t \\
&+ x_2{}' \cos 2\omega t + x_2{}'' \sin 2\omega t \\
&\qquad + x_3{}' \cos 3\omega t + x_3{}'' \sin 3\omega t \\
&\qquad\qquad + \cdots \qquad\qquad + \cdots
\end{aligned} \tag{2}
$$

Such a series is called a Fourier series and the $x_n{}'$ and $x_n{}''$ are called the Fourier coefficients of the series. The power of (2) is that given $x(t)$ we can readily determine the Fourier coefficients. Multiplying both sides of (2) by $\cos n\omega t$ and integrating over one period, we find for $n > 0$:

$$x_n' = \frac{2}{T} \int_{-\frac{1}{2}T}^{+\frac{1}{2}T} x(t) \cos n\omega t \, dt \tag{3}$$

and for $n = 0$

$$x_0 = \frac{1}{T} \int_{-\frac{1}{2}T}^{\frac{1}{2}T} x(t) \, dt \tag{4}$$

Multiplying by $\sin n\omega t$ and integrating we obtain:

$$x_n'' = \frac{2}{T} \int_{-\frac{1}{2}T}^{+\frac{1}{2}T} x(t) \sin n\omega t \, dt \tag{5}$$

In this way we can express $x(t)$ as an infinite series. We may combine (3) and (5) into a single expression by defining a complex Fourier amplitude for $x > 0$:

$$\underline{x}_n = \tfrac{1}{2}(x_n' + ix_n'') \tag{6}$$

and for $x = 0$

$$\underline{x}_0 = x_0$$

Writing $\cos n\omega t + i \sin n\omega t = e^{in\omega t}$ we have, combining (3) and (5),

$$\blacktriangleright \qquad \underline{x}_n = \frac{1}{T} \int_{-\frac{1}{2}T}^{+\frac{1}{2}T} x(t')e^{in\omega t'} \, dt' \quad \text{for} \quad n \geq 0 \tag{7}$$

It is still possible to write $x(t)$ in terms of a Fourier series, but we must take only the real part:

$$x(t) = \underline{x}_0 + 2\text{Re} \sum_{n=1}^{n=\infty} \underline{x}_n e^{-in\omega t} \tag{8}$$

An alternate way of writing (8) extends the definition of $\underline{x}_n$ to negative n for which from (7):

$$\underline{x}_{-n} = \underline{x}_n^* \tag{9}$$

Writing $x(t)$ as a sum over *all* n we obtain:

$$x(t) = \sum_{n=-\infty}^{n=+\infty} \underline{x}_n e^{-in\omega t} \tag{10}$$

1 Fourier Integral. Although the above discussion was developed for periodic phenomena, it can be extended to the discussion of aperiodic phenomena. This is done by letting the period T go to infinity. We define a new frequency $\omega = 2\pi n/T$ and a frequency increment $\Delta\omega = 2\pi/T$. We replace $\underline{x}_n$ by $\overline{x}(\omega)\,\Delta\omega$. Then in the limit that $T \to \infty$ **(7)** becomes:

$$\blacktriangleright \qquad \overline{x}(\omega) = \frac{1}{2\pi} \int_{-\infty}^{+\infty} x(t)e^{i\omega t}\, dt \qquad (11)$$

and **(9)** becomes:

$$\blacktriangleright \qquad x(t) = \int_{-\infty}^{+\infty} \overline{x}(\omega)e^{-i\omega t}\, d\omega \qquad (12)$$

The integral transforms **(11)** and **(12)** may be extended to any conjugate quantities with each quantity termed the *Fourier transform* of the conjugate quantity. For example, some function $f(x)$ may be written as a Fourier integral:

$$f(x) = \int_{\infty}^{+\infty} \overline{f}(k)e^{ikx}\, dk \qquad (13)$$

with:

$$\overline{f}(k) = \frac{1}{2\pi} \int_{-\infty}^{+\infty} f(x)e^{-ikx}\, dx \qquad (14)$$

The expression $f(k)e^{ikx}$ represents a complex wave of *wavenumber k*.

Equations **13** and **14** may be extended to three dimensions by defining wavenumbers k_x, k_y, and k_z:

$$f(x,y,z) = \int_{-\infty}^{+\infty} e^{ik_x x}\, dk_x \int_{-\infty}^{+\infty} e^{ik_y y}\, dk_y \int_{-\infty}^{+\infty} f(k_x,k_y,k_z)e^{ik_z z}\, dk_z \qquad (15)$$

$$\overline{f}(k_x,k_y,k_z) = \frac{1}{(2\pi)^3} \int_{-\infty}^{+\infty} e^{-ik_x x}\, dx \int_{-\infty}^{+\infty} e^{-ik_y y}\, dy \int_{-\infty}^{+\infty} f(x,y,z)e^{-ik_z z}\, dz \qquad (16)$$

Equations **15** and **16** may be considerably simplified by using vector notation. The usual position vector is written as:

$$\mathbf{r} = x\hat{\mathbf{x}} + y\hat{\mathbf{y}} + z\hat{\mathbf{z}} \qquad (17)$$

The corresponding *wavevector* is written as:

$$\mathbf{k} = k_x\hat{\mathbf{x}} + k_y\hat{\mathbf{y}} + k_z\hat{\mathbf{z}} \qquad (18)$$

The volume element in ordinary space is written as dV and the volume element in "wavenumber space" is written as dV_k. Now (15) and (16) may be written in compact form:

$$\blacktriangleright \qquad f(\mathbf{r}) = \int \bar{f}(\mathbf{k}) e^{i\mathbf{k}\cdot\mathbf{r}} \, dV_k \qquad\qquad (19)$$

$$\blacktriangleright \qquad \bar{f}(\mathbf{k}) = \frac{1}{(2\pi)^3} \int f(\mathbf{r}) e^{-i\mathbf{k}\cdot\mathbf{r}} \, dV \qquad\qquad (20)$$

Exercises

1. Let $x(t)$ in (11) be the Dirac delta function $\delta(t)$. Show that $\bar{x}(\omega)$ is equal to $1/2\pi$.
2. Let $x(t)$ be the unit step function $f(t)$, which is equal to $-\frac{1}{2}$ for $x < 0$ and equal to $+\frac{1}{2}$ for $x > 0$. Show that $\mathbf{x}(\omega)$ is given by

$$x(\omega) = -\frac{1}{2\pi i \omega}$$

Verify that $x(t)$ as given by (12) is an odd function of t.
3. Use the result of Exercise 1 to obtain the relation:

$$\int_{-\infty}^{\infty} e^{-i\omega t} \, d\omega = 2\pi \, \delta(t)$$

4. Show that (9) leads for real ω to:

$$\bar{x}(-\omega) = \bar{x}^*(\omega)$$

Show that this same result is obtained by requiring $x(t)$ real in (12).
5. Let $x(t) = 1$ for $-T < t < T$ and zero outside these limits. Obtain the result:

$$\bar{x}(\omega) = \frac{1}{2\pi} \int_{-T}^{T} e^{i\omega t} \, dt = \frac{1}{\pi} \frac{\sin \omega T}{\omega}$$

Using the result of Exercise 3 obtain for the Dirac delta function:

$$\lim_{T \to \infty} \frac{\sin \omega T}{\omega} = \pi \, \delta(\omega)$$

6. Let $x(t) = 1/2T$ for $-T < t < T$ and zero outside these limits. Obtain the result:

$$\bar{x}(\omega) = \frac{1}{2\pi} \frac{\sin \omega T}{\omega T}$$

In the limit that T goes to zero, show that $x(t)$ approaches $\delta(t)$ and $\bar{x}(\omega)$ approaches $\frac{1}{2}\pi$. In the opposite limit that T goes to infinity use the result of Exercise 5 to show that $x(t)$ approaches $\frac{1}{2}T$ and $\bar{x}(\omega)$ approaches $\delta(t)/2T$.

7. Let $x(t)$ be Heaviside unit step function $\sigma(t)$, which is equal to 0 for $x < 0$ and equal to 1 for $x > 0$. Use the results of Exercises 2 and 3 to obtain the Fourier transform of $\sigma(t)$:

$$\bar{\sigma}(\omega) = -\frac{1}{2\pi i\omega} + \tfrac{1}{2}\,\delta(\omega)$$

8. A function $x(t)$ is equal to $\sin \omega_0 t$ for *all* time. Use the result of Exercise 3 to obtain:

$$\bar{x}(\omega) = \frac{1}{2i}\,[\delta(\omega + \omega_0) - \delta(\omega - \omega_0)]$$

9. A function $x(t)$ is equal to $\cos \omega_0 t$ for *all* time. Use the result of Exercise 3 to write

$$x(\omega) = \tfrac{1}{2}[\delta(\omega + \omega_0) + \delta(\omega - \omega_0)]$$

10. A function is equal to $\sin \omega_0 t$ for $t > 0$ and equal to zero for $t < 0$. Obtain for the Fourier transform:

$$\bar{x}(\omega) = -\frac{1}{2\pi}\frac{\omega_0}{\omega^2 - \omega_0{}^2} + \frac{1}{4i}\,[\delta(\omega + \omega_0) - \delta(\omega - \omega_0)]$$

Use the Heaviside step function.

11. A function is equal to $\cos \omega_0 t$ for $t > 0$ and equal to zero for $t < 0$. Obtain for the Fourier transform:

$$\bar{x}(\omega) = -\frac{1}{2\pi i}\frac{\omega}{\omega^2 - \omega_0{}^2} + \tfrac{1}{4}[\delta(\omega + \omega_0) + \delta(\omega - \omega_0)]$$

Use the Heaviside step function.

12. Obtain the Fourier transform of the gaussian:

$$x(t) = e^{-t^2/2\tau^2}$$

and show that it is also Gaussian.

13. A function of position *and* time is expanded as a Fourier integral:

$$f(\mathbf{r},t) = \int_{-\infty}^{\infty} d\omega \int_{V_k} \bar{f}(\mathbf{k},\omega)e^{i(\mathbf{k}\cdot\mathbf{r} - \omega t)}\, dV_k$$

Obtain the expression for the Fourier transform of $f(\mathbf{r},t)$:

$$\bar{f}(\mathbf{k},\omega) = \frac{1}{(2\pi)^4} \int_{-\infty}^{\infty} dt \int_{V} f(\mathbf{r},t) e^{-i(\mathbf{k}\cdot\mathbf{r}-\omega t)} \, dV$$

---

2 [**Magnetic Flux Penetration**].[1] As an example of the use of the Fourier integral, we discuss the penetration of magnetic flux into a medium characterized by a dispersion relation:

$$k = k(\omega) \simeq k_0(\omega/\omega_0)^\beta \tag{21}$$

We apply at the surface a magnetic field:

$$B(t) = \int_{-\infty}^{\infty} B(\omega) e^{-i\omega t} \, d\omega \tag{22}$$

Then the field within the medium will be

$$B(z,t) = \int_{-\infty}^{\infty} \bar{B}(\omega) e^{i(kz-\omega t)} \, d\omega \tag{23}$$

From Faraday's law **(9-114)** the electric field within the medium is of the form:

$$E(z,t) = -\int \frac{\omega}{k} \bar{B}(\omega) e^{i(kz-\omega t)} \, d\omega \tag{24}$$

In particular, at the surface the electric field is given by:

$$E(0,t) = -\int \frac{\omega}{k} \bar{B}(\omega) e^{-i\omega t} \, d\omega \tag{25}$$

Now, we define the depth of flux penetration as

$$\delta(t) = \frac{\Phi}{B(0,t)} = \frac{1}{B(0,t)} \int_{0}^{\infty} B(z,t) \, dz \tag{26}$$

From **(9-111)** we may write, for the field at the surface,

$$E(0,t) = \frac{d}{dt} \Phi = \frac{d}{dt} [\delta(t)B(0,t)] \tag{27}$$

---

[1] This discussion follows an argument introduced by A. B. Pippard in *Low Temperature Physics*, edited by C. DeWitt, B. Dreyfus, and P. G. deGennes, Gordon and Breach, New York, 1962.

From **(25)** we have for the depth to which flux penetrates:

$$\delta(t) = \frac{1}{B(0,t)} \int_{-\infty}^{t} E(0,t)\, dt = \frac{1}{B(0,t)} \int_{-\infty}^{\infty} \bar{B}(\omega) e^{-i\omega t} \frac{d\omega}{ik} \tag{28}$$

Now, if $B(0,t)$ is a step function we have, from Exercise 2,

$$\bar{B}(\omega) = -\frac{B(0,t)}{2\pi i \omega} \tag{29}$$

Using the approximate dispersion relation of **(21)** we write for the time dependence of the depth of flux penetration:

$$\delta(t) = \frac{(\omega_0 t)^\beta}{\pi k_0} \int_0^{\infty} \frac{\cos x\, dx}{x^{(1+\beta)}} \tag{30}$$

COMPLEX VARIABLES

In the following sections we discuss explicitly the integration of equations like **(12)** and **(13)**. To perform an integral of the form of **(12)** we extend ω into the complex plane:

$$\underline{\omega} = \omega' + i\omega'' \tag{31}$$

as shown in Figure 1a. The integral **(11)** is then represented by the contour shown in Figure 1b along the real $\underline{\omega}$ axis.

3 **Cauchy-Riemann Conditions.** We must now examine the properties of the complex function of a complex variable $\bar{x}(\underline{\omega})$. We assume that $\bar{x}(\underline{\omega})$ is an analytic function of ω, that is, its derivatives with respect to both ω' and ω'' are defined at all but a few

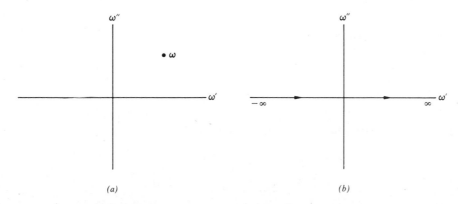

(a)   (b)

Figure 1 Complex plane. (a) A frequency in the complex plane has coordinate ω' on the horizontal axis and coordinate ω'' on the vertical axis. (b) An integral over real frequency is represented by a contour along the horizontal axis.

discrete number of points or lines in the complex plane where $\bar{x}(\underline{\omega})$ is singular. At other than singular points we write:

$$\bar{x}(\underline{\omega} + d\underline{\omega}) = \bar{x}(\underline{\omega}) + \frac{\partial}{\partial \omega'}\bar{x}(\underline{\omega})\, d\omega' + \frac{\partial}{\partial \omega''}\bar{x}(\underline{\omega})\, d\omega'' \qquad (32)$$

Note the similarity between (32) and (B-30), which suggests that we may think of ω' and ω'' as components of a vector in the complex plane.

We introduce the derivative of $\bar{x}(\underline{\omega})$ with respect to $\underline{\omega}$, which is the analog of the gradient (B-27):

$$d\bar{x}(\underline{\omega}) = \frac{d}{d\underline{\omega}}\,\bar{x}(\underline{\omega})\, d\underline{\omega} \qquad (33)$$

Comparing (33) with (32) we obtain

$$\frac{d}{d\underline{\omega}}\,\bar{x}(\underline{\omega}) = \frac{\partial}{\partial \omega'}\,\bar{x}(\underline{\omega}) = -i\frac{\partial}{\partial \omega''}\,\bar{x}(\underline{\omega}) \qquad (34)$$

Now if we write $\bar{x}(\underline{\omega})$ explicitly in terms of real and imaginary parts:

$$\bar{x}(\underline{\omega}) = x'(\underline{\omega}) + ix''(\underline{\omega}) \qquad (35)$$

we obtain from (34) the relations:

$$\frac{\partial}{\partial \omega'}\,x'(\underline{\omega}) - \frac{\partial}{\partial \omega''}\,x''(\underline{\omega}) = 0 \qquad (36)$$

$$\frac{\partial}{\partial \omega''}\,x'(\underline{\omega}) + \frac{\partial}{\partial \omega'}\,x''(\underline{\omega}) = 0 \qquad (37)$$

The equations (36) and (37) apart from a change in sign are the analogs of the divergence (B-43) and curl (B-57). The equations are called the *Cauchy-Riemann conditions* and are the complex equivalents of the statement that a function has no sources of flux or circulation.

Exercises

14. The real and imaginary axes may be identified with rectangular unit vectors $\hat{\mathbf{x}}$ and $\hat{\mathbf{y}}$. In this way $x(\underline{\omega})$ and $\underline{\omega}$ are identified with the vectors:

$$\mathbf{x} = \hat{\mathbf{x}}x'(\underline{\omega}) + \hat{\mathbf{y}}x''(\underline{\omega})$$

$$\boldsymbol{\omega} = \hat{\mathbf{x}}\omega' + \hat{\mathbf{y}}\omega''$$

Obtain the identity

$$\bar{x}(\underline{\omega})\underline{\omega}^* = \mathbf{x}\cdot\boldsymbol{\omega} - i\hat{\mathbf{z}}\cdot\mathbf{x}\times\boldsymbol{\omega}$$

15. Following Exercise 4, we introduce the gradient operators with respect to ω as:

$$\mathbf{V}_{\underline{\omega}} = \hat{\mathbf{x}}\frac{\partial}{\partial\omega'} + \hat{\mathbf{y}}\frac{\partial}{\partial\omega''}$$

$$\mathbf{V}_{\underline{\omega}^*} = \hat{\mathbf{x}}\frac{\partial}{\partial\omega'} - \hat{\mathbf{y}}\frac{\partial}{\partial\omega''}$$

show that the Cauchy-Riemann conditions are equivalent to:

$$\mathbf{V}_{\underline{\omega}^*} \cdot \mathbf{x} = 0$$

$$\mathbf{V}_{\underline{\omega}^*} \times \mathbf{x} = 0$$

16. Show that the function $\overline{r}(\underline{\omega}) = \underline{\omega}$ satisfies the Cauchy-Riemann conditions while the function $\overline{x}(\underline{\omega}) = \underline{\omega}^*$ does not.

17. Show that $\overline{x}(\omega)$ is an analytic solution of Laplace's equation:

$$\left(\frac{\partial^2}{\partial\omega'^2} + \frac{\partial^2}{\partial\omega''^2}\right)\overline{x}(\underline{\omega}) = 0$$

4   Cauchy's Integral Theorem. The circulation of $x(\underline{\omega})$ about a contour in the complex plane as shown in Figure 2*a* is defined by analogy with **(B-50)** as:

$$\Gamma = \oint_C \overline{x}(\underline{\omega})\,d\underline{\omega} \tag{38}$$

As shown in Figure 2*b* we may divide the area enclosed by the contour into increments:

$$dS = d\omega'\,d\omega'' \tag{39}$$

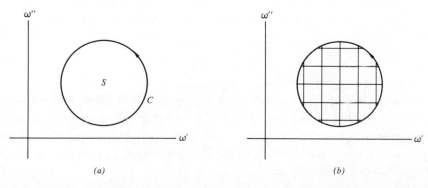

(a)   (b)

Figure 2   Circulation. (*a*) The circulation of a function of frequency in the complex plane is represented by the integral of the function over a closed contour in the plane. (*b*) The area enclosed by the contour may be divided into increments of area $d\omega'\,d\omega''$.

for which we can write, for the increment of circulation

$$d\Gamma = \left[\overline{x}(\underline{\omega}) + \frac{\partial \overline{x}}{\partial \omega'} \, d\omega' \right] i \, d\omega'' + \left[\overline{x}(\underline{\omega}) + \frac{\partial \overline{x}}{\partial \omega''} \, d\omega'' \right](-d\omega')$$

$$+ \, \overline{x}(\underline{\omega})(-i \, d\omega'') + \overline{x}(\underline{\omega}) \, d\omega'$$

$$= i \left(\frac{\partial \overline{x}}{\partial \omega'} + i \frac{\partial \overline{x}}{\partial \omega''} \right) dS \tag{40}$$

From **(38)** and **(40)** we obtain the complex equivalent of Stokes's theorem:

$$\oint_C \overline{x}(\underline{\omega}) \, d\underline{\omega} = i \int_S \left(\frac{\partial \overline{x}}{\partial \omega'} + i \frac{\partial \overline{x}}{\partial \omega''} \right) dS \tag{41}$$

But by **(34)** the integrand of the surface integral is identically zero from which we have the Cauchy integral theorem:

$$\oint_C x(\underline{\omega}) \, d\underline{\omega} = 0 \tag{42}$$

for $\overline{x}(\underline{\omega})$ analytic on S.

Exercise

17. Show that the following integral vanishes for $x(\underline{\omega})$ analytic and n not equal to one:

$$\oint_C \frac{\overline{x}(\underline{\omega})}{\underline{\omega}^n} \, d\underline{\omega}$$

5   Calculus of Residues. We wish to obtain the circulation of $\overline{x}/\underline{\omega}$:

$$\Gamma = \oint_C \frac{\overline{x}(\underline{\omega})}{\underline{\omega}} \, d\omega \tag{43}$$

where $\mathbf{x}(\underline{\omega})$ is analytic on S as shown in Figure 3a.

Now the function $\overline{x}/\underline{\omega}$ is analytic on S *except* at $\underline{\omega} = 0$ where the denominator introduces a singularity. The point at $\underline{\omega} = 0$ is called a simple pole of the function $\overline{x}/\underline{\omega}$. Thus, we may not apply the Cauchy integral theorem **(42)** to the integral over C. We may however apply the theorem to the integral over C_0 as shown in Figure 3b since the pole at $\underline{\omega} = 0$ is excluded:

$$\oint_{C_0} \frac{\overline{x}(\underline{\omega})}{\underline{\omega}} \, d\underline{\omega} = 0 \tag{44}$$

For $\bar{x}(\underline{\omega})$ single-valued around $\underline{\omega} = 0$ and continuous across the real axis, the contributions to C_0 along the real axis cancel as the two paths approach each other. In this limit we have the two contours shown in Figure 3c for which we are able to write

$$\oint_C \frac{\bar{x}(\underline{\omega})}{\underline{\omega}} \, d\underline{\omega} + \oint_{C_1} \frac{\bar{x}(\underline{\omega})}{\underline{\omega}} \, d\underline{\omega} = 0 \qquad (45)$$

To evaluate the integral over C_1 we let the contour be a circle with

$$\underline{\omega} = re^{i\theta} \qquad d\underline{\omega} = re^{i\theta} i \, d\theta \qquad (46)$$

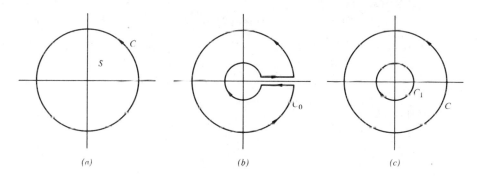

(a)   (b)   (c)

Figure 3 Residues. (a) We wish the circulation on the contour C of the function given in (43). (b) For the function to be analytic over the enclosed area, we must exclude the origin by integrating over the contour C_0. (c) The contour over C_0 is equivalent in the limit to a pair of contours, the original contour C and C_1.

We have then for the circulation of C_1:

$$\Gamma_1 = -\oint_{C_1} \frac{\bar{x}(\underline{\omega})}{\underline{\omega}} \, d\underline{\omega} = \int_0^{2\pi} \bar{x}(\underline{\omega}) i \, d\theta \qquad (47)$$

where the minus sign in (47) arises because the circulation is defined for a counter-clockwise contour and C_1 is clockwise. Letting the radius of the circle shrink to zero we obtain for (47)

$$\Gamma_1 = 2\pi i \bar{x}(0) \qquad (48)$$

and from (45) for the integral over C:

$$\oint_C \frac{\bar{x}(\underline{\omega})}{\underline{\omega}} \, d\underline{\omega} = 2\pi i \bar{x}(0) \qquad (49)$$

Exercises

---

18. Obtain the following result where C includes a simple pole at ω_1:

$$\oint_C \frac{\bar{x}(\omega)}{\omega - \omega_1}\, d\omega = 2\pi i \bar{x}(\omega_1)$$

19. Obtain the following result when C includes simple poles at ω_1 and ω_2:

$$\oint_C \frac{\bar{x}(\omega)\, d\omega}{(\omega - \omega_1)(\omega - \omega_2)} = 2\pi i\, \frac{\bar{x}(\omega_1) - \bar{x}(\omega_2)}{\omega_1 - \omega_2}$$

20. Show that the integral half way around a simple pole is equal to half the integral completely around the pole.

21. Use the calculus of residues to evaluate the integral

$$\int_0^{2\pi} \frac{d\theta}{1 + a \cos \theta} = \frac{2\pi}{(1 - a^2)^{1/2}}$$

22. Obtain the result:

$$\int_{-\infty}^{\infty} \frac{\sin \theta}{\theta}\, d\theta = \pi$$

Discuss carefully the contribution over the path that closes the contour.

23. Show by contour integration with ω'' finite:

$$\int_{-\infty}^{\infty} e^{-\omega^2}\, d\omega' = \sqrt{\pi}$$

---

6 Fourier Transforms. We have from **(12)** the integral for the Fourier transform of $x(t)$:

$$x(t) = \int_{-\infty}^{\infty} \bar{x}(\omega) e^{-i\omega t}\, d\omega \tag{50}$$

where $\bar{x}(\omega)$ is the Fourier transform of $x(t)$. We now examine the contour integral:

$$\Gamma = \oint_C \bar{x}(\omega) e^{-i\omega t}\, d\omega \tag{51}$$

where the contour C is shown in Figure 4.

In the limit that R goes to infinity, the integral over the curved part of the contour goes to zero under the conditions of *Jordan's lemma*:

  1. The function $\bar{x}(\omega)$ must go to zero on the contour in the limit of ω large.

2. For t negative the integral must be closed over the positive half plane while for t positive the integral must be closed over the negative half plane. In this way the factor $e^{\omega''t}$ is exponentially decreasing for ω'' large.

Under these conditions the integral over C is equal to the integral on the real axis, and we may write:

$$x(t) = \oint_C x(\underline{\omega})e^{-i\omega t}\, d\underline{\omega} \tag{52}$$

As an example we take $x(t)$ exponentially decreasing:

$$x(t) = 0 \text{ for } t < 0 \qquad x(t) = e^{-t/\tau} \text{ for } t > 0 \tag{53}$$

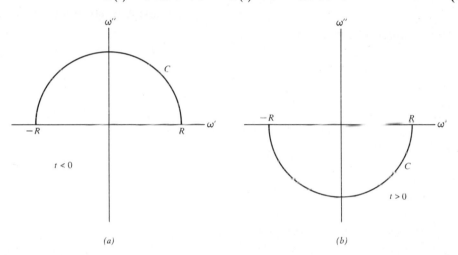

(a)   (b)

Figure 4 Fourier transform. The Fourier transform is taken over a closed contour that includes the real frequency axis. (*a*) At negative times the contour must be closed over the positive half-frequency plane. (*b*) At positive times the contour must be closed over the negative half-frequency plane.

From **(11)** we obtain for the Fourier transform of $x(t)$:

$$\bar{x}(\omega) = \frac{1}{2\pi} \int_{-\infty}^{\infty} x(t)e^{i\omega t}\, dt = \frac{1}{2\pi} \int_{0}^{\infty} e^{i(\omega + i/\tau)t}\, dt \tag{54}$$

Performing the integration we obtain:

$$\bar{x}(\omega) = -\frac{1}{2\pi i}\frac{1}{\omega + i/\tau} \tag{55}$$

Now to obtain the Fourier transform of **(55)** we write from **(52)**:

$$x(t) = -\frac{1}{2\pi i}\oint_C \frac{1}{\underline{\omega} + i/\tau} e^{-i\omega t}\, d\underline{\omega} \tag{56}$$

The integrand of (56) has a simple pole $\omega = -i/\tau$, which is in the lower half of the complex plane. At negative times the contour must be closed over the positive half plane as in Figure 4a. Since there are no poles in the positive half plane we obtain:

$$x(t) = 0 \text{ for } t < 0 \tag{57}$$

as given by (53). For positive times the contour is closed over the negative half plane. Since the direction of integration is clockwise rather than counterclockwise we obtain $-2\pi i$ times the residue at the pole:

$$x(t) = -\frac{1}{2\pi i}(-2\pi i)e^{-i(-i/\tau)t} = e^{-t/\tau} \tag{58}$$

which is in agreement with (53) for positive times.

Exercises

24. Taking $\bar{x}(\omega) = -\frac{1}{2}\pi i\omega$ as given by Exercise 2 for the unit step function, obtain $x(t)$. The contour on the real axis must exclude $\omega = 0$.

25. Let $f(t)$ be the Lorentz function

$$f(t) = \frac{1}{\pi}\frac{\tau}{t^2 + \tau^2}$$

Obtain the Fourier transform $\bar{f}(\omega)$.

PERIODIC STRUCTURES

We now discuss the representation by a Fourier series of a periodic function of position. The most general cell for three-dimensional periodicity is a parallelepiped of sides a, b, and c as shown in Figure 5a. The volume of the cell (Section A-6) is $V_c = \mathbf{a} \cdot \mathbf{b} \times \mathbf{c}$. All space may be filled with the structure shown in Figure 5a by displacing the parallelepiped by:

$$\mathbf{R} = m\mathbf{a} + n\mathbf{b} + p\mathbf{c} \tag{59}$$

If V_c is taken to be the smallest possible cell, then any two equivalent points will be connected by a vector of the form of (59) where m, n, and p are integers. Crystallographers refer to $\mathbf{a}$, $\mathbf{b}$, and $\mathbf{c}$ as primitive translation vectors and to V_c as the primitive cell. The vectors $\mathbf{R}$ establish a lattice of points and $\mathbf{R}$ is called a lattice vector.

Our discussion of Fourier series suggests that we might expect to write $f(\mathbf{r})$ as a sum over some discrete set of wave-vectors:

$$f(\mathbf{r}) = \sum_{\mathbf{k}} f_{\mathbf{k}} e^{i\mathbf{k} \cdot \mathbf{r}} \tag{60}$$

If $f(\mathbf{r})$ is to be a periodic function of position, we require:

$$f(\mathbf{r}) = f(\mathbf{r} + \mathbf{R}) \tag{61}$$

for all $\mathbf{R}$. Substituting **(60)** into **(61)**, we have:

$$\sum_{\mathbf{k}} f_{\mathbf{k}} e^{i\mathbf{k}\cdot\mathbf{r}}(1 - e^{i\mathbf{k}\cdot\mathbf{R}}) = 0 \tag{62}$$

for all $\mathbf{R}$ and for any position $\mathbf{r}$ in the cell. The only way that **(62)** can be satisfied is for each term in the series to vanish identically, which implies

$$\mathbf{k} \cdot \mathbf{R} = 2\pi N \tag{63}$$

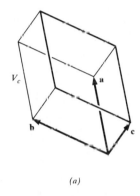 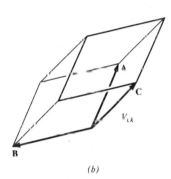

                       (*a*)   (*b*)

Figure 5 Primitive cell. (*a*) The most general cell for three-dimensional periodicity is a parallelepiped. (*b*) The vectors **A**, **B**, and **C** reciprocal to **a**, **b**, and **c** also determine a primitive cell, in the reciprocal lattice.

where N is some integer. As we will see in the next section, this condition establishes a discrete set of vectors $\mathbf{G}$ for the expansion of $f(\mathbf{r})$.

7 Reciprocal Lattice. We now develop a set of vectors **A**, **B**, and **C**, which are reciprocal to **a**, **b**, and **c**. We will show that any vector:

$$\mathbf{G} = h\mathbf{A} + k\mathbf{B} + l\mathbf{C} \tag{64}$$

satisfies **(63)**.

    The reciprocal vectors **A**, **B**, and **C** are required to satisfy the following conditions:

$$
\begin{array}{lll}
\mathbf{A} \cdot \mathbf{a} = 2\pi & \mathbf{B} \cdot \mathbf{a} = 0 & \mathbf{C} \cdot \mathbf{a} = 0 \\[2mm]
\mathbf{A} \cdot \mathbf{b} = 0 & \mathbf{B} \cdot \mathbf{b} = 2\pi & \mathbf{C} \cdot \mathbf{b} = 0 \\[2mm]
\mathbf{A} \cdot \mathbf{c} = 0 & \mathbf{B} \cdot \mathbf{c} = 0 & \mathbf{C} \cdot \mathbf{c} = 2\pi
\end{array}
\tag{65}
$$

These equations may be solved to give:

$$\mathbf{A} = 2\pi\,\frac{\mathbf{b} \times \mathbf{c}}{\mathbf{a} \cdot \mathbf{b} \times \mathbf{c}} \qquad \mathbf{B} = 2\pi\,\frac{\mathbf{c} \times \mathbf{a}}{\mathbf{a} \cdot \mathbf{b} \times \mathbf{c}} \qquad \mathbf{C} = 2\pi\,\frac{\mathbf{a} \times \mathbf{b}}{\mathbf{a} \cdot \mathbf{b} \times \mathbf{c}} \tag{66}$$

With this choice for **A**, **B**, and **C** we find from **(59)**, **(64)**, and **(66)**:

$$\mathbf{G} \cdot \mathbf{R} = (h\mathbf{A} + k\mathbf{B} + l\mathbf{C}) \cdot (m\mathbf{a} + n\mathbf{b} + p\mathbf{c}) = 2\pi(hm + kn + lp) = 2\pi N \tag{67}$$

The vectors **A**, **B**, and **C** are called the primitive translation vectors of the reciprocal lattice and **G** is called a reciprocal lattice vector. The vectors also determine a primitive cell of volume $V_{ck} = \mathbf{A} \cdot \mathbf{B} \times \mathbf{C}$ as shown in Figure 5b.

We may now expand $f(\mathbf{r})$ in vectors of the reciprocal lattice:

▶
$$f(\mathbf{r}) = \sum_{\mathbf{G}'} f_{\mathbf{G}'}\, e^{i\mathbf{G}' \cdot \mathbf{r}} \tag{68}$$

To determine $f_{\mathbf{G}}$ if $f(\mathbf{r})$ is known, we multiply both sides of **(68)** by $e^{-i\mathbf{G} \cdot \mathbf{r}}$ for some **G** and integrate over the cell. All terms but one vanish and we obtain:

▶
$$f_{\mathbf{G}} = \frac{1}{V_c} \int_{V_c} f(\mathbf{r}) e^{-i\mathbf{G} \cdot \mathbf{r}}\, dV \tag{69}$$

Had we started with vectors **A**, **B**, and **C** we might have used **(65)** to determine **a**, **b**, and **c**. In place of **(66)** we would have found

$$\mathbf{a} = 2\pi\,\frac{\mathbf{B} \times \mathbf{C}}{\mathbf{A} \cdot \mathbf{B} \times \mathbf{C}} \qquad \mathbf{b} = 2\pi\,\frac{\mathbf{C} \times \mathbf{A}}{\mathbf{A} \cdot \mathbf{B} \times \mathbf{C}} \qquad \mathbf{c} = 2\pi\,\frac{\mathbf{A} \times \mathbf{B}}{\mathbf{A} \cdot \mathbf{B} \times \mathbf{C}} \tag{70}$$

Thus, the vectors **a**, **b**, and **c** are reciprocal to **A**, **B**, and **C** as well and the relationship is a mutual one. Had we begun with **(68)** for some set of vectors **A**, **B**, and **C**, we could from **(70)** have determined the periodicity of $f(\mathbf{r})$.

Exercises

26. Obtain the lattice vectors of the body-centered-cubic (bcc) lattice. Find the set of vectors reciprocal to the lattice vectors and construct the reciprocal lattice.
27. Obtain the lattice vectors of the face-centered-cubic (fcc) lattice. Find the set of vectors reciprocal to the lattice vectors and construct the reciprocal lattice.

SUMMARY

1. A periodic function $x(t)$ may be written as a Fourier series:

$$x(t) = \sum_{n=-\infty}^{n=+\infty} \underline{x}_n e^{-in\omega t} \tag{10}$$

where $\underline{x}_n$ is given by

$$\underline{x}_n = \frac{1}{T} \int_{-\frac{1}{2}T}^{+\frac{1}{2}T} x(t) e^{in\omega t} \, dt \tag{7}$$

2. An aperiodic function $x(t)$ may be written as a Fourier integral:

$$x(t) = \int_{-\infty}^{+\infty} \mathbf{x}(\omega) e^{-i\omega t} \, dt \tag{12}$$

with

$$\bar{x}(\omega) = \frac{1}{2\pi} \int_{-\infty}^{+\infty} x(t) e^{i\omega t} \, dt \tag{11}$$

3. An aperiodic function of position $f(\mathbf{r})$ may be written as a Fourier integral:

$$f(\mathbf{r}) = \int \mathbf{f}(\mathbf{k}) e^{i\mathbf{k}\cdot\mathbf{r}} \, dV_k \tag{19}$$

with

$$\bar{f}(\mathbf{k}) = \frac{1}{(2\pi)^3} \int f(\mathbf{r}) e^{-i\mathbf{k}\cdot\mathbf{r}} \, dV \tag{20}$$

4. For $\bar{x}(\underline{\omega})$ an analytic function of complex $\underline{\omega}$ the Cauchy-Riemann conditions yield:

$$\frac{\partial}{\partial\omega'} x'(\underline{\omega}) - \frac{\partial}{\partial\omega''} x''(\underline{\omega}) = 0 \tag{36}$$

$$\frac{\partial}{\partial\omega''} x'(\underline{\omega}) + \frac{\partial}{\partial\omega'} x''(\underline{\omega}) = 0 \tag{37}$$

5. For $\bar{x}(\underline{\omega})$ analytic on S the integral of $\bar{x}(\underline{\omega})$ on the contour bounding S is equal to zero:

$$\oint_C \bar{x}(\underline{\omega}) \, d\underline{\omega} = 0 \tag{42}$$

6. The integral function with a simple pole on S is $2\pi i$ times the residue at the pole:

$$\oint_C \frac{\bar{x}(\omega)}{\omega}\, d\omega = 2\pi i x(0) \tag{49}$$

For the pole away from $x = 0$ the residue is evaluated at the pole. For more than one pole the integral equals $2\pi i$ times the sum of the residues.

7. The residue theorem is generally useful in obtaining for example a function $x(t)$ when its Fourier transform $\bar{x}(\omega)$ is known. One must be careful to close the contour over the positive half plane for negative times and over the negative half plane for positive times.

8. A periodic function of position $f(\mathbf{r})$ may be written as a Fourier series:

$$f(\mathbf{r}) = \sum_{\mathbf{G}} f_{\mathbf{G}}\, e^{i\mathbf{G}\cdot\mathbf{r}} \tag{68}$$

with

$$f_{\mathbf{G}} = \frac{1}{V_c} \int_{V_c} f(\mathbf{r}) e^{-i\mathbf{G}\cdot\mathbf{r}}\, dV \tag{69}$$

The vectors $\mathbf{G}$ are reciprocal to lattice vectors $\mathbf{R}$ with $\mathbf{G}\cdot\mathbf{R} = 2\pi N$.

REFERENCES

Arfken, G., *Mathematical Methods for Physicists*, Academic, 1970.
Bell, R. J., *Introductory Fourier Transform Spectroscopy*, Academic, 1972.
Champeney, D. C., *Fourier Transforms and Their Physical Applications*, Academic, 1973.
Churchill, R. V., *Fourier Series and Boundary Value Problems*, McGraw-Hill, 1963.
Hildebrand, F. B., *Advanced Calculus for Applications*, Prentice-Hall, 1962.
Kittel, C., *Introduction to Solid State Physics*, Fifth Edition, Wiley, 1976.
Seeley, R., *An Introduction to Fourier Series and Integrals*, W. A. Benjamin, 1966.
Tolstov, G., *Fourier Series*, Dover, 1976.

J DISPERSION RELATIONS

For any linear system with a driving force that we may write as

$$E(t) = E \cos \omega t \tag{1}$$

we expect a response:

$$p(t) = \alpha' E \cos \omega t + \alpha'' E \sin \omega t \tag{2}$$

We develop here a remarkable integral relation connecting α' and α'', first obtained in 1927 by H. A. Kramers. This work extended the observation of Kronig a year earlier that if α'' is known at all frequencies, one may write α' at any frequency. We first give the Kronig spectroscopic theory and then present the more general discussions of Kramers.

1 **Spectroscopic Theory.** For an oscillating dipole we have an equation of the form:

$$m \frac{d^2 p}{dt^2} + \frac{2}{\tau} \frac{dp}{dt} + m\omega_0^2 p = q^2 E e^{-i\omega t} \tag{3}$$

The complex polarizability is given by:

$$\alpha(\omega) = \frac{p}{E} = \frac{q^2/m}{(\omega_0^2 - \omega^2) - 2i\omega/\tau} \tag{4}$$

and the real and imaginary parts are then given by:

$$\alpha'(\omega) = \frac{q^2}{m} \frac{\omega_0^2 - \omega^2}{(\omega_0^2 - \omega^2)^2 + 4(\omega/\tau)^2} \tag{5}$$

$$\alpha''(\omega) = \frac{q^2}{m} \frac{2\omega/\tau}{(\omega_0^2 - \omega^2)^2 + 4(\omega/\tau)^2} \tag{6}$$

These quantities are sketched in Figure 1. In the limit that $\omega\tau$ is large we may approximate **(5)** and **(6)** by

$$\alpha'(\omega) \simeq \frac{q^2}{2\omega_0\, m} \frac{\omega_0 - \omega}{(\omega_0 - \omega)^2 + 1/\tau^2} \tag{7}$$

$$\alpha''(\omega) \simeq \frac{q^2}{2\omega_0\, m} \frac{1/\tau}{(\omega_0 - \omega)^2 + 1/\tau^2} \tag{8}$$

The integrated absorption is given by:

$$\int_0^\infty \alpha''(\omega)\, d\omega \simeq \frac{q^2}{2\omega_0\, m} \int_{-\infty}^\infty \frac{dx}{x^2 + 1} = \frac{\pi}{2} \frac{q^2}{\omega_0\, m} \tag{9}$$

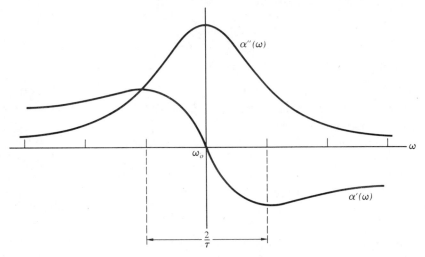

Figure 1 Complex polarizability. The real part of the polarizability α' is odd about the resonance ω_0. The imaginary part of the polarizability α'' is even.

where we have taken $x = (\omega - \omega_0)\tau$, and we observe that the integrated absorption is independent of τ. The lower limit of the second integral is strictly $-\omega_0\tau$, which we have let go to $-\infty$ for τ large. In the limit that the relaxation time τ becomes *very* long compared with $1/\omega$ we may approximate **(7)** by:

$$\alpha'(\omega) \simeq \frac{q^2/m}{\omega_0{}^2 - \omega^2} \tag{10}$$

In this limit $\alpha''(\omega)$ is zero everywhere except at $\omega = \omega_0$ where it has an integrable singularity, permitting us to write:

$$\alpha''(\omega) = \frac{\pi}{2} \frac{q^2}{\omega_0\, m}\, \delta(\omega - \omega_0) \tag{11}$$

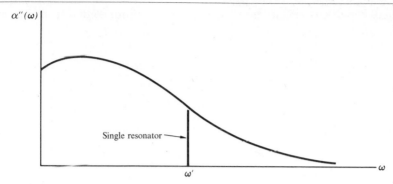

Figure 2 Energy loss. As pointed out by Kronig, a general energy loss may be associated with a distribution of delta-function oscillators.

where $\delta(\omega - \omega_0)$ is the Dirac delta function and has the property:

$$\int_{\omega_1}^{\omega_2} \delta(\omega - \omega_0) \, d\omega = 1 \tag{12}$$

as long as $\omega = \omega_0$ is included within the range ω_1 to ω_2. Otherwise the integral is zero.

Kronig pointed out that for some general loss as a function of frequency as shown in Figure 2, we may imagine that at each frequency there is a delta function oscillator of strength $(2/\pi)\alpha''(\omega)$, permitting us to write from **(10)**:

$$\alpha'(\omega) = \frac{2}{\pi} \int_0^\infty \frac{\alpha''(\omega')}{\omega'^2 - \omega^2} \, \omega' \, d\omega' \tag{13}$$

Exercises

1. Find the poles of $\alpha(\omega)$ in **(4)** and show that they are located in the lower-half complex plane.
2. Obtain **(13)** by eliminating q^2/m between **(10)** and **(11)** for a single oscillator and then superposing many such oscillators.

2 **Sum Rules and Pulse Response.** As a special case of **(13)** we have, for the static polarizability,

$$\alpha'(0) = \frac{2}{\pi} \int_0^\infty \alpha''(\omega') \frac{d\omega'}{\omega'} \tag{14}$$

A second special case is for ω high compared with any frequency for which the losses are important. In this limit we obtain

$$\alpha'(\omega) \simeq -\frac{2}{\pi\omega^2} \int_0^\infty \alpha''(\omega')\omega' \, d\omega' \tag{15}$$

At sufficiently high frequencies any oscillator must be limited only by its mass and we expect

$$\alpha'(\omega) = -\frac{q^2}{m\omega^2} \tag{16}$$

Comparing **(15)** and **(16)** we have the sum rule:

$$\int_0^\infty \alpha''(\omega')\omega' \, d\omega' = \frac{\pi q^2}{2m} \tag{17}$$

We now obtain an expression for the response to a pulse in the electric field

$$E(t) = \mathscr{E} \, \delta(t) \tag{18}$$

Writing for the polarization as a function of time:

$$p(t) = \int_{-\infty}^\infty \bar{p}(\omega)e^{-i\omega t} \, d\omega \tag{19}$$

with

$$\bar{p}(\omega) = \underline{\alpha}(\omega)\bar{E}(\omega) \tag{20}$$

and with

$$\bar{E}(\omega) = \frac{1}{2\pi} \int E(t)e^{i\omega t} \, dt = \frac{1}{2\pi} \mathscr{E} \tag{21}$$

we obtain finally for the response to a pulse:

$$p(t) = \frac{1}{2\pi} \mathscr{E} \int_{-\infty}^\infty \underline{\alpha}(\omega)e^{-i\omega t} \, d\omega \tag{22}$$

We evaluate **(22)** by extending ω into the complex plane and closing the contour before the pulse over the positive half plane as shown in Figure 3*a* and after the pulse over the lower half plane as shown in Figure 3*b*.

Since the response of the system must be identically zero before the arrival of the pulse there may be no poles of $\underline{\alpha}(\underline{\omega})$ in the upper half plane. The response of the system after the pulse is then determined by poles in the lower half plane. The transient time dependence associated with these poles *must* be a decaying exponential since ω'' is negative.

3 **Dispersion Relations for the Permittivity.** Following the arguments of Kramers we now develop the relationship between the real part of the complex permittivity $\epsilon'(\omega)$ and the imaginary part $\epsilon''(\omega)$ for any system that satisfies the following conditions:

1. The system must be linear. For the case of the permittivity this means that, for a driving field,

$$\underline{E}(t) = Ee^{-i\omega t} \tag{23}$$

the displacement will be given by

$$\underline{D}(t) = \underline{\epsilon}(\omega)\underline{E}(t) \tag{24}$$

where $\underline{\epsilon}(\omega)$ is a complex quantity independent of the magnitude of $\underline{E}(t)$.

2. The system must be in steady state, that is, $\underline{\epsilon}(\omega)$ may not be a function of time.

3. The system must respond according to ordinary causality. The response to a stimulus must *follow* the stimulus. This excludes the advanced solutions of the wave equation for example and allows only the retarded solutions.

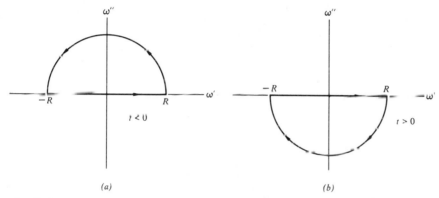

(a)   (b)

Figure 3 Pulse response. The response to a pulse is obtained from an integration of the polarizability over a closed contour that includes the real frequency axis. (a) At negative times the contour is closed over the positive half-frequency plane. (b) At positive times the contour is closed over the negative half-frequency plane.

4. The system must be in stable equilibrium so that the application of a *small* applied field will not produce a major change in the system. In particular, the response to a pulse can be an exponentially decaying polarization but not an exponentially increasing one.

In order to obtain the dispersion relations we first extend $\underline{\epsilon}(\omega)$, which is normally defined for positive frequencies only, to negative frequencies by requiring **(I-9)**:

$$\underline{\epsilon}(-\omega) = \underline{\epsilon}^*(+\omega) \tag{25}$$

We next extend ω into the complex plane, where the positive half of the plane represents a response of the form:

$$e^{-i\underline{\omega}t} = e^{-i(\omega' + i\omega'')t} = e^{\omega''t}e^{-i\omega't} \tag{26}$$

which is exponentially increasing and disallowed by our assumptions above. Conversely, the lower half plane represents exponentially decreasing solutions, which are allowed.

We take the contour integral of the function:

$$\frac{\epsilon(\omega)}{\omega - \omega} \tag{27}$$

where ω is a frequency on the positive real axis. We close the contour over the upper half plane, as shown in Figure 4, excluding $\omega' = \omega$ on a contour of radius r. Since the response of the system to a pulse is at frequencies for which $\epsilon(\omega)$ is singular and since there may be no singularities in the upper half plane, we have the result for the contour shown in Figure 4:

$$\oint \frac{\epsilon(\omega)}{\omega - \omega}\, d\omega = 0 \tag{28}$$

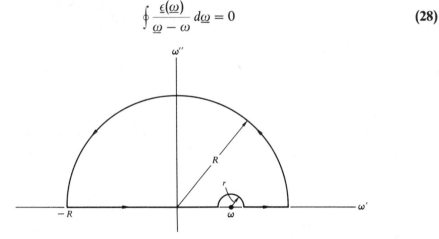

Figure 4   Dispersion relations. The dispersion relations connecting the real and imaginary parts of a response function are obtained by integrating over the contour shown.

The integral **(28)** may be broken up into four segments, which are to be integrated separately:

1. The segment along the real axis from $-R$ to $\omega - r$ is written as

$$\int_{-R}^{\omega - r} \frac{\epsilon(\omega')}{\omega' - \omega}\, d\omega' \tag{29}$$

2. The segment from $\omega - r$ to $\omega + r$ is taken over the contour $\omega = \omega + re^{i\theta}$ with $d\omega$ given by $re^{i\theta}i\, d\theta = (\omega - \omega)i\, d\theta$:

$$\int_{\omega - r}^{\omega + r} \frac{\epsilon(\omega)}{\omega - \omega}\, d\omega = \int_{\pi}^{0} \epsilon(\omega)i\, d\theta \tag{30}$$

3. The integral from $\omega + r$ to R is the real integral:

$$\int_{\omega + r}^{R} \frac{\epsilon(\omega')}{\omega' - \omega}\, d\omega' . \tag{31}$$

4. The final segment is over a contour of radius R with $\omega = Re^{i\theta}$:

$$\int_{-R}^{R} \frac{\epsilon(\omega)}{\omega - \omega} \, d\omega = \int_{0}^{\pi} \frac{\epsilon(\omega)}{1 - \omega/\omega} \, i \, d\theta \tag{32}$$

In the limit that r goes to zero and R to infinity, segments 1 and 3 may be combined to constitute the *principal value* of the integral:

$$\int_{-\infty}^{\infty} \frac{\epsilon(\omega')}{\omega' - \omega} \, d\omega' \tag{33}$$

The second segment becomes simply $-i\pi\epsilon(\omega)$. The fourth segment is $i\pi\epsilon_0$. The permittivity at infinite frequency can not include any contribution from the material medium since these terms must all be limited by inertia.

Combining terms we have finally:

$$\epsilon(\omega) = \epsilon_0 + \frac{1}{i\pi} \int_{-\infty}^{\infty} \frac{\epsilon(\omega')}{\omega' - \omega} \, d\omega' \tag{34}$$

Equating separately the real and imaginary parts of **(34)** we obtain two equations:

$$\epsilon'(\omega) = \epsilon_0 + \frac{1}{\pi} \int_{-\infty}^{\infty} \frac{\epsilon''(\omega')}{\omega' - \omega} \, d\omega' \tag{35}$$

$$\epsilon''(\omega) = -\frac{1}{\pi} \int_{-\infty}^{\infty} \frac{\epsilon'(\omega')}{\omega' - \omega} \, d\omega' \tag{36}$$

Using **(25)** we may rewrite **(35)** and **(36)** over positive real frequencies only:

$$\epsilon'(\omega) = \epsilon_0 + \frac{2}{\pi} \int_{0}^{\infty} \frac{\epsilon''(\omega')}{\omega'^2 - \omega^2} \, \omega' \, d\omega' \tag{37}$$

$$\epsilon''(\omega) = -\frac{2\omega}{\pi} \int_{0}^{\infty} \frac{\epsilon'(\omega')}{\omega'^2 - \omega^2} \, d\omega' \tag{38}$$

We observe that **(37)** has the same form as **(13)** while **(38)** is a new relationship not directly obtainable by the earlier approach. These relations **(37)** and **(38)** are called the *Kramers-Kronig relations*. An important use is in the experimental determination of $\epsilon'(\omega)$ and $\epsilon''(\omega)$ where the phenomenon being investigated depends on both the real and the imaginary part of the permittivity in a way such that they are not directly separable.

4 **[Exponential Increase.]** We investigate in this section the response of a dielectric medium to an electric field of the form:

$$E(t) = Ee^{t/\tau} \tag{39}$$

where τ is a real positive time. We expect, for the displacement,

$$D(t) = \epsilon(\tau)E(t) \tag{40}$$

From the fact that we have been able to write $\underline{\epsilon}(\omega)$ as an analytic function in the complex plane, we may regard (40) as corresponding to (24) with

$$\underline{\omega} = i\omega'' = i/\tau \tag{41}$$

To obtain $\epsilon(\tau)$ we repeat the discussion of Section 3, but instead of (27) we perform a contour integration of

$$\frac{\epsilon(\underline{\omega})}{\underline{\omega} - i/\tau} \tag{42}$$

where i/τ is a pole on the positive imaginary axis as shown in Figure 5.

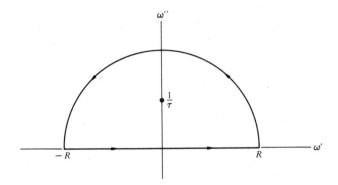

Figure 5   Exponential increase. The response to an exponentially increasing field is obtained by integrating (42) over the contour shown.

We obtain by Section I-5:

$$\oint \frac{\epsilon(\underline{\omega})}{\underline{\omega} - i/\tau} \, d\underline{\omega} = 2\pi i \underline{\epsilon}(\tau) \tag{43}$$

Taking $\epsilon(\omega)$ at infinite frequency to be ϵ_0 we write:

$$\epsilon(\tau) = \tfrac{1}{2}\epsilon_0 + \frac{1}{2\pi i} \int_{-\infty}^{\infty} \frac{\epsilon(\omega)}{\omega - i/\tau} \, d\omega \tag{44}$$

Using (25) we obtain:

$$\epsilon(\tau) = \tfrac{1}{2}\epsilon_0 + \frac{1}{\pi\tau} \int_0^{\infty} \frac{\epsilon'(\omega) \, d\omega}{\omega^2 + 1/\tau^2} + \frac{1}{\pi} \int_0^{\infty} \frac{\epsilon''(\omega)\omega \, d\omega}{\omega^2 + 1/\tau^2} \tag{45}$$

Exercises

2. An oscillator satisfies the differential equation

$$m\frac{d^2x}{dt^2} + \frac{m}{\tau_0}\frac{dx}{dt} + m\omega_0{}^2x = eE(t)$$

Show that for $E(t) = Ee^{i\omega t}$ the permittivity of a medium with n oscillators per unit volume is given by

$$\underline{\epsilon}(\omega) = \epsilon_0\left(1 - \frac{\omega_p{}^2}{\omega(\omega + i/\tau_0) - \omega_0{}^2}\right)$$

where $\omega_p{}^2 = ne^2/\epsilon_0 m$ is the square of the plasma frequency.

3. Show that if the system described in Exercise 1 is driven by a field

$$E(t) = Ee^{t/\tau}$$

the permittivity is given by

$$\epsilon(\iota) = \epsilon_0\left(1 + \frac{\omega_p{}^2\tau^2}{1 + \tau/\tau_0 + \omega_0{}^2\tau^2}\right)$$

4. Substitute the result of Exercise 2 into **(36)** to obtain the result of Exercise 3. Use the *residue theorem* (Section I-5) to evaluate the integral. Note that poles of $\underline{\epsilon}(\omega)$ are in the lower half of the complex plane.

5 [**Dispersion Relations for the Reflectivity.**] As shown in Section 13-3, the reflection coefficient of a surface is a complex quantity:

$$\underline{r} = \rho e^{i\theta} \tag{46}$$

where for normal incidence we have

$$r = \frac{Z_2 - Z_1}{Z_1 + Z_2} \tag{47}$$

and where Z_1 and Z_2 are the intrinsic impedances of the two media:

$$Z_1 = (\mu_1/\epsilon_1)^{1/2} \qquad Z_2 = (\mu_2/\epsilon_2)^{1/2} \tag{48}$$

The intrinsic impedances Z_1 and Z_2 are functions of ω through the dependence of the complex permittivity $\underline{\epsilon}(\omega)$ on frequency. Thus ρ and θ are functions of frequency.

Most reflectivity experiments do not yield phase information. All that is available is the reflectivity

$$R(\omega) = |\underline{r}|^2 = \rho^2 \tag{49}$$

Although it is certainly possible to write $R(\omega)$ in terms of $\epsilon'(\omega)$ and $\epsilon''(\omega)$ the expression is complicated, and one can hardly expect in practice to be able to extract the real and imaginary parts of the permittivity separately through the use of (37) and (38).

What is done instead is to develop dispersion relations for $\underline{r}$ by means of which it is possible to determine $\theta(\omega)$. Thus, it is much easier to extract $\epsilon'(\omega)$ and $\epsilon''(\omega)$ when used together with $\rho(\omega) = [R(\omega)]^{1/2}$. We observe that (46) describes a linear relation between incoming and outgoing wave fields and satisfies the criteria imposed on the permittivity. Since we wish to determine the phase angle θ, we work with the complex quantity:

$$\ln \underline{r} = \ln \rho + i\theta = \tfrac{1}{2} \ln R(\omega) + i\theta(\omega) \tag{50}$$

Using (38) we may write immediately:

$$\theta(\omega) = \theta(\infty) - \frac{\omega}{\pi} \int_0^\infty \frac{\ln R(\omega')}{\omega'^2 - \omega^2} \, d\omega' \tag{51}$$

We may simplify (51) by integrating by parts where we use:

$$\int \frac{2\omega \, d\omega'}{\omega'^2 - \omega^2} = \ln \left| \frac{\omega' - \omega}{\omega' + \omega} \right| \tag{52}$$

We obtain then for (51):

$$\theta(\omega) = \theta(\infty) - \frac{1}{2\pi} \ln R(\omega') \ln \left| \frac{\omega' - \omega}{\omega' + \omega} \right| \Big]_0^\infty$$

$$+ \frac{1}{2\pi} \int_0^\infty \ln \left| \frac{\omega' - \omega}{\omega' + \omega} \right| \frac{d}{d\omega'} \ln R(\omega') \, d\omega' \tag{53}$$

The first term is zero at both limits because of the argument of the logarithm and only the second term remains.

Where the reflection coefficient is a slowly varying function of frequency it may be removed from the integral and we have the simple approximate form:

$$\theta(\omega) \simeq \theta(\infty) - \frac{\pi}{4} \frac{d \ln R(\omega)}{d \ln \omega} \tag{54}$$

where we have used the tabulated integral:

$$\int_0^\infty \ln \left| \frac{\omega' - \omega}{\omega' + \omega} \right| \frac{d\omega'}{\omega'} = -\frac{\pi^2}{2} \tag{55}$$

Note from **(54)** that if $R(\omega)$ is dropping with frequency as $1/\omega$, the phase shift is $45°$; a $1/\omega^2$ drop in $R(\omega)$ corresponds to a $90°$ phase shift. This approximate relationship between phase shift and power response is widely used in network theory.

If $\theta(\omega)$ is known, it is possible to eliminate $\epsilon''(\omega)$ between the expressions for θ and for ρ. We are left, then, with the computational problem of obtaining $\epsilon'(\omega)$ at each frequency given θ and ρ at that frequency. Given $\epsilon'(\omega)$, one may use $\theta(\omega)$ to determine $\epsilon''(\omega)$.

REFERENCE

Kittel, C., *Introduction to Solid State Physics*, Fifth Edition, Wiley, 1976.

K UNITS, STANDARDS, AND SYMBOLS

In this appendix we discuss briefly the International System of Units (SI), which is used in this text, tabulate the principal physical constants in this system of units, and give the conversion factors between SI and the CGS system of units. We also list the symbols used in this text together with the quantities represented and the appropriate SI units.

INTERNATIONAL SYSTEM OF UNITS

The International System of Units (designated SI in all languages) is recommended by members of the General Conference of Weights and Measures for all scientific, technical, practical, and teaching purposes. This system of units has evolved from the metre (m) as the unit of length, the kilogram (kg) as the unit of mass, and the second (s) as the unit of time. Four additional *base* units have been added, the ampere (A) as the unit of electric current, the kelvin (K) as the unit of thermodynamic temperature, the candela (cd) as the unit of luminous intensity, and the mole (mol) as the unit of amount of substance. It is the recommendation of the General Conference that this system for scientific purposes largely replace the CGS system, which is based on the centimetre (cm) as the unit of length, the gram (g) as the unit of mass, and the second (s) as the unit of time. In the CGS system the units of current, temperature, luminous intensity, and the amount of substance are *derived* from the *base* units. Although there is a substantial body of physicists who prefer a system of units with only mass, length, and time as *base* units, the fact is that the SI system does have many advantages over the CGS system for technical and practical purposes. It is coming into wider use in teaching, and it is probably only a matter of time before the SI system largely replaces the CGS system for scientific purposes. Although we may feel with most physicists that the CGS system is more elegant, there is considerable advantage in working in a system of units that is widely used and understood. Principally for this reason, the present text employs the SI system of units.

1 **Definitions of SI Base Units.** The SI system is based on seven well-defined units that by convention are regarded as dimensionally independent:

1. The metre is the length equal to 1650763.73 wavelengths in vacuum of the radiation corresponding to the transition between the levels $2p_{10}$ and $5d_5$ of the krypton-86 atom.

2. The kilogram is the unit of mass; it is equal to the mass of the international prototype of the kilogram. (This international prototype made of platinum-iridium is kept at the International Bureau of Weights and Measures, Paris.)

3. The second is the duration of 9192631770 periods of the radiation corresponding to the transition between the two hyperfine levels of the ground state of the cesium-133 atom.

4. The ampere is that constant current that, if maintained in two straight parallel conductors of infinite length, of negligible circular cross section, and placed 1 metre apart in vacuum, would produce between these conductors a force equal to 2×10^{-7} Newton per metre of length.

5. The kelvin, unit of thermodynamic temperature, is the fraction $\frac{1}{273.16}$ of the thermodynamic temperature of the triple point of water. (The temperature in degrees Celsius is numerically equal to the thermodynamic temperature less 273.15 kelvin.)

6. The mole is the amount of substance of a system which contains as many elementary entities as there are atoms in 0.012 kilogram of carbon-12.

7. The candela is the luminous intensity of a source equivalent to the monochromatic radiation of $4\pi/683$ watts at a wavelength of 555 nm. A uniform point source with an intensity of 1 candela emits 1 lumen into a solid angle of 1 steradian.

2   Derived Units. The second class of SI units contains derived units—units that can be formed by combining base units according to the algebraic relations linking the

Table 1   SI Derived Units with Special Names

Quantity	Name	Symbol	SI Base Units	SI Derived Units
Capacitance	farad	F	$m^{-2} \cdot kg^{-1} \cdot s^4 \cdot A^2$	C/V
Conductance	siemens	S	$m^{-2} \cdot kg^{-1} \cdot s^3 \cdot A^2$	A/V
Electric charge	coulomb	C	$A \cdot s$	
Electric potential	volt	V	$m^2 \cdot kg \cdot s^{-3} \cdot A^{-1}$	W/A
Energy	joule	J	$m^2 \cdot kg \cdot s^{-2}$	$N \cdot m$
Force	newton	N	$m \cdot kg \cdot s^{-2}$	
Frequency	hertz	Hz	s^{-1}	
Inductance	henry	H	$m^2 \cdot kg \cdot s^{-2} \cdot A^{-2}$	Wb/A
Magnetic flux	weber	Wb	$m^2 \cdot kg \cdot s^{-2} \cdot A^{-1}$	$V \cdot s$
Magnetic flux density	tesla	T	$kg \cdot s^{-2} \cdot A^{-1}$	Wb/m^2
Power	watt	W	$m^2 \cdot kg \cdot s^{-3}$	J/s
Pressure	pascal	Pa	$m^{-1} \cdot kg \cdot s^{-2}$	
Resistance	ohm	Ω	$m^2 \cdot kg \cdot s^{-3} \cdot A^{-2}$	V/A

corresponding quantities. Several of these algebraic expressions in terms of base units have been given special names and symbols that may themselves be used to express other derived units in a simpler way than in terms of the base units.

For example, the unit of magnetic field strength H is the ampere per metre, given the symbol A/m. In Table 1 we give a number of derived units that have been given special names.

The *theoretical* values of the electric units are based on the ampere and are defined as follows:

1. The volt is the difference of electric potential between two points of a conducting wire carrying a constant current of 1 ampere, when the power dissipated between these points is equal to 1 watt.

2. The ohm is the resistance between two points of a conductor when a constant potential difference of 1 volt, applied to these points, produces in the conductor a current of 1 ampere, the conductor not being the seat of any electromotive force.

3. The coulomb is the quantity of electricity carried in 1 second by a current of 1 ampere.

4. The farad is the capacitance of a capacitor between the plates of which there appears a potential difference of 1 volt when it is charged by a quantity of electricity of 1 coulomb.

5. The henry is the inductance of a closed circuit in which an electromotive force of 1 volt is produced when the electric current in the circuit varies uniformly at the rate of 1 ampere per second.

6. The weber is that magnetic flux that, linking a circuit of one turn, would produce in it an electromotive force of 1 volt if it were reduced to zero at a uniform rate in 1 second.

In practice the ohm is linked to the farad and the henry by impedance measurements at a known frequency and where the inductance and capacitance are calculated. The volt is then deduced from the ampere, which is determined from the force between coils, and the ohm. Secondary standards of resistance, capacitance, and electromotive force are used to make available the results of absolute measurements. These secondary standards may in turn be checked by the Josephson effect in the case of the volt and by the gyromagnetic ratio of the proton for the ampere.

In addition, other derived units are expressed *by means* of units with special names. We give a few examples in Table 2.

In addition there are certain units that are used with the International System and whose values in SI units must be obtained experimentally. Among these are:

Table 2 SI Derived Units Expressed by Means of Special Names

Quantity	SI Derived Units	SI Base Units
Electric flux density	C/m^2	$A \cdot s \cdot m^{-2}$
Electric field strength	V/m	$m \cdot kg \cdot s^{-3} \cdot A^{-1}$
Permittivity	F/m	$m^{-3} \cdot kg^{-1} \cdot s^4 \cdot A^2$
Permeability	H/m	$m \cdot kg \cdot s^{-2} \cdot A^{-2}$

1. The electronvolt (eV), which is the kinetic energy acquired by an electron in passing through a potential difference of 1 volt in vacuum: $1 \text{ eV} = 1.60219 \times 10^{-19}$ J approximately.

2. The atomic mass unit (u), which is equal to the fraction $\frac{1}{12}$ of the mass of an atom of the nuclide C-12: $1 \ u = 1.66053 \times 10^{-27}$ kg approximately.

3 SI Prefixes. A series of names and symbols have been adopted to form decimal multiples and submultiples of SI units. These prefixes are given in Table 3.

Table 3 SI Prefixes

Factor	Prefix	Symbol	Factor	Prefix	Symbol
10^{12}	tera	T	10^{-1}	deci	d
10^{9}	giga	G	10^{-2}	centi	c
10^{6}	mega	M	10^{-3}	milli	m
10^{3}	kilo	k	10^{-6}	micro	μ
10^{2}	hecto	h	10^{-9}	nano	n
10^{1}	deka	da	10^{-12}	pico	p
			10^{-15}	femto	f
			10^{-18}	atto	a

STANDARDS

We list in this part representative physical constants together with their value in SI units and useful numerical ratios.

4 Physical Constants. We list in Tables 4 and 5 selected fundamental physical constants compiled by E. R. Cohen and B. N. Taylor under the auspices of the CODATA Task

Table 4 Primary Physical Constants

Quantity	Symbol	Numerical Value
Speed of light in vacuum	c	2.9979246×10^{8} m/s
Permeability of vacuum	μ_0	$4\pi \times 10^{-7}$ H/m
Permittivity of vacuum, $1/\mu_0 c^2$	ϵ_0	$8.8541877 \times 10^{-12}$ F/m
Elementary charge	e	$1.6021892 \times 10^{-19}$ C
Planck constant	h	6.626176×10^{-34} J · s
$h/2\pi$	$\hbar$	$1.0545887 \times 10^{-34}$ J · s
Avogadro constant	N_A	6.022045×10^{23} mol^{-1}
Electron rest mass	m_e	9.109534×10^{-31} kg
Proton rest mass	m_p	$1.6726485 \times 10^{-27}$ kg
Molar gas constant	R	8.31441 J · mol^{-1} · K^{-1}
Gravitational constant	G	6.6720×10^{-11} m^3 · s^{-2} · kg^{-1}

TABLE 5   Secondary Physical Constants

Quantity	Symbol	Numerical Value
Fine structure constant, $e^2/4\pi\epsilon_0\,\hbar c$	α	7.2973506×10^{-3}
	α^{-1}	137.03604
Magnetic flux quantum, $h/2e$	Φ_0	$2.0678506 \times 10^{-15}$ Wb
Bohr radius, $4\pi\epsilon_0\,\hbar^2/m_e e^2$	a_0	$5.2917706 \times 10^{-11}$ m
Rydberg constant, $\alpha/4\pi a_0$	R_∞	1.097373177×10^7 m^{-1}
Electron Compton wavelength, $h/m_e c = 2\pi\alpha a_0$	λ_C	$2.4263089 \times 10^{-12}$ m
Classical electron radius, $\alpha^2 a_0$	r_e	$2.8179380 \times 10^{-15}$ m^2
Thomson cross section, $(8\pi/3)r_e{}^2$	σ_e	$0.6652448 \times 10^{-28}$ m^2
Bohr magneton, $e\hbar/2m_e$	μ_B	9.274078×10^{-24} J T^{-1}
Nuclear magneton, $e\hbar/2m_p$	μ_N	5.050824×10^{-27} J $\cdot$ T^{-1}
Boltzmann constant, R/N_A	k	1.380662×10^{-23} J $\cdot$ K^{-1}

Group on Fundamental Constants and published in *J. Phys. Chem. Ref. Data*, **2**, 663 (1973). These values depend on a highly accurate determination of e/h from the ac Josephson effect in superconductors.[1]

CGS SYSTEM OF UNITS

5   CGS Gaussian Units. The CGS system is based on three units, the centimeter (cm), the gram (g), and the second (s), whose relation to the corresponding SI units is indicated by the indicated prefixes. The unit of force is the dyne (dyn) and is equal to 10^{-5} newton. The unit of energy is the erg (erg) and is equal to 10^{-7} joule.

In comparing electromagnetic units in the two systems a problem arises because in SI the ampere is a base unit, while in CGS the unit of charge is a derived unit. Thus in CGS Gaussian Coulomb's law **(1-2)** is written as:

$$F = \frac{qq_1}{r_{01}{}^2} \tag{1}$$

The unit of charge is the electrostatic unit (esu) or statcoulomb and is that charge that, if placed 1 centimeter from a similar charge, would experience a force of 1 dyne.

Comparing **(1)** with **(1-2)** and working through the units we obtain the result that 1 esu unit of charge corresponds to $1/(2.997925 \times 10^9)$ coulomb approximately.

The electric field in CGS units is similarly the force per unit charge:

$$E = \frac{F}{q} = \frac{q_1}{r_{01}{}^2} \tag{2}$$

[1] B. N. Taylor, W. H. Parker, and D. N. Langenberg, *Rev. Mod. Phys.*, **41**, 375 (1969). Reprinted by Academic Press, 1969.

and potential is work per unit charge:

$$\phi = \frac{q_1}{r_{01}} \tag{3}$$

The CGS unit of potential is the esu or statvolt. Comparing (3) with (1-19) we find that 1 statvolt corresponds to approximately 299.7925 volts.

The electrostatic unit of current or statampere is that current that carries in 1 second a quantity of charge equal to 1 esu. The force per centimeter between straight parallel conductors is written in CGS units as

$$\frac{F}{l} = \frac{2}{c^2} \frac{II_1}{r_{01}} \tag{4}$$

The magnetic field is given in CGS Gaussian units by

$$B = \frac{cF}{Il} = \frac{2}{c} \frac{I_1}{r_{01}} \tag{5}$$

The CGS unit of magnetic flux density B is the gauss (Gs). Comparing (5) with (6-65) and working through the units we find that 1 gauss corresponds to 10^{-4} tesla. Notice that this correspondence does not depend on the determination of the speed of light. The CGS unit of magnetic flux is the maxwell (Mx) and 1 maxwell corresponds to 10^{-8} weber.

In CGS units the permeability of vacuum is equal to 1 so that the magnetic field strength H is given by the same expression as (5):

$$H = \frac{2}{c} \frac{I_1}{r_{01}} \tag{6}$$

The CGS unit of magnetic field strength is the oersted (Oe). Comparing (6) with (10-14) we find that 1 oersted corresponds to $(1000/4\pi)$ amperes per metre.

This discussion may be summarized by writing the Maxwell equations in the two systems of units together with the Lorentz force law:

	SI Units	CGS Gaussian Units	
ME I	$\nabla \cdot \mathbf{E} = \rho/\epsilon_0$	$\nabla \cdot \mathbf{E} = 4\pi\rho$	(7)
ME II	$\nabla \times \mathbf{E} = -\partial \mathbf{B}/\partial t$	$\nabla \times \mathbf{E} = -(1/c)\,\partial \mathbf{B}/\partial t$	(8)
ME III	$\nabla \times \mathbf{B} = \mu_0(\mathbf{j} + \epsilon_0\,\partial \mathbf{E}/\partial t)$	$\nabla \times \mathbf{B} = (1/c)(4\pi\mathbf{j} + \partial \mathbf{E}/\partial t)$	(9)
ME IV	$\nabla \cdot \mathbf{B} = 0$	$\nabla \cdot \mathbf{B} = 0$	(10)
	$\mathbf{F} = q(\mathbf{E} + \mathbf{v} \times \mathbf{B})$	$\mathbf{F} = q[\mathbf{E} + (1/c)\mathbf{v} \times \mathbf{B}]$	(11)

6 CGS Electrostatic and Electromagnetic Units. In addition to the CGS *Gaussian* system, which we have described here, it is possible to develop other systems of electrical and magnetic units that are compatible with the CGS system. In the CGS *electrostatic* system electric field and charge are defined as in **(2)** while the magnetic field B is defined as:

$$B = \frac{F}{Il} = \frac{2}{c^2} \frac{I_1}{r_{01}}$$ (12)

rather than as cF/Il as in **(5)**.

The CGS *electromagnetic* system is based on an *absolute* definition of the unit of current as in SI:

$$\frac{F}{l} = \frac{2I_1}{r_{01}}$$ (13)

The electromagnetic unit (emu) of current is the abampere. The magnetic field is defined as

$$B = \frac{F}{Il} = \frac{2I_1}{r_{01}}$$ (14)

By defining the electromagnetic unit of charge, the abcoulomb, as the charge transferred in 1 second by a current of 1 abampere we are led to the following expression for the electric field, which is the force per unit charge:

$$E = \frac{F}{q} = c^2 \frac{q_1}{r_{01}^2}$$ (15)

The Gaussian system, by introducing a factor of c into the force law for currents combines *electrostatic* units of charge, current, and electric field with *electromagnetic* units of magnetic field.

7 Conversion Table. In Table 6 we express the definitions of selected CGS Gaussian units of measure as numerical multiples of SI units and give multiplying factors for converting *numbers* and selected units to corresponding new numbers and SI units. An asterisk follows each number that expresses an exact definition. Numbers not followed by an asterisk are only approximate representations of definitions, or are the result of physical measurement. Conversion factors are listed alphabetically.

It should be noted that the unit of inductance in the CGS Gaussian system is the same as in the CGS *electrostatic* system. This is most easily seen by recalling that the energy of an inductor is $\frac{1}{2}LI^2$ and that the unit of electric current in the CGS Gaussian system is the statampere.

Table 6   Conversion Factors

To Convert from	To	Multiply by
Dyne	newton	1.00×10^{-5}*
Erg	joule	1.00×10^{-7}*
Gauss	tesla	1.00×10^{-4}*
Maxwell	weber	1.00×10^{-8}*
Oersted	ampere per metre	7.9577472×10^{1}
Statampere	ampere	3.335640×10^{-10}
Statcoulomb	coulomb	3.335640×10^{-10}
Statfarad	farad	1.112650×10^{-12}
Stathenry	henry	8.987554×10^{11}
Statmho	siemens	1.112650×10^{-12}
Statohm	ogm	8.987554×10^{11}
Statvolt	volt	2.997925×10^{2}

SYMBOLS

8   Lower-Case Roman

Symbol	Quantity	SI units
c	velocity of light	metre per second
e	charge of the electron	coulomb
g	momentum density **(11-107)**	kilogram per square metre second
g	charge of the magnetic monopole **(D-30)**	ampere metre
g_0	quantum of magnetic charge **(11-53)**	ampere metre
h	Planck constant	joule per hertz
$\hbar$	Planck constant divided by 2π	joule second per radian
i	electric surface current density	ampere per metre
ī	magnetic surface current density	ampere per second
j	electric volume current density **(1-143)**	ampere per square metre
j̇	magnetic charge current **(11-56)**	ampere per metre second
k	wavevector	1 per metre
m	mass	kilogram
p	electric dipole moment momentum **(11-113)**	coulomb metre kilogram metre per second
q	electric charge	coulomb
$\bar{q}$	magnetic charge **(11-54)**	ampere metre
s	speed	metre per second
t	time	second
u	energy density	joule per cubic metre

v	velocity	metre per second
v_g	group velocity **(12-92)**	metre per second
v_p	phase velocity **(12-67)**	metre per second
w	work density	joule per cubic metre

9   Upper-Case Roman

Symbol	Quantity	SI Units
A	vector potential; the **curl** of vector potential is the magnetic field **B (11-172)**	tesla metre
B	magnetic field B; defined from the force on a moving charge **(11-1)**	tesla
C	capacitance **(4-76)**	farad
D	electric displacement; the divergence of **D** is the *external* charge density **(4-7)**	coulomb per square metre
E	electric field; defined from the force on a moving charge **(11-1)**	volt per metre
F	force	newton
G	conductance	siemens
H	magnetic field H, the curl of **H** is the *external* current density **(10-5)**	ampere per metre
I	electric current	ampere
J	charge flow **(1-141)**	ampere metre
L	inductance **(10-67)**	henry
L	generalized angular momentum **(7-6)**, **(8-84)**, **(G-27)**	kilogram square metre per second
M	magnetization density **(6-142)**	ampere per metre
P	pressure **(4-75)**, **(9-199)**	watt newton per square metre
P	polarization density **(1-130)**	coulomb per square metre
R	electrical resistance **(5-25)**	ohm
S	area	square metre
U	energy	joule
V	volume	cubic metre
	potential difference **(5-24)**	volt
W	work	joule
X	reactance	ohm
Y	admittance	siemens
Z	impedance, wave impedance	ohm
Z_0	intrinsic impedance of vacuum **(11-66)**	ohm

10   Sans Serif

Symbol	Quantity	SI Units
T	general tensor quantity	
Q	electric quadrupole moment **(1-76)**	coulomb square metre
$\overline{\text{Q}}$	magnetic quadrupole moment (Exercise 6-31)	ampere cubic metre
N	depolarization tensor **(1-167)**	1
	demagnetization tensor **(6-166)**, **(7-62)**	1

11   Upper-Case Greek

Symbol	Quantity	SI Units
Γ	circulation **(B-50)**	
Λ	irrotational vector **(B-45)**	
Π	Poynting vector **(11-88)**	watt per square metre
Φ	flux of the magnetic field B **(9-76)**	weber
Ψ	flux of the electric field **(1-30)**	volt metre
Ω	solenoidal vector **(B-59)**	

12   Lower-Case Greek

Symbol	Quantity	SI Units
α	atomic polarizability **(3-12)**	farad square metre
β	velocity relative to c	1
γ	$(1 - v^2/c^2)^{-(1/2)}$ **(6-7)**	1
	magnetomechanical ratio **(7-42)**	radian per tesla second
ϵ	permittivity **(4-11)**	farad per metre
ϵ_0	permittivity of vacuum **(1-2)**	farad per metre
θ	polar angle	radian
λ	linear charge density **(6-26)**	coulomb per metre
λ_C	carrier mean free path	metre
λ_D	Debye length **(5-95)**	metre
λ_L	London screening length **(8-87)**	metre
μ	permeability **(10-8)**	henry per metre
μ_0	permeability of vacuum **(6-34)**	henry per metre
v	frequency	hertz
ρ	volume electric charge density	coulomb per cubic metre
	electrical resistivity	ohm metre
$\overline{\rho}$	volume magnetic charge density **(11-55)**	ampere per square metre
σ	surface electric charge density	coulomb per square metre

σ	electrical conductivity **(5-10)**	siemens per metre
$\bar{\sigma}$	surface magnetic charge density	ampere per metre
τ	relaxation time	second
τ	torque	newton metre
ϕ	electrical potential	volt
χ	electric susceptibility **(3-65)**	1
	magnetic susceptibility **(7-23)**	1
ψ	gauge **(8-95)**	weber
ω	angular velocity	radian per second

13 Script

Symbol	Quantity	SI Units
$\mathscr{D}$	circulation of D **(4-9)**	coulomb per metre
$\mathscr{E}$	electromotive force **(9-111)**	volt
$\mathscr{F}$	motional electromotive force **(9-109)**	volt
$\mathscr{H}$	magnetomotive force (Problem 10-1)	ampere
ℓ	length	metre
$\mathscr{L}$	Langevin function **(7-34)**	1

14 Subscripts

clec	*electrical* as applying to forces, energy **(11-68)**, work **(9-154)**
eff	*effective* as applying to effective fields
eln	*electronic* contribution **(12-241)**
exch	quantum mechanical *exchange* **(7-58)**
ext	*external* as applying to charge **(4-7)**, energy **(4-66)** and **(10-60)**, work **(2-4)** and **(10-43)**, currents **(10-2)**, and fields **(1-114)** and **(9-127)**
field	referring to the electromagnetic field **(11-121)**
ind	*induced* as applying to charge **(4-4)** and current **(10-3)**
int	*internal* as applying to energy **(4-67)** and **(10-61)**
ion	*ionic* contribution **(5-157)**
mag	*magnetic* as applying to energy **(11-69)** and force **(9-201)**
mech	*mechanical* as applying to forces **(2-52)** and **(9-200)**, work **(9-153)**
D	referring to the displacement D
H	referring to the magnetic field H **(11-45)**
L	longitudinal **(5-158)**
	London **(8-105)**
	Larmor **(7-27)**
M	referring to magnetization **(6-151)**, **(6-152)**, **(6-164)**, and **(6-165)**
P	referring to polarization **(1-137)** and **(1-140)**
T	transverse **(5-159)**

REFERENCES

Astin, A. V., "Standards of Measurement," *Scientific American* **218** (6), 50 (1968).

Fundamental Physical Constants, NBS Special Publication 398, August 1974.

Page, C. H., "Definitions of Electromagnetic Field Quantities," *Am. J. Phys.* **42**, 490 (1974).

Page, C. H. and P. Vigoureux, eds., *The International System of Units (SI)*, NBS Special Publication 330, 1974.

INDEX